A TEXTBOOK OF
WATER POWER
ENGINEERING

A TEXTBOOK OF WATER POWER ENGINEERING

(Including Dams Engineering, Hydrology and Fluid Power Engineering)

[For the Students of B.E./B.Tech. Civil Engg., Institution of Engineers (India), U.P.S.C. Exam. & Practising Engineers]

R.K. SHARMA
Consultant, Water Resources

T.K. SHARMA
B.E.

S Chand And Company Limited
(ISO 9001 Certified Company)

S Chand And Company Limited

(ISO 9001 Certified Company)

Head Office: D-92, Sector–2, Noida – 201301, U.P. (India), Ph. 91-120-4682700

Registered Office: A-27, 2nd Floor, Mohan Co-operative Industrial Estate, New Delhi – 110 044, Phone: 011-49731800

www.**schandpublishing.com;** e-mail: **info@schandpublishing.com**

Marketing Offices:

Chennai	:	Ph: 23632120; chennai@schandpublishing.com
Guwahati	:	Ph: 2738811, 2735640; guwahati@schandpublishing.com
Hyderabad	:	Ph: 40186018; hyderabad@schandpublishing.com
Jalandhar	:	Ph: 4645630; jalandhar@schandpublishing.com
Kolkata	:	Ph: 23357458, 23353914; kolkata@schandpublishing.com
Lucknow	:	Ph: 4003633; lucknow@schandpublishing.com
Mumbai	:	Ph: 25000297; mumbai@schandpublishing.com
Patna	:	Ph: 4011400; patna@schandpublishing.com

First Edition 2003
Reprints 2008, 2010, 2011, 2013
Reprint 2014 (Twice), 2022

Reprint 2026

ISBN: 978-81-219-2230-2 **Product Code:** H3WPE46CIVE10ENAA03O

PRINTED IN INDIA

By Shri Shyamlal Printing Press Private Limited, Plot 20/4, Site-IV, Industrial Area Sahibabad, Ghaziabad – 201 010 and Published by S Chand And Company Limited, A-27, 2nd Floor, Mohan Co-operative Industrial Estate, New Delhi – 110 044.

PREFACE

The present book of **Water Power Engineering** aims at serving as a text book on the principles and practices of **hydro power engineering** and its associated disciplines *viz* **fluid power engineering, dam engineering** and **hydrology**. In a single volume, the book covers in full conformity, the syllabus of various Indian universities, IITs, UPSC examinations and Section B paper of the Institution of Engineers (India) on the subject in the most acceptable form. This will go a long way in providing a comprehensive text book to the undergraduate and post graduate engineering students and to the practising engineers with a veritable source of reference about the latest trends at the national level.

Hydro Power development has come of age in India. The country held centenary celebrations to commemorate the completion of 100 years of water power development in 1997. In the last five decades, the water power engineering has graduated from borrowed to indigenous expertise and state-of-the-art systems, equipment and turn-key plants tailored to be environmentally clean and highly cost-effective which has benefited several developing countries as well.

That there is a very great-need for a comprehensive book on the subject reflecting the level and achievements of the Indian specialists in hydro power development has been only too obvious to the authors for several years. The present volume is a culmination of the efforts to fulfil the long felt need. The attempt here is to synthesise the principles and practices of the water power engineering in a systematic and easy to understand manner. Every effort has been made to include the maximum available material and the latest developments on the subject with a view to comprehensively cover all the relevant topics. The text is profusely illustrated with figures, tables and solved examples so that the students get well-versed with the art of materializing upon those principles. Typical theoretical and objective type questions have been added to provide a thorough understanding and deep insight of the subject to the students and imparting them the feel of the type of the questions expected in the examination. Both MKS and SI units have been adopted.

Environment being at the centrestage of the development of hydro power has been given due importance. The last decade of the twentieth century has witnessed a major onslaughts of the environmentalists on the very construction of storage based power projects. The new millennium students and engineers should thus be better equipped to give proper direction to the preparation of project reports, planning and execution of the power projects with a view to ensure their timely clearance and to avoid time and cost over run.

In keeping with the vastness of the subject, the book running into twenty five chapters has been divided into four parts.

Part - I **Water Power Engineering**
Part - II **Dams Engineering**
Part - III **Hydrology**
Part - IV **Fluid Power Engineering**

Due emphasis has been placed, where it lies, on storage based, run-of-river, small, micro and pumped storage power developments. The important component of the modern electric supply system for frequency control, peak labbing, as an emergency standby and to meet persisting peak shortage and in view of high cost of installation of new generation capacity and the opposition from the so-called environmentalists.

In the presentation of this work the authors owe a great deal to authors and publishers of various books, journals and Central Board of Irrigation and Power Publications and Bureau of

Indian Standards Specifications. References are invited to the specifications for further details, which are available for sale with Bureau of Indian Standards, New Delhi and their regional and branch offices. The authors hereby record their sense of gratitude to them all. Although appropriate references have been given in the bibliography yet any omissions is unintentional.

The authors are indebted to Er. R.S.Goel, Director, Environment Management , Central Water Commission, Government of India for inspiration and material help.

The authors are grateful to Dr. C.D. Thatte, Secretary General, International Commission on Irrigation and Drainage, New Delhi, Er. Yogendra Prasad, Chairman and Managing Director, NHPC Er. P.L. Diwan, Chairman-cum-Managing Director, WAPCOS, Er. B.S.K. Naidu, Director REPSO-Winrock International, Er. A.M. Singal, Chief Engineer (retd.) Haryana Irrigation Department, Er. G.D. Gupta, Chief Engineer, Bhakra Dam, Er. S.K. Gupta, Chief Engineer, Rajasthan (Irrigation), Er. K.V.S. Thakur, Er. Rabindera Panday, Addl. Superintending Engineer, Secretary Er. M.K. Sharma, Er. R.K. Sharma "Rajan", Shri Laxmi Narain, Shri Raj Kumar and Sh. Roshan Lal Mankotia. Thanks are also due to M/s BHEL, M/s Jyoti Limited, Vadodara, M/s, Kvavrner Energy, New Delhi and M/s Kirloskar Brothers Ltd., Pune for the various illustrations included in the book highlighting their proprietory, quality products.

A word of praise for forbearance also goes to our spouses Mrs Janak Dulari Sharma and Mrs Neeru Sharma respectively. We are also thankful to Savita Sharma, Ashish, Amit, Mohit and Akhil Sharma.

The authors also express their gratitude to Shri Ravindra Kumar Gupta, Managing Director, Shri Navin Joshi, General Manager (S & M), Shri R.S. Saxena, Advisor, S. Chand & Co. Ltd. New Delhi and Shri P.J. Singh, Manager, Jalandhar Branch for their untiring efforts in bringing out the book in an elegant form, at a low price apart from its speedy publication.

Suggestions for the improvement of the book and pointing out any error in the text will be gratefully acknowledged for incorporation in the next edition.

R.K. SHARMA
T.K. SHARMA

CONTENTS

Part I – WATER POWER DEVELOPMENT

PART–I
WATER POWER DEVELOPMENT

- Energy
- Environment and Water Power Development
- Planning for Water Power Development
- Water Power Development
- Storage Power Development
- Run-of-River Power Development
- Pumped Storage Power Development
- Small Hydro Power Development
- Hydro Power Plants
- Hydraulic Turbines
- Hydraulic Valves and Gates
- Water Conductor System
- Penstocks and Pressure Shafts
- Hydraulic Transients and Surge Tanks
- Transmission Lines
- Hydraulic Tunnels

1

ENERGY

1.1. INTRODUCTION

Energy is one of the four fundamental requirements-food, water, energy and environment-which confronts the humanity. Energy is critical in the developing countries not only for social development and human welfare but also as a catalyst for overall development in the New Millennium. Energy in the form of electricity provides the basic infrastructural input to the national economic growth. The per capita electric consumption is considered as an index of a nation's progress and development. India's present per capita consumption around 500 kWh/year is far less than the world average of 2000 kWh/year as also that in the developed countries like USA 10,500, Sweden 12,500 and Norway 21,000. The exponentially growing population and the pace of development, urbanisation, energy intensive industries, rise in standard of living, modernisation etc. have resulted in steep decline in the availability of power, the demand is growing at an annual rate of 8 percent. This necessitates capacity addition of about 8000 MW per year during the Nineth Plan and 12,000 MW during the Tenth Plan. This is a formidable challenge. India has planned to produce 10,000 MW of power from renewable energy sources by 2012 AD. The country has a great potential to tap non-conventional sources of power; presently 3000 MW of power is being produced from wind, solar and micro hydel schemes.

1.2. SOURCES OF ENERGY

The three main sources of energy are water, fossil fuels and nuclear, briefly as under:

1. Renewable Energy

The primary resource is water. In addition, there are non-conventional and renewable sources like wind, tidal, solar and geothermal energies representing an insignificant contribution. At the beginning of the New Millennium, India can be proud of the great achievement it has made in this area and the strides in science and technology right from 70s when the country had to face a severe energy crunch due to steep price hike in imported oil which gave fillip to exploring alternative sources of energy still untapped but available in plenty within the country. The country is now fast emerging as one of the world's leader in the non-conventional sources of energy and is exporting technical expertise, services and equipment to developing countries. Switzerland has been benefited with technical expertise in the manufacture of biomass gasifier.

The conventional hydro power generation uses a renewable energy source and currently supplies about 20 per cent of the electricity generated worldwide. It plays an important role in a variety of long term scenarios for sustainable development.

2. Non-renewable Energies

Fossil Fuels: These include coal, oil, lignite, natural gas.

Nuclear Energy: The country has rich reserve of thorium and uranium for nuclear power plants, and these would help in the long run in meeting the rising power demand in the country.

3. Commercial and Non-Commercial Energies

Another classification of the energy sources is commercial and non-commercial, as given in Table 1.1.

Table 1.1 Energy sources.

Commercial	*Non-commercial*
Thermal power-coal, lignite	Animal dung
Hydro power	Agricultural waste
Nuclear power	Fuel wood
Hydrocarbons–oil, natural gas and biogas	

The share of each energy source logically may depend on the resources, their cost and constraints relating to their impact on the environment.

Hydroelectricity energy continues to be developed at a substantiable rate but it may only represent 3% of the consumption of primary energy on global basis and is not in competition with other sources of energy such as fossil, nuclear and solar.

The USA and Europe, which represent 20% of the world population, consume 62% of the prime energy–they themselves produce only 53%–and the same proportion of electricity. The less developed countries consume 35 times less electricity than the industrialised countries.

1.3. CLASSIFICATION OF ENERGY

Energy is total power consumed over a certain period, measured in kilowatt-hours (kWh). One kilowatt-hour, termed as unit, is equal to energy consumed when power is utilised at the rate of one kilowatt for one hour.

The energy is classified as conventional energy and non-conventional energy, briefly as under:

1. Conventional Energy

The conventional energy is hydro, thermal and nuclear, although thermal generation dominates in the country.

2. Non-conventional Energy

The non-conventional energy has entered into the present scenario of energy problem primarily because of high cost of conventional fuel and their gradual depletion and above all due to non-exploitation of the 78% of the country's hydro power potential. Nevertheless, the matter is not bounded within the limitation of this country but has assumed global dimension. The non-conventional energy is solar, wind, gobar gas, bio-mass etc.

The pattern of consumption of conventional energy is showing downward trend with increase in non-conventional form. The pricing and availability are main criteria. The trend is a significant one indicating adoptability of human society to the new form for the better, since renewable source otherwise is inexhaustible.

A brief resume of the various types of energy (electricity) is as follows. Hydro power has been dealt with in detail separately.

1.4. THERMAL POWER

Coal fired thermal power plants generate energy in the form of heat by the process of combustion; coal is burnt to release heat which in turn is utilised to convert water into high pressure steam to turn the turbines and generate electricity.

First thermal plant of 1000 kW was commissioned in the country in Kolkata in 1899. This was followed by the installation of regional plants at a large number of towns and cities in the first half of the twentieth century. The installed capacity in 1947 was 854 MW, 1260 MW in 1951 (including diesel) which rose to 60,122 MW (including steam and gas) in March 1997. Till 1969 all the thermal sets in the range of 30 to 60 MW were imported ones. At present the indigenous sets range from 200/210/250 MW and 500 MW. The count.y celebrated in 1999 completion of 100 years of the thermal power development. The thermal generation has by now emerged as the principal conventional power generating source in the country constituting 76 percent of the total capacity. The operating availability at present is of the order of 78 percent of the installed capacity and plant load factor is more than 60 percent.

The predominantly coal based thermal plants require large and continuous supply of coal. They consume on an average 0.74 kg/kWh of coal and 11.24 ml/kWh of oil. The high consumption is due to inferior quality of coal, frequent start-ups of units due to forced outages and backing down of thermals due to poor hydro-thermal mix.

With more and more thermal plants coming up, the consumption of coal is enormous; 240 million tonne which is expected to rise to 407 million tonne annually by 2006-7. At the present rate of consumption, the coal deposits of 211 billion tonne (out of which about 180 billion tonne are of non-cooking variety which fuel the power stations) may last hardly two centuries.

Thermal plants are, however, suitable at pit-heads to minimise transportation of coal and its burden on the railways and to reduce cost of generation with a stronger generation base.

The thermal power generation has many implications in terms of environmental problems of disposal of ash–about 75 million tonne are currently being produced, economics of coal transportation to distant power stations, thermal efficiency and performance of boilers and cost of electricity generated. The coal fired plants employ pulverised coal firing. The high ash content and erosive nature of ash leads to fast wear-out of grinding components of the coal pulverisers and erosion of coal conveying pipes, the burners and the heat transfer surfaces in the boilers. The high ash content in the flue gases and low conductivity of the ash also leads to the problems related to collection of fly ash into the electrostatic precipitators. The power plants also entail release of waste heat released to the atmosphere from the stack with flue gas and heat rejected by cooling tower through evaporative cooling of circulating water and through cooling tower blow down water.

Advantages

(*i*) Shorter gestation period, (*ii*) Located near the load centre and as such long transmission lines and transmission losses are avoided, (*iii*) Low initial cost compared to that of hydropower which requires huge investment on the construction of dam and appurtenant works, besides the power plant, and (*iv*) High capacity rating machines.

Disadvantages

(*i*) The speed of output variation of a large capacity, high efficiency thermal is generally low; 3 to 5 MW/min. in a 500 MW unit, (*ii*) Thermal loss occurs when starting and stopping and several hours are required for restarting; at times there are unstable conditions of output declining to less than 50% of the unit capacity, (*iii*) Serious pollution and ecological hazards are associated with it, (*iv*) Complicated thermal equipment, (*v*) Poor industrial relations in generating stations and complicated operation and maintanence of thermal plants, (*vi*) Emissions to the atmosphere of greenhouse gas which may lead to important changes and also have consequences on the hydrology and hence on water supply and agriculture and on the sea level, (*vii*) Long distance transportation of coal and high cost of mining, and (*viii*) Limited operational life of thermal plants; hardly 25 years.

1.5. NUCLEAR POWER

Power generation is an important use of nuclear energy. The first commercial nuclear power plant was commissioned in U.K in 1956 and in USA in 1957. India's first nuclear power plant was operated at Tarapur in 1969 which was optimistically viewed as opening of a major new source of economical and clean electricity supply. A nuclear plant, like a thermal plant, uses steam to generate electricity but not by burning coal but is fueled by uranium, an ore mined from the earth and refined. In simple terms, the process is to use the energy from the fission of the atom which releases energy so intense that a uranium fuel pellet of the size of a large pencil eraser can produce as much heat as a tonne of good quality coal. This heat is removed by suitable coolants which in turn exchange the energy picked by the coolant to convert water into steam, which run turbines to generate electric power. After turning the generator, the steam is converted back into water in a condenser system and is returned to the steam generators where the cycle starts again.

The nuclear fission is accompanied by release of ionising radiation. The nuclear plant concentrates great amounts of radioactivity in one place, creating a potential threat to the surrounding environment if accidentally released from the plant. However, there are many physical barriers besides backup equipment systems that go to work any time the normal operation of the plant is disrupted. Moreover, the entire plant is monitored and supervised from a control room by highly trained operators. A most controversial issue related to nuclear plant is how to dispose of used fuel and other radioactive materials. About a third of the fuel assemblies in a reactor have to be replaced every year. After the fuel is removed, it is stored in water pools. Most of the radioactivity decays rapidly, but a significant fraction of the high-level radioactivity remains in the fuel and can stay active for many thousands of years. Fig. 1.1 shows cross section of reactor building.

The nuclear power plants handle large amount of cooling water for condensers and other auxiliaries as also for the heat transport system between the reactor and the steam generator. However, the boiler needs are very low compared to cooling requirement.

Advanced countries now depend fully on nuclear power. In France 80 percent of national power generation is by nuclear, 40 percent in Sweden and 30 percent in Japan. India has been lagging behind and the share of nuclear power is rather low at 2.21 percent. Nuclear power is more scientific, economical, reliable and ecologically more acceptable and increasingly important for sustainable development. Tarapur power plant generates electricity at a nominal 83 paise a unit, the lowest cost per unit. With the plant load factor increasing, the cost of nuclear power is expected to be much less.

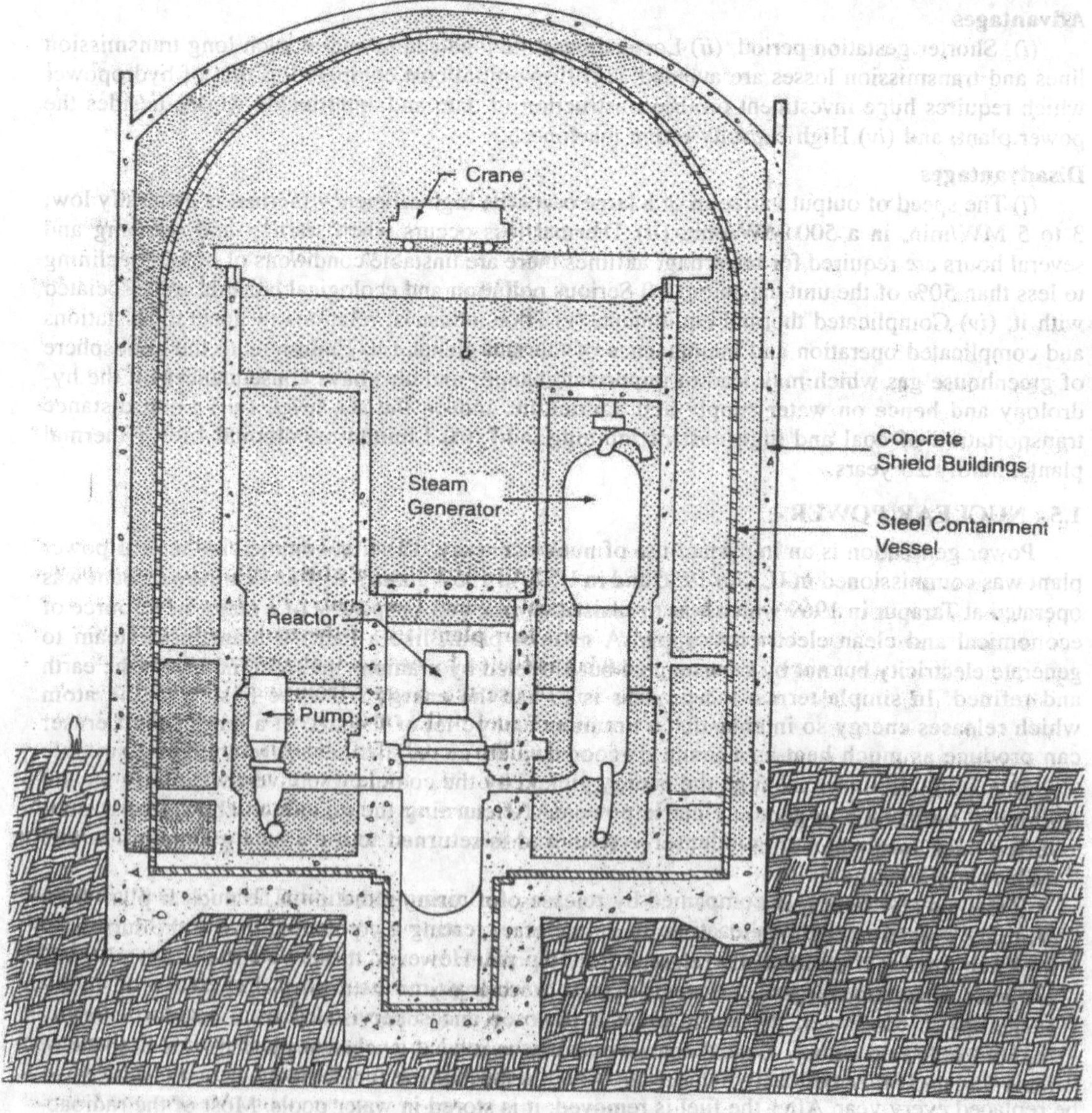

Fig. 1.1 Cross section of reactor building.

1.6. TIDAL POWER

Tidal power is generated by harnessing energy in the tides of oceans i.e. tidal range or the difference in head during the rise in water level during the tidal period and falling during the ebb period. Sea water rises and falls twice a day each full cycle occupying about 12 hours 25 minutes. The tide can be appreciably high. The quantum and intensity of generation of tidal power can be predicted as tidal timings are published in advance, called the tide tables.

Tidal power is generated by the flow of water from higher to lower levels through turbines. The principal components of a tidal power plant are power house, dykes (embankments) to form basin or basins and sluice ways from the basin to the sea and vice versa (Fig. 1.2). During high tide the gates are opened and water from sea enters the basin and fills it and in the process turn the turbines to generate power. During falling or ebb tide, water flows back to the sea turning the turbines and generating electricity. The same set of turbines may produce electricity in high and low tides if designed to generate power in either direction. The power generation is not constant for 24 hours, rather in a regular interval of 4 hours, 4 times a day in a specific pattern. The sites suitable for tidal power generation are essentially to have sufficient tidal volume and the available mean head difference.

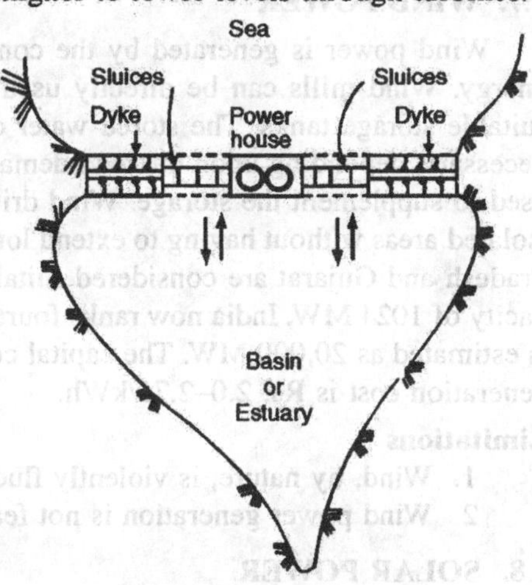

Fig. 1.2 Tidal power plant

The installation capacity of the plant is fixed by economic consideration rather than by available flow.

The first project of 240 MW power generation from tidal energy was constructed in France at Rance Estuary in 1967. In India, tidal power generation is envisaged in gulf of Cambay near Bhavnagar (5000-6000 MW) where the tidal range is of the order of 11-12 m, besides that in Gulf of Kachchh (900 MW) and Sunderbans (3000-4000 MW). Total tidal power potential is estimated as 10,000 MW.

The turbines may be of steel or fibreglass. The steel turbines are costly and render the electricity prohibitively expensive. The helical turbine made of fibreglass (Fig. 1.3) is cheaper to develop and run. Unlike hydraulic turbines at dams which are driven by high pressure water, the helical turbine uses the low pressure, free flow action of the ocean to generate power. While passing over the blades water spins the turbine on an axle perpendicular to the flow, driving an electricity generator. The light weight machine spins very quickly in relatively slow-moving sea water.

Electric generator

Fig. 1.3 Helical fiberglass turbine.

Advantages

(*i*) Power generation is pollution free, (*ii*) Power is firm as the output can be predicted in advance, (*iii*) Power generation is not susceptible to vagaries of nature such as failure of monsoon. There are no wet or dry seasons for any summer, winter and monsoon on the availability of tidal energy. The inequalities due to spring and neap tides are known or can be accurately predicted sufficiently in advance, (*iv*) It is everlasting, (*v*) It does not involve any extra submergence or uprooting of population, and (*vi*) There is no loss of forest cover and dislocation to wild life.

Disadvantages

(*i*) High initial cost, (*ii*) Low head as tidal range is limited, (*iii*) Variation in the power produced both in time and magnitude. While power consumption is based on solar cycle, the power production is as per the lunar cycle, (*iv*) The turbines have to work on a wide range of head variation resulting in low efficiency, (*v*) Equipment is susceptible to corrosion with the corrosive sea water, and (*vi*) Fishing and navigation may be affected.

1.7. WIND POWER

Wind power is generated by the conversion of energy available in the wind into electrical energy. Wind mills can be directly used to pump water to a height where it can be stored in suitable storage tanks. The stored water can be released to drive the micro hydel set, whenever necessary, depending upon the load demand. The water pumped during light load periods can be used to supplement the storage. Wind driven electric generators help in electrification of remote isolated areas without having to extend long, uneconomic transmission lines. Tamil Nadu, Andhra Pradesh and Gujarat are considered suitable for wind power generation. With a wind power capacity of 1024 MW, India now ranks fourth in the world. The wind power potential of the country is estimated as 20,000 MW. The capital cost of wind energy is about Rs. 50 million/MW and the generation cost is Rs. 2.0–2.75/kWh.

Limitations

1. Wind, by nature, is violently fluctuating in its speed as well as in its duration.
2. Wind power generation is not feasible in calm localities.

1.8. SOLAR POWER

Energy from the giant power house in the sky, the sun, offers a promising source of power generation in the 21st century. India receives 5000 trillion kWh solar energy annually, sufficient to generate 20 MW solar power per sq. km of land area. Research is being conducted for providing low-capacity solar-powered electrical generators for specific consumers and the other for large-scale solar power stations.

Solar photovoltaic and solar-thermal technologies are finding ready acceptance for a variety of industrial and thermal applications as well as in the rural unelectrified areas. Solar photovoltaic is capital intensive and rate as low capacity device. Solar power in the solar-thermal application has developed to a stage of commercial operation. The country has the world's largest deployment of solar photovoltaics comprising about 5,00,000 pv systems aggregating to 35 MW and covering over 30 different applications. The capital cost of solar power is Rs. 30 million/MW and the generation cost is Rs. 10–12/kWh.

1.9. GAS TURBINE GENERATION

In view of the need for accelerated capacity addition to tide over the problem of power shortage, emphasis is, of late, on the installation of gas turbine generators, both open cycle and combined cycle.

At the end of March, 1996 there were 23 gas turbine (GT) stations with a total capacity of 5064 MW. Over and above these, there were mobile GT units of 81 MW in Assam.

1.10. GEOTHERMAL POWER

India's new emphasis on multidimensional development of non-conventional energy may accelerate rapid utilisation of available geothermal fields. The country's geo-thermal fields are of intermediate to low grade potential temperature exceeding 100° to 150°C at 2-4 km depth whereas deep seated reservoir temperatures are between 100° to 250°C compared to successful countries like New Zealand, USA, Itlay, Mexico and Japan. The total stored heat potential in the country is about 40×10^{18} cal, which is equivalent to 27.6 billion barrels of petroleum. The total geothermal power potential assessed by the geoscience organisations is of the order of 2,000 to 10,000 MW. The most promising geothermal areas are Puga-chhumathnng, Manikaran and Tapoban in Himalayas, Konkan, Cambay and Bombay offshore, Taptapani (Orissa), Gondwanic grabens, and volcanic areas of Andaman-Nicobar chain.

Deep boreholes can be successfully utilised for geothermal power generation by hydrofracturing and using commercially available single bore hole geothermal energy extraction system technology.

1.11. BIOGAS

It is a clean, cheap and non-polluting fuel produced from organic material like cattle dung, human waste and different types of biomass. Biogas plant produces quality organic fertilizer besides other social benefits such as reduction in felling of trees for firewood, easing the drudgery of women, decline in incidence of eye and lung diseases, improvement in environmental sanitation, and employment to rural artisans. However, biogas is only a medium term policy dependent on weather and water besides other factors. Its production falls with a fall in temperature and water availability which is one of the prime requirements of the biogas plants.

Biomass, a potential source of renewable energy for use as solid, liquid and gaseous fuel, includes animal waste, forest waste, crop residues, firewood trees and shrubs besides aquatic biomass covering aquatic plants, sea weeds, micro-algae, water hyacinth and other marine floating plants. Biomass research centres located in six different agro-climatic regions are developing new plant species. The two major features of biomass utilisation are biomass briquetting and biomass gasification. Under the former, agricultural are estimated at 145 million tonnes besides forest residues could be converted into briquettes to generate 14,000 MW of power. Under the gasification programme, 12 designs of biomass gasifiers are developed generating thermal energy for industrial applications, for water pumping and for power generation. A 500-kW capacity biogas gasifier system for power generation has been installed in Gausaba, Sunderbans island. About 2.75 million biogas plants are in use in the country.

The capital cost of biomass power is about Rs. 40 million/MW and generation cost is about Rs. 1.75–2.0/kWh.

1.12 ENERGY CONSERVATION

Energy saved is the energy produced – a very wise saying – is very apt. in the current scenario of acute electricity shortage in the country. A watt of electricity saved is equal to 2.3 watts of generated power. Furthermore, it costs about Rs. 5 crore and 5 years to create 1 MW of new generation capacity with additional expenditure on transmission and distribution system but the same amount spent on energy conservation can provide 2-3 MW of avoidable generation capacity and yields results in 1 year. In USA, for instance, power companies prefer giving incentives to consumers to use energy saving devices rather than enhance generating capacity.

Energy conservation is a matter of pure economics and is a vital factor to salvage the situation of acute power shortage the nation is plagued with. The spiralling fossil fuel costs, dwindling reserves and the ever increasing demand for electric power is causing depletion of scarce non-renewable energy sources such as coal, gas and fossil fuels etc. besides polluting the environment.

The economic growth of a country, developed or underdeveloped, is essentially related to the energy availability which in turn is dependent on the energy efficiency and conservation.

Electricity, the most widely used energy input in India, has lowest efficiency and the highest percentage of losses and wastes. Energy efficiency improvement and conservation, briefly as under, can bring about a huge saving in energy:

1. Power system

(*i*) *Generation Efficiency*: The huge gap between demand and supply is due to low generation efficiency of the power stations. The average power load factor across the country is about 60 per cent. The low operational efficiency is mainly due to the poor operation and maintenance of power stations. The eco-friendly renovation/uprating of the existing machines at 20 per cent of new capacity addition, modernisation and life extension exercises can help improve generation efficiency.

(*ii*) *Transmission and Distribution Losses*: The huge transmission and distribution losses put avoidable pressure on the existing infrastructure. The high losses are due to non-commercial losses or power theft. The technical losses may be due to low voltage transmission, inadequate distribution network, sparsely distributed loads over rural areas, poor maintenance of the transmission lines and equipment. With better management and control, the losses can be reduced appreciably at minuscule investment.

(*iii*) *Improvement in Power Factor*: The loads in the distribution systems are mostly inductive and draw a huge amount of reactive power from the grid. It adversely affects the power factor of the system and as the power factor falls, the distribution losses in the grid increase exponentially. Reactive power consumption and better load management can solve this problem. Installation of shunt capacitor banks near the load centres are a viable and cheap option, considering the incremental benefits as a result of reduction in line losses.

(*iv*) *Energy Auditing*: The energy auditing is essentially required to be taken up in order to zero in on the potential energy conservation areas, especially at the micro level where industry-wise energy accounting can go a long way in identifying "energy expensive" activities. About 10 to 15 percent saving in energy consumption can be effected by better housekeeping and retro-fitting measures.

(*v*) *Hydro*: *Thermal Mix*: The current hydro thermal mix of 25 : 75 needs to be improved to bring in more of the cheap hydro power for long-term sustainability and energy conservation.

(*vi*) *Tariff Adjustment*: Tariff adjustment is an important measure extensively used in the West to induce energy-efficient habits in the consumers. The tariff at all times should penalise inefficient utilisation of energy. First and foremost, tariff needs to be rationlised to reflect the true cost of generation and transmission and distribution plus some returns for the intermediaries.

Tariff variants could also be used to flatten the daily load curves. Time-of-day tariff could be one such technique. This tariff structure could involve high rates at peak hours, thereby discouraging wasteful energy use during such times.

2. Consumption system

The largest end users of the electricity in the country are as given in Table 1.2 and the effort at conservation can be focused accordingly :

Table 1.2 End users and electricity consumption

End user	*Percentage consumption*
Industrial motors	38
Agricultural pumps	30
Lighting	20
Domestic appliances	12

(*i*) *Industrial Consumption*: Electric motors account for two-third of industrial electricity consumption. Substantial amount of electricity can be saved by using energy efficient motors, properly sized and placed as close to load as possible, run motors/equipment in a staggered manner so as not to affect the process and at the same time keeping the maximum demand within limits and by using transformers with less no-load losses and operating them at high efficiency in the range of 75-80% of their rated capacity etc.

Shortage of power in the country is being made up through power capitive, using imported diesel, by industry to the extent of 25% consumption is at a very heavy cost to the nation, not only a heavy drain on foreign exchange but also makes a big dent in the revenue of state electricity boards which earn mainly from the industrial sector. Besides, through captive power evasion of taxes is very easy and this invisible loss may be very colossal.

(*ii*) *Agricultural consumption*: Agricultural in USA accounts for only 10 per cent of the total power which is 30 per cent in India. More than 12 million pumping sets in agriculture sector consume 6000 million kWh of power, highly subsidised or provided free through populous political trends. This is the biggest potential area for energy conservation because at present there is no incentive for farmers to conserve energy. A saving of about 30 per cent in power consumption can be effected by (*i*) Using high efficiency pumps with proper alignment of the pump and motor, (*ii*) Providing shunt capacitors on the motor terminals to keep voltage conditions stable and avoiding damage to the motor, (*iii*) Improving operational and maintenance practices, and (*iv*) Providing metered subsidised supply instead of free to encourage conservation of power.

(*iii*) *Lighting consumption*: The domestic lighting accounts for about 20 percent of the total electric supply and essentially it is another critical area for energy conservation. Use of tubular fluorescent lamps and mercury vapour lamps consume 80 percent less electricity than ordinary incandescent bulbs and last 10 times longer. Power factor corrected fluorescent tubes need to be provided in all new installations. The conventional ballast has a power factor of 0.5 as against 0.95 of the power factor corrected electronic chokes. Bombay alone has about a million lamps and if replaced by CFL, would save 10 MW. A comparison of light output between ordinary bulbs and compact fluorescent lamps (CFL) is as under:

Bulb (watt)	40	60	75	100	125
CFL (watt)	9	11	15	20	23

The high initial cost difference is paid back in a very short period of time besides conservation in power consumption and saving in replacement labour. Likewise, the rivals constituting two-third of the population need to be educated to switch over to the new energy-efficient lighting equipment to consence electricity.

3. Non-Conventional Energy Development

Small hydro, solar PV, bio-mass wind energy and cogeneration power plants-popular in sugar producing areas of the country-developed at an accelerated pace will go a long way in easing pressure on the limited fossil fuel reserves.

4. Multiple Time Zones

To reduce the gap for peaking power many countries have multiple time zones; USA 4, Russia 9, to make use of sunlight. Tata Research Institute has estimated a saving of Rs 400 crore a year with multiple time zones.

Realising the need for better quality power for the overall economic development, unbundling of the transmission and distribution in a time bound manner, 100% electrification of the villages and energy auditing at all levels are essential for the modernisation, upgradation and conservation of the electricity.

QUESTIONS

1. Discuss renewable and non-renewable energy sources.
2. Give a brief resume of conventional and non-conventional energies.
3. Give a brief account of hydro, thermal and nuclear power.
4. Discuss tidal power with emphasis on its merits and demerits.

2

ENVIRONMENT AND WATER POWER DEVELOPMENT

2.1. INTRODUCTION

Global environment consists of four main components viz., (*i*) land, (*ii*) water, (*iii*) air and (*v*) plant and animals including human beings. Another external component of the environment is the perennial energy that it receives from the sun. Thus the environment is the totality of all extrinsic and intrinsic factors–both physical and biotic-that effect the life and behaviour of all living things including man.

Water is a major component of that environment in which man occupies the centre-stage. Man is an integral part of nature and his interaction with the fragile ecological systems which surround him is important for his continual survival. Functional relationship exists between the environment, its ecology and man. Ecology is the study of inter-relationship between living things and the environment.

Environmental science represents the integrated essence, which is holistic and not reductionist, to look at the process of development and its effect on environment. This integrated essence of environmental science encompasses chemistry, biology, geology, meteorology with the disciplines associated with human development like economics and social science. The global dimensions and the implications of environmental issues are global warming, ozone depletion, ever-increasing threat of pollution and desertification etc.

2.2. CONCERN ABOUT ENVIRONMENT

Concern about environment dates to vedic period. The Atharva Veda says in a hymn to Mother Earth "what of Thee 1 dig out, let that quickly grow over. Let me not hit Thy or Thy heart." The recent concern arises from the environmental degradation that the country has suffered so far due to human pressure and over-exploitation of resources. There have been a considerable degree of deforestation, depletion of wildlife, degradation of land, soil erosion and pollution of the air and water. Taking note of the general concern for environment enshrined in the Directive Principles of our Constitution, Government of India has set up a Ministry of Environment and Forests for examining water resources and power projects from the environmental angle.

The term "sustainable development" was brought into common use by the World Commission on Environment and Development (1987). It was defined as the development which meets the needs of the present generation without compromising the needs of future generation. One basic premise for sustainable development is the recognition that environment and development are complementary and interdependent and in long run mutually reinforcing.

In the present scenario of economic reforms and liberalisation, concern for environment is rapidly increasing. The development of hydropower world-wide has progressively responded to valid criticism namely displacement of population, actual rate of sedimentation higher than that anticipated in design, watershed management for soil conservation, microclimate changes, endangered flora and fauna and protection of bio-diversity, insufficient provision for social and economic needs of those displaced. It brings out that environmental implifications of hydel projects are subjected to impact assessments to a satisfactory standards.

2.3. ENVIRONMENTAL PLANNING AND MANAGEMENT

1. Environmental planning

Environmental planning is absolutely essential to achieve long-term sustainable development, economic and social development. Short term gains, at the cost of natural environment, may be tempting but can end up only in disaster. The real cost of growth would be very high when irreparable damage is caused to fragile ecology. The various aims of environmental planning are (*i*) Protect environment, (*ii*) Assume safe, healthy, productive, aesthetically and culturally harmonious surroundings, (*iii*) Attain widest range of beneficial uses through sustained development with minimum environmental degradation, (*iv*) Prevention of long-term environmental side effects by incorporating mitigative measures so that the remedies do not become unimaginable and prohibitive in the long run, (*v*) Preserve important historic, cultural and natural aspects of national heritage, maintain environmental diversity, and (*vi*) Enhance quality of renewable resources, recycle depletable resources.

Long term planning can take care of all environmental dimensions like minimum flows in the rivers, sustainable water and land-use management, afforestation, and better soil management i.e. watershed treatment programme to mitigate the adverse effects of deforestation, resettlement and rehabilitation of the population displaced due to project and even eliminate adverse impacts and enhance positive impacts in the interest of sustainable development.

2. Environmental Management

The environmental management encompasses (*i*) Environmental impact assessment (EIA), (*ii*) Environmental and Forest clearance, (*iii*) Environment management plan, (*iv*) Rehabilitation and resettlement, (*v*) Sedimentation, and (*vi*) Watershed treatment.

The environmental study concurrently with the feasibility study is very essential to assess and analyse the positive and negative impacts of the scheme qualitatively and quantitatively. The selection is made which may have substantial positive environmental impacts and limited adverse impact, also stipulating the implementation of environment management and mointoring programme. The environment impact assessment (EIA) is all the more necessary in the case of schemes with large capital outlay. A balanced approach for the socio-economic needs and the environmental concerns should be adequately assessed, weighted and accommodated in the project approved and taken up for execution after securing a number of statutory and non-statutory clearances from central and state governments.

The Ministry of Environment and Forests has made submission of Environment Impact Assessment statement and Environment Management Plan mandatory, without which no project is cleared for investment.

2.4. ENVIRONMENT AND HYDROPOWER DEVELOPMENT

The legacy of environmental issues of the 20th century has essentially forced out into the new millennium. Environment conducive development in the 21st century is a major challenge for the nation. An increasing constraining factor in hydropower development has been the conflict between the expanding electric energy demands and the concern for environmental effects of dams, reservoirs, tunnels etc. associated with the water power projects. The negative impacts on the environment of hydroelectric plants are sometimes over emphasised. They generally come from minority groups in industrialised countries-North America, Europe-where the hydroelectric potential has largely been exploited. Hydro electrical potential is wealth. Its development, with the strong backing of the state, contributes to the well-being of the population. The works performed are then part of its heritage.

1. Conventional Environmental Concerns

Conventional hydropower generation uses a renewable energy source and constitutes about 22 percent of the country's annual output of electricity and about 20 percent of electricity generated worldwide. To provide a still more significant contribution to sustained development keeping in view that 78 percent of the country's potential yet remains to be tapped, the hydropower development must address a variety of environmental concerns.

Hydropower plays an important role in a variety of long-term scenarios for sustainable development. The hydroelectric projects not only provide electricity but also multi-purpose benefits like inland navigation, reservoir fishery, incidental irrigation, recreation, flood control etc.

The hydropower is recognised as the cheapest and reliable supply of power in India; its economic benevolence has of late been enhanced due to steep rise in energy costs from fossil fuels and imported oil. In both short term and the long term, further development of conventional hydroelectric energy generation can increase clean, sustainable energy production and make an important contribution to the reduction of greenhouse gas emissions. Hydropower produces no carbon dioxide, sulphur dioxide, nitrous oxides or other air emissions besides no solid or liquid wastes disposal problem. Some contend that new large dams may be arguably the worst electricity option in terms of damage to ecosystems per unit of electricity. But it is not factual.

The environmental impacts of hydro power development projects are related differently to storage projects and small hydro schemes, briefly as under :

Storage Power Projects

Major impacts on environment due to hydel projects in the developing countries like India include submergence, land loss and change in use, dislocation of native/tribal people and their forced resettlement away from their area, potential for outbreaks of water-borne diseases, siltation, reduction of fisheries, micro-climatic changes and intensification of regional water rights conflicts. The negative effects on ecosystems include modification or destruction of aquatic and terrestrial habitat, interruption of daily and seasonal stream flows, alteration of seasonal temperature patterns, reduction in sediments downstream, diruption to the natural flow of organic materials and nutrients through the aquatic ecosystem, and a consequent decrease of biodiversity. The impoundments and releases from hydropower facilities may have adverse impact on the aquatic life upstream, downstream and fish passing through the turbines, besides others mentioned below:

(*i*) **Reservoir:**

Water storage reservoirs have to be created to ward off shortage of water in the event of monsoon failure as also to meet the demand during dry season. The present live storage, about 174 billion cu m is nearly 10% of the total water potential available in the country. The construction of large storages in the Himalayan belt ranging from Jammu and Kashmir in the west to Arunachal Pradesh in the North-East is important to achieve the additional generation of hydropower by the end of the Eleventh Plan (2007-12). The environmental aspects of the reservoirs are briefly as under:

The stilling of river flow in the reservoir does bring about changes in the ecosystem e.g. temperature stratification, variation in nutrient content and dissolved oxygen at different levels etc. This may have adverse effect on aquatic and riparian life which originally existed in the project area. Life forms new to the area may invade to the disadvantage of original life forms. Pollution from settlements and siltation also affect water quality. However, impoundment of water in hydro-storage reservoirs has beneficial effects on water quality due to settlement of suspended materials thereby decreasing turbidity. The construction of a dam undoubtedly would result in the change of environment but it will not be correct to presume that the construction of a large dam will result in ecological disaster. Supreme Court of India in a recent judgement (October, 2000) has ruled that construction of big dams cannot be equated with setting up of polluting industries as far as their effect on the environment is concerned. The construction of a dam undoubtedly would result in the change of environment but it will not be correct to presume that the construction of a large dam will result in ecological disaster.

(*a*) **Submergence:** The submergence of the area and the creation of large water body, dislocation of human habitats, micro-climatic changes, alteration in flora and fauna, and loss of timber and species diversity are the major ecological impacts of a reservoir. On long term basis, reservoir environment may be better than degraded forest land.

(*b*) **Water Quality:** Reservoir water quality is affected depending on retention period, reservoir soil and area flooded if large compared to annual volume of water passed, especially where the reservoir area is covered by dense tropical forest, a very large amount of biomass is submerged. The problem tends to be more serious in reservoirs in moist tropical regions than in dry or temperate regions. Pre-impoundment clearing of the reservoir area is an important preventive measure, besides pre-treatment of sewage before outfalling into reservoir.

Hydropower is effected by water quality which includes reduction in oxygen content and gas release (methane, sulphuretted oxygen). Large amount of oxygen is consumed by anaerobic decomposition of inundated vegetation and produces noxious gases that are toxic to aquatic life and harmful to power plant machinery.

Water quality is also deteriorated by water and bottom sediments becoming enriched with nutrients; the process called 'eutrophication' is accelerated by the wide spread use of fertilizers, detergents and growing quantities of waste water. The excess nutrients produce a bloom of plankton, algae and weed growth which when die and rot use up the available oxygen at depth.

(*ii*) Fish Production

Free passage of migratory fish to and from spawning grounds is disturbed due to the obstruction caused by a dam. Slow water flow can lead to thermal stratification with warm water on top and cold water underneath. As cold water is not exposed to the surface, it loses oxygen and becomes uninhabitable for fish. A continuous release of a small quantity of water from deep reservoir layers at the foot of dam creates a constant inflow of oxygenated water to the deeper layers.

Stones and soils contain naturally occurring mercury and other metals. When land is inundated, mercury dissolves in the water and passes into fish and other species and thence to human being, thus fish lead to mercury poisoning.

Stagnant water promotes water hyacinth at the cost of native species. The proliferation of water weeds affects water quality as also cause water loss through evapotranspiration. Air bubbles are forced into the water when cascading over a spillway. These bubbles can be absorbed into fish tissue, ultimately killing them.

Dams often block fish migration resulting in fisheries loss and reduced fish diversity. However, fresh water reservoirs provide significant opportunities for the development of sport and commercial pisciculture like that at Bhakra and Pong dams which are recognised as fish bowl of the country, besides providing employment to the people living along the periphery of the reservoir.

(*iii*) Sedimentation and Erosion

Sediments are trapped in the reservoir usually at an accelerated rate than the estimated volume of natural silt transported by river at the design stage causing early termination of the life of reservoir and interference with turbines and other hydropower machinery (Fig. 2.1). The storage is also impaired by erosion of reservoir banks due to shore exposure, wind velocity and duration, reservoir water level fluctuations, human activities etc. causing land slips, landslides in steep valleys due to reservoir filling bringing excessive sediments into reservoir. Nevertheless, the sedimentation has positive aspect too. The sedimentation caused by erosion of sediments at water's

Fig. 2.1 Silting in pong reservoir

edge increases the capacity of reservoir's active pool because the material so eroded deposits in dead storage zone (Fig. 2.2.).

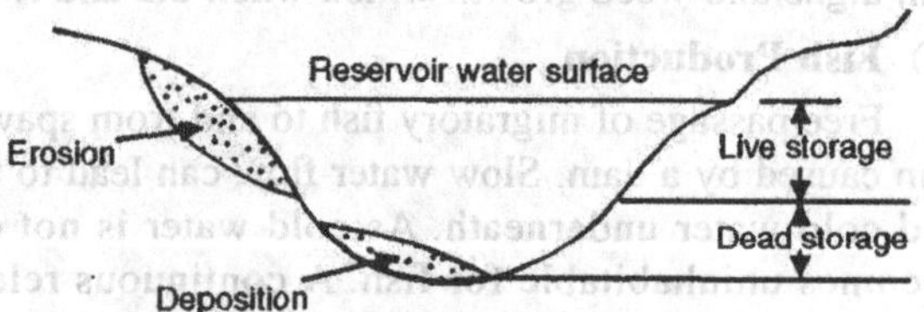

Fig. 2.2 Shoreline erosion and sedimentation.

The soil erosion and hence sedimentation is also accelerated due to cultivation on marginal land, unplanned and illegal logging resulting in denudation etc.

The power generated by hydel schemes can obviate the need of felling of trees and help restoring the environment. Each MW of power brings about enormous amount of increase in agriculture and industries, promotes forests by saving fuel wood and thus creates environmental balance.

(*iv*) Flora and Fauna

The flora and fauna surviving in the catchment area of the reservoir may be disturbed. The adverse effects are removal of feeding areas, loss of habitat, limitation of mobility, loss of wild life, impact on endangered species and reduction of biological diversity. These effects are considerably reduced by proper planning, rather improved. Brindraban and Shalimar gardens, Ukai and Periyar wild life resorts etc. are direct by products of various hydro related projects.

The reservoirs attract a steady stream of resident and migratory birds. During winter the water bodies are important oasis for winged tourists. In addition, efforts are made by the project to intensify the protection measures and also to improve the habitat along and below the reservoirs besides setting up of bird sanctuaries.

The treatment of degraded forests and the development of compensatory afforestation not only compensate for the submergence of forests in the catchment area but also greatly enhance the environmental quality of the affected area.

(*v*) Reservoir Induced Seismicity

The risk of reservoir seismity is no longer contentious and environmental issue because the design takes care of likely impact of reservoir loading on the geological faults by specifying the safe rate of the reservoir filling to ensure that the induced stresses and strains are within the acceptable limits. The modern techniques in design and construction specifications ensure considerable strength to withstand severe earthquakes. The reservoir induced seismicity (RIS) is found to decrease with time and terminate after a couple of years. Geological formations like sandstone, claystone rock formations which can absorb stress and deform without fracturing as at Bhakra and Pong dams do not favour accumulation of stresses and, therefore, preclude occurrence of RIS.

(vi) **Climate**

Large reservoirs affect the local climate by creating new microclimate with modified or increased rainfall pattern. The observations at Bhakra and Pong reservoirs have shown that there is a gradual but marked change in the microclimate around the reservoir area. The maximum atmospheric temperature has become lower in summer months and risen in the winter months. Likewise, the minimum temperatures are higher almost throughout the year after the formation of the reservoir. Even in Thar desert of Rajasthan, the command area of Pong dam, the development of greenery has improved microclimate and ecology of the area.

(vii) **Flood Protection**

The storage hydroelectric plants serve as regulatory factors for the flow of streams and rivers and, therefore, act as protection against devastation caused by floods in the downstream area.

(viii) **Rehabilitation and Resettlement**

Submergence is caused by storage power plants which constitute less than 40 percent of the total number of hydropower schemes identified. However, storage plants have become synonymous with submergence and displacement of the population from the submerged area as also land acquired for appurtenant civil works which has become a very emotional issue as it leads to social, occupational problems apart from cultural, psychological and personal hardships generally owing to resettlement in area where climate, environment and customs are entirely different. With a view to minimise them, the areas selected for the resettlement should, as far as possible, be under similar environmental, climatic, topographic and occupational pattern. The special characteristics, life style, cultural norms of tribals and adivasis should necessarily be preserved. Not only their cultural values are preserved, they should be better off in the resettled area as a result of humanitarian consideration in compensation, liberal grants and socio-cultural specific resettlement housing units and other medical, educational, communication and occupational amenities. Presently, the most important and emotive environmental issue in the country is the submergence of land and consequent population displacement.

In the revised policy, the Government of India has related land acquisition to rehabilitation and resettlement. The rehabilitation is essentially to provide services and facilities which will restore to communities, families and individuals their normal living standards while at the same time encouraging adjustment to the drastic changes. They should be able to enjoy the standard of living comparable with people benefiting from the project. The resettlement has to be consistent with the desire of each community to preserve its own identity and culture.

2. New Environmental Concerns

The adverse environmental impacts of conventional hydropower generation can be reduced and hence its acceptance as a source of renewable energy with an important role in sustainable development can be enhanced through new technologies which are emerging the world over.

The major new environmental concerns are improvement of dissolved oxygen levels and the provision of minimum flows to protect aquatic habitat in tailwaters and greenhouse gas emissions from large reservoirs. The typical environmental parameters include water tempertures, incoming DO values, downstream DO values, total dissolved gas, multiple differential pressures and corresponding air flow rates, oxygen flow rates (where appropriate).

The five key indicators used by TVA (USA) to monitor and rate overall ecological health of a reservoir are as given in Table 2.1.

Table 2.1 Key indicators

Sl. No.	*Key Indicator*	*Good rating means*
1.	Dissolved oxygen	Enough oxygen throughout the water column to support a healthy population of fish and other aquatic life.
2.	Algae	Enough nitrogen and phosphorus to stimulate the growth of algae (the beginning of the reservoir's food web), but not so much that levels become excessive.
3.	Sediment quality	Reservoir bottom is not polluted by chemicals.
4.	Bottom life	A variety of animals like worms, insects and snails live on reservoir bottom.
5.	Fish	There are a large number and variety of healthy fish.

(*i*) Increasing Dissolved Oxygen

Improved hydro turbine designs help to enhance DO concentration in releases from Francis-type turbines. Most cost-effective auto-venting or self aerating technologies use the low pressures created by flows through turbines to induce additional air flows. At Norris project (USA), turbine aeration is typically initiated in July, when the DO level monitored upstream from the turbines begins to drop till November when cold, dense surface water promotes enough vertical mixing to reduce the thermal stratification. Various combinations of auto-venting technology options are used, throughout the low DO season, based on the head, power and required DO uptake. Specially shaped geometries for turbine components ensure enhanced low pressures at appropriate locations, allowing the air to be drawn into an efficiently absorbed bubble cloud as a natural consequence of the design and minimising power losses due to the aeration. The new turbines also show significant reduction in both cavitation and vibration. With careful monitoring of operating conditions and environmental conditions, individual operator judgement and appropriate optimisation software are utilised to balance the energy needs and the environmental needs.

(*ii*) Green house Effect

The hydroelectric reservoirs are considered to emit large quantities of greenhouse gases, (GHGs) mainly CO_2 and CH_4 but these compare favourably with emissions by fossil-fuelled electricity generation. The green house gases are produced during bacterial decomposition of flooded peat and forest biomass. The amount emitted is positively related to the area flooded. World Commission on Dams in its report has stated that the greenhouse gases are emitted for decades from all reservoirs in the boreal and tropic regions for which measurements have been made. This is in contrast to the wide spread assumption that such emissions are zero. A number of factors influence the greenhouse gases emissions from reservoirs including the inflow of carbon from the surrounding catchment area but it can not be directly explained by the volume of submerged biomass or its carbon content. The report reveals that dam intervenes in, and modifies, the natural carbon cycle within the catchment. The dam serves to store this carbon and interrupt its natural flow to the sea or to the atmosphere. However, it is recognised that while dams may produce net greenhouse gas emissions, the time scale over which gases are stored and emitted and the quantity of gas released varies considerably for each reservoir. The commission appreciated that the release of GHGs should not detract from how dams compare to the other options available to any country.

Luc Gagnon, an advisor on climate change for Hydro-Quebec in Canada, has emphasised that despite these findings, hydropower is still a powerful tool to reduce GHG emissions.

(*iii*) Increasing Fish Passage Survival

Provision of minimum flows in the river channel below the dam and fish-friendly operation of turbines help in development of suitable varieties of fish in the downstream river channel and long canal system. Certain species of fish live on the weed growth generally found in canals, the introduction of which may check weed growth.

(*iv*) Managing Sediments

The degree to which silt, sand and other solids pose challenges to the power plants varies according to the location and configuration of the facility. Generally sedimentation limits operating effectiveness and increases maintenance costs. The methods range from major dredging plans that seek to minimise sediment problems while optimising efficiency. The sediment management plans include measures that enhance environmental and other non-power benefits of hydro projects, mainly management of watershed area.

In conclusion, it is possible to harmonise the implementation of a hydro plant with conservation of the environment and with regard for people affected by a project.

2.5. SMALL HYDROPOWER SCHEMES

Compared to storage based hydel projects, small hydropower schemes are considered to be less destructive to aquatic life, and cause less damage to the general environment and other aesthetic factors. The small hydropower schemes have relatively small associated reservoirs for which available storage space or detention time is of little concern. Some of these reservoirs may be small regulating reservoirs with cycle times measured in hours or days instead of months or years. Others of these reservoirs may have little or no active pool such as in run-of-river plants.

The small hydropower is distinct from the conventional high and medium hydel projects in that it is simple in layout and mostly does not interfere with the existing regime of the flowing water They involve minimal submergence, rehabilitation and resettlement, deforestation etc. and have practically no adverse effect on the environment and ecology of the area. The schemes with capital outlay up to Rs. 50 crore are exempted from obtaining environmental clearance from Ministry of Environments and Forest and Pollution control board.

The environmental impacts may be classified as under:

1. Due to Location: It includes impacts on account of rehabilitation and resettlement of the population uprooted, encroachment into genetic resources, wild life parks, and forests, besides soil erosion.

2. Due to Project Design: It covers impacts due to seismicity, water quality and safety of structures.

3. During construction: It includes impacts due to soil erosion and consequent sedimentation, as a result of increased silt load stemming from the construction sites.

4. During Operation and Maintenance: It may include impacts due to variations in flows, first filling of the reservoir and its development during first few years and in operation phase, water quality, sedimentation, waterlogging and salinity, if applicable.

QUESTIONS

1. Give an account of environmental impacts of water power development projects.
2. Discuss sustainable development and the role water power projects play in promoting it.
3. Briefly mention various concerns for the development of hydropower.
4. What do you understand by environmental planning and management of hydropower projects in the context of present scenario of economic reforms and liberalisation?
5. What is meant by environment impact assessment and environmental analysis of hydropower projects?
6. Small hydro projects are considered to have minimal adverse environmental impacts compared to storage type projects, why?
7. Give a detailed account of the conventional and new modern environmental concerns about hydropower development projects.

3

PLANNING FOR WATER POWER DEVELOPMENT

3.1. INTRODUCTION

Planning for water power development, on a long term basis, is imperative in view of the important considerations of capital intensive, long gestation period, resources constraints, uncertainties of future and the increasing criticality of electrical energy as a vital input to the country's economic, social, agricultural and industrial development. The long term planning also helps in merging facilities created for meeting the short term requirements in the long term perspective plan and results in avoiding redundant constructions and premature obsolescence.

In view of the uncertainties in the load demands and the generation programme in a long time frame, some times resort has to be made to carry out long term system planning studies for specific load levels without pegging them to any particular time frame.

3.2. OBJECTIVES OF PLANNING

The prime objective of the planning is to derive maximum benefit from the available water with least tinkering with the environment. It is, therefore, imperative to adopt state-of-the-art approach in planning, analysis, design, construction and operation of the water power projects. Another important objective is the long term planning to meet the challenges of rapid rise in power demand with reasonable levels of reliability and at the lowest possible cost to the consumer. The various factors considered are as under:

1. **Acceptable level of reliability.**
2. **Options for power development including their technological, economic, environmental and ecological aspects as well as the global trend in fuel availability and pricing of fuel.**
3. **Demand and supply side management options.**
4. **Policies of the government.**

3.3. PLANNING FOR WATER POWER DEVELOPMENT

Planning for water power development is very complex in view of the competing demand and various complexities involved during investigation, planning, cost estimation and design because of the complications of topography, environmental, rehabilitation and resettlement etc. Various guidelines, manuals and Bureau of Indian Standards are available in this regard.

1. Water-Demand Planning

The hydropower essentially involves water supply and head (or fall), the planning has to be for the optimum development of these two elements. The determination of the amount of stream flow and its variation at a power site is fundamental for water power development, the estimation of head is easier, rather it is related to the availability of water vis-a-vis storage needed to meet

the demand. The power from a plant is limited by the available water supply, head characteristics and storage which in turn limit the capacity of the installed equipment.

The hydro plant may be run-of-river or storage based, in either case certain basic hydro-meteorological data are needed for analysis, dealt with separately.

Storage reservoirs are planned to conserve water, to regulate outflow and control floods. For hydropower generation, as for irrigation, domestic and industrial uses, assured water supply is needed. In reservoir designing, the critical period is the minimum flow period during the driest year of the record but essentially covering a period of several successive dry years to determine storage requirements and the hold over storage to meet the demands in the periods of critical low flows as also to meet future projected demands based on population or load growth. The reservoir capacity also bears an important relationship to carrying capacity of the downstream river/stream channel due to peak releases, their sudden changing pattern and long sustained flows. The storage may be single purpose, as for hydroelectric or of multipurpose concept.

In single purpose reservoir, planning for power is a simple relationship among the available water, power load demand and the volume of storage to be provided.

In multipurpose reservoir, it is more complex which has to cater for seasonal distribution of inflow, its regulation and varying demand for the various intended purposes. It is still more complex in the case of seasonal distribution of flow to provide additional storage for power in non-flood season. Certain losses, inevitable as they are, like evaporation, seepage and sedimentation have to be estimated and taken into consideration.

No matter how well a plant is planned, designed and constructed, its operation to meet the intended power demands is severely hampered if water supply to it is inadequate or overestimated.

The economics of reservoir development is reckoned on the basis of the cost attaining a given objective and the return on the investment, that is the ratio of benefits to the cost. It appears to be a simple principle of economics but is complicated in actual application due to evaluation of various benefits of the multi-objectives. The hydropower benefits are direct and easy to evaluate.

2. Power Demand Planning

Planning for the power demand forecasts is very necessary. The demand for electrical power is directly related to rate of growth of national economy, level of industrialisation, level of mechanised agriculture and its pumped water requirement for irrigation; agriculture draws power heavily mainly in kharif season to keep paddy fields under 5 to 7 cm of irrigation water, level of literacy and general standard of living and level of electrification, domestic as well as railways etc. The studies done indicate that for every point increase in national economy, the corresponding rise in the power demand would be 1.8. With projected 6 to 7 percent growth rate of the national economy, the corresponding increase in the power demand would be of the order of 8 percent per annum. The power planning without load forecast has no meaning. In the country Chief Electricity Authority conducts power surveys at regular intervals. The forecast of the electric power demand for the country as a whole is also done by the Central Electricity Authority.

Supply and shortage: It includes information on (*i*) Comparison of present demand with available supply at various locations, (*ii*) Surplussing and shortage during various periods of the year, and (*iii*) Comparison of future demand with available supply.

Power Generation Studies

It includes information on (*i*) Power supply position; (*ii*) Future needs and proposals for generating additional power, and (*iii*) Projected water requirement for additional power generation.

Scheduled Maintenance: The scheduling of annual maintenance of generating units is a very important aspect for generation planning studies. There are two basic approaches to scheduling namely (*i*) To achieve constant reserve margin, and (*ii*) Constant reliability. The difference between the two arises out of the varying forced outage rates of the units depending on the time, unit size, and muturity. However, the aim is to achieve the minimum outage of thermal plants when the need for its service is maximum. The staggering of the maintenance is, therefore, dependent on factors like thermal-hydro mix, occurrence of seasonal peak load etc. The effect of plant maintenance can be considered by increasing the load by an amount corresponding to plant maintenance outage.

3. Low, Medium and High Head Stations

The salient features of low, medium and high head plants having bearing on power planning are briefly as mentioned below:

Plant features	*Low head plants*	*Medium head plants*	*High head plants*
Head	< 30 m	30–300 m	over 300 m
Topography	Flat Land (ocassionally Hilly)	Hilly (occasionally mountainous)	Mountainous
Relative share of discharge	Large	Moderate	Small
Structures for control of water	Barrage with gates	Barrage or dam	High dam
Water conductor system	Diversion canal or run-of the river	Generally canal	Diversion tunnel, penstock etc.
Character of pondage or storage	None or daily pondage	Daily or weekly pondage	Annual & even carry over
Character of foundation	Loose to clay soil	Rock	Rock
Turbine-type	Kaplan, fixed blade propeller, tubular, bulb, high speed Francis	Francis-Kaplan	Impulse or low speed Francis.
Relative size of equipment	Large size turbine	Medium size	Small size
Character of energy generation	Fluctuating	Slightly fluctuating or occasionally continuous	Continuous as per load requirement
Production cost	High	Moderate	Low
Supply of Load	Base load plants run in co-operation with other types	Base load plants (occasionally Peakload)	Peakload plant or occasionally run independent
Arrangement for P.H. structure	Machine hall and sub-structure in one unit	Arrangement of either low head type or high head type	Separate power house structure.

4. Automation of Plant

The increased size of the hydro generating equipment and integrated optimisation and control of power grids, demand for high-performance and reliable control systems has necessitated digital control of various equipment with communication interface in-between and other generating equipment at the same or a remote site. The digital control system helps in achieving total automation of the plant and reduces Mean Time Between Failure (MTBF) in the plant by providing more information to the operator in fault diagnosis, operation of the plant as also identifying in advance the likely faults to occur. Fig. 3.1 shows typical levels of hierarchy of controls for a hydropower plant.

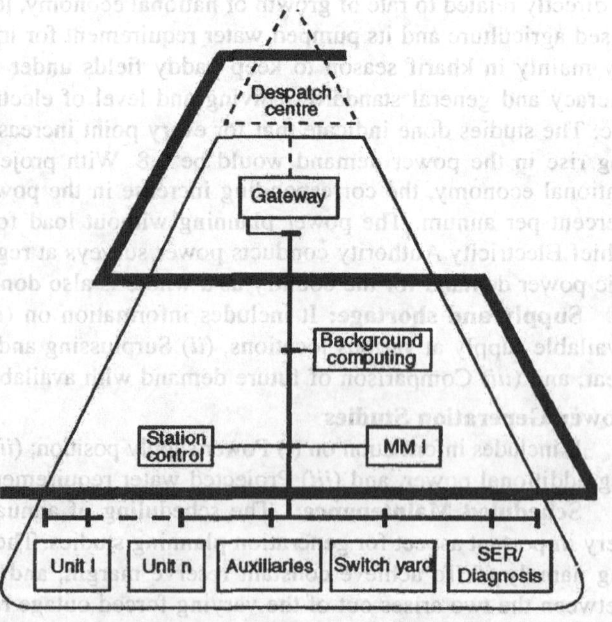

Fig. 3.1 Hierarchy of controls

A typical power plant digital control divided into different levels such as (*i*) plant/station control, (*ii*) unit control, (*iii*) functional group control and (*iv*) drive control, as developed by BHEL is shown in Fig. 3.2.

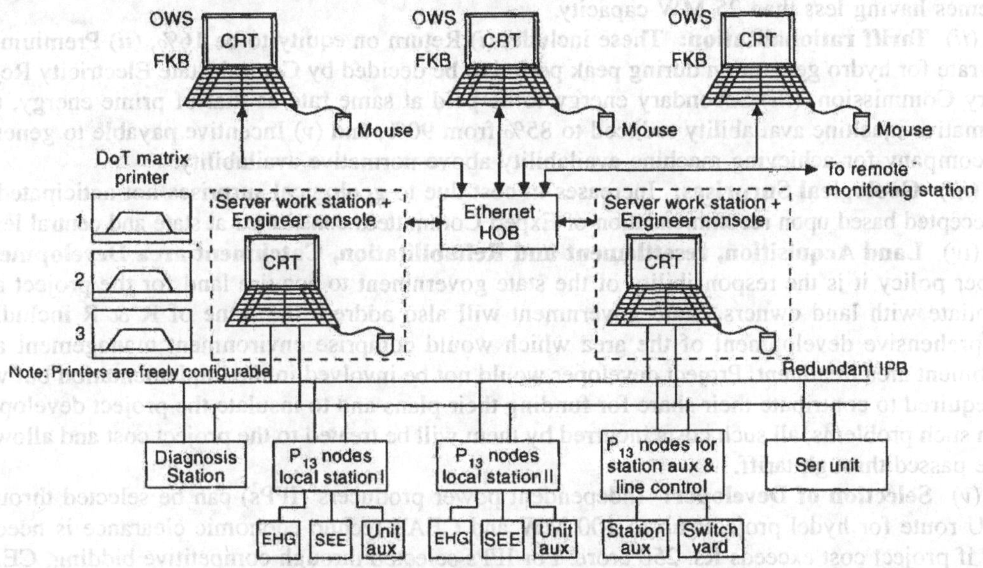

Fig. 3.2 Digital controls

For the realisation of these control functions in a cost-effective manner, the hardware is divided into two groups. The first group comprising computers, function key boards, printers etc, is dedicated to plant level control. At the lower level, a microprocessor-based controller ensures unit, functional group and drive controls. While the equipment provides centralised control at the unit level, to achieve the required hierarchical decentralisation, dedicated microprocessor modules are provided for unit or programme control, functional group and drive control respectively, which communicate among themselves over a common data bus in the panel.

The various advantages accruing from automation of the power plants are (*i*) Enhanced safety, (*ii*) Increased availability, (*iii*) Rapid start-up and shutdown of machines, (*iv*) Operation at high efficiency and (*v*) Ease of operation and reduced risk.

5. Hydropower-policy

The Indian growth in the power development has not been able to keep pace with the accelerated power demand. However, in response to the new economic realities, the government of India formulated a policy in 1991 to encourage greater investment by the private sector in the power development. To accelerate waterpower development a policy on hydropower development was announced in 1998 which was followed by a mega power policy.

The hydropower policy envisages benefits to the investors by appropriately adjusting the normative levels for incentives, further simplification of procedures and reassurances of returns on investments. The objectives of the policy in brief are (*i*) Exploitation of vast hydroelectric potential at a faster pace, (*ii*) Strengthening of the role of public sector undertakings/state electricity boards for taking up new hydel projects, and (*iii*) Increasing private investment.

6. Policy Initiatives

(*i*) **Planning for Faster Exploitation:** These include (*i*) Creating a National Power Development Fund through levy of cess on the power consumed for accelerated hydropower development, (*ii*) Enacting studies for integrated basin-wise development of hydro potential, prioritising and sequencing of hydropower projects to optimally harness hydro potential of each river basin, (*iii*) Vigorous measures for survey and investigation of promising sites on an advanced scientific basis with the aim to prepare bankable detailed project report of international standard, (*iv*) Funding support to survey and investigation projects from National Power Development Fund, (*v*) Advance action on 13 projects to be initiated during 9th Plan to reap their benefits in 10th Plan i.e. by 2007, and (*vi*) Proposal for special financial incentives for development of small hydropower schemes having less than 25 MW capacity.

(*ii*) **Tariff rationalisation:** These include, (*i*) Return on equity to be 16%, (*ii*) Premium on sale rate for hydro generation during peak period to be decided by Central/State Electricity Regulatory Commission, (*iii*) Secondary energy to be paid at same rate as that of prime energy, (*iv*) Normative machine availability reduced to 85% from 90%, and (*v*) Incentive payable to generating company for achieving machine availability above normative availability.

(*iii*) **Geological Surprises:** Increases in cost due to geological surprises not anticipated to be accepted based upon recommendation of Expert Committees constituted at state and central level.

(*iv*) **Land Acquisition, resettlement and Rehabilitation, Catchment area Development:** As per policy it is the responsibility of the state government to acquire land for the project and negotiate with land owners. State government will also address the issue of R & R including comprehensive development of the area which would comprise environment management and catchment area treatment. Project developer would not be involved in the implementation but will be required to contribute their share for funding their plans and to insulate the project developers from such problems, all such costs incurred by them will be treated to the project cost and allowed to be passed through tariff.

(*v*) **Selection of Developer:** Independent power producers (IPPs) can be selected through MOU route for hydel projects up to 100 MW and CEA's techno economic clearance is needed only if project cost exceeds Rs. 250 crore. For IPPs selected through competitive bidding, CEA's clearance is not needed for projects costing up to Rs. 100 crore.

(*vi*) **Transfer of clearances:** Simplified procedure for transfer of clearances already accorded to non-starting hydro projects in state sector in favour of central public sector undertaking/ independent power producer/joint venture of CPSU and IPP has been put in place.

(*vii*) Promotion of hydroelectric projects through joint ventures.

7. Reducing T & D Losses

The emphasis need to be laid from generation for meeting the demand to supply side by drastically reducing transmission and distribution losses. A rupee saved is a rupee earned.

8. Renovation and Upgradation

About 20,000 MW can come through renovation and upgradation of the existing plants. At a cost 20% of new capacity addition.

9. Superior Quality Power

The country needs power of superior quality so that a hike in tariff can be legitimately justified.

3.4. INVESTIGATIONS AND STUDIES FOR WATER POWER DEVELOPMENT

The power projects are necessarily required to be planned on the basis of long term hydrological data so as to cover wet as well as dry years cycles of rainfall. The longer the period of data more realistic would be the estimation of runoff. In the country, observed flow data are available for short-term period since river gauging on most of the rivers was started only recently in a systematic manner. Long term river flow data has, therefore, to be developed from the observed hydrological data. For the purpose correlation has to be established between observed rainfall and runoff data. Basic meteorological data such as rainfall, temperature, evaporation etc. are continuously available in the country since 1891 for a fairly good network of stations. This long term data can be used to estimate hydrological parameters through various techniques and methodologies developed. Such generation of synthetic data is an accepted national and international practice. The Bureau of Indian Standards too recommends generation of runoff data from the observed rainfall data. Satellite photographs can be used to estimate areal extent of seasonal snow cover over the Himalayan catchments which receive large amount of precipitation in the form of snow during winter and spring months. The river discharge of Himalayan snow fed rivers of a unit area is approximately twice that of peninsular rivers of South India, mainly due to perennial contributions from snow melting and glacial drainage.

1. Essential Data

The essential stream flow data required for water power studies of a storage project are as follows:

1. Hydrological Data: (*i*) Daily, weekly or monthly flow data over a long period of time for plant capacity and estimated output which is a function of the average flow as well as its distribution during a year as revealed by flow-duration curves. The curve for power site forms the basis for the determination of plant capacity and available power at all times, (*ii*) Maximum or flood flows for impoundment, spillway capacity and safety of the power plant. Unit hydrograph is a useful tool for it, (*iii*) Minimum or low water flows to compute primary or dependable power availability and size of auxiliary, if used. The minimum (dry) period flow without storage is determined after a period of drought has prevailed for some time. The low flow analysis is best done by probability studies. The minimum flow of a stream can be increased substantially by storage. Usually, a daily record of flow of say 10 years or more provides for most of the above, with perhaps the exceptions of the extremes of high and low flows which may be predicted with fair accuracy for a 10-year record of flow by taking into account precipitation and flow data of the nearby streams, (*iv*) Hydropower development consists of changing the space and time availability of waters or their quality to suit the intended needs. The essential hydrological aspects are briefly as under:

A reasonably accurate assessment of available water at the plant site requires a sufficiently long sequence of data depending on the type of pondage or storage, type of development and variability of inputs. The longer the period of simulation, the more is the confidence about the overall performance of the plant.

Table 3.1 indicates the minimum length of data required to arrive at a minimum flow with 90% dependability for design purposes.

Table 3.1 Minimum length of data

Type of development	*Minimum length of data*
Run-off-river (diversion)	10
Within the year storage	25
Over the year storage	40
Complex system involving combination of above	Dependent upon the predominant element.

It is also desirable to assess from the highest flood marks (levels) at the site to arrive at design flood or level for which the civil structures need to be designed.

(*v*) **Design flood:** The design flood is the flood hydrograph or instantaneous peak discharge adopted for the design of the hydropower project and its structures. The flood characteristics, frequencies and potentialities of the contributing drainage area above the site, social, economic and other considerations influence the selection of design flood. The design criteria for safety in fixing spillway capacities indicated in IS: 11223-1985 is given in Table 3.2.

Table 3.2 Design criteria for spillway capacity

Classification	*Gross storage (Mm^3)*	*Hydraulic head (m)*	*Inflow design flood*
Small	0.5 – 10	7.5 – 12	100 year flood
Intermediate	10 – 60	12 – 30	Standard project flood
Large	> 60	> 30	Probable maximum flood
Barrage	Standard project flood or 100-year flood, whichever is larger.		

Depending upon the availability of data or specific requirement either the hydrometeorological approach or the frequency analysis can be adopted to determine the design flood.

2. Hydrometeorological Approach

At waterpower development sites long term surface flow data are generally not available and resort has, therefore, to be made to make use of flow data of the adjoining streams or even rainfall data. The study of maximum rainfall, of 1 day or so, is of fundamental importance in flood probabilities analysis and safety of waterpower development. The storm data given by recording gauges is considered because hourly rate of rainfall are not given by non-recording gauges or a large storm giving 1-day rainfall may be spread over two successive days. The stream flow or runoff for waterpower development consists of surface flow as well as percolation. In determining maximum flow, excessive rainfall or a combination of large rainfall and snow melting, characteristics of the drainage area, path of stream etc. are considered. The high floods generally occur during monsoon as a result of excessive rainfall when the soil is also wet.

The rainfall frequency is at best an estimate of the average time interval of occurrence based on past data without indicating when a rainfall equal to or greater than any of this value may actually occur. However, it helps in deciding factor of safety to be considered in the design of civil structures of the hydropower project.

The essential parameters considered are as under:

(*i*) **Design storm:** (*i*) Inventory of storms in and around the basin, (*ii*) Storm analysis – depth duration, depth-area-duration, temporal distribution, (*iii*) Storm transposability, (*iv*) Moisture adjustment–moisture maximisation, barrier correction etc., and (*v*) Transposition with adjustments and deciding time and area distribution.

(*ii*) **Estimation of effective rainfall:** The effective rainfall that is the net rainfall contributing to the runoff is determined by separating the losses by ø index method etc.

(*iii*) **Catchment response function:** The effective rainfall falling over a basin travels and reaches the point of discharge measurement on the river. The attenuation and translation effects of the basin are reflected through a suitable response function. Unit hydrograph method is generally used to represent the catchment response function in flood estimation and prediction.

(*iv*) **Frequency Analysis:** When sufficiently long series of annual flood peaks is available, it can be used for flood frequency analysis to determine floods of different return periods.

2 Investigations and Surveys

For proper harnessing and utilisation of India's water power wealth through dams and reservoirs, certain basic studies are necessary. The hydropower projects involve many complexities during investigation and planning due to complications of topography, geology, hydrology, seismicity, environmental and other features. The investigations and surveys of the potential hydropower sites is essentially thorough, systematic and on advanced scientific basis so that bankable project reports are prepared which go a long way in the construction of techno-economically viable and environment-friendly hydel projects while avoiding time and cost over runs.

1. Topographical Surveys

(*i*) **Dam and Reservoir:** Major and medium hydel projects require construction of dam and reservoir, the topographical surveys needed in this regard are indicated in chapter "Dams and Reservoirs".

(*ii*) **Barrage:** Diversion structures like a barrage or a weir are required for run-of-river schemes, the topographic survey essential for which comprises, (*i*) An index plan of entire catchment area upstream of the barrage site indicating gauge-discharge sites, raingauge sites, road-railway crossings etc, (*ii*) Contour plan of the area around the barrage site with contour interval 0.5 m up to an elevation of at least 2.5 m above HFL besides cross sections of the river at 2 km intervals up to pondage effect on upstream, (*iii*) Longitudinal section of river. For model studies, cross sections are needed at close intervals, say 100 m up to 2 km upstream and 1 km downstream of the proposed barrage or weir site.

(*iii*) **Power channel:** Longitudinal sections covering power channel, penstocks and tailrace channel is prepared. It gives levels at 50 m intervals and indicate log of boreholes at points recommended by geologist of Geological Survey of India. Cross sections are also taken at 100 m intervals extending sufficiently on both sides.

(*iv*) **Tunnels:** For tunnels, longitudinal section along tunnel alignment and contour plan with 50m intervals covering about 100 m on either side of the alignment and also up to contours corresponding to tunnel grade, if applicable, indicating location of adits. More details in the chapter "Tunnelling".

(*v*) **Surge Tank:** Contour plan of surge tank area at 5 m contour interval.

(*vi*) **Underground Power House:** Contour plan of 1 in 4000 scale showing contour intervals at 15 m covering locations of access tunnels, tailrace tunnel and switch yard etc.

2. Geological and Foundation Investigations

Geological sections to show known and interpreted sub-surface conditions is prepared. Seismic conditions of the region are investigated with reference to the geological map of the vicinity. The intake structure and power house building are designed taking into consideration seismicity of the area.

(*i*) **Barrage/Weir:** Exploratory holes spaced 150 m apart are put along the alignment of the barrage to determine the soil profile up to a depth of 30 m below average bed level on permeable strata; 3 m in case sheet rock is met earlier. Samples of the soil obtained are tested for properties such as mechanical analysis, atterberg limits, proctor compaction, permeability, triaxial shear tests and solubility tests in soils with high soluble content.

(*ii*) **Water Conductor System:** The longitudinal section of the water conductor system from intake structure to tailrace is given. In power channel portion, trial pits to adequate depth be put in at suitable intervals so as to indicate all changes of strata.

For tunnel, a geological section of the tunnel alignment and surge shaft and drill holes adequate to establish cover of rock and important geological features is given.

(*iii*) **Power Station:** 2 to 4 drill hole details for the power house area may be given. The number, location, spacing and depth is decided in consultation with geologist. The survey plan may cover an area sufficient to include switchyard and alternative plant layouts with contours at 5 m intervals. Low water level, maximum observed flood level, rock outcrops, sand shoals etc. at the power house site are indicated therein. Indications of slides and their locations and the possibility of rockfalls in the hill slopes is explored. Ground water level at the plant site is determined.

For underground power house, a geological section of power house, pressure tunnel, access tunnels and switchyard and sufficient drill holes/drifts to establish their geological features.

(*iv*) **Sediment studies:** These are undertaken for the design of storage dam or diversion structure, refer details in the chapter "Dams and Reservoirs".

3. Final Investigations and Studies

The final investigations and studies include (*i*) Determination of relative merits of the finally selected out of two or more prospective sites investigated for the formulation of project report, (*ii*) Type of dam or diversion structure, (*iii*) Nature of foundations, topography of the site, (*iv*) Water flow availability vis-a-vis pondage or storage needed, which in turn determine the cost of dam and other structures including safety aspects, (*v*) Final location of the dam or diversion structure, intake, spillway, headrace and tailrace conduits, surge tank, power house, workshops, stores, administrative and field offices, residential colonies, rail-roads, sources of materials, borrow areas etc., (*vi*) Land acquisition of a reservoir, if any, dam, workshops, colonies etc. (*vii*) Environmental aspects, and (*viii*) Railroad and highway/village roads to be realigned or laid for the carriage of heavy equipment etc.

3.5. INSTITUTIONAL FRAMEWORK

1. Connected with Hydrological and Environmental Data

Accurate and adequate hydrological, and climatological data, very essential for the formulation of water power development schemes/projects, is normally available with the sources given as under:

Data	*Source*
Precipitation and climatological data	IMD and state government departments.
Stream flow data, quality, water availability, utilisation studies and sedimentation data	Central Water Commission, Rashtriya Barh Ayog (National flood commission), state government irrigation departments.
Evaporation, transpiration and infiltiration data	Department of Agriculture, Government of India, State irrigation and agriculture departments/universities.
Topographic maps	Survey of India.
Geological maps	Geological survey of India.

Data	*Source*
Environmental data	Ministry of Environment and Forests, Government of India, State Forest and wild life departments.
Surface water developments	Ministry of Water Resources.
Power related data	Ministry of Energy and Central Electricity Authority.

2. Decision Making and Implementation of Power Schemes

A number of organisations, departments and institutions, as under, are connected with decision making and speedy implementation of hydroelectric projects to protect the interests of people and society. The institutional mechanism obviously requires a decision making process in which each of the participant in it provides inputs as a specialist. These inputs are then synthesised by specialist bodies like Central Electricity Authority and Central Water Commission, Government of India.

Organisation	Role
1. Central Electricity Authority	Location and optimisation of project design to national power resources.
2. Central Water Commission 3. Geological Survey of India	Location of the project and to determine issues of water with the aim of maximising the use of water.
4. Ministries — Union Ministry of power — Union ministry of food, agriculture and irrigation — State ministries of irrigation and power and state electricity boards	In addition, issues relating to generating competition and reducing cost of projects, preventing interstate conflicts without delays, arrange funds, implementation and to evolve the commercial mechanism to earn return on investments, development funds for regional and social uplift should be decided to be investigated by the central and state governments to become part of commercial enterprise.
— Union ministry of environment and forests — Central pollution control board — State pollution control boards	Finding environment and pollution protection measures which are practically feasible and compatible with the project location and design.
5. Multilateral Funding Agencies	Funding of project.
6. Non-government organisations 7. Social organisations and pressure groups 8. Various other departments 9. District administration and forest department	To speed up the process of rehabilitation and resettlement including administrative issues of land acquisition, settlement of project affected families besides economical and effective afforestation programmes.

3.6. DEFINITIONS

The two components for hydropower generation are head (H) and discharge (Q).

Available or Gross Head: In storage plant, it is the fall from reservoir level at the dam to water level in the tailrace. Precisely at any hydropower plant installation it is the difference between headrace level and tailrace water elevation measured at the same time, at the upstream and downstream ends of the power generating system. It is represented by H (Fig. 3.3). It is not constant but varies as the river discharge varies, the rise in reservoir level being different from that at the tailrace. The backwater effect in the tailrace results in lesser amount of available head and consequently power is lessened not only due to head reduction but also by lower wheel efficiency. In planning hydropower development, the backwater effect in tailrace is duly consi-

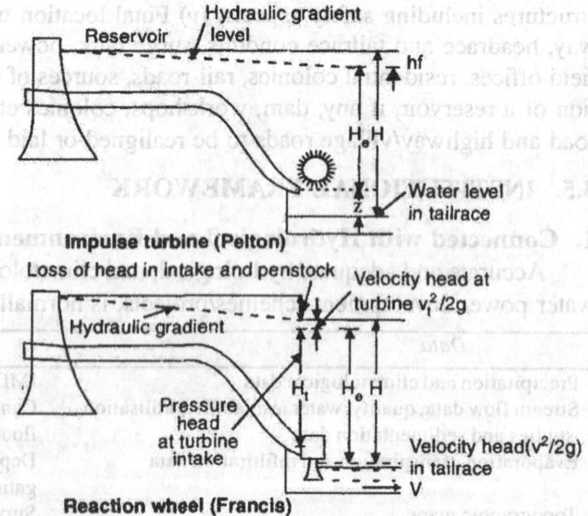

Fig. 3.3 Heads acting on impulse and reaction tubines

dered. The gross head available at any river stage is determined by plotting curves of river or reservoir water elevation above the dam, as the case may be and at the tailrace site. The available head varies with the draft from pondage or storage.

In run-of-river plant, it is the difference in water level at the point of diversion of water to power plant and the water level at the point where it is returned to rivulet or stream. In pondage scheme the decrease of head due to draft from pondage is duly considered. Normally the mean daily elevation of the pond which is the average of maximum and minimum levels is considered. However, to be precise graphical study of flow, load curve and height of pond is done to get true weighted average of the head during the day.

In reservoir power development, there are wide fluctuations in head water level (or reservoir level) and as such the average head is determined from hydrograph studies of the use of water from the reservoir. The reservoir-depletion curve provides the basis for the determination of available average head.

Operating Head: It is the head available for turbine for power generation. Alternatively it is the simultaneous difference in elevation of the water surface at the entry to penstocks in the high head stations, or forebay in low head plants and the water surface at the tailrace (just outside the draft tube), allowance being made for velocity heads.

Net or Effective Head: It means gross head minus the losses in friction, trashrack and entrance losses, losses in bends in penstocks, if any, unrecovered velocity head in the draft tube etc. Usually they account for 2 to 5 percent of the gross head.

These may be assumed as given in Table 3.3 for preliminary studies:

Table 3.3 Loss of head in power plant

Gross head (m)	7.5	15	30	60	150	300
Loss of head (m)	0.3 – 0.6	0.5 – 1.0	0.6 – 1.5	1.25 – 2.5	3 – 6	6-12

The net head applicable to the type of turbine (Fig. 3.3) is as under:

Reaction turbine, $h = H - V_d^2/2g - h_f$

Pelton or impulse turbine, $h = H - Z - h_f$

where, H is gross head, h is net head, V_d is exit velocity, Z is height of pelton wheel exit above tailrace water level, and h_f is head loss in penstock.

Maximum Head ($H_{max.}$): It is the gross head resulting from the difference in elevation between the maximum forebay level and tailrace level without spillway discharge and with one unit operating at no load speed. Under this condition, hydraulic losses are negligible and may be disregarded.

Minimum Head ($H_{min.}$): It is the net head resulting from the difference in elevation between the minimum forebay (or MDDL) and the tailrace level with all turbines operating at full gate.

Weighted Average Head ($H_{av.}$): It is the net head determined from reservoir operation calculations, which will produce the same amount of power in kilowatts between that head and maximum head as is developed between the same head and minimum head. This weighted average head usually lies approximately two-third of the total range of head above minimum.

Design Head (H_d): It is the net head at which peak efficiency is attained while operating at rated speed. This head should preferably approximate the weighted average head but must be so selected that the maximum and minimum heads are not beyond the permissible operating range of the turbine. This is the head which determines the basic dimensions of the turbine and, therefore, of the power plant.

Rated Head (H_r): It is the net head at which the full gate output of the turbine produces the generator rated output in kilowatts. The turbine name plate rating is usually given at this head.

$$H_r = H_{max} (P.F/1.15)^{2/3} = 0.91\, H_{max} (P.F)^{2/3}$$

where, P.F is the generator power factor. The factor 1.15 is for a generator having an overload capability of 115 percent of its rated capacity. H_{max} is the net head resulting from difference in elevation between maximum forebay level and tailrace level without units operating at full gate.

Critical Head (H_{cr}): It is the net head at which the full gate output of the turbine produces the permissible overload on the generator at unity power factor (usually 115 percent of generator K_{VA} rating). This head produces the maximum discharge through the turbine.

Range of Head: The head variation is restricted to the following percent to ensure a normal runner life and a reasonable minimum maintenance.

Type of turbine	*Permissible operating head %*
Francis	65–125
Fixed blade propeller	90–110
Adjustable blade propeller	65–125

The impulse type of turbine, working under extremely high heads, operates at a very nearly constant head and the problem of variable head is seldom important.

Rated Discharge

For hydropower plant design, mainly from generation point of view, the rated discharge capacity is considered as the average discharge or the discharge available for 50 percent of the time i.e. plant is designed to operate at its full capacity at least 50 percent of the year. As per recent trend for the evaluation of rated discharge for generation of power from run-of-river schemes or those primarily for irrigation purposes, the rated discharge capacity is fixed in the range of discharge which exceeds at least 15–30 percent of the time. It involves higher initial cost of the power plant.

3.7 ESTIMATION OF AVAILABLE WATER POWER

Hydropower is generated with the fall of water through a head. The requirement of water for power generation is determined as under:

Let a discharge of Q cumecs falls through a net height H metres

Rate of work done = QWH kg-m/s

Hose power generated = $(QWH)/75 = 13.33\ QH$, taking $w = 1000$ kg/m³ and metric $H.P = 75$ kg m/s.

Actual H.P. generated = $13.33\ QH\ \eta$, η being the overall efficiency of the system i.e. of generator, turbines etc. (refer Table 3.4)

Table 3.4 Typical overall plant efficiency (η)

Item	*Turbines*	*Generators*	*Speed increasers*	*Step-up transformer*	*Transmission Lines*	*Step-down Transformer*	*Overall plant Water-wire*
Av. η		88	96	97	98	98	75
Range	83–90	95–97	96–98	98–99	>97	98–99	70–85

$\therefore$ Electrical energy = $13.33\ QH\ \eta \times 0.736$ kilowatts, taking 1 metric H.P. = 736 watts = 0.736 kilowatts.

= $9.8\ QH\ \eta$ kilowatts

= $9.8\ QH \times 0.8$, taking efficiency η as 80%

$\simeq 8\ QH$.

3.8. POWER DURATION CURVE

The flow duration curve representing a long period data can be used for the determination of duration of floods. The annual flow data is arranged in descending order of discharges. The plotting position (P_p) of any discharge (Q) is given by $P_p = \left(\frac{m}{N+1}\right) \times 100$. Where, m is rank of discharge, N is total number of records, and P_p is percentage probability of the flow magnitude being equalled or exceeded. This flow duration curve obtained by plotting Q and P_p is further used for the estimation of the amount of power that can be produced under average conditions by the conversion of discharge scale to power units corresponding to the available head. The curve has shape similar to that of flow duration curve but indicates the relative duration of different powers available from the water flowing at the site. The area under the curve represents the average power yield from the plant.

The capacity estimates for firm power are made from the average dry year's flow duration curve and the secondary capacity is determined by considering normal year's or wet year's flow. A better approach is to use the entire recorded data and to summarise it in a single curve by total period method and calendar year method, described hereinafter.

The available power from a run-of-river plant wherein head is presumed to be constant at an average value is represented by a power-duration curve which exactly corresponds to a flow-duration curve. However, for a precise power-duration curve the head corresponding to any discharge is required. The average monthly power from the plant is determined by the following formula:

$$P = 9.8\, QH\, \eta \text{ kW}$$

The power duration curve is then drawn by arranging the monthly power values corresponding to monthly discharges in ascending order and following the same procedure as for flow-duration curve.

Storage is a very important factor in operating a hydropower plant. The reservoir size required to satisfy the specified flow demand is accomplished by mass curve analysis (refer "Dam and Reservoirs").

(*i*) *Total period Method*: In this method, the entire available record of say 10 years is used for drawing the flow duration curve to produce 120 values of monthly average. These are tabulated in ascending order starting with the lowest value (the driest month) and ending with the highest value (wettest month) and the flow duration curve is then drawn making use of these values.

(*ii*) *Calendar year Method*: In the method, each year's average monthly values are tabulated in ascending order. Then the average flow values corresponding to the driest month, second driest month and so on up to wettest month are determined by taking arithmetic mean of all values of the same rank. The flow-duration curve is then plotted with these average values. Such a curve has only 12 points.

The total period method gives more correct results than the calendar year method which averages out extreme value.

Limitations

(*i*) The considerable variation in the monthly values are not reflected in the curve as it is drawn from mean monthly values, (*ii*) It does not represent the values in the natural sequence of occurrence, (*iii*) It does not indicate whether the lowest values occurred in consecutive periods or scattered throughout the period of data.

3.9. STORAGE AND PONDAGE

The power generation has essentially to conform to the pattern of demand which implies that water is made available to power plant in the required amount. The flow rate in a river or stream varies (on account of various factors natural such as rainfall and/or snowmelt and artificial due to pondage of water at power plant upstream) which certainly does not correspond to the variation in power demand. The water supply and demand thus need to be evened. Moreover, as electricity can not be stored, resort has to be made to create storage or pondage to provide regulated flow to the power plant for continuous power generation or as intended.

1. Storage

Storage is defined as impounding of sufficient amount of surplus wet season flow in a reservoir to equalise monthly, seasonal or yearly fluctuations in river flow by the construction of a dam. The storage may cater for within the year or over the year demand. Storage based is the most efficient water power development.

Operation of Storage: The operation of storage reservoir is designed to yield maximum value or benefits from the storage capacity. The principles of reservoir planning viz., seasonal inflow, varying demands, power load and the greatest benefits determine the pattern of releases made from the reservoir. The operation of the reservoir can be classified into the following categories:

1. Insurance Method: In this method minimum needed outflow is made during dry period through regulated releases and available storage conserved to increase the dependable flow during future dry period and as such envisages carry over storage.

2. Within the year Use: Also called the yearly or annual use method. In this method use of practically all the available storage is made each year without regard to maintaining minimum regulated flow. There is no effort to carry over any considerable amount of stored water for the next season. The water is released as and when the inflow from the stream or river falls short of the minimum demand to operate power plant at full capacity. The drawback of the method is that the reservoir may dry up in drought years. Storage reservoirs for power development are operated by this method especially where safeguard of primary power is possible by interconnection with grid.

3. Over the year Use: Also called safe-yield or best dependable flow method. It is an improved and modern concept of the above two methods. It implies to so regulate the operation of reservoir that it is capable to provide a dependable minimum flow and at the same time optimum utilisation of the stored water every year. The reservoir operation is governed by the rule curves based on the previous record of the reservoir operation, making best use of the stored water (except in extremely wet years of infrequent intervals), the withdrawal of water from the reservoir capacity is not to exceed safe yield. The stored water may be held over more than one season or several seasons so as to ensure always availability of optimum amount of storage to meet the demand. In this way, depending on regulation, only about 50% or less of the storage capacity is made use of in an average year and as such more waste of water over the spillway and a less total available power output in kilowatt hours per year, although it gives a maximum of primary power from the stored water. The storage use may be up to nearly full capacity under favourable conditions of use and with relatively small reservoir capacity, taking into account losses from the storage because of leakage or irregular load etc.

The storage reservoir may be at the power plant site which simplifies to regulation of flow obtainable from and at the reservoir. The mass curve method of studying the effect of storage is very convenient. Where storage reservoir is at some distance upstream of the plant and an uncontrolled drainage area intervenes, usually a simple assumption is made with regard to the total amount of storage and its effective location. The operation is planned for the centre of gravity of the power system. The hydrograph method with accompanying reservoir-depletion curve is usually adopted.

There may be two or more power plants, some distance apart on the river or served by the reservoir. Since equal regulation can not be obtained at all downstream plants, the regulation is determined with due regard to (*i*) The relative size of the plants, (*ii*) Plant for which regulation of flow is to be made, and (*iii*) A system which ensures maximum power output from the combined plants. The point of regulation of flow is generally the centre of gravity of power for the system based on either wheel capacities or developed head and drainage areas, the available flow is determined by hydrograph study and corresponding flow at the various power plant site is determined.

Where two or more reservoirs exist, the regulation becomes complicated with some of the plants between the reservoirs. The hydrograph study is usually done considering the entire drainage area above the lowest reservoir to be largely controlled by the total reservoir capacity.

2. Pondage

Pondage is defined as the holding back (storing) a relatively small amount of water and release later to equalise daily or weekly fluctuations in inflow or to facilitate irregular hourly use of water by the turbines in accordance with the variations in the load demand; the hourly and daily load variations being usually considerable with 24 hour use of power. Thus sufficient pondage is required to (*i*) Equalise fluctuations in river flow to avoid irregular use of water, and (*ii*) Fluctuations in load under usual operating conditions. The pondage is usually accomplished through the construction of a diversion structure across the river or stream and of a capacity sufficient to regulate the flow during the day or the week ends. Pondage is an important factor especially in the operation of hydropower plants where inflow may experience irregularity due to pondage at upstream plant (s).

The storage and pondage are, however, relative terms. The storage is related to lerge reservoirs to equalise monthly, seasonal or yearly fluctuations in the inflow from a river, whereas the latter is associated with small and medium power plants with limited pondage to equalise hourly, daily or weekly fluctuations in streams or rivulets. In terms of capacity alone, an amount of storage which on a river might suffice only for pondage may be sufficient as storage on a small stream.

At hydropower plants pondage is required to be provided usually under the following site layouts:

1. Hydropower plant is located some distance below a storage reservoir and water is supplied to the plant through a power channel which in turn gets supply from a diversion barrage or weir constructed in the river channel downstream of the dam to pick up the released water from the dam.

A regulating reservoir termed as pond is created upstream of the barrage as the releases from the dam may not necessarily correspond with the water requirements to take care of the variation in the daily or weekly power load demand at the power plant.

2. Power plant is located on a rivulet or stream and pondage is needed to regulate the natural perennial inflow of the stream to accord with the variations in daily or weekly power demand as in the case of run-of-river plants. It is accomplished by the construction of a weir or barrage on the stream.

3. Power plant is located on a canal fall, remote from a barrage or weir (at head), on a power channel and pondage in the shape of a forebay or enlarged channel section is provided just upstream of the power plant to serve as a regulating reservoir to take care of the variation in power load demand.

4. Power Plant is located considerably away from the diversion dam and water is supplied through a power channel or pressure tunnel. A balancing reservoir is provided nearest to the plant to take care of variations in hourly, or daily power load demand as also as a peaking station when water is stored during off-peak hours to run the turbines during peak hours of the same day.

With lower load factors firm capacity of water power can be increased by pondage so that a run-of-river plant may show a primary power several times the 24-hr minimum power.

Pondage Factor

Pondage factor, also termed as concentration factor, expresses the flow or power available during the working day in terms of the 24-hour flow or power.

For example, consider a stream constantly discharges into a pond with 8-hour daily use of water, 6 days in the week, if water is ponded during the remaining 16 hours of the day, the pondage factor would be

$$\frac{7\times24}{6\times8} = 3.5$$

Pondage factor for 8 hours, 6 days allowing 10% as wastage = 3.5 × 0.9 = 3.15.

The pondage capacity required is to hold 16/24 times the 24 hour flow to be filled every night/day and drawn during 8 hours next day.

In case Sunday flow is excluded, the pondage factor for 8 hours, 6 week days would be

$$\frac{6\times24}{6\times8} = 3.0$$

$$\text{Pondage capacity} = \frac{16+24}{24} = \frac{40}{24} \text{ times the 24-hour flow.}$$

To summarise, the pondage capacity required for night time (8 hours) ponding is about two-third of the average day load. Fluctuation in load with 24-hour use of power requires a pondage capacity varying from one-third of the average load for a 30 percent load factor.

Storage Vs Pondage

A comparison of storage and pondage, as under, shall be of interest:

Storage	*Pondage*
Water holding back (storage) for long term requirements, within the year or over the year.	Water holding back (pondage) is for a very limited time–hourly, daily or weekly cycle.
Holding back of excess flow available in wet months for continuous release during dry season.	Pondage is of daily normal flow. Flood water is not conserved.
Regulation of flow over a long cycle of years	Regulation of flow is diurnal
Equalises monthly, seasonal or yearly fluctuations in river flow.	Equalises hourly, daily or weekly fluctuations in rivulet/stream flow.
Flow equalisation at different times of the year means large amount of storage and creation of high head.	Facilitates irregular hourly use of water by turbines during different hours of 24-hour day compatible with load demand, without serious fluctuations in head.
Water impoundment is done in a reservoir	The water ponding area is called pond area.
The regulatory structure is a dam.	The regulatory structure is a barrage or a weir.
Maximum primary power from storage but less total available power, kilowatt-hours of output per year, if storage is for over the year use.	Kilowatt-hours of output are not increased but increases primary horse power capacity and enables a large load to be carried in the day hours.
Power output may be continuous in base load plant or depending on load if working as peaking station.	Power output is more or less concentrated into the day hours.
Primary power is related to outflow from storage which may be many times more than minimum river flow depending on amount of storage.	Primary power is mainly related to minimum flow, besides on available pondage.

3.10. LOAD STUDIES

1. Terms relating to water power

Firm Power: Also called primary power. It is the minimum continuous power which can be generated for 100 percent of time. It corresponds to the minimum stream flow with due consideration of the effects of pondage and load factor (Fig. 3.4). It may also be defined as the portion of the total installed capacity which performs the same function on that part of the load curve to which it is assigned as is performed by an alternative steam plant.

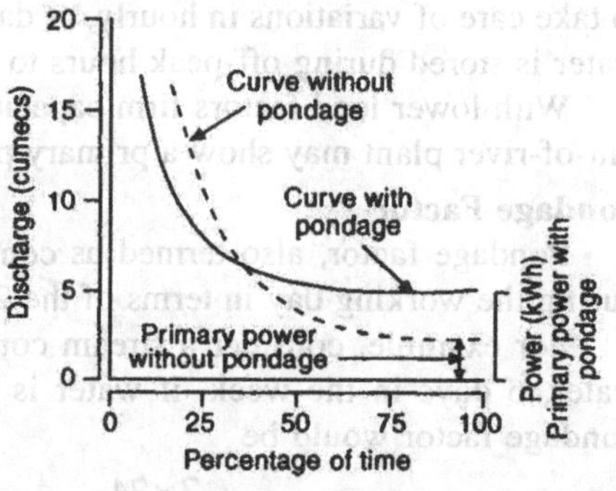

Fig. 3.4 Firm power

In the case of storage reservoir it corresponds to the minimum outflow at the minimum head, the condition usually met with just before the filling period.

Firm power = $9.81\ Q\ H\eta \times 10^{-3}$ MW

where, Q is the average inflow (cumecs) during the lean months of 90% dependable year. H is average net head (m) during the period, η is combined efficiency of turbine and generator (%).

Annual firm energy benefit = Firm power (MW) × 870×10^{-3} GWH.

Secondary or Surplus Power: It is the additional power which can be generated at a plant in excess of the primary, depending upon the spare capacity provided over the constant load, is known as secondary power (Fig. 3.5). It is limited by the generating capacity of the plant, the head and the availability of water in excess of

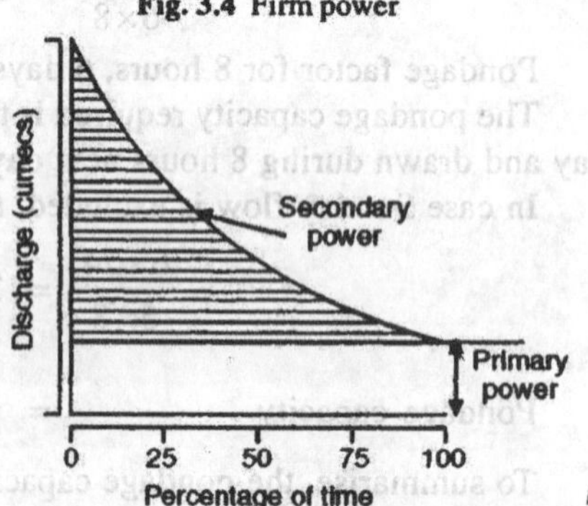

Fig. 3.5 Primary and secondary power

that corresponding to the firm power and as such is comparatively less valuable. A certain amount of secondary power may approximate firm power in being available a large percentage of time, whereas other amounts may be available for shorter periods.

In run-of-river plants which are generally designed for Q_{90} (the discharge available at 90% time), the firm power would be available for 90% of the time. When the available discharge is more than Q_{90}, extra available power is called as secondary power.

A comparison of firm power and secondary power, as under shall, be of interest :

Firm power	*Secondary Power*
Minimum continuous power generated for 100 % of time	Additional power over and above the firm power.
Generation corresponds to minimum stream flow with due consideration of pondage and load factor	Generation depends on availability of water in excess of that corresponding to firm power.
Corresponds to minimum outflow at minimum head, condition usually met with just before the filling period.	Corresponds to excess flow available at higher head, condition met with after filling of reservoir.

2. Terms Relating to Plant Capacity

Plant Capacity: Also termed as installed capacity. It is the total generating capacity (MW or kW) of all the turbine-generator units installed in the power plant. It is generally given in horse power for a given load, discharge and speed at the best efficiency.

Dependable capacity: It is the load carrying capacity of the power plant with respect to the characteristic of the load to be supplied during a specified time interval and the period. It is determined by the power factor, capability and the load on the power plant.

Optimum Installed Capacity: To determine optimum installed capacity, a number of alternatives of installed capacities are considered starting with the load factor (LF) of about 40% down to about 15%. Total energy generation during 90% dependable year is computed for each of the alternative of installed capacity based on average inflows in 10-days time intervals.

Output of Plant: It is the power produced during a period of time at a variable rate, the maximum of which is equal to or less than the plant capacity.

Reserve Capacity: The electric power demand varies from moment to moment and the available generation has to meet the fluctuations without in any way affecting the equality of service. Besides, adequate reserve generation capacity is necessary to take care of possible forced or partial outages of the plant and planned maintenance so that the desired reliability of the system in meeting demand at any time is maintained. Spinning reserve generally takes care of the former situation while the latter is taken care of by what is called as cold reserve. The proportion of spinning reserve in a system depends upon the estimated time required to take care of the fluctuations and as such hydro sets are ideally suited for the same. Another factor essentially to be considered is the power demand of thermal station auxiliaries. It is recognised that the same as percentage of plant output is higher at part load than at full load and it is usually taken as 10% of rated output for planning purposes. Thus the total installed capacity of a system has to be higher than the anticipated peak load. The capital intensive power plants necessitate to arrive at a balance between the effects of power shortage and investment costs in adding generating capacity. The cost considerations also necessitate a desired reliability level and the fixation of reserve necessary to obtain this desired reliability level.

In a single plant serving a power system, usually one unit is allocated for reserve capacity. With the present interconnected system with power grid, some reserve capacity is required to cater for expected peak load as also to take care of maintenance shut downs and breakdowns, besides for secondary power generation when excess supply is available in wet season.

Water power is adapted to provide reserve capacity on demand because a stand by unit can become operative in a few minutes; usually 3 to 4 minutes as against 30 minutes minimum required for steam units.

Methods to Determine Reserve Requirements

The generation reserve requirement is determined by the following two methods:

1. Deterministic Methods

(*i*) *Percentage Reserve Method*: It is the earliest and simplest method. In this method, installed reserve is taken as a fixed percentage of the annual system peak load.

(*ii*) *Contingency Method*: In this method, the system reserve is taken as equal to one (preferably the largest) of the following:

(*a*) A fixed percentage of annual peak load.

(*b*) The capacity of the largest unit.

(*c*) The capacity of the largest two units.

(*d*) The capacity of the largest unit plus a fixed megawatt value.

These methods take no cognisance of the daily, monthly and seasonal load patterns or to the characteristics of generating equipment peculiar to individual system such as unit availabilities and the mix of unit type and sizes and as such are unable to measure the system reliability in a quantitative manner. The use of computers and probabilistic methods are useful tools in this regard.

2. Probabilistic Methods

These methods evaluate the risk in meeting the system load requirements by inter-relating in some way the load and capacity models developed for the particular system and time period under study. The four basic methods to assess the generation reliability are as under:

(*i*) *Loss of Energy probability (LOEP) Method*: In the method, the ratio of energy not served to the total energy requirement is used as an index.

(*ii*) *Frequency and Duration (FD) Method*: It takes into account the factors of frequency and duration of capacity shortages which are not considered in the method of loss of load probability. The method enables production of expected number of days between shortages of capacity and the average number of hours that such a shortage may occur.

(*iii*) *Expected value of Demand Not Served (DNS) Method*: It is the widely accepted method because of its obvious physical significance. It is derived after calculating LOLP and quantifies the value of demand not served.

(*iv*) *Loss of Load probability (LOLP) Method*: Based on load characteristics the load models incorporating the probability of occurrences of load is built up. This then is combined with the probability of occurrences of forced outages. The loss of load probability corresponding to outage of a particular capacity is equal to the product of its probability of occurrence and the probable time the load is likely to exceed the remaining generating capacity. This figure is taken as an index of reliability in the units of days per year. The reliability index followed by several North American and European systems is 1 day in 10 years but in India a reliability index corresponding to the loss of load probability of 1 day per year is considered.

The Development of Generation Reserve Model: The capacity model and load model are merged together to give a generation reserve model which depicts the occurrence of surplus capacity. The meshing of capacity model with load model is the basis for evaluating the risk of capacity not being able to satisfy the load requirement.

3. Terms Relating to Load

Load: It is the amount of power delivered or received at a given point at any instant.

Average Load: It is the total load produced divided by the number of hours in the day. It is expressed in kilowatt hours (kWh) i.e. 1 kW of power delivered for 1 hour. It is also the hypothetical constant load over a given period of time that would produce the same energy output as the actual loading produced.

Peak Load: It is the maximum instantaneous load or a maximum average load over a specified period of time. The maximum average load is generally used.

Absolute Peak: It is the highest salient of a load curve during a specified period, usually a year.

Potential peak period: It is the period in which a reasonable possibility is considered to exist of an absolute peak developing.

Load factor: It is an index of load characteristics. It is the ratio of the average load over a certain period to the peak load during the same period. The load factor may be daily, monthly or annual depending on the period chosen such as one day, month or year. It may apply to a plant or to a system. The load factor of an industrial area is higher than that for domestic area, the power demand by industry being more uniform. A high load factor is indicative of the better utilisation of the installed capacity and consequently the unit generating cost is less. A low load factor means that the generating capacity is not being fully utilised and as such the generation cost is high.

Hydro plants being basically peak load stations are designed to take variations in load and for that they must have sufficient regulating storage so as to be able to work on low load factor. Operating on 50 percent load factor needs sufficient storage to utilise twice the inflow for half the time. Likewise, at 25 percent load factor, the plant should be able to utilise four times the inflow for a quarter of the time. The lower the load factor, the larger the storage required.

Capacity factor: It is a measure of plant use. It is the ratio of the average load to the plant capacity. The capacity factor may be for a day, month, year or any given period of time. When peak load just equals the plant capacity, the capacity factor and load factor are the same. With maximum demand less than the plant capacity, the capacity factor is greater or less than the load factor, depending on load factor itself.

Utilisation factor: It is a measure of plant use as affected by water supply. It is the ratio of water actually utilised for power generation in the plant to that available in the river. With constant head, utilisation factor is also the ratio of power utilised to the power available. The utilisation factor equals capacity factor if there is always sufficient water to run the plant. A shortage of water results in reduction in the output and may either increase or decrease the utilisation factor according to the plant load factor.

Plant factor: Also termed as capacity factor or plant use factor. It is the ratio of the average output to the installed capacity of the power plant. The capacity factor is identical with load factor when the maximum load just equals the plant capacity.

i.e. $$\text{Capacity factor} = \frac{\text{Maximum load}}{\text{capacity}} \times \text{load factor.}$$

However, load factor is different if the plant is not used to its full capacity. The plant factor as an index of economic viability of a plant should be more than that indicated below:

Plant	*Percent*
Peaking	25
Multipurpose storage	50
Run-of-river	60

$$\therefore \quad \text{Plant factor} = \frac{\text{Average output or load}}{\text{Installed capacity}}$$

$$= \frac{\text{Average load}}{1.1 \times \text{peak load}}, \text{ taking reserve capacity as 10 percent of the peak load.}$$

Power factor: It is the ratio of actual power (in kilowatts, kW) to the apparent power (in kilo volt amperes, KVA). The actual power generated depends on the power factor as the electrical machines are generally rated in KVA.

3.11. LOAD DURATION CURVE

The variation of load demands and other factors such as load factor, capacity factor, plant use factor etc., are studied for a particular duration by means of daily, weekly or monthly load curves.

For longer duration, like that of a year, plotting of daily load curves and using them for calculations on yearly basis is very difficult. Furthermore, the variation of daily load curves is a complicated one as it varies for different seasons and under varying demands etc. Fig. 3.6 shows typical daily load curve of Eastern region of the country. It shows that load remains less than 80% of peak for more than 75% of the time. Thus requirement of peak may be taken care of by peaking station or by load shedding, the latter being economical.

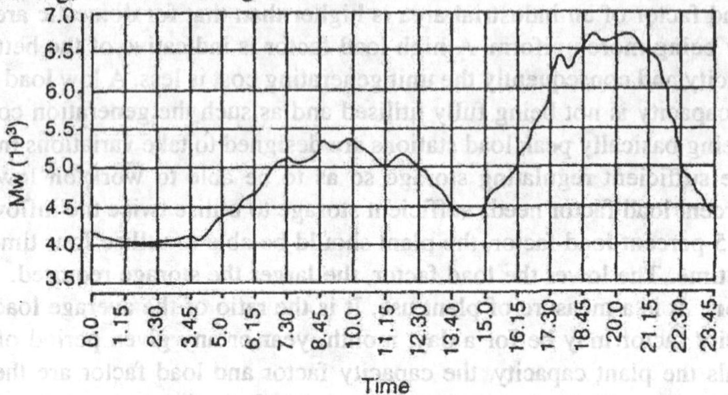

Fig. 3.6 Typical daily load curve of eastern region.

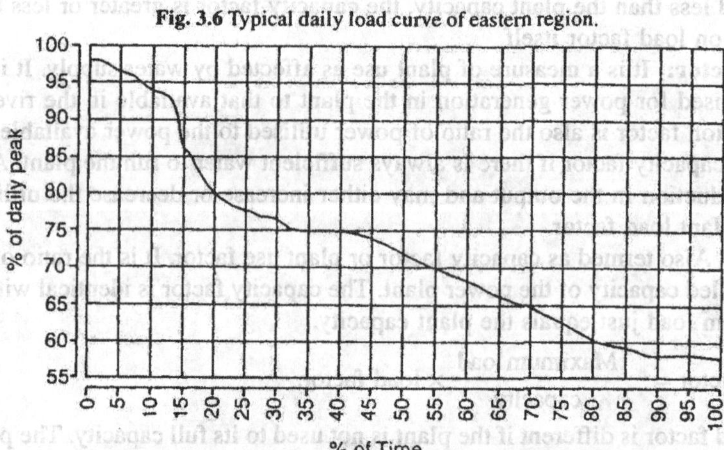

Fig. 3.6(*a*) Typical daily load duration curve of eastern region.

A simple technique lies in drawing annual load duration curves from the daily load curves by splitting the daily load curve into many vertical strips and then arranging them in descending order alongwith the time during which they occurred. Thus a plotting of loads and the percentage of time during which these loads or higher occurred is prepared and is called load-duration curve. The load duration curve is, as such, a rearrangement of daily load curve with loads set up in descending order of magnitude (Fig. 3.6*a*). The curve may be constructed for any duration of time. The areas under the load duration curve and corresponding load curve are equal and measure kW-hr of energy for that period. The annual load factor is given by the ratio of the area under the curve to the area of the rectangle corresponding to the peak demand occurring during the year.

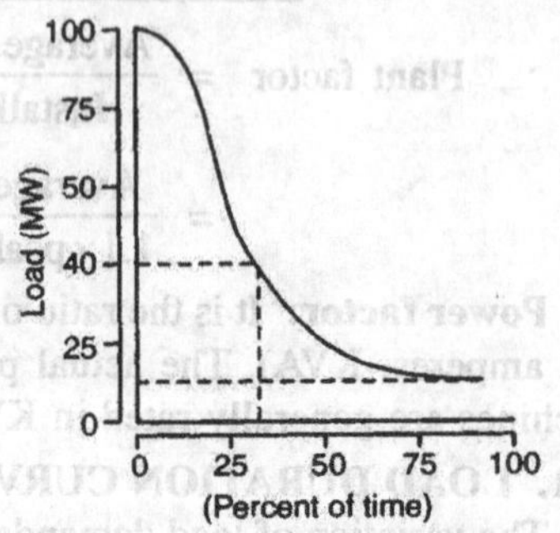

Fig. 3.6(*b*) Load-duration curve.

Refering to Fig. 3.6(*b*) which shows a load-duration curve for a year, it is evident that a load of 40 MW occurs for 30 percent of the time or say 2628 hours in a year. A minimum load of 25 MW occurs for 100 percent of the time or 8760 hours (365 days) in a year.

3.12. VARIATIONS IN LOAD FACTOR

The load factor varies with the type of power development, whether run-of-river plant or storage plant, as mentioned below:

1. Load Factor of Run-of-River Plants

These plant can be operated as base load plants on the rivulets/streams which have substantial and steady minimum flow. The plant size is such that the discharge requirement at full load does not generally exceeds the minimum stream flow. The installed capacity is low and load factor is high if run as an interconnected station. The size of thermal plant can be substantially reduced or altogether eliminated from the integrated system. It thus conserves coal otherwise burnt to run thermal plant for base load requirement. With pondage, the plant can be planned exclusively as a peak load station or may cover a greater range of the load curve including the peak part. The load factor may be as low as 20 to 30 percent.

With varying flow in the stream, these plants can operate on base of the load curve when the flow is steady, gradually changing position as flow decreases and operating entirely on the peak of the load curve under the minimum flow conditions. The generation has to vary according to the load requirements and the power plant should be able to cater for the peak load.

The annual load factor is less than the weekly or 10 day load factors which are less than the daily load factor.

To elaborate the load factor in run-of-river plant as a base station and peak load station, consider a plant which can generate 15 MW of power. The peak load the station can take, as such, is 15 MW. During non-peak load period if load factor is 0.6, power developed is 15 × 0.6 = 9 MW which means the stream inflow can not be utilised fully and has to be wasted.

In case pondage is possible to store water in the non-peak periods, this water can be utilised to generate more power during the peak hours. With the utilisation of the entire water, except evaporation loss etc., 15 MW will become the average power, instead of being the peak. The peak load the plant can take will be 25 MW (i.e., 15 ÷ 0.6) thereby substantially increasing the power output of the plant.

Pondage: The pondage required is determined by the study of load curve for one day. From the load curves, the surplus water available at various periods of the day are computed and the total maximum accumulation gives the capacity of the pondage required (Fig. 3.7). The cost of providing the pondage has to be commensurate with the additional benefits to decide the feasibility of the pondage.

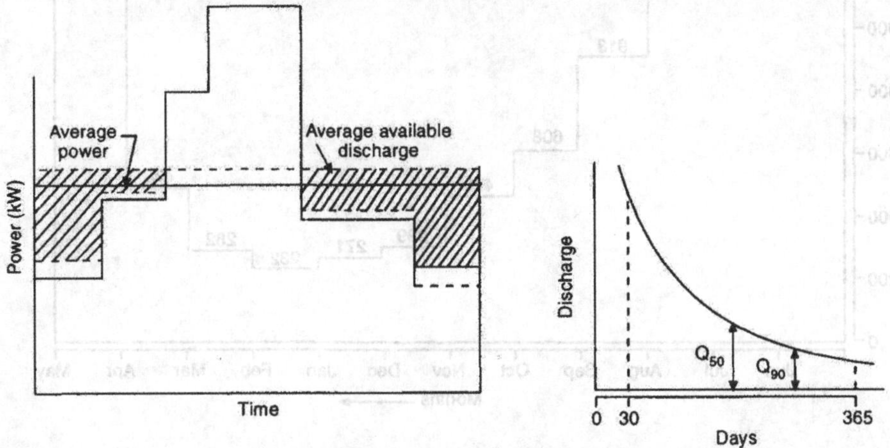

Fig. 3.7 Pondage required

The storage consideration can be done on the weekly basis when generator capacity of the storage will be needed. This will, however, increase the total units generated, and the peak power since in a provision for daily load factor, the water has to be wasted during the days of slack demands such as saturdays and sundays.

2. Load Factor of Storage Plant

The storage based plant gives continuous power which for best utilisation is equal to the average load connected to the station. The peak load corresponds to the average load divided by the load factor. The load factor in this case is taken as the yearly load factor. The station has to be provided with enough installation to meet the peak demand if the plant is an isolated one and takes its own peak. Where the plant is required to run as base load station in a grid, the installation needs be enough only to assure the constant supply of firm power. In case it is a peak station only, it is then run for short period only when much higher installations are needed.

The statutory variation of voltage variation is ± 6 percent of the declared voltage at consumer's terminals.

3.13. POWER SYSTEM LOAD

The power system load is normally met by the various generating sources, as under:

1. **Base Load:** Hydropower without daily/weekly pondage, irrigation controlled without regulating pondage, nuclear and coal fired thermal plants, combined cycle gas turbines (CCGT).

2. **Intermediate Load:** Storage based hydropower plants, CCGT and also sometimes by less efficient old thermal power plants.

3. **Peak Load:** Storage based hydropower plants, run-of-river with daily/weekly pondage, pumped storage, gas turbine plants etc.

The pattern of energy and rate of a typical run-of-river hydroelectric project and with diurnal storage during high (July), intermediate (October) and low flow period (February) is as given in Fig. 3.8 and 3.8*a*.

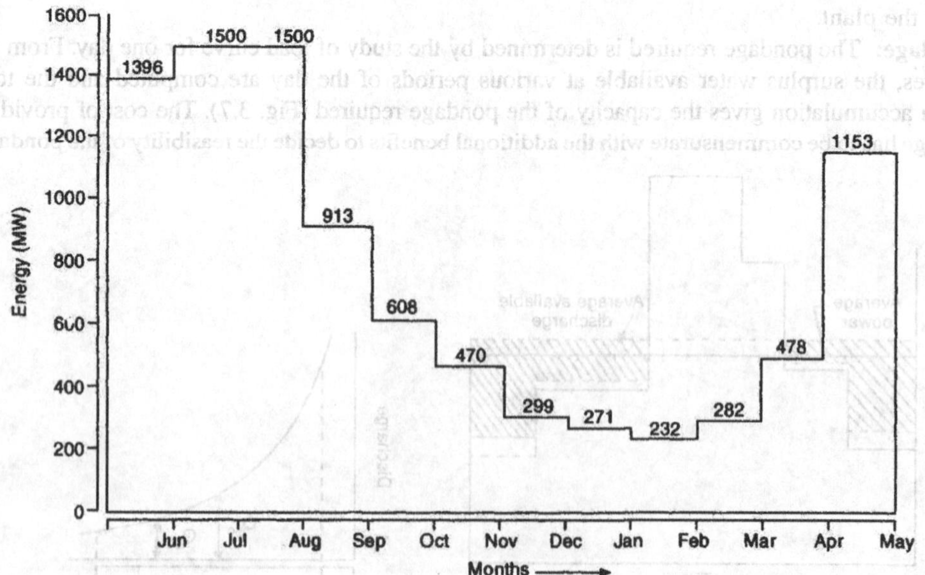

Fig. 3.8 Energy generation pattern from a hydro plant.

Typical Load Curve

The application of power plants for a typical load curve is shown in Fig. 3.8*a*.

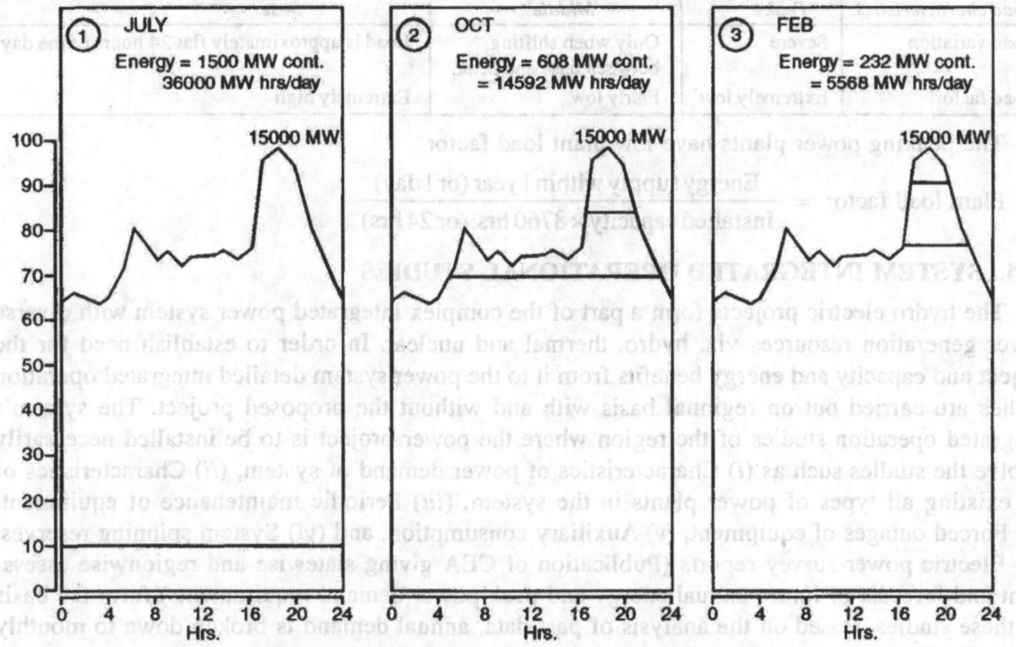

Fig. 3.8*a* Operation of hydropower station in daily load curve

The output of a hydropower station (Fig. 3.9) varies rapidly at around 8, 12, 13, 16 and 17 hours, whereas that of thermal plant does not change as much. The thermal power stations are usually operated for base load to minimise fuel cost. On the other hand, a hydropower station fed by a reservoir or regulating pond responds quickly to load variations and mainly takes care of variable load.

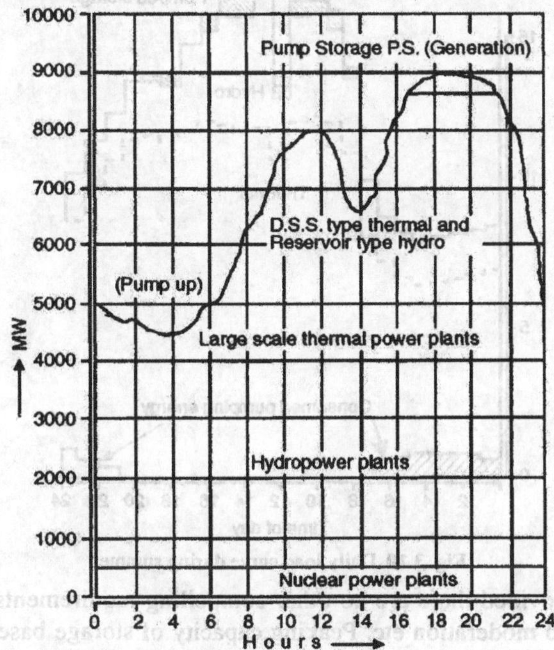

Fig. 3.9 Typical load Curve

When the load is divided into three zones of peak, middle and base zones, the features of the respective zones are as under:

Load characteristics	*Peak*	*Middle*	*Base*
Load variation	Severe	Only when shifting between base and peak.	Load is approximately flat 24 hours of the day
Load factor	Extremely low	Fairly low	Extremely high

The peaking power plants have low plant load factor

$$\text{Plant load factor} = \frac{\text{Energy supply within 1 year (or 1 day)}}{\text{Installed capacity} \times 8760 \text{ hrs. (or 24 hrs)}}$$

3.14. SYSTEM INTEGRATED OPERATIONAL STUDIES

The hydro electric projects form a part of the complex integrated power system with diverse power generation resources viz. hydro, thermal and nuclear. In order to establish need for the project and capacity and energy benefits from it to the power system detailed integrated operation studies are carried out on regional basis with and without the proposed project. The system's integrated operation studies of the region where the power project is to be installed necessarily involve the studies such as (*i*) Characteristics of power demand of system, (*ii*) Characteristics of the existing all types of power plants in the system, (*iii*) Periodic maintenance of equipment, (*iv*) Forced outages of equipment, (*v*) Auxiliary consumption, and (*vi*) System spinning reserves.

Electric power survey reports (Publication of CEA giving statewise and regionwise assessment and forecast of future annual energy and peak power demand requirements) form the basis for these studies. Based on the analysis of past data, annual demand is broken down to monthly and daily figures.

A run-of-river power plant with diurnal pondage may operate as base load station during high flow period. Some plants would assume the role of a peaking station during low flow period. A storage based plant can be operated in a more flexible manner to suit the requirements of peak

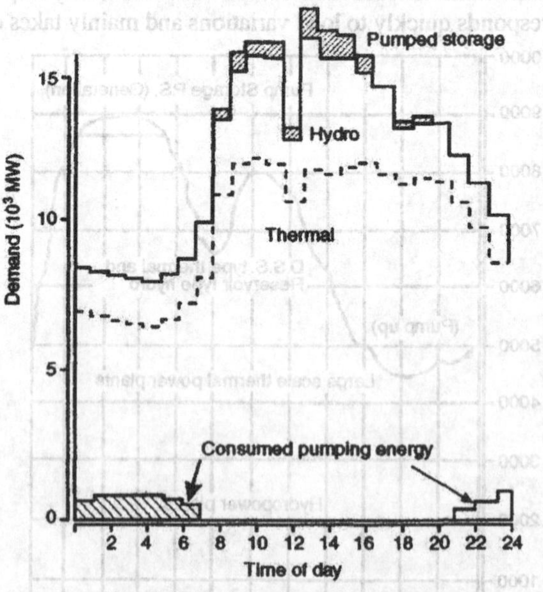

Fig. 3.10 Daily load curve during summer.

demand and energy, provided there are no other compelling requirements/constraints on storage such as irrigation, flood moderation etc. Peaking capacity of storage based plant increases when reservoir level is high during monsoon and reduces as the reservoir level goes low. Nuclear (at

85% PLF) and coal fired thermal plant (at 60% PLF) are preferred for base load operation. Gas turbines and combined cycle gas turbines, diesel and fuel oil fed turbines can be started and stopped quickly and are suitable for meeting peaking demands.

3.15. LOAD PREDICTION

Power sector is highly capital intensive and entire planning of generation, transmission and distribution follows an axiomatic approach. Based on load forecasting, the various aspects which influence the load curve are changing consumption pattern etc. and peak, lean as well as energy forecast based on time series analysis.

The load prediction, load forecasting or forecasting power demand for diverse uses like domestic, agricultural, industrial etc., has essentially a relation with the exponentially growing rate of population and pace of development. Like water flow forecasting, it too is of three types, as given in Table 3.5.

Table 3.5 Load prediction

Type of prediction	*Design period (years)*	*Purpose*
Short term	5	For areas of deficit or surplus power
Medium term	5–10	For planning development of generation and transmission of power
Long term	20	For perspective planning for the power generation capacity of the country.

The hydropower projects have a usual gestation period of 10–15 years and as such they have necessarily to cater for the load demands of next 15 years minimum. Since the hydropower development is usually in stages, the first stage corresponds to the immediate demand, and the balance potential in the second and third stages of the project.

The future power requirements are essentially based on the rate of consumption in the past years and variations in this rate as a result of the pace of development in the past vis-a-vis planned or forecasted for the future, technological changes etc. Predicting the load for the various forms of energy on the long term basis is very difficult for a country of India's size and on the march of development, population and diversity.

The peak load variation in a day is approximately as given in Table 3.6. The typical peak load variation curve in a year in chandigarh, with the highest per capita consumption (1978 kWh) in the country, is depicted in Fig. 3.11.

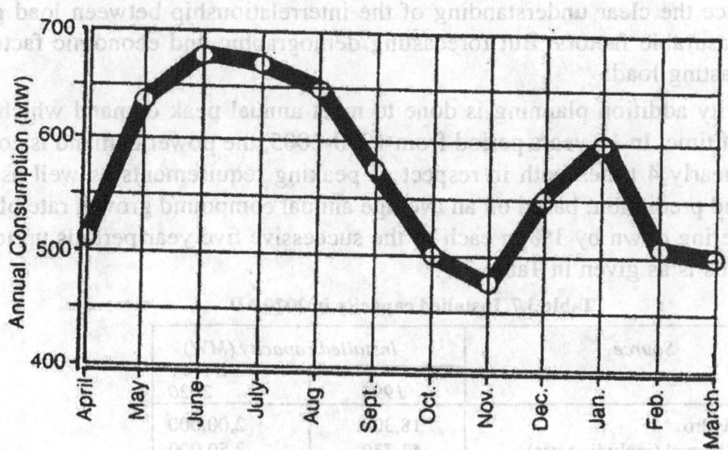

Fig. 3.11 Typical annual load curve

Table 3.6 Peak load variation

Use	*Time of peak load*	*Remarks*
Mechanised agriculture (irrigation)	Mid-day	Operation of pumping at the height of irrigation season.
Domestic commercial, industrial, manufacturing and street lighting	Late afternoon	Traditional time of peak load
Air conditioning	Hot summer	Increased use of air conditioners in metropolitan cities.

The generation capacity corresponds to the load for peak day of the year, whereas the power requirements of peak week or peak month indicate the amount of power storage in the form of fuel or water.

Methods of Load Prediction

The various methods of load prediction are (*i*) Past rate of consumption, (*ii*) Linkage of power development and national economy, and (*iii*) Empirical formulae.

The past rate of all-India consumption of energy in the country has maintained an average annual growth rate of 8 percent. The pattern of electricity consumption has under-gone changes. Industrial consumption has come down from 60% (in past two decades) to about 30% due to energy conservation strategies and captive generation adopted in industries. There is rise in domestic consumption from 10% to 20% on all India basis due to rise in standard of living. Agriculture energy consumption too has doubled, from 15% to 30%. With national economy growth aimed at 6 percent, the corresponding increase in power demand per annum may be of the order of 10 to 11 percent. The demand doubles every 10 years and as such generation too needs to be doubled just to maintain the same per capita consumption. Since per capita consumption is rising rapidly, the generating capacity has to keep pace with it which means generation has necessarily to multiply manifold.

Proper load forecast is the basis of any power planning. The load prediction is directly related to (*i*) Rise in general standard of living, (*ii*) Increase in literacy, (*iii*) Irrigation requirements, (*iv*) Mechanised agriculture, (*v*) Pace of industrialisation, (*vi*) Population growth rate, (*vii*) Electrification of villages and railway lines, (*viii*) Rate of growth of national economy, (*ix*) Weather which too has influence on load requirements. The economic correlation techniques of forecasting relate system loads to various demographic and economic factors. The approach has the merit that it takes into cognisance the clear understanding of the interrelationship between load growth patterns and other measurable factors. But forecasting demographic and economic factors is more difficult than forecasting load.

In India, capacity addition planning is done to meet annual peak demand which lasts for a very small period of time. In 15 years period from 1990-2005, the power demand is considered to multiply itself by nearly 4 times both in respect of peaking requirements as well as energy requirements. The load prediction, based on an average annual compound growth rate of 9% during 1990-2005 and tapering down by 1% in each of the successive five year periods up to 2020 AD, for the peak demands is as given in Table 3.7.

Table 3.7 Installed capacity in 2020 AD.

Source	*Installed capacity (MW)*	
	1990	*2020*
Hydro	18,308	2,00,000
Thermal (including gas)	43,730	2,50,000
Nuclear	1,565	30,000
Non-conventional sources	15	20,000
Total	63,618	5,00,000

Empirical formula

Belgium formula $E = K \times M^{0.6} \times 2^{0.465t}$

where, E is electricity consumption, k is adjustment factor, M is index of manufacture of production and t is time for which consumption is to be projected. The formula is in use in Belgium.

Stochastic Modeling: Forecasting is an apocryphal exercise, as different models with different assumptions are likely to come up with different results, each with their set of assumptions and justification. The set of forecasts that come closest with the actual may be treated as the best ones.

3.16. MARKET REQUIREMENTS OF POWER

The market characteristics effect the sale price and value of power when developed. A power development has two essential characteristics; natural which effect the cost of development and regional which effect the use or market for power. The power development investment hinges on pattern of load demand, the time frame for meeting the energy need and economics of the power generation. The value of a power development too remote from a market up to an amount that represents the capitalised value of saving in cost of power afforded by its use at the market that is available.

The market demand of power in the country for agriculture, domestic, industrialisation far outstrips the available capacity. Average annual load factor of agricultural supplies is over 15% in states with heavy agricultural use. It is obvious that full provision for simultaneous supply of agricultural demand and the rest of the demands including industry, domestic, commercial etc., is not made. It is, therefore, inherently assumed that the transitory agricultural demands will be met by temporary restrictions on other supplies which is also an infinite element. Rainy day in paddy cultivation may bring down demand. Electricity is the most preferred form of commercial energy and its consumption is growing at an average annual compound rate of 10%. Electricity provides the basic infrastructure input to the national economic growth. It is in fact the life line of developing Indian civilisation. The country's market energy requirements are thus enormous.

A brief resume of valuing a power development by three methods is as under:

1. **Real Estate Method:** The method involves fixing a price per horse power of capacity with a reasonable assumed wheel installation, based upon the sale price per horse power of waterpower development of similar character in the vicinity.

2. **Net Income Method:** It is the estimation of net income from hydro plant which is the difference between annual cost of waterpower and the prevailing wholesale price for power in the locality and capitalising this net income at say 10 percent.

3. **Steam Power Comparison Method:** It is the method of comparing the cost of hydropower with that of steam power generation in the locality capitalising the difference in favour of hydropower.

3.17. INSTALLED CAPACITY

The wheel capacity to be installed at a power plant is dictated by economics. The best capacity is that which shows maximum net return in value of power produced besides the consideration of various other factors such as increase in wheel capacity not only increases its cost but also of the generators and power house foundations as well, apart from character of load, load factor, spare units etc. In modern plants, connected to grid, with full development a wheel capacity corresponding to 20 to 40 percent of the available flow is provided. It may be equal to a flow available for 5 percent of the time or less for a peaking plant.

From a flow-duration curve for the power site and the net available head, the available power is computed for any given wheel capacity by planimetering or scaling the area under the flow duration curve up to the line of wheel-discharge capacity, thereby getting the average yearly flow to be utilised.

1. Run-of-River Power Plants

These plants are planned based on 90% yield dependability criteria which is worked out from the inflow data available. Assuming that the inflows data is available for N number of years, 90% dependable year would be 0.9 (N + 1)th year when years are ranked in descending order with respect to total inflows during lean periods.

Although individual run-of-river plants are generally planned on the basis of 90% availability i.e., discharge available for at least 330 days in a year, yet 75% availability also serves well and is adopted in grid system development.

In run-of-river plants without pondage, number of generating units is decided keeping in view the varying discharge and limitations proposed by turbine operating characteristics.

At canal head power plant on Sardar Sarovar project, the units are to operate between a maximum head of 48 m and minimum head of 15.5 m, the design head being about 30.5 m. From the salient parameters for unit capacities of 75 MW, 60 MW and 50 MW sets (Table 3.8), it is seen that adopting unit size of 50 MW to 60 MW would be better and more economical.

Table 3.8 Salient details of unit capacities.

Details	*Capacity (MW)*		
	75	60	50
Runner dia (m)	7.35	5.64	5.2
Setting of C.L. of distributor below minimum pond level (m)	2.5	1.8	1.4
Bottom of draft tube below C.L. of distributor (m)	20.6	14.7	13.5
Spacing of units (m)	32	24	23
Dia. of penstock (m)	9.1	6.7	6.7
RPM of machine	88.24	115.4	125

It is concluded to opt for smaller size units of 50 MW or 60 MW instead of 75 MW. However, from the point of view of height, flexibility and better utilisation of the unit at lower heads, unit size of 50 MW has been preferred.

2. Storage Power Plants

The installed capacity of storage plants operated on annual basis, 90% dependable hydrological year, is determined as in the case of run-of-river plants considering long term water power studies, ensuring no failure for 0.9 (N + 1) years, when N represents number of years for which the hydrological data is available. 90% dependable output in case of carry over storages is higher than the output which is possible on the basis of utilisation of inflows available annually.

The power output for different heights and storages is worked out to determine optimum storage parameters as well as installed capacities. The firm power is obtained by detailed operation tables.

As regards secondary power in reservoir plants, the inflow continues to be high for a long time during wet years and during floods when the reservoir level is higher than full reservoir level, extra power can be generated which is available as secondary power. The secondary power being in addition to the firm benefits is available at comparatively cheaper rates.

Power from a plant is limited by the capacity of the installed equipment, available water supply, head characteristics and storage.

3.18. BENEFITS EVALUATION OF INSTALLED CAPACITIES

The benefits due to the project with different possible installed capacities are evaluated to determine the optimum installed capacity. For the purpose, costs of generation from alternative sources such as coal based thermal, gas turbines and combined cycle gas turbines (CCGT) are computed and the least cost of energy from these options is adopted for evaluating the benefits of the installation considering the corresponding load factor of operation. Benefits of the seasonal energy (i.e., total annual energy–firm annual energy), which can be absorbed in the system, is evaluated with the average incremental cost of fuel consumed in the coal based thermal plant for generating the same energy.

Alternative for installed capacity which provides maximum net benefit (B–C) and assures incremental $\Delta B/\Delta C$ ratio higher than unity is considered optimum installation. Where, C is the present worth of the capital cost and operation and maintenance expenses during the life (50 years) of hydroelectric project under consideration and B is the present worth of benefits which is the value of firm energy of the project under consideration, evaluated at the cost of energy generation of alternative thermal/GT/CCGT plants plus the value of secondary energy evaluated at the cost of resultant fuel saved at fuel based generating station (s).

3.19. SIZE AND NUMBER OF UNITS

The decision about size and number of units to be installed in a power station is interdependent on each other. For a total installed capacity, the total cost of power plant is increased in direct proportion to the number of units, the cost of turbines, generators etc. per kW too increases but not in direct proportion to it. The choice of number of units depends upon total firm power potential and its periodical variation, plant location and use, whether isolated or operated within a grid system. Basically the number of generating units are kept as less as feasible consistent with safety and economy in operation, because the cost of installation increases with an increase in number of generating units in the power plant. The transportability of the components of generating units is also the other basic consideration in deciding the size of the units. The various other factors influencing the choice, number and size of units are as under:

1. For a given total plant capacity, minimum number of units of largest feasible capacity are selected.
2. Smaller the number of units lesser is the cost of civil structures and plant.
3. Large capacity units not only reduce the number of units and hence total plant cost but larger capacity turbines as well as generators also have better efficiencies.
4. There have to be diurnal and annual variations, sometimes substantial in the load on a power station. Number of units are to be so fixed that the total plant efficiency for the whole year may be the maximum. In case of substantial load variation, multi-unit plants are preferred from economy consideration as more units work an full load and only some unit/units work under variable load.
5. Transportability of the components of generating units is the basis in deciding the size of the units. There may be limitations to the size of a single machine by consideration of size of components vis-a-vis transportation capacity available in Indian Railways. Size of castings, especially for runner and inlet valve door which require transportation as a single piece may be limited, otherwise split construction has to be opted for these components which impairs strength and is uneconomical. For example a split runner of Francis turbine may cost 25-30% more than its equivalent integrally cast variant.

It may be concluded that for the supply of electric power in a stable manner to meet large increases in power demand, large capacity power stations are imperative. In planning large scale power plants, it is more economical to increase unit capacity rather than number of units, and for this purpose, the two options of increasing discharge or increasing head are conceivable, but generally it is more economical to make discharge small and head large than vice versa. In this way, the reservoir capacity for the same length of operating time becomes small and as a consequence a small dam with reduced size of water conductor system and headrace and tailrace while by increase in speed of turbine-generator the structure can be made compact with advantage.

The selection of number and type of units is illustrated by an example at appendix-I.

QUESTIONS

1. What are the various objectives of planning for waterpower development?
2. Give an account of planning for waterpower development on long term and short term basis.
3. Mention various components of planning for waterpower development.
4. Give a brief account of salient features of low, medium and high head plants with regard to head, topography, discharge, storage/pondage, generation cost etc.
5. Briefly discuss hydropower policy formulated by the Government of India to accelerate waterpower development.
6. What are the various investigations and studies required to be carried out for the development of storage power development.
7. Estimate available waterpower at a site for the given discharge and head.
8. Discuss power duration curve and its role for the assessment of amount of hydropower available at a site.
9. Why is storage at a power plant necessary? How does storage differ from pondage and in which respects?
10. What are the various load studies carried out for waterpower development?
11. Discuss firm power and secondary power. In what respects the two differ?
12. Describe reserve capacity. Why is it necessary and how determined?
13. What is load duration curve? Illustrate with the aid of a sketch.
14. Discuss load prediction and the various methods for its determination.
15. What do you understand by installed capacity of a power plant? What are the various considerations in fixing installed capacity with regard to size and number of units?

4

WATER POWER DEVELOPMENT

4.1. INTRODUCTION

The human endeavour to harness water for the generation of electricity is time old. Electricity generated with water is usually referred to as Hydropower. The term "hydro" is derived from the Greek word "Hudor" meaning water. Hydropower is the motive energy contained in water which can be converted into electricity through hydropower plants.

Water is needed to run a hydroelectric plant. The level of water in the reservoir created behind the dam is higher than the power plant. When water is released from the reservoir into the penstocks it rushes down into the power house to strike the blades of turbines with such force that they begin to spin. The spinning turbine is connected by a shaft to the generator. Inside the generator is a large electromagnet which spins inside a coil of wire producing electricity. The electricity from the generator is carried to the switchyard where the voltage is boosted for transmission over long distance power lines. Water is released through the base of the turbine and joins the river downstream of the dam.

Hydropower generation has assumed all the more importance in recognition of its distinct and clear economic and environmental advantages over thermal being inexhaustible, perpetual and renewal energy source, more reliable because of its non-dependence on costly and at time unreliable and low quality supply of coal, imported oil etc., non-polluting, high conversion efficiency, flexibility in operation, lower cost of generation as well as operation and maintenance, immune to inflation and threat of shortage of coal due to long distance transportation and strikes etc. It has inherent ability for instantaneous starting and stopping, almost instantaneous load acceptance and rejection. It is ideal for peak load demand. The fuel cost being zero, the cost of generation declines over time unlike thermal power and in the long run it turns out to be economical. Irrigation, flood control, drinking and industrial water supply, development of pisciculture, recreation etc., are the additional benefits in the context of multipurpose river valley project. Presently the hydropower provides the high-value peaking capacity to the power grid, operating with thermal and nuclear plants as base loads.

4.2. HISTORY OF HYDROPOWER DEVELOPMENT

Hydropower generation is an improved variant of primitive water wheel for grinding cereals. As hydroelectric power it emerged in USA in 1882, followed by Sweden and Japan. In India, a hydropower plant of 130 kW installed capacity was commissioned in 1897 at Sidrapong in Darjeeling (West Bengal) and followed by 4.5 MW plant at Sivasamudram (Karnataka) in 1902. By 1915 about 70 percent of the country's demand for electricity was being met with from hydroelectric stations. During the period between two world wars, a number of hydel power plants such as 48 MW at Jogindernagar (H.P.), 17.4 MW Ganga canal power plant (U.P.), 38.75 MW Pykara and 30 MW Metter (Chennai), 13.5 MW Pallivasal (Andhra Pradesh) were commissioned,. From installed capacity of 1362 MW, out of which hydropower was 508 MW in 1947, the pace of growth has been rapid in post-independence era. The notable giant multipurpose projects commissioned in the period are Damodar Valley (Bihar and Bengal), Bhakra Nangal Project (Himachal and Punjab), Hirakud dam (Orissa), Chambal Valley project (Madhya Pradesh and Rajasthan), Tungabhadrra project (Andhra Pradesh and Mysore), Ramganga Project (Rajasthan), Pong dam (Himachal Pradesh) and Beas Sutlej link project (Himachal Pradesh). From an installed capacity of 2000 MW at the beginning of Five year plans, it grew to almost 30,000 MW by the end of Fifth five year plan, out of which hydropower was about 12000 MW. By the end of 7th Plan (March 1990) it rose to 18,300 MW. Only 881 MW of hydro capacity was added during the two annual plans (1990–92). During 8th plan, installed capacity of 38,000 MW hydropower was planned but the achievement was 16,500 MW. The Task Force on 9th Five year plan has recommended hydropower capacity addition of 10,515 MW. The hydel installed capacity by end 2001 is 25,574 MW, out of total capacity of 1,02,907 MW. (Fig. 4.1).

4.3. ADVANTAGES OF HYDROPOWER

Hydropower has numerous inherent advantages, as under, which makes it the most preferred form of environmentally bengin renewable source of energy, and above all, on full development, it will be a factor in providing energy independence for the country.

The thermal and nuclear power plants are given precedence simply because of their shorter gestation period. While their development is no doubt imperative, hydel power generation deserves priority because it utilises a renewable natural resource which is constantly recycled in nature and is perennially available. It also enables conservation of limited reserve of non-renewable fossil fuel. The several technical merits of hydropower are that due to flexibility in operation, hydropower plants are adaptable to meeting peaking loads, load variations and frequency control and spinning reserve duty thereby lending stability and reliability to the power system as a whole. In the context of ozone depletion and global warming it is looked upon as the energy source of the future which is clean. It offers many advantages such as:

(*i*) Renewable source of energy thereby saves scarce fuel reserves, (*ii*) Economical source of power, (*iii*) Non-polluting and hence environment friendly, (*iv*) Reliable energy source with approximately 90 percent availability, (*v*) Low generation cost compared with other energy sources and cheapest in the long run due to zero fuel component and as such free from inflationery effects, (*vi*) Indigenous, inexhaustible, perpetual and renewable energy resource. Hydropower plants make use of water but do not consume it, (*vii*) Helps in management and regulation of water resources, storage based plants are often of multipurpose concept with incidental benefits of flood control, irrigation, public water supply, navigation, fishing, tourism etc., (*viii*) Provide employment opportunities being labour intensive, (*ix*) Lead to development of remote areas where these are generally located, (*x*) Long span of life; plant in service since 1897, (*xi*) Afford a degree of independence from costly and at times unreliable supply of imported fuel, (*xii*) Technically more reliable than many thermal plants. Stabilisation period is 2 to 3 months against 6 to 12 months for thermal plants, (*xiii*) Plant equipment is simple and less complicated than thermal, nuclear or even diesel power generating equipment, (*xiv*) Low operation and maintenance cost, (*xv*) Better service operation flexibility, the operation of the plant can be matched with load requirements i.e., they can operate both as base load and peaking station, (*xvi*) Possible to build power plants of high capacity, (*xvii*) Quick to start and stop, picking up and instantaneous load acceptance/rejection make it suitable to meet peak demand and enhance system reliability and stability, (*xviii*) Speed of output variation, depending on unit capacity, is high; 10-15 MW/s, (*xix*) Peaking operation enables optimum utilisation of thermal capacity in the system, (*xx*) Reliable source of energy with high availability factor and minimum maintenance needs, (*xxi*) Socio-economic benefits being located usually in remote areas, (*xxii*) Practically immune to labour problems; much less labour strength is needed for operation, (*xxiii*) Its proven technology is accessible for all countries, (*xxiv*) Foreign exchange requirement for the manufacture of hydroelectric equipment is much less compared to nuclear and thermal, (*xxv*) Higher efficiency, 95% to 98%, compared to 35% thermal, 42% gas in combined cycle and 28 to 30% in open cycle.

Limitations

(*i*) Susceptible to vagaries of nature such as drought, (*ii*) Longer construction period which means a disadvantageously freezing of the capital invested, (*iii*) Breakdown in hydropower equipment may result not only in proportionate reduction in power generation but also, particularly in the case of run-of-river plants, in letting the precious water run waste, (*iv*) High initial cost, (*v*) Long term flow data is essential for proper assessment of irrigation and power benefits, (*vi*) Loss of large land due to submergence in the reservoir so created, (*vii*) Non-availability of suitable sites for the construction of dam, (*viii*) Displacement of large population from reservoir area and their rehabilitation, (*ix*) Environmental aspect; reservoir *vs* river ecology, and (*x*) High cost of transmission system for remote sites.

4.4. HYDRO VS THERMAL POWER

The relative usefulness of two sources of power is determined in consideration of availability of raw materials and their distance from the plant, unit cost of energy, load factor, dependable capacity, unit rate, cost benefit and other parameters over a long period. A comparison of hydro and thermal power is as under:

Hydel power	*Thermal power*
Environmentally clean	Non eco-friendly
Cheapest energy source; about one third to one-half of that of thermal power.	Costly energy source.
Water is a renewable source. Production of power does not consume water.	Coal consumed in generating power is a non-replenishable source and is being fast consumed.
Maximum financial return against small fraction of total investment on multipurpose project.	All the investment on power development only, and indirectly on coal mining.
Offers possibility of storing surplus energy in the form of pumped storage generation.	Storage of electricity is not possible.
Plants are robust requiring minimum maintenance. Outages are relatively less.	Delicate structures with high maintenance cost. More outages.
Lesser operation and maintenance staff requirement.	Large operation and maintenance staff requirement.
Overhaul and maintenance need shut-down of very short durations.	Require longer shut-downs for overhaul and maintenance.
Water transportation is a natural process under gravity.	Requires quality coal besides its transportation from mines.
Water resource is almost uniformly distributed offering power development in large regions of the country.	Coal reserves are concentrated in certain geographical locations only. At locations away from coal pits it is costly, time consuming and at times unreliable due to strikes etc.
Pollution free generation of power.	Pollution and ash disposal problems.
Plant equipment is simple. Operation is also simple.	Complicated plant equipment and generation.
High efficiency. Highest efficiency prime movers; turbine efficiency 90-95%.	Low efficiency, thermal turbines efficiency 30-40%.
Non-consumable small quantity of water is required for cooling system.	Large amount of water supply is consumed for production of steam.
High extra cost on dam and appurtenant works.	Construction of dam and appurtenant works is not involved.
Serious rehabilitation and resettlement problem of the population uprooted from large reservoir area.	Limited R & R problem.
Reliable energy source with approximately 90% availability.	Low reliability energy source.
Greater life expectancy, 50-100 years	Low life expectancy, 25 years.
Minimum stabilisation period of 2 to 3 months.	Longer stabilisation period of 6 to 12 months.
Smaller operating work force make them immune from labour problems.	Large operating work force besides that engaged in transportation and mining of coal susceptible to labour problems.

Hydel power	*Thermal power*
Numerous incidental benefits of flood control, navigation, pisciculture, recreation facilities etc.	Single purpose venture.
Low capacity rating machines.	High capacity rating machines.
Longer gestation period.	Shorter gestation period.
High service operational flexibility, enables better matching of operation of power plant with load requirement.	Lesser operational flexibility.
Generating sets can run and pick up load in almost no time to start and to stop.	Considerable time is required to run and pick up load.
Operation is dependent on water which in turn is susceptible to vagaries of nature.	Operation is dependent on the supply and carriage of coal by rail—a means of communication susceptible to shortage and strikes etc.
Sites at isolated and inaccessible locations away from load centre necessitating long transmission lines.	Located usually near the load centre, the essential requirement being availability of coal and water in adequate quantities.
High cost on long transmission lines to carry the energy as also high transmission losses.	Short transmission lines and hence less transmission losses.
Large area is submerged in the reservoir which is required to be acquired besides that for dam and appurtenant works.	Land required for the power plant is less but large additional area is required for disposal of ash and mining of coal.
Land acquisition in hilly areas is cheap.	Land acquired is in plain area and very costly.
Speed of output variation depending on unit capacity is high (10 to 15 MW/s).	Speed of output variation is slow (3 to 5 MW/min. in 500 MW units).
Starting and stopping is in 2-3 minutes, even less in modern plants.	The equivalent period is 20-30 minutes even in 'warmed up' condition.
Output varies rapidly.	Output does not change as much.
Mainly take care of variable load.	Operated for base load to minimise fuel cost.
Foreign exchange requirements for manufacture of hydro-electric equipment is minimal.	Foreign exchange requirement is more compared to hydro.
Improvement in micro-climate; becomes humid and cool.	Deterioration in micro-climate.
Submergence of flora and fauna. Barrier to fish life.	Dislocation to flora and fauna is negligible.
Impact zone of storage plant limited to area of submergence.	Serous health hazard zone around the plant. Even ground water is polluted by fly ash disposal.
Storage dams and reservoir are tourist spots.	Living nearby plants is hazardous due to air, water pollution and emission of gases.

4.5. HYDROPOWER POTENTIAL AND ITS DEVELOPMENT

The global hydropower potential economically exploitable is 15,000 billion units (kWh) annually and that of India is about 4% of it. The country ranks 5th in the world.

There are 12 major river basins, as well as 22 west and 24 east-flowing medium sized rivers. The rivers have been classified into 6 river systems for the assessment of their hydro potential.

The country's total hydropower potential has been assessed as 84,044 MW at 60 percent load factor, which is more than the oil potential of the Arab countries. (Table 4.1). The region-wise potential at 60 percent load factor is as given in Table. 4.2. It is equivalent to total installed capacity of about 1,50,000 MW which will yield an annual energy of 442 billion units of energy. With seasonal energy, the total energy potential is assessed to be 600 billion units per year.

Table 4.1 Distribution of Hydropower potential (MW)

Conventional						
Basin/river	*Potential at 60% L.F. (MW)*	*Number of schemes*	*Potential already developed upto 31.3.99 (MW)*	*Potential development from ongoing projects (MW)*	*Total (MW)*	*% of potential developed & under development*
Indus	19,988	190	2,936	1,373	4,309	22
Ganga	10,715	142	1,860	1,138	2,998	28
Central Indian rivers	2,740	53	634	1,528	2,162	79
West flowing rivers	6,149	94	3,518	684	4,202	68
East flowing rivers	9,532	140	3,663	699	4,362	46
Brahmaputra	34,920	226	453	367	820	2
All India	84,044	845	13,064	5,749	18,853	22, 25% up to 2002

Firm power potential is 50,426 MW. Theoretical potential is 3,01,117 MW out of which economical potential is 84,044 MW.

Table 4.2 Region-wise hydro potential

Region	*North-east*	*North*	*South*	*West*	*East*
Potential (MW)	31,857	30,155	10,763	5,679	5,590
Percentage	37	36	13	7	7

The basin wise and schemewise break up of the potential is as given in Table 4.3.

Table 4.3 Hydro potential

Name of river/basin	*Number of schemes identified*			*Potential at 60% load factor*		*Installed capacity*
	Storage	*R-O-R*	*Total*	*(MW)*	*%*	*(MW)*
Indus basin	23	167	190	19,988	23.8	33,842
Ganga basin	35	107	142	10,715	12.8	20,711
Central Indian river system	38	15	53	2,740	3.3	4,152
West flowing rivers of southern India	69	25	94	6,149	7.3	9,418
East flowing rivers of southern India	90	50	140	9,532	11.3	14,511
Brahmaputra basin	76	150	226	34,920	41.6	66,065
Total	331	514	845	84,044	100.0	1,48,699

About 78% of the total hydropower potential (64,000 MW) is in the Himalayan region, comprising the basins of the Ganga, the Indus and the Brahmaputra. The Brahmaputra river alone has maximum potential (34,920 MW), 41.55% of the total potential but the least development hardly 2 percent. Some of the major projects envisaged therein include Dihang (20,000 MW), subansiri (4800 MW) and Tipaimuk (1500 MW). Among the states, 32% of the potential exists in Arunachal Pradesh. Himachal Pradesh with 14% of the country's potential ranks second.

Potential of conventional as well as non-conventional waterpower is as given in Table 4.4.

Table 4.4 Total hydropower potential

Conventional hydro (MW)	*Non-conventional hydro (MW)*			
	Pumped storage	*Small hydro*	*Tidal energy*	*Total*
84,044	96,524	10,000	10,000	1,16,524

The hydropower potential is not only impressive in national as well as international context, but is of greater importance in as much as it is continually renewable, non-polluting resource, available free of cost and provides cheapest energy. Hydropower is also an important part of multipurpose projects.

The potential of a hydroelectric project represents the energy content of a scheme and is expressed at 60 percent load factor which has been considered common basis for operating a system involving various modes of generation.

It transpires from the Table 4.1 that about 75 percent of the hydropower potential still remains to be tapped.

4.6. PROJECTED INSTALLED CAPACITY

The Fourth National Power Plan (1997) assumed a compound growth rate of 7 percent in the installed generation capacity for the country during the period 1997-2012. The commission made estimates of future installed capacity, assuming compound rate of 7 percent per year up to the year 2025 and 6 percent during the period 2025 to 2050. The source-mix assumption has been made keeping in view the known potential, capital cost, gestation period, economic and environmental implications relating to hydropower, nuclear energy and other renewable sources of energy. For sensitivity analysis, production capacities for low and high scenarios have been taken as 95 percent and 105 percent of projected activity.

According to 16th electric power survey, the need-based addition during, forthcoming 10th and 11th plans is 55,000 MW and 52,000 MW respectively.

According to projections made by Central Electricity Authority, total installed capacity at the end of Eleventh Plan (2007-12) will be 2,40,000 MW; the hydro component being 74,000 MW which is nearly three times of the present installed capacity. The projected energy related water requirement (evaporation plus consumptive use) in the year 2025 will be 71 billion cubic metre with respect to the corresponding present (2000 AD) demand of 27 billion cu. m and 150 b. cu. m in 2050.

Projects Under Construction/To be implemented

During the 8th Five-year plan (1992-97), against the proposal to add 9282 MW of hydro capacity, the actual achievement was only 2602 MW. The decline in implementing hydropower has resulted in sub-optimal use of existing capacity, and an increase in the overall cost of generation. Emphasis is, therefore, being placed on adding a substantial amount of hydro capacity over the next decades.

Table 4.5 shows some of the major schemes which have either been sanctioned, or are already under construction.

Table 4.5 Large projects under construction or sanctioned.

Project	*State*	*No. and capacity of units (MW)*	*Total capacity (MW)*
Nathpa Jhakri	Himachal Pradesh	6 × 250	1500
Dul Hasti	Jammu & Kashmir	3 × 130	390
Uri	Jammu & Kashmir	4 × 120	480
Tehri Stage I	Uttar Pradesh	4 × 250	1000
Dhaulganga	Uttar Pradesh	4 × 70	280
Lakhwar Vyasi	Uttar Pradesh	3 × 100	300
Sardar Sarovar	Gujarat	6 × 200	1200
Ghatgarh	Maharashtra	2 × 125	250
Koyna Stage IV	Maharashtra	4 × 250	1000
Srisailam LBPH	Andhra Pradesh	6 × 150	900
Gangawai (Bedhi)	Karnataka	2 × 105	210
Koel Karo	Bihar	4 × 172.5	690
Puralia (PS)	West Bengal	4 × 225	900
Pykara Ultimate	Tamil Nadu	3 × 50	150

It is anticipated that during the 9th Five-year plan (1997-2002) a total of about 17,410 MW of hydro capacity will be added in the various regions as given in Table 4.6.

Table 4.6 Anticipated capacity addition

Region	*North*	*West*	*South*	*East*	*Northeast*
Capacity addition (MW)	8,412	4,914	2,521	900	657

Of this, it is likely that 5530 MW will be implemented by the so-called Central Sector, which means directly by Government of India Undertakings; 11,090 MW will be implemented by the State Sector, which means various state electricity boards and power corporations; and 790 MW will be undertaken by private sector agencies.

Table 4.7 shows some of the principal schemes which are expected to be implemented over the next ten years.

Table 4.7 Some schemes to be implemented over the next 10 years.

Project	*State*	*No. and capacity of units (MW)*	*Total capacity (MW)*
Chamera Stage II	Himachal Pradesh	3 × 100	300
Baspa II	Himachal Pradesh	3 × 100	300
Baglihar	Jammu & Kashmir	3 × 150	450
Tehri Stage II	Uttar Pradesh	4 × 250	1000
Vishnu Prayag	Uttar Pradesh	4 × 100	400
Lohari Nagpala	Uttar Pradesh	3 × 94	282
Teesta Stage II	Sikkim	6 × 200	1200
Teesta Stage IV	Sikkim	3 × 170	510

Hydropower development in India is often hampered by limited financial resources, complex environmental clearance procedures and long gestation period. A number of incentives are now being offered for private sector participation. Formalities for obtaining environmental clearance are also being streamlined with a view to overcoming delays. These measures will give a major thrust to hydropower development over the next ten years, which should accelerate its development and reduce to a great extent the present imbalance in the hydro-thermal mix.

4.7. HYDRO: THERMAL MIX

A certain proportion of hydropower is essential in the regional power generation grid because of its inherent advantage in providing peak energy during peak hours.

In a power grid, some power plants such as thermal and nuclear are run as based-load stations to give their best when they are fully loaded, and essentially some other like hydro, including pumped storage cater for peak load. In power production mix of hydro, thermal and nuclear, the two considerations are percentage of each type of source depending on availability of the natural resource and operation of various plants to meet the demand of load with overall economy in view. System analysis studies are done to determine the optimum mix. The mix in terms of storage hydro, run-of-river hydropower, thermal, nuclear etc., is important as without hydel back up the overall cost of meeting the power demand is very high besides for the conservation of the fast depleting coal reserves of the country. It is estimated that every 1000 MW of hydro capacity conserves normally 3 to 4 million tonnes of coal per annum. The pumped storage power development can meet peak requirements to a very large extent resulting in saving in new installations of fossil fuels, and other modes of power generation.

Water's role in the energy sector was identified at a very early stage of planning in the country. It was acknowledged that hydropower was the cheapest, perpetual and eco-friendly of all available power generation sources. The share of hydropower was 33% at the start of First Five Year Plan in 1951, rose to 46% by the end of Third Plan with a maximum of 50.62% in 1963 which in 2002 has dipped to 25% (Fig. 4.1).

Hydro share in India vis-a-vis other countries is dismal as indicated in Table 4.8.

Table 4.8 Hydro share (%)

Country	*India*	*Norway*	*Brazil*	*New Zealand*	*Austria*	*Canada*	*Sweden*
Share	24	99	93	74	69	68	64

The principal factors responsible for the dismal share of hydroelectric plants in the overall installed generating capacity in the country are unrealistic planning, poor implementation and ineffective monitoring besides the following :

1. Uneven distribution of hydroelectric resources, among the political divisions.

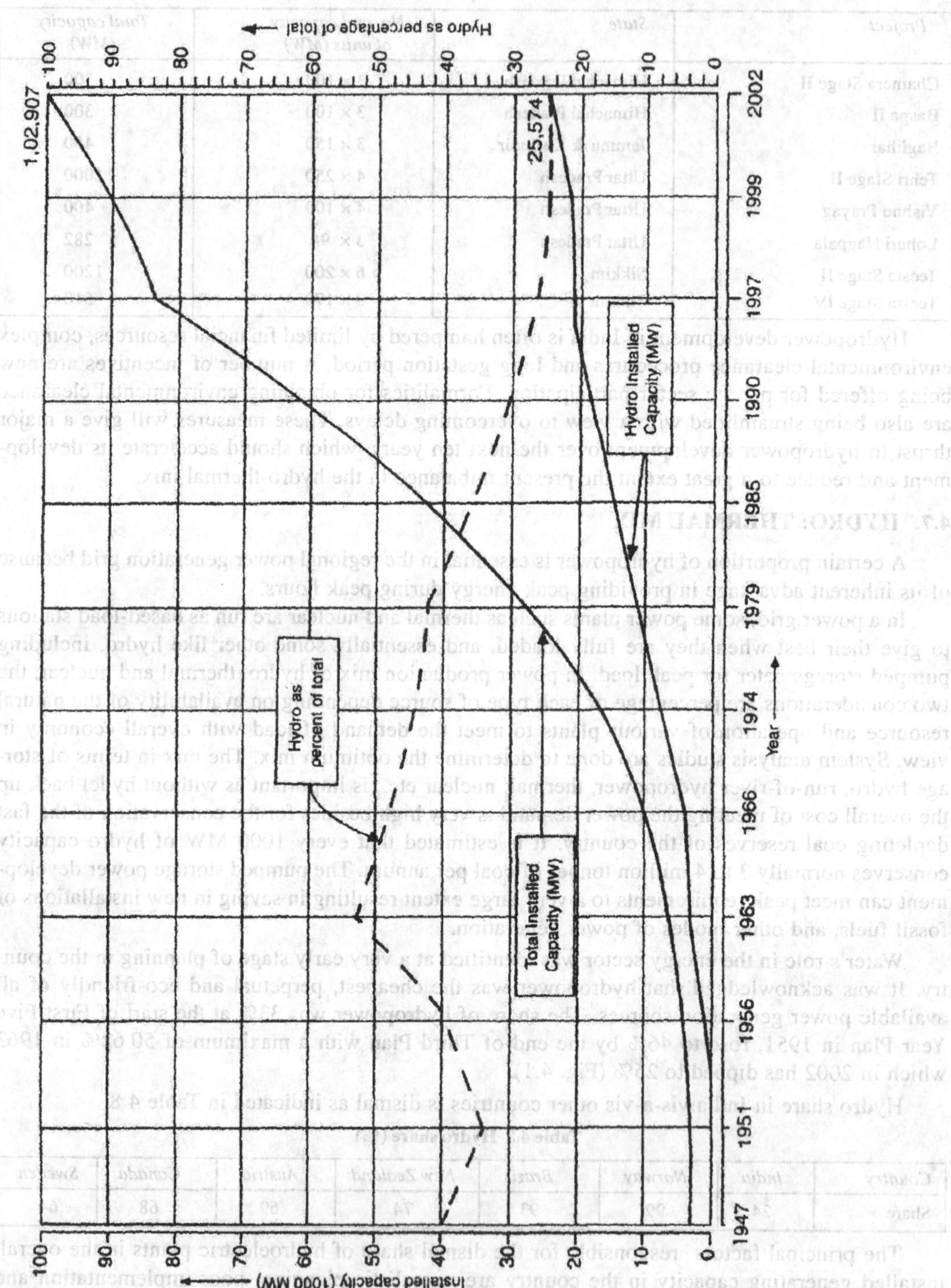

Fig. 4.1 Installed capacity and hydro share (Source CEA)

2. Concentration of hydro resources in the states having neither the demand for power nor capability for large hydro development.
3. Environmental constraints in terms of deforestation, submergence and rehabilitation.
4. Geological uncertainties leading to slowing down of the civil works construction.
5. Socio-economic impediments due to remoteness of sites far away from the urban centres.
6. Attitudinal impediments in terms of choices for shorter gestation schemes, and
7. Financial constraints.

The main cause for the gradual but steady decline in the hydro share is long gestation period of hydropower projects which has made planners to shift to thermal power generation to tide over the energy crunch the country is facing.

Adverse Effects of Poor Hydro: thermal Mix

The operation of thermal plants at a very low load, during off-peak period, adversely effect the efficiency of the system, the plant life and fuel consumption. Hence the imperative need for hydropowers to take care of peak load, whereas thermal is eminently suitable for base load.

Optimal hydrothermal mix is related to the basic energy sources of water and coal; optimisation of the consumption of coal and water use as also that the system should have a mix of plant characteristics capable of meeting economically the fluctuations in the load demand.

For an efficient and optimal operation of a power system, an optimal hydro : thermal mix of 40:60 is considered as most desirable for economic operation of the system and environmentally sustainable development. Likewise, ideal hydro-thermal-nuclear mix for healthy sustainable development is 41 : 56 : 13. The adverse effects of poor hydro: thermal mix are (*i*) Load carrying capacity of the grid is constrained, (*ii*) Poor efficiency of the grid, (*iii*) Frequent load shedding, (*iv*) Fluctuations in voltage and frequency, (*v*) Backing down of thermal plants, (*vi*) More atmospheric pollution and environmental degradation, (*vii*) Early exhaustion of coal reserves, and (*viii*) Consumers made to pay more.

The economy of different power sources is reflected in the price of the electricity. The countries which have a large proportion of hydropower in their systems have the lowest tariffs.

To achieve the ideal mix of 40:60, rapid increase in hydropower generation is imperative; the un-exploited hydro potential to the extent of 75% promises to be a significant amount in the future energy mix.

4.8. CONSTRAINTS IN HYDROPOWER DEVELOPMENT

Development of hydropower is beset with many problems. Most of the dam sites are in difficult terrain and geology, on interstate rivers involving joint construction projects, with problems of equitable sharing of waters, sharing of costs and benefits, lack of common rehabilitation and resettlement policy, issue of land for land, environmental concerns, loss of forest cover, besides dislocation of tribal communities, etc. During the last decade, several dam projects have been slowed down because of agitations, repeated fasts into death, "Jal-Samarpan" threats etc. The main disadvantages attributed to hydroelectric power projects are high investment, longer gestation period and dependence on nature, etc. etc. briefly as under:

1. Longer gestation: The longer gestation period attributed to the hydropower is on account of the inclusion of the period of investigation and surveys, whereas the gestation for the activities associated with thermal plants such as exploration and development of coal mines etc. is not taken into account. Hydel projects can now be completed within 5–7 years as in other countries, as also like thermal plants by advance planning, state-of-the art technology, better administrative and modern management techniques, standardisation of design of project components. Interim benefits of generation may even be derived before raising the dam to full designed height.

86 MW Malana Hydro Proejct (Himachal Pradesh) has been completed in 2001 in a record period of two-and-half years which otherwise was scheduled to be completed in five years.

2. Poor Planning: The continued shift in the power development programme to thermal-power stations tentamounts to dismal failure in planning in power sector to provide the most economical hydro energy.

3. Financial Constraint: The states where bulk of hydropower potential is located have neither the resources to develop major hydropower projects nor have power demand. Paucity of funds and poor management of the available resources are responsible for slow development of hydropower. The funding policy has resulted in the long term handicap for the hydro projects which are consequently suffering from lack of resources.

4. Environmental and Forest Clearance: It is an important statutory clearance for the construction of a project. The over emphasis on environmental consideration by concerned agencies have adversely affected securing final clearance for taking up construction. Identification of non-forest land, its acquisition and preparation and implementation of afforestation plans, compensatory afforestation etc., contribute to considerable delay in obtaining forest clearance for hydro projects.

5. Higher initial investment: Hydropower plants are considered to require relatively large capital investment on dam and appurtenant works in addition to the power plant itself but thermal plants too need huge expenditure on (*i*) Development of coal mines, (*ii*) Ash disposal, (*iii*) Rehabilitation of ecological environment disturbed in the mining area, (*iv*) Investment on rolling stock, new rail lines, signalling facilities etc., (*v*) Auxiliary consumption of a thermal plant is 10-12% and, therefore, capital investment per kW of the useful thermal power thus becomes 10-12% more.

According to a survey, the investment on a thermal plant is about double that of the hydro equivalent. Furthermore, the average cost of the thermal power generation is about three times that of the hydel power.

6. Delays in clearance and implementation: A number of statutory and non-statutory clearances are required for hydropower projects from central and state governments. The lag time involved in obtaining the clearances before start of construction and/or even stoppage later on have been a nightmare for the development of hydropower. Development has also been delayed due to lack of planning/uncertainity about the implementing agencies.

7. Central government guarantee: No hydropower project proposal is complete unless proper power purchase agreement for the financial institutions to lend the money for the project is backed up by the central government guarantee for sales realisation is tied up before hand. This is not done in all cases and a review is required and hence delay.

8. Rehabilitation and Resettlement: It is one of the most sensitive, complex and human issue. Storage hydropower projects involve submergence of large areas necessitating involuntary shifting of the habitat to new areas and inherent human nature to oppose such a change in habitat; occupation, custom and culture being different in alien land of resettlement, the liberal R & R policy notwithstanding. The poor record of resettlement of project affected people is an organisational failure-common in every field in the country.

9. Interstate disputes: Most of the perennial rivers, in India, flow through one or more states and the lean period supply gets reduced substantially. A number of projects have common river systems between adjoining states. The interstate disputes whether the state on upper reach can store water or not and sharing of the water of many rivers are very intricate to resolve and finally land with supreme court resulting in delay and keeping the proposal for the project in abeyance. The total installed capacity of such projects is about 4000 MW.

10. International dimensions: The largest part of the country's water resources is contributed by transboundary rivers which rise from, or flow into or through neighbouring countries viz., Nepal, Bhutan, Tibet/China, Myanmar, Bangladesh and Pakistan. Optimum development of the hydropower potential is constrained taking cognisance of this geo-political reality. There are varied problems and issues that usually arise in working out such cooperation as a result of global concern for environment, bio-diversity, emission, R & R of displaced persons as also the possible effects on such cooperation by multi-national corporations.

11. Uneven Distribution of Potential: A paradoxical situation exists in that the potential awaiting exploitation is concentrated in northern and north eastern regions of the country which neither have demand nor have adequate financial resources to develop it.

12. Socio-economic impediments: The hydropower potential is lying locked due to remoteness of sites in inaccessible areas, the better one having already been tapped.

13. Inadequate Investigation: Proper investigation and planning, one of the most important activity in a hydropower project, helps in advance planning of the construction activities according to the hydrological and geological conditions of a particular project thereby avoiding delays in execution and escalation in construction cost.

14. Geological surprises: In spite of extensive investigations using new techniques, unexpected geological conditions are found greatly in the Himalayas and the neighbouring mountains than in peninsular India resulting in cost and time over-runs due to time taken in finding engineering and technological solutions. The construction of large tunnels have encountered geological uncertainties like poor geological strata, land slides, rock slides etc., causing delay in completion and escalation in the cost of construction.

15. Project Sharing: Hydel power potential is in the mountainous states which do not need that much of power to consume. Locating the buyer for the surplus power and/or having a partner in the project construction and finalising memo of understanding causes delay in taking up such projects.

16. Attitudinal impediments: There may be attitudinal impediments in terms of choices for shorter gestation schemes.

17. Land acquisition: Acquisition of land rests with the state revenue department. It is a lengthy and time consuming process, more so if located in other state which not only upsets construction schedule and commissioning of the power plant but also results in cost over-run. Generally hydro projects suffer delays of 2 to 5 years due to the land acquisition problems which have many facets.

18. Time and cost over-runs: Located as they are in remote and inaccessible areas besides other constraints/uncertainities during execution, there have often been delays in completion of hydropower projects leading to cost and time over-runs.

19. Law and Order: Slow progress and even stoppage of the construction work results from disturbed law and order conditions affected by insurgent and militant elements in the region where a hydropower project is to be constructed.

20. Small vs large dam: The controversy of small vs large dams is detrimental to the development of hydropower. Every dam is site-specific. A number of small dams cannot check floods or generate electricity as a large dam can.

21. Dependence on nature: It is true to some extent but it has to be considered that the hydro generation is less only in the case of continued dry spell and it is for such contingencies that adequate back up thermal capacity is required as base load generation.

22. Contractual Problems: The contractual problems arise due to dearth of competent and resourceful contractors with modern equipment. The private sector is hesitant to venture for mega hydro projects. Managerial problems too arise as a result of introduction of contracting system in the execution of hydro development projects.

23. Reservoir induced seismicity: The impounding of huge mass of water in a reservoir is apprehended to trigger reservoir induced seismicity and progress of construction has been retarded/stopped in a number of projects in the country, although there is no convincing relation established between reservoir impounding and reservoir induced seismicity.

24. Reservoir siltation: The reservoirs trap sediments and thereby get their useful life reduced. Hydro plants do not cause any soil erosion as such but sedimentation is on account of the developmental activities like spurt in the construction of new roads and habitations, unscientific mining, shifting cultivation, deforestation due to creation of new agricultural lands etc.

25. Loss of Forest cover: The storage plants cause some loss of forest cover but the same is very insignificant percentage of the economic benefits to a large command area.

26. Natural calamities: Natural calamities like unprecedented floods and flash floods add to time and cost over runs.

4.9. ACCELERATED HYDROPOWER DEVELOPMENT

Hydropower projects known for long gestation period can now be constructed expeditiously by modifying the conventional construction techniques. The minimum period for different activities from the start of concreting to the commencement of generation can be reduced to 3 years (refer appendix-II) besides the following:

1. Flow through diversion tunnels and difference in head between inlet and outlet be utilised to run tubular/bulb turbines.
2. Water conductor system and power house construction be started alongwith diversion tunnels so that power house starts spinning after cofferdam construction.
3. Erection bay placed in middle, instead of at the end, would isolate few units from others under construction.
4. Stage development may be resorted to where dam height and mass is large. Two tier intakes, tunnel spillway for the first stage and mobile runners may be resorted to. Diversion tunnel of first stage may be converted into tunnel spillway for this stage which will act as diversion tunnel for the second stage. The power generation can, however, be continued during the progress of the work.

4.10. CLASSIFICATION OF HYDROPOWER DEVELOPMENT

The development of hydropower has varied classifications and types. The type of development depends upon the general layout arrangement of the whole scheme from the offtake of water to its discharge outside, while the classification of power plant is limited within the walls of the power house itself i.e., structural characteristics. However, the broad classification is as under:

1. Classification Based on Function

Run-of-River Plants: The run-of-river plants generate electricity with the stream flow as it comes; normally there being little or no pondage (Fig. 4.2) or that provided is only for taking care of diurnal fluctuations (Fig. 4.2*a*) such as at a weir or barrage constructed across a rivulet or stream. The full output of such a plant is considered as firm. The power plants on canal falls also fall in this category.

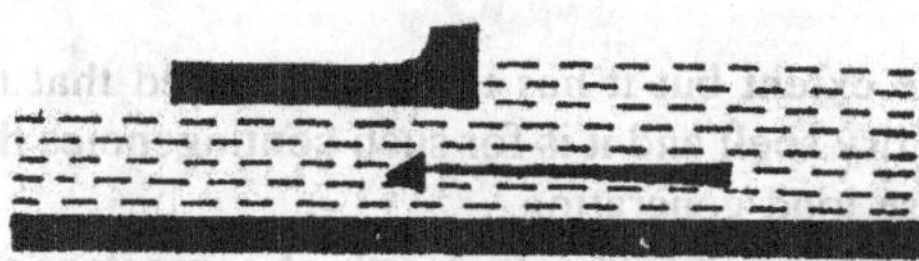

Fig. 4.2 Run of river without pondage.

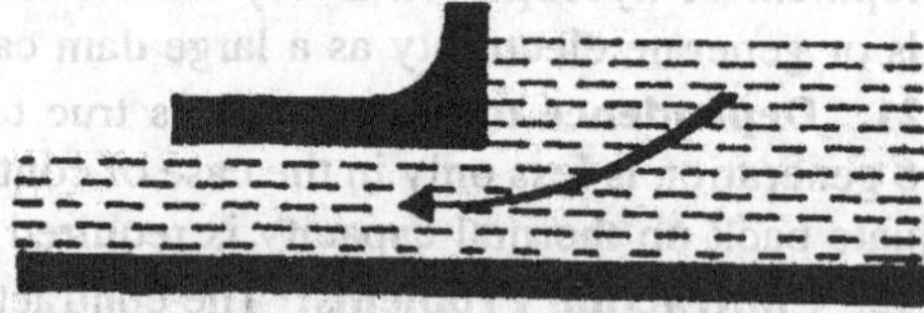

Fig. 4.2(a) Run of a river with pondage.

Storage Plants: Storage plants are necessarily associated with mountainous regions. In storage plants substantial reservoir storage is provided by the construction of a dam across a river to create high head. (Fig. 4.3) Storage is done during wet season when flow is in excess of demand to maintain continuous hydel power generation. The demand is met from the run-of-river when the flow is in excess of demand and from storage during the low supply period. These storage plants are designed so that the estimated yield for reliability is 90 percent. The power output is largely firm. Such a plant can be used as base load as well as peak load plant. Storage plants are designed for

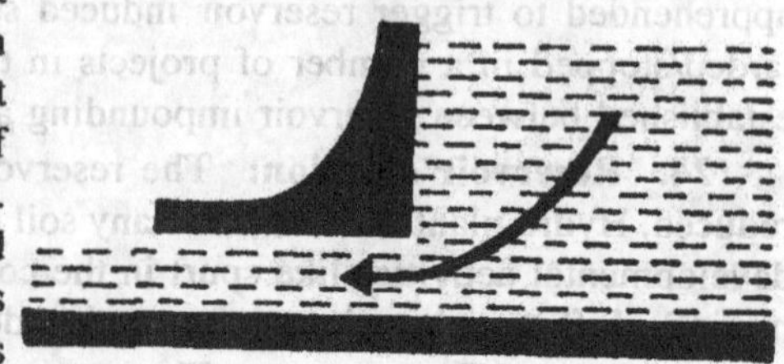

Fig. 4.3 Storage plant

small as well as large heads. Accordingly they are classified into high, medium and low head power stations.

The biggest power plant in the world Dehadu Bansri, is to come up in India (Arunachal Pradesh) with an installed capacity of 22,000 MW at an estimated cost of Rs. 110,000 crore (Rs. 5 crore/MW).

Cascade Plants: In cascade plant, available water is used more than once in a series of hydel power plants constructed to utilise the available drop in the head (Fig. 4.4). Such hydel stations may either have separate storage reservoir or a common upstream reservoir or can simply be a run-of-river cascade with no significant storage.

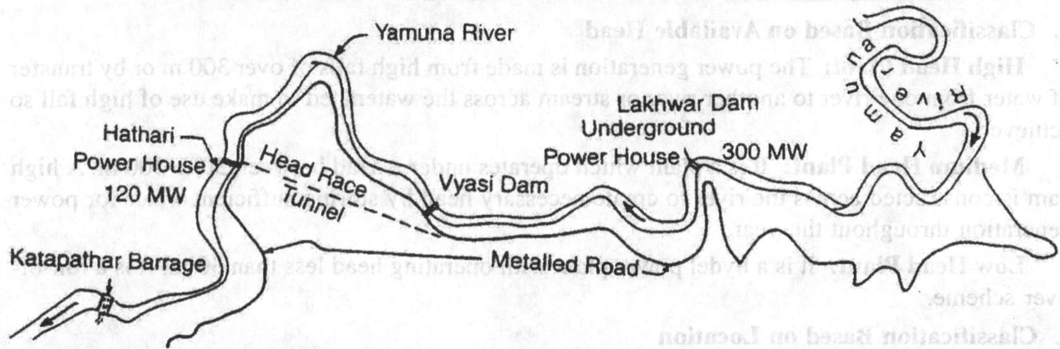

Fig. 4.4 Cascade power plants

Pumped Storage Plant: It is a peak head plant in which power is generated during the period of peak load from the water stored in an upper reservoir. The water is pumped back from a lower reservoir into the upper reservoir during the periods of low energy demand, utilising the power from the system. Thus same water is reused. Usually such plants follow diurnal cycles but some may follow seasonal cycles. The machine used as a turbine during generating electricity is used as a pump during off-peak load hours to pump water back into the upper reservoir. Similarly the machine used as a generator is used as a motor (Fig. 4.5.). It embodies the advantages of improved load factor of the power system as also economy in the cost of the power plant.

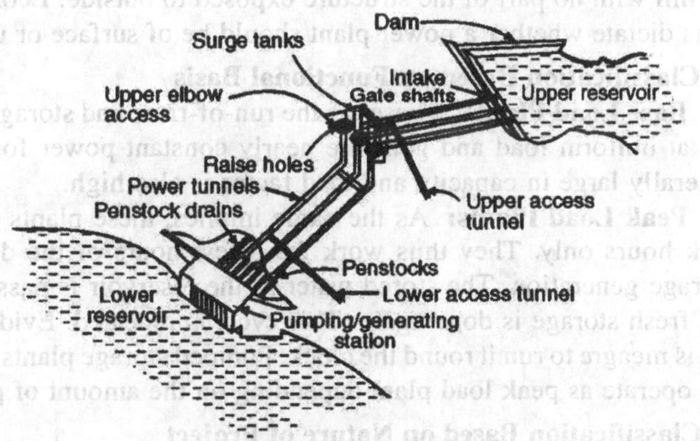

Fig. 4.5 Pumped storage plant

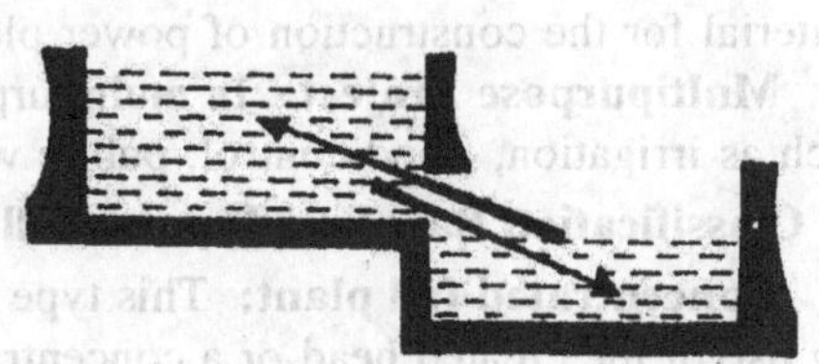

Fig. 4.5(a) Illustration of pumped storage plant

Tidal Power Plant: Water at the time of high tide when stored in a basin at a high level can be made to fall into sea during low tide through turbines to produce tidal power. The installation capacity of the plant is fixed by economic consideration rather than by available flow. Tidal power generator is, however, to be justified by the minimum reliable peaking power it generates as an alternative to a stand by thermal plant of similar capacity.

2. Classification Based on Plant Capacity

In India, the unit rating based on generation capacity is as given in Table 4.9.

Table 4.9 Unit rating

Type of plant	*Unit size*	*Total installed capacity*
Micro	1 to 100 kW	up to 100 kW
Mini	101 to 1000 kW	up to 2000 kW
Small	1001 kW to 5 MW	up to 25 MW
Medium	15 to 50 MW	up to 250 MW
High	above 50 MW	above 250 MW

3. Classification Based on Available Head

High Head Plant: The power generation is made from high falls of over 300 m or by transfer of water from one river to another river or stream across the watershed to make use of high fall so achieved.

Medium Head Plant: It is a plant which operates under a head between 30 to 300 m. A high dam is constructed across the river to create necessary head by storing sufficient water for power generation throughout the year.

Low Head Plant: It is a hydel power plant with operating head less than 30 m. It is a run-of-river scheme.

4. Classification Based on Location

Surface Power Plant: Also termed as overground i.e., located over the ground surface.

Underground Power Plant: It is a power plant built underground i.e., located in a cavity in the hill with no part of the structure exposed to outside. Economic considerations and site conditions dictate whether a power plant should be of surface or underground type.

5. Classification Based on Functional Basis

Base Load Plants: These are the run-of-river and storage power plants which are designed to run at uniform load and generate nearly constant power for long periods. Base load plants are generally large in capacity and load factor is also high.

Peak Load Plants: As the name implies, these plants run at their peaking capacity during peak hours only. They thus work for a few hours of the day, when the load is more than the average generation. The stored water in the reservoir is passed down in the running of the plant and fresh storage is done again. This cycle is repeated. Evidently supply in the river at the plant site is meagre to run it round the clock. Pumped storage plants and run-of-river plants with pondage can operate as peak load plant depending on the amount of pondage.

6. Classification Based on Nature of Project

Power Project: In mountainous regions, hydroelectric stations are constructed only for power generation. Long transmission lines are required for connecting to the load centres in plains besides high cost on construction of rail/road communication for the transportation of men and material for the construction of power plant.

Multipurpose Project: In multipurpose project, power is generated besides other functions such as irrigation, flood control, public water supply etc.

7. Classification Based on Type of Fall

Concentrated fall plant: This type of plant is located close to the dam or the barrage making use of the created head or a concentrated fall.

Divided fall plant: Depending on topography, the power house is located a considerable distance downstream of the dam or a barrage on a feeding canal or tunnel to make use of the head created at the dam or barrage plus due to steep fall available in ground surface; the bed slope in the canal or tunnel being mild as compared to steep bed slope in the river.

8. Classification Based on Physical Characteristics

Indoor plant: It is the plant wherein generators are placed in a machine hall having indoor crane.

Semi-outdoor Plant: In this plant the generator room is fully enclosed with a relatively low roof and the gantry crane, characterised by the outdoor stations, transfers the equipment.

Outdoor Plant: In this plant, the generators are housed in individual cubicals or recessed in the roof deck, there being no generator room.

9. Classification Based on Interconnection

Isolated Plant: It is the plant so located as to meet the local, isolated power demand.

Interconnected Plant: It is the plant the output of which is interconnected with other plants as also with the grid for stability and reliability of the system.

10. Classification Based on Location and Topography

Located in Hills: Since bulk of hydropower is available in rivers in mountainous or hilly areas, the major hydroelectric projects are located there and may or may not be associated with storage reservoirs. They require long transmission lines through difficult terrain and at a high cost to convey power to the load centres in plains.

Located in Plain: The hydel power plant in plains is associated with a barrage or a weir with little pondage. The transmission lines are of short length.

4.11. WATER REQUIREMENT FOR POWER DEVELOPMENT

Power sector uses substantial quantity of water. Normally, water is drawn for thermal and nuclear power stations from rivers and canals and the cooling water system could either be a direct cooling system for condenser i.e., without cooling towers where substantial quantity of water is returned back to the source, or an indirect cooling system with cooling towers which are a part of a closed circuit system and in which case all the make-up water drawn from the source ultimately becomes consumptive. The nuclear power stations handle much larger quantities of cooling water for condensers and other auxiliaries as also for the heat transport system between the reactor and the steam generator. The boiler water needs are very low compared to cooling needs. The consumptive water requirement is about 0.20 cumec per 100 MW. The water requirement for a typical coal-fired power station using cooling towers is indicated in Table 4.10.

Table 4.10 Water use by Thermal plant

Type of use	*Consumption*	
	m³/hr/MW	*Mm³/yr/100MW*
Consumptive	4.48	3.92
Non-consumptive	1.32	1.15

Hydropower stations, in general, do not have any consumptive use except for evaporation from reservoirs. It is considered incidental to the irrigation releases in a multipurpose project. Yet it may cause a restraint on the reservoir operation because at occasions releases are to be made to generate power even when demand of water for irrigation is not there to that extent.

Water requirement for energy/power sector has been estimated for high and low water scenarios, as given in Table 4.11:

Table 4.11 Water demand for energy/power

Demand scenario	*Water demand (M ha m)*		
	2010	2025	2050
Low demand	1.8	1.9	6.3
High demand	3.3	6.3	7.0

80 percent of the water requirement is expected to be met from surface water sources and balance 20 percent from ground water sources.

4.12. ECONOMICS OF HYDROPOWER DEVELOPMENT

From the Environmental Impact Assessment statement of a project all the environmental and displacement aspects are 'costs' (in addition to the direct financial costs) and the balancing of these costs against the 'benefits' which the project will bring (increased industrial and agricultural activities made possible by hydroelectric power, increased agricultural production resulting from irrigation, their multiplier effects, etc.) is a cost-benefit analysis.

The initial investment for the construction of hydroelectric power plant is substantial but it has to be viewed in the context that the life span of these plants is relatively long (over 100 years) and as such it is a part of sustainable character of this energy resource. A quite large initial investment is required for investigation and studies to ensure technical and economical feasibility of the project and arranging finance for execution.

The economics of power generation through non-renewal sources of power–fossil fuels-changes with the exploitation of every successive ounce of such fuels. The renewable sources of hydro and tidal offer the only option in the long run. With the status of technologies developed for exploitation of this renewal resource of energy and substantial unexploited potential, coupled with a number of other major technical and economic advantages by virtue of non-polluting nature, high conversion efficiency, flexibility in operation, relatively low cost of generation, operation and maintenance, longer life of equipment etc., hydropower development is economically attractive.

The hydropower constitutes the cheapest source of power generation in India even after the enormous increases in the cost of the equipment and the implementation of civil works. The economic advantages of hydropower have been enhanced in the recent years with the steep increases in the energy costs from fossil fuels and the rapid approaching limits to the exploitable resources of such fuels.

The economy of the different power sources is reflected in the price of the electricity. The countries which have a large proportion of hydropower in their systems have the lowest tariffs.

The apparent initial investment of a thermal plant tends to be more or less the same as that of a hydropower plant, the real investment taking into cognisance auxiliary factors and infrastructural costs may be nearly double that of hydropower schemes. Hydropower is actually cheaper, more than half that of thermal alternative.

The British engineers at one time were not willing to build Bhakra dam in view of the poor demand of power. Thanks to the imaginative and futuristic vision of Late Pt. Jawahar Lal Nehru, Dr. K.L. Rao and Dr. A.N. Khosla that Bhakhra dam and its twin Beas project are the lifeline of the northern region. The remarkable social, economic and industrial development in the Punjab, Haryana, Rajasthan, U.T Chandigarh and even Delhi is due to the tremendous amount of water and power released from these monuments of prosperity to the posterity.

4.13. ECONOMIC VALUE OF HYDROPOWER

The economic value of hydropower is usually compared with that of thermal power, the latter being the largest kind of power in general use. The various factors which govern the relative economy of hydropower development may be summarised as under:

1. Site specific

It includes consideration of topographical, geological and hydrological conditions obtainable at a specific site of plant for water and power development. Favourable topography and easy accessibility means less cost on the construction of power plant, residential colony and connecting roads resulting in economy in power development. A site at which waterpower can be developed at a low cost is the high value one.

Topographical conditions determine the design of dam and its cost as also, water conductor system, whether canal, penstock or tunnel or a combination thereof.

Geological conditions particularly the rock foundation availability determine the type of dam.

Hydrological conditions that is the relation of head to discharge is important because greater the head compared to discharge, the economical is the power development as higher discharge means larger and costlier civil structures. High head plants are economical.

The hydrology of the watershed area, slope of the river bed and storage/pondage potential and its economy and acceptability from environmental angle, possibility of use of discharge at cascading falls and several power plants, dependability of the water availability vis-a-vis primary and firm power output, nature of floods in the river/stream too influence the cost of power development at a site. The cold climate and ice presence, where applicable, means costly development besides the problems in operation and maintenance.

The power plant located nearest to the demand centre is economical because of low cost of transmission lines and vice versa. A power plant at a promising site and economical in cost may work out to be costlier if transmission lines are long and hence add considerably to overall cost of power development at the site.

2. Power Demand

The power demand should be commensurate with generation capacity and other means of power viz., thermal which is costlier. Load factor should be high. Cost of power plant and equipment varies immensely as the load factor.

The canal power development is essentially costlier owing to relatively low head, low load factor being isolated plant and poor load characteristics, small capacity and large number of generating units. But transmission and distribution losses are less, theft of power can be controlled because power generated is supplied in the nearby areas and as such it is competitive in cost.

3. Cost and Availability

The cost per unit of electricity supplied to the consumer and its reliability of availability are other factors.

In essence, economical waterpower development and low cost power produced, apart from the favourable site conditions and demand characteristics, has to have essentially the following salient features:

1. High head and small discharge.
2. Dependable flow with favourable storage facility.
3. Promising dam site, good foundations, narrow gorge, minimum of submergence and rehabilitation and resettlement problem, availability of land to resettle the oustees, if any, minimum cost of acquisition of submerged land, habitations and properties, minimal dislocation to wildlife.
4. Favourable location for water transmission system particularly penstocks and spillway.
5. Large capacity generating units with minimum number.
6. Short transmission lines to the demand centre.
7. High load factor.
8. Eco-friendly power development.

Economic Analysis

The economic justification for taking up a hydro power project is influenced by various factors such as pattern of load demand, time schedule for completion to meet the energy demand as also the economics of power generation from alternative sources. With expanding industrialisation, growing population and rising standard of living etc., the energy demand outstrips the available capacity and as a consequence bulk of the capital investment is allocated to the energy sources which require least time to create the capacity. This goes against the immediate interests of hydropower development and long-term interest of the nation's power systems.

In the computation of economics of a waterpower project, there are two components:

1. Capital Cost: It includes the capital outlay on the construction of dam/barrage, water conductor system, penstocks, power house, switch yard, tramission lines, residential complex, workshops, acquisition of land and properties, rehabilitation and resettlement of oustees etc. The dam and appurtenant civil structures usually constitute major portion of the plant cost except in low head plants being location-specific. However, the cost of power house and its equipment constitute a more definite and smaller part of the plant cost.

The capital cost of a hydro plant may be twice or more than thermal plant, resulting in greater fixed charges. However, maintenance and labour costs are very less, there being no element of fuel cost. But hydro plants are usually located in mountainous regions and as a result the power generated has to be transmitted by a long distance to load centre at a high cost. But life of hydro plants is very high compared to thermal plants and hence the economy on this account.

The recovery or gain factor of primary energy power plants (Tables 4.12 and 4.13) is the maximum for hydropower. The recovery factor, F_G, is given by:

$$F_G = \frac{E_{produced}}{E_{direct} + E_{indirect}}$$

where, F_G = recovery or gain factor of primary energy.
$E_{produced}$ = produced energy by power plant during the life time.
E_{direct} = expense of direct energy (non-renewable) during the lifetime.
$E_{indirect}$ = expense of indirect energy (non-renewable) during the lifetime.

Table 4.12 Recovery/gain factors of hydropower plant

Type of installation	*Recovery or gain factor*	*Technical life (years)*
High head storage	280	80
High head run-of-river	221	80
Low head run-of-river	170–180	80

Compared to this, the recovery gain factor of non-renewable energy source power plant is as given in Table 4.13.

Table 4.13 Recovery/gain factors of non-renewal energy plant

Type of power plant	*Recovery or gain factor*	*Technical life (years)*
Coal fired cycle	0.97	37.5
Nuclear (pressurised water)	0.81	40

If the long operational lives of hydro plants, reliability and reduced green house gases emissions are taken into account, hydro is often the most attractive power option in regions with sufficient water and head (fall) resources. If full account is taken of all the costs of other options, and multipurpose hydro projects are considered (with the costs for non-hydro benefits borne by the direct beneficiaries and not by electricity consumers), possibly with construction in stages, the capital cost (investment burden) may be significantly reduced.

2. Total Annual Charges: It includes fixed annual charges consisting of (*i*) annual interest on capital cost, (*ii*) annual depreciation charges; 2 to 4% of capital cost, (*iii*) annual insurance and tax charges; 1-2% of capital cost. The total fixed charges are, however, plant-specific and constitute most of the cost of hydropower, especially large plants. Thus tentatively annual charges = Rs. 0.13 C.

Assume total installed capacity is E kW, 60% is annual load factor, 10% of peak load is reserve capacity and S is the sale price of 1 kWh,

$$\text{Annual generated energy} = \frac{E}{1.1} \times 0.6 \times 365 \times 24 = 4780\ E \text{ kWh}$$

$$\text{Generation cost} = \frac{\text{Annual total charges}}{\text{Annually generated energy}} = \frac{0.13C}{4780E} \text{ (Rs/kWh)}$$

Annual revenue from energy sale = 4780 *ES* rupees

$$\text{Benefit cost ratio} = \frac{\text{Annual revenue}}{\text{Annual charges}} = \frac{4780\,ES}{0.13\,C}$$

Alternative Method

Capital recovery factor $(CRF) = [i(1 + i)]/[(1 + i)^n - 1]$

where, i is annual interest, say 12%, n is life of plant, say 30 years.

Annual cost of plant, $C_a = C_p \times CRF$

Annual maintenance of plant, $C_m = 0.015 \times C_p$; Assumed 1.5% of C_p

Total investment, C, (Rs.) per year, $C = C_a + C_m$

$$\text{Annual benefit, } B, \text{ (Rs)} = \sum_{i=1}^{12} 9.81 \times 24 \times 30\eta\, QHM$$

where, C_p is capital cost (Rs.) Q is monthly discharge (cumecs) for 90% dependable (dry) year, μ is tarrif (Rs/kWh). Say Rs. 1.

Benefit cost ratio $= B_{50}/C$

where, B_{50} is benefit with 50% dependable (average year) discharge.

The benefit-cost ratio should be more than 1 for a hydro project's financial justification for taking up for construction. The primary function of economic analysis of a hydroelectric project is to determine the size of the hydro plant at a particular site which will return maximum revenue.

The economy of hydropower development has other considerations too, briefly as under:

1. Natural and Regional Considerations

The relative economy of waterpower development also depends on the nature of the region of the country. The areas with abundance of hydropower potential and availability of fall and sustained river/stream flow and optimum opportunities for storage cost less in power development, just as thermal power may be less costlier near coal heads.

In India, Indus river offers large power potential (19,988 MW), good sites and storage opportunities but as high as 78% has yet to be developed. Likewise, Ganga and Brahmaputra rivers too afford large power potential. Himachal Pradesh is a State endowed with water power potential of the order of 21,000 MW which is one-fourth of the total identified hydro potential in the country as per latest estimates. The hydro plants improve operational reliability, stability of the power system and also help in optimising the operational economy and meeting the peak load requirements.

A certain amount of hydropower is essential in the regional power generation grid because of its inherent advantage in providing peaking energy during peak hours. Pumped storage power generation having a large potential of about 96,224 MW in the country can meet peak requirements to a very large extent resulting in economy due to saving in new installation of fossil fuel based and other modes of power generation.

2. Environmental Considerations

Keeping in view the global energy and environmental perspectives, the hydropower appears to be the best compromise between the energy needs and the environmental conservation. The hydropower has very small share in the environmental degradation and as such in overall perspective means development at the least environmental cost. Hydro plants also serve as regulatory factors for the flow of streams and rivers and, therefore, act as protection against floods, besides the benefits of irrigation where applicable, pisciculture etc, discussed in detail separately.

In assessing the economic feasibility of a hydro plant, various factors considered are that the capacity is firm, energy output will be absorbed in the system and transmission liability are taken into account.

4. 14. COST OF WATER POWER

In the selection of suitable power plant alternatives, the two basic cost parameters are (*i*) installation cost (Rs/kW installed), and (*ii*) generation cost (Rs/kwh generated).

1. Installation Cost: It is the capital cost (investment) of creating a power capacity at a certain point of time at a particular site. The installation cost varies over a very wider range. The total cost of plant may not bear any direct relation to the amount of installation. However, pertinent portions of the total cost such as intake, water conductor system, power plant and equipment are approximately proportional to the installed capacity of the plant and is called "increment cost installation". Thus the cost of power production at a plant can be expressed in terms of installed capacity. The installed capacity of the waterpower in the country roughly varies from 30 to 50 million Rs/MW.

The investment per kw generally tends to reduce with higher and higher unit sizes plants. In the medium range of 50-200 MW, hydro plants may be cheaper than thermal depending on location. However, small capacity units (15-50 MW) are costlier.

The relation between head and the increment cost installation is not that easy to find. The increment cost installation per unit for high head plants may or may not be low compared to low head plants. Rather some low head plants may have low increment unit cost.

As far as energy is concerned, hydro plants are cheaper, their running cost being negligible. On an average, cost of hydro energy is nearly half that of thermal.

The capital investment on certain hydro plants is relatively high, the best sites having already been developed, and construction is difficult at the remaining sites with consequent delay in completion due to geological surprises, rehabilitation and resettlement problems. The implicit costs of delayed execution together with increased construction works, implies that the real cost of hydro development is higher. The initial investment is substantial but life span of these plants is relatively long, about 100 years. This is part of the sustainable character of hydro power.

2. Generation Cost: Also termed as total annual cost. It is the cost incurred to generate power from the capacity created which includes interest on investment, taxes, depreciation besides annual operation and maintenance cost. It indicates the investment choice which not only represents the investment recovery obligations as also the economy of the power generated. It is more comprehensive, since it combines investment liability as well as the operating cost, and more relevant to the ultimate consumer who has to pay on the basis of Rs./kWh (consumed). It is competitive and stable kWh cost, as there is no combustible cost.

3. High Load Factor: A high load factor reduces the overall unit rate of the power generated because for a given maximum demand, the number of units generated is more corresponding to a high load factor.

4.15. TOTAL ANNUAL COST OF A HYDRO PROJECT

The total annual cost of a waterpower project consists of three elements:

(*i*) *Fixed charges*: It includes (*a*) fixed charges on plant, (*b*) interest, taxes and insurance, and (*c*) depreciation and obsolescence.

(*ii*) *Operation and Maintenance Costs*: It includes operating costs, fuel cost (nil in the case of hydro power), supervisory, labour, maintenance, repairs and miscellaneous expenses. The annual operation and maintenance cost is roughly proportional to the capacity of the plant and the number of units installed. It also varies with the pay structure and strength of the supervisory and maintenance staff. Automatic plants have less operating cost. The annual maintenance cost is usually taken as 1.5% of capital cost.

(*iii*) *Transmission Cost*: It covers the cost of transmission facilities to connect the power generated to the system load. Where in a hydropower, an alternative thermal plant needs as large an annual cost for additional transmission as the waterpower, there is no transmission liability against the hydro.

In conclusion fixed charges constitute most of the cost of hydropower, especially large plants. The operating charges consisting mostly of wages and small repairs are very less. A large plant may require more staff but with a smaller proportionate cost per kilowatt hour due to its higher capacity.

The other influencing factor on total cost is capacity factor. A low capacity factor means a larger and expensive power plant, since waterway capacity is fixed by average head, rather than the maximum head. For the usual form of load curve, its cost is substantially independent of capacity factor. But the cost of power plant and equipment varies approximately inversely with the capacity factor.

4.16. OPERATION AND MAINTENANCE OF HYDRO PLANTS

The essence of hydropower development is to provide a significant contribution to sustainable overall development of the Nation through clean, sustainable energy production at minimum operation and maintenance cost and address environmental responsibilities. However, the dismal state of operation of the power sector in India, as per Chairman and Managing Director of Power Finance Corporation is that it is suffering from one-third shortage syndrome. On a long term basis, only two-third of the power generation capacity planned gets actually implemented; of the generated power only about two-third is supplied to the consumer (the balance being lost in transmission and distribution), out of the power supplied only two-third is generally billed; and again, about two-third of the billing is collected from the consumer.

1. Operation: The objective of the operation is to match the power demand with the supply and as closely as possible. It is generally complex and reiterative and the use of computers may simplify and improve it. The complexity varies depending on the scope for manipulating power supplies to meet the demand. The functional operation includes:

1. Water supply availability in wet and dry seasons vis-a-vis power generation (supply),
2. Power demand estimate,
3. Matching supply and demand.

Water Availability: The source of hydropower is water. Its availability should be determined (forecast and hold over) on a periodic basis by carrying out adequate water power studies for computing plant capacity so that the power generation is phased accordingly. Since hydro uses water–the free fuel–little attention is paid to operating efficiency compared to thermal plants wherein fuel used is costly and as such greater importance is given to fuel efficiency. The hydro plant essentially operates under a balancing of competing and at times conflicting demands on water such as power generation, irrigation, drinking water supply, flood control, industrial, water quality, navigation, if applicable, etc. There may be constraints also due to cavitation or vibration limits. Operation within the restricted zones may be permissible under certain circumstances but the economic benefits must be weighed against the increased maintenance costs and the future value of the wasted water. The grid operation should ensure complete avoidance of spilling over of water in hydro plants for overall economy of the integrated power system.

Many power plants utilise a reservoir, often a part of a multipurpose scheme, to store water during periods of high river flow for subsequent use during periods of low flow. The dam and reservoir are the edifice which must be operated under specific rules and procedures formulated during planning, design and continually reviewed in operational phases.

Emergency Procedures: Emergency procedures should be developed to take care of unprecedented floods, due to cloud-bursts, and drought besides for all facilities for which failure or malfunctioning could be apprehended. Liaison with meteorological department to get timely and precise weather forecasts is very essential.

Communication: An efficient communication system is essential for the prompt flow of information required for operation within the system.

Power Demand: The power demand is essentially estimated, rather it is a part of the water-power studies essentially carried out at planning and design stages of the power scheme. The power demand is continually reviewed in consideration of water short situations and increasing social, economic, agricultural and industrial requirements.

Matching Supply-Demand: Procedures to be adopted to match supply with demand should be defined and documented in operation and maintenance manual of the plant. Method of power distribution e.g., on demand or pre-arranged demand under a grid system or under continuous flow are periodically reviewed.

Operational Availability: It is possible to achieve 100 percent plant availability; in India the average lies in the range 85-88 percent. This can be improved through planned and preventive maintenance as well as renovation and modernisation, wherever necessary.

Power System Management: The power system management requires systemisation in respect of (*i*) Power generation, (*ii*) Power demand, and (*iii*) Power transport including regulation thereof. The basic tenents which form the core of power system operation management are: (*i*) Maintain sufficient reserve in spinning plant to be sure to pick-up load demand rise in case of probable outage of generating unit, (*ii*) Keep sufficient back-up reserves to bring in, in case of schedule needs, (*iii*) Watch system security, implying transmission line flows and parameters to ensure that tripping of generation or of one transmission circuits will not upset the system either in power flow or stability, and (*iv*) Keep above system reliability and security requirements as primary schedule plant operation so as to get maximum saving.

MAINTENANCE

Essentially details of strategies, standards, procedures for maintenance should be followed as delineated in the operation and maintenance manual of the power plant. The most viable function of the maintenance system is generation, transmission and distribution of power. Sustained success of the function depends on the efficient operation of the maintenance system.

1. Maintenance Policy: The maintenance policy may include:

(*i*) Degree to which preventive maintenance as opposed to identification and resolution of problems on an adhoc basis is to be relied upon;

(*ii*) Appropriateness of deferring maintenance on facilities for modernisation;

(*iii*) Approach to modernisation/uprating of the plants during performance of maintenance activities;

(*iv*) The relationship between system maintenance and modernisation.

The maintenance and repairs problems associated with the plant and its machines and equipment, hydraulic structures etc., are important due to age, natural attrition and design and construction inadequacies etc.

2. Conventional Maintenance: The conventional maintenance is to perform maintenance periodically according to a schedule based on tradition, experience of the maintenance personnel, statics and budget or some combination of these factors. The latest maintenance technique is based on the maintenance effects from varying operating conditions. A functional relationship is established between the life for a component of a generating unit (or an entire unit) and the generalised stress (bearing vibrations, output from cavitation monitors and inefficiency) encountered in the operating environment of the equipment. The measured parameters are used to assess the additional maintenance costs produced by variable operating conditions. The maintenance cost methodology is a simple approximation but it serves to reinforce the strong link between present operating conditions and future maintenance costs. This operation-based maintenance concept is important for hydro participation in auxiliary services markets. This methodology also provides the basis for a quantitative assessment of cavitation and vibration data to determine the appropriate minimum operation limits and maximum load limits for an individual turbine's operation or to evaluate a unit for hydro modernisation.

Annual Maintenance: The duration of planned annual maintenance depends on different conditions of operation and the resultant extent of damages. There is scope for improvement in optimising downtime by resorting to: (*a*) Modular concept of maintenance as in aircraft industry, (*b*) The forced outages can be reduced through a planned preventive maintenance schedule relevant to a power plant, and (*c*) The design of turbines can be tailored to suit the excessive silt load in inflow during monsoons resulting in loss of stainless steel from underwater components of the plant.

3. Modern Concepts of Maintenance

(*i*) *Environmental Performance*: The conventional approach is to react to environmental pressures. The innovative designs and technologies for environmentally advanced turbines and control systems focus on improving levels of dissolved oxygen (DO) in turbine discharges and increasing fish passage survival by optimising environmental performance. The performance related parameters (unit status, power, flow rate, head water level, tail water level, gate opening, efficiency, cavitation level) are monitored or computed.

(*ii*) *Improving DO Levels*: Innovative designs and technologies for environmentally advanced turbines and control systems focus on improving levels of dissolved oxygen (DO) in turbine discharges.

(*iii*) *Fish Friendly Operation*: Optimised environmental performance also includes increasing fish passage survival, provision of minimum flows to protect aquatic habitat in tail waters. The fish friendly operation may differ somewhat from the point of maximum efficiency. Safe passage module, using a patented technique, operates turbines at its most fish-friendly point when fish are detected near the turbine's intake. The fish survival means optimisation of revenue when fish are present and optimising the hydraulic and economic performance when fish are not present. Scale model tests of new turbines reveal hydraulic performance improvements as well.

(*iv*) *Optimum Load Management*: In view of the peaking power shortage, load management or flattening of system load curves may be adopted to contain the peak demand. It envisages reducing the peak hour load demand and creating additional loads during off-peak hours by transfer of peak load to average load hours during the day.

(*v*) *Specialised Maintenance and Training*: The mechanical and electrical machines and equipment are generally specialised and manufacturer's comprehensive specific instructions for maintenance are required to be followed together with specialised training for maintenance personnel. Adequate supply of commonly used spare parts should always be kept on hand to ensure continuous operation.

4. Maintenance of hydraulic structures

Dam and Reservoir: The general maintenance problems to be addressed to are sedimentation and siltation, water quality, bank erosion and instability etc. The work plan includes monitoring watershed management, control of polluting sources, bank protection etc.

The dams usually associated with hydropower development are concrete, earth, rockfill or a combination of these materials. The maintenance monitoring parameters are stress and strain, instability, abnormal seepage, concrete deterioration, abnormal floods and danger of overtopping, inadequate capacity of downstream channel to cater for high releases. The work plan includes concrete refurbishment, maintenance of gates and hoists, seepage control, foundation grouting riprap repairs, maintenance of control facilities and implementation of dam safety procedures.

Power Channel: The power channels are generally lined. The maintenance aspects are silting, reduction in carrying capacity due to deterioration in coefficient of roughness of lining, sealants, underdrainage. The work plan includes monitoring silt removal, lining repair and seepage control etc. An important aspect in the operation and maintenance of a power channel is to safeguard lining against excessive drawdown which is essentially related to power-load and deterioration in the carrying capacity of the canal with passage of time due to increase in coefficient of rugosity of the weathered lined surface.

Control Structures: These include spillway, barrage or weir, tunnels, power outlets, irrigation outlets, intakes, etc. Periodical inspection and prompt maintenance as embodied in operation and maintenance manual need to be strictly followed so as to ensure operational efficiency of these structures.

QUESTIONS

1. What is hydropower development? Give a brief account of perspective power development vis-a-vis power demand in the country.
2. Hydropower plants bestow many benefits, what are their limiations, discuss.
3. Discuss hydro and thermal power generation in comparative form.
4. What is the hydropower potential of the country, conventional as well as non-conventional?
5. What is the present status of development of hydropower vis-a-vis demand?
6. Discuss the statement that hardly one-fourth of hydropower potential has been tapped so far in the country.
7. What are the various factors considered in planning hydropower development?
8. What is hydro thermal mix? Why has it come down? What is the ideal mix?
9. Narrate the principal factors responsible for poor hydro-thermal mix. What are its adverse effect on nation's economy?
10. Briefly discuss the various constraints in optimum development of hydropower potential.
11. Discuss various types of hydropower plants. Why run-of-river plants are preferred these days over storage based plants? Draw a neat sketch of typical run-of-river power plant.
12. Hydropower has non-consumptive water use, discuss.
13. Give relative economics of a hydropower plant and a thermal power plant.
    1. (*a*) Compare advantage and disadvantage of three major types of power development.
       (*b*) Explain how hydropower stations are good for taking care of peak loads. (*P.U. 1999*)
    2. (*a*) What are base load and peak load plants? For what type of load conditions hydroelectric power is very much suitable. (*P.U. 1999*)
14. Discuss power system management. How can it ensure maximum reliability, stability and economy.

5

STORAGE POWER DEVELOPMENT

5.1. INTRODUCTION

The hydropower potential of the country is essentially storage based. The terrain and the economics of the power generation necessitate construction of large and high dams for tapping the bulk of the available hydro potential in Brahmaputra-Ganga system besides Beas-Sutlej system. The Rashtriya Barh Ayog (1980), after elaborate studies, has strongly recommended "there is a need of storage in various forms, which would even out the flow and also conserve water for use during the dry period". Furthermore, the perspective studies prove that even the absolute minimum demands of drinking, industrial and cooling water requirements for thermal and nuclear power plants in 2010 AD can not be met without speedy construction of large dams.

Big vs Small Dams: The notion that big projects are necessarily divisible into smaller component projects is mistaken. Larger projects often sustain smaller ones. The big and small projects are not competitive and as such ideally operated as integrated and conjunctive systems. Shallow storage causes proportionately greater loss of land area due to submergence besides higher evaporation losses for equal amount of storage in addition to higher cost. It is also difficult to find large number of alternative sites for medium projects even if such an alternative is preferred over deeper storage. The large water storage projects are essentially better alternatives wherever the parameters such as the volume of water flow, geological and topographical considerations and regional requirements indicate such possibilities especially in view of large power potential awaiting development. The dynamic nature of the environmental effects which seem adverse to environment at the time of construction generally tend to stabilise and become favourable.

The growing energy needs of the country can be met with only through tapping of the vast hydro potential by storage power projects, the long gestation period, non-availability of easy storage sites, environmental impacts of reservoir related submergence and rehabilitation and resettlement, resources crunch etc., notwithstanding. National policy on resettlement and rehabilitation of persons or families adversely affected or displaced on account of compulsory acquisition of land will go a long way in addressing this emotive environmental issue.

5.2. COMPONENTS OF STORAGE POWER DEVELOPMENT

The storage hydropower development is site-specific, essentially of various types and so are the components as given in Fig. 5.1.

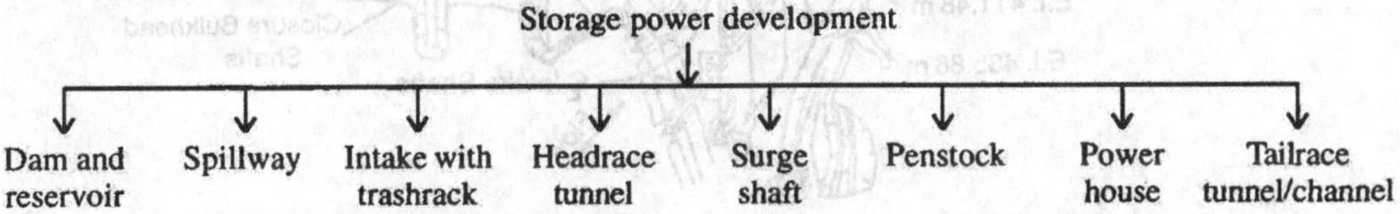

Fig. 5.1. Typical components of a storage power development.

These components in the scheme of power generation are illustrated in Fig. 5.2.

Fig. 5.2. Storage power development

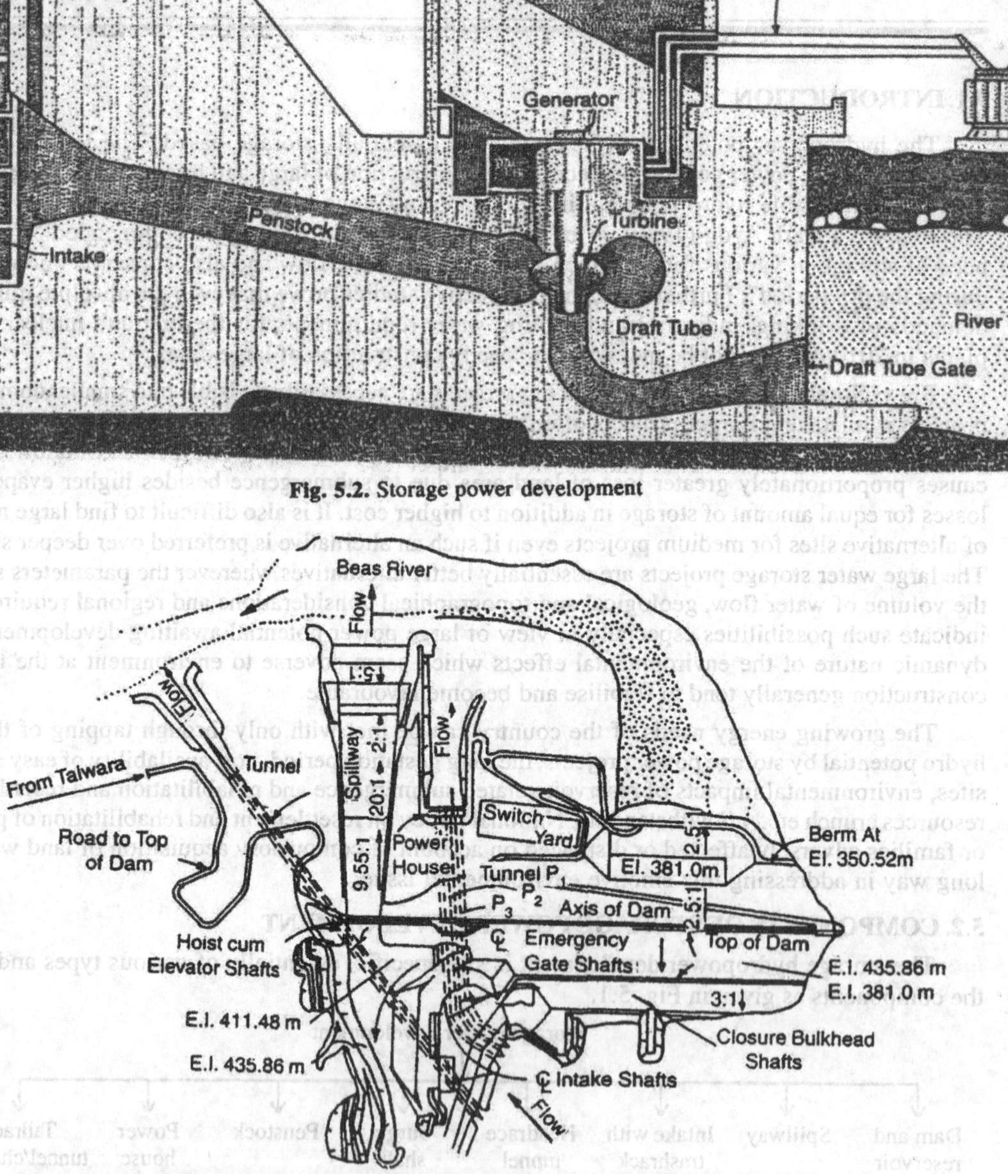

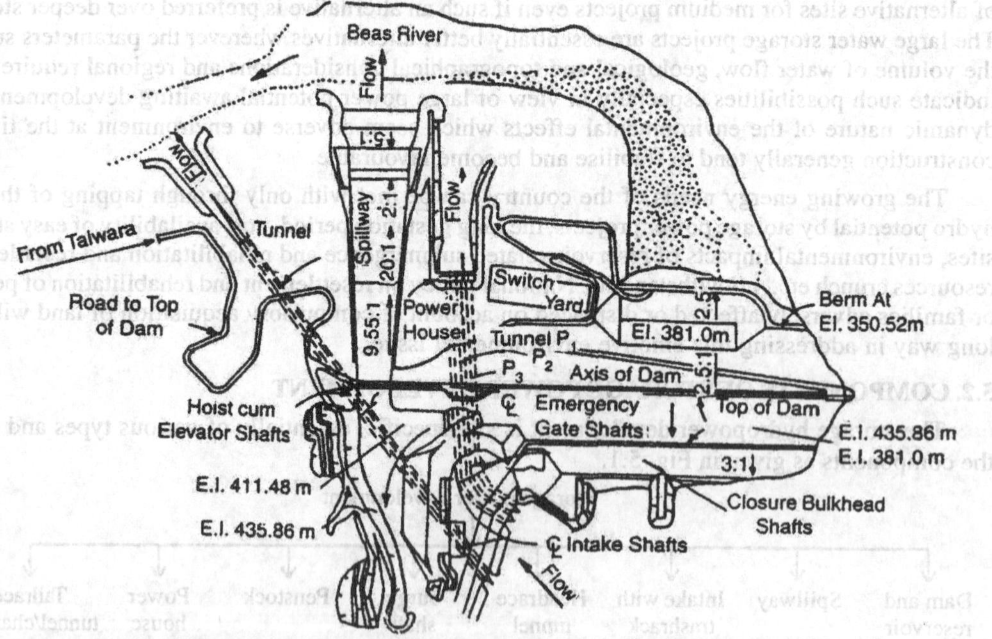

Fig. 5.3. Shows the general layout of multipurpose storage Pong dam and power plant.

The salient features of a storage power development are briefly as under:

1. Dam and Reservoir: A dam is essentially constructed to create head and impound wet season inflows in the form of a storage reservoir to ensure sustained release of water for development of hydropower. The storage may be annual or even carry over from one year to the next. Boulder dam reservoir on Colorado river (USA) has such a large capacity that the entire flow of the river for two years can be stored. The type of dam largely dictates the nature of components of the power development.

2. Intake: It is the structure meant for drawing water from the reservoir to feed turbines. It essentially consists of trashracks, rakes to clear the trashracks and penstock regulating gates with their hoist mechanisms. The intake design essentially depends on the category of dam viz., low, medium or high head. The type of intake is also related to the type of dam viz., concrete, masonry or earthen dam. In concrete or masonry dams, the intake structure is located either at the toe when operating head is low or in the body of the dam itself when operating head is medium or high (Fig. 5.2). In earthen dams, the intake is a sloping intake or tower type intake. In high head installations, the intake, located as it is at a distance from the dam is a tower type semi-circular, circular, rectangular or inclined structure.

3. Water Conductor System: The typical water conductor system in a storage reservoir project, discussed separately, is shown in Fig. 5.2.

4. Surge Tank: Penstocks are subjected to water hammer pressures due to fluctuations in load. The short length penstocks are designed to take up this extra pressure but long penstocks are usually provided with a surge tank to absorb the water hammer pressure and to provide water to meet sudden load increases. The surge tank is thus a sizeable water receptacle interposed between power plant and high pressure penstock on the downstream side and low pressure tunnel and reservoir on the upstream side. Its main function is to intercept and dampen high internal pressure generated in the high pressure penstock due to water hammer and not allow it to transmit into the low pressure tunnel upstream. Besides, it serves as a storage basin i.e., to absorb the excess discharge when load on the turbine decreases and the rejected water flows into the surge tank and thereby raising its level to absorb the flow and when the load increases it instantly supplies the extra needed water to the turbine.

5. Penstock: It carries water from the intake structure to the power plant. It essentially consists of large diameter steel pipes to carry water under pressure. The diameter of the penstock is determined from economic diameter criterion. The penstocks may be buried underground, exposed pipes supported on pedestals or embedded in concrete in tunnel (Fig. 5.3). The long penstocks are usually

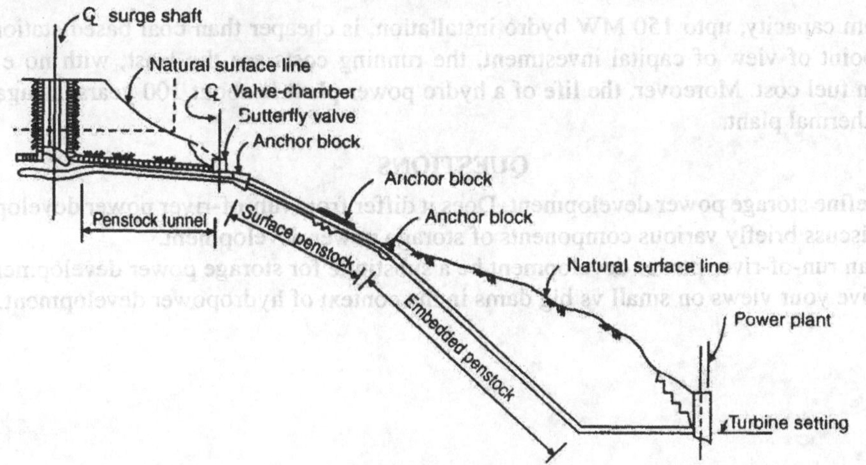

Fig. 5.4. Types of penstocks.

branched at the lower end to feed several turbines. Gate control is provided at the entrance in the intake to permit inspection and repairs of penstocks. Downstream from the gate, an air vent and stand pipe connecting the top of the penstock with open air is provided to facilitate entry of air into the penstock when the gates are closed and water drawn off through turbine as otherwise serious negative pressures may be exerted on the penstocks tending to collapse them.

In concrete dams, a separate turbine is fed by each penstock which is embedded inside the non-overflow sections of the dam. In earth and rockfill dams, penstocks are generally taken out from diversion tunnels. Sometimes the penstocks may be laid in trench excavated below the earth and rockfill dam and buried in concrete. In underground power house, as at Tehri Dam (U.P.), the water is led though pressure shafts.

6. Scroll casing: The water from penstock is delivered to the scroll casing which surrounds the turbine runner and guide mechanism. It is a spiral shaped steel or concrete conduit to distribute water evenly around the guide vanes and at the same time avoid formation of eddies. The concrete scroll casing is used usually in low head units.

7. Turbines: Turbines are the machines which convert hydraulic energy into mechanical energy which in turn is used in running an electric generator. In general, a turbine consists of a wheel, called runner, which is provided with specially designed curved blades or vanes so arranged that water flows over them. The high velocity water strikes the runner and makes it to rotate. The rotation is passed on to generator which is directly coupled to the shaft of the turbine. The generator armature so rotated produces hydropower.

8. Power Plant: The power plant is a multistoreyed structure wherein the hydroelectric power is generated. It consists of substructure, intermediate structure and superstructure. The power house may be a surface structure or underground structure. The underground power plant is housed in a large cavity excavated into a mountain whereon surface space for the structure is not available.

9. Tailrace: The tailrace which carries the outflow from the draft tubes to the river may be an open channel or a closed conduit (tunnel) running free and/or under pressure. The draft tubes may discharge directly into the tailrace or through a manifold as in the case of Innertkirchen plant (Switzerland). In tailrace tunnel of considerable length it is essential to provide a tailrace surge tank. The tailrace channel is aligned to meet the river tangentially or as smoothly as possible and the fluctuations in the tail water level due to roller or ski jump bucket is the minimum. The channel should have neither silting nor scouring tendency. It is so designed that minimum tail water is ensured at all times. There should not be bed load entry from the river bed into the tailrace channel.

5.3. ECONOMIC ASPECTS

Medium capacity, upto 150 MW hydro installation, is cheaper than coal based stations even from the point of view of capital investment; the running costs are the least, with no effect of inflation on fuel cost. Moreover, the life of a hydro power plant is about 100 years, as against 25 years of a thermal plant.

QUESTIONS

1. Define storage power development. Does it differ from run-of-river power development ?
2. Discuss briefly various components of storage power development.
3. Can run-of-river power development be a substitute for storage power development.
4. Give your views on small vs big dams in the context of hydropower development.

6

RUN-OF-RIVER POWER DEVELOPMENT

6.1. INTRODUCTION

This type of power development has generation limited to the instantaneous discharge of the rivulet/stream but the minimum flow should not dwindle to such an extent that the generation is not beneficially fitted into the load pattern. A variant of this type of development is that in which some pondage is provided to take care of diurnal load fluctuations or weekly cycle. The pondage enhances usefulness of the plant and enables generation of additional power during peak demand periods as a peaking station, may cover a greater range of the load curve including the peak part.

The run-of-river power plants are seldom exactly similar, rather are site-specific with regard to topography of the stream, site characteristics, available water flow and head etc. Usually advantage is taken of the drop in the river bed by cutting across the band or the loop of the river (Fig. 6.1) or cutting across the ridge and dropping the tail water into the adjacent river/stream (Fig. 6.1).

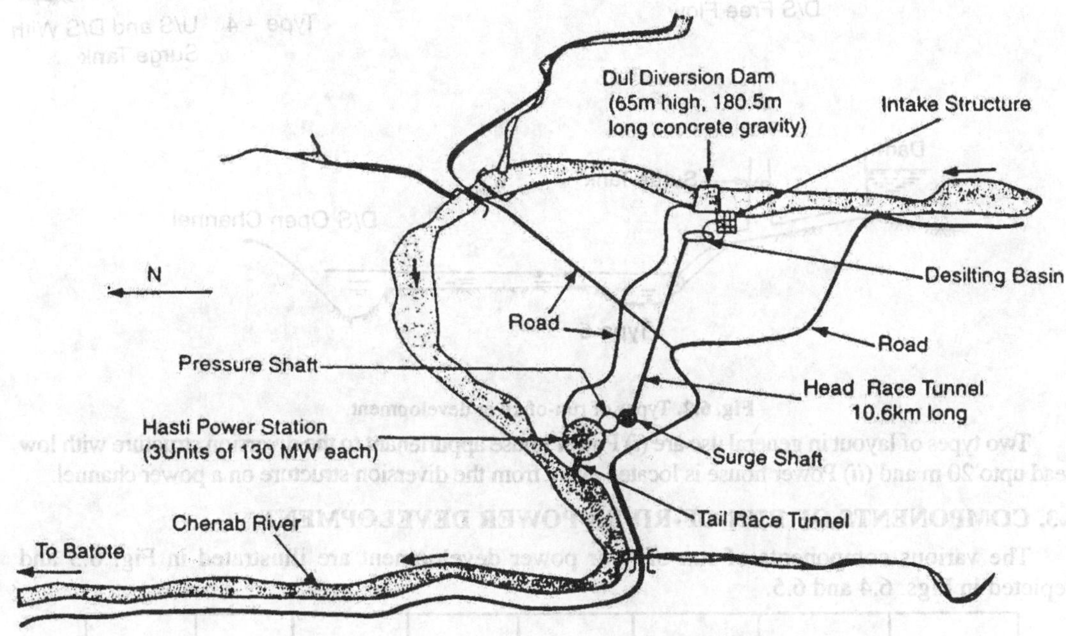

Fig. 6.1. Layout plan of Dulhasti hydroelectric project.

6.2. TYPES OF PLANTS

The various types of run-of-river power development are shown in Fig. 6.2 and briefly as under:

Type I : U/s surge tank and D/s free flow
Type II : U/s surge tank and D/s pressure flow
Type III : U/s without surge tank and D/s free flow

Type IV : U/s and D/s with surge tank
Type V : D/s open channel

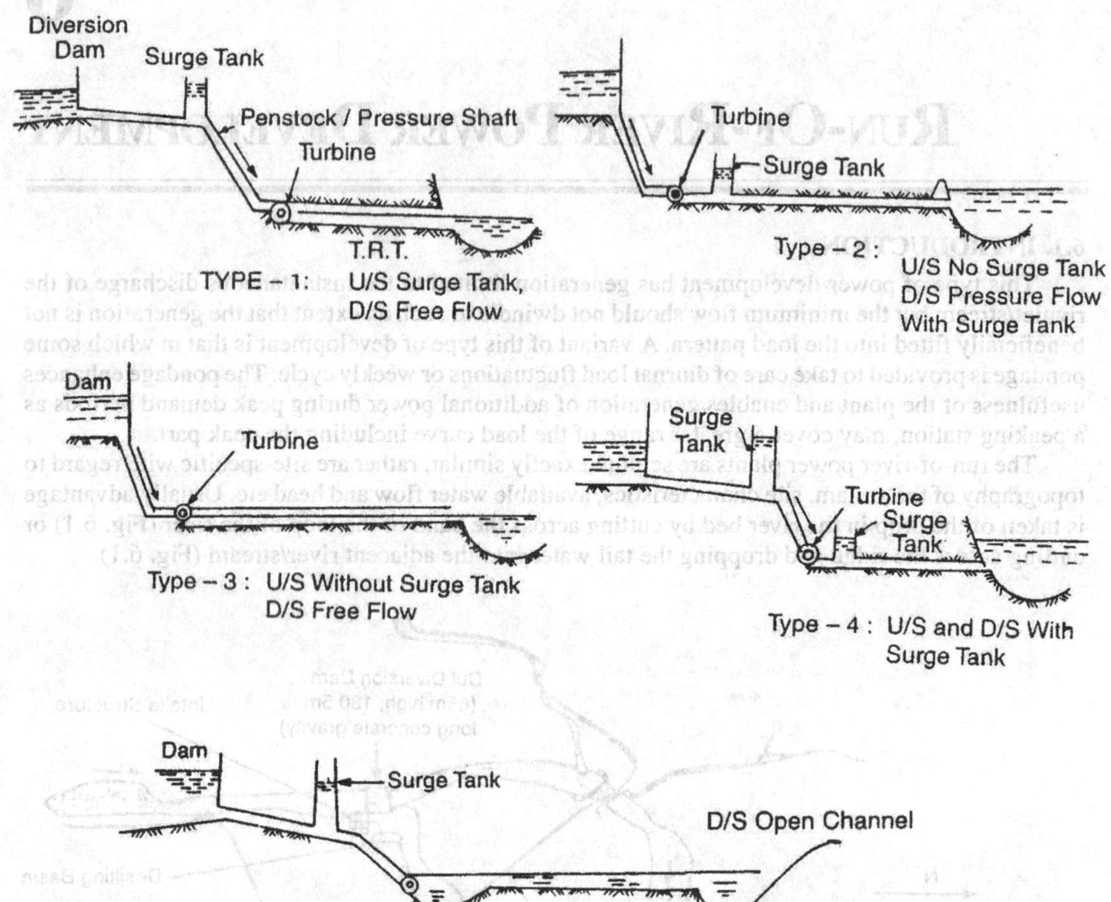

Diversion Dam
Surge Tank
Penstock / Pressure Shaft
Turbine
T.R.T.
TYPE – 1 : U/S Surge Tank, D/S Free Flow

Turbine
Surge Tank
Type – 2 : U/S No Surge Tank D/S Pressure Flow With Surge Tank

Dam
Turbine
Type – 3 : U/S Without Surge Tank D/S Free Flow

Surge Tank
Turbine
Surge Tank
Type – 4 : U/S and D/S With Surge Tank

Dam
Surge Tank
D/S Open Channel
Type-5

Fig. 6.2. Types of run-of-river development.

Two types of layout in general use are (*i*) Power house appurtenant to the diversion structure with low head upto 20 m and (*ii*) Power house is located away from the diversion structure on a power channel.

6.3. COMPONENTS OF RUN-OF-RIVER POWER DEVELOPMENT

The various components of run-of-river power development are illustrated in Fig. 6.3 and depicted in Figs. 6.4 and 6.5.

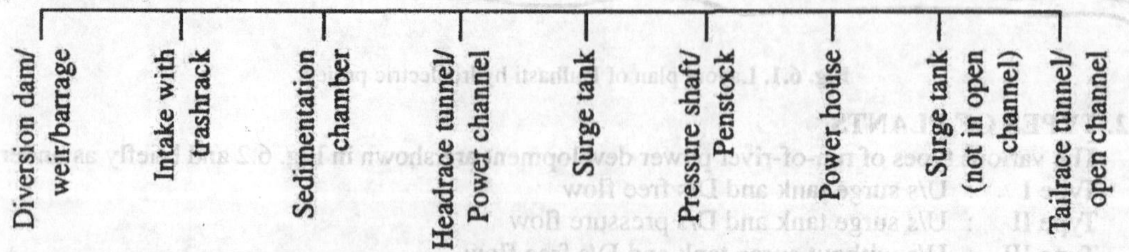

Fig. 6.3. Components of run-of-river power development.

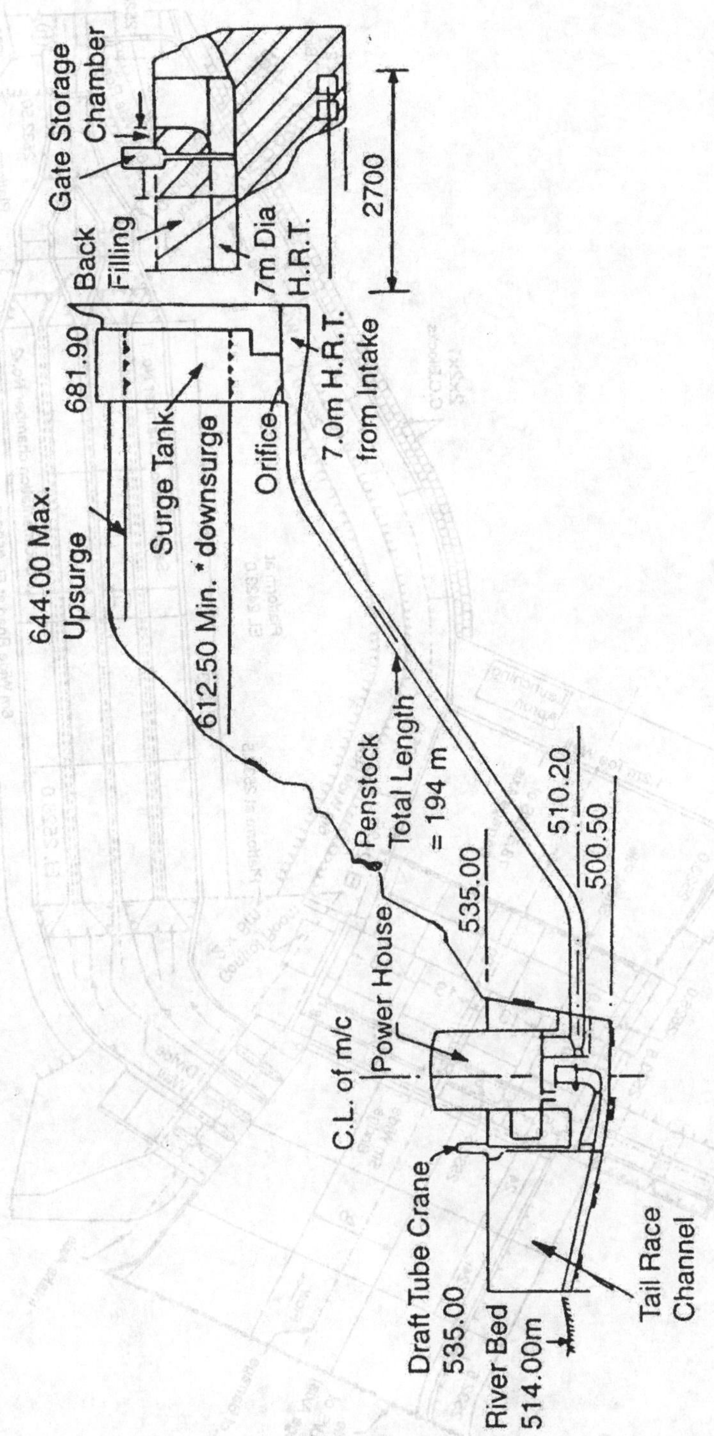

Fig. 6.4. Components of diversion dam development.

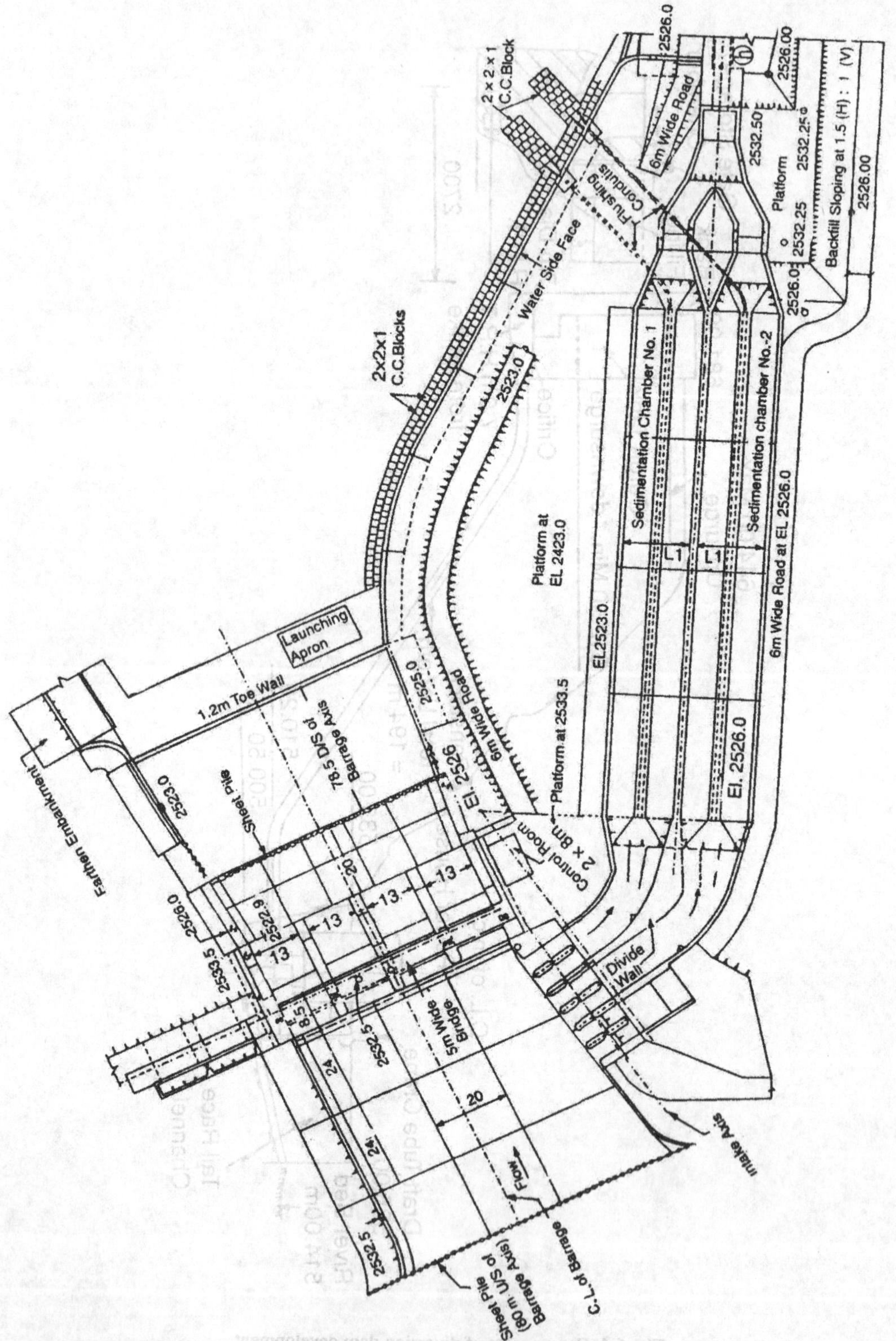

Fig. 6.5. Components of weir type development.

1. Diversion Structure: The diversion structure consists of a relatively small dam (Fig. 6.4) or barrage/weir (Fig. 6.5) constructed across a mountain stream to divert the river water into the intake which in turn feeds the power house through a water conductor system. The location of the diversion structure is usually governed by the topographical and geological considerations. The river/stream width is essentially narrow and foundation and abutments geologically sound. It is relatively easily approachable and provides adequate working space for undertaking construction.

The diversion structure is likely to be silted up to the crest within a short period and as such no appreciable storage is envisaged, being available only above the crest of the structure and up to the top of gates, unless dredging arrangement is provided as at Pandoh diversion dam on Beas river in Beas-Sutlej link project in Himachal Pradesh. The pondage normally provided is to meet the diurnal variations and for peaking purposes. The power generation mostly depends on the run-of-river flow.

2. Intake Structure: Water is conveyed from river/stream to the water conductor system of the power plant through intake structure. Since the streams, especially in the Himalayan region, carry large sediment load, large quantity of sediment may enter the intake and damage the tunnel lining, the turbine blades and auxiliaries. The intake structure in run-of-river schemes as such is located on the outer curve of the rivulet so that it may draw comparatively lesser amount of sediments. The crest of intake is kept sufficiently higher than the anticipated silted bed at the intake location and suitable arrangement, usually sedimentation chamber, is made to exclude the sediments before being allowed through the water conductor system to the power house. Suitable trashrack is provided at the intake and/or exit end of the power channel that is at the forebay portion, wherefrom the penstocks offtake, to avoid trash entering the penstocks. Fig. 6.5 shows intake of a typical run-of-river power plant.

3. Water Conductor System: The layout of water conductor system is governed by the topography and geology of the area and overall economy. Fig. 6.6. shows typical layout of water conductor system of a run-of-river plant on Baspa river (Himachal Pradesh) with installed capacity

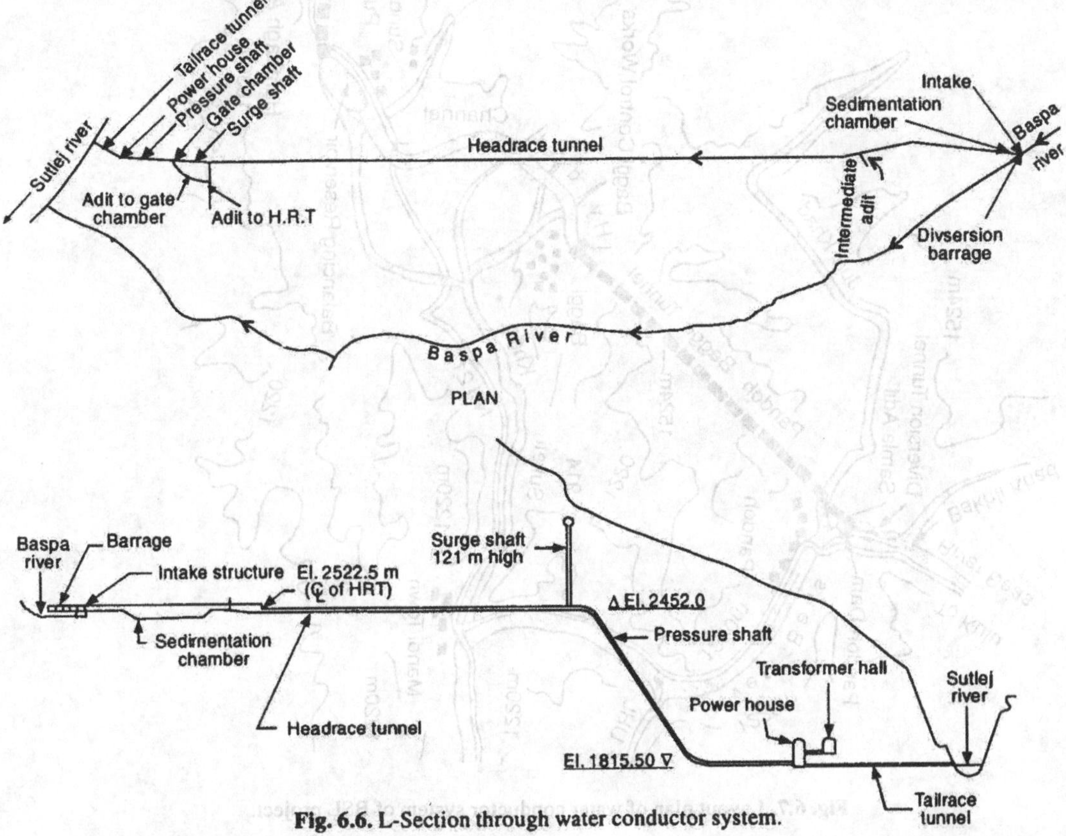

Fig. 6.6. L-Section through water conductor system.

of 3 × 100 MW and utilising a head of 716 m between the barrage and the power house.

The layout of water conductor system of the Beas Sutlej link project (Himachal Pradesh) comprising 61m high diversion dam (Pandoh dam), 13 km long tunnel, 11.8 km hydel channel, balancing reservoir and 12.35 km headrace tunnel to generate 990 MW of power at Dehar power plant through a fall of 320 m into Sutlej river upstream of Bhakra dam is shown in Fig. 6.7 and Fig. 6.8.

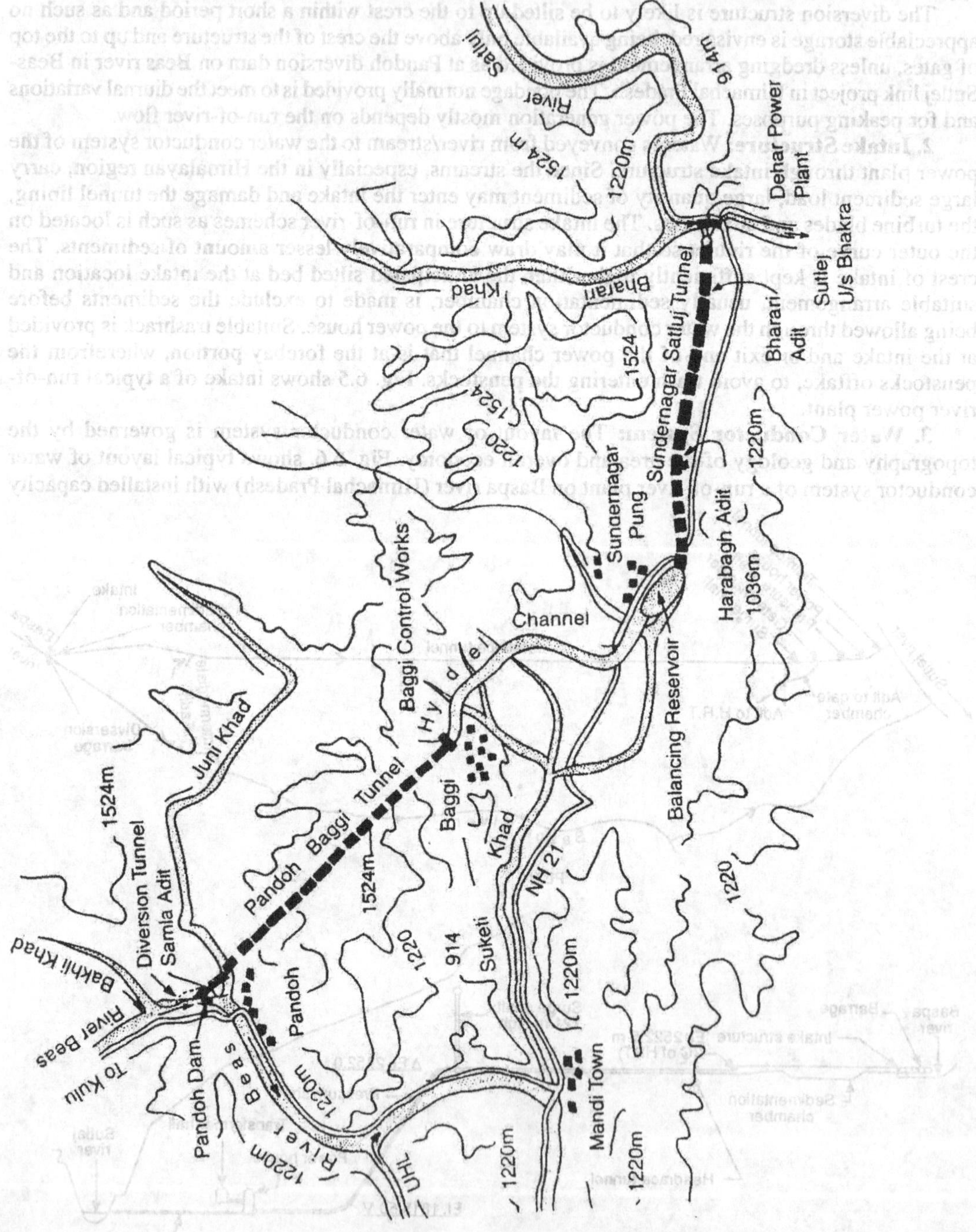

Fig. 6.7. Layout plan of water conductor system of BSL project.

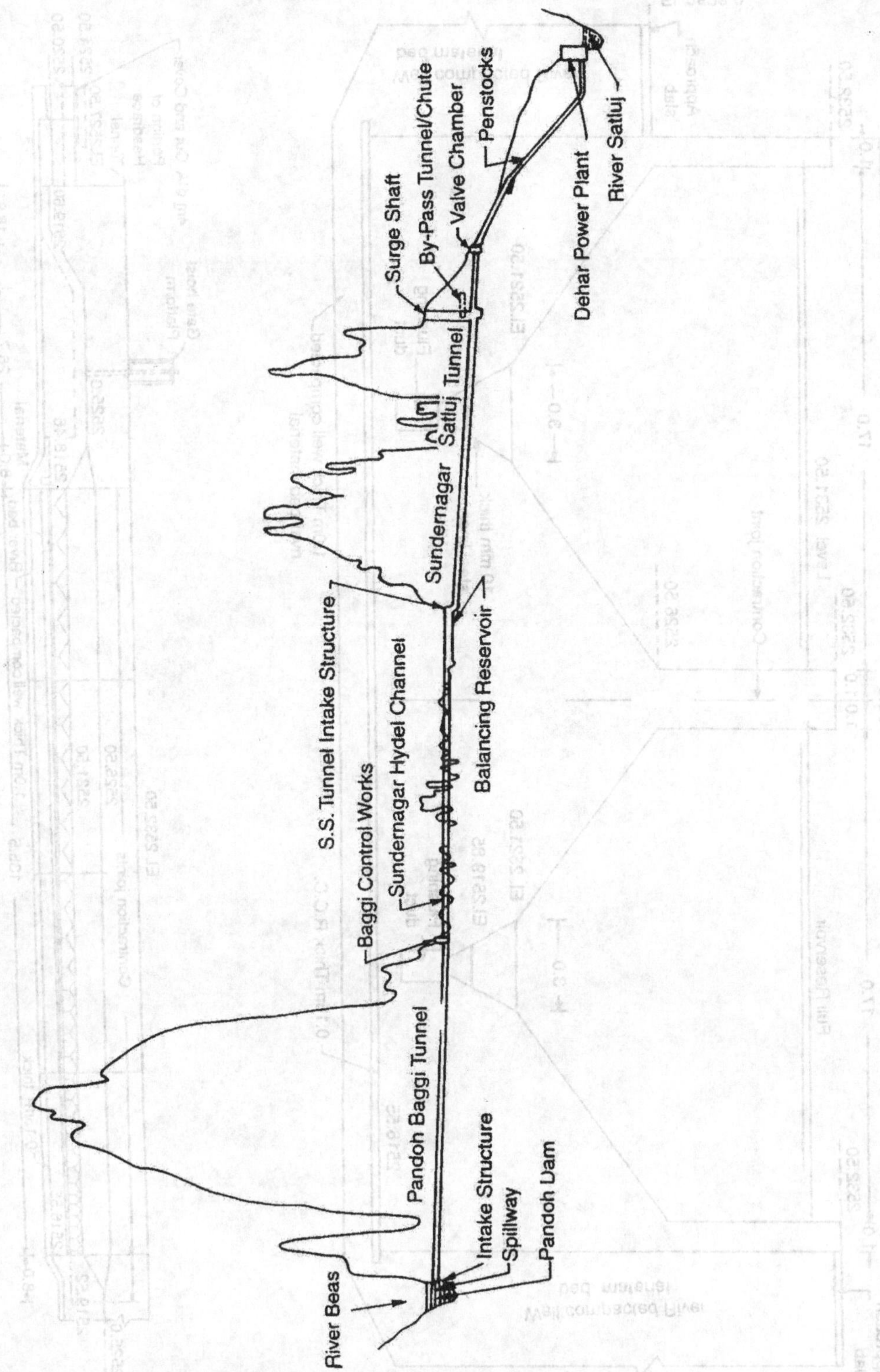

Fig. 6.8. L-section of water conductor system of BSL project.

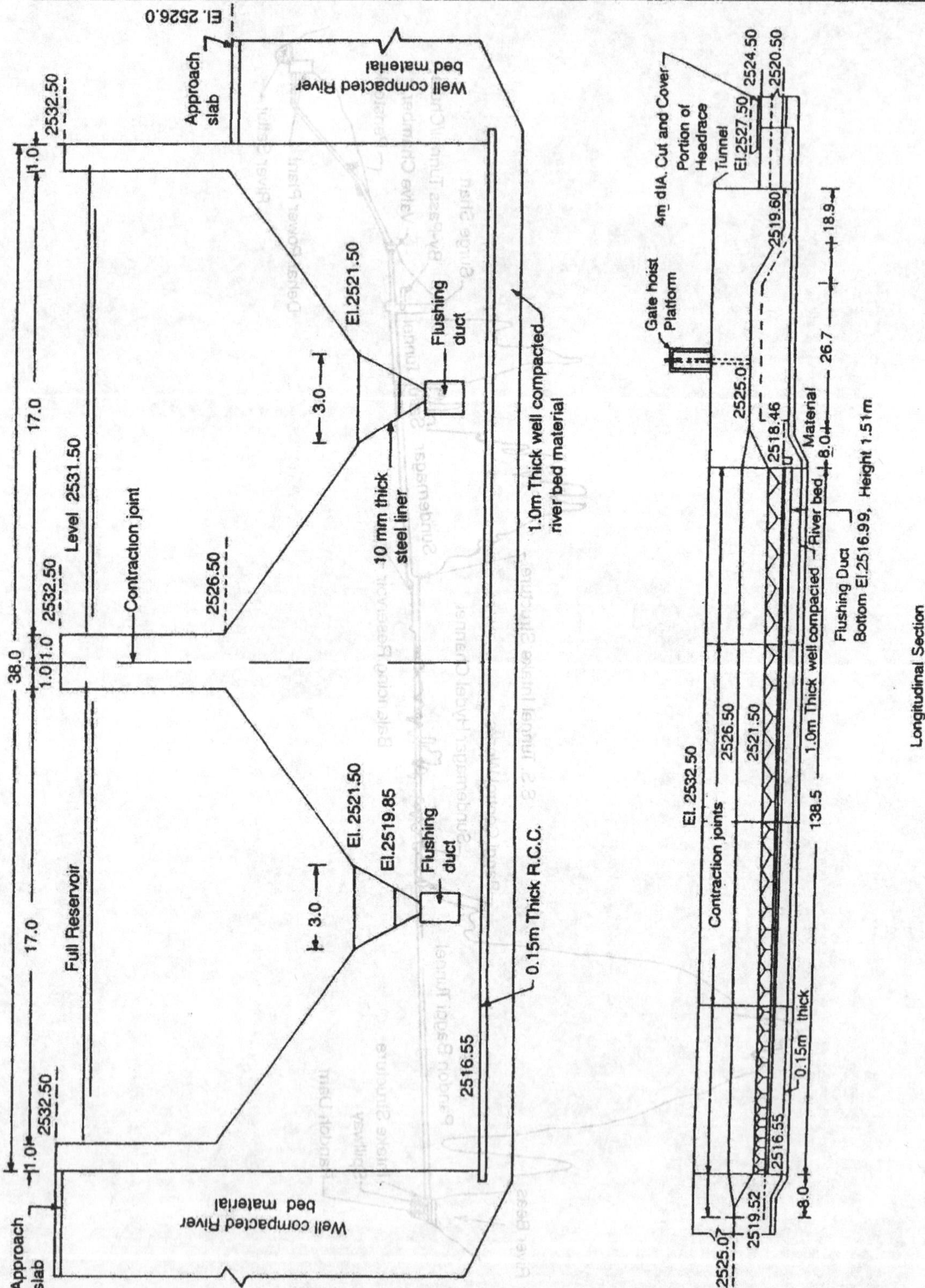

Fig. 6.9. L-section through sedimentation chamber.

Fig. 6.10. Cross section of sedimentation chamber.

4. Sedimentation Chamber: Usually a sedimentation chamber is introduced at the head of headrace tunnel to exclude sediments from travelling down into the water conductor system. Typical sedimentation chamber as provided at Baspa-II hydroelectric project (Himachal Pradesh) consists of two similar compartments of size 138.5 m long × 17 m wide (Fig. 6.9). The depth of the chambers up to a level from where the hoppers start is 11m. Flushing conduits located at the bottom of each compartment (Fig. 6.10) is connected to the chamber through a number of holes aligned in a line at the centre of the compartments. In all 32 holes of different sizes have been provided to connect 1.0 m wide duct. The depth of the duct has been so varied as to provide accelerating velocities varying from 3 to 4.3 m/s in the duct. Each duct is to carry 6.5 cumecs for flushing the sediment.

5. Headrace Tunnel: In low head power plants, in general, there is an open power channel where the discharge is large and penstock (or pipe) will be too large and uneconomical. In high head plants it is a pressure conduit which is normally concrete lined to prevent seepage losses and reduce head loss or it may be a combination of open channel, cut and cover conduit and tunnel as at Beas Sutlej Link Project (Figs. 6.7 and 6.8.). In good and sound rocks, unlined headrace tunnel can be provided but usually lined one. In some cases it is considered desirable and economical to increase the size of the tunnel to reduce the head losses. Velocities up to 6 to 7 m/s are usually allowed in the tunnel, more precisely economical diameter based on studies of power loss vis-a-vis saving in the cost of tunnel by reducing its size is adopted. The maximum size of the tunnel is governed by the geology.

Usually an intermediate adit in headrace tunnel and another near the surge shaft end is provided.

(a) *Pressure Shaft:* The pressure shaft is an inclined conduit (Fig. 6.11) constituting an alternative to headrace tunnel. It may branch off directly from the surge tank (Chibro underground power house) or downstream of the surge tank (Khodri power house). The layout of the shaft is such that its length is the minimum consistent with overall economy. The orientation of the shaft may vary from vertical to an inclination of about 50° which is nearly the limiting angle for self mucking during excavation from below using rise climber etc. The advantages of pressure shaft is the shorter excavation length.

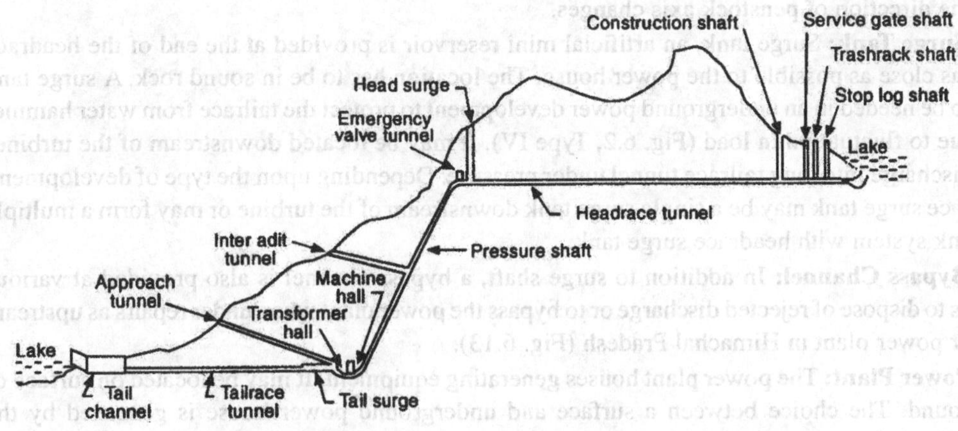

Fig. 6.11. Schematic layout of K.H.E.P. (Stage-IV).

The excavation of pressure shaft is done from top downwards in full section by using controlled blasting technique with support system of shotcreting and rock bolting. The excavation from top downwards is preferred over from bottom upwards on account of the advantages such as (*i*) Easy access to the working face, (*ii*) Due to horizontal face and direction of progress downwards drilling is easier and the accuracy in drilling is better, (*iii*) Scaling can be done thoroughly, reducing considerably the possibility of accidents or rock falls, (*iv*) Ventilation is better as gases being lighter are carried to the top, (*v*) Convenient to make ancilliary facilities viz. compressed air line, lighting and water supply up to the working face, and (*vi*) Work can commence without waiting for procurement of special imported equipment.

(*c*) ***Air Cushioned Surge Chamber:*** Developed in Norway at Duria Power house, air cushioned surge chamber is economical to replace the conventional horizontal low pressure head race tunnel, surge shaft and pressure shaft system with one inclined pressure tunnel running straight from the intake to the power station with an enclosed air-cushion surge chamber located close to the power house (Fig. 6.12). Extensive instrumentation and monitoring system is provided to have a continuous check on the water level and air volume/pressure. The air cushion chamber provides a better and smooth hydraulic functioning technically as the air-cushion functions as a shock absorber to ensure hydraulic stability. However, it requires constant elaborate observation and maintenance system, whereas a conventional surge shaft, though not so efficient hydraulically, does not require elaborate maintenance and operational observance. The layout with air-cushion chamber comprises tunnel system with practically no steeply inclined/vertical shafts as required for conventional layout. It generally expedites construction and effects economy in cost.

6. Penstocks: Penstocks are large diameter steel pipes to carry water under pressure from storage reservoir or power channel to the turbines. More precisely penstocks connect water conduction from forebay to the scroll case of a turbine. They are subjected to water hammer pressures due to fluctuations in load. Short length penstocks are designed to take this extra pressure. However, in long penstocks an arrangement for the dissipation of water hammer pressure, usually a surge shaft or pressure relief valve, is provided. The penstocks are usually branched at the lower end to run each turbine. The intake of the penstocks at forebay is kept at a level low enough to provide necessary water seal under low reservoir condition. The entry to the penstocks has to be smooth to ensure minimum head loss. The penstocks are either buried, underground, embedded in concrete in tunnels used earlier as diversion tunnels or kept exposed. Exposed penstocks are suitably anchored by embedding into anchor blocks to hold down penstocks at some regular intervals or at locations where the direction of penstock axis changes.

7. Surge Tank: Surge tank, an artificial mini reservoir is provided at the end of the headrace tunnel, as close as possible to the power house. The location has to be in sound rock. A surge tank may also be needed in an underground power development to protect the tailrace from water hammer effect due to fluctuation in load (Fig. 6.2, Type IV). It may be located downstream of the turbines which discharge into long tailrace tunnel under pressure. Depending upon the type of development, the tailrace surge tank may be a single surge tank downstream of the turbine or may form a multiple surge tank system with headrace surge tank.

8. Bypass Channel: In addition to surge shaft, a bypass channel is also provided at various locations to dispose of rejected discharge or to bypass the power plant when under repairs as upstream of Dehar power plant in Himachal Pradesh (Fig. 6.13).

9. Power Plant: The power plant houses generating equipment. It may be located on surface or underground. The choice between a surface and underground power house is governed by the topography and geological conditions. In case there is no terrace available for the location of surface power plant and rock is sound and massive, an underground power house is preferred. If the rock is not sound and is likely to need heavy steel supports for the power house cavity and other underground works, it may be desirable to go in for a surface or semi-underground power house. A surface power house is the obvious choice in case a sufficiently wide terrace is available to accommodate the power house structure.

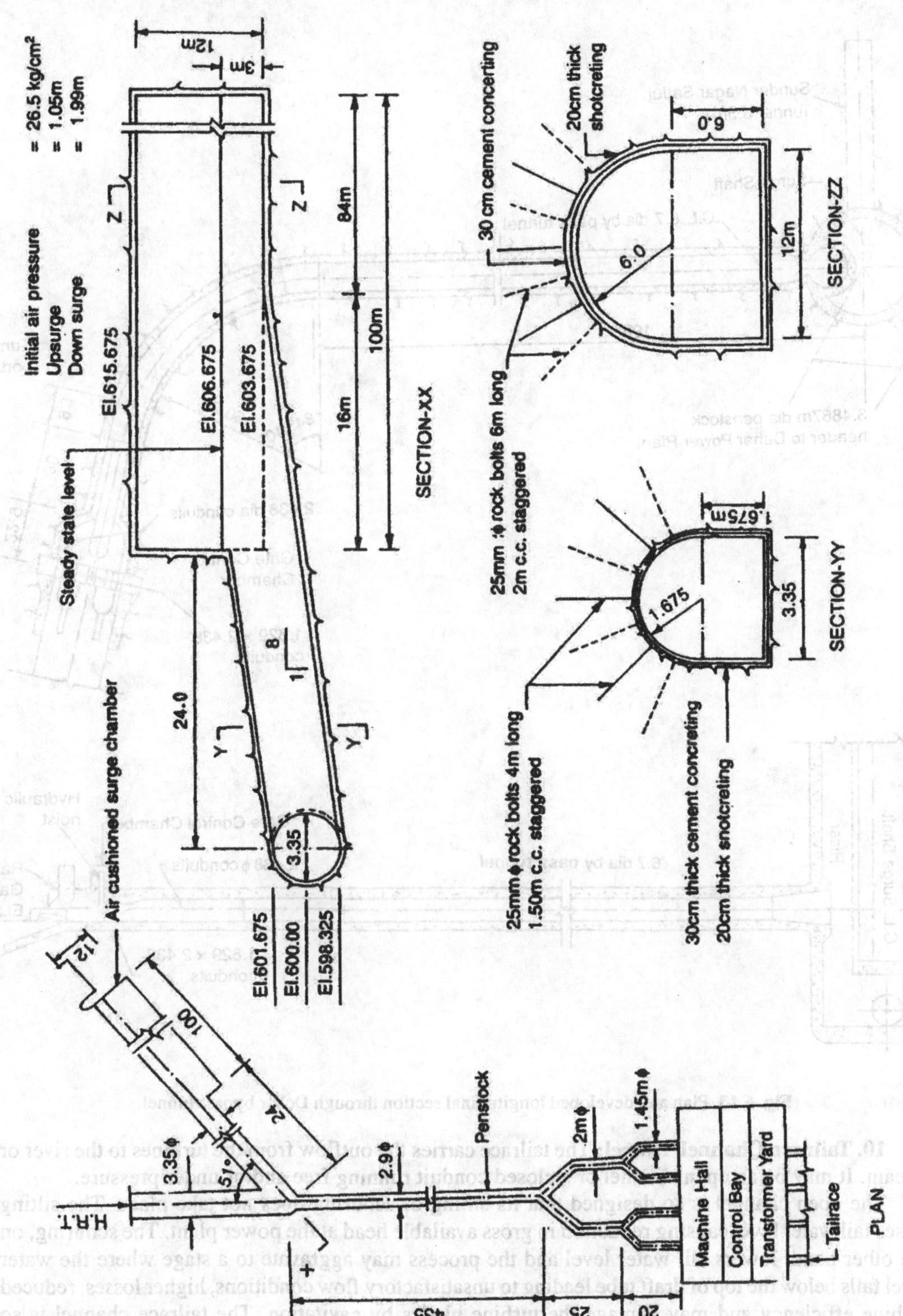

Fig. 6.12. Air cushioned surge chamber.

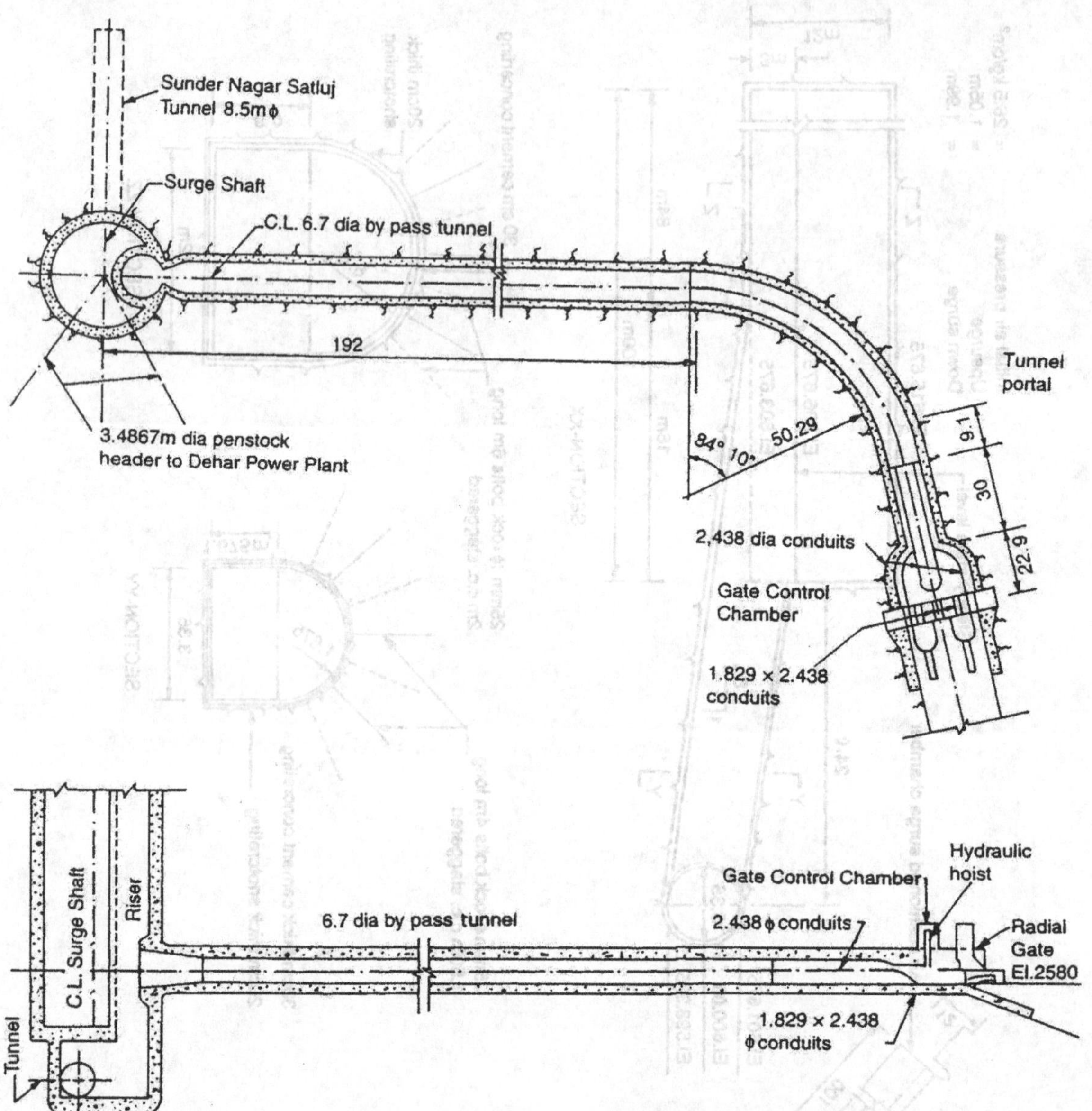

Fig. 6.13. Plan and developed longitudinal section through Dehar bypass tunnel.

10. Tailrace Channel/Tunnel: The tailrace carries the outflow from the turbines to the river or stream. It may be an open channel or a closed conduit running free and/or under pressure.

The open channel is so designed that its silting or scouring does not take place. The silting raises tail water level causing reduction in gross available head at the power plant. The scouring, on the other hand, lowers tail water level and the process may aggravate to a stage where the water level falls below the top of draft tube leading to unsatisfactory flow conditions, higher losses, reduced turbine efficiency and may damage the turbine blades by cavitation. The tailrace channel is so designed that minimum tail water level is ensured at all times. The tailrace is so aligned that it meets the river tangentially or as smoothly as possible and fluctuation in the tail water level due to roller

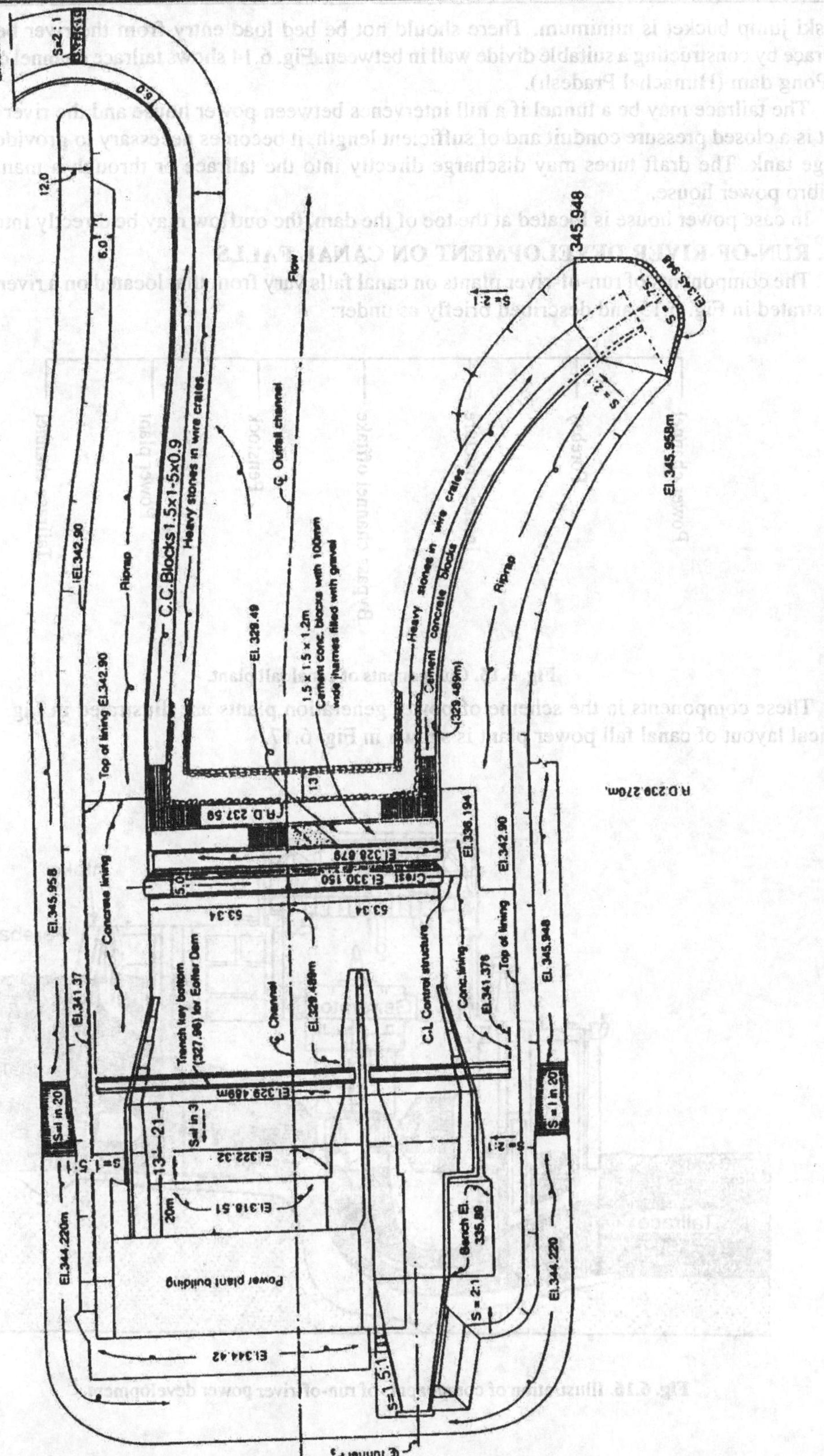

Fig. 6.14. Tailrace channel at Pong dam.

or ski jump bucket is minimum. There should not be bed load entry from the river bed into the tailrace by constructing a suitable divide wall in between. Fig. 6.14 shows tailrace channel constructed at Pong dam (Himachal Pradesh).

The tailrace may be a tunnel if a hill intervenes between power house and the river or stream. If it is a closed pressure conduit and of sufficient length, it becomes necessary to provide a tailrace surge tank. The draft tubes may discharge directly into the tailrace or through a manifold as at Chibro power house.

In case power house is located at the toe of the dam, the outflow may be directly into the river.

6.4. RUN-OF-RIVER DEVELOPMENT ON CANAL FALLS

The components of run-of-river plants on canal falls vary from that located on a river/stream as illustrated in Fig. 6.15 and described briefly as under:

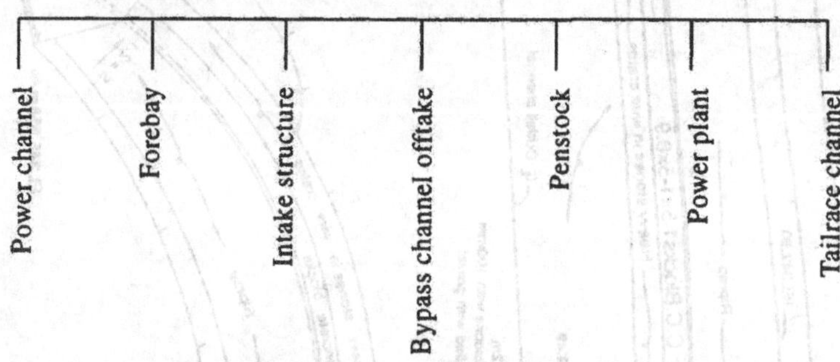

Fig. 6.15. Components of canal fall plant.

These components in the scheme of power generation plants are illustrated in Fig. 6.16. The typical layout of canal fall power plant is shown in Fig. 6.17.

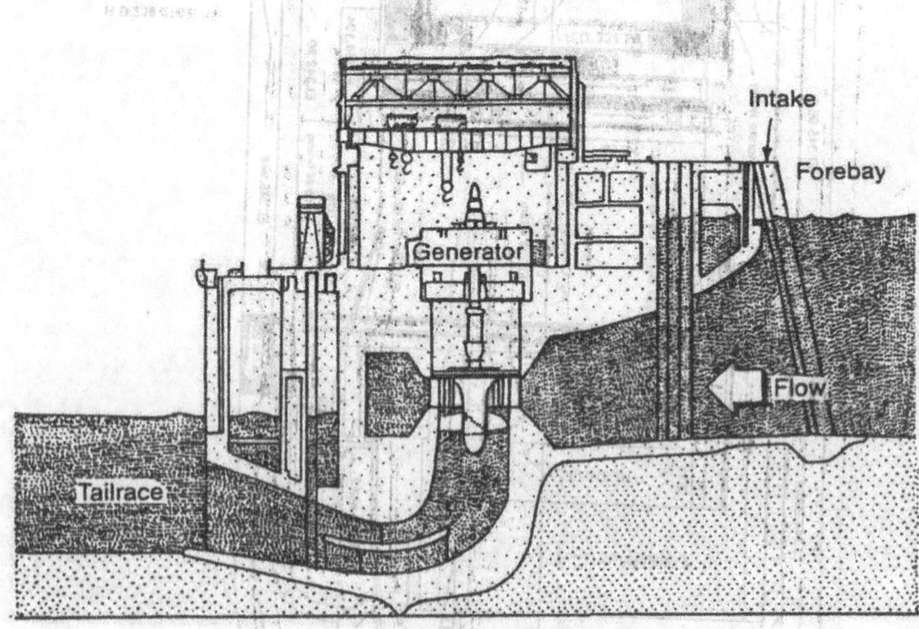

Fig. 6.16. Illustration of components of run-of-river power development.

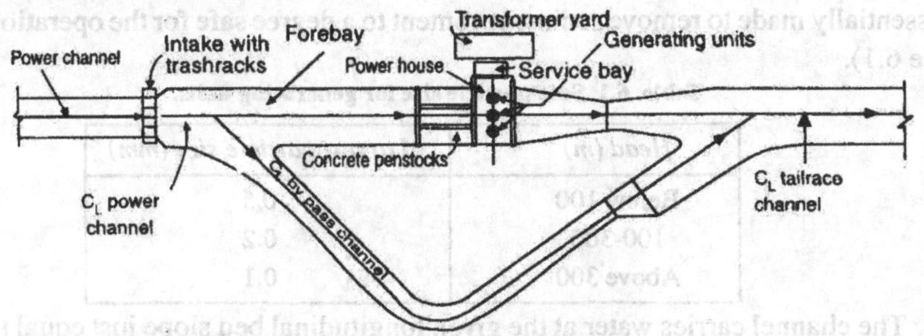

Fig. 6.17. Typical layout of canal power plant.

11. Power Channel: A hydel channel is an important component of a hydroelectric project. Oftenly the water carried by it is also used for irrigation in downstream areas. As such the essential requirements which it must satisfy are (*i*) The section is structurally strong and durable to admit all time flow in the channel because closures for repairs, if any, are rare and also they are of short duration only, interrupting power generation for brief periods, (*ii*) The section is hydraulically efficient with relatively flatter slope for the same discharge capacity, (*iii*) The section is strong enough to take care of the phenomenon of surges due to variation of discharge on account of load demand or rejection.

Free flow channel is the preferred form of water conductor system up to forebay upstream of penstock in low head power plants depending upon topography, sub-surface strata, soil characteristics, texture, composition and permeability, water availability, sub-soil water level, drainage facilities of the area, quantity and gradation of sediment load, permissible sediment grade through water turbines, discharge requirements for irrigation, seismicity of the area, available construction materials etc.

Alignment: The alignment of the channel has to be the shortest or follow the contour of the side hill from the dam to forebay with due consideration of economy. Continuous high embankments are avoided.

Channel Design:* The design of the channel depends upon its hydraulic characteristics; whether upstream and downstream water levels are fixed or fluctuate considerably with the change in load. Three conditions of operations normally adopted are:

1. Constant discharge in channel and constant water levels upstream and downstream with bypass arrangements from upstream to downstream.
2. Channels with balancing reservoir to take care of diurnal fluctuations with reduced channel capacity upstream of the reservoir.
3. Lock operation of the channel, in which the channel is used similar to a lock and the discharges in channel fluctuate with the load of the generating stations.

Capacity: The discharge capacity is fixed based on water power studies made for arriving at the installed capacity of the power plant. These studies include flow-duration curve and mass curve for the available discharge or the storage volume or pondage required vis-a-vis load demand and power plant functioning–whether base load or peaking load plant etc. In stage development plant, the channel capacity is fixed for the ultimate stage of the power development as permissible by economic studies. A spillway or bypass channel is usually provided to prevent too great rise of water level in the channel.

Sediment: The quantity of sediment that can be permitted depends on the type of turbine, its head, the size and mineralogical content of the sediment. Provision of silt excluder and/or silt

*For details refer "Irigation Engineering' by the authors.

ejector is essentially made to remove sediment content to a degree safe for the operation of generating units (Table 6.1).

Table. 6.1. Safe particle size for generating units.

Head (m)	*Largest particle size (mm)*
Below 100	0.5
100-300	0.2
Above 300	0.1

Grade: The channel carries water at the given longitudinal bed slope just equal to that required to overcome the friction and other losses at velocity at which the water is flowing. The canal grade is fixed in consideration of the probable draft of pond and as permitted by Manning's formula $\left(V = \frac{1}{n} . R^{2/3} S^{1/2}\right)$ for the permissible velocity in the channel; The minimum velocity is limited to 2 m/s. Value of n usually adopted in concrete or brick lined channel is 0.018. At Narmada canal, the bed slope provided is 1 in 12,500.

Cross Section: The cross section of the channel, bed slope etc., are designed on the basis of economic studies considering optimum cost of construction and cost of energy lost due to head loss in friction. The best section is computed by determining the cost for varying depths of the canal based on a reasonable percentage allowance which includes fixed charges on account of interest, taxes, maintenance and depreciation etc. The loss of head is determined for the average discharge corresponding to the product of plant capacity and capacity which in turn depends on its load factor. The total annual cost for each assumed depth of the channel is its cost plus value of power lost. A curve is drawn of annual cost as ordinate and depth as abscissa and the minimum point on the curve or depth giving the minimum cost is determined (Fig. 6.18). The design parameters for a trapezoidal lined section are as given in Table 6.2 (refer example 6.1).

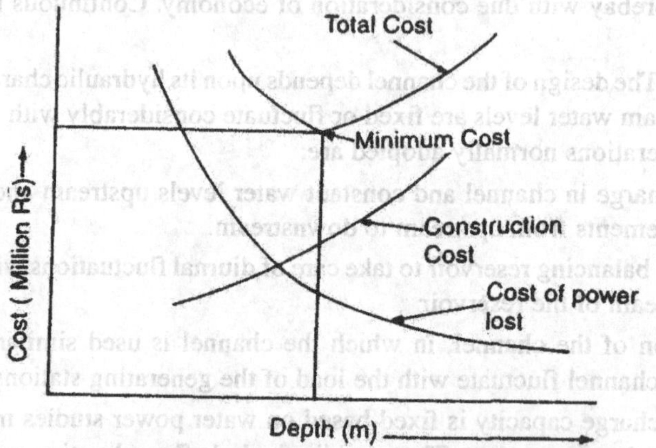

Fig. 6.18. Minimum cost canal depth.

Table 6.2. Design parameters for trapzoidal section.

Parameter	*Side slope 1:1*	*1.25:1*	*1.5:1*	*2:1*
Sectional area (A)	$BD + 1.785\ D^2$	$BD + 1.925\ D^2$	$BD + 2.088\ D^2$	$BD + 2.463\ D^2$
Wetted perimeter (D)	$B + 3.57\ D$	$B + 3.85\ D$	$B + 4.176\ D$	$B + 4.926\ D$
Hydraulic mean depth (R)	←	A/P	→	
Velocity (V)	←	$\frac{1}{n} R^{2/3} S^{1/2}$ (m/s)	→	
Angle	45°	38°-40′	33°-42′	26°-30′
Tangent length	0.4142 D	0.35 D	0.30 D	0.235 D

The channel is usually rectangular, trapezoidal in section and essentially lined to reduce head loss to the minimum and corresponding more operating head made available to the plant, minimum cross section, stable section at maximum velocity and lower operation and maintenance cost. The best hydraulic section may not necessarily be the economical one (Fig. 6.19).

Side Slope: The side slopes of the channel section are designed to suit drawdown conditions. Fluctuations in water level occur depending on the pattern of load demand and extent of storage provided in balancing storage reservoir, if any. The effect of drawdown and rate at which it occurs govern the stability of the side slopes and hence lining of the channel. At Narmada canal (Gujarat) the drawdown stipulated is 15 cm in one hour, 30 cm in two hours and 65 cm in 24 hours. The side slope may vary from 1:1 to 2:1. In Sundernagar hydel channel at Beas Sutlej Link Project in Himachal Pradesh it is 1.5:1 for a head discharge of 255 cumecs, bed slope 1 in 6666 and 6.26 m water depth. At Narmada canal, the biggest in the world, with 1133 cumecs capacity, the side slope provided is 2:1.

Free Board: Sufficient free board is provided to avoid overtopping of water on channel sides, usually 1.0 m. About 25 percent extra free board is provided to avoid reduction in power generation due to higher coefficient of roughness observed in subsequent years, the surge phenomenon due to load rejection, load acceptance, rate of closure of units or acceptance of load and size of forebay or surge basin on the channel.

The maximum surge height due to load rejection is determined from the following formula, as per IS: 7916:1992.

1. For abrupt closure, $h_{max} = \sqrt{k^2 + 2kh}$.
2. For gradual closure within the period required for the first wave to travel twice the channel length, $h_{max} = \frac{k}{2} + V\sqrt{h/g}$

where,

h_{max} is maximum wave height (m), K or $V^2/2g$ is velocity head, V is mean velocity of flow and h is effective depth i.e., area of cross section/top width.

In the event of incomplete dissipation of energy upstream of power channel, provision for wave suppression in the head reach of the channel is made.

Underdrainage: Suitable underdrainage system is provided to protect the lining where the channel is running in terrain with very high ground water level and is subjected to high seasonal variations or where the soil is sufficiently impervious to present free drainage of seepage or leakage from the channel.

12. Forebay. It is a small balancing storage at the head of penstock to absorb the short interval variations of intakes of water into turbines in accordance with the fluctuating loads. Power demand varies and so varies the discharge required by turbines. An arrangement, named forebay, is provided to absorb the short interval variations of inflow water into turbines by maintaining a storage basin in front of the intake to temporarily store the water rejected by the plant due to reduced power load during off-peak hours and to supply water for initial increments of an increasing load on sudden increased demand while water in the canal is being accelerated. It thus serves as a safety valve by releasing the water pressure built up in penstock caused by water hammer action in the event of sudden rejection of water by turbines. Where plant is situated at the base of a dam, the reservoir constitutes the forebay. It gets supply from the reservoir through headrace channel and in turn feeds the penstock.

The layout of forebay is so selected that it is founded on rock or on firm soil. The lining in the forebay portion is essentially reinforced to take care of water waves associated with water rejection or withdrawal of water by the turbine (s).

When a hydel channel conveys water to the turbine, the canal itself serves as a forebay. The canal section is enlarged to cater for the flow variations (Fig. 6.17). When penstock directly takes water from reservoir, the later itself serves as forebay.

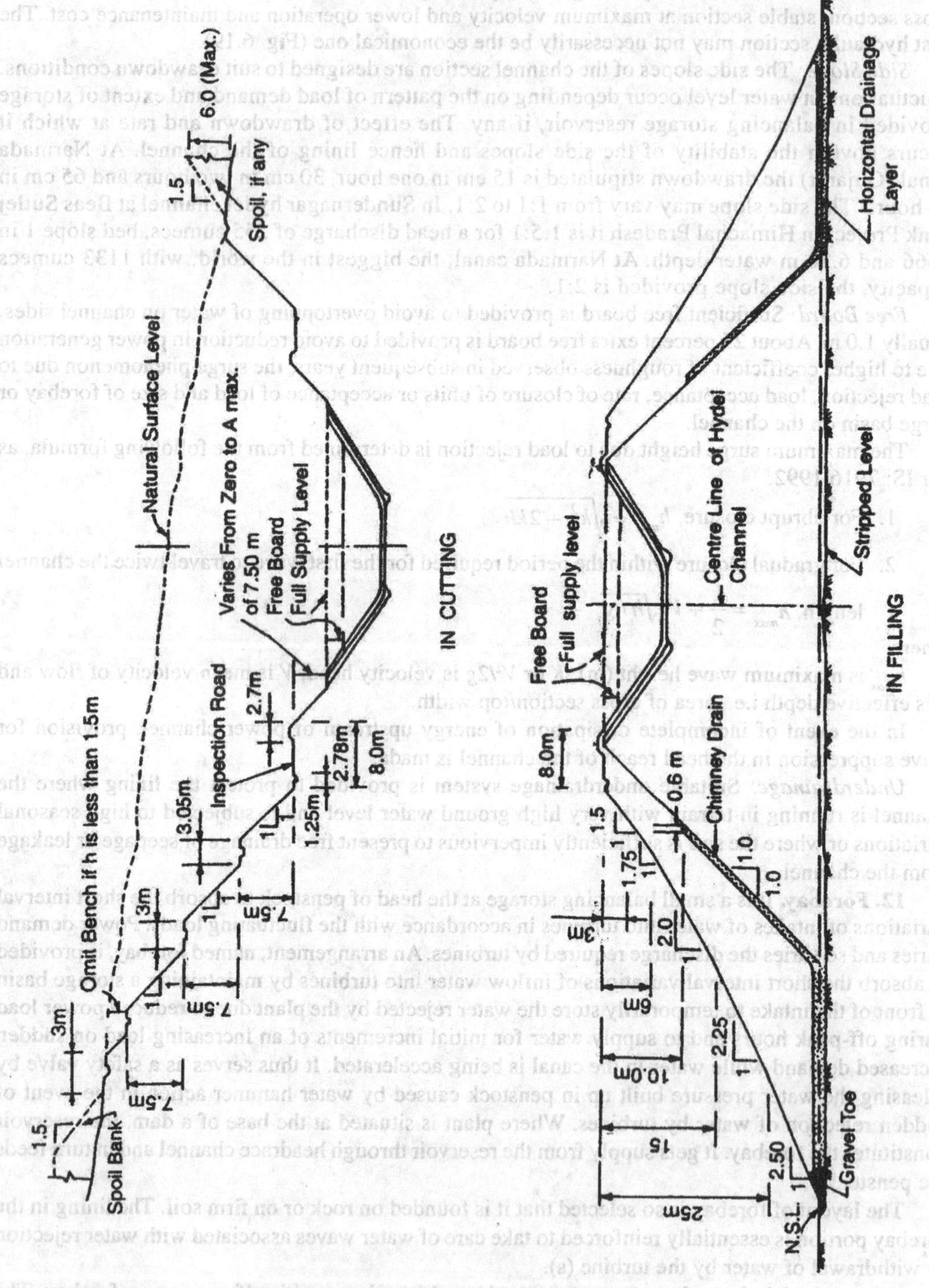

Fig. 6.19. Typical cross section of a power channel.

13. Intake: The adjunct to the forebay is intake structure to divert water to penstocks. Suitable trashracks are provided to avoid trash entering the penstock.

14. Bypass Channel: It is a channel with crest in the forebay provided for diverting the flow from headrace to the tailrace by-passing the power plant. Its function is to dispose of the water in the event of sudden stoppage of the power house. The gates in the channel are regulated automatically, activated hydraulically by water level in the forebay.

SOLVED EXAMPLES

Example 6.1. *Design a lined power channel to convey a discharge of 250 cumecs to a power plant, side slope is 1.25:1, bed slope is 1 in 6666 , Manning's n = 0.018.*

Solution.

In the trapezoidal lined section, assume a velocity of 1.75 m/s.

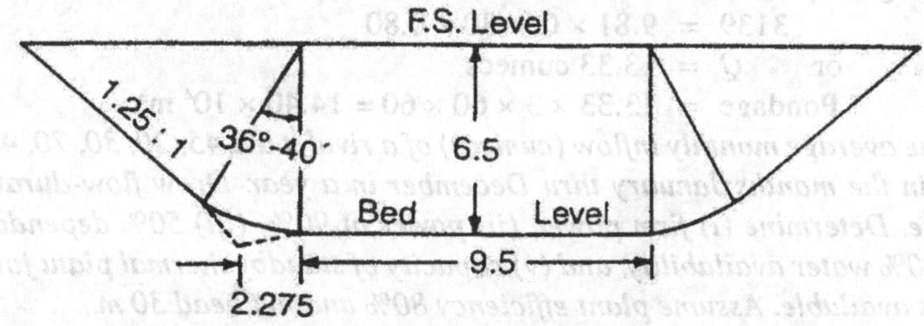

Fig. 6.20. Lined section.

$\therefore$ Section area, $A = Q/V = 250/1.75 = 142.85\ \text{m}^2$

Velocity $\quad V = \dfrac{1}{n} R^{2/3} S^{1/2}$

$$1.75 = \frac{1}{0.018} R^{2/3} (1/6666)^{1/2}$$

or $\quad R = 4.14$ m

Also $\quad R = A/P$

or $\quad P = 142.85/4.14 = 34.3$ m

For 1.25:1 side slope

$$A = BD + 1.925\,D^2$$
$$P = B + 3.85\,D$$

Putting values of A and P.

$$142.85 = BD + 1.925\,D^2 \qquad (i)$$
$$34.30 = B + 3.85\,D \qquad (ii)$$

Multiply (*ii*) by D and deduct (*i*) from it, we get

$$1.925\,D^2 - 34.3\,D + 142.85 = 0$$

or $\quad D = \dfrac{34.3 \pm \sqrt{34.3^2 - 4 \times 1.925 \times 142.85}}{2 \times 1.925}$

$\therefore \quad D = 6.5$ m

From (*ii*) $\quad 34.3 = B + 3.85 \times 6.5$

or $\quad B = 9.5$ m

Lacey's $\quad f = 2.5\,V^2/R$

$\quad = 2.5\,(1.75)^2/4.14 = 1.85$

Example 6.2. *A run-of-river plant on a stream has inflow of 20 cumecs and net head of 30 m with provision for pondage to meet daily peak demand with a load factor of 60%. Determine the power* *generation capacity of the plant at 80% overall efficiency, (b) The plant runs as peaking station for 3 hours and balance period in the day for average load. What amount of pondage is needed?*

Solution:

Average power generation, $P = 9.81\,QH\eta$

$= 9.81 \times 20 \times 30 \times 0.80 = 4709$ kW

$$\text{Peak load} = \frac{4709}{0.60} = 7{,}848 \text{ kW}$$

$\therefore$ Maximum power generation capacity

$= 7{,}848$ kW.

Additional load to be generated from storage for peaking

$= 7{,}848 - 4709$

$= 3{,}139$ kW

Additional discharge required

$$3139 = 9.81 \times Q \times 30 \times 0.80$$

or $Q = 13.33$ cumecs

Pondage $= 13.33 \times 3 \times 60 \times 60 = 14.40 \times 10^4$ m^3.

EXAMPLE 6.3. *The average monthly inflow (cumecs) of a rivulet are 45, 20, 30, 70, 40, 10, 100, 120, 80, 65, 55, 50 in the months January thru December in a year. Draw flow-duration curve, power-duration curve. Determine (i) firm power, (ii) power at 90%, (iii) 50% dependability, (iv) storage required at 50% water availability, and (v) capacity of standby thermal plant for the same, the storage being not available. Assume plant efficiency 80% and net head 30 m.*

Solution: Computation for flow-duration curve

Discharge (cumecs)	10	20	30	40	45	50	55	65	70	80	100	120
Power (10^3 kW)	2.35	4.71	7.06	9.42	10.59	11.77	12.95	15.30	16.48	18.84	23.54	28.25
% of time	100	91.7	83.3	75	66.7	58.3	50	41.7	33.3	25	16.7	8.3

Flow-duration and power-duration curves are shown in Fig. 6.21.

(*i*) Firm power* $= 9.81 \times Q \times H \times \eta$ $= 9.81 \times 10 \times 30 \times 0.80 = 2.35 \times 10^3$ kW

(*ii*) Power at 90% dependability $= 9.81 \times 23 \times 30 \times 0.80 = 5.42 \times 10^3$ kW

(*iii*) Power at 50% dependability $= 9.81 \times 55 \times 30 \times 0.80 = 12.95 \times 10^3$ kW

(*iv*) Storage required, area of hatched portion $= 112.5$ cumec – months $= 2.95 \times 10^8$ m^3, assuming a month of 30.4 days

(*v*) Capacity of thermal plant $= (55 - 10) \times 9.81 \times 30 \times 0.80$
$= 10{,}594.8$ kW

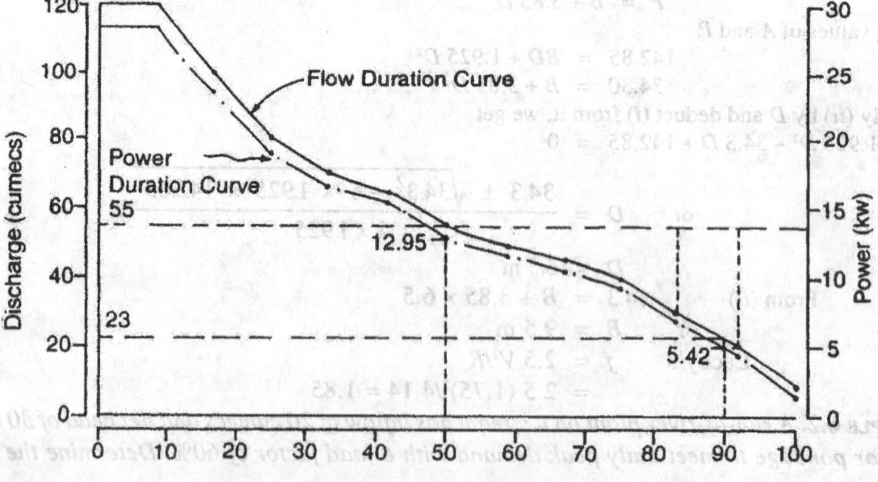

Fig. 6.21. Flow Duration and Power Duration Curves.

QUESTIONS

1. **Explain with sketch a typical arrangement of a power house on canal fall. Show the superstructure and sub-structure when the turbine used is a Kaplan type. *(P.U.)***
2. **Describe various components of a run-of-river power development getting supply from (*i*) a diversion dam, and (*ii*) barrage.**
3. **Why desilting chambers are needed in a run-of-river power development on a stream, illustrate with a sketch?**
4. **How do various components of a run-of-river development on a stream differ from that of a canal fall power plant and why?**
5. **Give sketches of various components of surface and underground run-of-river power developments.**
6. **What do you understand by headrace tunnel and pressure shaft? Where and why latter is used?**
7. **A canal carries a discharge of 120 cumecs and in a fall the head available is 20 m. If the efficiency is 88%, find the power that can be developed. *(AMIE)***

   Hint: Firm Power $= 9.81\, QH\eta$

   $= 9.81 \times 120 \times 20 \times 0.88$

   $= 20{,}719$ kW.
8. **Explain with sketches a typical arrangement of a power house on a river. Show the superstructure and substructure when the turbine is a Kaplan type.**
9. **What are the advantages and disadvantages of an underground power plant? *(AMIE)***
10. **How will you work out the general requirement of water for a hydel power project? Explain with suitable examples? *(AMIE)***
11. **How will you estimate the water requirement for a hydropower project knowing distribution of electrical load? What improvements will be achieved if hydroelectric plants will be used to generate power for peak loads only? *(AMIE)***
12. **Write a note on water requirements of a hydel power project. *(P.U.)***
13. **How is the water requirement for hydel project calculated? *(AMIE)***
14. **What do you understand by flow-duration curve and power-duration curve?**
15. **What is flow-duration curve? How is it prepared?**
16. **State briefly storage and pondage. What is the essential difference between the two, illustrate with a sketch?**
17. **Give a comparative description of storage and pondage.**

7

PUMPED STORAGE POWER DEVELOPMENT

7.1. INTRODUCTION

Electricity is not amenable to direct storage, the primary aim being to generate power to the tune of demand in the grid at all the times. To meet the demand, with a given quality of service, the generating mechanism may include peaking facilities offering a rapid response to load fluctuation, whether cyclical or random. Pumped storage power plants are the indirect way of storing of electrical energy, by utilising the surplus energy available in off peak hours usually at night or on weekend days, through thermal, nuclear etc., in the form of 'water potential' by pumping water from lower reservoir to the upper reservoir and releasing the same during peak demand hours, usually during day time, to generate power as in a conventional hydropower station. Water thus forms an excellent medium for the generation and accumulation of electrical energy. The pumped storage power plants can meet peak requirements to a very large extent resulting in saving in new installation of fossil fuels and other modes of power generation.

The various purposes of a pumped storage plant as such are (*i*) To improve power factor of the network, (*ii*) To control network frequency, (*iii*) To serve as emergency reserve, (*iv*) To meet weekly and seasonal peak load demands, and (*v*) To cover daily peak power demands and to improve capacity factor of base-load thermal plants.

7.2. ESSENTIAL REQUIREMENTS OF PSP PLANT

A PSPP essentially requires two reservoirs, upstream and downstream, with a high difference in water levels between the two (Fig. 7.1). The higher head results in (*i*) The reduced size of the water conductor system for a given power output and (*ii*) Increase in the speed of the generating/pumping unit which in turn reduces the physical size and cost of the generating/pumping unit. The lowest

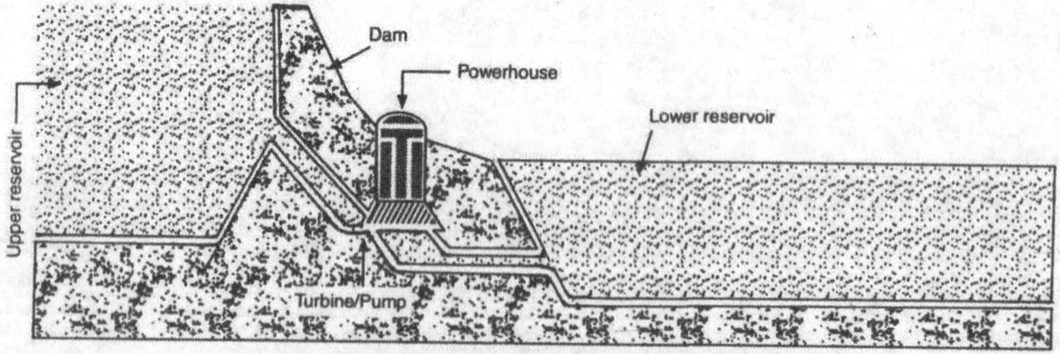

Fig. 7.1. Essential elements of a pumped storage plant.

limit of head is considered around 200 m. However, the creation of the upper and lower reservoirs is not compulsive. The pilot plant of 30 MW pumped storage power plant in Japan uses sea water as the lower reservoir. Likewise the Mount Hope 2000 MW pumped storage power plant in USA utilises an underground reservoir at the site of a former magnetite iron mine.

7.3. NECESSITY FOR PUMPED STORAGE POWER DEVELOPMENT

The pumped storage power development is needed urgently in the country to tide over the present peaking power shortage of more than 11 percent, besides a variety of other compulsive factors on account of which these plants constitute a better option compared to the conventional hydropower and gas turbine power etc. The necessity for pumped storage power plants can be summarised as (*i*) The conventional hydropower and gas turbines can meet the demand for peaking power but they are not economical when operated for low load factor in respect of investment and energy cost. A levelled load curve is essential in any electrical system for economical operation of the generating units. Some units are needed to be operated for short periods to meet the peaking power. Such power systems are necessarily to have the characteristics of quick start and economical operations at lower load factors, (*ii*) In some regions, like Orissa, Bihar and Tamil Nadu almost the entire hydropower potential has since been exploited or the available resources are very limited and located far away from the load centre, (*iii*) With declining hydro-thermal mix and restrained development of hydropower due to a variety of reasons, storage of water may be affected restricting the system capability to meet the peak demand only through hydropower projects, (*iv*) The utility system needs to be improved to bring about capacity benefit and energy benefit (fuel cost saving) which a pumped storage plant can provide, (*v*) Some means are needed for absorbing and storing energy utilising cheap power produced from base load plants during off peak hours, (*vi*) There is an imperative need to generate cheaper power during peak hours to replace costly gas turbine power, (*vii*) Quality of power needs improvement for frequency regulation and voltage control, (*viii*) Reliable power system is needed to cover forced outages of thermal power plants, and (*ix*) Environmentalists have diminished opportunities for conventional hydropower development.

7.4. ADVANTAGES OF PUMPED STORAGE POWER

(*i*) Simple, premium peaking power. Supplies high value energy by absorbing low value excess energy during off-peak period that otherwise goes as waste, (*ii*) Absorbing and storing surplus energy from base load plant (s) during off-peak hours and generating cheaper power (stored in the form of head of water) during peak hours. Replacing costly means, (*iii*) Very short start-up time, a few minutes, (*iv*) Excellent capability to follow load, 30 percent per minute, significantly higher than coal or gas turbine power plants, (*v*) Stabilising the energy system load by absorbing the surplus energy for pumping water during off-peak hours, smoothening the load curve and enabling continuous operation of efficient base load system, thereby improving grid stability, (*vi*) Ideally suited for frequency regulation and voltage control thus providing quality power, (*vii*) Reliable power system ideally suited to cover forced outages of thermal power plants, (*viii*) Help in optimum utilisation of installed capacity in thermal and nuclear plants by facilitating their running at full load capacity, (*ix*) Improve flexibility of performance of the system, meeting rapid changes in load demand etc. leading to improved efficiency in operation of all modes of supply sources, (*x*) Reservoirs are small compared to hydropower plants with consequent economy in the cost of diversion structures and lesser amount of displacement of population, submergence of land etc., (*xi*) Unlike thermal, no emission of particulate and gases and water temperature rise, (*xii*) Flexibility in siting and as such can be so located that environmental impacts are minimum, (*xiii*) Improve plant load factor of base load plants, (*xiv*) Ideally suited when there is a significant difference between peak and off peak demands.

Limitations

The pumped storage plants need energy for pumping. They generally have the economic advantage when operated as a power source for short peak period and emergencies. Care has to be exercised in the selection of the site of the plant to keep investment per MW low. There is a conversion loss of about 23% in the system because of losses in conversion equipment.

7.5. HISTORY OF PUMPED STORAGE POWER DEVELOPMENT

The concept of pumping water to augment power was first developed in Italy in 1892. The PSPP were at first of the type of storing surplus river water of the high-water season and using it in the low-water season but now with the purpose of daily and weekly operation for efficient operation of electric power systems. The first hydro plant making use of the augmented storage for power generation was established in 1909 in Switzerland. In U.K. first PSPP was installed in 1920 in Scotland. The first reversible pump-turbine was installed in 1931 in Germany. In USA, Tennessee Valley Authority installed a reversible pump-turbine, Francis type, in 1955. In Japan, the PSPP generation is more than 44 percent of its hydropower capacity.

In India, the earliest PSP plant was Nagarjuna Sagar power station in Andhra Pradesh (7 units of 100 MW each + 1 conventional unit of 110 MW). Later on, Kadana in Gujarat (4 × 60 MW) was followed by Kadamparai in Tamil Nadu (4 × 100 MW) in 1988. Of late, the development of pumped storage has attracted much attention in view of the important role these plants can play to even out energy generation from base load and in meeting fluctuations in power demands and system contingencies. The pumped storage stations as a means of meeting peak power requirement in large power supply system with different characteristics are not only becoming desirable but economical as the peak demands are rising fast.

7.6. POWER POTENTIAL OF PUMPED STORAGE POWER DEVELOPMENT

The Central Electricity Authority in 1987 identified 56 sites for pumped storage schemes with a total installation capacity of about 94,000 MW (Table 7.1). Of late, the pumped storage power plants have evinced great interest on account of the possibilities of (*i*) Partial conversion of existing hydropower plants to pumped storage plants, and (*ii*) Development of deep underground plants.

The pumped storage potential in the country, presently assessed, is 96524 MW.

Table 7.1. PSPP potential

Region	Power potential (MW)
West	38,220
East	25,985
South	16,650
North	13,065

6 pumped storage plants with total installed capacity of 1284 MW are in operation. Schemes worth 2620 MW are under execution. The Kachchh project has installed capacity of 900 MW.

7.7. CLASSIFICATION OF PUMPED STORAGE POWER DEVELOPMENT

The pumped storage power development is classified, as described below:

1. Classification Based on Inflow

(*i*) **Pure pumped storage plant:** Also called recirculating type. In this type there is a negligible flow in the upper or lower reservoirs, water stored at either the upper or lower reservoir is circulated in one cycle of operation without utilising river runoff except in small measure that is required to make up for losses due to evaporation and seepage etc. The essential requirements are upper and lower reservoirs and a site to secure a high head. The generating capacity depends upon the amount of water pumped from the lower reservoir to the upper one. Kadamparai PSPP belongs to this type.

The plant may be based on daily regulation or weekly regulation. In daily regulation scheme, the capacity of the upper reservoir is sufficient to meet daily peaking requirements while in the weekly regulation it is more. Additional pumping energy available during weekends is utilised to pump additional water for power generation during week days.

(ii) **Mixed type :** Also called multiple use type. It is an ordinary reservoir type or regulating pondage type power plant which incorporates pumped storage features that is utilises river runoff flowing into the upper reservoir provided with pump up facilities. The site selection requirements are the same as for a high head hydropower plant. The amount of inflow allows motive power cost for pump-up to be saved so that it may be considered as pure pumped storage with construction cost made cheaper by the amount of the above mentioned savings capitalised. Precisely, the upper reservoir has substantial amount of inflow and accordingly the operation of the plant is of mixed type. During high inflow season, the plant operates only in generation mode. However, during lean period, pumped storage operation is resorted to as at Nagarjuna Sagar and Kadana plants.

2. Classification Based on Reservoir Capacity

(i) **Daily regulation type:** In this plant surplus supply capacity is pumped during period of light load and power generation is done during peak hours that is operated on a daily cycle of pumping and generation.

(ii) **Weekly regulation type:** It is the plant wherein the pumping is restricted to during light load week ends, Sundays and holidays, for regulation not only weekly fluctuations in load but also supplementary operations for times such as of problems in thermal generation. The peak durations are about 6 to 12 hours. Evidently the reservoir capacity is slightly larger than in daily regulation type.

(iii) **Seasonal regulation type:** In this plant pumping is done during light demand season and generation in high demand season. This type is used in a system where hydropower is the main source and generally surplus power is available in high-water season for pumping. Large reservoir hydropower stations with pump-up facilities also belong to this type, but in such cases they are more often operated on seasonal regulation as reservoir type hydro plants in general and as pumped storage in particular. It is possible for seasonal regulation to be done but emphasis is placed on supplementary operation as daily peak supply capability and supply capability for time zones when fluctuations in power demand are great and this type is often used in systems with large proportions of thermal power.

3. Classification Based on Equipment Type

It is based on the relative arrangements of turbines and pumps.

(i) **Four unit type:** In the initial stages of development there were four different units namely pump, motor, generator and turbine which lent it the name of four unit installation. The pump and motor coupled together were completely independent of turbine and generator coupled together which afforded a complete independence of operation. It required more spacious power plant with consequent high cost.

(ii) **Tandem type:** Also termed as three unit system. The motor for driving the pump and the generator are possessed in common and the pump, generator-motor and turbine are coupled on a single axis. When the turbine is run, the unit operates as a generator and when the pump is operated, the same unit operates as a motor. These are of two types— horizontal shaft and vertical shaft. The vertical shaft setting is more compact but requires a higher power plant structure. Presently, reversible pump-turbines are generally used because of economic consideration.

(iii) **Reversible type:** Also called two-unit system. A pump-turbine and generator-motor are provided, the equipment used as a motor and pump during pump-up, and as a turbine and generator during power generation with revolutions in the reverse direction from that during pump-up. Since

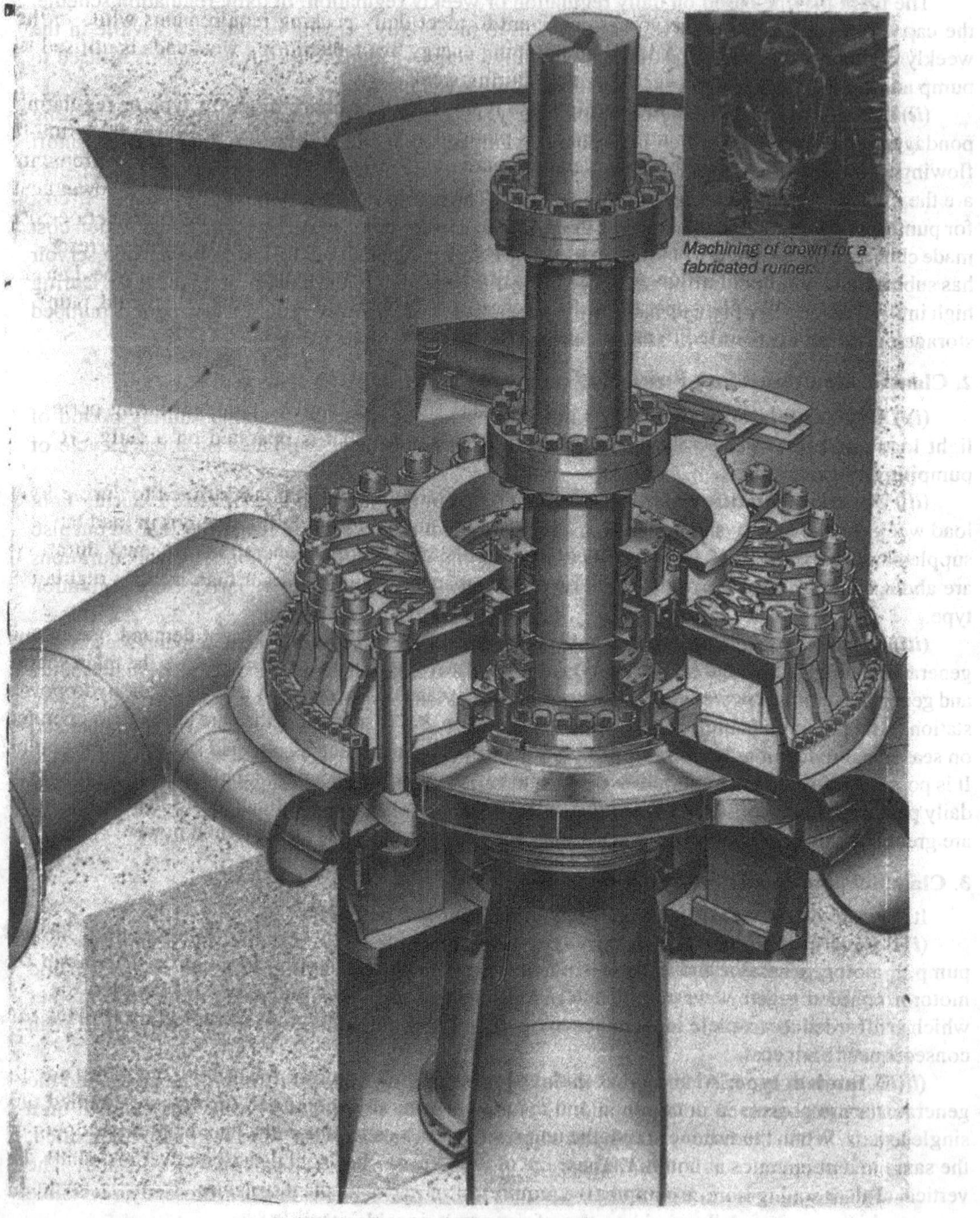

Fig. 7.2. Typical high head reversible pump-turbine.

single machines are pump-cum-turbine and motor-cum-generator, it is difficult for both to be used for maximum performances and as such has the drawback that comparatively more time is required for switching between pump-up and power generation but the construction cost as a whole including appurtenant equipment and the plant building is low. The latest trend is to use reversible type pump-turbine (Fig. 7.2). The efficiency tends to decline with high head greater than 400 m but this can be ignored compared with the merit obtained through adoption of larger-capacity units. With the advances made in reversible pump-turbines it is now possible for high head sites of 500-600 m to be developed.

The reversible pump-turbines have to perform well both as a pump for pumping and as a turbine for generation of required power at a synchronous speed. These are best suited as they are compact, economical and efficient and hence are universally adopted for modern pump storage plants. These types of machines are built for higher heads and larger capacities. Machines of 600 m head (single stage) and capacity per unit as high as 300 MW are in operation while machines of 800 m head (single stage) are being developed. It is desirable in design to make use of the pump and turbine characteristics of the machine in such a way that these operate as far as possible in the best efficiency zone either as a pump or as a turbine and are less prone to adverse cavitation conditions. The hydraulic design of this type of machine is more difficult than a conventional machine as the range of satisfactory operation is much more limited.

Whereas all reaction turbines may be used for reversible pumping and generating mode, but Francis turbines are generally employed. The reversible Francis turbine is essentially a centrifugal pump, suitably modified to perform as a turbine when operating in the opposite direction.

Comparison of Tandem and Reversible Types

Tandem type	*Reversible type*
Simple in operation, high cost, more space is required.	Intricate in operation, economical, compact and less space is required.
Suitable for high heads; multistage centrifugal pumps can be used.	Limitation of head; generally a single-stage arrangement is used.
Pump and turbine being separate units can be optimally designed to operate at maximum efficiency.	Constraint in design. Single machine designed as a pump-cum-turbine and motor-cum-generator has lesser efficiency.
Duration of rotation is generally kept identical; only coupling or decoupling the pump with generator is involved.	Machine rotation has to be reversed to change from one operating mode to the other.
Time required for switching between pump-up and power generation is less.	Time required for switching between pump-up and power generation is comparatively more.
Cavitation problem is not severe as multistage pumps are used.	More cavitaton problem.

7.8. DEVELOPMENT PLAN FOR PSP DEVELOPMENT

The various parameters to be considered in the formulation of development plan for pumped storage plant are briefly as under:

(*i*) Maximum Power, Discharge (*Q*) and Head (*H*): For supplying electric power in a stable manner to meet large increases in power demand, large capacity power stations are needed. In planning a large station it is more economical to increase the capacity rather than number of units and for this purpose the two directions of increasing Q or increasing H are conceivable, but generally it is economical to make Q small and H larger than to make Q large and H less. A high head means small Q and hence lesser reservoir capacity and small dam for the same length of operating time. The cross sectional area of water conductor structures such as headrace and tailrace is reduced, while by increase in speed of the turbine-generator the structure can be made compact with economy.

(*ii*) Water Conductor Length and Head: Topography plays an important role in this consideration. With a view to reduce unit construction cost of a pure pumped storage plant, the essential consideration from the standpoint of topography is that the length of the water conductor system between two reservoirs be kept as short as possible and making head high. In multiple-use pumped storage plant, the head may not be high. It is desirable from economic considerations that the ratio of length of water conductor and the head between upper and lower reservoirs be as small as possible and in general it is 2 to 10.

***(iii)* Unit Capacity:** The rotating speed of the turbine-generator is determined from the head and power discharge and unit capacity is selected giving consideration to engineering limits to manufacturing and limits in transportation.

***(iv)* Power Transmission and Transformation:** For the transmission voltage, transmission line route and step-up system consideration is given to economy and reproductibility. The transmission cost of power transmission facilities is roughly proportional to distance so that it is essential that the location of the power station is close to the load area.

***(v)* Dam Height and Reservoir Capacity:** The optimum dam height and hence reservoir capacity is selected from the considerations of topography and geology, peak duration times and economics. In case reservoir capacity is small, the critical daily production (electric energy produced in one day) is small and it will not be possible to operate at critical output for the operating time required from the standpoint of demand and supply (peak duration time) so that latency of output occurs, and kW-value decreases as a result. However, the construction cost is low. On the other hand, in case the reservoir capacity is large, the critical daily production is large and the degree of latency is small or output will become apparent for effective output to be increased and kW-value of the site increases, but construction cost of making large capacity reservoir is increased.

It thus becomes important when considering the scale for economic analysis to be made of the increment production cost for extending reservoir capacity by 1 hour and the effect of increasing kW-value to grasp the economic nature of the site and to gain an approximate measure of the optimum scale.

***(vi)* Reservoirs:** As upper and lower reservoirs are essentially required, and when an existing reservoir or a natural lake can be utilised the economics will be improved by that much.

7.9. COMPONENTS OF PUMPED STORAGE POWER DEVELOPMENT

The essential components of a pumped storage power development (Fig. 7.3) are briefly, as under:

1. Upper Reservoir: The requirements of upper reservoir are the same as for general hydropower but earthfill dams generally adapted for pumped storage have additionally to be more stable due to rapid and frequent drawdown conditions associated with P.S.P. plants. The upper reservoir on hill top is preferred and is designed considering topography, geology and rapid drawdown conditions.

2. Lower Reservoir: Lower reservoir is generally developed on natural streams and even existing reservoirs meant for flood control, power generation, and other purposes can be utilised. In the case of pure pumped storage, water to cover evaporation, leakage and seepage losses must be supplemented from inflow to the low reservoir.

3. Conduits: These are in general the same as in hydropower plant except that here water flow is essentially in both directions. The intake and outlet at time of power generation will respectively become outlet and inlet at the time of pump-up. This is of importance in designing manifolds and bifurcations. The water conduit structures must be designed for smooth flow of water in both directions. Trashracks are provided both at the intake as well as at the end of draft tubes to avoid entry of trash during pump-up stage in off-peak hours.

4. Surge Tank: The requirement of surge tank is there but for a purpose different from that of a hydropower station in that pump tripping may occur during pump-up and minimum water level is reached due to this. Water flows from bottom to top during pump-up, and when power supply is cut off, the water will continue to flow in that direction for a short while due to inertia, but after that it will flow from top to bottom. The pump will accordingly start to rotate in reverse (turbine- direction rotation), and in case the valve is not closed, unrestricted speed of the turbine is reached. This is the stage when the surge tank will reach the minimum water level and designing is done based on this. The tailrace surge tank can be omitted even in cases of tailrace length of several hundred metres.

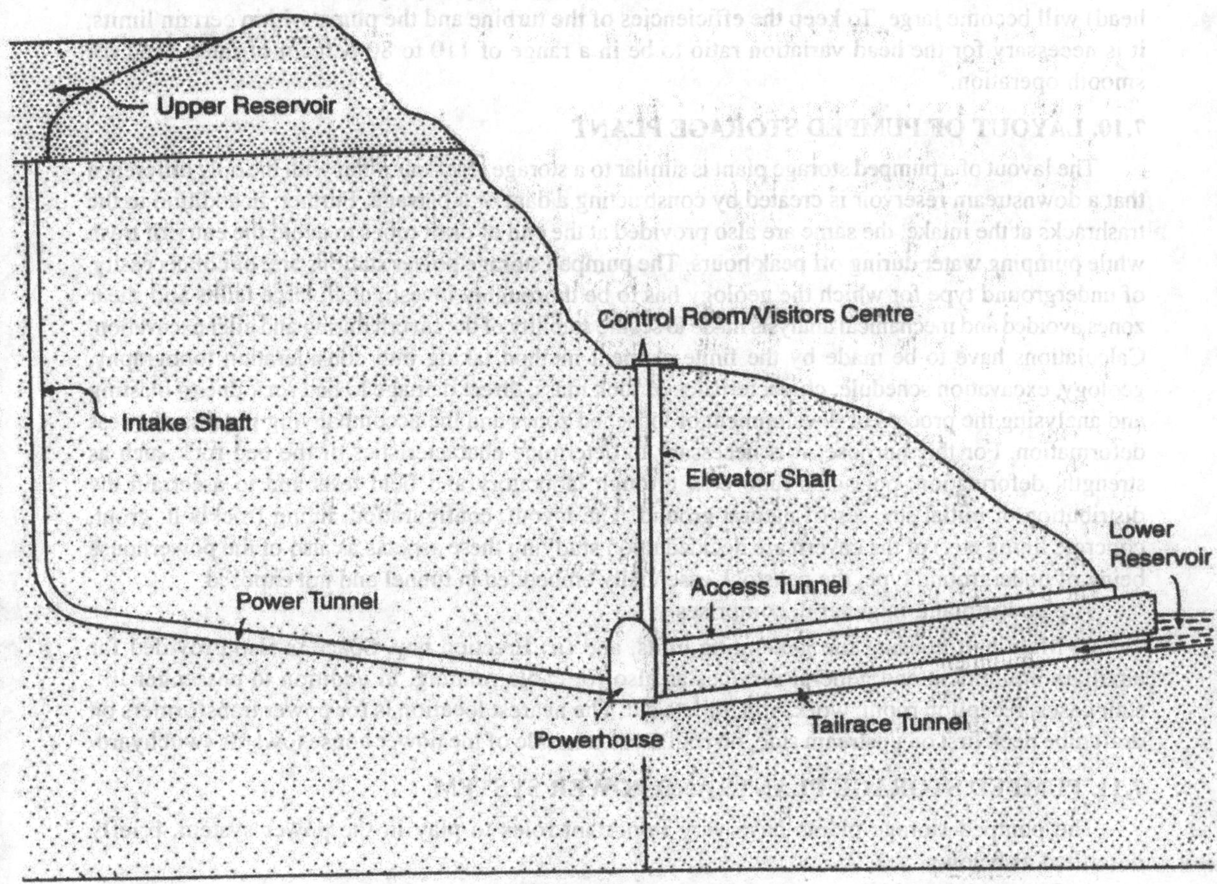

Fig. 7.3. Bad Creek (USA) pumped storage power plant.

5. Penstock and Power House: With a view to prevent cavitation during pumping operation, it becomes necessary to provide draft head by lowering runners below tail water level. With a pump-turbine of high head and large capacity, negative draft head is very high necessitating pumped storage plant of underground type. The cavern for underground plant has to be of a large size than conventional hydropower station and with increased output the cavern cross section becomes even larger. The penstock is of necessity embedded in tunnel and not to be exposed. The inside diameter of the penstock is selected for minimum loss of head considering inview the economy. The flow velocity in the penstock is generally 2.5 to 5 m/s with a run-off-river type but in high head pumped storage the maximum velocity is about 10 to 12 m/s. It is to ensure that in the former case head loss is made small by reducing velocity and in the latter case, since the utility factor is low there may be some amount of head loss, it would be normal for low velocity to be increased while penstock diameter and pipe thickness are held down. Even then design internal pressure will be increased and pipe thickness increased as head is made higher. However, penstock is generaly of high tensile steel and welded and part of pressure is borne by surrounding rock with rock cover of 0.4 to 0.5 H.

6. Draft Head: The draft head is determined from the characteristics of the pump-turbine and is increased with higher head. The construction cost is increased in case draft head is higher and consequent higher design water pressure of the penstock/tailrace. It is desirable for a turbine of small draft head and favourable cavitation characteristics to be developed. With a high head pump-turbine, restrictions related to head variation ratio (maximum pumping head/rated head minimum head) will become large. To keep the efficiencies of the turbine and the pump within certain limits, it is necessary for the head variation ratio to be in a range of 110 to 80 percent of rated head for smooth operation.

7.10. LAYOUT OF PUMPED STORAGE PLANT

The layout of a pumped storage plant is similar to a storage reservoir plant with the only difference that a downstream reservoir is created by constructing a dam or a barrage. Further, in addition to the trashracks at the intake, the same are also provided at the end of draft tubes to avoid the entry of trash while pumping water during off peak hours. The pumped storage power stations are, out of necessity, of underground type for which the geology has to be thoroughly investigated, large faults and shear zones avoided and mechanical analysis made to secure stability of the cavern during and after excavation. Calculations have to be made by the finite element method taking into consideration topography, geology, excavation schedule, characteristics of rock mass, direct damage to bed rock due to blasting and analysing the process of development of loosened zones and the accompanying process of stress deformation. For this purpose, it is necessary to determine characteristics of the bed rock such as strength, deformation, poisson's ratio etc., through laboratory and field tests and to ascertain the distribution of initial stresses of natural ground. The layout, configuration, lining (rockbolt, grout, concrete lining etc.) of the cavern are decided upon studying these aspects as also of the powerhouse being of underground type, the penstock essentially embedded in tunnel and not exposed.

The power plant, in general, comprises:

(*i*) Main hall housing the generating units, and (*ii*) Erection bay. Space is also provided for auxiliary equipment and general services as also for cable running, in addition to passenger lift, stair cases, reception room, workshop, toilets etc. The nearest location is for power transformers on draft tube deck first on upstream side, on roof or on one side of the power house towards switchyard.

7.11. PUMPED STORAGE PLANTS AND POWER SYSTEM

The pumped storage plants have very important roles to play in the power system, briefly described as below:

1. Conversion to high value peak energy: The PSPP is essentially to convert low value off-peak energy into high value on peak energy (Fig. 7.4).

The typical load duration curve of 900 MW Purulia pumped storage plant (West Bengal) for peak supply capability of eastern region and maintaining schedule frequency through load following operation in the early morning, in the evening and at peak hours when load rises up in eastern

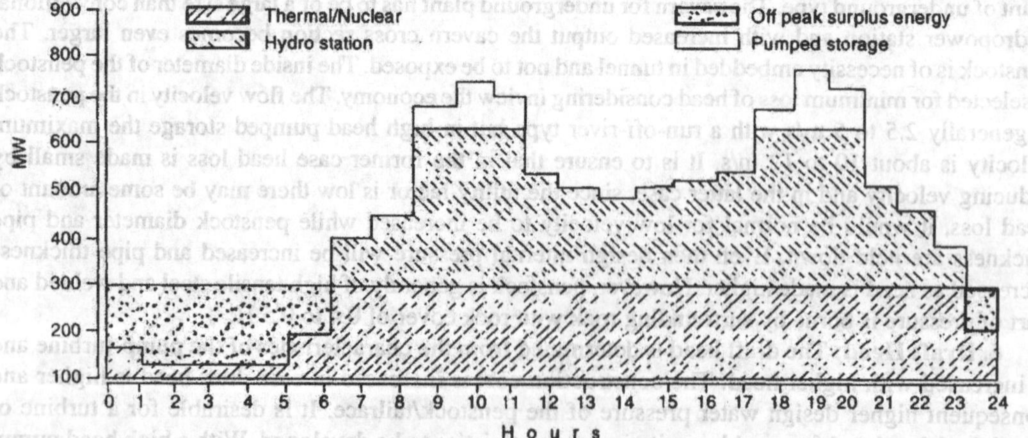

Fig. 7.4. Typical generation scheduling.

region is shown in Fig. 7.5. Pump up operation daily through off-peak energy (middle of the night, lunch), will increase midnight load to make possible high efficiency operation of thermal power plant. When reservoir water level of a hydro storage plant has dropped, PSPP will make it possible to increase the reservoir water level in the storage hydro plant by suspending or limiting power generation from these plants. The plant factor for the hydro storage projects in a dry month is 20 percent and from this the equivalent peak duration of the PPSP is about 6 hours taking account of energy for pumping.

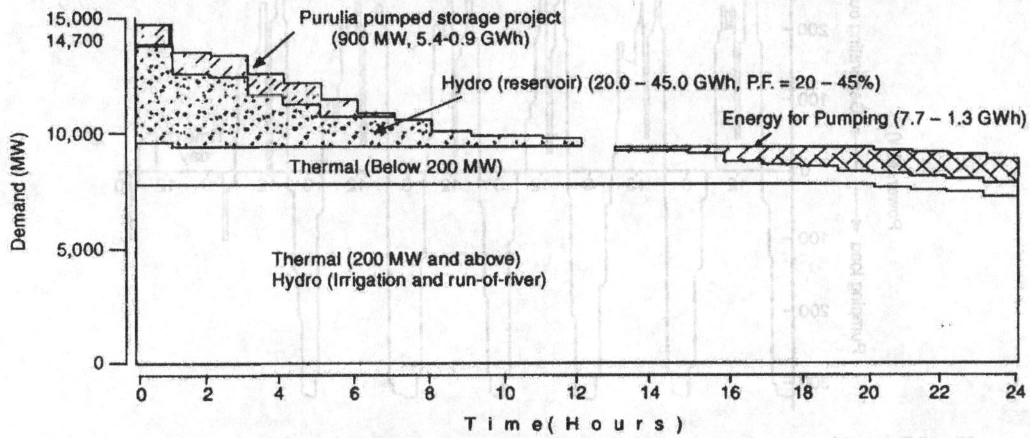

Fig. 7.5. Typical load duration curve of peak day (Eastern region).

2. Generating Additional Power Demand: By meeting additional power demand in peaking hours, PSPP helps in generating additional power demands during the off-peak hours (Fig. 7.6) which results in further increasing the utilisation of thermal and nuclear installed capacities. The approximate increase in power demand in off-peak hours is given by the following relation:

$$D_1 - D_2 + A\,(B/C) \quad \text{MW}$$

where, D_1 is increased power demand at any instant, D_2 is original power demand (without PSPP) at the same instant, A is the peak demand met by pumped storage storage, B is the original demand (D_2)—minimum power demand during the day, and C is original maximum demand–minimum demand.

3. Improved Load Factor: The essential function of PSPP is to improve the load factor of the base-load stations. The thermal plants and especially the nuclear plants are most cost-effective and trouble free when operated constantly as close as possible to the full load capacity for base load.

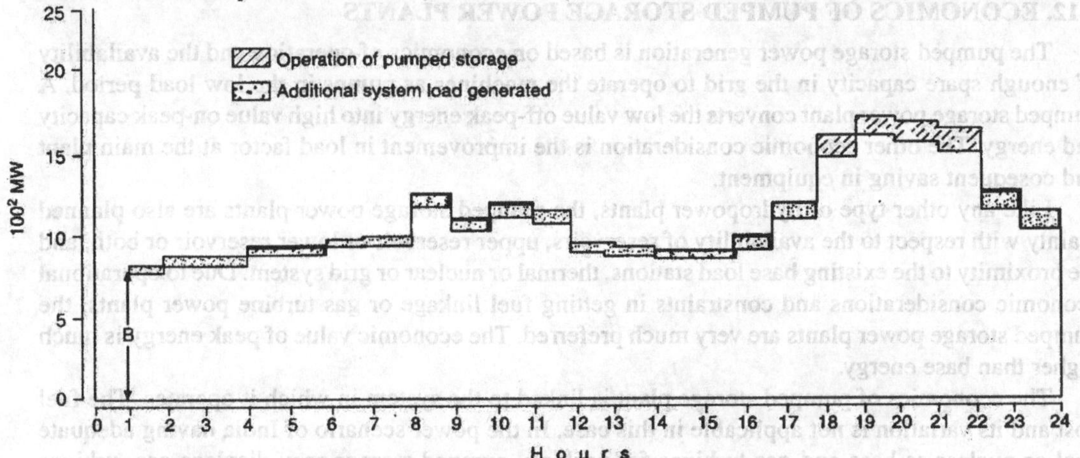

Fig. 7.6. Role of pumped storage in the system.

A typical daily load curve of 300 MW power station at Ikehara power station (Japan) is indicated in Fig 7.7 where power demand in the middle of the night declines to around 50% of peak load. Thus the same water is reused. Usually such plants follow diurnal cycles but some may follow seasonal cycles.

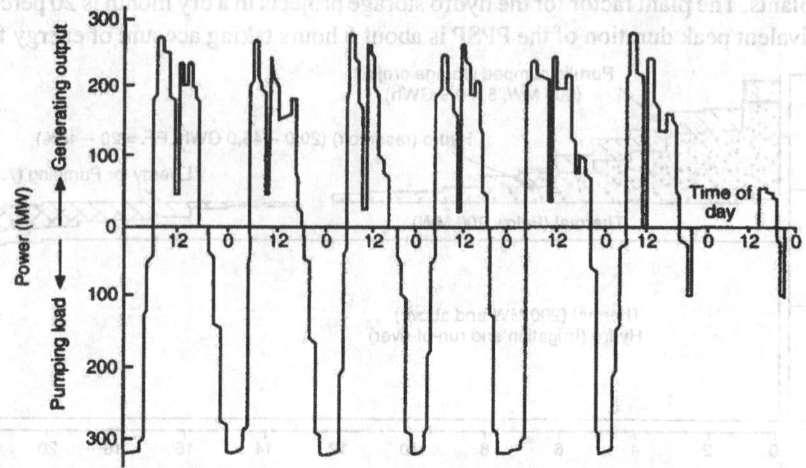

Fig. 7.7. Weekly operating record of pumped storage power station from Ikehara power station

The main function of a pumped storage power plant is to utilise the surplus power of the base load station during non-peaking periods, when the demand is below the output of the base load, and generate power to meet peak load demand. The essential requirement is upper reservoir for pondage preferably at a sufficiently high elevation to provide high head for power generation, a lower reservoir and proximity to the existing base load stations— thermal, nuclear or grid system. Due to operational and economic considerations, the pumped storage plants with low cost plant and equipment are preferred. A pumped storage power plant helps the utility system resulting in capacity benefit and energy benefit (fuel cost saving).

The peaking shortages exist in the country and are likely to continue for some more time. The pumped storage plants are a better option compared to the conventional hydropower and gas turbine power in view of several technical advantages. In view of above, the pumped storage power development need to be taken up on a large scale at suitable locations.

7.12. ECONOMICS OF PUMPED STORAGE POWER PLANTS

The pumped storage power generation is based on economics of operation and the availability of enough spare capacity in the grid to operate the machines as pumps in the low load period. A pumped storage power plant converts the low value off-peak energy into high value on-peak capacity and energy. The other economic consideration is the improvement in load factor at the main plant and cosequent saving in equipment.

Like any other type of hydropower plants, the pumped storage power plants are also planned mainly with respect to the availability of reservoirs, upper reservoir or lower reservoir or both, and the proximity to the existing base load stations, thermal or nuclear or grid system. Due to operational economic considerations and constraints in getting fuel linkage or gas turbine power plants, the pumped storage power plants are very much preferred. The economic value of peak energy is much higher than base energy.

The economics of pumped storage plant is linked to the system in which it operates. The fuel cost and its variation is not applicable in this case. In the power scenario of India having adequate coal or nuclear as base and gas turbines for peaking, pumped storage may displace gas turbines from the considerations of economy and conserving oil. The saving in start up and shut down of pumped storage plants are significant when compared to thermal plants. But pumped storage has high investment cost and long return period compared to gas turbine. Uprating a hydro plant by adding reversible units decreases investment cost and construction period, thereby improving the economics of pumped storage plants. But pumped storage plant involves emptying and filling cycles which can affect stability of the dam and may even necessitate raising the dam height to augment reservoir capacity.

Economic Evaluation

1. Size of reservoir: The economic evaluation criteria is the cost of generation of energy which should be comparable to cost of generation from alternative options. A simple method of determining the cost of generation is to work out annual cost and annual energy keeping in view the contribution of the PSPP in the system as a whole. The cost of a pumped storage plant consists of fixed cost (amortization of interest on construction cost, personnel cost, repair cost, general administrative expenses, various taxes and levies etc.) and variable costs (fuel cost and others). As regards power generation, the operating hours are short compared to conventional hydro, so that the weight of variable costs in overall cost is small while since the general part of variable costs is fuel cost, the variable cost per kWh, may be considered to be practically unchanged according to power station. As such with a view to reduce expenses of pumped storage power station, it is necessary to reduce fixed costs, in other words, economise on the construction cost.

On the other hand, in the way of values of pumped storage hydro, there are kW-value as supply against demand and kWh-value as energy production. The weight of kWh value in overall value is small because of the short operating time of pumped storage power generation. The kW-value is affected mostly by reservoir capacity and peak duration required from the point of view of demand. As regards the latter, it is determined readily if size of demand, shape of daily load curve and hydro, thermal and nuclear facilities are definite. Accordingly, the degree at which the size of reservoir capacity is set greatly influences the economics.

2. Generation Features: A pure pumped storage plant can be constructed provided a suitable head and a site where a small capacity reservoir can be created are available. When supply capability in unnecessary, pump-up is not done and when necessary if some amount of high cost is allowed, a constant supply capability can be obtained at all times by using surplus thermal as motive power for pump-up and the degrees of freedom and sureness as supply capability are much higher than hydro in general.

From start point of operation, such a plant is a hydel power station but from the energy point of view it is a low efficiency thermal having a quick response property. It requires a few minutes from start to full capacity operation while the output following ability under load variations is very great compared with thermal power generation. Thus from operation aspect , it is like a hydro in general but since it requires motive power for pump-up, a fuel cost is required similar to thermal. The thermal efficiency in this case is as follows and it is equivalent to a low efficiency thermal as seen from the economic point of view of power generation.

Thermal efficiency of P.S. power generation = **Additional firing efficiency of pump-up × overall pump-up efficiency**

Say = 40% × 70% = 28%

Economic Analysis

C/V Technique: The C/V ratio indicates which is more economical comparing the size of *C* and *V* of equal effect and reliability. Where, *C* is cost of pumped storage power plant which includes fixed capital cost plus the cost towards pumping during non-peak period, and *V* is the cost of alternative gas turbine power plant which includes fixed capital cost plus variable cost towards fuel.

The time for development of pumped storage hydro is to be the point in time when $C/V<1$ giving consideration for fitting in of demand, and the limit to development in the point when the economics of gas turbine plant and pumped storage hydro are equivalent that is the point when $C/V = 1$.

The pumped storage plants are economical compared to gas turbine power plants. That may have led to planning to establish pumped storage power plants at least to offset peak load demand.

The economic evaluation of a hydropower project is done by comparing its total costs converted to the present worth with those of an alternative thermal power plant considered most suitable for performing the same power generation as that by the hydel power project. The various methods of economic evaluation are briefly as given below:

1. Cost Estimate in Real Terms: In this method all terms are expressed in real terms of the latest year excluding any future inflation. The annual costs in respect of operation and maintenance, fuel cost for generation of gas turbine, fuel cost for pumping up for storage to upper reservoir of pumped storage power plant are compared.

2. Equalising Discount Rate Method: Also called economic internal rate of return method. A discount rate is calculated which equalises the total costs incurred from the beginning of the construction to the end of service life or a given period of the proposed pumped storage power project to the corresponding total costs of the alternative gas turbine power project, both converted to the present worth. If the equalising discount rate thus calculated is larger than the social rate of discount should reflect the opportunity cost of the capital, generally 8%, the proposed hydropower project is considered to be economical.

3. Simple Sensitivity Analysis : In this method the annual costs of pumped storage power plant and the ultimate gas turbine power project is increased by a factor to account for variation in process of coal used for thermal power base load station and diesel oil used by gas turbine.

4. Benefit-cost Ratio: In this method, the benefit-cost ratio of the pumped storage power plant is compared with the alternative gas turbine power plant. The benefit is the amount obtained by multiplying the annual energy generated units by the ruling tariff rate. The cost is the amount of the annual cost and investment costs.

7.13. COST OF POWER GENERATED

A simple method of determining cost of generation in the case of PSPP is to work out annual cost and annual energy, taking into cognisance the contribution of the scheme in the system as a whole. The annual cost may include (*i*) Annual interest on the capitalised cost, (*ii*) Maintenance cost, (*iii*) Depreciation, (*iv*) Cost of pumping energy at marginal cost of generation at thermal/nuclear plants (fuel cost), and (*v*) Cost of energy generation (at marginal cost) at thermal/nuclear plants to meet additional power demand generated during off-peak hours.

The annual energy may comprise (*i*) Energy generation by PSPP during peak hours. and (*ii*) Energy generation by thermal/nuclear plants to meet the additional demands generated during off-peak hours.

$$\text{Cost of energy generation} = \frac{\text{Annual cost}}{\text{Annual energy}}$$

The economic evaluation of PSPP on the above basis indicates that the cost of energy generation compares favourably to that at alternate modes of energy.

QUESTIONS

1. What are pumped storage plants? Explain with suitable sketches their working and state the advantages. (*P.U. 1999*)
2. Discuss briefly the various essential components of a pumped storage power plant.
3. Mention the various parameters considered in the formulation of development plan for PSPP.
4. Why of late there is a shift towards the development of pumped storage power in the country? What purposes they serve?
5. Give a brief account of (a) Necessity for the development of pumped storage power in the

country, (b) Mention advantages and limitations of the development of pumped storage power.

6. **How will you classify pumped storage power development? How does it differ from the conventional hydropower development?**
7. **What essential role the PSP plants play in the grid power system?**
8. **PSP plants convert low value off-peak energy into high value on peak energy, discuss with the help of a sketch.**
9. **(a) What are base load and peak load plants? For what type of conditions hydroelectric power is very much suitable?**
   (b) What are pumped storage plants? Explain with suitable sketches their working and state the advantages. ***(AMIE)***
10. **Explain how hydropower stations are good for taking care of peak loads.** ***(AMIE)***

8

SMALL HYDROPOWER DEVELOPMENT

8.1. INTRODUCTION

Small hydro, one of the earliest energy sources, has recently caught the countrywide attention despite a history of 100 years behind it, mainly because it is eco-friendly, has a short gestation period. The economic viability can be improved by using cost-effective designs and new technology based turbines, controls, equipment and materials. In the present scenario of energy starvation, danger of global warming and depletion of fossil fuels and their rising prices, the small hydro has assumed significant complimentary role in the hydropower development, as means of environmental protection and socio-economic development of remote rural areas.

The small hydro plants on global basis, (potential 50,000 MW), cover station sizes of 5 to 50 MW installed capacity but in India the upper limit of station size has recently been increased from 15 to 25 MW which is indicative of the importance of exploiting hydroelectric resources to serve the nation's electrical energy needs.

The small hydro plants are specially adapted to serve remote hilly areas on independent basis to save large investment needed for extending transmission and distribution for low consumption centres from the power grid supplies by harnessing hydro potential of small rivers and streams besides canal falls utilising the releases made therein for irrigation and other purposes through construction of small, mini and micro hydel stations. Kerala State has started work on 18 small hydel projects with Chinese assistance to attain a total installed capacity of 108 MW. The Chinese are considered to be the pioneers in micro-power technology.

8.2 ADVANTAGES OF SMALL HYDROPOWER

Of all the Non-conventional, renewable energy sources, small hydro power represents highest density resource and stands in the first place in generation of electricity from such sources throughout the world. Its advantages are:

(*i*) Only source for scattered remote rural areas where it is not economically prudent to extend the grid supply, (*ii*) Low investment, (*iii*) Least interference with water flow regime, (*iv*) Minimal and less time consuming investigations, (*v*) Short gestation period up to 3 years, (*vi*) Negligible submergence, (*vii*) Lesser requirement of land acquisition and hence minimal community displacement and rehabilitation problem and hence avoids local resistance, (*viii*) Least impact on flora and fauna, bio-diversity and ecosystem, (*ix*) Least forest land disturbance, (*x*) Simple construction and small size equipment installation, (*xi*) Generally does not impinge on the rights of the downstream population on the stream on account of small scale of operation, (*xii*) Improve the socio-economic conditions in rural areas, retard migration to cities for electricity based employment enterprises, (*xiii*) Non-polluting energy source, (*xiv*) Fairly constant cost of generation; inflation free, (*xv*) Low operation and maintenance cost and special adaptability to remote control operation; operation made simple by microprocessor, (*xvi*) Reliability and flexibility, (*xvii*) Attractive in remote mountain areas and in plains through development of canal falls, (*xviii*) Provides impetus to set up small scale industries and lift irrigation in remote areas, (*xix*) Non-polluting energy source, (*xx*) Environmentally bengin, (*xxi*) Operationally flexible, (*xxii*) Suitable for peaking support to local grid as well as stand alone applications in isolated remote areas, and (*xxiii*) Quicker electricity and economic returns.

Limitations

(*i*) Essentially of insignificant capacity addition, (*ii*) Costlier, high per kW cost than large projects as also may submerge proportionately larger land areas for the equal amount of storage and generation, (*iii*) Steep gradients in river beds and sedimentation problem limit their efficacy, and (*iv*) Sustainability of the development is dependent on the capacity of the local village population to pay for the high cost generated electricity.

Advantages of Canal Power Plant

(*i*) Availability of firm power as hydel canals are run almost round the year except for annual repairs, (*ii*) Power generated is very near to the load centres and hence minimum length of transmission lines. Transmission losses are also minimal, (*iii*) Little adverse impact on environmental quality, (*iv*) Energy generated does not draw on depleting energy supplies like oil, coal or gas, (*v*) Instant availability and high degree of reliability, (*vi*) Low maintenance, (*vii*) Low head turbines used are more efficient, cavitation is low.

8.3. CLASSIFICATION OF SMALL HYDROPOWER

1. Classification Based on Unit Rating

In India, the unit rating based on generation capacity is as given in Table 8.1.

Table. 8.1. Unit rating

Type	*Station capacity (kw)*
Micro	Up to 100 kW, with capacity of individual unit ranging from a few kW to 100 kW.
Mini	Up to 2000 kW, with capacity of individual unit ranging from 10 to 1000 kW.
Small	Up to 25 MW, with individual unit capacity ranging from 1001 to 5000 kW.

2. Classification Based on Head

The hydraulic head under which the turbine operates is classified as given in Table 8.2.

Table. 8.2. Classification of head.

Type of power plant	*Head (m)*
Medium/high head	Above 40
Low head	Less than 40
Ultra low head	Below 3

8.4. SMALL HYDRO POTENTIAL

As per recent assessment, the total small hydro potential of the country is 10,000 MW in terms of installed capacity from various schemes which include those already in operation, under construction and identified, as also 564 MW from canal falls over 3m. The small hydro installations presently aggregate to about 500 MW.

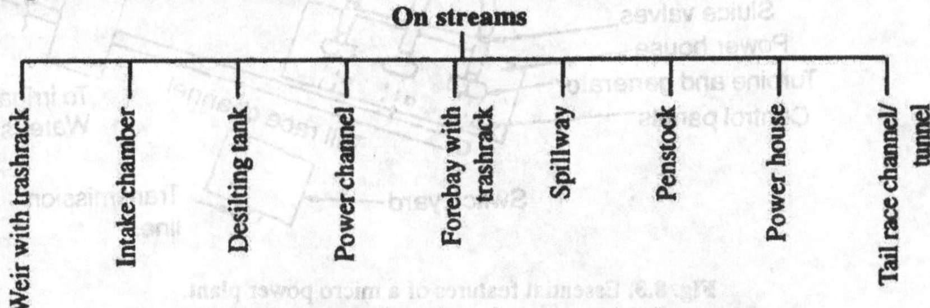

Fig. 8.1. Components of small hydro power development.

8.5. COMPONENTS OF SMALL HYDROPOWER DEVELOPMENT

The various components of small hydropower developments on streams and canal falls are depicted in Fig. 8.1. and 8.2 hereinafter.

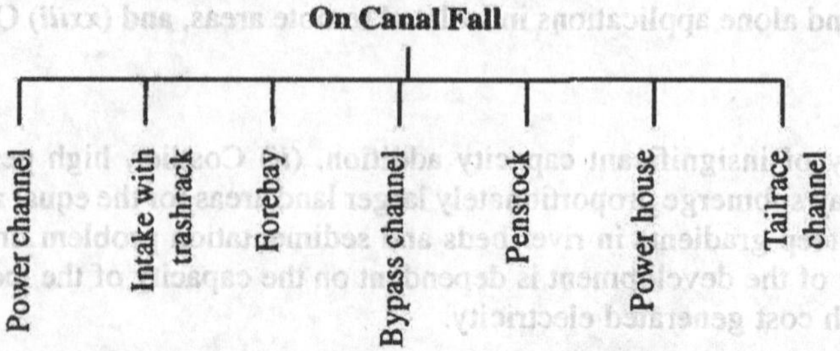

Fig. 8.2. Components of small hydropower development.

Fig. 8.3 illustrates essential features of a micro power plant on a rivulet /stream.

The salient features of the components have been discussed under run-of-river power development. In remote areas and streams with very steep bed slope and carrying huge silt load, instead of gates a trench type weir is constructed. In this case the power draft is comparatively small and this type of weir is generally adopted for micro hydel schemes or to supplement the discharge.

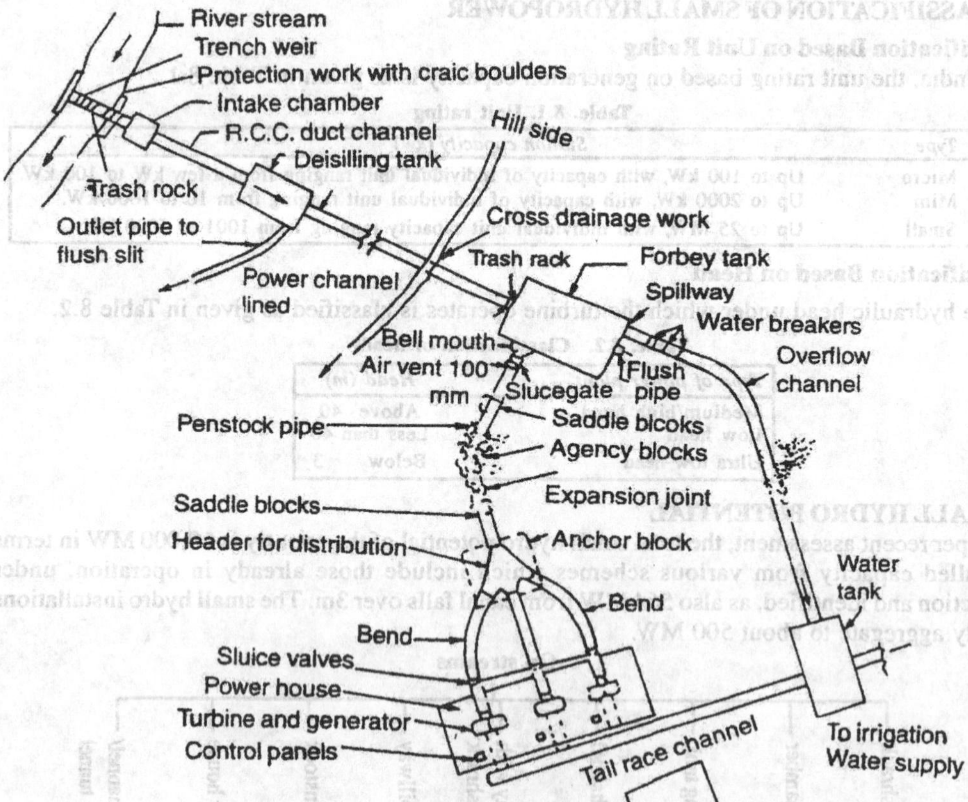

Fig. 8.3. Essential features of a micro power plant.

8.6. FORMULATION OF DEVELOPMENT SCHEMES

1. Investigation and studies: To identify the sites for the development of small hydro schemes, the profiles of hilly streams and small rivers are studied on survey of India maps of scale 1:25,000 (wherever available), 1:50,000 and 1:2,50,000. The catchment area and drop in the bed elevation of small rivers/hilly streams from initial reaches to their confluence with the main rivers are examined to identify potential sites. From economic considerations the schemes with open channel as water conductor system, short length of penstocks and surface type power house are considered. In some cases, however, where open channel is not possible but otherwise the scheme features are attractive, the provision of small tunnels as water conductor may be considered.

The schemes are essentially run-of-river type. During investigation of these schemes, suitable location for balancing reservoir has to be identified to serve for peaking purposes. For the rivers which are only rainfed, the possibility of developing storages would also need to be examined at investigation stage to improve utility of such schemes. The schemes which are located away from inhabitated areas and which would require long transmission lines to bring power to the load centres may not be economically feasible.

Micro hydel projects require least geological data to establish feasibility. The basic data needed is type of strata whether boulderly, rocky, soft or hard rock.

2. Hydrology of Rivulets/streams. Gauge and discharge data at identified sites is generally not available in most of the cases. The monthly/10-daily discharge data at the identified sites may, therefore, be worked out based on the gauge-discharge data available at other locations on the river/stream under consideration or the data available for adjacent river/stream on catchment area proportion basis.

For calculating firm discharge, lean period inflows in hydrological years are computed and arranged in descending order. From the series 75% dependable year is identified. The firm discharge is the average of the lean period inflows in a 75% dependable year.

3. Firm Power, Installed Capacity and Unit size: The firm power at the identified sites is worked out by the following formula:

$$\text{Firm Power} = 8\,QH \quad \text{kW or } (LPS \times H)/130$$

where, Q = firm discharge (cumecs), and H = head (m), LPS is discharge in lt/s.

The possibility of peaking operation during non-monsoon months is kept in view in arriving at installed capacity. Other factors considered for fixing the installed capacity are monthly pattern of power output, load factor of the system etc. The units are usually small in size, discussed hereinafter.

8.7. CHOICE OF UNITS

A micro hydropower plant may have a Francis, Pelton or a Turgo impulse unit for power generation. However, the turgo turbines have better performance over Francis turbines in hilly areas due to lesser wear and tear of runner blades and lesser shut down. However, it is a combination of head and discharge that determine the type of units. In Himalayan rivulets/streams, the heads available are generally medium/high but discharges are low for which Francis/Pelton turbines are suitable. In peninsular India, heads are low but discharges high for which tubular/bulb units are used. Future developments are likely to concentrate on configurations that minimise the cost of turbines which are usually large due to large quantity of discharge that must be conveyed through the turbines. The use of low efficiency cross flow turbines may be restricted only for micro range and remote areas.

The turbines have to be tailor-made for each specific site, based on the requirements of head and discharge. Nevertheless, there is need to standardise them to suit various combination of heads and output. As recommended by CEA, the kW rating of the generating units is chosen from the following standard ratings:

10	25	50	100	200	250	350	500	750	1000
1250	1500	2000	2500	3000	3500	4000	4500	and 5000 kW.	

In any type, a few larger units cost less than more smaller units for the same output. The number of units provided in a power house may vary from two to four. It is also prudent to have units of the same size and type in various locations within an administrative zone to the extent possible to cut down on the spares and for easier inter-changeability. The technical knowhow in the remote areas is also a limiting factor. The small off-period loads suggest choice of smaller units to prevent unnecessary wear on the machine, though Turgo impulse turbines are better from efficiency consideration in this respect. These are considered to have better part load efficiency.

Rated Turbine Output $P_r = \frac{H.Q.\gamma}{1000} \eta_t \eta_s$ (kW)

where, H is rated head (m), Q is rated discharge (cumecs), γ is specific weight of water (N/m³), η_t and η_s are rated turbine efficiency and efficiency of speed increaser, if used, respectively.

8.8. CLASSIFICATION OF TURBINES

The turbines for small, mini and micro power plants are classified as illustrated below:

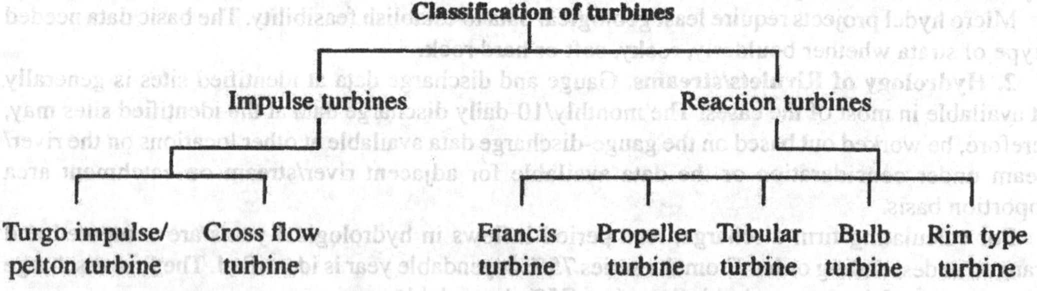

1. Impulse Turbines

(i) Turgo Impulse/Pelton Turbines: It has one or more free jets discharging into an aerated space and impinging on the buckets of a runner. In Turgo impulse, the jet impinges on several buckets continuously, whereas only single bucket per jet is effective at any instant in Pelton wheel. These may be mounted horizontally or vertically.

(ii) Cross Flow Turbine: Also called Banki turbine. It is an impulse turbine with partial air admission.

2. Reaction Turbines

(i) Francis Turbine: It has a runner with fixed vanes to which the water enters the turbine in a radial direction with respect to the shaft and is discharged in an axial direction (Fig. 8.4). Steel plate/concrete water supply case or open flume are used. These turbines may be mounted with horizontal or vertical shafts.

(ii) Propeller Turbine: It has a runner with blades in which water passes through the runner in an axial direction with respect to the shaft. The pitch of the blades may be fixed or movable. The conventional propeller of Kaplan (variable pitch blade) turbines are mounted on a vertical, horizontal or slant shaft.

(iii) Tubular Turbines: These are inclined axial flow turbines with fixed blades or adjustable blade propeller. These are adapted to low head plants up to 15 m with water passage formed directly in the concrete mass of a low dam as also for tidal power plants (Fig. 8.5). The conventional scroll casing and draft tubes are not required; instead a straight conical horizontal tube extending from head water to tail water is provided which lends it the name tubular turbine. The generator may be located inside the turbine wherein it is fixed annularly to the tips of the propeller fixed blade band, the stator located in the walls of the tube. It is called peripheral tubular turbine. Gears, couplings and generator shaft are not required.

It is horizontal or slant mounted with propeller runner. The generator is located outside of the water passage way. These are equipped with fixed or variable pitch runners and with or without gate assemblies. The tubular turbines have given place to bulb turbines in the present day low head plants.

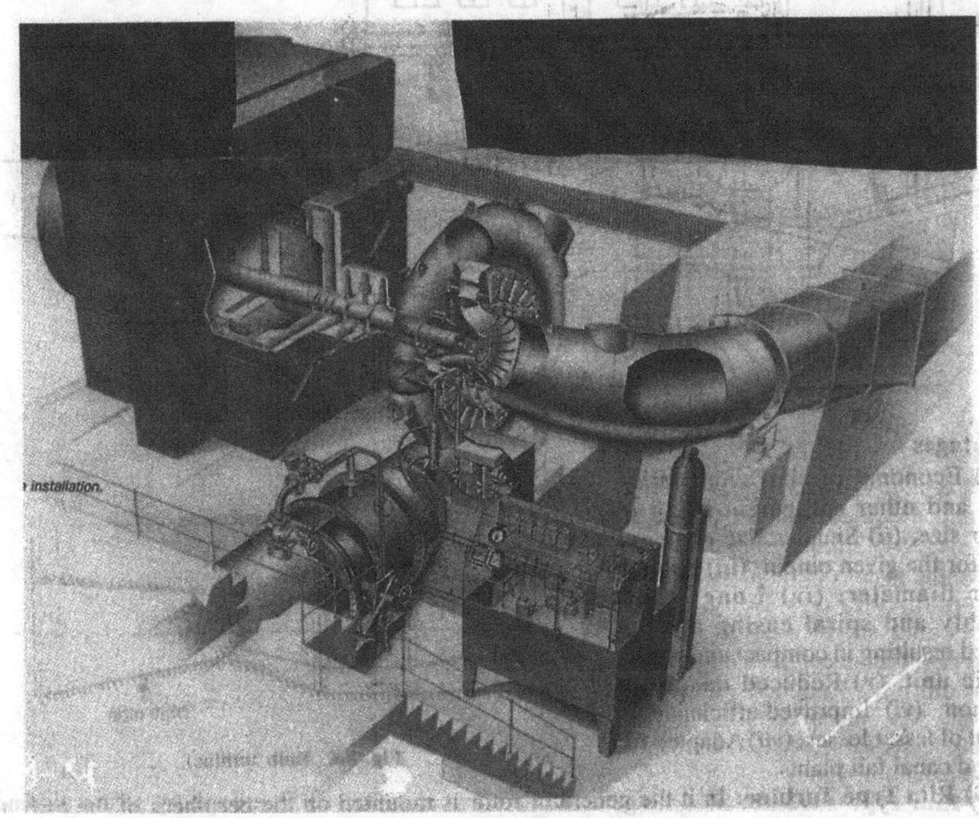

Fig. 8.4. Small scale Francis turbine installation.

(v) Bulb Turbine: It is a variant of Kaplan turbine with horizontal arrangement and having generator located inside bulb over which water flows while approaching to runner. It is axial flow turbine suitable for low head run-of-river and tidal power plants. For units less than 5 MW capacity, it consists of a conical guide apparatus which can actuate the movement of conical guide in the inclined flow passages. In bulb turbine, the generator is housed in an enclosed bulb-shaped casing which is installed in the middle of the passage of flow. The bulb is water tight and streamlined structure which forms the outer boundary of the water passage from the reservoir and together with propeller turbine forms an integral unit. Its draft tube is a straight conical flaring tube (Fig.8.6).

The two arrangements of the bulb vis-a-vis runner are; bulb either upstream or downstream of the runner. The upstream bulb arrangement is commonly used. Two spiders consisting of streamlined columns support the bulb which in turn supports the runner and guide vanes outside the bulb and alternation inside it. The generator is air cooled and bulb is at double the atmospheric pressure.

In downstream arrangement the bulb is placed just after the runner which houses the alternator. Another bulb, provided upstream, houses the radial and thrust bearings, seals etc.

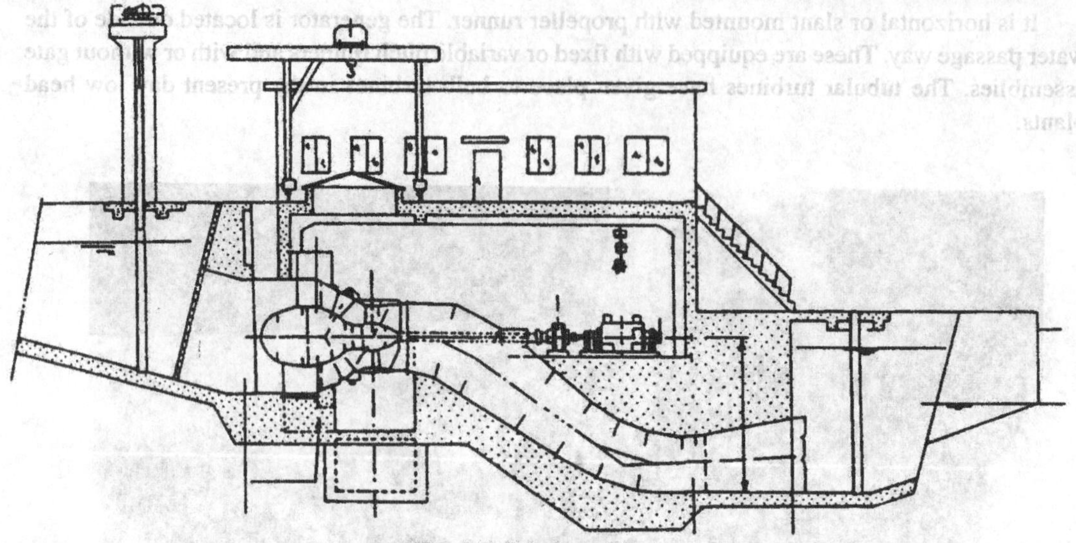

Fig. 8.5. Tubular turbine.
(Courtesy M/s Kirloskar Brothers Ltd., Pune).

Advantages

(*i*) Economy in civil engineering works and other equipment being of smaller size, (*ii*) Smaller size of power house for the given output, (*iii*) Smaller runner diameter, (*iv*) Long shaft-assembly and spiral casing are not required resulting in compact and space-specific unit, (*v*) Reduced danger of cavitation, (*vi*) Improved efficiency on account of lesser losses, (*vii*) Adapted for tidal and canal fall plants.

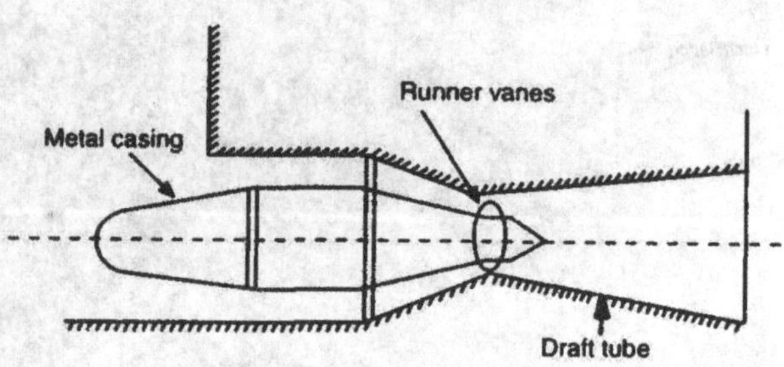

Fig. 8.6 Bulb turbine.

(vi) Rim Type Turbine: In it the generator rotor is mounted on the periphery of the turbine runner blades. It is called as straight flow turbine also. It has a small bulb upstream which houses the bearing. It is a horizontal unit.

Portable Micro Hydel

An inovative concept that has been developed and is under implementation is the installation of light weight portable micro hydel set of 5 to 15 kW in the Himalayan region.

8.9. SELECTION OF TURBINE

The selection of hydraulic turbines is made on techno-economic considerations, on the recommendation of standard manufacturers, besides the following factors:

1. Specific Speed: It is an important criteria for the selection of type of turbine. The range of speed and specific speed for various types of turbines is given by the equation:

$$\frac{n}{n_s} = \frac{H_r^{5/4}}{\sqrt{P_r}}$$

where, n is rotational speed of turbine (r.p.m.), n_s, is specific speed of turbine, P_r is rated turbine output (m.H.P., equivalent to 0.9863 H.P.) at full gate opening.

2. Maximum Head: It is an important criteria for the selection of type of turbine. Normal range of maximum net head with respect to power output for the selection of the type of turbine is indicated in Fig. 8.7. In overlapping head ranges more detailed analysis is carried out to optimise costs for selecting exact type of turbine.

3. Head Variation: Turbine performance is ideal at design head. The turbine efficiency falls at head higher and lower than the design head. Normal range of head variations for various types of turbines is shown in Table 8.1.

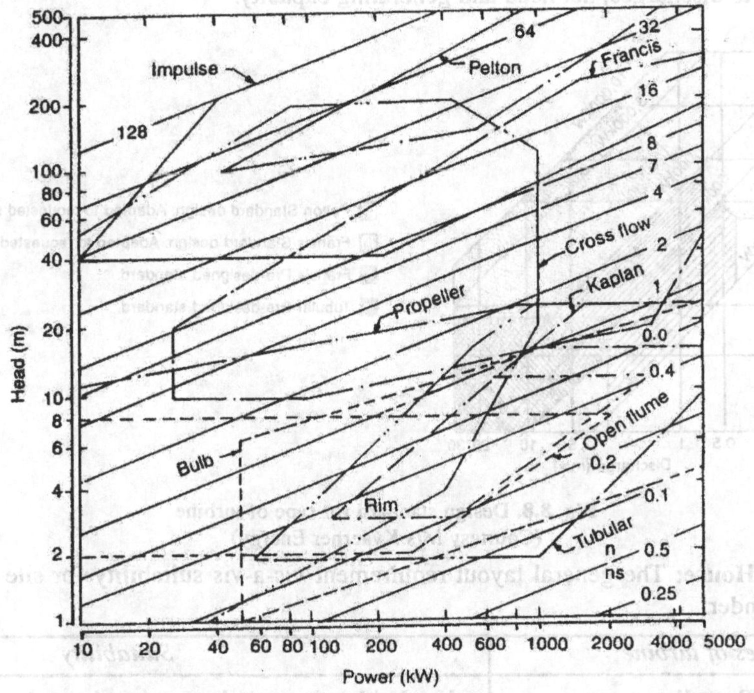

Fig. 8.7. Turbines vs head (IS: 12800-part 3-1991).

4. Load Variation: Turbine efficiency varies with load. Necessity of operating turbine at part loads influence choice of turbines in the overlapping head ranges. Minimum load up to which the turbine may be operated without undue cavitation and vibration is dictated by the type of turbine as given in Table 8.1.

Table 8.1. Turbine performance characteristics (IS: 12800 - Part 3 - 1991).

Sl. No.	*Type of turbine*	*Head application (m)*		*Head variation (%)*		*Load variation (%)*	
		Min	*Max*	*Min*	*Max*	*Min*	*Max*
1.	**Impulse**						
	Turgo turbine	**40**	**200**	**90**	**110**	**40**	**115**
	Pelton	**100**	**500**	**90**	**110**	**40**	**115**
	Cross flow	**1**	**200**	**60**	**125**	**30**	**115**
2.	**Reaction (Mixed Flow)**						
	Francis-horizontal	**10**	**250**	**65**	**125**	**60**	**115**
	Francis-vertical	**10**	**250**	**65**	**125**	**60**	**115**
	Francis-open flume	**2**	**8**	**90**	**110**	**50**	**115**
3.	**Reaction (Axial Flow)**						
	Vertical fixed blade propeller	**2**	**25**	**85**	**100**	**80**	**115**
	Vertical adjustable propeller (Kaplan)	**16**	**40**	**65**	**125**	**40**	**115**
	Tubular with adjustable blades and fixed gates (Horizontal Kaplan)	**2**	**25**	**75**	**115**	**75**	**115**
	Tubular fixed blade with wicket gates (Horizontal propeller)	**2**	**25**	**85**	**110**	**85**	**115**
	Bulb (Horizontal propeller)	**2**	**25**	**50**	**140**	**30**	**115**
	Rim (Horizontal propeller)	**2**	**25**	**65**	**125**	**85**	**115**

5. Efficiency: IS: 12800 (Part 3) - 1991 gives efficiency curves for various load variations in respect of each type of turbine.

6. Discharge: Fig. 8.8. is an illustration for the selection of various types of turbines corresponding to discharges, net head and generating capacity.

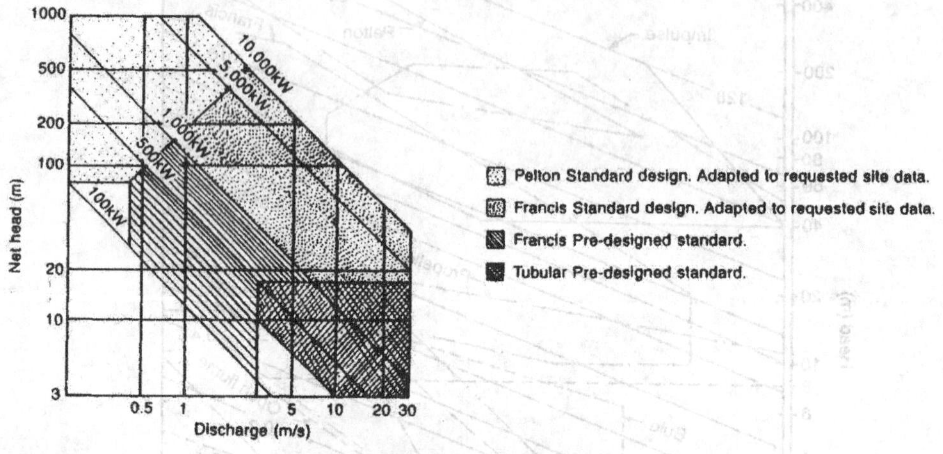

Fig. 8.8. Design standard for type of turbine (Courtesy M/s Kvaerner Energy)

7. Power House: The general layout requirement vis-a-vis suitability for site conditions is sumarised as under:

Types of turbine	*Suitability*
Vertical/Propeller	Used either in an indoor or outdoor plant depending on site conditions.
Propeller	Adapted to canal development.
Horizontal Propeller Francis	For small turbines, there may be a cost advantage in using Francis type with a horizontal shaft. The arrangement is simpler than those for a vertical shaft unit.
Open flume configuration	A Francis turbine is used in flume or canal at an existing drop or vertical discontinuity in the flume or canal. Penstocks are not used.
Tubular	It is easily adapted to a canal installation. Normally generator is housed within a building. However, it is feasible to have the major erection or overhaul areas outdoors.

8.10. ECONOMICS OF SMALL HYDROPOWER DEVELOPMENT

The major cost of a small hydropower development is the turbine-generator unit. The capital cost of small hydro installations varies from 30 to 60 million Rs/MW. The cost of generation is of the order of 1.0 to 2.0 Rs/kWh as estimated by Renewable energy development agency, Ltd, New Delhi.

The rural economy is essentially agriculture based with limited industrial or commercial activities. As such it is important to consider the relationship between the cost of generation of electricity and its tariff to the income of the locals. Evidently it calls for generating paying capacity of the village population by encouraging local electricity based industries. At the same time innovative technology and machinery may be brought into use to make the power generation cost effective. Once the rural consumer is able to generate surplus income from electricity based activities, it will lead to sustainable economic development.

Small hydro plants are not economically attractive but rapidly rising prices of imported oil, constituting a big drain on the foreign exchange, has made it inevitable to exploit whatever hydro sources are available. Usually small hydro is suitable for remote hilly areas on independent basis or on canal falls in plains, there being no cost involved in extending long transmission lines to reach such villages for electrification. They find application only in the systems where existing diesel or coal fired plants can be relied upon to produce the necessary back up capacity.

SOLVED EXAMPLE

Example 8.1. *The runoff data of a river are 240, 200, 100, 300, 250, 150, 700, 500, 350, 275, 125, 325, 280, 400, 375, 390, 320, 140, 130 cumecs. Determine 75% dependable discharge for power generation.*

Solution: The runoff data in descending order are:

Sl. No.	1	2	3	4	5	6	7	8	9	10	11	12	13	14	15	16	17	18	19
Discharge	700	500	400	390	375	350	325	320	300	280	275	250	240	200	150	140	130	125	100

Number of observations = 19

75% dependable value = $(N + 1) \times 0.75$

$= (19 + 1) \times 0.75$

i.e., 15th value from top, which is 150 cumecs.

QUESTIONS

1. What are the advantages and disadvantages of small hydropower?
2. Define small hydropower. How will you classify it? Why is rating of generating units done?
3. Differentiate between a small hydropower development on a stream and a canal fall.
4. Explain with sketch typical arrangement of power development on a stream.
5. Illustrate components of small hydro development on a stream and a canal fall. If these differ, why?
6. What is the prescribed average lean period inflow dependable year? Does it differ from high head power development? **(Ans. 75% and 90% Respectively)**
7. What is small hydropower potential of the country? Small hydropower is called eco-friendly, why? **(Ans. 10,000 MW)**
8. Small hydropower is considered to overcome most of the disadvantages of storage power development; what these are?

9

HYDROPOWER PLANTS

9.1. INTRODUCTION

Powerhouse or power plant is a structural complex, usually multistoreyed, to house the essential equipment which needs protection from weather such as generators, switchboard, low tension switching equipment, bus bar, service shop, office etc.; the transformers and high tension switching equipment is generally kept outdoors near the powerhouse. As such it constitutes an important part of an hydroelectric development and has essentially to be simple, efficient on functional basis for construction, operation and maintenance, apart from aesthetic considerations of harmonising with the surrounding. The basic dimensions of the powerhouse are fixed on the consideration of the arrangement of the component features of the plant which in turn is determined by the electrical, mechanical and hydraulic requirements. The reinforced concrete framed structure is the usual choice for reasons of durability and economy.

A powerhouse generally comprises main powerhouse structure housing generating units and having either separate or combined generator and turbine rooms, erection bay and service bay.

9.2. POWER PLANT STRUCTURE

Structurally, the powerhouse consists of three parts; superstructure, intermediate structure and substructure (Fig. 9.1).

Superstructure: The part of the structure lying above the generator floor is called the superstructure. It houses operation related equipment such as generators and governors, exciters and the auxiliary equipment for ventilation and cooling, overhead travelling crane besides control room, engineer's room, electrical and mechanical shops for light jobs, air duct, compressor room, service bay, stores, reception room etc. The vertical shaft turbines are placed immediately below the generators as against the horizontal shaft turbines which are placed on the generator floor alongside the generators. The generating units are placed in a row at right angle to the direction of water flow through the power plant.

Intermediate Structure: It extends from generator floor down to top of draft tube. It houses turbines including their casing, governor, gate valves and as such called turbine floor. The turbine casing may be made of concrete in low head units but consists of steel plates embedded in concrete in high head units. The galleries for the auxiliary machines and the governor servo-motors system are also located here. It has access, through stairs, with generator floor on top and to the intermediate structure parts.

Substructure: It provides most of the mass needed for stability including water passages and rooms as well as galleries for mechanical and electrical equipment and services. Access galleries to draft tube and spiral case are also provided. The substructure in the usually used reaction turbines is the concrete block extending from foundation to the turbine level which is integrally cast while concreting water bringing scroll casing and water exit draft tube. A basement floor is usually accommodated in it.

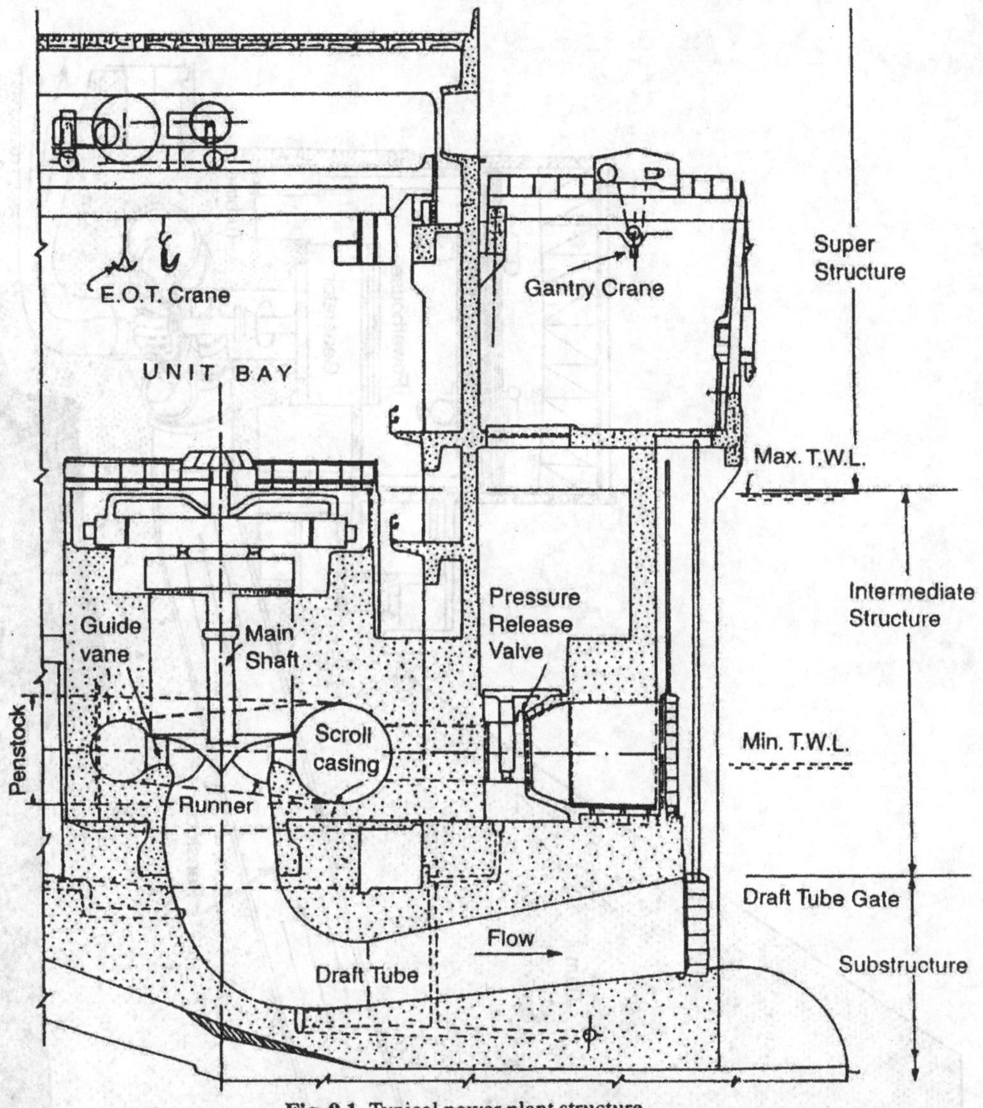

Fig. 9.1. Typical power plant structure.

The powerhouse complex may be located at surface, semi-underground or underground largely dictated by topography, site conditions and economic considerations.

9.3. LAYOUT OF HYDROPOWER PLANTS

The powerhouse location depends upon the overall project development, type of dam, foundation conditions, location of spillway when dam is located adjacent to powerhouse, valley width, river channel condition below the dam, location of switchyard and transformer and accessibility.

The layout of a hydropower plant envisages positioning of the various components of the plant to ensure optimum use of the available space for its efficient and convenient erection, operation and maintenance.

Powerhouse: The powerhouse is positioned at the toe of the concrete or masonry dam where suitable rock to lay foundation is available (Fig. 9.2). Each turbine is fed by a separate penstock which is embedded inside the non-overflow section of the dam. The power house is separated from the dam by expansion joints. With a view to minimise the fluctuations in the tail water level, especially

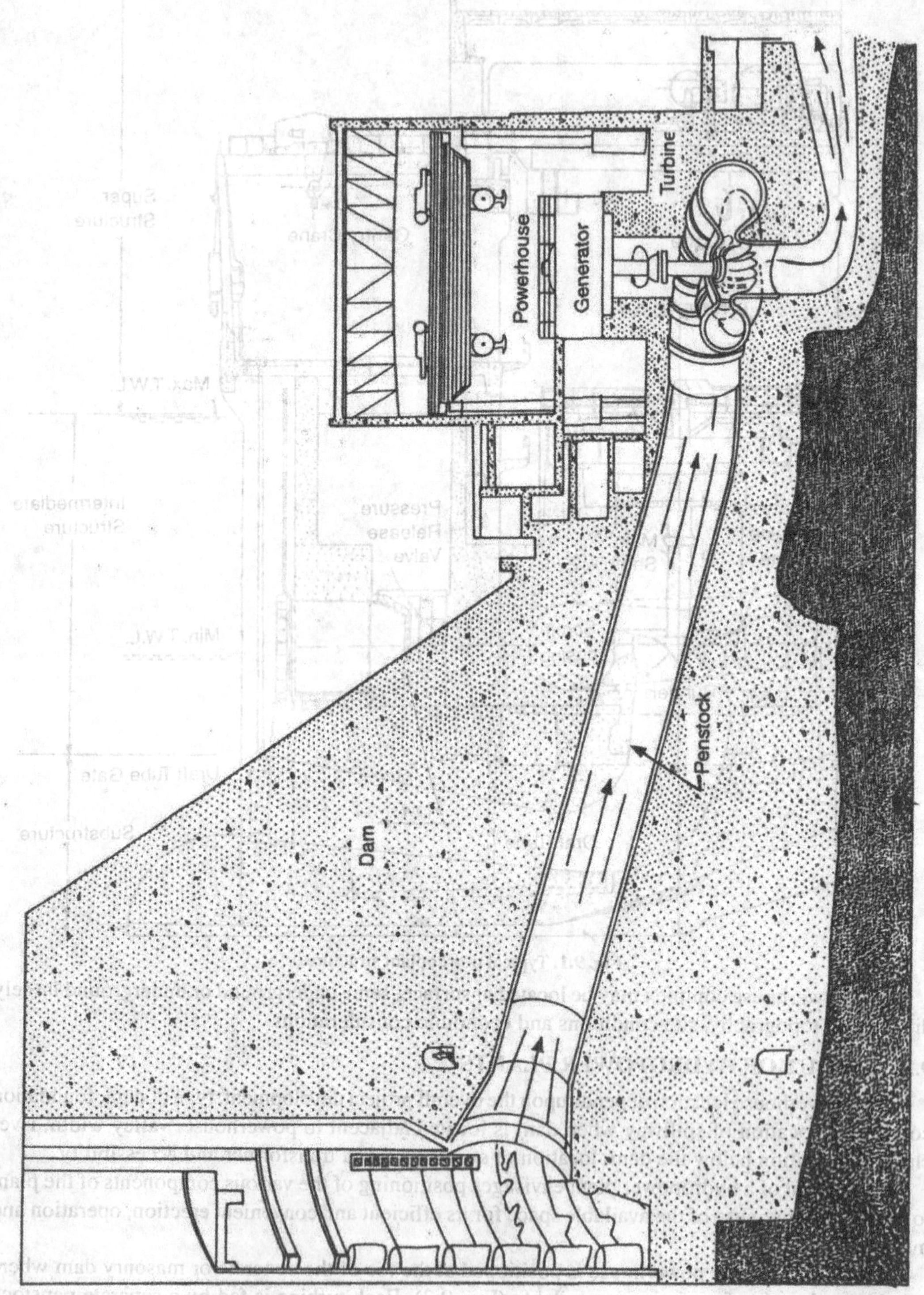

Fig. 9.2. Powerhouse and dam.

due to ski jump trajectory, the powerhouse may be located further downstream and fed through a tunnel branching into individual penstocks near the powerhouse.

The powerhouse may be located underground, led through pressure shafts or pressure tunnels with surge tank. It is economical to lay the power house partly underground and partly on surface or to change the orientation by aligning it parallel to the river to reduce excavation and consequently slope stabilisation problems. The powerhouse may be located below the ski jump bucket itself.

In the case of earth and rockfill dams, the powerhouse is separated from the dam founded on suitable location and fed by penstocks generally taken out from a tunnel earlier used as diversion tunnel (Fig. 13.3). Sometimes the penstock may be laid in a trench excavated below the dam and buried in concrete.

The layout of the three principal component systems of a powerhouse is briefly as under:

1. Hydraulic System: It includes penstock, inlet valve, turbine-runner, guide apparatus, spiral casing, draft tube and its gates besides auxiliaries for monitoring, controlling and operating them.

(*i*) *Main Inlet Valve:* It may be butterfly valve for head up to 200 m or spherical type if the head is more. The hydraulic mechanism to operate the valve comprising high pressure compressed air system for operating the hydraulic mechanism is placed on the turbine floor.

(*ii*) *Turbine:* The dimensions of machine hall depend upon the type of turbine; vertical or horizontal turbine. The horizontal turbine requires larger surface area, reduced roof height and reduced number of floors as generator and turbine are on the same floor but the reverse is the case with vertical turbine which needs lesser surface area but the height of the roof is increased to have clear travelling space between E.O.T. crane and various machines keeping in view the length of the longest part of the unit. The turbine floor also houses the panels for cooling water heat exchangers of generator and thrust bearing. The hydraulic mechanism to operate the regulating ring of the turbine guide vanes and the compressed air system to operate this hydraulic mechanism is also located on the turbine floor or alternatively compressed air system is located in the adjoining hall of the control room near to the turbine floor level.

(*iii*) *Draft Tube:* The outline of draft tube is generally determined by the turbine manufacturer to suit the turbine operating requirements. The physical requirements are spacing and setting of the units, depth of foundations and elevation of tailrace. Normally the draft tube gate is controlled from a floor at the elevation of machine hall or control room. The operating mechanism for the working of the hoist of the draft tube is located on this floor.

2. Power System: It includes generator, neutral grounding system, bus duct, transformer, switchyard and connected auxiliaries to operate, monitor and control.

(*i*) *Generator:* It is installed in the unit barrel above the turbine in a vertical type turbine, and its shaft is coupled to the turbine above the turbine top cover. Unit control panels and excitation transformers are installed outside the generator barrel and on the generator floor. Unit auxiliary transformer and excitation transformer are connected to the bus duct at convenient points between generator barrel and generator transformer. From generator transformers HV bushings connection is taken to the switchyard bus bars via sets of high voltage isolators and circuit breakers. From the bus bar outgoing connection to associated transmission line is taken through another set of isolators and circuit breakers. In surface power plant, the generator, transformer and switchyard are located outside the powerhouse complex.

Unit control panels and excitation control panels are located on the floor of machine hall by the side of respective unit barrels whereas switchyard control panels are located in control room. The control room also houses remote control panels of units alongwith its auxiliaries. By the side of the control room or on a floor adjacent to it in the control room building are provided PLCC room, computer room, LT supply control panels, battery charger and DC distribution board, air conditioning equipment, fire fighting control panels.

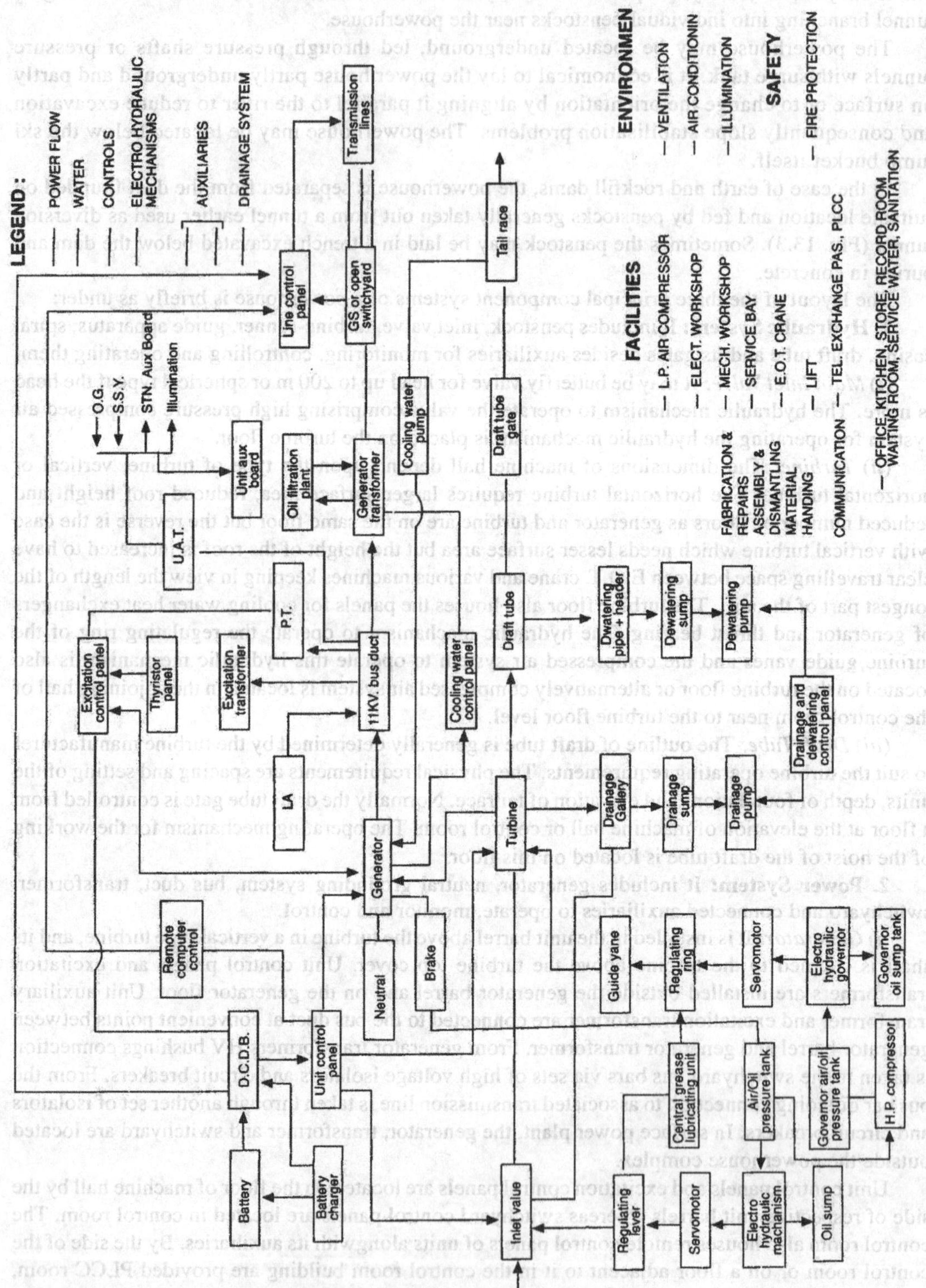

Fig. 9.3. Schematic layout of hydroelectric powerhouse.

EOT crane of adequate capacity is also provided in the machine hall for handling, shifting, assembling, dismantling and maintenance of various components in service bay. A service bay is located at the end of the generator room, at the same floor elevation and of length equal to at least one generator bay.

Cooling water system and pumps and heat exchanger etc., are provided for each unit.

Low pressure compressed air system is also installed either on turbine floor or generator floor.

Transformer oil and lubrication oil treatment plant is also installed near the transformer. Mechanical workshop is also provided as also an electrical laboratory for carrying out various tests during erection and maintenance. Likewise, store, office room and a record room are also provided in control room wing alongwith pantry, kitchen and other facilities.

An elevator is provided in a powerhouse having more than three floor levels.

Openings with hatches are provided at suitable locations on all the floors for access of EOT crane for erection and maintenance at lower floors.

Fig. 9.3 shows schematic layout of an hydroelectric powerhouse.

9.4. TYPES OF POWERHOUSES

The types of powerhouses may be classified based on location and structural characteristics, briefly as under:

1. Classification Based on Location

(*i*) *Surface Power House:* It is the best choice when sufficient area is available to accommodate the powerhouse within economical and convenient excavation. The three types of surface powerhouses, depending on superstructure, are outdoor, semi-outdoor and indoor types. The choice between them depends on (*i*) Level of approach to the plant, (*ii*) Height of tail water level with respect to generator floor, (*iii*) Seismicity of the area, and (*iv*) Severity of climate variations at the site. While deciding about the type of powerhouse the merits and demerits of indoor machine hall and of the semi-outdoor arrangement only are compared.

The surface Pong power plant on Beas river in Himachal Pradesh has six generating units (Fig. 9.4) and service bay alongwith its extension. The unit bays and service bay are each 18 m wide, and service bay extension 7.5 m, each bay is separated from the adjoining bay by a 12 mm thick expansion joint. Each unit bay, 22.75 m long, houses one 60 MW generating unit, consisting of 66.67 MVA umbrella type generator suitable for coupling to a Francis turbine having rated output of 86,200 mhp at a speed of 166.7 rpm under net head of 65.5 m. In unit bays and service bay galleries are provided running longitudinally as well as transversely for the installation of required auxiliaries and service facilities at various floors which are easily accessible by means of stairs, passenger elevator and hatch openings. The compressors for the supply of compressed air required for synchronous condenser operation are provided in an underground room adjoining service bay extension with its top at approach road level. Fig 9.4. shows the typical transverse section of unit bays.

(*ii*) *Semi-Underground Power House:* The surface power house with setting of turbines below the minimum tail water level may involve substantial excavation and then backfilling with concrete to facilitate construction of high retaining walls for protection against floods and as such semi-underground shaft power house is the obvious choice (Fig. 9.6.). In this type, vertical shafts are driven in rock for housing part of draft tube, spiral casing, turbines and generators. The construction work for substructure and superstructure can also be taken up simultaneously resulting in reduced construction time. Approach road to service bay may also sometimes involve substantial excavation. For reasons of economy, an unloading bay is provided in the powerhouse at higher elevation with the provision of an additional smaller crane to lower the equipment into service/erection bay.

(*iii*) *Submersible Powerhouse:* It is the type of power plant which is incorporated in the body of spillway beneath the crest. The head water elevation is maintained with the help of vertical lift crest gates. It has the advantage of economy because separate powerhouse structure is avoided in this arrangement.

Fig. 9.4. Generator house in Pong power plant.

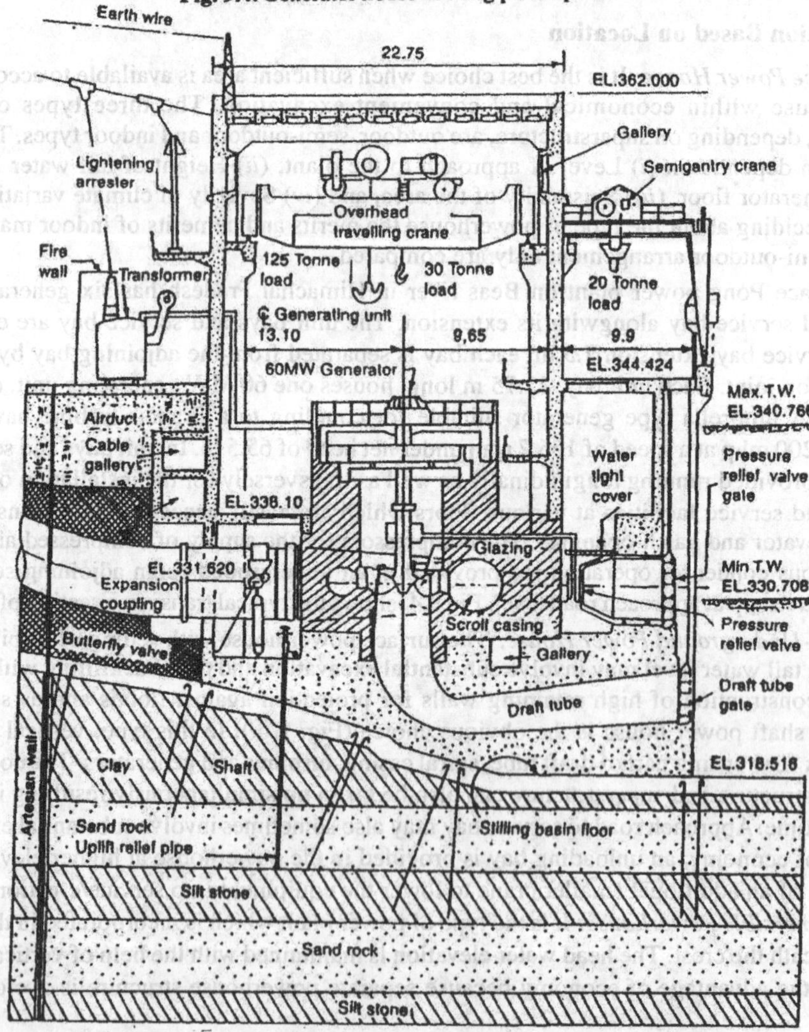

Fig. 9.5. Pong power plant — transverse section of unit bays.

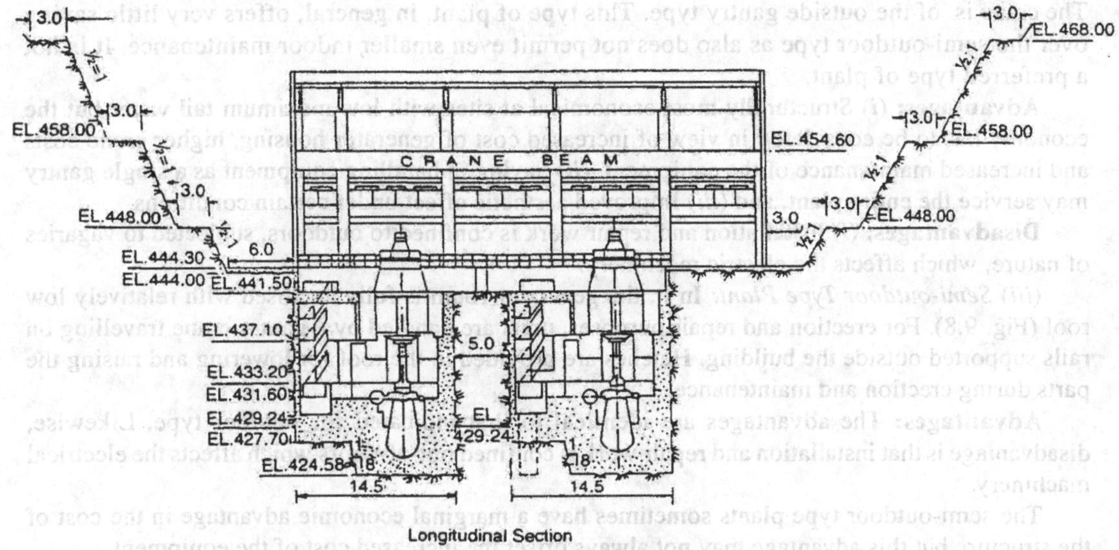

Fig. 9.6. Semi-Underground type -Giri Bata power house.

(*iv*) *Underground Powerhouse:* It is the power plant of recent origin favoured by the two basic considerations of security and economy depending on site conditions (Fig. 9.17). The subject matter has been dealt with in detail separately.

2. Classification Based on Structural Characteristics

(*i*) *Indoor Type Plant:* In this type of powerhouse, the generator room is fully enclosed and has an indoor crane for handling of equipment (Fig. 9.5.). The height of generator room is more or less fixed by head room requirements. The crane beam, in this arrangement, transmits large eccentric loads to the columns which raises its cost high. The higher machine hall does not blend well with the landscape. The indoor type plant affords great protection from weather and facilitates operation and maintenance of equipment.

(*ii*) *Outdoor Type Plant:* In the outdoor type plant, all electrical and hydraulic equipment are in metal enclosure or cubicles instead of a conventional superstructure in the indoor type (Fig. 9.7).

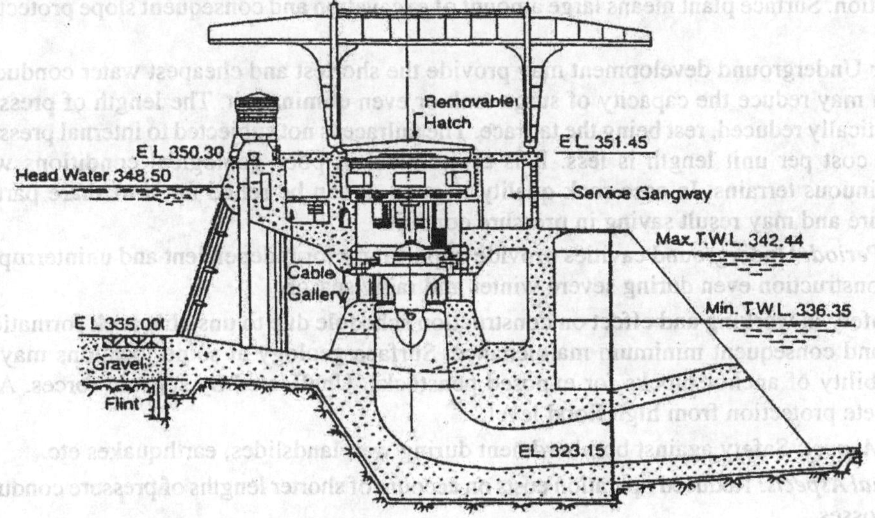

Fig. 9.7. Out door type powerhouse.

The crane is of the outside gantry type. This type of plant, in general, offers very little saving over the semi-outdoor type as also does not permit even smaller indoor maintenance. It is not a preferred type of plant.

Advantages: (*i*) Structurally most economical at sites with low maximum tail water but the economy has to be considered in view of increased cost of generator housing, higher crane costs and increased maintenance of the equipment, (*ii*) Saving in handling equipment as a single gantry may service the entire plant, and (*iii*) Improved aesthetic effect under certain conditions.

Disadvantages: (*i*) Installation and repair work is confined to outdoors, subjected to vagaries of nature, which affects the electric machinery.

(*iii*) *Semi-outdoor Type Plant:* In it, the generator room is fully enclosed with relatively low roof (Fig. 9.8). For erection and repair purposes, units are handled by a gantry crane travelling on rails supported outside the building. Hatches are provided in the roof for lowering and raising the parts during erection and maintenance.

Advantages: The advantages are identical to that indicated for outdoor type. Likewise, disadvantage is that installation and repair work is confined out-of-doors which affects the electrical machinery.

The semi-outdoor type plants sometimes have a marginal economic advantage in the cost of the structure but this advantage may not always offset the increased cost of the equipment.

9.5. UNDERGROUND POWERHOUSE

The concept of underground powerhouse had origin in Switzerland in 1897 but partly within rock. The first fully underground station came up in 1911 in Sweden. The second world war gave impetus to the underground plants mainly as a measure of immunity against air attacks. Thereafter there was spurt in the construction of this type of power plants in Canada, UK, USA, Japan, USSR and in India in post independence era. The first to come up was Manthon powerhouse in 1952-58, followed by Koyna power station, Idikki hydel plant, Chibro plant, Salal hydel project, Chamera project, Baspa II hydoelectric project, Tehri dam etc.

Advantages of Underground Plant

An underground power house is preferred over a conventional surface plant due to a variety of considerations, given as under:

Topography: Non-availability of sufficient level land area at the location of the plant at the required elevation. Surface plant means large amount of excavation and consequent slope protection measures.

Economy: Underground development may provide the shortest and cheapest water conductor system, which may reduce the capacity of surge tank or even eliminate it. The length of pressure conduit is drastically reduced, rest being the tailrace. The tailrace is not subjected to internal pressure and hence its cost per unit length is less. It is economical in poor geological conditions with complicated sinuous terrains. In case rock quality is good, it can be relied upon to share part of internal pressure and may result saving in pressure conduit.

Working Period: Underground cavities provide opportunity for independent and uninterrupted working for construction even during severe winter and rainy season.

Safety: Safety in working and effect on construction schedule due to unstable rock formations is controlled and consequent minimum maintenance. Surface geology at some locations may be adverse to stability of anchor blocks for exposed penstocks. Unaffected by seismic forces. Also there is complete protection from high flood level.

Strategic Aspect: Safety against bombardment during war, landslides, earthquakes etc.

Operational Aspects: Reduced operation costs on account of shorter lengths of pressure conduits, smaller head losses.

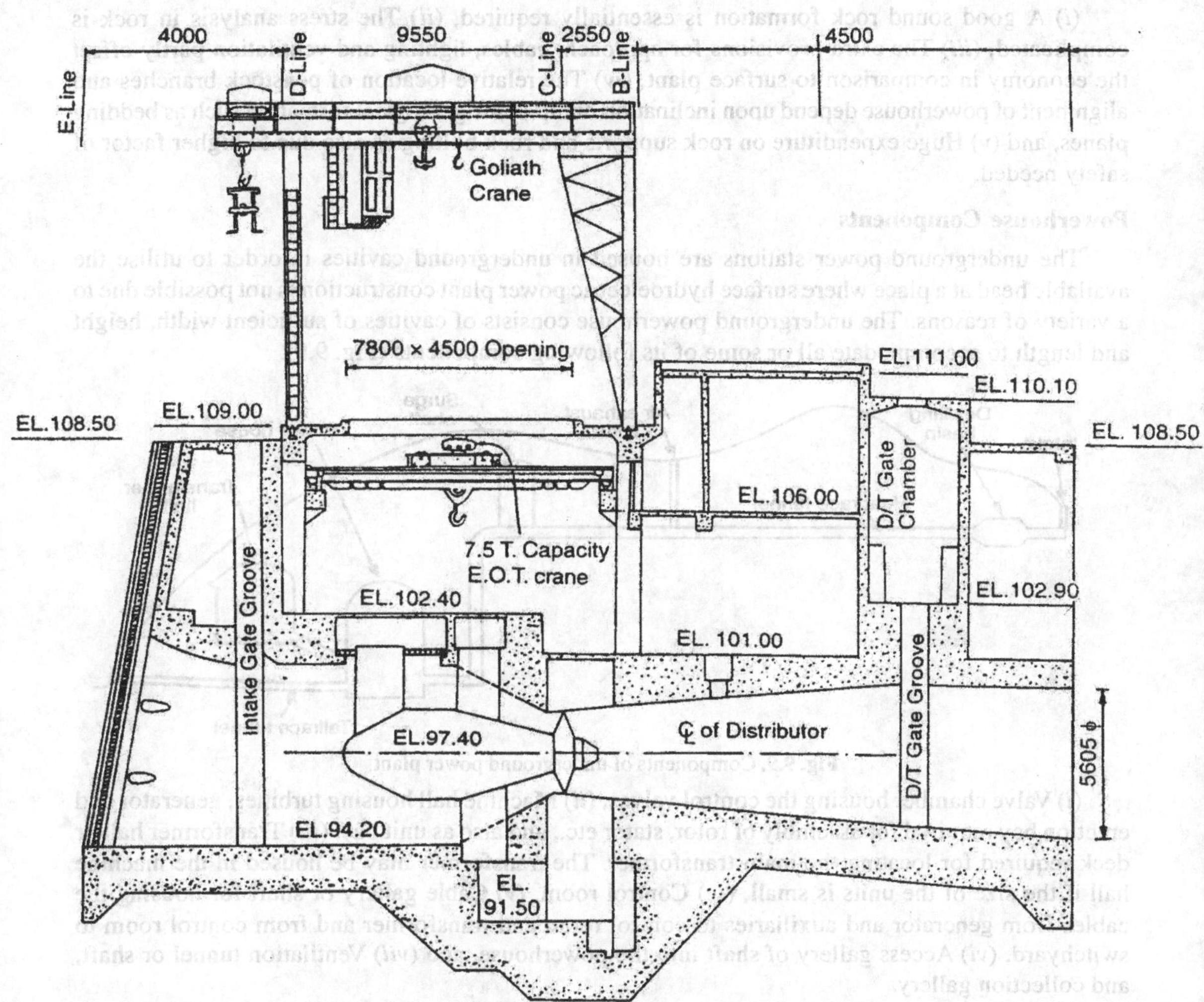

Fig. 9.8. Semi-outdoor type powerhouse (Bulb turbine).

Simple governing conditions: **Reduced length of penstock means reduced water hammer and consequently simplified governing conditions.**

Aesthetic: **The underground development does not interfere with the natural landscape features and hence no deforestation for power plant construction and consequently lesser effect on environment.**

Operation and Maintenance: **The regular operation and maintenance cost is less.**

Disadvantages

(*i*) Expensive excavation of large size cavities, (*ii*) Access tunnels and additional galleries have to be provided which add to the length of tunnelling, longer construction period and consequent higher cost, (*iii*) Special ventilation and air conditioning arrangement and equipment has to be provided at additional cost, (*iv*) Higher lighting charges, and (*vi*) In the case of long tailrace tunnels, downstream surge tank may have to be provided.

Limitations

(*i*) A good sound rock formation is essentially required, (*ii*) The stress analysis in rock is complicated, (*iii*) The extra provisions for approach, cables, lighting and ventilation partly offset the economy in comparison to surface plant, (*iv*) The relative location of penstock branches and alignment of powerhouse depend upon inclination or dip of the geological formation such as bedding planes, and (*v*) Huge expenditure on rock supports and rock bolting as also due to higher factor of safety needed.

Powerhouse Components

The underground power stations are housed in underground cavities in order to utilise the available head at a place where surface hydroelectric power plant construction is not possible due to a variety of reasons. The underground powerhouse consists of cavities of sufficient width, height and length to accommodate all or some of its following components (Fig. 9.9).

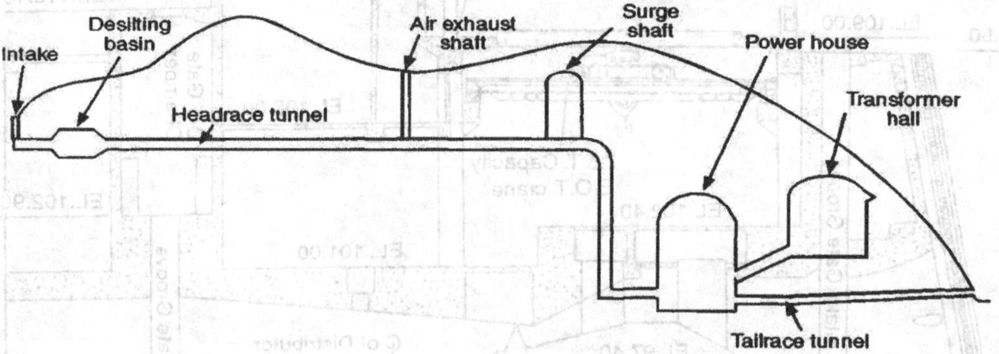

Fig. 9.9. Components of underground power plant.

(*i*) Valve chamber housing the control valves, (*ii*) Machine hall housing turbines, generator and erection bay required for assembly of rotor, stator etc., and also as unit bay, (*iii*) Transformer hall or deck required for locating the main transformer. The transformer may be housed in the machine hall if the size of the units is small, (*iv*) Control room, (*v*) Cable gallery or shaft for housing the cables from generator and auxiliaries to control room and transformer and from control room to switchyard, (*vi*) Access gallery of shaft into the powerhouse, and (*vii*) Ventilation tunnel or shaft, and collection gallery.

Sometimes, more cavern is required to serve as a surge tank on the upstream side if the power plant is located deep below the ground level. Another cavity, nearer to the transformer yard, is required for switchyard in case it is not economical to locate it overground. The switchyard cavity may be of smaller size in view of the advent of gas insulated switchgears. Even generator transformer and switchyard may be located in a separate cavern adjacent to machine hall. From underground generator transformer or gas insulated switchgears power is transmitted through E.H.V. oil filled cable or XLPE cables to the outdoor transmission gallery.

9.6. TYPES OF UNDERGROUND POWER PLANTS

The five types of underground powerhouses are briefly as under:

1. Free Level Tailrace without a Surge Tank

In this type, the tailrace tunnel is long but is not under pressure in steady and unsteady conditions (Fig. 9.10.). A downstream surge tank is avoided. Pelton turbine is usually used as it does not interfere with the flow in the tailrace tunnel.

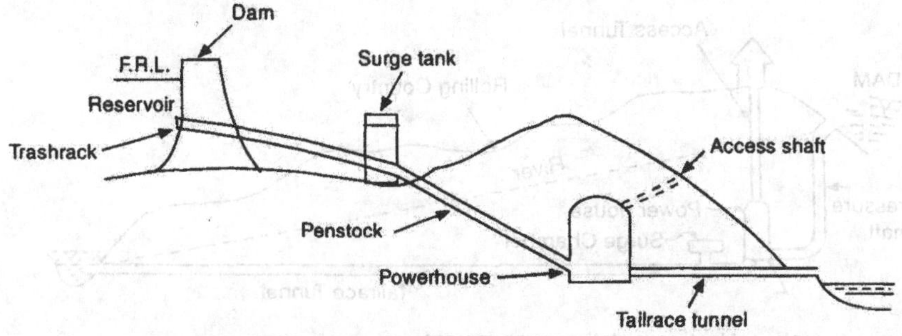

Fig. 9.10. Free level tailrace tunnel without a downstream surge tank.

2. Upstream Station Arrangement

In this type, called Swedish type, powerhouse is located close to the intake. A vertical pressure shaft extending down directly from the intake to the power plant and surge tank at the entrance to the tailrace pressure tunnel is provided (Fig. 9.11*a*). The surge tank is located close by to the draft tube because of the development of negative water hammer waves in the draft tube. It is adapted to continuously and mildly rolling terrain which does not permit construction of horizontal pressure tunnel in the upstream. Valve chamber and conduit valves before the plant are not necessitated, the pressure shaft being too short. A vertical access shaft is provided to facilitate entry into the powerhouse for operation as also for the transportation of machinery and equipment. This type of plant is suitable with reaction turbine. The intake has automatic mechanism for water control to the turbine. Maithon powerhouse (DVC) is of this type in India.

3. Downstream Plant or Alpine Type

In this type of plant, water from intake is carried through a long horizontal pressure tunnel, terminating in a surge tank wherefrom a steep pressure shaft leads water to the power plant (Fig. 9.11*b*). Individual control valves are required for turbines. These valves are located in main cavern or in a separate valve gallery. Access galleries to the surge tank and power plant are provided. Usually the tailrace tunnel is of short length. In the case of long tailrace tunnel, a surge tank has to be provided to take care of occurrence of surges in the tail water. This type of plant is suitable for a rugged terrain and high head fall. Koyna I stage (240 MW) belongs to this type.

4. Intermediate Station Arrangement

This arrangement, as the name implies, has power station at an intermediate location selected from economy point of view with long pressure tunnels constituting the headrace tunnel and tailrace tunnel respectively as also individual surge tanks in them (Fig. 9.11*c*). Individual control valves are required for the turbines. In impulse turbine, there is no interference between the tailrace and headrace level and the size of both the surge tanks is not large. In reaction turbine, the oscillations in the tailrace and headrace interfere with each other necessitating larger surge tanks.

Chibro power plant in Yamuna hydel scheme belongs to this type.

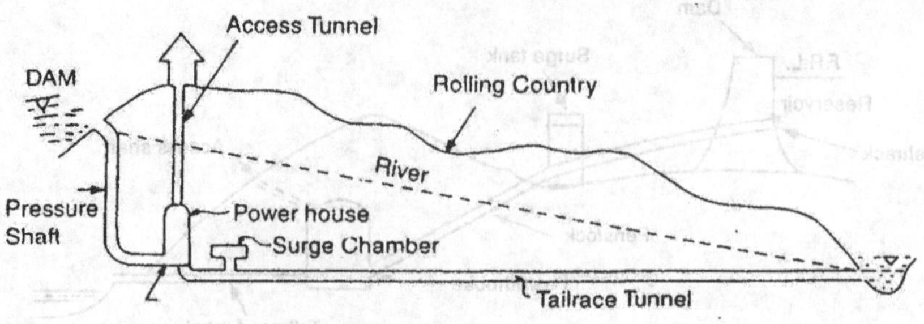

(a) Upstream station arrangement (Swedish type of development)

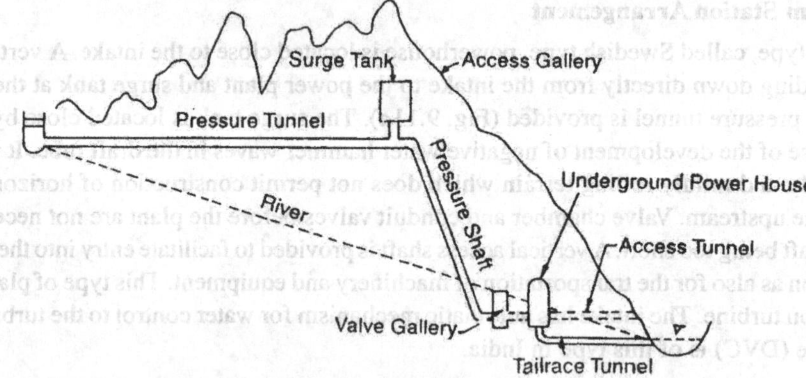

(b) Down stream station arrangement (Swiss type of development)

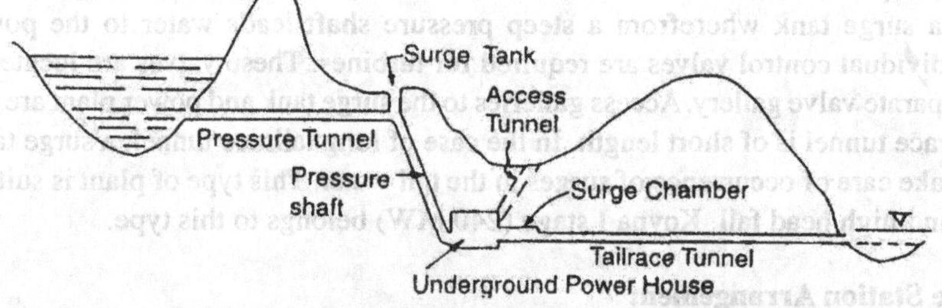

(c) Intermediate station arrangement

Fig. 9.11. Types of underground development.

5. Station Arrangement without Surge Tank

In this arrangement, surge tanks, both upstream and downstream are eliminated, the length of upstream pressure shaft and tailrace tunnel being of short length (Fig. 9.12). Water hammer effect and danger of cavitation in the turbine are taken care of. The cavitation is avoided by adopting the proper turbine axis level above the axis level of the draft tube. The transformer is located in a transformer room adjacent to the power house.

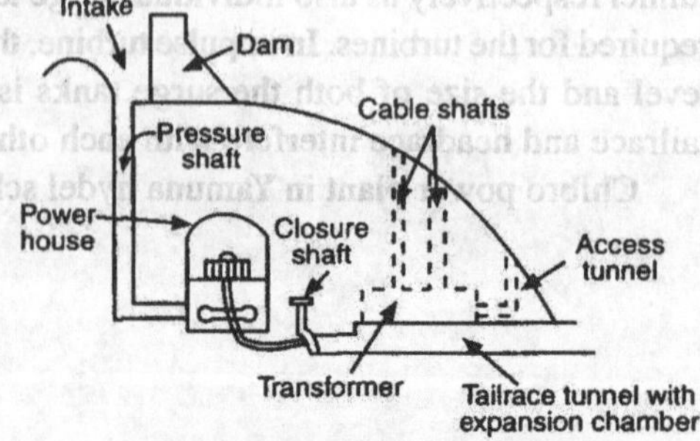

Fig. 9.12. Powerhouse without surge tank.

9.7. ALIGNMENT AND LAYOUT OF CAVITIES

Layout

An underground power house has a number of cavities to locate its various components within the rockmass which has necessarily to be massive hard and least jointed.

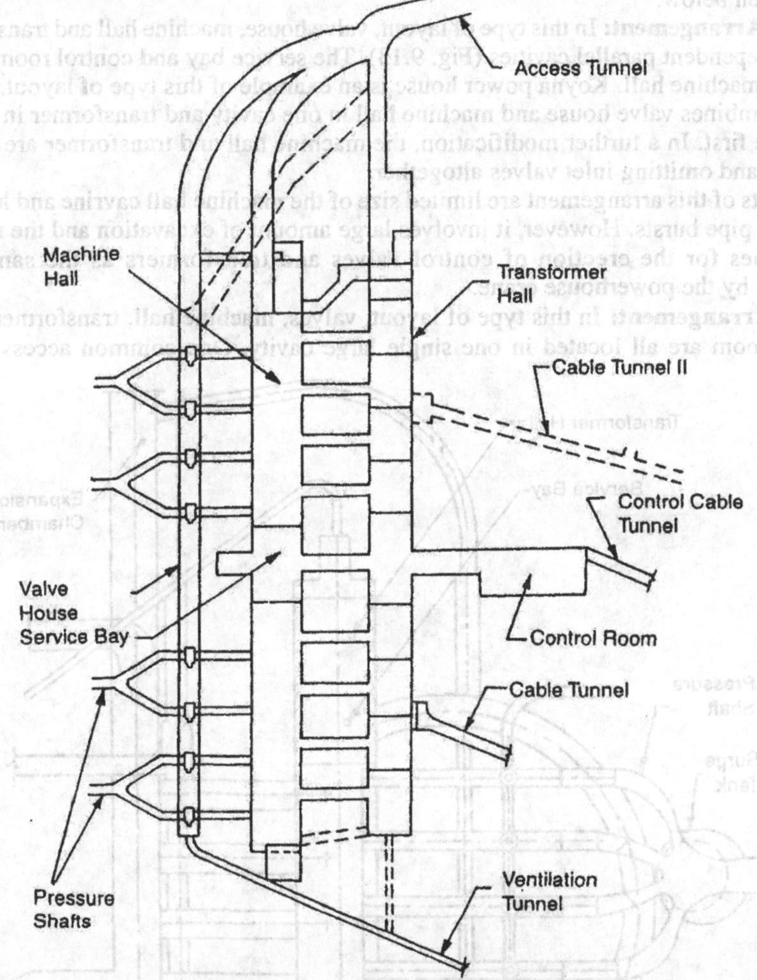

Fig. 9.13. Parallel hall arrangement powerhouse.

The largest cavern and involving difficult excavation is machine hall. The cavern is aligned normal to the rock strike to utilise to the greatest extent the rock arching action. Where the joints, slip planes and seams are numerous, the machine hall may be aligned on the basis of an optimum compromise between the direction of the ruling strike and the direction of such features to ensure that inferior quality rock is confined to the shortest dimension of the cavern.

A number of layouts of cavities is possible in the power plant depending upon geology and the rock strata, approach and access facilities and arrangement of auxiliary openings as under:

1. Control valves may be housed in (*i*) a separate cavity parallel to the machine hall on upstream side, (*ii*) main cavity or (*iii*) separate cavity skew to the machine hall.

2. Service bay may be located (*i*) between units in the centre of the cavity, or (*ii*) on one side of the machine hall.

3. Transformer may be placed in (*i*) a separate cavity parallel to the machine hall, (*ii*) a separate cavity at right angles to the machine hall, (*iii*) machine hall with lateral cavities on either side, (*iv*) main cavity along the wall of the machine hall, or (*v*) machine hall in continuation of service bay.

4. Control room may be located in (*i*) an independent cavity, (*ii*) main cavity and located centrally between units but separate from the units, or (*iii*) main cavity on either side of the machine hall.

Classification of Layout of cavities

The layout of the cavities, on the basis of the arrangement indicated above, is classified, briefly as given below:

Parallel Arrangement: In this type of layout, valve house, machine hall and transformer hall are located in independent parallel cavities (Fig. 9.13). The service bay and control room are located in line with the machine hall. Koyna power house is an example of this type of layout. A variation of this layout combines valve house and machine hall in one cavity and transformer in another cavity parallel to the first. In a further modification, the machine hall and transformer are located in two parallel halls and omitting inlet valves altogether.

The merits of this arrangement are limited size of the machine hall cavrine and higher safety in fire hazard or pipe bursts. However, it involves large amount of excavation and the requirement of separate cranes for the erection of control valves and transformers as the same can not be accomplished by the powerhouse crane.

Linear Arrangement: In this type of layout, valves, machine hall, transformers, service bay and control room are all located in one single large cavity. One common access serves all the

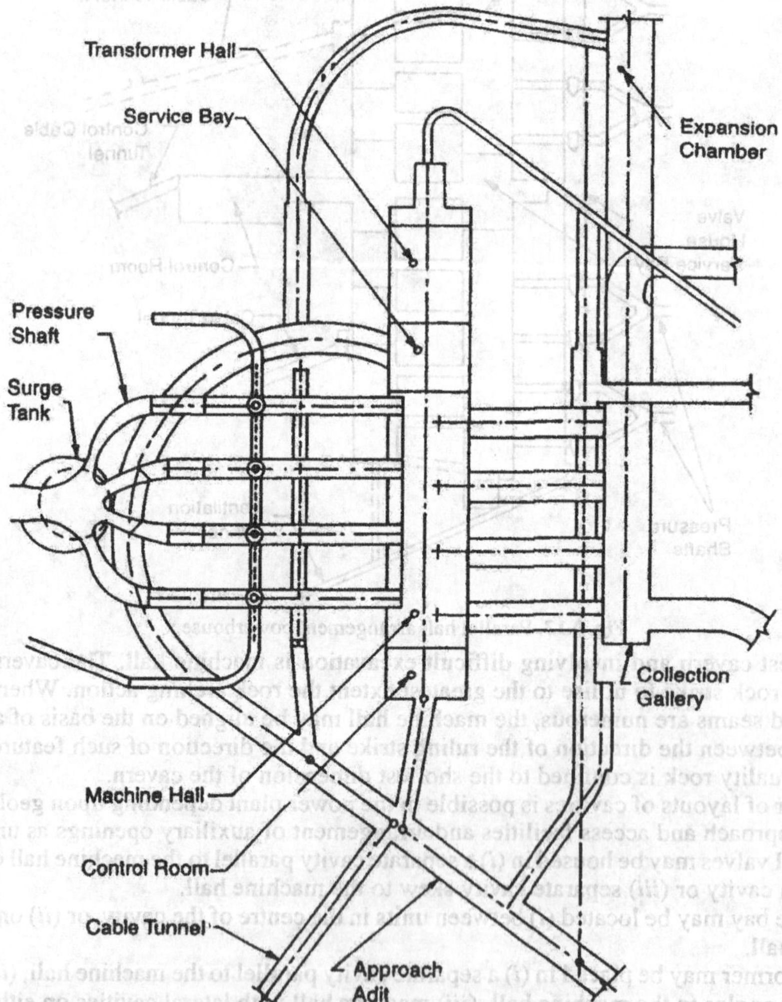

Fig. 9.14. Linear arrangement power house.

component units and a single gantry crane installed for the generating units can be used for handling the transformer also.

The arrangement has the merit that (*i*) Gantry crane meant for generating units serves for handling the transformers as well, (*ii*) A common access serves all the component units, (*iii*) Separate bus bar galleries are not required as low tension cables from the units are taken to the transformers along the machine hall walls, and (*iv*) Limits the span in the caverin.

Composite Arrangement: It combines the characteristics of the parallel and linear arrangements. In this type, control valves, generating unit and transformers are all located in the machine hall cavity; the valves are located on the upstream side and the transformers on the downstream side of the units (Fig. 9.15).

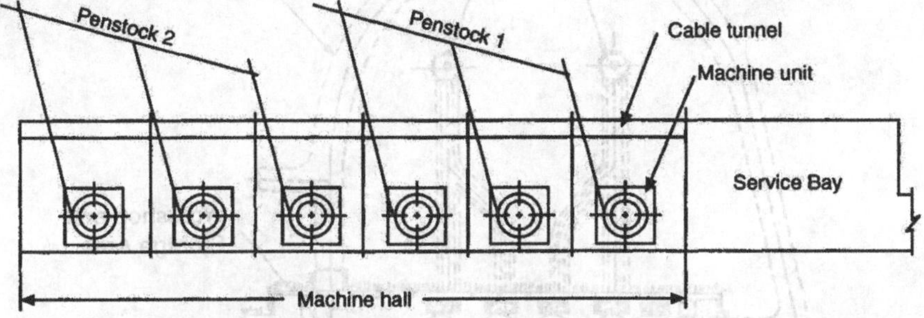

Fig. 9.15. Composite arrangement power station.

The advantages are (*i*) The power house crane serves all the purposes, (*ii*) One access gallery meets the requirement, and (*iii*) Reduces the length of the low tension cables to the minimum.

The arrangement requires a cavity of comparatively large span, possible only in good geological condition. A variant of this type locates the valve house separately, parallel to the machine hall and transformer hall caverin.

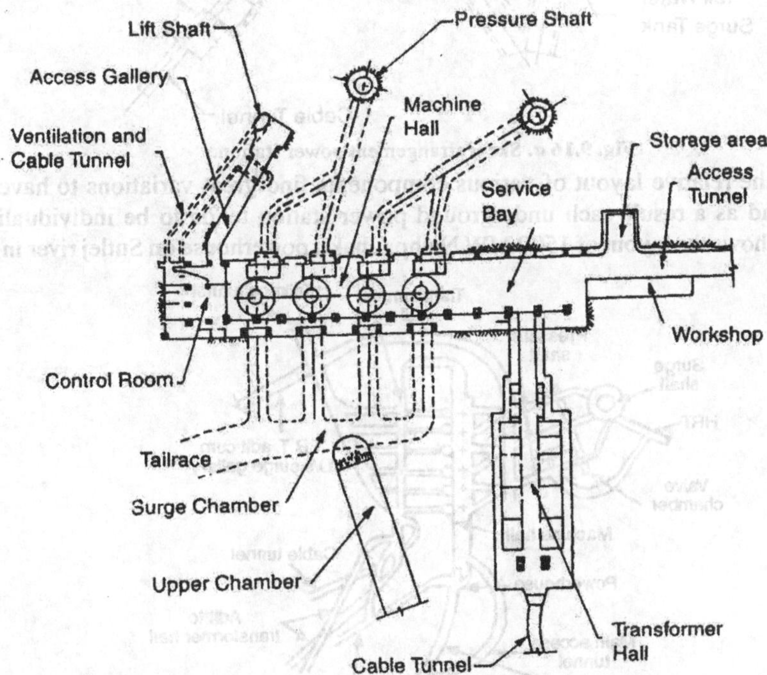

Fig. 9.16. Perpendicular arrangement power station.

Perpendicular and Skew Arrangement: These types of layout are slight modifications of the types mentioned above in that the transformers are located with certain variation at right angle or skew to the machine hall (Figs. 9.16 and 9.16*a*). The transformer hall approach is through the service bay and adjacent to it. The powerhouse crane serves the handling of trasnformer also.

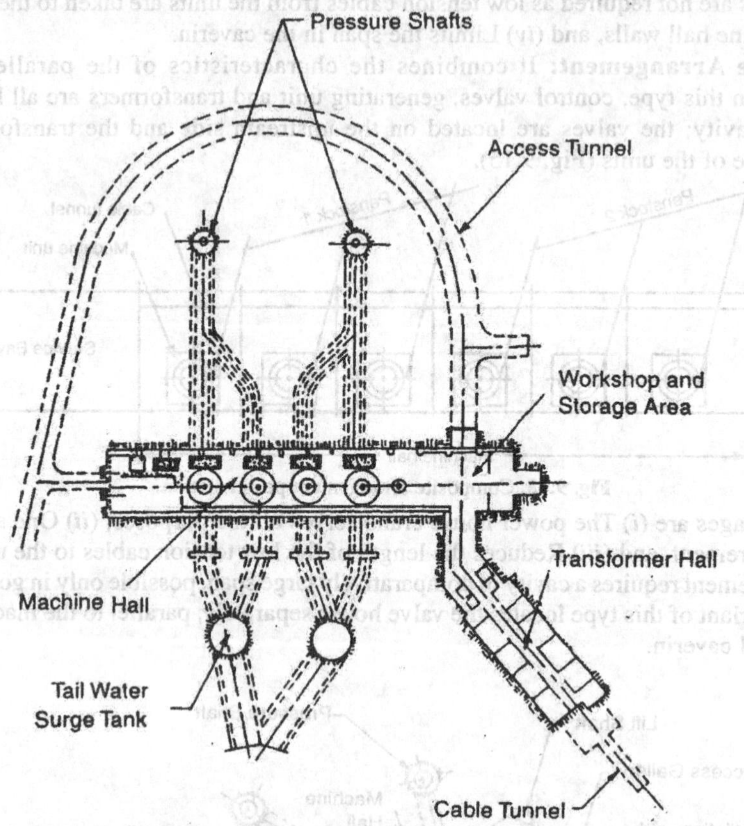

Fig. 9.16 *a*. Skew arrangement power station.

In practice, the relative layout of various components find great variations to have cavity of minimum size and as a result each underground power station tends to be individualistic in its layout. Fig. 9.17 shows the layout of 1500 MW Nathpa Jhakri powerhouse on Sutlej river in Himachal Pradesh.

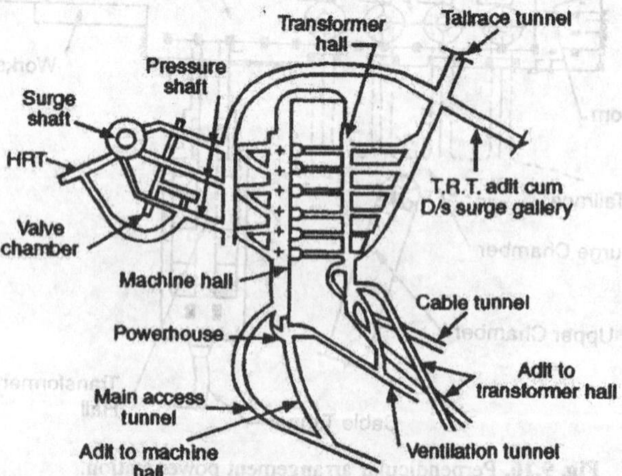

Fig. 9.17. Layout of Nathpa Jhakri powerhouse.

Typical layout of 1020 MW powerhouse complex of Tala hydroelectric project (Bhutan) is shown in Fig. 9.18 to illustrate that layout of each power house is essentially site-specific.

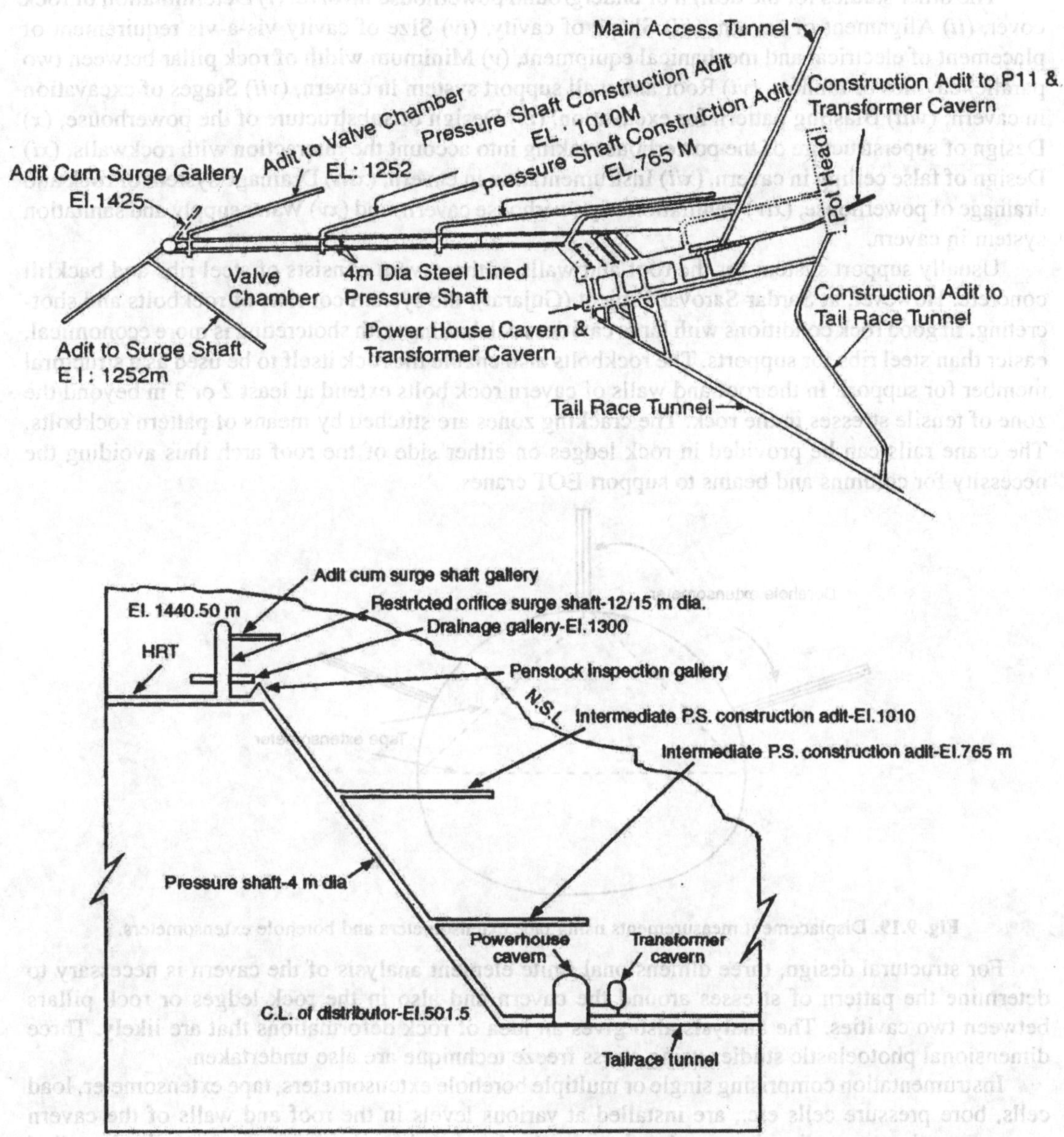

Fig. 9.18. Powerhouse complex of Tala hydroelectric project (Bhutan).

9.8. INVESTIGATIONS AND STUDIES

For underground powerhouse, investigations made are topographical, geological and geotechnical for the collection of various data required for planning and design. The type of rock formation and its engineering properties such as C, ϕ, unconfined compressive strength, modulus of elasticity, poissons ratio, density, permeability, etc., are determined. Details of type of joints, their orientation and location, properties of infill material if present, are also collected. A system of drifts of size 1.5 × 2.5 m are also excavated along the length and across the width of cavern to study geological and engineering properties of rock. K_0 value i.e., the ratio of horizontal and vertical in situ stresses in the rock is also determined.

The other studies for the design of underground powerhouse involve: (*i*) Determination of rock cover, (*ii*) Alignment of cavern, (*iii*) Shape of cavity, (*iv*) Size of cavity vis-a-vis requirement of placement of electrical and mechanical equipment, (*v*) Minimum width of rock pillar between two parallel cavities or tunnels, (*vi*) Roof and wall support system in cavern, (*vii*) Stages of excavation in cavern, (*viii*) Blasting pattern for excavation, (*ix*) Design of substructure of the powerhouse, (*x*) Design of superstructure of the powerhouse taking into account the interaction with rockwalls, (*xi*) Design of false ceiling in cavern, (*xii*) Instrumentation in cavern, (*xiii*) Drainage system of rock and drainage of powerhouse, (*xiv*) Ventilation of powerhouse cavern, and (*xv*) Water supply and sanitation system in cavern.

Usually support system for the roof and walls of the cavern consists of steel ribs and backfill concrete. However, at Sardar Sarovar Project (Gujarat), the system consists of rockbolts and shotcreting. In good rock conditions with large cavities rock bolting with shotcreting is more economical, easier than steel ribs for supports. The rockbolts also enable the rock itself to be used as a structural member for support. In the roof and walls of cavern rock bolts extend at least 2 or 3 m beyond the zone of tensile stresses in the rock. The cracking zones are stitched by means of pattern rockbolts. The crane rails can be provided in rock ledges on either side of the roof arch thus avoiding the necessity for columns and beams to support EOT cranes.

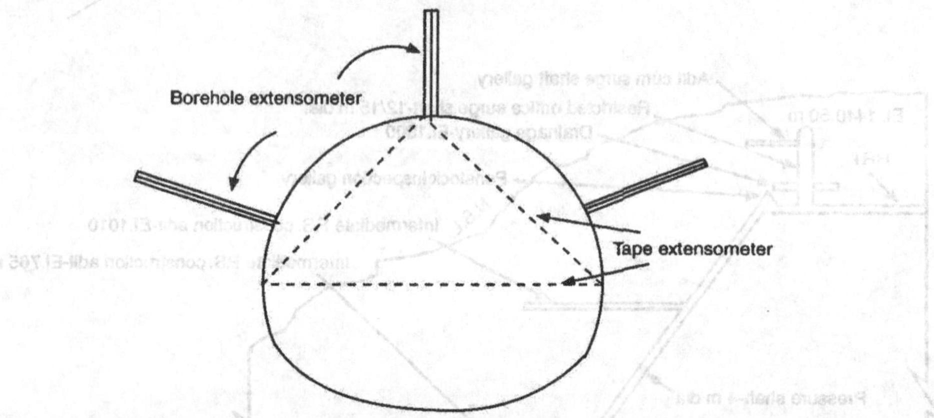

Fig. 9.19. Displacement measurements using tape extensometers and borehole extensometers.

For structural design, three dimensional finite element analysis of the cavern is necessary to determine the pattern of stresses around the cavern and also in the rock ledges or rock pillars between two cavities. The analysis also gives an idea of rock deformations that are likely. Three dimensional photoelastic studies using stress freeze technique are also undertaken.

Instrumentation comprising single or multiple borehole extensometers, tape extensometer, load cells, bore pressure cells etc., are installed at various levels in the roof and walls of the cavern as soon as the excavation is completed up to the levels where instruments are to be installed (Fig. 9.20). The installation coincides with that of support placement. Observations taken at regular intervals manually or by using data loggers for automatic recording and printing of data can be useful for monitoring the rock deformations and convergence of the cavity. If the rock deformations are found to be higher than 3 to 5 times the theoretical deformations then the support system has to be redesigned to limit the deformations.

A system of drainage galleries surrounding the cavern is provided to drain the ground water to avoid saturation of rock mass. Grouting holes are provided at regular spacings from the gallery towards the top and bottom of the gallery to enable grouting of the rock mass.

Sardar Sarovar Powerhouse: The powerhouse cavern here is the largest underground power house cavern in the country. The powerhouse cavern is located about 157.5 m downstream from the reference axis of the dam. The cavern is sited mainly in basalt formation intruded by dolerite dykes and silts. The cavern is 23 m wide, 210 m long and 58 m high and D-shaped at the roof. The roof of the powerhouse has a rise of 6 m and radius of 14.0 m.

The support system of powerhouse roof consists of 6 m long 25 mm dia, 14/12 tonne pre-tensioned expansion shell type solid/hollow rockbolts spaced at 1.75 m centre to centre staggered, 2 layers of 38 mm each shotcrete with welded wiremesh reinforcement in between. Drainage holes of 75 mm dia, 12 m long are provided in the roof of powerhouse. Additional treatment is by way of two layers each of 38 mm thick shotcrete with one layer of welded mesh. Hollow groutable rockbolts of 25 mm diameter, 6 m long at 1 m centre to centre staggered are provided in entire width of roof. The section of powerhouse cavern is shown in Fig. 9.21.

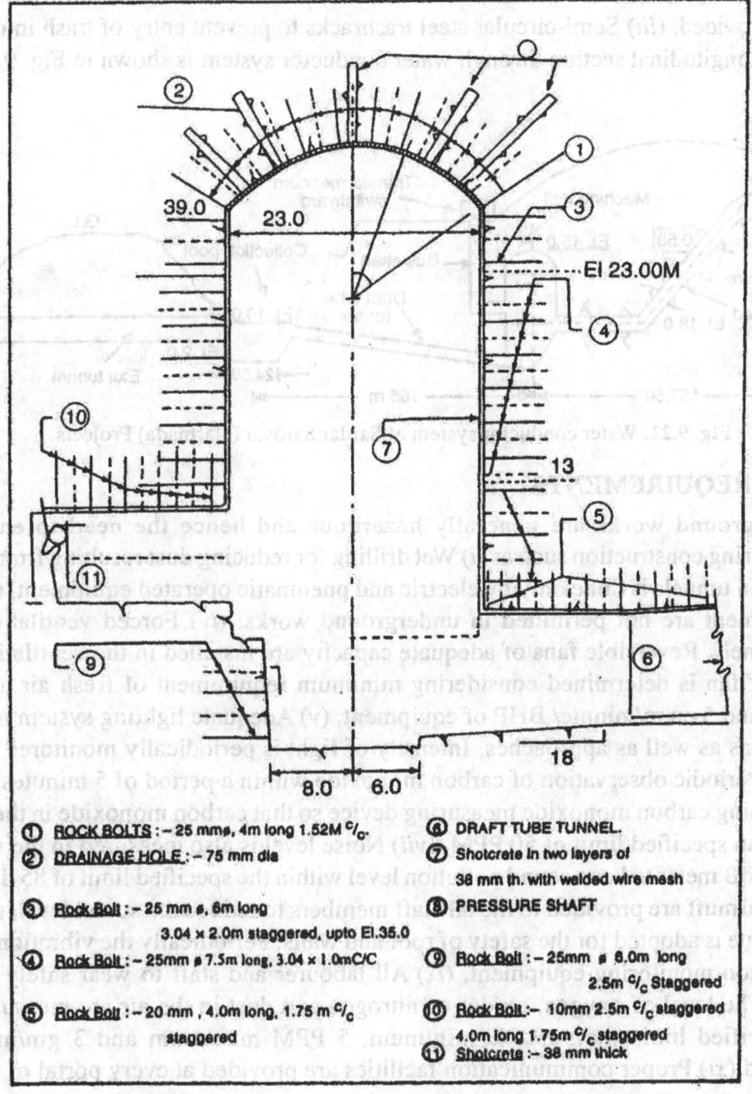

Fig. 9.20. Support system at Sardar Sarovar Project.

The power units are spaced at 25 m centre to centre. The design discharge through turbine in generation and pumping modes are 220 cumec and 233 cumec respectively. The turbine generator sets are of reversible type.

The powerhouse is initially to operate as a conventional unit but after 45 years when full irigation will be developed it will generate power during monsoon and in order to retain the peaking capacity afforded by the units, they will work as pump-turbine units and as such equipped with six reversible turbine units each of 200 MW capacity necessitating construction of a weir downstream for reverse mode operation.

The water conductor system at underground power house of Sardar Sarovar (Narmada) Project comprises (*i*) Intake; independent for each unit through 7.62 m diameter penstock housed in main dam, with bell mouth entry, (*ii*) Penstock gate provided in a gate shaft. Intake gates to be operated by means of a hoist. One set of stoplogs at the intake before bellmouth for maintenance and repairs to the penstock gate. As cover over the intake is not adequate antivortex device in the form of perforated breast wall is provided, (*iii*) Semi-circular steel trashracks to prevent entry of trash into the intake is provided . The longitudinal section through water conductor system is shown in Fig. 9.22.

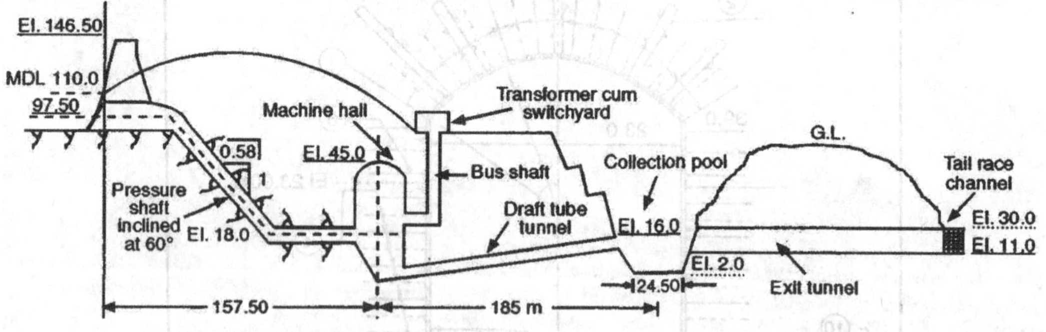

Fig. 9.21. Water conductor system of Sardar Sarovar (Narmada) Projects.

9.9. SAFETY REQUIREMENTS

The underground works are generally hazardous and hence the need to enforce safety requirements during construction such as (*i*) Wet drilling for reducing dust resulting from dry drilling, (*ii*) Excavation in tunnels is done only by electric and pneumatic operated equipment, (*iii*) Gasoline powered equipment are not permitted in underground works, (*iv*) Forced ventilation system is provided in tunnels. Reversible fans of adequate capacity are installed in the ventilation duct line. The capacity of fan is determined considering minimum requirement of fresh air at 4.25 cu m/ minute/person and 5 cu m/minute/ BHP of equipment, (*v*) Adequate lighting system is provided in the working areas as well as approaches. Intensity of light is periodically monitored by using the lux meter, (*vi*) Periodic observation of carbon monoxide within a period of 5 minutes after blast is monitored by using carbon monoxide measuring device so that carbon monoxide in the air does not exceed more than specified limit of 50 PPM, (*vii*) Noise level is also measured in the working area with the help of dB meter to keep sound pollution level within the specified limit of 85 dB. Protective measures like earmuff are provided to the all staff members to reduce the sound level, (*viii*) Control blasting technique is adopted for the safety of roof and walls. Periodically the vibration is measured by using vibration monitoring equipment, (*ix*) All labourer and staff to wear safety helmets and gumboots, (*x*) The level of oxygen, oxides of nitrogen and dust in the air are measured and kept within the specified limits i.e., 19.5% minimum, 5 PPM maximum and 3 gm/m^3 maximum respectively, and (*xi*) Proper communication facilities are provided at every portal of tunnel (s).

9.10 SIZING OF A POWERHOUSE

1. General Arrangement

A powerhouse in general comprises (*i*) Main powerhouse structure housing generator units and having either separate or combined generator and turbine rooms, and (*ii*) Erection bay.

1. Powerhouse: The main feature of a powerhouse is generator room about which other bays are grouped. The generator room is normally located in a separate bay. The space required for the unit determine the location and spacing of the transverse contraction joints. The width of the generator room provides for a minimum 3 m wide passage between the generators and one power plant wall. The height of the generator room is more or less fixed by (*i*) Head room required for dismantling and/or moving major items of generator and turbine, (*ii*) Location of crane rails due to erection bay requirements, (*iii*) Crane clearance requirements, and (*iv*) Type of roof frame.

The elevation of turbine room floor is such as to permit a minimum requirement of 1.0 m of concrete over a steel spiral case or a minimum roof thickness of 1.25 m for a semi-spiral concrete casing.

The generator and turbine room, if not combined, are kept apart in consideration of the size of equipment to be handled in the turbine room, the head room between platform in the turbine pit and the generator room floor construction.

2. Erection Bay: Generally erection bay is located at the end of the generator room at the same floor elevation and of the same length as of at least one generator bay. It is considered an extension of the generator room.

Dimensioning of Powerhouse

The overall dimensions of a power house are governed by (*i*) Overall dimensions of the turbine, draft tube and scroll casing, (*ii*) Overall dimensions of the generator, (*iii*) Number of units in the powerhouse, and (*iv*) Size of erection bay. The provision for inlet valve, erection of rotor and undertaking of transformers is so made that the space required is minimum and is consistent with the operational and maintenance requirements. IS: 12800 (Part I): 1993 specified overall dimensions of the powerhouse are briefly as under:

1. *Length of Powerhouse:* The unit spacing, erection bay length and the length required for the E.O.T. crane to handle the last unit govern the length of the powerhouse.

The unit spacing or the centre lines of the successive units is determined by drawing a plan indicating the overall dimensions of spiral casing, draft tube and hydro generator (plus 0.5 to 1.5 m to its inner diameter) with respect to the vertical axis of the machine, plus a clearance of 1.5 to 2.0 m on either side of the extremities to determine the unit spacing.

These clearances take cognisance of 2.0 to 2.5 m concrete thickness on either side of the concrete scroll casing and 1.0 to 1.5 m in the case of fully embedded steel scroll cases.

The erection bay length is taken as 1.0 to 1.5 times the unit bay size in accordance with the erection requirements.

The total length, L, of powerhouse thus amounts to:

$$L = N_o \times (\text{unit spacing}) + L_S + K.$$

where, N_o is number of units, L_S is erection bay length and K is the length required for E.O.T. crane to handle the last unit, usually 3.0 to 5.0 m depending upon the number and size of E.O.T. crane.

2. *Width of Powerhouse:* The internal width of the powerhouse superstructure is determined by drawing a plan indicating the overall dimensions of the spiral casing and the hydrogenerator with respect to the vertical axis of the machine to which is added 2.0 to 2.5 m on the downstream extremities to locate the superstructure columns. On the upstream side, provision is made for (*i*) 1.5 – 2.0 m clearance for concrete of the upstream scroll casing, (*ii*) 1.5 – 2.0 m wide gallery for approaching draft tube manhole, (*iii*) Valve pit of appropriate size, in case the main inlet valve is accommodated in the powerhouse, (*iv*) 1.5 – 2.0 m clearance for pressure relief valve in the scroll case, if required, and (*v*) Any space required, over the above, considering layout of essential equipment and operational requirements. The column width may be added to compute the outer width of the powerhouse.

3. *Height of Powerhouse:* (*i*) Fig. 9.23 indicates typical cross section through generating unit of a powerhouse. The height H_1 is determined from the bottom of draft tube to the centre line of the spiral casing. The thickness of concrete below draft tube may be assumed as 1.0 – 2.0 m.

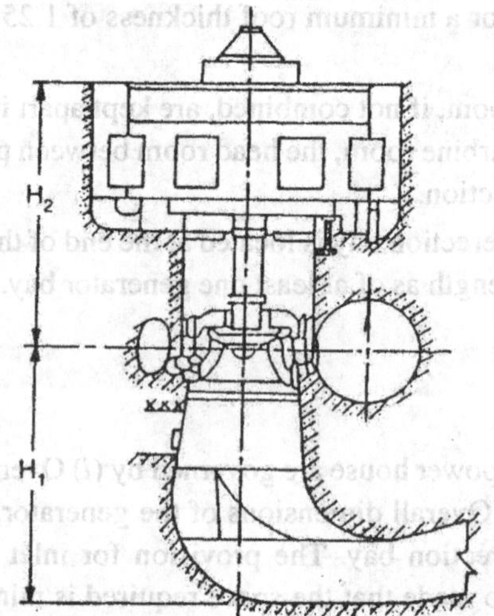

Fig. 9.22. Cross section through generating unit.

(*ii*) The height of power house from the centre line of spiral casing up to top of the generator, H_2, (Fig. 9.23) is computed as under :

$$H_2 = L_f + h_j + K$$

where, L_f is length of stator frame which is determined by adding about 1.5 m to the length of stator core, h_j is the height of load bearing bracket and K is taken as 5.5 to 7.0 depending on the size of the machine.

(*iii*) The height of machine hall above top bracket of the generator depends on E.O.T. crane hook level and the corresponding E.O.T. crane rail level plus the clearance required between the ceiling and top of the crane. The height of the ceiling above E.O.T. crane hook is usually 4 to 6.5 m depending on the width of powerhouse superstructure and capacity of E.O.T. crane. The height of service bay floor from where the equipment is to be handled is also considered in fixing the height of the powerhouse.

Fig. 9.24 shows pong power plant building (6 × 66 MW).

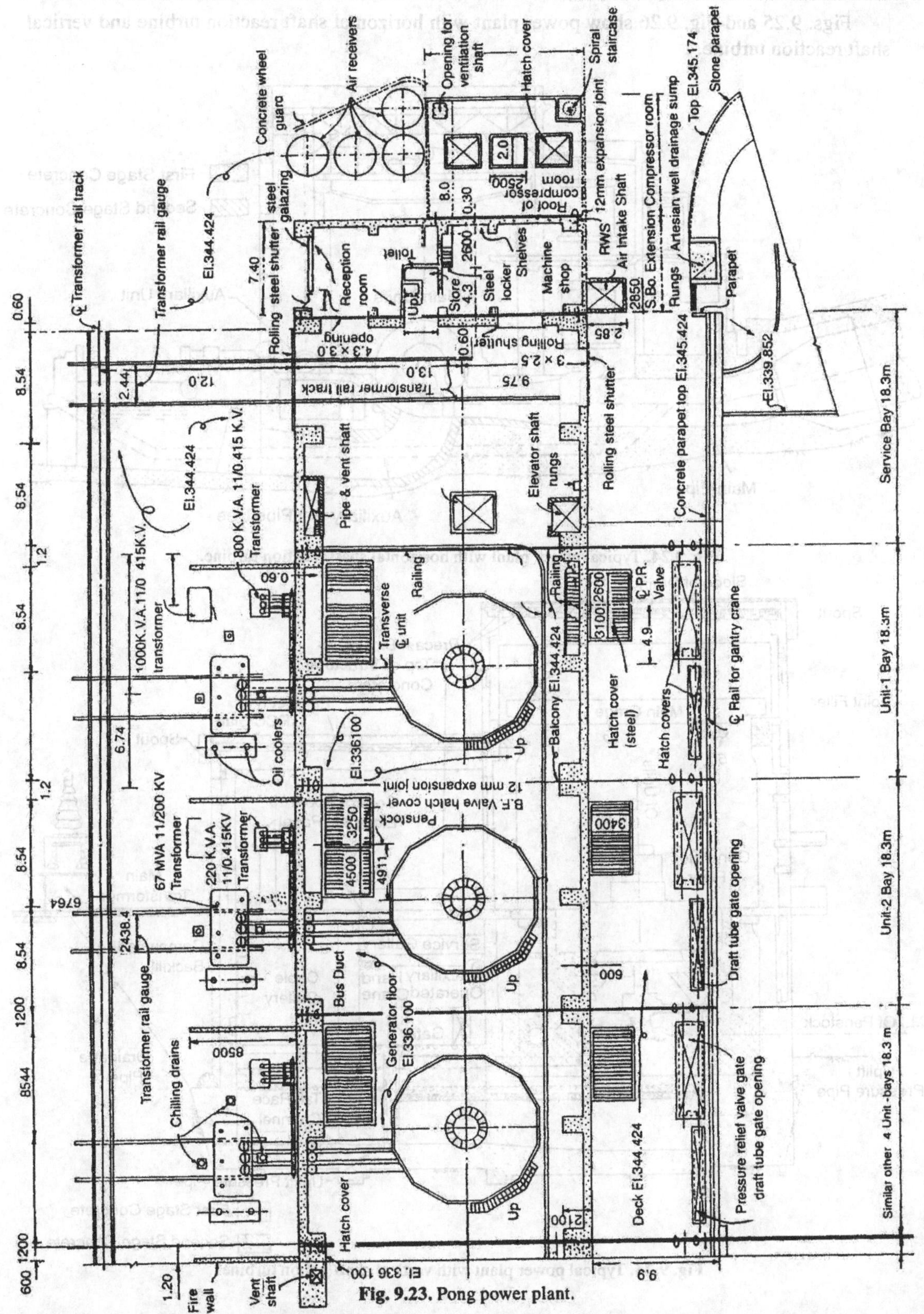

Fig. 9.23. Pong power plant.

Figs. 9.25 and Fig. 9.26 show power plant with horizontal shaft reaction turbine and vertical shaft reaction turbine.

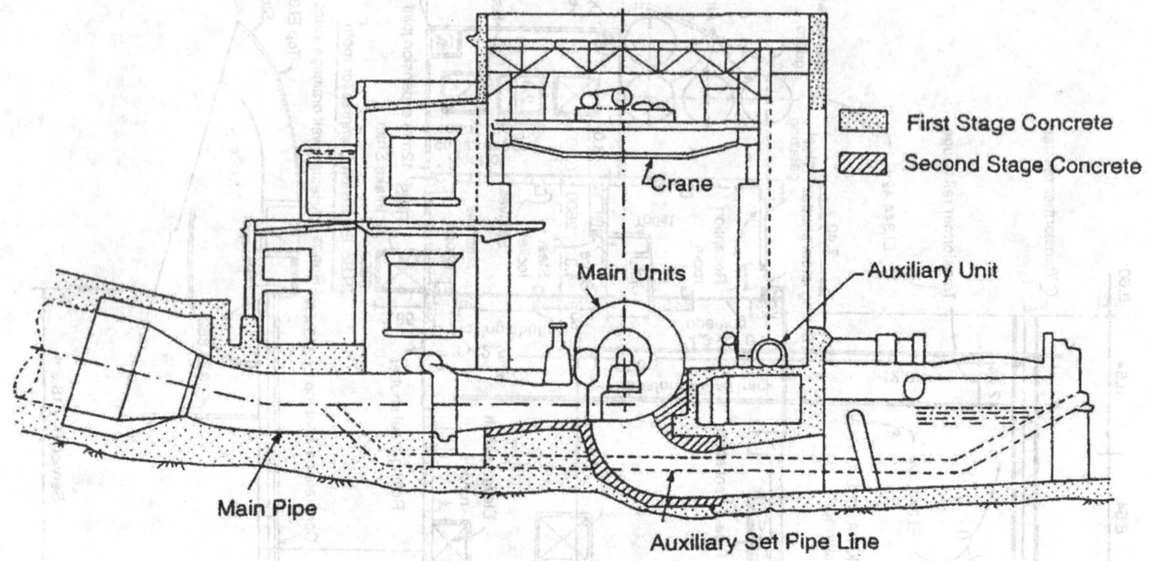

Fig. 9.24. Typical power plant with horizontal shaft reaction turbine.

Slope of Drain
Spout
Precast Blocks
Top of Structural Concrete
Joint Filler
Main Crane
5f
Of Units
Parapet
RCC Slab
Spout
Control Room
Electrical Panel
Generator Floor
Main Transformer
Service Gallery
Compacted Backfill
Auxiliary Hand Operated Crane
Cable Gallery
C.L. Of Penstock
Gate
Uplift Pressure Pipe
Drainage Pipe
Tail Race Channel
Uplift Pressure Pipe
First Stage Concrete
Second Stage Concrete

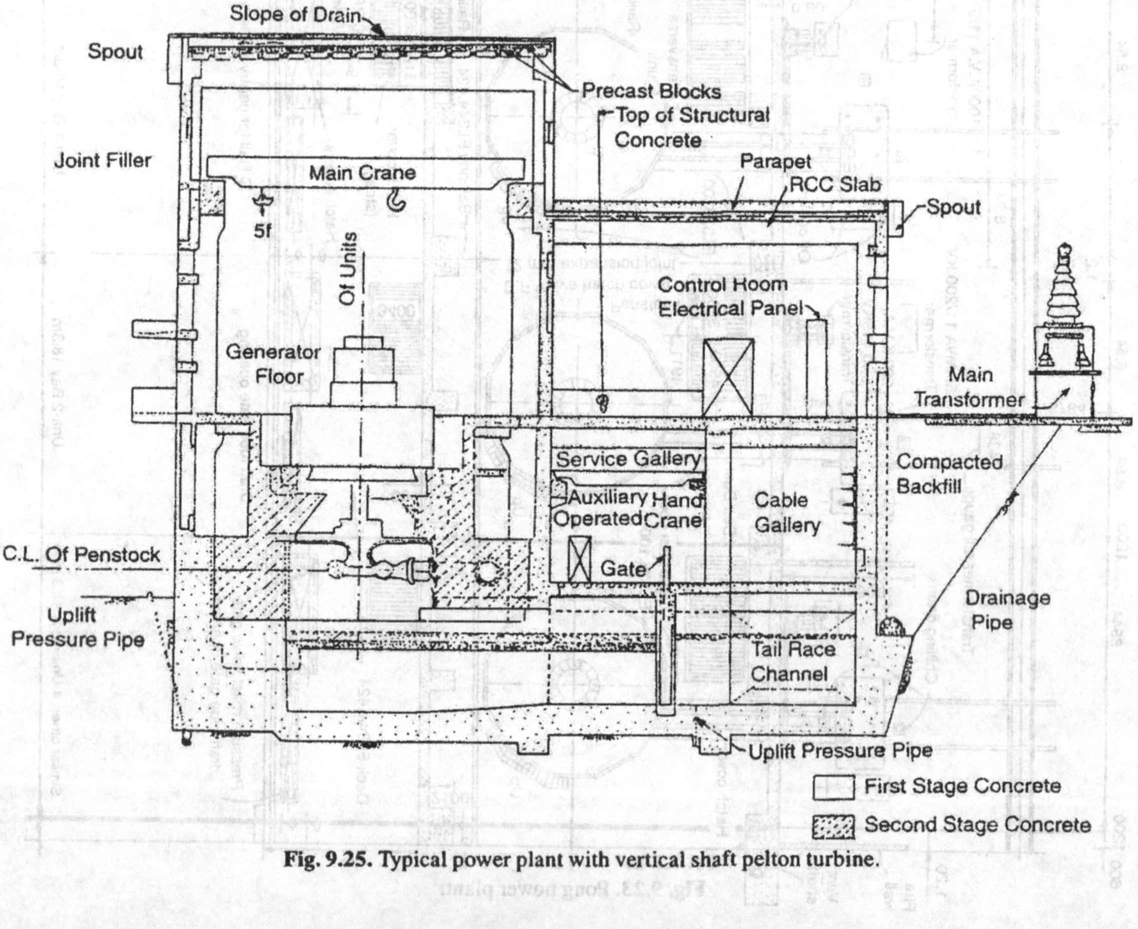

Fig. 9.25. Typical power plant with vertical shaft pelton turbine.

9.11. JOINTS IN HYDROPOWER PLANTS

Power plant structures essentially require joints (*i*) To facilitate construction work, (*ii*) To allow for expansion, (*iii*) To ensure against misalignment of machinery that would otherwise result from structural distortion, (*iv*) To prevent development of cracks, (*v*) To safeguard against differential settlement, and (*vi*) Most importantly to sectionalise its different parts in order to separate them so as not to transmit vibratory stresses or thrusts during operation of the plant.

The various types of joints are briefly as under:

1. Construction : These are produced by the placement of fresh concrete against the surface of hardened concrete already placed necessitated by the construction sequence dictated by the availability of hydromechanical and hydroelectrical as also construction equipment. These may be vertical or horizontal. After thoroughly cleaning the lower layer of concrete, 5 mm thick layer of cement mortar of the same mix as of the concrete is placed over the surface before laying fresh concrete. The reinforcement is essentially continued across these joints and good bond between the two placement of concrete is ensured. Where heavy shear caused by lateral loads is to be resisted by the joint, shear keys are also provided (Fig. 9.27). For walls more than 2 m thick one or more keys are provided (Fig. 9.28).

W
b = 4A
A
New concrete
Water stop
Exterior surface
A = W/100
a = W/2 − A = b
C.L. of key
Old concrete

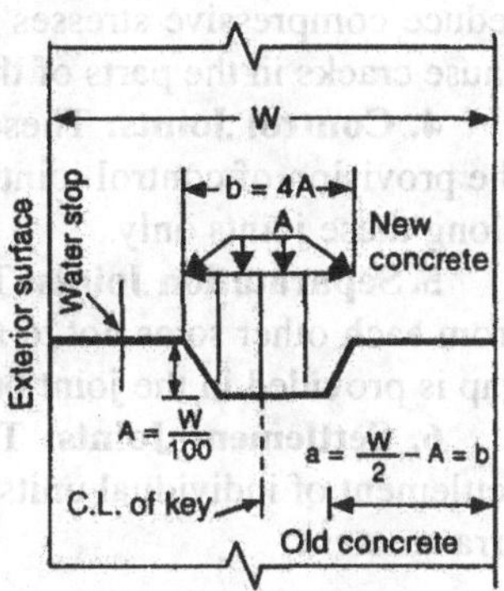

Fig. 9.26. Shear key.

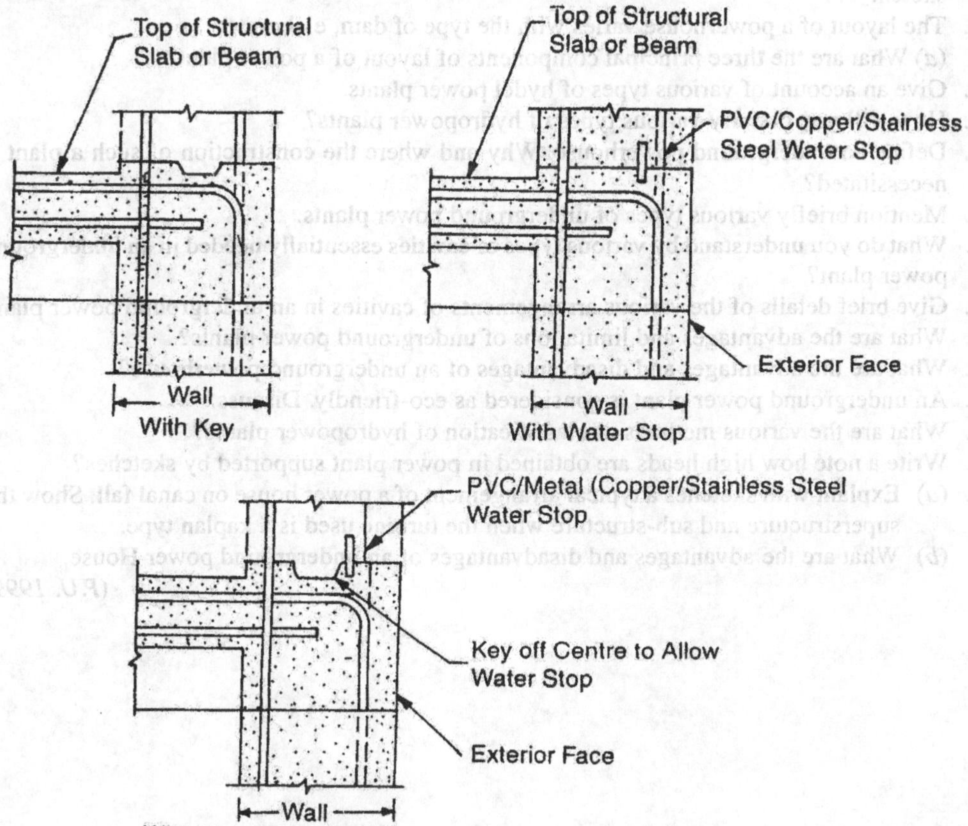

Top of Structural Slab or Beam
Wall
With Key
Top of Structural Slab or Beam
PVC/Copper/Stainless Steel Water Stop
Exterior Face
Wall
With Water Stop
PVC/Metal (Copper/Stainless Steel Water Stop
Key off Centre to Allow Water Stop
Exterior Face
Wall
With Water Stop and Key

Fig. 9.27. Horizontal construction joint in walls.

2. Construction Joints: These are provided to take care of tensile stresses due to shrinkage and where temperature variations are small and there is no likelihood of expansion. A contraction joint also serves as a construction joint in so far as break in pouring of concrete is concerned but neither reinforcement is continued across it, nor bond is allowed to be developed between the joint faces to ensure structural discontinuity.

3. Expansion Joints: These are the joints usually 12 to 25 mm thick left between the concrete joint faces to permit expansion. These are essentially provided to completely eliminate or substantially reduce compressive stresses in concrete resulting from thermal expansion which may otherwise cause cracks in the parts of the structure. The joint is filled with expansion joint filler.

4. Control Joints: These are provided at locations where cracking can not be avoided. With the provision of control joints, planes of weakness are introduced so that the cracking takes place along these joints only.

5. Separatation Joints: These are the joints introduced to keep the units of the structure separate from each other so as not to transmit harmful vibrating stresses or thrusts. Like expansion joint, a gap is provided in the joint but it is not used by the expansion of members.

6. Settlement Joints: These are the joints, usually 40 mm wide, provided to take care of settlement of individual units of the structure. It is like an expansion joint but has different sealing arrangement.

QUESTIONS

1. Define powerhouse. Discuss the three parts of hydel power plant with the help of a neat sketch.
2. The layout of a powerhouse varies with the type of dam, elaborate:
   (*a*) What are the three principal components of layout of a power plant?
3. Give an account of various types of hydel power plants.
4. How will you classify various types of hydropower plants?
5. Define an underground powerhouse. Why and where the construction of such a plant is necessitated?
6. Mention briefly various types of underground power plants.
7. What do you understand by various types of cavities essentially needed in an underground power plant?
8. Give brief details of the various arrangements of cavities in an underground power plant.
9. What are the advantages and limitations of underground power plants?
10. What are the advantages and disadvantages of an underground powerhouse?
11. An underground power plant is considered as eco-friendly. Discuss.
12. What are the various methods of classification of hydropower plants?
13. Write a note how high heads are obtained in power plant supported by sketches?
14. (*a*) Explain with sketches a typical arrangement of a power house on canal fall. Show the superstructure and sub-structure when the turbine used is a kaplan type.
    (*b*) What are the advantages and disadvantages of an underground power House.

(*P.U. 1999*)

10

HYDRAULIC TURBINES

10.1 INTRODUCTION

Hydraulic turbines occupy a distinct place amongst the prime-movers used for power generation by virtue of their high efficiency. Turbines are the machines which convert hydraulic energy into mechanical energy which in turn is used in running an electric generator to produce electricity. In general, a turbine consists of a wheel called runner, provided with vanes or buckets along its periphery, connected to a shaft, a mechanism which controls the quantity of water and passages leading to and from the wheel. The high velocity water strikes on the vanes which deflect the water towards outlet and in doing so impart the energy to the runner and causes it to rotate about the shaft axis. The shaft of the runner is directly coupled to the generator shaft thereby the rotation is passed on to the generator. The generator armature so rotated produces hydropower.

10.2 CLASSIFICATION OF TURBINES

The hydraulic turbines are classified as under:

1. Classification Based on Hydraulic Action: The modern hydraulic turbines classified in respect of hydraulic action are (*i*) Impulse turbine, and (*ii*) Reaction turbine.

2. Classification Based on Direction of Flow: The types of turbines with respect to the direction of flow are given in Table 10.1.

Table 10.1 Type of turbines w.r.t. flow direction.

Flow direction	*Type of turbine*
Radial inward or mixed type	Francis
Tangential	Pelton
Axial	Propeller and Kaplan
Diagonal	Deriaz

3. Classification Based on Specific speed: The types of turbines on the basis of specific speed are as given in Table 10.2.

Table 10.2 Specific speed of turbines.

Type of turbine	*Specific speed*
Pelton	15–65
Francis	60–400
Kaplan	300–800
Deriaz	200–400

4. Classification Based on Head: The types of turbines on the basis of head are given in Table 10.3.

Table 10.3 Turbines for various heads.

Type of turbine	*Maximum net head (m)*
Pelton	Above 300
Francis	30–400; even up to 500–600
Deriaz	50–150
Kaplan	10–60
Bulb	3–20

10.3 TYPES OF TURBINES

Types of hydraulic turbines used for medium and large power plants are illustrated in Fig. 10.1.

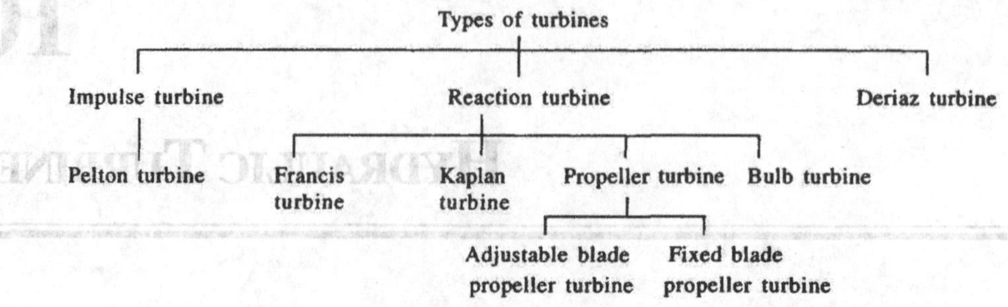

Fig. 10.1 Types of turbines

The two basic types of hydraulic turbines are (*i*) Impulse or velocity turbine, and (*ii*) Reaction or pressure turbine.

1. Impulse Turbine

Pelton is the single type of impulse turbine. In this type of turbine, all the potential energy of water is converted into kinetic energy, before it acts on the runner, by expanding through a nozzle. It has one or more free jets of water discharging into an aerated space and impinging on a series of split buckets mounted around the periphery of a wheel, a needle valve in the nozzle for controlling the rate of flow without any change in jet direction, thereby eliminating shock losses at the entrance to the bucket and maintaining sustained high efficiency at part load, a housing and a discharge passage. The jet is produced by passing the water brought by the penstocks through a contracting nozzle. In small turbines only one jet is used while in large turbines two or more jets impinge on different points around the wheel. It causes the wheel to revolve free in air and water is in contact with only a part of the wheel at a time.

The pressure of water is all along atmospheric and as such there is no difference of pressure in the water at the inlet to the runner and that discharging from the bucket to the tailrace. To guide the water discharged from the bucket to tailrace as also to prevent splashing a casing is provided. The wheel must not run full having been placed at the foot of the fall and above the tailrace.

The impulse turbines are made in both horizontal and vertical shaft configuration. These turbines are essentially low-speed turbines and are used for high heads 150 to 1000m and low discharges; the maximum head of over 1770 m has been utilised in Switzerland.

To cater for water hammer occurrence in the penstock in case the nozzle is suddenly closed, a bypass valve is provided with some nozzles which opens whenever the needle valve is closed quickly. Likewise, a jet deflector can be provided to deflect the jet away from the wheel in the case of drop in load.

The peak efficiency of a good impulse turbine is high; 85–90 percent, for which the bucket width is 3 to 4 times and wheel diameter 15 to 20 times of the jet diameter.

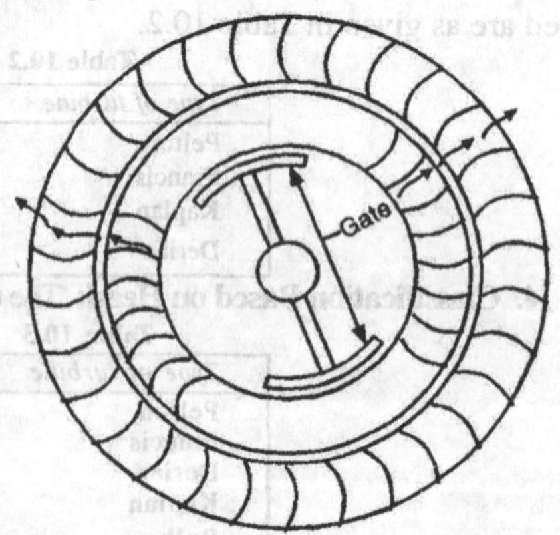

Fig. 10.2 Radial flow turbine.

Classification of Impulse Turbines

1. Radial Flow Turbine: The water flow may be inwards or outwards. The water enters the wheel through fixed guide blades. A centrifugal head is impressed by the revolving wheel as the water flows over the moving vanes which is converted to velocity head. In case the flow is inward, the relative velocity is reduced by the centrifugal force which renders easy control over the speed than the outward flow turbine. The wheel tends to adjust itself automatically to the load; any small increase of speed of the wheel due to load lightening decreases the flow through the wheel and hence decreased power.

2. Axial Flow Turbine: The conditions governing the radial flow impulse turbine apply to this type of turbine except that there is no centrifugal head impressed on the water. Girard and Pelton wheels are the principal types of the axial flow turbines.

2. Reaction Turbine

In this type of turbine, the flow takes place under pressure in a close chamber and atmospheric air has no access at any point. The water enters all around the periphery of the wheel under pressure, wheel passages completely full with water and flows over the vanes. In the process of passing over the vanes, the pressure head is converted into velocity head and is finally reduced to atmospheric pressure before leaving the wheel.

The reaction turbine utilises pressure energy as well as kinetic energy. As only a part of the available potential energy is converted into velocity head at the entrance to the runner, and the balance, a substantial portion, remains as pressure head, the water entering the turbine has pressure as well as kinetic energy and the wheel rotates under the action of both of these forces. It obtains rotative force by deflecting the flow of water in closed passages and utilising the consequent reaction. The water leaving the turbine also possesses some pressure as well as velocity head. The pressure at inlet to the turbine is much higher than that at the outlet. Since the water is under pressure, the wheel must run full and is entirely submerged below the tailrace. It may also discharge into the atmosphere or placed about 9 m above the foot of the fall and discharge into a suction or draft tube.

A reaction turbine has a water tight casing, a mechanism for controlling the quantity of water and for distributing it equally over the entire runner intake and a draft tube. The wheel may run in a vertical or horizontal position, more convenient in the horizontal position on account of more accessibility.

The principal types of reaction turbines, in general use, are (*i*) Francis turbine, (*ii*) Propeller turbine-adjustable blade and fixed-blade propeller turbines, and (*iii*) Kaplan turbines.

Classification of Reaction Turbines

The reaction turbines are classified according to the direction of flow in runner. The flow through the turbine may be radially inward, axial or mixed i.e. partly radial and partly mixed.

1. Outward Flow Turbine: It consists of a wheel in the shape of a cylindrical disc mounted on a shaft carrying blades around the perimeter (Fig. 10.3). The wheel is surrounded by a water-tight casing capable of running in vertical or horizontal position. The water enters into the wheel at the centre and passes through fixed radial guide blades into the moving blades causing them to rotate and leaves at the outer edge. The water in the wheel is under pressure and as such the wheel must run full.

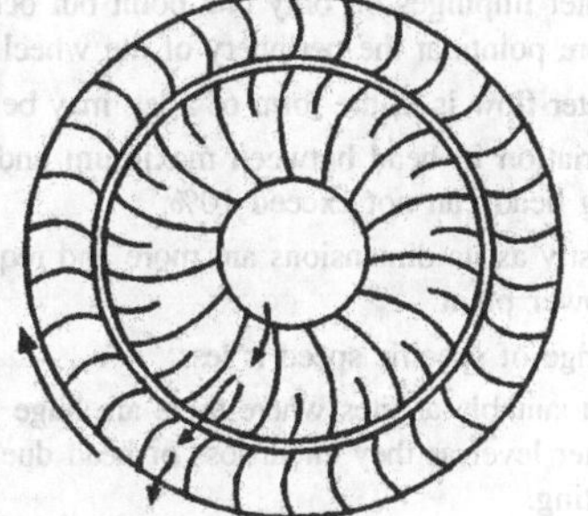

Fig. 10.3 Outward radial flow turbine.

The wheel my be submerged below tailrace or placed in a suction or draft tube above the foot of the fall, in which position it is convenient being more accessible.

The revolving wheel causes a contrifugal head to be impressed on the water passing through it which increases the relative velocity of the water and as a result tends to increase the quantity of flow passing through the wheel. The wheel tends to race due to increase in the centrifugal force if there is a slight increase in speed. The efficiency is increased by discharging the water radially with zero velocity of whirl at outlet.

2. Inward Flow Turbine: In this turbine, the water enters the wheel through the outer periphery and flows radially inward i.e. towards the centre and leaves it in a direction parallel to the axis. The fixed guide blades surround the revolving blades externally and the whole mechanism is surrounded by an outer casing. The revolving wheel causes a centrifugal head to be impressed on the water acting against the radial flow of the water and as a consequence any increase in the speed of the wheel tends to reduce the quantity of flow through the wheel which means reduced power but prevents racing which is its merit over the outward flow turbine. The efficiency of the turbine is the highest when the discharge is radial with the velocity of leaving water as small as possible. The wheel may be placed below the tailrace or in a suction tube above the foot of the fall.

The inward flow turbines have certain merits over the outward flow turbines, such as (*i*) The speed of the turbine is better regulated in that an increase in rotor speed tends to reduce the rate of flow through the rotor and accordingly reduces the power, (*ii*) The chances of eddy formation and pressure loss are reduced since the area of flow becomes gradually convergent, (*iii*) Better regulation is possible as the guide vanes can be located on the outer periphery of the runner, (*iv*) Less friction losses as water velocity over the vanes is reduced, and (*v*) Can be used for relatively high heads without increasing the speed of the turbine since centrifugal head supports appreciable part of the supply head.

3. Axial Flow Turbine: In axial flow turbine, water enters the wheel at the side and flows parallel to the axis (Fig. 10.4) and that is why also termed as parallel flow turbine.

4. Mixed Flow Turbine: It combines the characteristics of the inward radial flow and the axial flow turbines by curving the blades in two directions. The water flows into the wheel radially and leaves at the centre axially. However, the modern mixed flow turbines are inward flow turbines. The Francis and the Deriaz are the two mixed flow type turbines in use. The latter is similar to the former except that the blade angles can be changed according to the guide blade position.

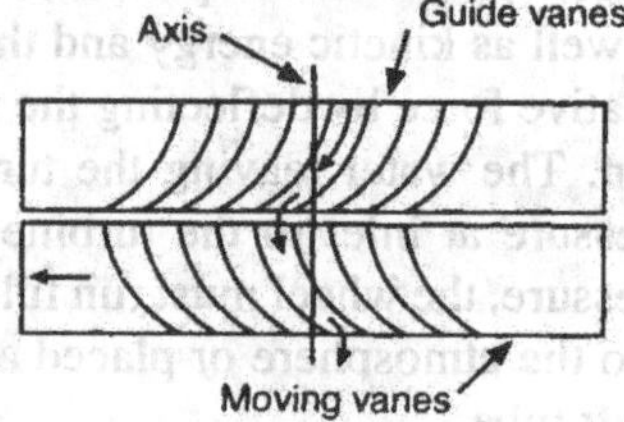

Fig. 10.4 Axial flow turbine.

10.4 IMPULSE VS REACTION TURBINES

A comparison of impulse and reaction turbines, as under, shall be of interest:

Impulse Turbine	*Reaction Turbine*
Free jet of water impinges on the runner which is exposed to atmospheric pressure.	Flow takes place under pressure in a closed chamber.
Water impinges on only one point but occasionally two or more points at the periphery of the wheel at any instant.	Water enters all around the circumference of the wheel.
Water flow is in the form of a jet, may be two or more.	The flow may be radially inward, axial or mixed.
Variation in head between maximum and minimum operating head can not exceed 10%.	Can better accommodate variations in operating heads.
Costly as its dimensions are more and require large size power plant.	Dimensions of generator, power plant etc are smaller and hence comparatively economical.
Range of specific speed is less.	Range of specific speed is more.
Not suitable at sites where there are huge variations in tail water level as they entail loss of head due to their higher setting.	Better adapted where the variations in tail water level are relatively large compared to total head.
Energy delivered to the turbine is wholly kinetic	It utilises pressure energy as well as kinetic energy.
Entire potential energy is converted into kinetic energy.	Only a portion of potential energy is converted into kinetic energy.
Used only for high head installations.	Comparatively low head wheels.
Efficiency changes very little with size.	Somewhat higher efficiency in larger sizes.
Low speed turbine.	High speed units.

Impulse Turbine	*Reaction Turbine*
Some nozzles are provided with a bypass valve to take care of water hammer occurrence in the penstocks.	A relief valve or a surge tank is generally required to avoid serious water hammer pressures.
Efficiency decreases fast with wear.	Efficiency does not decrease with wear as fast as in pelton units.
Less operating efficiency at full and half load.	Good operating efficiency especially between half and full load.
Turbine casing has no hydraulic function except to prevent flashing and guide discharge.	Wheel is essentially surrounded by a watertight casing.
Draft tube is not required.	Draft tube is an integral part of the turbine.
Low power is developed by a wheel of a given diameter compared to reaction turbine.	More power is developed by a wheel of the same size and operating under the same head as impulse turbine.
Flow can be regulated through the wheel without loss.	Flow difficult to regulate without loss.
Can work satisfactorily with moderately silt laden water as it can withstand abrasive action and yet maintain good efficiency.	More sensitive to abrasive effect of silt.
Wheel must not run full with water.	Wheel must run full with water.
Speed of the wheel is kept constant under varying loads by changing the jet discharge.	The quantity and direction of flow is controlled by adjustable vanes in accordance with the variation in load.
Made in both horizontal shaft and vertical shaft configuration.	Wheel may run in vertical or horizontal position.
Pressure atmospheric and the wheel must be placed at the foot of the fall and above the tailrace.	Wheel is under pressure greater than atmospheric and wheel may be entirely submerged below the tailrace or discharge into atmosphere.
No difference of pressure in water at the inlet and outlet.	Pressure varies throughout; higher at inlet than at outlet end.
Action of the turbine depends on a change in momentum of the water so that a dynamic force is exerted on the runner.	Action of the turbine is similar to that of the impulse turbine.

10.5 FRANCIS TURBINE

It essentially consists of a scroll casing, guide vanes, runner and draft tube forming its integral part (Fig. 10.5). The turbine has fixed blades runner, where the water under pressure enters the runner through a series of guide vanes, known as wicket gates, in radial direction and leaves the runner practically in axial direction. The contracting passages in the vanes convert pressure head to velocity head. The vanes being adjustable, their rotation actuated by governor control, the quantity and direction of flow is varied by changing the size of inlet of the turbine. When load drops rapidly, the governor operates to close the gate. The flow through the runner is firstly inward in the radial direction, gradually changing to axial. After passing past the runner, the water leaves through draft tube to outfall into tailrace. The effective head under which the turbine operates thus comprises the difference in total energy contained in the water immediately before its entrance into the turbine and the energy immediately after discharge from draft tube. The draft tube reduces the discharge velocity and hence the loss of head at exit and also allows the runner a setting higher than the tailrace without sacrificing the corresponding head. Usually the turbines are mounted on a vertical axis but on horizontal axis, also. The turbine is provided with a scroll casing but for heads less than 12 m open flume setting is adopted.

Advantages

(*i*) Operating head variation is easily controlled, (*ii*) Operating head is utilised even when the variation in tail water level is relatively large compared to total head, (*iii*) Ratio of maximum and minimum operating head may be even two, (*iv*) Size of runner, generator and power plant required is small and hence economical, and (*v*) Mechanical efficiency due to wear does not decrease as faster as with pelton turbine.

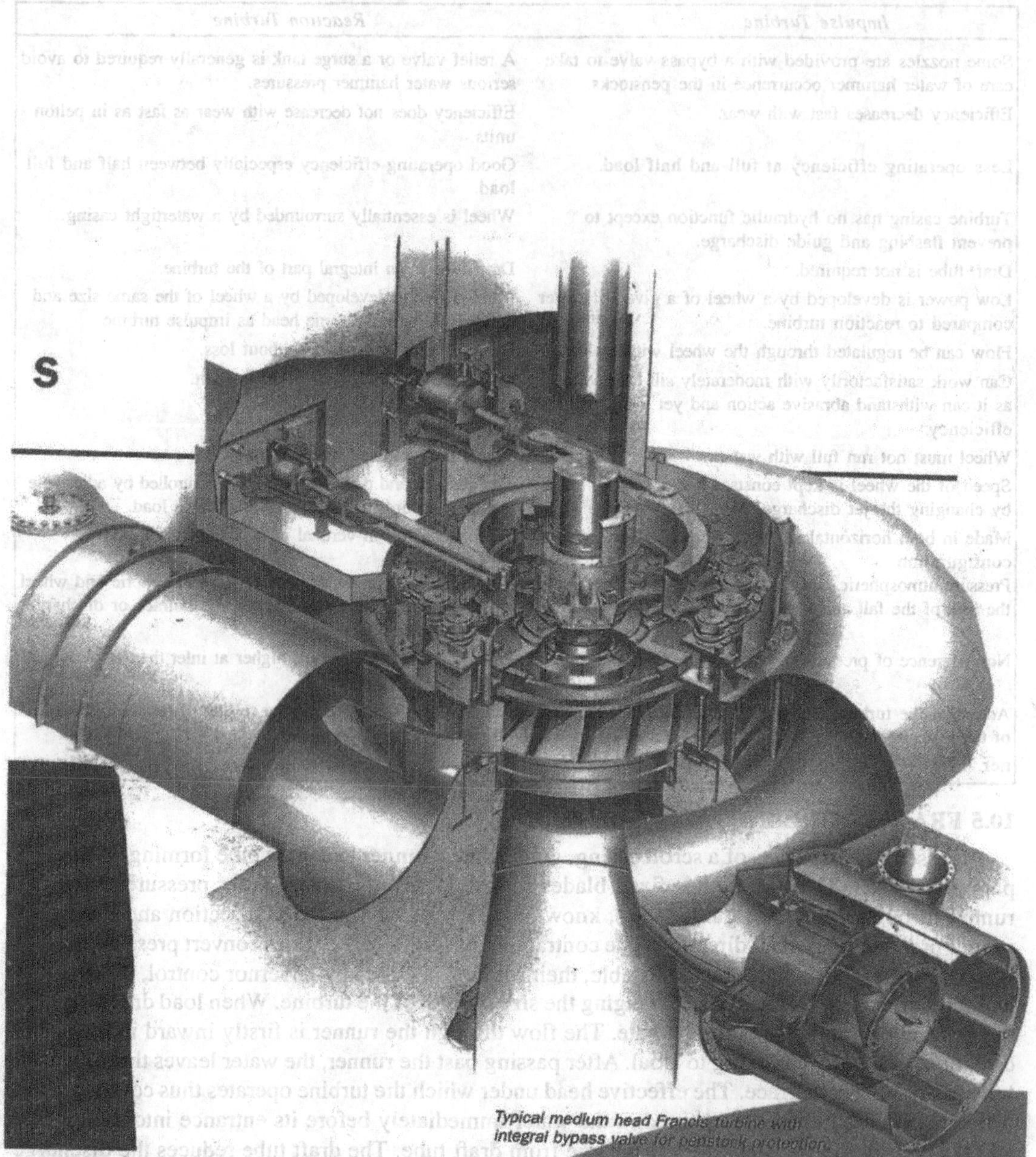

Fig. 10.5 Francis turbine.

Disadvantages

(*i*) Sediment in water flow rapidly wear high head turbines; rather the damage is more serious in small diameter machines than in large ones, (*ii*) Inspection and repair is tedious; worn out parts have to be replaced which is a time consuming job, (*iii*) More susceptible to water hammer effect, (*iv*) Cavitation problem is endless, and (*v*) Efficiency is reduced if run at below 50 percent load for a long time as also cavitation problem becomes serious.

10.6 KAPLAN TURBINE

It is an adjustable blade propeller type axial flow turbine suitable for low heads. The runner consists of a central hub on the periphery of which curved vanes are mounted usually 3 to 8 in number, depending on specific speed range. The runner is capable of adjustment automatically to runner-blade angle as the wicket gates are opened or closed, actuated hydraulically by the governor.

Each angular position of the runner blades corresponds to a certain definite gate opening. The hub which houses the mechanical device to automatically turn the blades, according to load, is of diameter dependent upon head. The adjustability of the blades maintains efficient conditions of flow without shock losses at entrance to the blades. The turbine has very high efficiency at part gate and maintains sustained efficiency at part load.

It is costlier than fixed blade turbine but may economise in number of units in a plant thereby compensating as to cost.

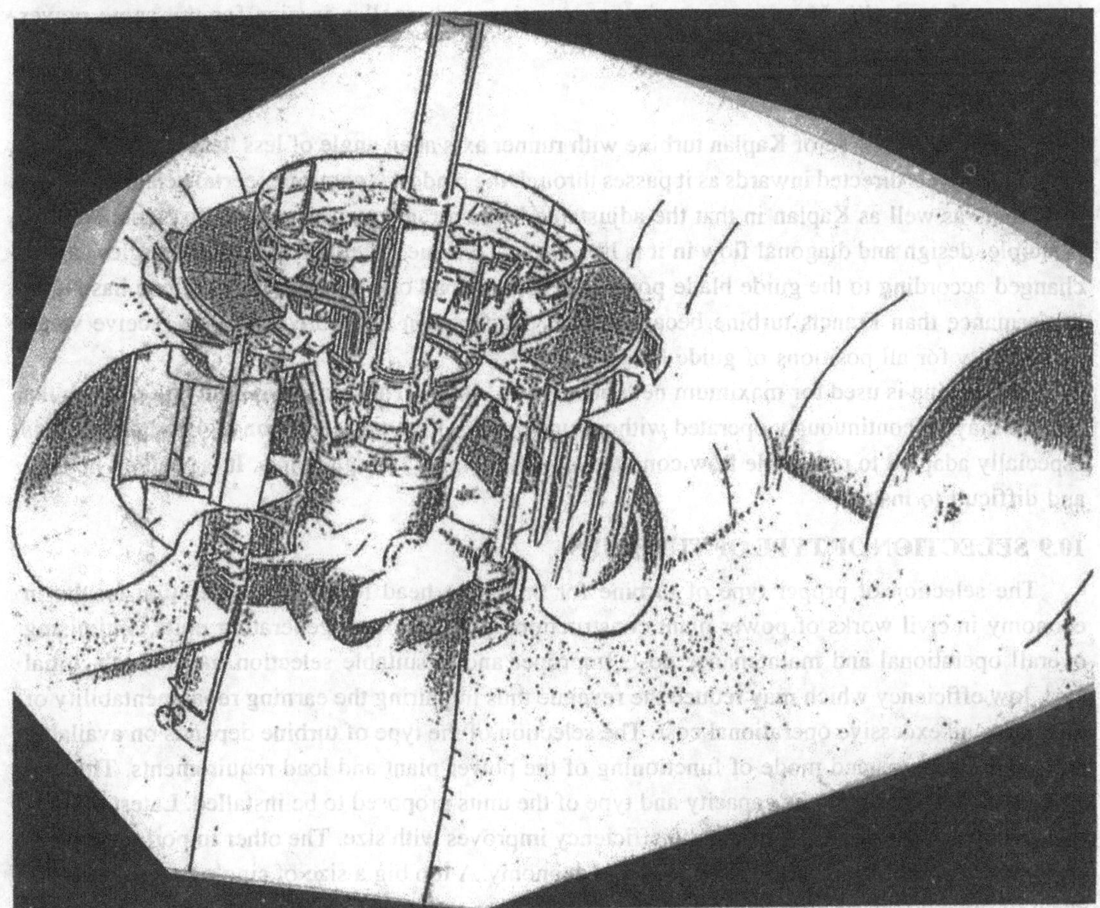

Fig. 10.6 Kaplan Turbine.

10.7 PROPELLER TURBINE

It is an axial flow turbine with its runner confined in a closed conduit system and has no exposed free surface. It is a very high speed runner and usually used for heads less than 20 m to avoid cavitation problems. The runner consisting of a central hub with curved vanes mounted on its periphery has 4 to 8 blades. In simple type, the blades are fixed in position and is not susceptible to clogging in operation. The relative velocity between the blades and the water is high and comparatively little change in this relative velocity as water passes through the runner.

Kaplan turbine belongs to propeller type but has adjustable blades and is costlier than fixed-blade wheel. Where turbines are operated at or close to the best gate, due to load conditions, this type of turbine is economical due to being lighter in weight and the higher speed necessitates low cost generator.

The disadvantage of this type of turbine is the rapid decrease of efficiency at part-gate and is essentially considered in determining size and number of units for a given load conditions.

Advantage

(*i*) Part load efficiency is high. The efficiency curve remains almost flat over the whole load range, (*ii*) Lesser frictional losses in passing through the blades on account of small number of blades used, and (*iii*) More compact in construction and smaller in size for the same power output.

10.8 DERIAZ TURBINE

It is a diagonal type of Kaplan turbine with runner axis at an angle of less than 90° to the shaft axis. The flow is directed inwards as it passes through the blades. It combines certain characteristics of Francis as well as Kaplan in that the adjustable blade arrangement is similar to Kaplan and in principle, design and diagonal flow in it is like Francis turbine except that the blade angles can be changed according to the guide blade position. For part load operation, Deriaz turbine has better performance than Francis turbine because blade position can be altered so as to receive water tangentially for all positions of guide blades.

The turbine is used for maximum net head of 50–150 m. The minimum load up to which the turbine may be continuously operated without undue cavitation and vibration is 40 percent and is especially adapted to reversible flow conditions as in pumped storage plants. It is costlier, heavier and difficult to install.

10.9 SELECTION OF TYPE OF TURBINES

The selection of proper type of turbine for the given head for power generation results in economy in civil works of power plant construction and erection of generating units, optimising overall operational and maintenance cost. Improper and unsuitable selection means high initial cost, low efficiency which may reduce the revenue thus impairing the earning repaymentability of the plant and excessive operational cost. The selection of the type of turbine depends on available head and discharge and mode of functioning of the power plant and load requirements. The two aspects in this regard are the capacity and type of the units proposed to be installed. Latest trend is to use bigger size turbines as hydraulic efficiency improves with size. The other important considerations are optimum working conditions and economy. A too big a size of single unit may, in the event of breakdown, result in considerable loss of power generation. A too small size turbine may be economical in so far as the cost of civil works and the plant are concerned but it means more number of units and hence on the whole uneconomical. The various considerations in the selection of type of units are as under:

1. Site Conditions: The techno-economic considerations of the generating equipment cost of power house and relative benefits of power generation determine the type of turbine specific to site conditions as given in Table 10.4 as per IS: 12837-1989:

Table 10.4 Criteria for turbine selection.

Type of turbine	*Head variation (percent of rated head)*	*Load variation (percent of rated output)*	*Specific speed (m. mhp)*	*Peak efficiency (%)*
Pelton	120–80	50–100	15–65	90
Francis	125–65	50–100	60–400	93
Deriaz	125–65	50–100	200–400	92
Kaplan	125–65	40–100	300–800	92
Propeller	110–90	90–100	300–800	92
Bulb	125–65	40–100	600–1200	92

2. Specific speed: It is a fundamental index for the selection of correct type of turbine for a particular set of conditions. It is defined as the speed (in r.p.m.) at which a turbine of homologous design would run, at its designed efficiency, if the runner were reduced to a size which would develop one metric horse power under one metre head

$$n_s = n\frac{\sqrt{P \times 1.358}}{H^{5/4}}, \quad \text{Also } N_s = \frac{n\sqrt{Pr}}{H^{5/4}}$$

Where, n_s is specific speed of turbine (r.p.m), n is rated speed of turbine (r.p.m.), P is turbine output (kW), and H is rated head (m), Pr is rated turbine output in m H.P at full gate opening, as 1 m. H.P = 0.736 kW (*i.e.* 1 kW = 1.358 HP).

The specific speed (N_s) being a fundamental index of turbine hydraulic properties forms a common basis of comparison of all types of turbines. Technical and economic requirements together with manufacturing and operational problems establish a relationship between the specific speed and design head of the type.

The present trend is to have higher and higher specific speeds to have smaller generator, higher efficiency and overall economy of power house. But mechanical design considerations, cavitation tendency, vibrations, drop in peak efficiency and increased ratio of runaway speed to normal speed etc. limit the economic feasible specific speed. In addition, greater speed requires deeper setting of units consequently increasing excavation and structural costs.

3. Head: Maximum net head acting on a turbine is an important consideration in the selection of type of turbine for a power plant. There is a range of maximum net head for which each type of turbine is suitable as given in Table 10.5.

Table 10.5 Types for heads (IS: 12837-1989)

Type of turbine	*Type of Head*	*Maximum net head (m)*
Pelton	High head	Above 300
Francis	Medium head	30–400m; even up to 500–600 m
Kaplan	Low head	10–60
Deriaz	Low head	50–150
Bulb, Propeller	Very low head	3–20

In overlapping head ranges more detailed analysis is carried out to optimise cost for selecting the exact type of turbine.

4. Head and Specific Speed: High specific speed for a given head and power output means economy as smaller size of turbines, generators and power house are required but there is some loss of efficiency. But the modern trend is for the selection of high specific speed machines leading to smaller and higher speed machines for the same design requirements and output.

Furthermore, at high heads more than 300 m for a given speed and power output, a low specific speed machine like impulse turbine is required. For low heads, propeller turbines with high specific speeds are the obvious choice. However, an impulse turbine is suitable for low head and small discharge plant but usually the size of impulse turbine is not desirable. For Francis turbine, the limiting head is about 300 m.

Two machines of identical geometric shape but of different size may have same specific speed and hence the emphasis on specific speed for the selection of the type of turbine. Fig. 10.8 shows the specific speeds in the range of best efficiency for various types of turbines.

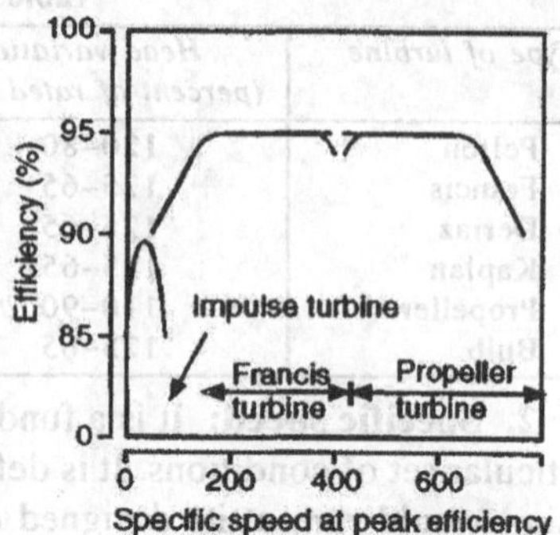

Fig. 10.7 Maximum turbine efficiency as a function of specific speed (After Moody).

Whereas higher the specific speed for a given head and power output, in general, the smaller is the size of turbine and generator and hence economy in installation, it also entails (*i*) Some loss of efficiency, (*ii*) Deeper setting of units consequently increase excavation and structural costs, (*iii*) Reduces the head range under which the turbine can operate satisfactorily, (*iv*) Too high a velocity into the draft tube resulting in high residual velocity head loss which may cause cavitation problem on runner blades, (*v*) Increased cost due to high strength requirements caused by the high speed. On the other hand, too low a specific speed for a low-head turbine of a given power output means increased cost of the generator due to slow speed. The low speed turbines are suitable for high heads and high speed types are adapted to low head plants.

Highest specific speed of turbine results in higher speed of rotation for generator with consequent reduction in cost of generator. This criteria is very important for the selection of turbines from cost consideration in the overlapping head ranges (Fig. 10.8).

5. Efficiency: The turbine efficiency varies with power output and head. Turbine performance is best at design head. Francis and Propeller turbines have higher fall of efficiency in comparison to Pelton, Kaplan and Bulb types. As such in overlapping head ranges selection of type of turbine is affected by head variation existing at site.

The choice of turbines in the overlapping head ranges is also influenced by the necessity of operating turbines at part loads for longer time because turbine efficiency varies with load as well. Fall of efficiency at part load for Francis and Propeller turbines is much steeper in comparison to that for Kaplan and Pelton types. As such in head ranges were both Kaplan and Francis are suitable, in case of requirement of large head and load variation, Kaplan turbine is superior to Francis from higher power generation consideration on account of better overall efficiency. Similarly, in the overlapping head ranges where both Francis and Pelton can be used, the latter is better in overall performance level when variation of load and head is higher.

6. Load: The turbine selection is also influenced by the variability of load. Under low demand, high capacity necessarily has low efficiency. For reasonably constant load, a single unit of requisite capacity may do. Two units may be needed if the load goes down considerably.

Minimum load up to which turbine may be continuously operated without undue cavitation and vibration is dictated by the type of turbine. Normal range for continuous operation for various types of turbines are given in Table 10.6 as per IS: 12837–1989:

Table 10.6 Minimum output for turbines.

Type	*Minimum output for continuous operation (%)*
Pelton	30–50
Francis	50
Kaplan	30–40
Bulb	30–40
Propeller	85
Deriaz	40

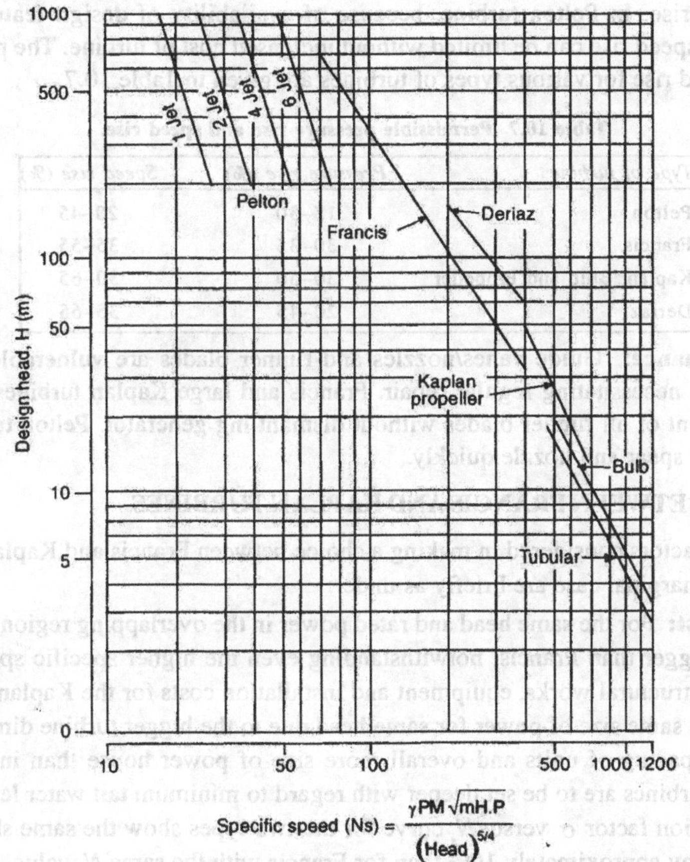

Fig. 10.8 Relation between head and specific speed (IS: 12837-1989)

7. Cavitation: Cavitation is an important consideration in the selection of a turbine for the given head and specific speed. It is on account of cavitation limit that high-speed turbines are not used for high-heads, but low-speed turbines can work under high heads, there being no apprehension of cavitation due to their favourable cavitation characteristics. In general, cavitation coefficient for Francis turbines is much less than that for Kaplan turbines.

8. Turbine Setting and Excavation Requirement: Setting of reaction turbine with reference to minimum tail water level is dictated by requirement from cavitational consideration. Further, in view of bent draft tube, excavation up to bottommost point of knee that is much deeper than runner centre line is required. Francis turbine requires relatively lesser submergence and excavation. Pelton turbine is installed above maximum tail water level which means minimum excavation cost.

9. Transportation Cost: It is a consideration of importance in the selection of large unit rating machines. Runner is one of the costliest and the most critical part which has to be handled in one piece from the consideration of cost, reliability and reduced site work as in the case of Pelton turbine. Large size Kaplan runner has lesser transportation problem because of the possibility of removing runner blades easily as they are fixed by bolts and keys. Francis turbine runner can be manufactured in two pieces from transportation limitation in size but it means higher cost.

10. Pressure Rise and Speed Rise Consideration: It is more important a consideration in high head machines as higher pressure rise affects the penstock cost substantially. Francis and Kaplan turbines incur extra cost for they require pressure relief valves in the long water conductor system to limit pressure rise. In Pelton turbine, because of availability of design feature of deflector, pressure rise and speed rise can be limited without increased cost of turbine. The permissible pressure rise and speed rise for various types of turbines are given in Table 10.7.

Table 10.7 Permissible pressure rise and speed rise.

Type of turbine	*Pressure rise (%)*	*Speed rise (%)*
Pelton	15–30	20–45
Francis	30–35	35–55
Kaplan, Bulb and Propeller	30–50	30–65
Deriaz	20–45	35–65

11. Maintenance: Guide vanes/nozzles and runner blades are vulnerable to damage by cavitation and silt necessitating regular repair. Francis and large Kaplan turbines can permit, by design, replacement of all runner blades without dismantling generator. Pelton turbines have the merit of replacing spear and nozzle quickly.

10.10 CHOICE BETWEEN FRANCIS AND KAPLAN TURBINES

The various factors considered in making a choice between Francis and Kaplan turbines in an overlapping and marginal case are briefly as under:

1. Initial Cost: For the same head and rated power in the overlapping region, Kaplan turbine dimensions are bigger than Francis, notwithstanding even the higher specific speed values. The rock excavation, structural works, equipment and installation costs for the Kaplan are higher than Francis unit of the same size of power for same head due to the bigger turbine dimensions, bigger centre to centre spacing of units and overall more size of power house than in the latter case. Further, Kaplan turbines are to be set deeper with regard to minimum tail water level. Since while the Thoma cavitation factor σ versus N_s curve for the two types show the same slope, average σ values are higher by approximately 10% than for Francis with the same N_s value.

2. Efficiency: Francis turbine has better peak efficiency at full load and nearby range than Kaplan. But in higher specific speed zone, Francis shows better part load efficiencies only than propeller but is not as good as Kaplan turbine. Efficiency at part load operation and requirement of operation at part load is briefed under that head.

3. Running and Maintenance Cost: The present trend is to favour Francis rather than Kaplan in the overlapping feasible region. Controlling, operating and maintenance of Kaplan is more troublesome and expensive than Francis turbine. The control gear of Kaplan blades is more complicated and requires additional maintenance compared to Francis type where this mechanism does not exist. Recent practice is to change the old Kaplan runners with Francis type in such overlapping regions.

4. Head Variation: In order to ensure a normal runner life with minimum of cavitation and to achieve reasonable efficiency levels, head variation is restricted as given in Table 10.4.

With the range of permitted head variation, Francis and Kaplan turbines both suit the requirement.

4. Part-Load Operation: Turbines have maximum efficiency for only one load, as determined by the study of the performance or characteristic curves of the turbines. In case these are to be operated at part loads, the turbines the efficiency of which do not fall appreciably with part loads are selected. The curve of efficiency vs power output for a given turbine operated at varying loads (varying gate openings) has different characteristic forms for various types of turbines and specific speeds (Fig. 10.9).

Part-gate efficiency of medium and low specific speed wheels, as are used for medium and high head plants, is better than that of high specific speed wheels used for low heads. Part-gate efficiency is more important in medium and high head plants as water is a precious commodity. The lighter loads can be handled efficiently in case one or more Kaplan adjustable blade wheel units are included in the installation.

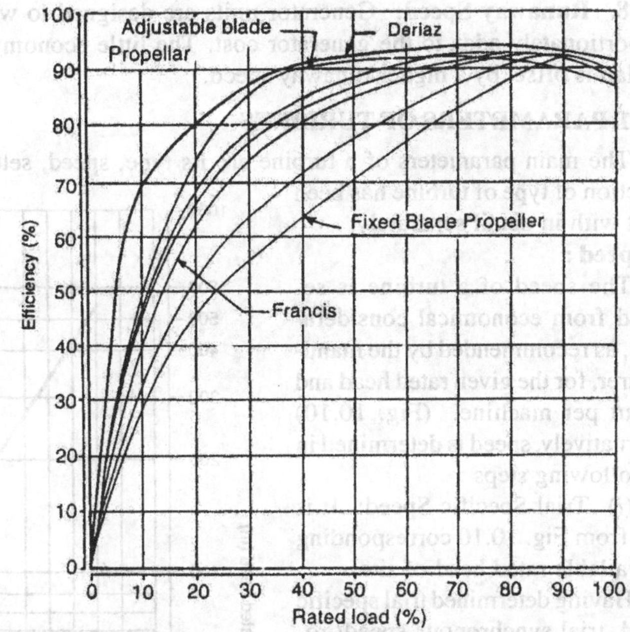

Fig. 10.9 Efficiency vs load for various types of hydraulic turbines.

In general, the efficiency at part-loads or overloads is less than normal and falls off at different rates depending on the type of turbine and specific speed of the wheel.

Kaplan and Pelton turbines give better performance than Francis and Propeller turbines at part load. Kaplan turbine gives high efficiency compared to Pelton over a considerable range of load. Low speed Pelton and Francis turbines give sustained efficiency with high part-gate opening and over-gate efficiencies are suitable for plants serving to regulate a power system and taking load variation. On the other hand, high speed Francis turbine has low part-load efficiencies and as such is suitable to operate within a narrow range of outputs. However, in low head run-of-river plants, with limited pondage and wherein water is surplused through spillway under reduced load condition, the low part-gate efficiencies are of little consideration but if high degree of regulation is desired, Kaplan turbine is more suitable than even low-speed Francis or Pelton turbines. Kaplan is also suitable to operate under reduced heads while maintaining good power output in low head plants with wide load variations.

The choice of turbine is also influenced by the variability of load if the head lies between 150 to 300 m or below 30 m. For higher range of heads Pelton turbine is preferable for part load operation than Francis turbine. Despite high initial cost, Kaplan turbine is preferred for part-load operation for heads below 30 m in comparison to propeller turbine.

5. Sediment Content: Sediment in water may also dictate suitability of the type of turbine. Francis turbine under high head and on a river carrying silt laden water is not suitable because its runner is damaged by erosive action of silt in water.

6. Hydraulic Thrust: The thrust bearing of an axial flow runner (Kaplan unit) is more heavily loaded than that of corresponding Francis runner and, therefore, Kaplan is expensive and thrust bearing always stressed.

7. Transportation of Castings: Kaplan unit becoming bigger will be expensive and difficult in transportation.

8. Runaway Speed: Generator units are designed to withstand full runaway speed which proportionately adds to the generator cost. The little economy due to higher operating speed of Kaplan is offset by a higher runaway speed.

10.11 PARAMETERS OF TURBINES

The main parameters of a turbine are its type, speed, setting, runner and spiral casing. The selection of type of turbine has been dealt with in detail separately.

1. Speed:

The speed of a turbine is selected from economical considerations, as recommended by the manufacturer, for the given rated head and output per machine. (Fig. 10.10) Alternatively, speed is determined in the following steps :

(*i*) Trial Specific Speed: It is read from Fig. 10.10 corresponding to available rated head of site.

Having determined trial specific speed, trial synchronous speed/ rotational speed n', is computed from the formula :

$$n' = \frac{n_s' H^{5/4}}{\sqrt{P \times 1.358}}$$

(*ii*) Rotational/synchronous/ rated speed (in r.p.m): It is determined from the formula:

Rated speed, $n = (60 \times f)/p$ (r.p.m). Where, f is frequency in cycles per second (In Indian power systems, frequency is 50 cycles/s), and p is number of pairs of poles.

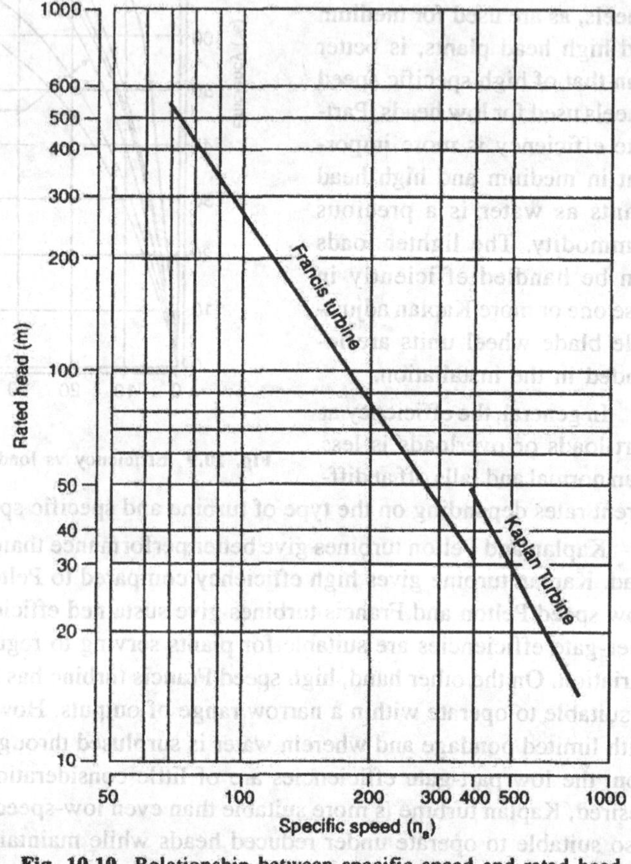

Fig. 10.10 Relationship between specific speed and rated head

Having determined the rated speed, the specific speed is determined from the formula:

$$n_s = \frac{n\sqrt{P \times 1.358}}{H^{5/4}}$$

where, n_s is specific speed of turbine (in r.p.m), n is rated speed of turbine (r.p.m), P is turbine output (kW), and H is rated head (m).

Further, selection of speed is governed by head variation, as given in Table 10.8 :

Table 10.8 Head Variation Vs Speed.

Head variation from design head (%)	*Speed chosen*
< 10	Next greater speed
> 10	Next lower speed

2. Turbine Setting:

The turbine setting, in a reaction turbine, with respect to minimum tail water level is fixed keeping in view the consideration of cavitation. The suction height of distributor centre line above minimum tail water level is computed from the formula:

$$H_s \leq H_b - \sigma H - H_v$$

where, H_s is section head (m), H_b is barometric pressure (m) of water column, H_v is vapour pressure; value of $H_b - H_v$ can be read from Fig. 10.11, for given altitude above mean sea level and given temperature, usually taken as 30°C, σ is Thoma's cavitation coefficient (Fig. 10.12).

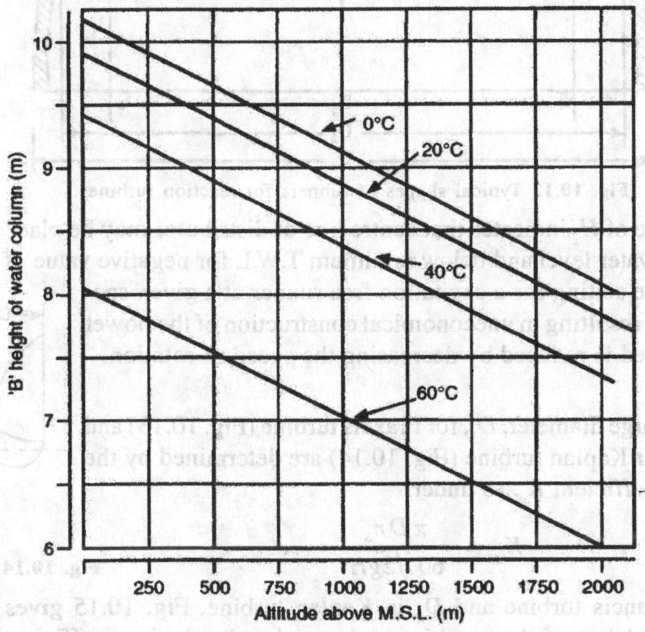

Fig. 10.11 Height of barometric water column at different temperatures of water and altitudes above sea level.

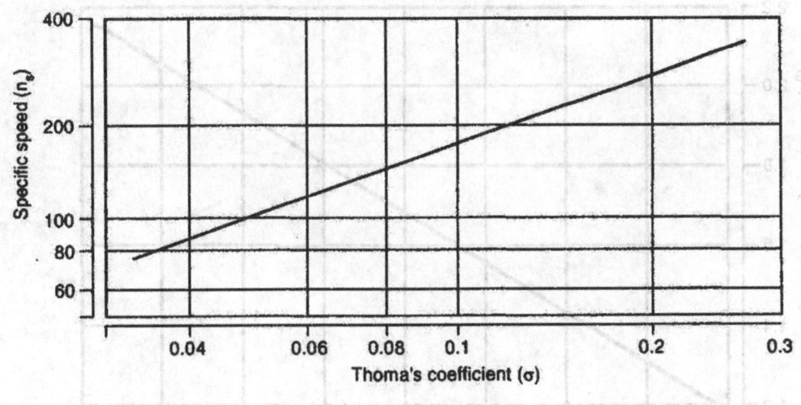

Fig. 10.12 Thoma's coefficient for Francis turbine.

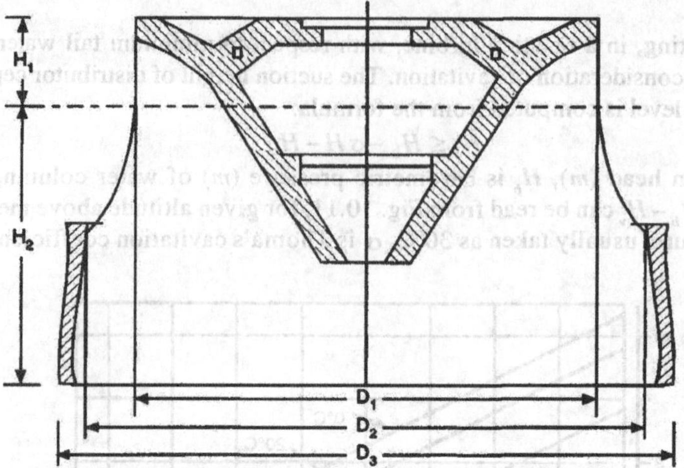

Fig. 10.13 Typical shapes of runners for reaction turbines.

The positive value of H_s indicates that centre line of distributor may be placed up to H_s metres above minimum tail water level and below minimum T.W.L for negative value of H_s.

In case the turbine setting for a cavitation free runner at a given specific speed is very low resulting in uneconomical construction of the power plant, the specific speed is reduced by decreasing the speed of rotation.

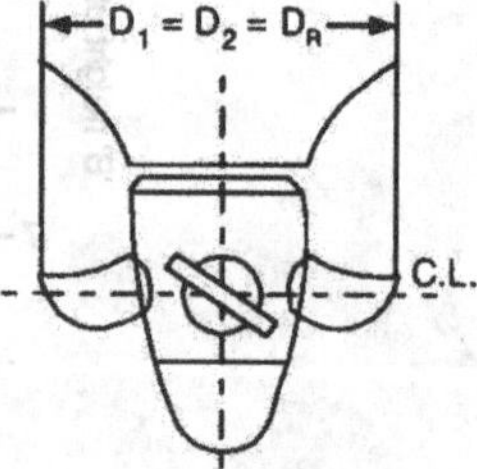

Fig. 10.14 Kaplan Turbine

3. Runner

The runner discharge diameter, D_3, for Francis turbine (Fig. 10.13) and runner diameter D_R for Kaplan turbine (Fig. 10.14) are determined by the peripheral velocity coefficient K_u, as under:

$$K_u = \frac{\pi D n_s}{60\sqrt{2gH}}$$

where, D is D_3 in Francis turbine and D_R in Kaplan turbine. Fig. 10.15 gives the relationship between specific speed (n_s) of the machine and peripheral velocity coefficient (k_u) for Kaplan turbine and Fig. 10.16 for Francis turbine.

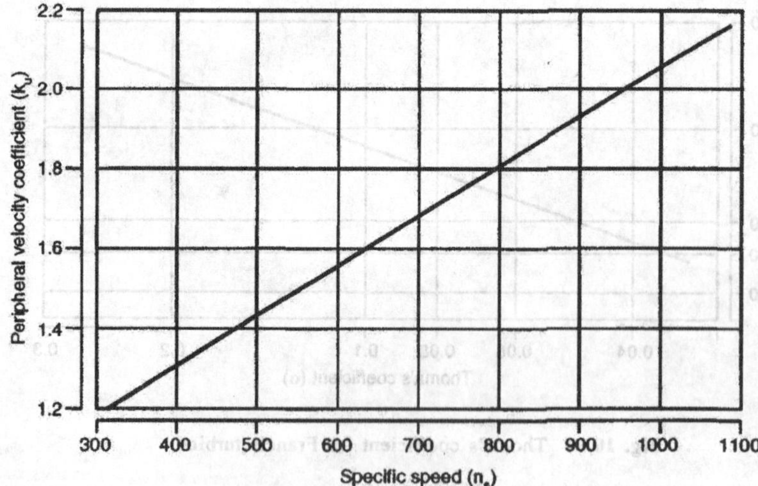

Fig. 10.15 Relationship between specific speed (n_s) and peripheral velocity (k_u) for Kaplan turbine.

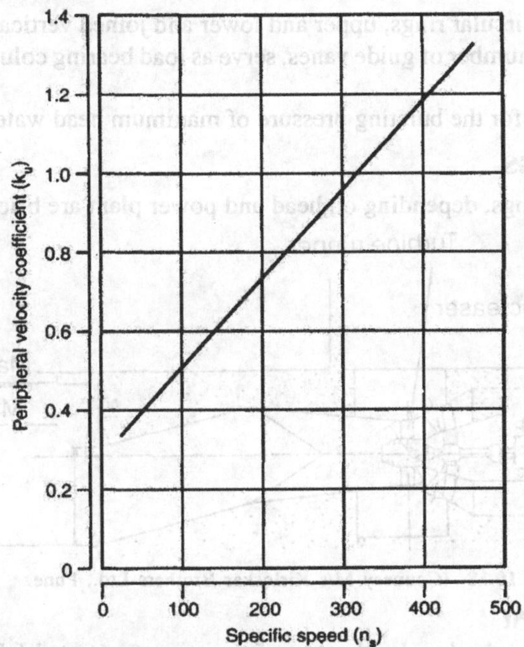

Fig. 10.16 Relation between specific speed (n_s) and K_u (Francis turbines)

The other runner dimensions of Francis turbine indicated in Fig. 10.13 are obtained with respect to the diameter D_3 and specific speed n_s from the curves shown in Fig. 10.17.

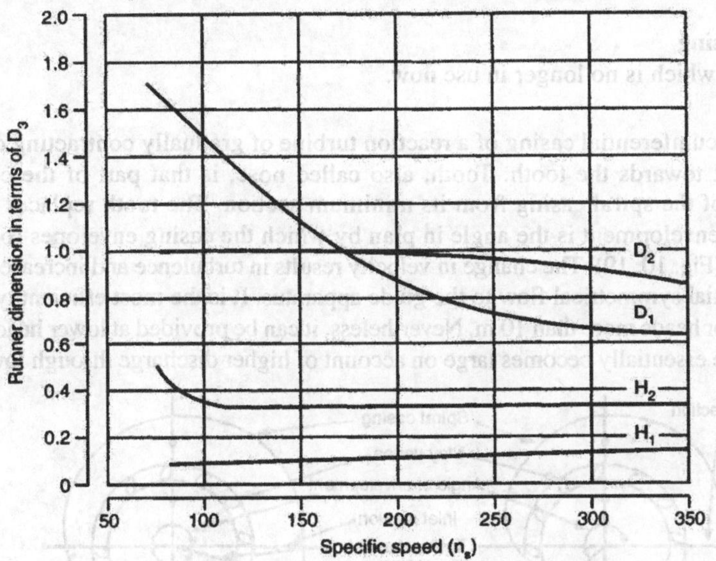

Fig. 10.17 Runner dimensions (Francis turbines)

10.12 TURBINE CASING

Turbine casing is the water passage surrounding the turbine. Water from the penstock is delivered to the casing which distributes it around the stay ring, wicket gates and runner. It is a winding duct reducing in cross section and height along its length with opening on the runner side to distribute water evenly around the guide vanes and at the same time avoiding eddies formation. Inside the casing speed ring is placed which transmits the load of runner to the foundation of the turbine. The

speed ring consists of two circular rings, upper and lower and joined vertically by stay vanes. The stay vanes, usually half the number of guide vanes, serve as load bearing columns as also of guiding water to the water vanes.

The casing is designed for the bursting pressure of maximum head water plus water hammer.

10.13 TYPES OF CASINGS

The three types of casings, depending on head and power plant are briefly as under:

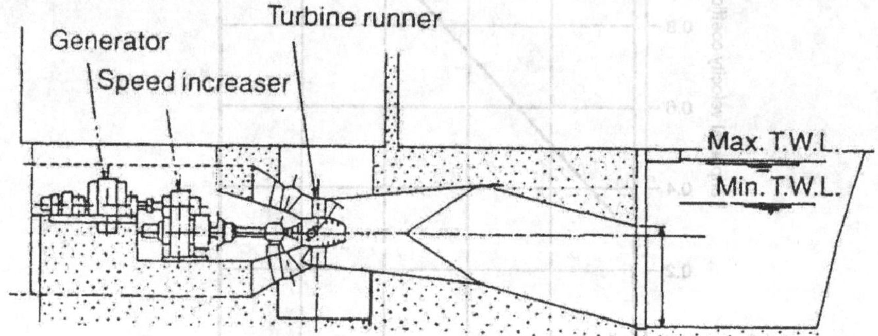

Fig. 10.18 (Counesy M/s Kirloskar Brothers Ltd., Pune.

1. Open Flume or Wheel Pit

For small installations under low heads, up to 5 m, on concentrated falls scroll casing is not required and instead open flume settings or simple concrete pit closed from top, usually at a level below minimum tail water level with an assured minimum cover over turbine is adopted. Water flows directly from it into the guides (Fig. 10.18). It is no longer in use as tubular turbines are now used for low heads.

2. Cylindrical Casing

It is the casing which is no longer in use now.

3. Spiral Casing

It is a fixed, circumferential casing of a reaction turbine of gradually contracting cross section from point to point towards the tooth. Tooth, also called nose, is that part of the casing which separates the inlet of the spiral casing from its minimum section. The tooth replaces the last stay vane. The angle of envelopment is the angle in plan by which the casing envelopes (or surrounds) the guide apparatus (Fig. 10. 19). The change in velocity results in turbulence and increased losses. The spiral provides an axial symmetrical flow to the guide apparatus. It is the most efficient type of casing and is widely used for heads more than 10 m. Nevertheless, it can be provided at lower heads with better efficiency but its size essentially becomes large on account of higher discharge through low head units.

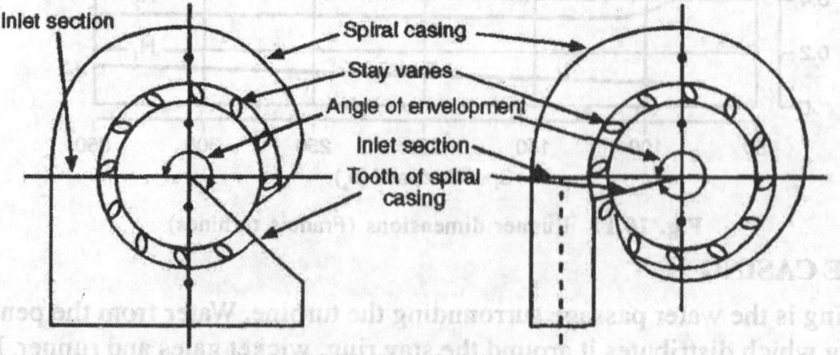

Fig. 10.19 Spiral Casing.

The scroll casing which has spiral sector 315° and above is termed as full spiral casing. The partial casing has spiral sector less than 315°.

The spiral casing provides a uniform flow to the wicket gates and on this account the cross section of the casing, as also its height, is essentially reduced along its length in proportion to the decreasing flow rate passing a given section with a view to maintain constant velocity. The cross sectional area is determined by computing discharge at any section by the relation:

$$q = \frac{Q\theta^\circ}{360^\circ} \quad \text{(cumecs)}$$

and working out area assuming mean velocity as $0.2\sqrt{2gH}$, where Q is flow through penstock.

Velocity in the inlet section of the spiral casing, taken as the design velocity, depends on the rated head and is determined from the formula:

$$V \leq K\sqrt{H}$$

where, V is design velocity (m/s), K is coefficient, value read from Fig. 10.20 for different heads and turbines and H is rated head (m)

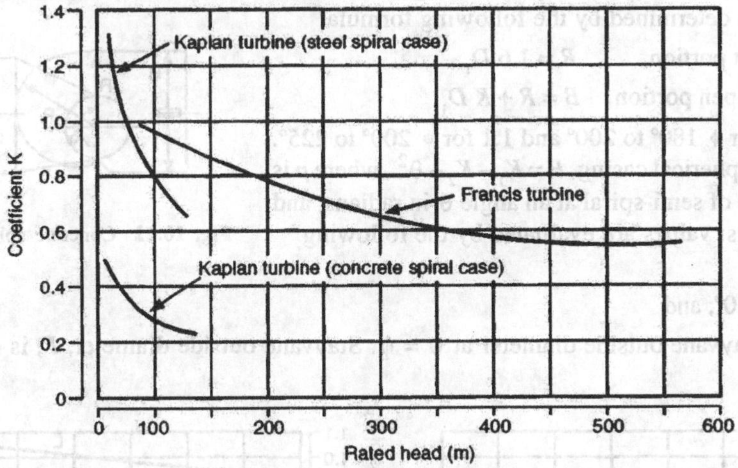

Fig. 10.20 Coefficient *K* vs rated heads.

10.14 CLASSIFICATION OF SPIRAL CASINGS

The spiral casings may be classified as illustrated below:

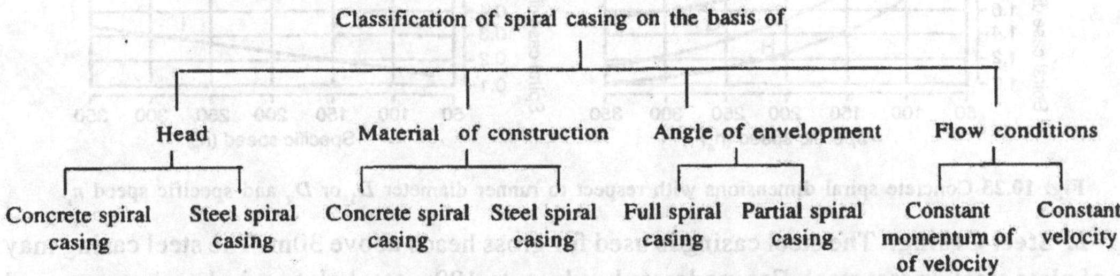

1. Concrete Spiral Casing: It is the casing generally used for low heads upto 40 m with low allowable velocities. When used for higher heads, thin steel lining is provided to cater for high velocities and water tightness, but the liner is not designed to take loads. The casing essentially has sufficient concrete thickness and reinforcement for the worst possible combination of loading conditions to ensure water tightness and freedom from undesirable vibrations. In order to keep block width within limits, the casings do not normally envelope the guide apparatus fully and the angle of envelopment used is generally from 180° to 225°. The casing is generally of trapezoidal section (Fig. 10.21) on account of easy construction compared to circular or elliptical shapes. Concreting of the casing is done in stages with suitable construction joints provided with water seals. As far as possible, simultaneous concreting the sections symmetrically placed about the centre line of the spiral casing is done to reduce chances of shifting of the stay ring.

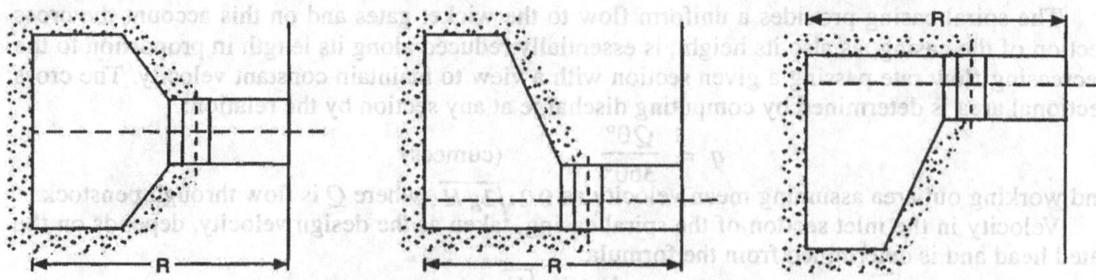

Fig. 10.21 Trapezoidal shapes of spiral casing.

Fig. 10.22 indicates concrete spiral casing. The salient dimensions can be determined by the following formula:

Radius of inlet portion, $R = 1.6\,D_1$

Width of the open portion, $B = R + K\,D_1$

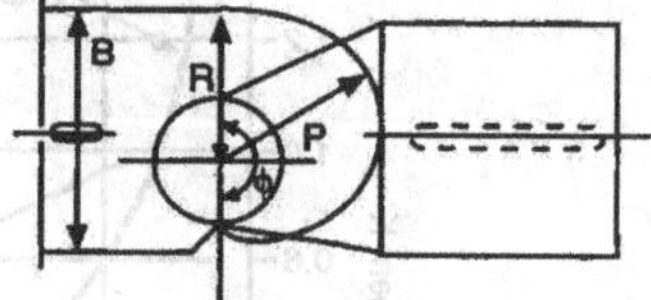

Fig. 10.22 Concrete spiral casing.

where, $K = 0.95$ for ϕ 180° to 200° and 1.1 for ϕ 200° to 225°. Equation for semispherical casing, $P = K_1 - K_2 - \theta^2$, where p is radius of curvature of semi-spiral at an angle θ in radians, and K_1, K_2 are constants: values are evaluated by the following conditions:

$P = R$ at $\theta = 0°$, and

$P = 0.5 \times$ stayvane outside diameter at $\theta = \phi$. Stayvane outside diameter, F, is determined from Fig. 10.23

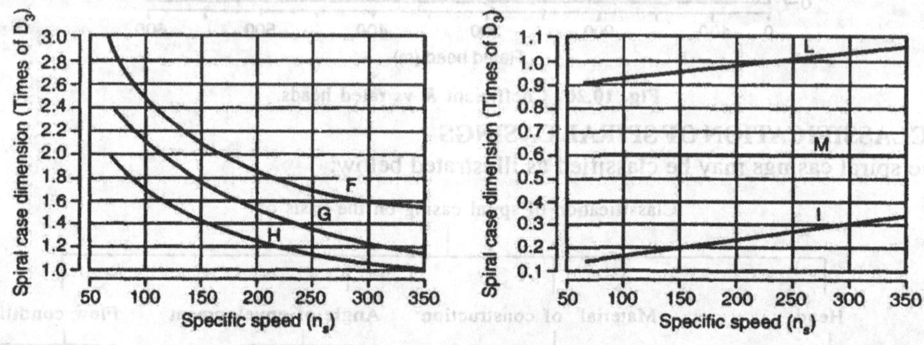

Fig. 10.23 Concrete spiral dimensions with respect to runner diameter D_3 or D_R and specific speed n_s.

2. Steel Casing: The steel casing is used for gross heads above 30m. The steel casing may be of plate steel or cast steel. For moderate heads up to 100m steel plate spiral casings are used and cast steel casings for higher heads. The steel casings are generally of circular or circular-cum-elliptical sections. The major dimensions of the casing indicated in Fig. 10.24 may be obtained as a function of n_s referred to runner diameter D_3 or D_R from the curves shown in Fig. 10.25 and 10.23. For large sizes, the casings are made in several segments, strutted properly to prevent loss of shape during transportation and are welded with the stay ring at site.

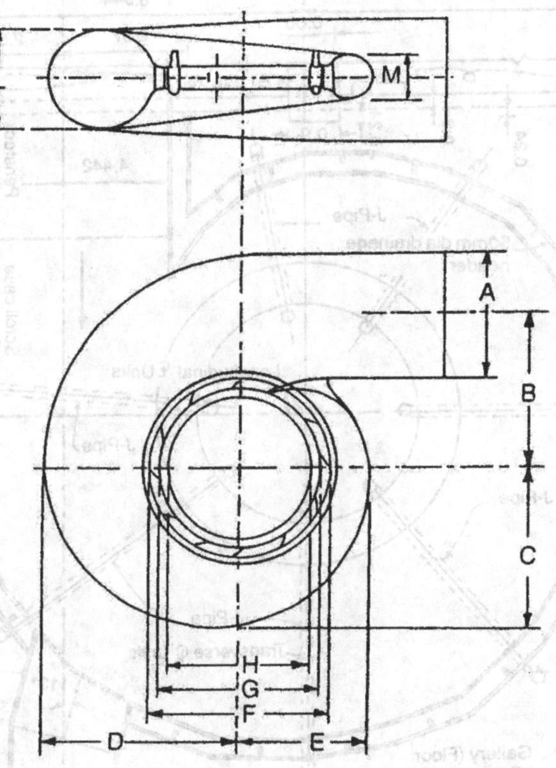

Fig. 10.24 Major dimensions of the spiral casing.

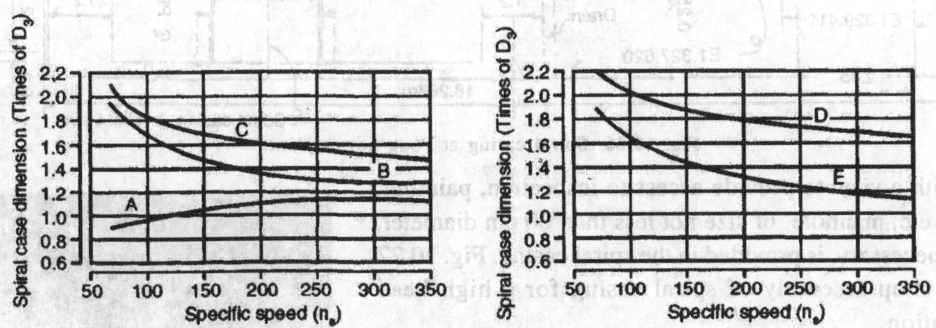

Fig. 10.25 Casing dimensions with respect to runner diameter D_3 or D_R and specific speed n_s.

The casing in a unit with vertical setting is installed as (*i*) Embedded in concrete, or (*ii*) Embedded in concrete with an elastic packing between the casing and the concrete on the upper half, or completely exposed and supported on spillar foundation, or half embedded. Fig. 10.26 shows scroll casing provided for 4.44 m diameter penstock at Pong power house.

The casing is designed for the maximum head including pressure rise due to water hammer in the casing. The casing may be divided into a number of sections and each section designed separately as a toroidal shell of uniform section identical to the given spiral casing section.

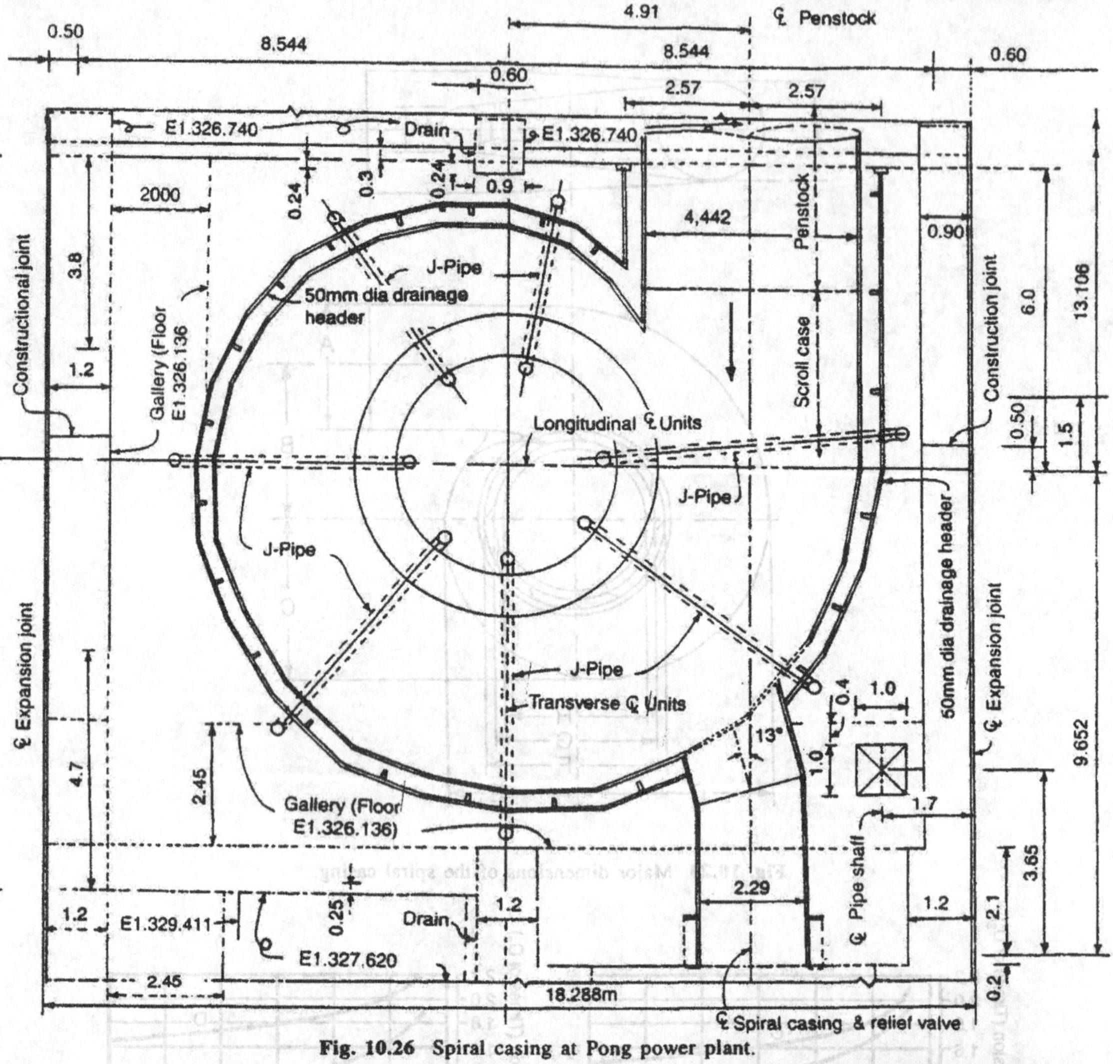

Fig. 10.26 Spiral casing at Pong power plant.

With a view to provide access to inspection, painting, repairs etc, manhole of size not less than 60 cm diameter, where necessary, is provided in the spiral casing. Fig. 10.27 shows shop assembly of spiral casing for a high head installation.

Fig. 10.27 Shop assembly of spiral casing and guide apparatus for a high head turbine installation.

10.15 DRAFT TUBES

The draft tube, an integral part of a reaction turbine, is the conduit connecting the exit from the runner to the tailrace to improve efficiency of the plant. It is not required for impulse turbines and the head from the nozzle to the tail water is necessarily lost.

The two functions of a draft tube are (*i*) It allows the setting of the pressure turbine and the connecting machinery above the tail water level under high-water or flood conditions of river without sacrificing the elevation difference, and (*ii*) To recover large part of the velocity head of water leaving the turbine and utilise it by increasing the total pressure difference on the runner.

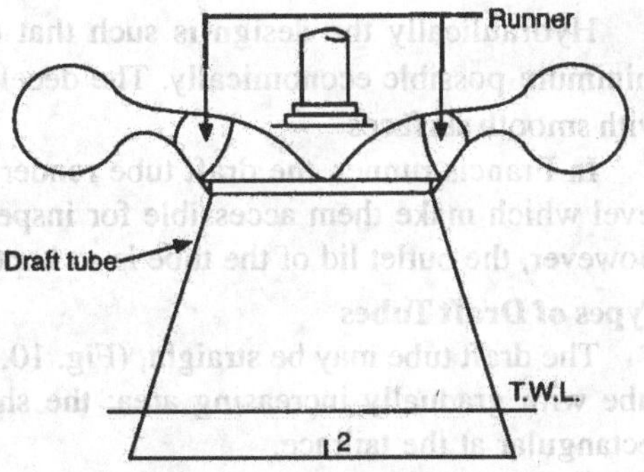

Fig. 10.28 Draft tube functioning.

The draft tube decelerates the flow gradually with a minimum of losses by means of a progressively increasing cross sectional area of the tube, the flare limited to 1 in 10, so that a large part of the residual kinetic energy remaining in the runner discharge is reconverted into effective head which is utilised through a reduction in the static pressure against which the runner discharges. Thus it functions to utilise the differential elevation between the runner exit and the tail water level, termed as static suction head, recovering as much as possible of the velocity head in the water leaving the runner. The energy recovered in the draft tube is substantial in high speed turbines.

The recovery of velocity head at runner outlet is achieved because of subatmospheric pressure developing at runner exit due to the introduction of draft tube. Referring to Fig. 10.28 and applying Bernaullis equation between 1 and 2:

$$\frac{P_1}{w} + h_s + \frac{V_1^2}{2g} = \frac{V_2^2}{2g} + h_{1-2}$$

or

$$\frac{P_1}{w} = -h_s - \frac{V_1^2 - V_2^2}{2g} + h_{1-2}$$

Draft tube efficiency,

$$\eta_{dt} = \frac{1}{\frac{V_1^2}{2g}} \left(\frac{V_1^2 - V_2^2}{2g} - h_{1-2} \right)$$

or

$$\frac{P_1}{w} = -h_s - \eta_{dt} \frac{V_1^2}{2g}$$

where, h_{1-2} is loss of head between the points 1 and 2. As P_1 is the pressure at the outlet of the runner, the turbine overpressure increases by the value of its negative value. The draft tube efficiency increases the overall efficiency of the turbine.

The layout of a draft tube is indicated by the turbine manufacturer to suit the turbine operating and physical requirements i.e. spacing and setting of units, depth of foundations and tailrace water level.

The dimensions of the draft tube depend on the specific speed, size and spacing of the unit. At the inlet, the draft tube diameter is fixed by the diameter of the runner. At the exit, the dimensions of the draft tube are fixed by the limits of velocity which varies from 0.1 $\sqrt{2gH}$ for high specific speed turbines to 0.02$\sqrt{2gH}$ (m/s) for low specific speeds. The cross sectional area of the draft tube may increase towards exit. The ratio of inlet to outlet diameter of the tube dictates its length to get a desirable flare of about 25 cm per metre length of the tube.

Hydraulically the design is such that the total losses in the draft tube and exit losses are minimum possible economically. The deceleration has to be gradual so that transition is gradual with smooth surfaces

In Francis runner, the draft tube renders it possible to set the wheel above normal tail water level which make them accessible for inspection without resorting to dewatering of the tailrace. However, the outlet lid of the tube is so kept that it is sealed under worst low water conditions.

Types of Draft Tubes

The draft tube may be straight, (Fig. 10.29) conical or with a circular section or elbow-shaped tube with gradually increasing area; the shape changing from circular at the runner section to rectangular at the tailrace.

1. Straight Conical: **It is the simplest type of draft tube (Fig. 10.30). It has the shape resembling frustrum of a cone, the vertex angle of the cone of divergence less than 10° so as to avoid separation of flow from the walls of the tube. But the length of the tube has to be such as to ensure exit velocity of about 1 m/s. Since the length required is very long in terms of throat diameter its use is limited to relatively smaller turbines. Large turbines entail excessive excavation.**

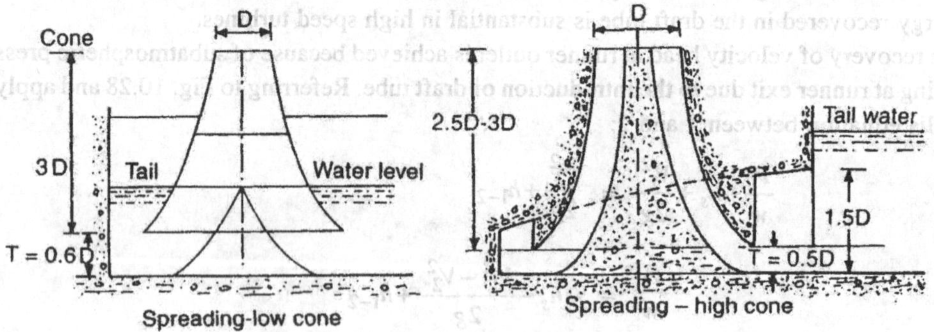

Fig. 10.29 Types of draft tubes

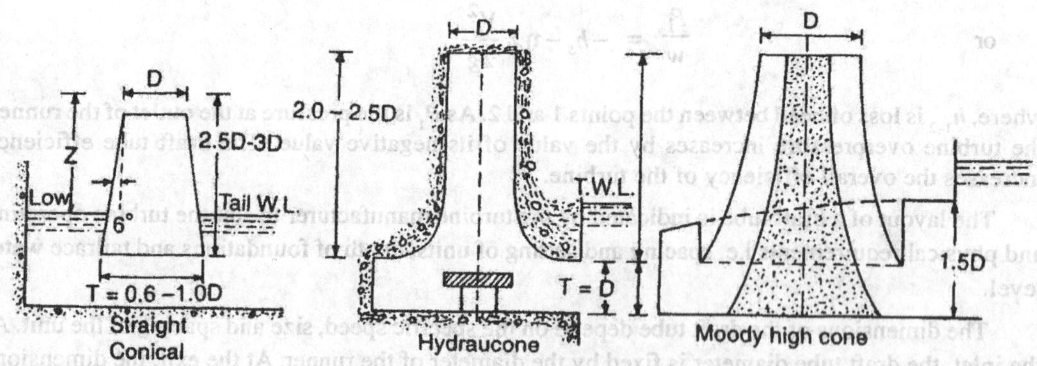

Fig. 10.30 Types of draft tubes.

The draft tube inlet is essentially set at a limited distance Z above the tail water level to ensure that no where within the turbine the absolute pressure drops to the vapour pressure of water.

It is the most efficient type of draft tube as about 90% of the energy is converted into potential head.

2. Hydraucone Type: It is a modification of conical tube by placing a horizontal baffle under the exit end of the pipe to reclaim some of the energy of whirl and reduce exit loss (Fig. 10.30). This type of draft tube provides a central solid cone to get a eddy-free flow. It utilises the hydraucone action of a water jet impinging on a plate, preferably concentric with the axis of the entering flow, and being deflected therefrom in all directions. The bottom of the tube is usually considerably lower than river or stream bed and is joined with a reverse slope of 5:1. Moody or spreading draft tube belongs to hydraucone type of draft tubes. It has the shape of a falling jet on a plate. It better utilises the whirl component of water entering the draft tube and has better efficiency as it intends to regain some of the energy lost in whril. The hydraucone effect is substantially obtained by an actual concrete cone constructed symmetrically in the tube in the shape of an inverted free vortex. The total length of the tube is less compared to the other types.

Elbow Type Draft Tube: It is an efficient type of draft tube and is generally preferred in the power plants where vertical draft tube involves excessive excavation (Fig. 10.31). It permits change of section from circular at the runner end to rectangular by introducing a transition section at the bend of the tube. A tail gate may be provided at the outlet end of the tube to facilitate complete dewatering of the turbines for repairs.

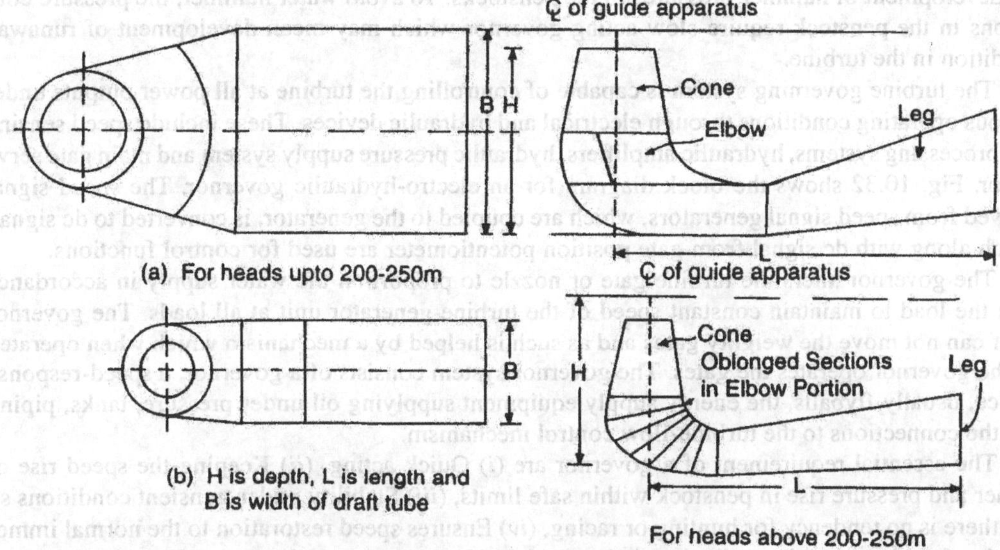

Fig. 10.31 Elbow draft tube.

The salient dimensions of the tubes (Fig. 10.30) as per IS: 5496: 1993 are in accordance with the values of *H*, *L* and *B*, as under:

(*i*) For inlet cone, the half angle of conicity is 6° to 10°. A bed slope of about 10:1 (vertical: horizontal) is provided subject to a maximum of 6:1.

(*ii*) The outlet end of the tube is so located that it remains submerged under all operating conditions of the unit at least by $V_e^2/2g$ or 30 cm whichever is greater, V_e is exit velocity. The slope of excavation connecting the tube floor at exit and bottom of the tailrace should not be steeper than 4:1.

(*iii*) The height of the tube at exit is usually 0.94 to 1.32 times the diameter of the runner, depending upon the specific speed of the turbine, the lower value being for lower specific speeds.

(*iv*) Depth of the tube is reckoned from the centre line of the guide apparatus. For Kaplan turbines normally a depth of 2.3 to 3.0 D is adopted depending upon specific speed of the turbine. For Francis turbines, a depth of 2.5 to 3.3 D is adopted.

(*v*) The length of the tube, measured from the tube exit, is normally 4 to 5 times the runner inlet diameter.

(*vi*) The basic width (excluding the pier of minimum width of 1.0 m) is 2.6 to 3.3 D.

Bulkhead gates are provided to close the draft tube against tail water pressure to facilitate maintenance, by a gantry operated from a suitable operating platform.

10.16 GOVERNING OF TURBINES

The water-wheels are essentially constant speed machines. The instantaneous variation in speed corresponding to variation in load is regulated by means of a governor. The turbines are essentially required to be so run that the frequency of generation is maintained constant by keeping the speed constant or limiting speed drop between no load and full load to less than 5 percent and adjusting the load on the machine accordingly. The load variation is accompanied by change in quantity of water passing through the turbine which means pressure changes in the water conveyance system. The maximum pressure rise results in an instantaneous closing of the turbine gates or of the valve located near the turbine and hence faster acting of the governor. Rapid variations in flow result in the development of hammer pressures in the penstocks. To avoid water hammer, the pressure conditions in the penstock require slow acting governor which may mean development of runaway condition in the turbine.

The turbine governing system is capable of controlling the turbine at all power outputs under various operating conditions through electrical and hydraulic devices. These include speed sensing and processing systems, hydraulic amplifiers, hydraulic pressure supply system and main gate servo motor. Fig. 10.32 shows the block diagram for an electro-hydraulic governor. The speed signal derived from speed signal generators, which are coupled to the generator, is converted to dc signal, which along with dc signal from gate position potentiometer are used for control functions.

The governor alters the turbine gate or nozzle to proportion the water supply in accordance with the load to maintain constant speed of the turbine-generator unit at all loads. The governor itself can not move the weighty gates and as such is helped by a mechanism which when operated by the governor operates the gates. The governor system consists of a governor, a speed-response device, usually flyballs, the energy supply equipment supplying oil under pressure, tanks, piping and the connections to the turbine-flow control mechanism.

The essential requirement of a governor are (*i*) Quick acting, (*ii*) Keeping the speed rise of runner and pressure rise in penstock within safe limits, (*iii*) Stability under transient conditions so that there is no tendency for hunting or racing, (*iv*) Ensures speed restoration to the normal immediately after alteration due to load variation, and (*v*) No serious water hammer effect should result.

Certain terms in the context of governor functioning are as under:

Sensitivity: It is the degree of speed change required for the turbine-gates to make a responsive move.

Stability: It is the ability of the governor to avoid any false or excessive movement of the turbine gates.

The sensitivity and stability are related rather are contradictory to each other. A highly sensitive governor loses its stability that is it becomes so stable that it does not respond to speed changes.

Compensation: A governor has a tendency to move turbine gates farther than necessary to return the speed to normal after a change of load and hence speed change due to momentum of the gate mechanism as also the time needed to accelerate or decelerate the water moving to the turbine. A governor should compensate for this tendency so that hunting or racing does not result. The compensating device consists of a dashpot, generally combined with springs, interposed in the relay mechanism and operated from the regulating cylinder.

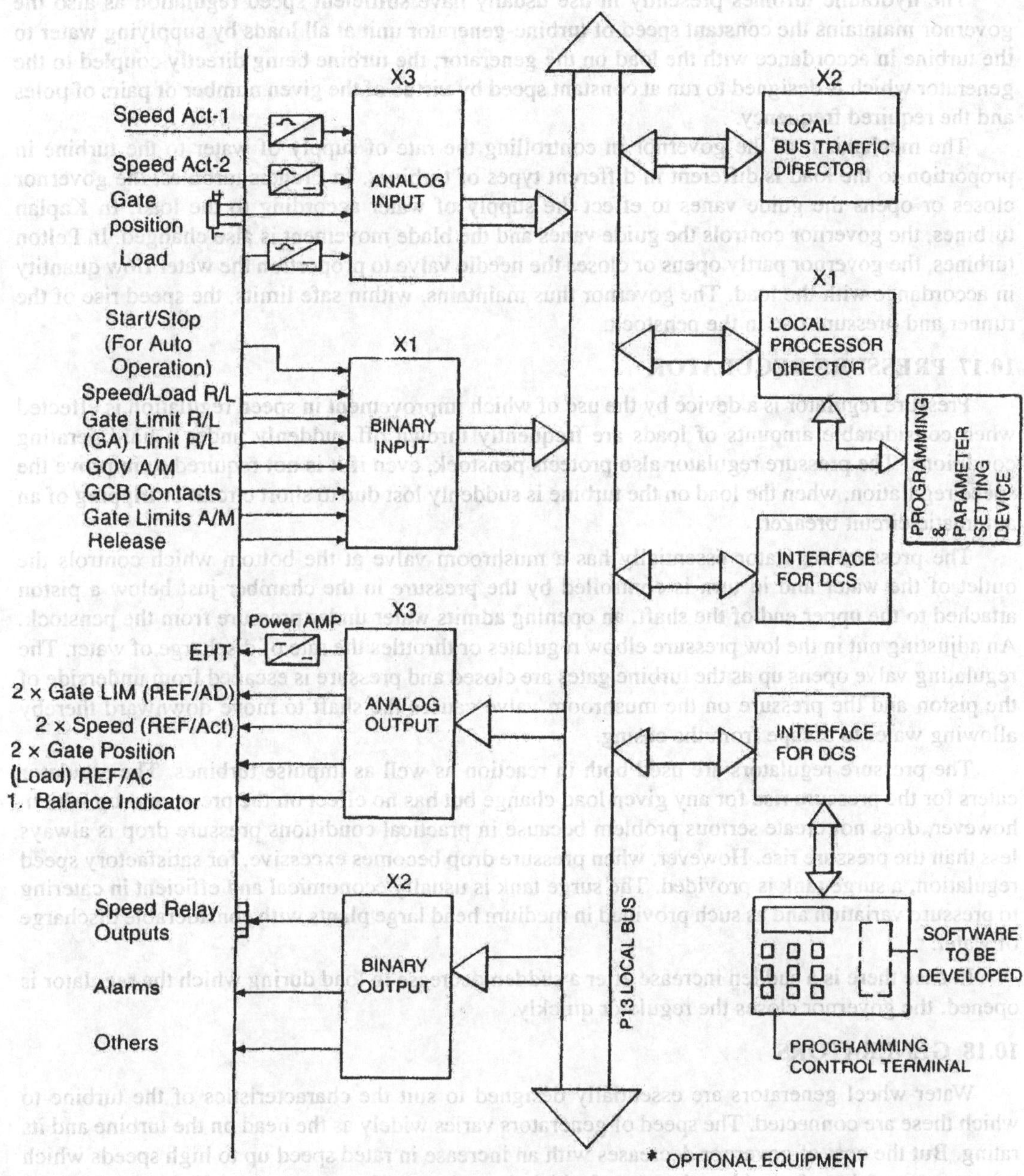

Fig. 10.32 Block diagram for EHG system
(Courtesy M/S BHEL, Bhopal)

The compensation as such may be defined as a modification of the gate motion introduced to prevent its overtravel and thereby produce stability. The compensation may be classified as primary and secondary. The temporarily retardation of the gate movement to prevent overtravel is termed as primary compensation. The secondary compensation is the gradual neutralisation of the primary compensation by means of yielding of the dashpot under the reaction of its spring, which takes place while the speed recovery is restoring the governor head to the position when it is in stable condition.

The hydraulic turbines presently in use usually have sufficient speed regulation as also the governor maintains the constant speed of turbine-generator unit at all loads by supplying water to the turbine in accordance with the load on the generator; the turbine being directly coupled to the generator which is designed to run at constant speed by virtue of the given number of pairs of poles and the required frequency.

The mechanism of the governor in controlling the rate of supply of water to the turbine in proportion to the load is different in different types of turbines. In Francis turbines, the governor closes or opens the guide vanes to effect the supply of water according to the load. In Kaplan turbines, the governor controls the guide vanes and the blade movement is also changed. In Pelton turbines, the governor partly opens or closes the needle valve to proportion the water flow quantity in accordance with the load. The governor thus maintains, within safe limits, the speed rise of the runner and pressure rise in the penstock.

10.17 PRESSURE REGULATOR

Pressure regulator is a device by the use of which improvement in speed regulation is effected when considerable amounts of loads are frequently thrown off suddenly under usual operating conditions. The pressure regulator also protects penstock, even if it is not required to improve the speed regulation, when the load on the turbine is suddenly lost due to short circuit or tripping of an automatic circuit breaker.

The pressure regulator essentially has a mushroom valve at the bottom which controls the outlet of the water and in turn is controlled by the pressure in the chamber just below a piston attached to the upper end of the shaft, an opening admits water under pressure from the penstock. An adjusting nut in the low pressure elbow regulates or throttles the rate of discharge of water. The regulating valve opens up as the turbine gates are closed and pressure is escaped from underside of the piston and the pressure on the mushroom valve causes the shaft to move downward thereby allowing water to escape from the casing.

The pressure regulators are used both in reaction as well as impulse turbines. The regulator caters for the pressure rise for any given load change but has no effect on the pressure drop which, however, does not create serious problem because in practical conditions pressure drop is always less than the pressure rise. However, when pressure drop becomes excessive, for satisfactory speed regulation, a surge tank is provided. The surge tank is usually economical and efficient in catering to pressure variation and as such provided in medium head large plants with considerable discharge of water.

In case there is a sudden increase after a sudden decrease in load during which the regulator is opened, the governor closes the regulator quickly.

10.18 GENERATORS

Water-wheel generators are essentially designed to suit the characteristics of the turbine to which these are connected. The speed of generators varies widely as the head on the turbine and its rating. But the cost of governor decreases with an increase in rated speed up to high speeds which is in contrast to that of a turbine, the cost of which decreases and becomes more efficient as the rated speed decreases up to moderately low speed. The generators are usually designed for full runaway speed of the turbine. The generators are generally large due to the comparatively slow speed of revolution and the requirement of large WR^2 for stability.

The modern generators are of vertical direct-connected type. The excitation current for the magnetic poles on the rotor is furnished by power derived from the network. The starting and shut down of the units, controlled by a switch from control room in the power plant, is automatic. The time required for a generator to attain normal speed is 3 to 8 seconds, depending upon the WR^2 of the machine. The run out time that is the time to come to rest from the operating speed is 5 to 10 minutes.

The speed of a generator is fixed by the number of poles and the frequency of the supply. Thus low speed means more number of poles and increased size of the generator and hence higher cost. Conversely, higher speed means reduced size and cost of the generator but requires high quality materials for rotor and also be safe for the turbine runaway speed.

The usual features of a generator are (*i*) A field or assembly of magnets arranged to produce a magnetic flux, (*ii*) An armature or assembly of electric conductors arranged across the path of magnetic flux. The field and armature are so mounted that a relative motion is produced between the magnetic flux and the electric conductors inducing in the conductors an electromotive force on the application of mechanical force. The polarity of the electromotive force induced in the armature is alternating as the field poles are arranged alternatively positive and negative around the periphery of the generator.

In hydro power stations, alternating current synchronous generators are used since the output is transmitted over long transmission lines after transformation to suitable transmission voltage. It is also used where the output is distributed locally.

The generator coupled as it is to the turbine has magnetic poles mounted on the rotor and the stator produces power. The carrying capacity of the generator depends on the current carry capacity of the stator coils without undue heating. The stator coils supporting metallic frames are cooled to temperature within safe limits. The large size generators are of completely enclosed type and surface type coolers cool the circulating air after passing through the station. Nevertheless, extra power can be produced for a short period but at a higher temperature. The usual generation voltage in hydro power units is 11 kv to 15 kv; low voltage generator requires conductors of high current carrying capacity.

The use of induction generator for grid connected installations has the advantages in maintenance, robustness and cost-effectiveness, even though it draws reactive power from the grid. For isolated micro units up to 100 kW the newly developed self-excited induction generators are preferable. With the development, use of frequency controlled thyristors, electronic load controllers for up to 250 kW unit are useful for isolated units.

10.19 CHARACTERISTIC CURVES FOR TURBINES

The characteristic curves for a turbine are the plot of efficiencies for full range of its running. These are useful for the study to obtain conditions for the maximum efficiency of a turbine. These are generally provided by the manufactures of the turbine. These curves are relevant to the particular turbine. The curves are drawn with reference to the points obtained by testing the turbine for the various gate openings and at various speeds.

The performance characteristics of the turbine such as power, speed and quantity expressed in unit power, unit speed and unit quantity are plotted to obtain these curves.

Unit Power: It is the power developed under 1 m head, $P = C_1 H^{3/2}$. The coefficient C_1 varies with the efficiency of the turbine *i.e.* with the gate opening and speed.

Unit Speed: It is the speed when the turbine is running under a head of 1 m, $n = C_2 H^{1/2}$. The coefficient C_2 varies with the conditions of the turbine running.

Unit Quantity: It is the volume of water passing under 1 m head, $Q = C_3 H^{1/2}$. The coefficient C_3 depends on the condition of turbine running. These parameters render it possible to compare the performance of the turbine under different heads.

The turbine is tested for a particular gate opening; speed and head are varied and quantity and BHP are measured. The values of efficiency, unit power and unit speed are computed for various speeds and heads at that gate opening from the results of testing. Then a curve is plotted for this particular gate opening; power and unit speed as ordinates and the efficiency for each point is noted on the curve at that point (Fig. 10.33). Likewise, the efficiencies for various gate openings are ascertained from repeat tests and plotted. The lines of equal efficiency are drawn by interpolation after examining the efficiencies noted on each point. Such a plot is a contour plan known as characteristic curves for the particular turbine. The following important observations can be made from these curves:

1. The speed of the turbine which at any gate opening is to give best efficiency for that gate opening.
2. The maximum efficiency of the turbine for all conditions and gate opening and speed which produce this maximum efficiency.

This is the normal condition of running the turbine.

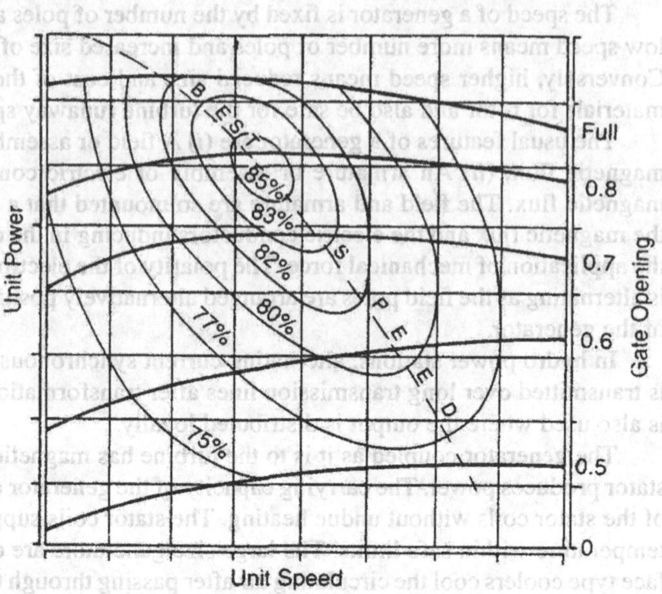

Fig. 10.33 Characteristic curves

10.20 MODEL TESTING OF TURBINES

Model is a small scale representation of the prototype. Model testing is a useful tool for the design of hydraulic machines. The performance of a prototype turbine is predicted by model testing of its small model similar to the actual on the basis of the principle of similarity. With the tests on the model, its horse power is measured under a known head and speed and measured discharge. The results of these tests are employed to compute the performance of the prototype turbine.

For similar turbines the parameters like V, v, V_r, V_f are proportional $\sqrt{H}$. Accordingly :

1. Head coefficient, $K_1 = H/n^2D^2$
2. Discharge coefficient, $K_2 = Q/nD^3$
3. Power coefficient, $K_3 = P/n^3D^5$

where, h is head (m), n^2 is specific speed (r.p.m), D is wheel diameter (m), Q is discharge (cumces) and P is power developed (kW).

SOLVED EXAMPLES

Example 10.1 *An inward flow reaction turbine has a flow of 0.5 cumec under a head of 20 m. The velocity of wheel periphery at inlet is 10 m/s. The outlet pipe is of 0.3 m diameter. The radial velocity of flow through the runner is the same as that in outlet pipe. Determine (i) vane angle at inlet, (ii) guide blade angle, and (iii) power generation in kW. Assume data not given.*

Solution.

Velocity of flow, $V_f = V_1 = \dfrac{0.5 \times 4}{\pi (0.3)^2} = 7.07$ m/s

$$\frac{V_w \cdot v}{g} = H - \frac{V_f^2}{2g}$$

$$\frac{V_w .10}{9.81} = 20 - \frac{(7.07)^2}{2 \times 9.81}$$

$$V_w = 17.12 \text{ m/s}$$

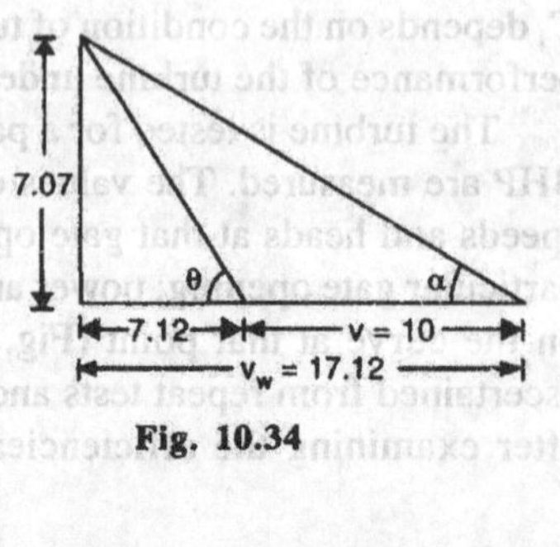

From Fig. 10.34, $\tan \alpha = 7.07/17.12 = 0.41$

Fig. 10.34

$$\alpha = 22.29^\circ$$

$$\text{Tan } \theta = 7.07/7.12$$

$$\theta = 44.80^\circ$$

Power output $= 9.81\, QH\eta$

$= 9.81 \times 0.50 \times 20 \times 0.85$ (Assume total $\eta = 0.85$)

$= 83.39$ kW.

Example 10.2 *Determine number of turbines and diameter of runner for a power plant having 20 cumecs inflow, 10 m head, turbine has efficiency 88%, speed 150 r.p.m. Assume specifc speed as 250 and speed ratio of runner as 0.85.*

Solution.

Power potential, $P = \dfrac{wQH\eta}{75} = \dfrac{1000 \times 20 \times 10 \times 0.88}{75} = 2347$ H.P.

Power developed by a turbine $N_s = \dfrac{N\sqrt{P}}{H^{5/4}}$

$$250 = \frac{150\sqrt{P}}{(10)^{5/4}}$$

or $P = 878$ H.P.

∴ No. of turbines required $= 2347/878 = 3$

Speed ratio, $0.85 = \dfrac{v}{\sqrt{2gH}} = \dfrac{1}{60} \times \dfrac{\pi DN}{\sqrt{2gH}}$

or $D = \dfrac{0.85 \times 60\sqrt{2 \times 9.81 \times 10}}{\pi \times 150}$

∴ Runner diameter, $D = 1.52$ m.

Example 10.3 *Determine type and number of turbines which can be installed at a stream with flow of 20 cumecs and fall 20 m. Assume overall efficiency 80%, speed 250 r.p.m. and $N_s = 600$.*

Solution.

Power potential $= \dfrac{20 \times 1{,}000 \times 20 \times 0.8}{75} = 4267$ H.P.

$$N_s = (N\sqrt{P})/H^{5/4}$$

$$600 = (250\sqrt{P})/(20)^{5/4}$$

or $P = 10{,}304$

Turbine type = Referring to Fig. 10.8 for a head of 20 m, $N_s = 600$, Kaplan propeller turbine is suitable.

Number of turbines $= 4267/10{,}304 \simeq 1$.

Examples 10.4 *A pelton wheel develops 5000 kW under 300 m head at 80% efficiency. Assuming coefficient of discharge 0.98, ratio of peripheral velocity of wheel to jet velocity 0.46, N_s 20, determine diameter of bucket circle, size of buckets.*

Solution.

Peripheral velocity, $V = 0.98\sqrt{2gH} = 0.98\sqrt{2 \times 9.81 \times 300} = 75.19$ m/s

Jet velocity, $v = 0.46\, V = 0.46 \times 75.19 = 34.59$ m/s.

Specific speed, $N_s = \dfrac{N\sqrt{P}}{H^{5/4}}$

or $$N = \frac{N_s H^{5/4}}{\sqrt{P \times 1.358}} = \frac{20(300)^{5/4}}{\sqrt{5000 \times 1.358}} = 303 \text{ r.p.m.}$$

For volume of water, $$P = Q\gamma H\eta$$

$$5000 = Q \times 9.81 \times 300 \times 0.80$$

$$Q = 2.12 \text{ cumecs}$$

$$v = \frac{\pi DN}{60}$$

or Runner diameter, $$D = \frac{60v}{\pi N} = \frac{60 \times 34.59}{\pi \times 303} = 2.18 \text{ m}$$

Area of jet $$= \frac{Q}{v} = \frac{2.12}{34.59} = 0.06 \text{ m}^2$$

Size of jet $$\frac{\pi d^2}{4} = 0.06$$

∴ Diameter of jet $$= 0.276 \text{ m.}$$

Example 10.5 *At a power plant a pelton turbine generates 15,000 kW at a head of 300 m, number of jets is 3. Assume overall efficiency 80%, exit blade angle 165°, velocity of water in jet is 95% of the theoretical velocity, coefficient of discharge is 0.98, speed ratio is 0.46. Determine (i) quantity of water in cumecs, (ii) size of jet.*

Solution.

$$P = 9.81\, QH\eta$$

$$15{,}000 = 9.81 \times Q \times 300 \times 0.8$$

$$Q = 6.37 \text{ cumecs}$$

Flow per nozzle $$= Q/3 = 2.12 \text{ cumecs.}$$

Jet velocity, $$V = 0.98\sqrt{2 \times 9.81 \times 300} = 75.19 \text{ m/s}$$

Size of jet, $$\frac{\pi d^2}{4} V \times 0.95 = 2.12$$

or $$d = \sqrt{\frac{2.12 \times 4}{\pi \times 75.19 \times 0.95}} \times 100 \simeq 20 \text{ cm.}$$

Example 10.6 *A kaplan turbine runner has an output of 10,000 kW against net head of 10 m, and overall efficiency of 0.85. The flow ratio is 0.67, speed ratio is 2.0 and diameter of boss is 1/3 that of the runner. Determine runner diameter, speed and specific speed of the turbine.*

Solution.

$$v_1 = 2.0\sqrt{2gH} = 2.0\sqrt{2 \times 9.81 \times 10} = 28.01 \text{ m/s}$$

$$V_{fl} = 0.67\sqrt{2gH} = 0.67\sqrt{2 \times 9.81 \times 10} = 9.38 \text{ m/s}$$

$$P = 9.81\, QH\eta$$

$$10{,}000 = 9.81 \times Q \times 10 \times 0.85$$

or $$Q = 119.93 \text{ cumecs}$$

Also $$Q = \frac{\pi}{4}(D^2 - d^2)V_{f1},$$ **D is runner dia. and d is a hub dia.**

$$119.93 = \frac{\pi}{4}\left(D^2 - \frac{1}{9}D^2\right)9.38$$

Runner dia. $$D = 2.14 \text{ m}$$

$$v_1 = \frac{\pi DN}{60}$$

$$28.01 = \frac{\pi \times 2.14 \times N}{60}$$

Speed, $N = 249.88$ r.p.m.

Specific speed, $N_s = \dfrac{249.88\sqrt{10{,}000 \times 1.358}}{10^{5/4}}$

$= 1637.50$ r.p.m.

Example 10.7 *(a) A 20,000 kW hydro power plant with 30 m head operates at 30% load factor as peaking station. Determine the discharge required if it is to work as a base load station. Assume plant efficiency as 80%.*

(b) What is the maximum load factor if the discharge available is 35 cumces?

Solution.

Average power generated as peaking station $= 20 \times 10^3 \times 0.30 = 6 \times 10^3$ kW

$$P = 9.81\, QH\eta$$

$$6 \times 10^3 = 9.81 \times Q \times 30 \times 0.80$$

or $Q = 25.48$ cumecs.

(*b*) Power developed with 35 cumecs discharge $P = 9.81 \times 35 \times 30 \times 0.80 = 8240$ kW

Maximum load factor $= \dfrac{\text{Average load}}{\text{peak load}}$

$$= \frac{6 \times 10^3 \text{ kW}}{8240} \times 100 \simeq 73\%.$$

Example 10.8 *A run-of-river plant for peak load demand of 8 hours a day has minimum inflow of 20 cumecs, net head 25 m and plant efficiency 85%. Determine firm power output of the plant,* (*i*) *without pondage and* (*ii*) with pondage. Assume a water loss of 10% on account of evaporation etc.

Solution.

(*i*) Firm output without pondage $= 9.81 \times 20 \times 25 \times 0.85 = 4.169 \times 10^3$ kW

(*ii*) With pondage; water stored in 16 hours when not in operation $= 20 \times 16 \times 60 \times 60 = 1.152 \times 10^6\ m^3$

Net water available from pondage when plant is operating for 8 hours, allowing 10% losses $= (1.152 \times 10^6) \times 0.9$
$= 1.037 \times 10^6\ m^3$

Discharge available from pondage $= (1.037 \times 10^6) / (8 \times 60 \times 60)$
$= 36$ cumecs

$\therefore$ Firm power output of plant with pondage $= 9.81 \times (20 + 36) \times 25 \times 0.85$
$= 11.674 \times 10^3$ kW

Example 10.9 *In a preliminary survey of possible hydro power station, the expert(s) surveyed water heads of 20,100 and 300 m at three different sites. What would be the suggested type (s) of turbine (s) to be installed at each site indicating their operating specific speed range (AMIE).*

Solution.

Referring to Table 10.5 & 10.4 the suggested type of turbine at each site is Bulb, Deriaz and Francis turbine with corresponding specific speed range of 600 – 1200, 200 – 400, 60 – 400 respectively.

Example 10.10 *A power plant can generate 100 × 10⁶ kwH of electricity in a year. When working as a peaking station it has load factor of 30%. What is the installed capacity of the plant. Also determine reserve capacity if plant capacity factor is 20%.*

Solution.

$$\text{Average load} = \frac{10\times10^6 \text{ kW}}{24\times365} = 11{,}416 \text{ kW}$$

$$\text{Load factor} = \frac{\text{average load}}{\text{peak load}}$$

$$0.30 = \frac{11{,}416}{\text{peak load}}$$

$\therefore$ Peak load $= 38{,}052$ kW

$$\text{Capacity factor} = \frac{\text{Average output}}{\text{Installed capacity}}$$

$$0.20 = \frac{11{,}416}{\text{Installed capacity}}$$

$\therefore$ Installed capacity $= 57080$ kW

$\therefore$ Reserve capacity $= 57{,}080 - 38{,}052$

$= 19{,}028$ kW.

Example 10.11 *A hydropower plant operates for 12 hours a day and six days a week with constant available inflow of 30 cumecs, net head 30 m and overall efficiency 80% and 10% water losses. Determine (i) The maximum constant power as can be obtained to cater for the given demand, (ii) minimum pondage capacity, and (iii) percent increase in power with pondage.*

Solution.

$$\text{Pondage factor (with 10\% losses)} = \frac{24\times7}{12\times6}\times0.90 = 2.10$$

Discharge available for 12 hours operation $= 2.10 \times 30 = 63$ cumecs

(*i*) Maximum constant power generated $= \dfrac{1000\times63\times30}{75}\times0.80 = 20{,}160$ HP.

(*ii*) Minimum pondage capacity $= \dfrac{12+24}{24} \times 30 \times 24 \times 60 \times 60$

$= 3.888$ Mm³

(*iii*) Power increase due to pondage $= \left(\dfrac{63-30}{30}\right)100 = 110\%$

Example 10.12 *A pelton turbine develops 14000 kW working under a head of 900 m and 600 r.p.m. Obtain the least diameter of the jet and the mean wheel diameter for an overall efficiency of 90%.*

(b) Determine jet ratio and water power. (AMIE)

Solution.

$$P = 9.81\, Q H\eta$$

$$14{,}000 = 9.81 \times Q \times 900 \times 0.90$$

$\therefore$ Discharge through wheel, $Q = 1.76$ cumecs

Velocity of jet, $V = C_v \sqrt{2g\times H}$, assuming $C_v = 0.98$

$$= 0.98\sqrt{2\times9.81\times900}$$

$$= 130.23 \text{ m/s}$$

Discharge of jet $= V \times \pi\, d^2/4$

$$1.76 = (130.23 \times \pi \times d^2/4)$$

or $d = 131$ mm

∴ Dia. of jet $d = 0.131$ m

Now, $v = 0.46\,V$, assuming speed ratio as 0.46

$$= 0.46 \times 130.23 = 59.91 \text{ m/s}$$

$$v = \frac{\pi DN}{60}$$

Mean wheel diameter, $D = \dfrac{60v}{\pi N} = \dfrac{60\times 59.91}{(22/7)600} = 1.91\,\text{m}$

(*b*) Jet ratio, D/d $= 1.91 / 0.131 = 14.5$

Water power $= wQH$ $= 9.81 \times 1.76 \times 900 = 15{,}539$ kW

Example 10.13 *Determine (a) minimum number of Francis turbines required for a hydro plant on a stream having 250 cumecs flow and 25 m head. The generator is directly coupled to the turbine which has specific speed of 250, efficiency of 85%. The frequency of generation is 50 cycles/s and number of poles used is 20.*

(b) What will be the minimum number of kaplan turbines with a specific speed of 700.

Solution.

The speed of turbine is synchronous with that of the generator, both being directly coupled.

$$N = \frac{120f}{20} = \frac{120\times 50}{20} = 300\,\text{r.p.m.}$$

Power potential, $P = \dfrac{wQH\eta}{75} = \dfrac{1000\times 250\times 25\times 0.85}{75} = 7.083 \times 10^4$ H.P.

(*a*) Power generated by one Francis turbine, $N_s = \dfrac{N\sqrt{P}}{H^{5/4}}$

$$250 = \frac{300\sqrt{P}}{(25)^{5/4}}$$

or $P = 2170$ H.P.

∴ No. of turbines required $= \dfrac{7.083\times 10^4}{2170} = 33$

(*b*) Kaplan turbine, $N_s = \dfrac{N\sqrt{P}}{H^{5/4}}$

$$700 = \frac{300\sqrt{P}}{(25)^{5/4}}$$

or $P = 1.701 \times 10^4$ H.P.

∴ No. of turbines required $= \dfrac{7.083\times 10^4}{1.701\times 10^4} = 5$

Example 10.14 *At a power plant of 3000 kW output, an inward mixed flow turbine works under a head of 20m at an overall efficiency of 85%. Speed of runner is 250 r.p.m. and hydraulic* *efficiency is 88%. Velocity of flow is 0.30 $\sqrt{2gH}$ and C_v is 0.98. Determine (i) guide blade angle, (ii) wheel inlet angle, and (iii) outer diameter of wheel.*

Solution.

$$V_1 = 0.98\sqrt{2gH} = 0.98\sqrt{2\times 9.81\times 20} = 19.41 \text{ m/s}$$

$$V_{f1} = 0.30\sqrt{2gH} = 0.30\sqrt{2\times 9.81\times 20} = 5.94 \text{ m/s}$$

Considering that there is no whirl loss at the runner outlet

$$\frac{V_{w1}V_1}{gH} = \frac{V_{w1}\times 19.41}{9.81\times 20} = \eta_h = 0.88 \text{ (given)}$$

$$\therefore \quad V_{w1} = 8.90 \text{ m/s}$$

$$V_{w1} \text{ Tan } \alpha_1 = V_{f1}/V_{w1} = 5.94/8.90 = 0.667$$

$$\therefore \quad \alpha_1 = 33°\ 40'$$

$$\text{Tan } \beta_1 = \frac{V_{f1}}{V_1 - V_{w1}} = \frac{5.94}{19.41 - 8.90} = 0.565$$

$$\beta_1 = 29°\ 30'$$

$$V_1 = (\pi D_1 N)/60$$

$$D_1 = 1.48 \text{ m}$$

$$P = 9.81\, QH\eta$$

$$3000 = 9.81 \times Q \times 20 \times 0.85$$

or $\quad Q = 17.99$ cumecs

Discharge at outlet = $(\pi D^2 V_{f1})/4$, assuming $V_{f1} = V_{f2}$

$$17.99 = (\pi D^2 \times 5.94)/4$$

$$D_2 = 1.96 \text{ m}$$

Fig. 10.35

Example 10.15 *A power plant with 3 units of 60,000 kW output runs under a head of 100 m at 100 r.p.m. with 90 percent efficiency. A model of the turbine is tested on a flume of 0.3 cumecs under a head of 5 m. Compute scale ratio of the model, its speed and power.*

Solution.

Power generated by a turbine = 60,000/3 = 20,000 kW

Now $\quad P = 9.81\, QH\eta$

$$20{,}000 = 9.81 \times Q \times 100 \times 0.90$$

or $\quad Q = 22.65$ cumecs.

Since dimensionless parameters of discharge coefficient, head coefficient and power coefficient of the prototype and model shall be the same.

Now
$$\frac{Q_p}{D_p^2\sqrt{H_p}} = \frac{Q_m}{D_m\sqrt{H_m}}$$

or
$$\frac{D_m}{D_p} = \left[\left(\frac{Q_m}{Q_p}\right)\left(\frac{H_p}{H_m}\right)^{1/2}\right]^{1/2} = \left[\left(\frac{0.3}{22.65}\right)\left(\frac{100}{5}\right)^{1/2}\right]^{1/2}$$

$\therefore$ Scale ratio = 0.243 = 1 : 4.115

Speed,
$$\eta_m = \eta_p\sqrt{\frac{H_m}{H_p}} \times \frac{D_p}{D_m} = 100\sqrt{\frac{5}{100}} \times \frac{1}{0.243} = 92.02 \text{ r.p.m.}$$

Power, $P_m = P_p\left(\frac{\eta_m}{n_p}\right)^3\left(\frac{D_m}{D_p}\right)^5 = 20{,}000\left(\frac{92.02}{100}\right)^3 (0.243)^5$

$= 13.20$ kW.

Example. 10.16. *The installed capacity of a plant is 108 MW at a minimum head of 47.85 m. Determine generating capacity at a rated head of 65.5 m. Assume ratio of turbine efficiency at rated head to minimum head as 0.84/0.80.*

Solution: Generating capacity at rated head $= \left(\frac{65.5}{47.85}\right)^{3/2} \times \frac{0.84}{0.80} \times 108 \simeq 180$ MW.

Example. 10.17. *What is specific speed of a turbine? Derive a relation for specific speed of turbine.* *(P.U. 1999)*

Solution: The equation for specific speed of a turbine is determined by assuming that all turbines are similar with their linear dimensions in proportion and blade angles are constant and by applying the principle of similarity.

Tangential velocity of wheel at inlet, $v = wD/2$

or $D \propto v/w$

But $w \propto n$

From inlet triangle, $v \propto V$ where V is absolute velocity of entering water.

Also $v \propto \sqrt{H}$

$\therefore$ $D \propto \sqrt{H}/n$ ...(*i*)

$b \propto D$ ...(*ii*)

Assuming linear dimensions of turbines to be similar

From (*i*) and (*ii*)

$b \propto \sqrt{H}/n$

From inlet triangle, $V_f = V$

or $V_f \propto \sqrt{H}$

Discharge passing through turbine = Radial area of flow × flow velocity

$= \pi Db \times V_f$ (*iii*)

Discharge/s $\propto \frac{\sqrt{H}}{n} \cdot \frac{\sqrt{H}}{n} \cdot \sqrt{H} = \frac{H^{3/2}}{n^2}$ ...(*iv*)

H.P. = WH/75

From equation (*iv*) $P \propto \frac{H^{3/2}}{n^2} \times H = \frac{H^{5/2}}{n^2}$

or $n \propto \frac{H^{5/4}}{\sqrt{P}}$

or $n = \frac{KH^{5/4}}{\sqrt{P}}$

Under specific speed condition *i.e.*, turbine developing 1 H.P. under 1 m of head, *k* is equal to *n* which is the specific speed n_s

Thus $n_s = \frac{(n\sqrt{P})}{H^{5/4}}$

QUESTIONS

1. Define specific speed of a turbine. Mention the range of specific speed for a pelton turbine.
2. Explain salient features of a Kaplan turbine with the help of sketches.
3. What do you understand by characteristic curves of a turbine? How are these plotted. Narrate their utility.
4. (*a*) What is cavitation? Explain how cavitation is a consideration for setting of turbines relative to tail water.

   (*b*) What is specific speed of a turbine? Derive a relation for specific speed of a turbine. (P.U. 1999)
5. For an impulse turbine obtain an expression for maximum hydraulic efficiency.

   A pelton turbine develops 14000 kW working under a head of 900 m, running at 600 r.p.m. Obtain the least diameter of the jet and the mean wheel diameter for an overall efficiency of 90%. (AMIE)
6. What are the characteristic curves for a Francies turbine? Draw these.
7. Describe with neat sketch the governing of a turbine. (AMIE)
8. A Kaplan turbine developing 40,000 H.P. under a head of hm has an overall efficiency of 90%. If velocity of flow be 10 m/s, flow ratio 0.6, obtain peripheral velocity for a speed ratio of 2.0. Obtain also the diameter and turbine speed, if hub diameter be 0.35 times the outer wheel diameter. (AMIE)
9. What is cavitation? Explain how cavitation is a consideration for setting of a turbine relative to tail water.
10. What is specific speed of a turbine? Derive a relation for specific speed of turbine.
11. A turbine develops 1000 kW under a head of 25m at 135 r.p.m. Obtain specific speed what would be its normal speed and power under a heat of 37.5 m. (AMIEW 2000)
12. A kaplan turbine developing 50000 HP under a head of hm has an overall efficiency of 90%. If velocity of flow be 12 m/s. Flow ratio 0.6, obtain peripheral velocity for a speed ratio of 2.0 obtain also the diameter and turbine speed, if hub diameter be 0.35 times the outer wheel diameter. (AMIEW 2000)

11

HYDRAULIC VALVES AND GATES

11.1 INTRODUCTION

The hydraulic valves are used in conduits, pipe lines and penstocks to regulate or control the flow through them. Distinguished from the gates, the valve construction is such that the closing member operates and remains within the water passage way and moves or rotates longitudinally to control the flow of water whereas in the gates the leaf or the closure member moves across the waterway from an external position to control the water flow and its construction is such that the closing member is completely removed from the water passage way when the gate is in fully open condition.

Valves Vs Gates

A comparison of hydraulic valves and gates, as under, shall be of interest:

Valve	*Gate*
The closing member operates and remains within the water passage.	The closing member is completely removed from the water passage under fully open condition.
It moves or rotates longitudinally to control the flow.	It moves across the water passage to control the water flow.
Used at strategic points in the course of a long pipe line.	Used at intake of conduits.
Either closed or opened under unbalanced hydraulic conditions except butterfly valve which operates under balanced hydraulic conditions.	Used under unbalanced and balanced conditions depending on the type and location of the gate.
Valves with concentric and fairly close-fitting parts that move upon each other are avoided in sandy, silty or carbonate water as they get choked and rendered inoperative.	Sliding and rotating types of gates are more suitable than valves for use in sandy, silty or carbonated water.
Even under fully opened condition, valves present obstacle to flow.	Under fully opened condition, gates present no obstacle to flow.

11.2 HYDRAULIC VALVES

The hydraulic valves are used at the strategic points in the course of a pipe line for effecting control and safe operation. In general, the valves are provided at two places; one at the upper end of the conduit after the surge tank and one at the lower end of the pipe line near the entry to the turbine. The present trend is to install the valve below surge tank in the case of pressure shafts in sound rock. The valve at top, if provided, is fitted with an automatic over-velocity tripling device which closes the pipe when velocity exceeds 20% of the normal. In penstocks, valves are provided for control of the flow into the penstocks for operation, inspection and repairs and are normally used only either in fully open or fully closed conditions. Even under fully opened condition, the valves present obstacle to flow in the form of disc thickness.

The design of the valve should be such that it is safe against the cavitation for successful operation. The valve may be operated hydraulically, electrically or manually.

The various considerations in the selection of valves for installation primarily depend on (*i*) Service conditions that is operating and effective static heads, (*ii*) Sediment quality and quantity in

the water flow, (*iii*) Dispersion such as where spray from free-discharge valves is objectionable, (*iv*) Maintenance such as ease and frequency of repairs, avoids cavitation, (*v*) Cost like initial cost and maintenance cost, (*vi*) Quantity of water to be discharged, (*vii*) Frequency of valve operation, (*viii*) Length of time the valve may be in service, (*ix*) Climatic conditions, (*x*) Control requirements, (*x*) Valve position in the conduit or pipe line, and (*xi*) Accessibility of valves and controls.

11.3 TYPES OF HYDRAULIC VALVES

Air vents and air valves are provided on the immediate downstream side of control gate or valve to facilitate connection with the atmosphere. The various types of valves are briefly discussed as under:

1. Air Vent

To prevent occurrence of excessive negative pressures in the penstock during emergency gate closure and to exhaust air during penstock filling operations, one or more air vents of suitable size are provided on just downstream from the control gate or valve at upper end of penstock to facilitate connection with the atmosphere. The air inlets serve the purpose of admitting the air into the penstock when the control gate or valve is closed and the penstock is drained thus avoiding collapse of the pipe due to vacuum or excessive negative pressure, as also the air admitted accelerates the draining action of the pipe. Similarly when the penstock is being filled up these vents facilitate the proper escape of the air from the pipe.

The air vent is designed for adequate capacity of air entry, maintaining a pressure of not less than ½ atmosphere in the penstock at the maximum rate of depletion of water from penstock under emergency closure conditions, as serious accidents and failures may result due to inadequate venting. The opening in the intake roof has essentially to be as close as practicable to the gate. The upper end of the air passage should be open to the atmosphere sufficiently above head water and so located that it is not readily accessible. Air vent in the penstock is not needed when gates sealing on the upstream side are sometimes used because sufficient air is introduced into the water passage through the opening between the downstream side of the gate and the concrete structure. It is safe to provide the air vent for full discharge of the penstock. The factors governing the size of the vents are the length, diameter, thickness, head of water and discharge in the penstock and strength of the penstock under external pressure viz. collapsible strength of the pipe.

Generally the air valves are provided in a penstock line always in pairs to ensure required degree of safety. The air valves must be protected against freezing by enclosing them in separate compartment and using electric heaters or by insulating the valve sufficiently to avoid freezing. The gate house is so arranged that air essentially enters freely at all times.

An air vent is also provided downstream of the intake gate. It is designed to admit air at the rate the turbine is discharging water under full gate conditions.

Formula for the design of air vent capacity is as under:

$$A = \frac{Q\sqrt{S}}{7{,}50{,}000\,C}\left(\frac{d}{t}\right)^{3/2}$$

where, A is area of air inlet (m^2), Q is flow of air through inlet (m^3/s), S is factor of safety against collapse of pipe, C is coefficient of discharge through air vent, generally 0.6, d is diameter of the steel pipe (mm) and t is thickness of steel pipe (mm).

2. Air-Inlet Valve

The penstock profile is usually such that its lower end has much steeper slope near the power plant than in the upper leg of its length (Fig. 11.1). On sudden load acceptance by the turbine, water is drawn out faster from the lower steeper part of the penstock than the upper flatter portion thereby tending to cause a collapsing

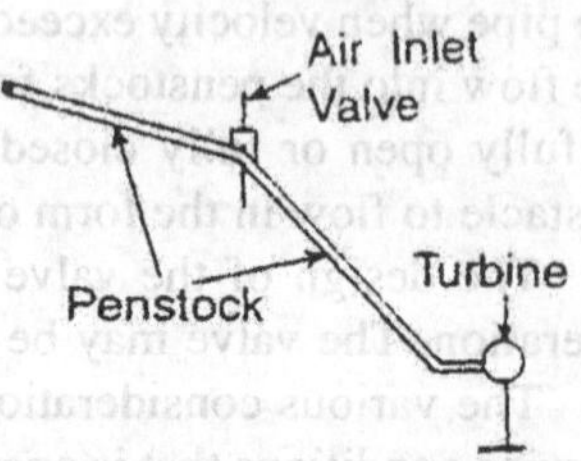

Fig. 11.1 Air inlet valve

pressure on the penstock at the point of change of grade due to the tendency of the separation of the column of water. To remedy this, either a surge tank at or near the part of grade change or an air-inlet valve (Fig. 11.2) is provided to permit entry of air into the penstock as needed. This air-inlet valve is a simple check valve set in the top of the penstock which normally remains shut by the penstock pressure so that the valve does not operate entailing wastage of water for slight pressure differences.

The air-inlet valve is essentially set in a protective covering or a manhole is provided to protect it from freezing and at the same time there is free access of air. Fig. 11.2 shows a typical air valve.

3. Needle Valve

Needle valve is essentially a valve having a circular orifice which is closed by a conical plunger. It comprises an outer globular casting containing a cylindrical inner mechanism consisting of a movable closed cylinder telescoping over a stationary cylinder. The cylindrical mechanism has needle points both at the upstream and downstream ends to guide the flow of water. The pressure for the operation of the valve is supplied by a compensating valve located below the body of the main valve. In a needle valve, the hydraulic forces acting on the needle are approximately balanced so that the force required to move the needle through its entire travel can readily be provided by mechanical operation.

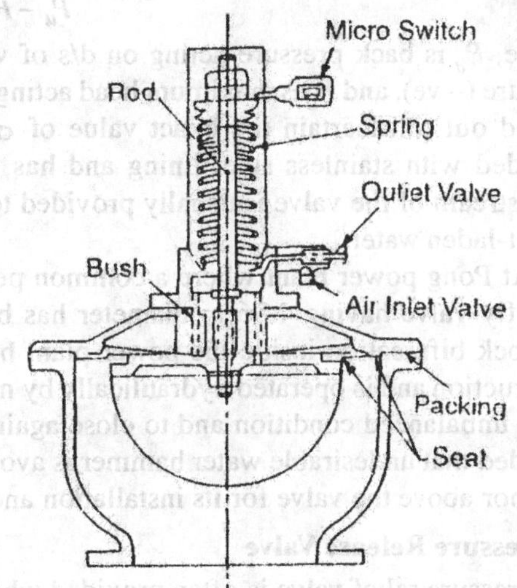

Fig. 11.2 Typical air valve

Needle valve is provided in the penstock for bypassing the regulated discharge whenever required. In outlet conduit it is normally installed at the downstream end discharging directly into the atmosphere.

The various types of needle valves are ensign valve, motor operated needle valve, balanced needle valve and internal differential needle valve.

4. Tube Valve

It is an improved version of the internal-differential needle valve wherein the water passages are similar but the downstream end of the needle is eliminated with a view to minimise cavitation. The shape of water passages is important and is so designed to give a gradual increasing velocity without abrupt changes within the valve. Tube valves are used in outlet works, located either in the mid portion of a contuit or at its downstream end.

5. Butterfly Valve

Butterfly valves are frequently used in large size penstocks as service or emergency gates with the provision of automatic closure in the event of power failure. These are installed immediately upstream from the turbine. For emergency closure, these are operated either fully open or fully closed. The valves are also used in outlet works as both shut off and regulating devices. Whenever there is more than one generating unit in a power station, each unit needs to have butterfly valve at the intake to each penstock to facilitate repair of the main inlet valve when required.

A butterfly valve consists of a cylindrical or conical body with a circular leaf mounted on a transverse shaft carried on two bearings, opposite to each other, in the valve body. The leaf rotates through 90° about a central axis perpendicular to the axis of the pipe so that it aligns itself in the direction of flow in open position and in opposite direction when closed. The valve is operated in balanced load condition during its operation of opening or closure for which the pipe on both sides of the valve has to be full of water which condition has limited its use up to 250 m of water head. For high heads and large valves, the leaf is operated by means of hydraulic cylinders or rotors. In closed condition some leakage takes place past the valve for which suitable adjustable sealing device is essentially provided to ensure no leakage at high heads. The leaf and body are so shaped to avoid abrupt changes in velocity. When used upstream and adjacent to a turbine, the water passage is so shaped to give a constant increase in velocity in the direction of flow.

Under certain conditions cavitation is caused if a non-dimensional critical cavitation index

$$\sigma = \frac{P_d - P_v}{P_u - P_d} \quad \text{exceeds 2.0 to 2.5.}$$

Where, P_d is back pressure acting on d/s of valve, P_v is vapour pressure of water relative to air pressure (– ve), and P_u is maximum head acting u/s of valve. However, model studies are essentially carried out to ascertain the exact value of σ. The cavitation is controlled if water passage is provided with stainless steel lining and has streamlined flow. Air vent for the injection of air downstream of the valve is usually provided to avoid cavitation. The butterfly valves are suitable for silt-laden water.

At Pong power plant where a common penstock is provided to serve two generating units, a butterfly valve having 4.56 m diameter has been provided in each unit bay, downstream of the penstock bifurcation inside the power plant building. The butterfly valve has a lattice type door construction and is operated hydraulically by means of a servomotor. The valve is designed to open under unbalanced condition and to close against full flow in an emergency. Its closing time is so provided that undesirable water hammer is avoided. A hatch opening, 3.55 × 7.75 m, is provided in the floor above the valve for its installation and maintenance by the main EOT crane.

6. Pressure Release Valve

Pressure relief valve is often provided where the length of penstock from reservoir to the turbine is appreciable but provision of a surge tank can be avoided. The valve diverts the water from runner when there is a sudden load rejection to reduce the effect of water hammer. Such an installation is necessitated when the time of governor action is shorter than that given by the relation:

$$\frac{\text{Length of penstock (ft)} \times \text{velocity (ft/ s)}}{10 \times \text{head (ft.)}}$$

Conventionally where the unregulated length i.e. length from the free water surface in the reservoir to the turbine is more than 5 times the maximum operating head, provision of surge shaft has to be considered. The condition may be waived off if pressure variation due to water hammer and speed rise can be taken care of suitably by other means.

Since a surge shaft is efficient if located near the power house, at Pong power plant it was not feasible due to topographical constraint, besides being excessively high and hence expensive. Moreover, provision of surge shaft was undesirable in the poor rock and highly seismic area. Accordingly 2.286 m diameter pressure release valves of 100% capacity have been provided in each unit bay (Fig. 11.3) taking off from the scroll case and connected by means of suitable links with the servomotor operating the turbine guide vanes and open on the downstream of power house into the tailrace. The valves operate when the wicket gates of the turbine close in the case of shut down. Thus water is bypassed and the pressure rise is limited to 23 percent of the maximum static head.

11.4 GATES

The gate, in general, is a structural steel frame consisting of end vertical girders with properly spaced horizontal girders between them, the spacing depends on the design water pressure and the gate dimensions. Skin plate is provided on the upstream or downstream side of the frame which in turn is held by secure welding or riveting. The gate is designed for the hydrostatic and hydrodynamic forces taking into consideration forces arising from wave effects, seismic loads and ice formation, if applicable.

In hydro power generation, the gates finds use in spillway, intake, pressure release valve, where applicable and draft tube, discussed briefly as under :

11.5 SPILLWAY GATES

Controlled spillways are provided with means to control the outflow from the reservoir which lend them the name "*Gated Spillways*". Opposed to it, uncontrolled or ungated spillways are the simplest form of spillways which are not provided with gates but automatically release water whenever the reservoir rises above the crest level. The storage is made against the raised crest and because of permanent obstruction in the waterway the rise in reservoir level is much more than that in gated spillway.

Gate is a welded or riveted steel frame consisting of end vertical girders with horizontal girders spaced between them depending on the design water pressure and dimensions of the gate and provided with a skin plate on the upstream side (*i*) To hold water against it, (*ii*) To protect the steel frame from corrosion by reservoir water, damage by ice or heavy debris, (*iii*) To minimize downpull, and (*iv*) To facilitate maintenance. The skin plate may be on downstream side, in the case of fixed wheel gate, or in exceptional cases on both upstream and downstream sides.

Types of Spillway Gates

1. Radial Gates

Also called tainter gates. These gates are extensively used as spillway gates and in certain cases as outlet gates in a dam. A radial gate, as the name implies, is a curved gate in the shape of a portion of a cylinder rotating about a horizontal axis. Normally the water is against the convex side of the gate but sometimes water load is applied on the concave side of the gate also. The upstream skin plate is bent to an arc with convex surface of the arc on upstream side, attached to the suitably spaced stiffeners either horizontal or vertical or both supported by radial arms. If horizontal stiffeners are used, these are supported by suitably spaced vertical diaphragms which are connected together by horizontal girders transferring the load to the two end vertical diaphragms. The end beams are supported by radial arms, emanating from the trunnion hubs located at the axis of the skin plate cylinder. If vertical stiffeners are used, these are supported by suitably spaced horizontal girders which are supported by radial arms. The arms transmit the water load to trunnion anchorage girder. Suitable seals are provided along the curved ends of the gate and along the bottom. Each gate is pivoted on two trunnion assemblies anchored in the downstream portion of the spillway piers (Fig. 11.3) such that the gate is capable to rotate about the fixed horizonatal axis. Fig. 11.4 shows radial gate on a barrage.

The radial gates are widely used as spillway crest gates and adapt itself easily to automatic control.

2. Drum Gates

Also termed as *sector gates.* These gates (Fig. 11.5a and b) are completely enclosed on all the three faces and at ends to form a water tight vessel. It resembles the quadrant of a circle formed by skin plate on the upstream attached to internal bearings. The gate rotates about the hinge generally provided at the upper edge so that the buoyant force aids in its lifting. A recess is made in the crest of the spillway to accommodate the gate in lowered position; the skin plate surface serving as a lid over the recess conforming to the shape of the ogee crest.

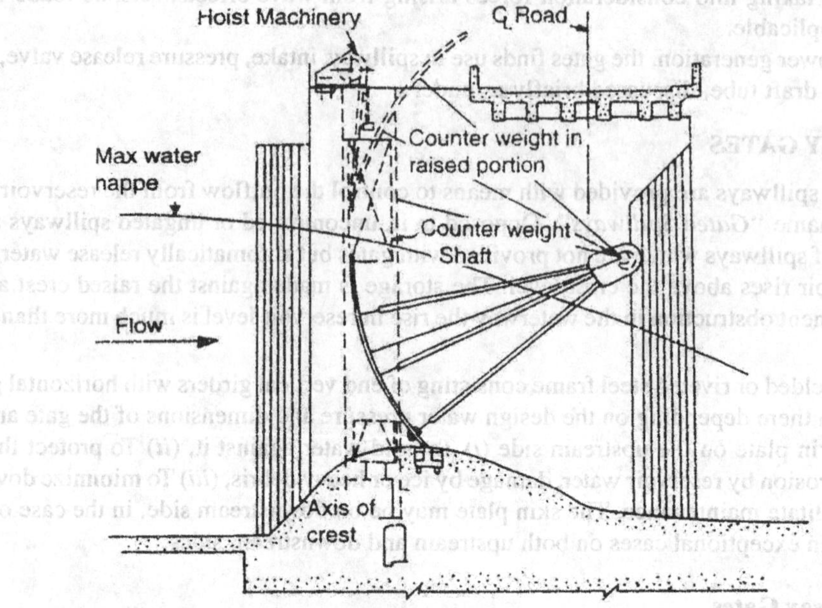

Fig. 11.3 Radial gate on a dam spillway.

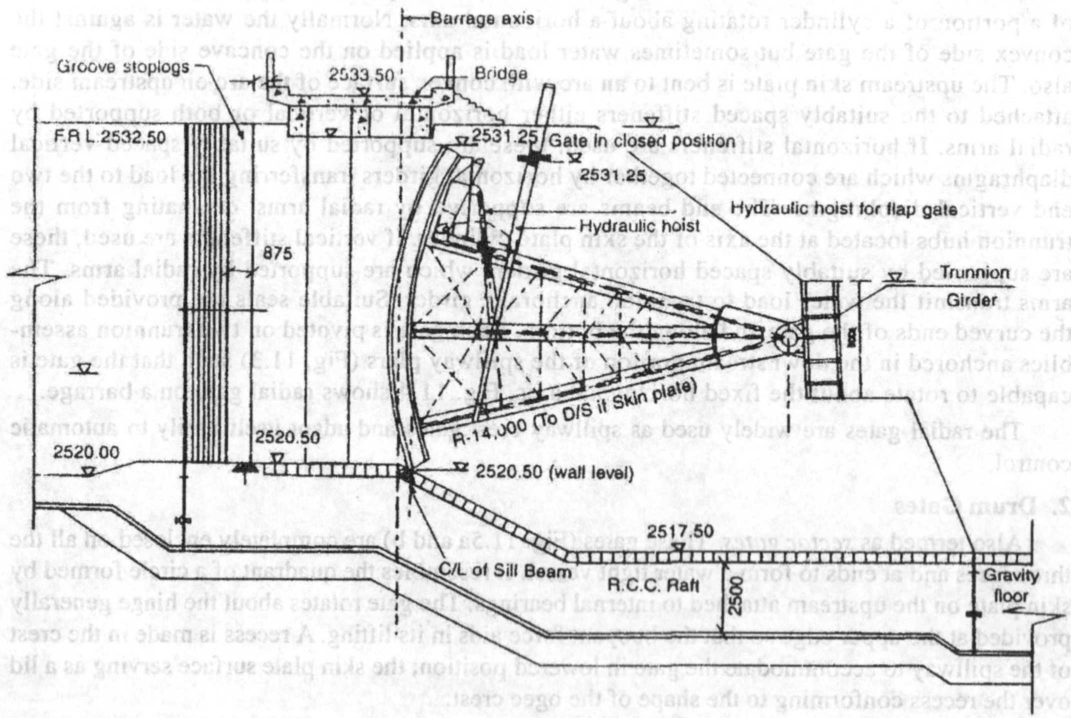

Fig. 11.4 Baspa barrage with radial gate

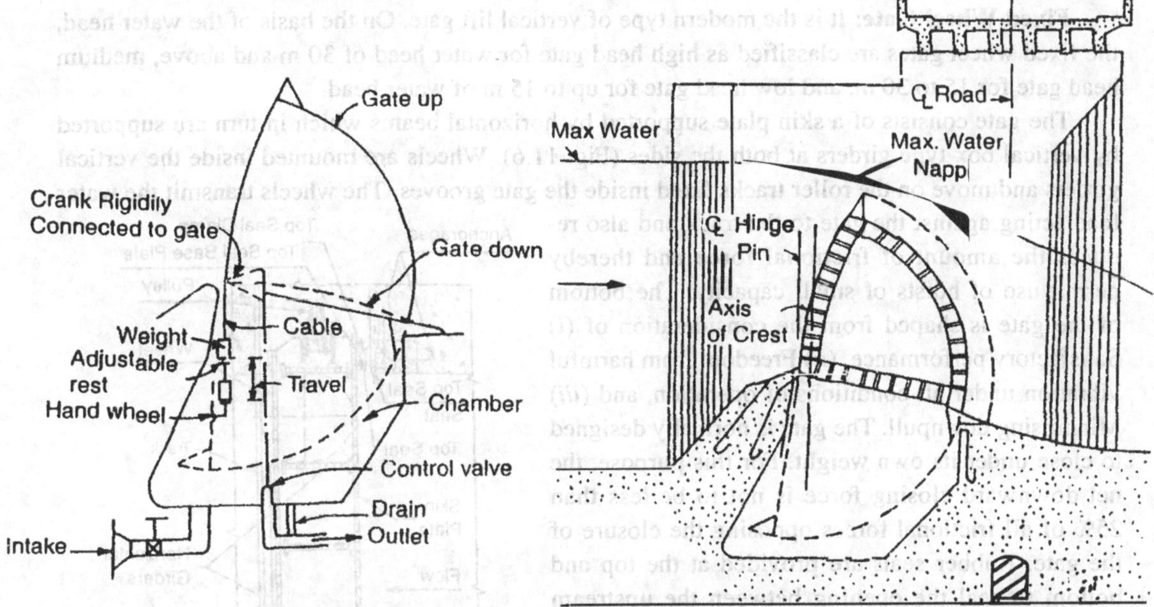

Fig. 11.5 (*a*) Drum gate. Fig. 11.5 (*b*) Diagramatic layout of drum gate.

The gates are usually operated by water pressure. When the water is admitted through the inlet pipe, controlled by hand operated sluice, into the recess the hollow drum gate is forced upwards to the raised position. When surface regulation of reservoir is to be done by drum gate, the gate is lifted by control equipment housed inside the piers. These controls are fairly complicated.

Drum gates are well adapted for long spans of 40 m or more and under medium heights of 10m or so for surface regulation of reservoir but are not suitable for low dams having low structures on account of the deep excavation required and the liability of the flooding of the recess in which the gate floats.

Disadvantages: (*i*) Gates are very costly, (*ii*) Gate in lowered position is liable to clogging, and (*iii*) Controls required for operation of the gate are complex in operation and are costly.

3. Vertical Lift Gates

Earlier to the development of radial gates, vertical lift gates were commonly used. A vertical lift gate is a rectangular steel frame consisting of skin plate on the upstream supported by suitably spaced beams and girders. The gate spans horizontaly between the grooves provided in piers and abutments and move vertically in its own plane. The grooves termed as guide grooves are lined with channel section of suitable size with a view to provide smooth bearing surface. Large friction forces are developed between the gate and the downstream guides due to hydrostatic force caused by the water against the gate and the lifting mechanism operated from the top must be able to overcome these high frictional forces. The frictional forces can be reduced by placing cylindrical rollers between the bearing surfaces of the gates and guide grooves. Vertical lift gates are simple to install and do not require lengthy piers as for radial gates. The gates are designed as wheeled gates rather than slide gates to enable lowering by gravity and for lower hoisting capacity requirement. The common types of vertical lift gates are discussed below:

Slide Gate: The slide gate, as the name implies, operates vertically by sliding inside gate frame. The gate leaf is usually made of cast steel or of welded construction of structural steel. It consists of an upstream skin plate with all the sides closed except the downstream side. The leaf is adequately stiffened by horizontal and vertical ribs. The gate frame is made of cast steel or alternatively fabricated from structural steel.

Fixed Wheel Gate: It is the modern type of vertical lift gate. On the basis of the water head, the fixed wheel gates are classified as high head gate for water head of 30 m and above, medium head gate for 15 to 30 m, and low head gate for up to 15 m of water head.

The gate consists of a skin plate supported by horizontal beams which in turn are supported by vertical box type girders at both the sides (Fig. 11.6). Wheels are mounted inside the vertical girders and move on the roller tracks fixed inside the gate grooves. The wheels transmit the water load acting against the gate to the track and also reduces the amount of frictional force and thereby permit use of hoists of small capacity. The bottom of the gate is shaped from the consideration of (*i*) Satisfactory performance, (*ii*) Freedom from harmful vibration under all conditions of operation, and (*iii*) Minimising downpull. The gate is normally designed to close under its own weight. For this purpose, the net downward closing force is not to be less than 25% of all frictional forces opposing the closure of the gate. Rubber seals are provided at the top and bottom to seal the opening between the upstream skin plate of the gate and the sides of the pier grooves. The type of rubber seal depends on the class of gate. IS:4622-1992 specifies the structural design requirements of fixed wheel gates.

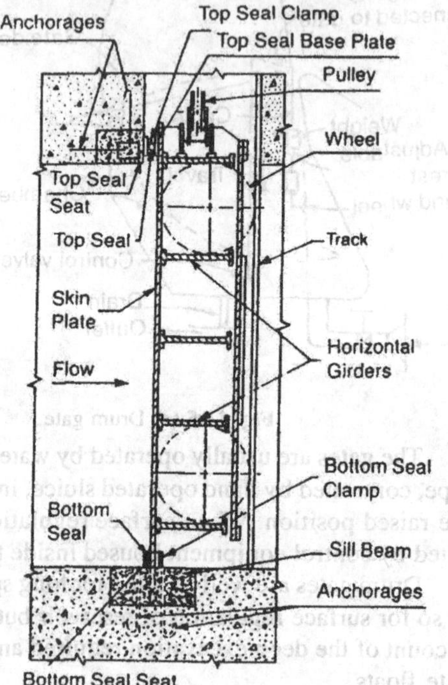

Fig. 11.6 Fixed wheel gate

Disadvantages: (*i*) Low coefficient of discharge. Gate grooves come in the waterway with consequent decrease in the coefficient of discharge necessitating increased net spillway width, (*ii*) Operational difficulties as the gate is liable to get jammed during operation due to extraneous matter getting wedged between the gate and slots, (*iii*) Regulation not possible as it is less suitable at partial openings and is more liable to vibration, and (*iv*) Maintenance difficult, as for the replacement of seals, the gates have to be taken out of the groove which is a time consuming and costly operation.

Roller Gate: Also called *stoney gate.* It is the oldest type of vertical lift gate. Roller gate is a steel cylinder spacing across the piers to the height of the opening. The gate slides on roller trains placed between the gate and roller track (Fig. 11.7). The rollers are neither attached to the gate nor to the guide groove but are placed independent of the gate and guide groove to roll vertically between the two when the gate is moved. The arrangement eliminates axle friction and only roller friction is developed and the entire water load is transmitted from a bearing strip on the gate to the roller track on the guide. The gate moves on the circumference of the roller and as such its speed is always twice the speed of roller trains. Roller gate is suitable for long span and moderate height.

Disadvantages: (*i*) Old type of gate; presently used at barrages only, and (*ii*) Not suitable for operation in rivers carrying large quantity of silt and debris.

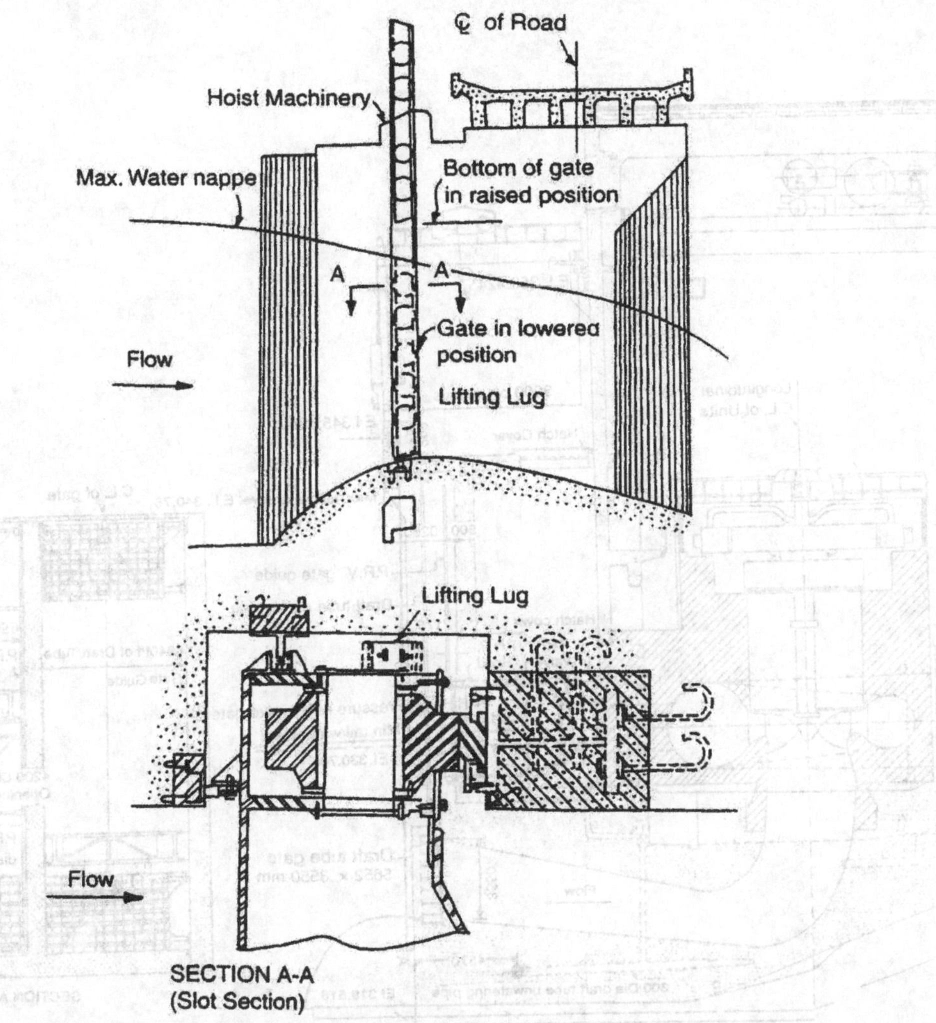

Fig. 11.7 Roller gate.

3. Tainter Gate

It consists of a skin plate formed to a segment of a cylinder with vertical elements of circular arcs and supported by a framework of horizontal or vertical purlins and stiffeners. The purlines are supported by vertical or longitudinal guides from which two or more radial struts converge downstream to horizontal pins anchored in the piers and carry the full thrust of water load. Since the skin plate is made concentric to the pin, the resultant of the water pressure passes through the pin and as such there is no moment to be overcome in hoisting the gate. The hoisting load consists of weight of gate, friction between the side seals and piers and minimal force necessary to overcome the moment of the frictional resistance at the pins. Anchorage to cater for half the water load of each gate is provided as near each face of the pier as possible. Music note seal is provided. The gates are hoisted by two chains or cables attached to the upstream face near the bottom by a shackle and extending to the drum of a hoist overhead.

Tainter gates are *(i)* Simple and economical for moderate heights, *(ii)* Suitable where span requirements are not too great and where objectionable submergence by high water does not occur, *(iii)* Reasonably water tight, *(iv)* Suitable for severe ice conditions. However, they are not suited to passage of floating material unless fully open which means wastage of water.

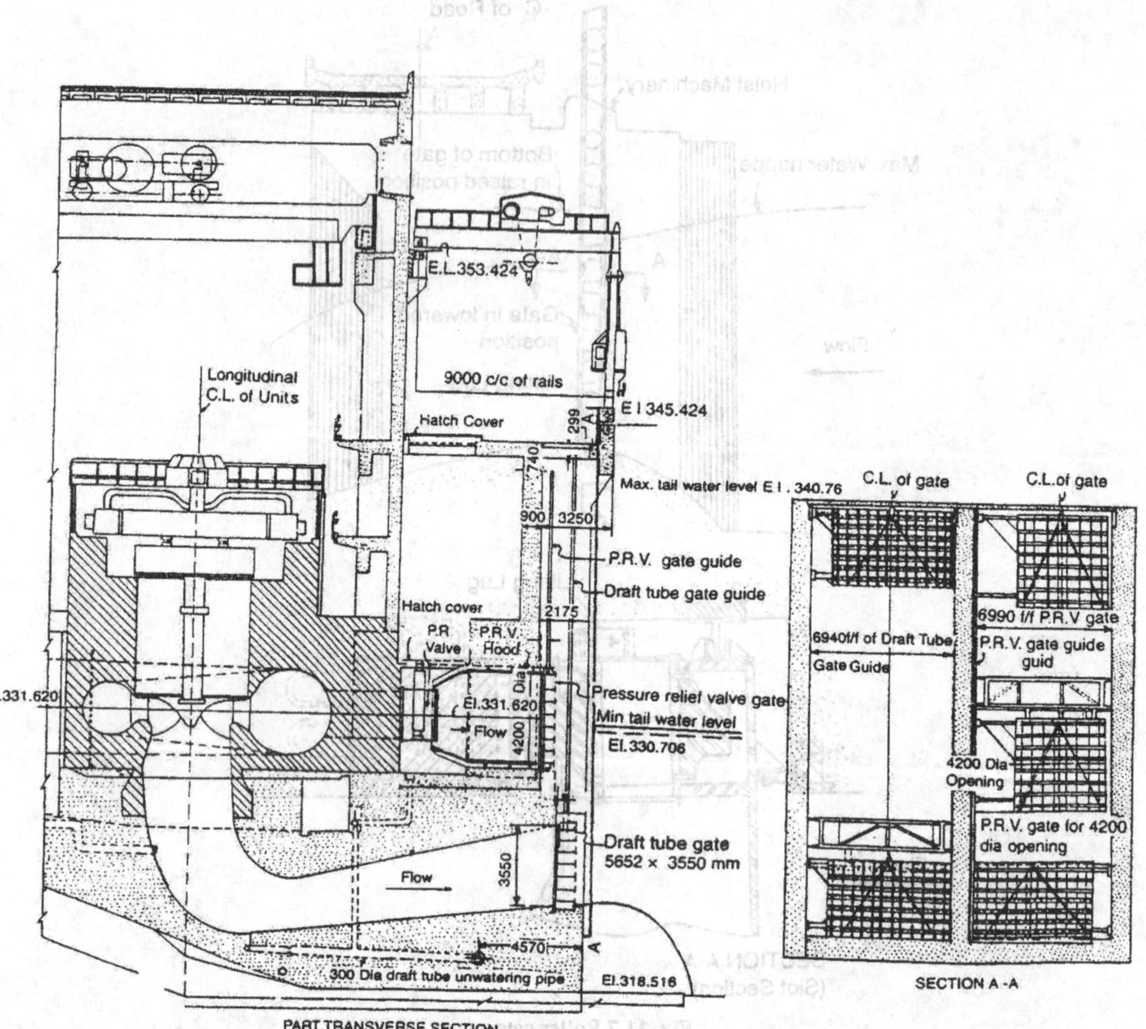

Fig. 11.8 Draft tube gate and PRV gate at pong power house.

4. Draft Tube Gate

Draft tube gates are provided to isolate the draft tubes from the tailrace channel or tunnel for closing them for dewatering, whenever required, for inspection, repairs and maintenance of generating units. Draft tube of each unit at Pong Power plant, has two identical openings each of 5.652 × 3.55 m size separated by an intermediate pier. As such two gates for closing the draft tube of one unit are provided (Fig. 11.8). Due to the provision of an opening for pressure relief valve in each unit, both the draft tubes for the two openings of one unit are not identical and as such particular gate is used for closing a particular opening. The gate is flat sliding type. It is a welded steel construction having 20 mm thick skin plate on the upstream side, supported on horizontal T-griders spaced at 510 mm centre to centre. The horizontal girders are further supported by vertical stiffeners and end plates.

11.6. GATE INSTALLATION IN PENSTOCKS

Pressure Release Valve Gate: The PRV gate is provided for inspection and maintenance of PRV and its pit. At Pong power house, four PRV gates, in each unit, have been provided to cater for the need of all the six units. The gate is designed to be raised and lowered under balanced head conditions and to withstand maximum tailrace water head of 12 m. The gate is flat sliding type. The gate is welded steel structure having 20 mm thick skin plate on the upstream side, supported by horizontal girders spaced at 570 mm centre to centre, which in turn are supported by 16 mm thick 150 mm deep vertical stiffeners spaced at 570 mm to take up water load (Fig. 11.8).

Penstocks are usually equipped with head gates at the forebay which can be closed to permit repair of the penstocks. An air-inlet valve downstream from the gate is necessary to prevent collapse of the pipe immediately after the head gate is closed.

QUESTIONS

1. Differentiate between a valve and a gate.
2. Why and where hydraulic valves are provided in penstocks?
3. What are the salient purposes served by hydraulic valves in penstocks? What are the various considerations in the selection of the type of valve for installation?
4. Discuss briefly various types of hydraulic valves used in penstocks.
5. Define gate. What are the various types of gates used in penstocks and spillways?
6. Mention briefly the salient features of radial gates. Why are they generally preferred for spillway control?

12

Water Conductor System

12.1 INTRODUCTION

Water conductor system, the life line of a hydro power project, is designed to carry silt-free water from the source, after desilting, to the turbines of a power house. The essential requirements of a water conductor system are (*i*) "Direct-route" as far as possible which may involve open channel, tunnel or a combination of the both. Generally a power canal is the usual choice for carrying large mount of water instead of a pipe which may be too large and uneconomical. A tunnel too is essentially a very costly form of water conductor system but is usually inevitable in mountainous areas to meet the requirement of shortest alignment, (*ii*) Entail minimum loss of head enroute to the plant, (*iii*) Safe means of water conveyance, and (*iv*) Economical water delivery to the plant at minimum construction and maintenance cost. The system may consist of (*i*) intake, (*ii*) open channel/cut and cover duct, (*iii*) Tunnel-either free or pressure flow, (*iv*) penstock, and (*v*) tail race tunnel or channel.

The water conveyance system varies with the type of water power development whether run-of-river plant, storage power plant or canal power plant, apart from topography, geology and site conditions and a brief resume thereof has been given in the respective type of power developments. It is, however, emphasised that the water conductor system is planned and designed carefully to ensure efficient functioning of the power plant.

12.2 ALIGNMENT

The various considerations in the finalisation of the alignment of a water conductor system are briefly as under:

1. It entails minimum loss of head so as to permit use of maximum head for power generation.
2. It affords maximum safety.
3. It has safe hill slopes, adequate vertical and horizontal rock cover.
4. Safe hydraulic gradient over its length.
5. It permits easy and expeditious construction.
6. Minimum construction cost.
7. Minimum maintenance cost.

12.3 INTAKE

Intake works draw the desired quantity of water with minimum of silt load to feed the water conductor system. The intakes are of varying types depending on the type of plant i.e. storage plant or run-of-river plant. In the intake structure housed in dam body, as in a concrete dam, the centre line of the penstock opening is necessarily kept 0.8 times the height of opening below minimum drawdown level of the reservoir, as per USBR design criteria, so as to avoid formation of vortices in front of the intake of the penstocks. However, if flow conditions are oblique to the intake, the water cover over the penstock opening, if insufficient, and penstock intake level can not be depressed, anti-vortex devices such as parallel fin type, perforated breast wall and RCC louvered grid may have to be provided. At intake structure of Sardar Sarovar dam, perforated breast wall has been preferred because of ease of design and construction and overall economy. At pong reservoir a floating grid has been provided (Fig. 12.4).

The intake gate slot is enclosed in a structure designed to guide the water into the rectangular opening without side contraction. Where gates are located in a gate shaft, suitable transition from circular to rectangular gate slot is provided. An air vent downstream of the intake gate is provided to admit air at the rate the turbine is discharging water under full gate conditions (Fig. 12.4).

12.4 LOCATION OF INTAKES

The various factors influencing the choice of location of an intake structure are (*i*) Type of edifice that is storage reservoir or run-of-river scheme, (*ii*) Location and type of dam/weir, (*iii*) Type of water conductor system that is canal or tunnel, and (*iv*) Topographical features of the river. Intakes are so located and their withdrawal level is so designed as to draw quality water that is clean and free from sediments so as not to injure the power generating equipment, discussed as under:

1. *Quality water.* Location of intake is required to be such as to draw the best quality water from the reservoir. Depth of water at intake is important. Quality of water varies at different levels in a reservoir and it is necessary to draw water from different elevations of the reservoir at different seasons of the year for which multiple-level intakes are frequently provided.

2. *Optimum water utilisation.* Intake is located in the deepest part of the impounding reservoir to enable full utilisation of the capacity of the reservoir and to protect intake from sediments in the reservoir. In reservoirs with wide variations in the water surface elevation, the intake is better located at the lowest stage so that one inlet is always submerged and operative to draw supply and minimum operating head is always available.

3. *Minimum silt entry.* In order to prevent silt from the reservoir being carried into the outlet system, location of intake at low points or pockets in the reservoir is avoided. Bottom sediments are kept out of the intake if the lowest entrance ports are sufficiently above the reservoir bed level.

4. *Wave action.* The intake structure is placed sufficiently removed from the currents that might threaten the safety of the structure. Waves exert pressure on the superstructure of the intake. Waves may also stir up the bottom sediments if the intake is not so located deep in the water below the water surface. Locations where winds may drift debris into the intake are avoided.

5. *Ice action.* Ice may push down and clog ports. Ice troubles are reduced in case entrance velocities are kept between 7 to 10 cm/s. Low velocities do not transport ice and hold the entry of leaves and debris into the intake as also fish can escape from being drawn into by intake currents.

6. *Type of dam.* Usual location of intake in an earthen dam is in deep water near the upstream toe of the dam while in concrete or masonry dam, the intake may be in the dam structure itself.

7. *Sewage pollution.* The intake is located to avoid sewage pollution. Locations where stagnant water or algae can accumulate are unsuitable for intake structure. Navigation adds the hazard of pollution from passing ships. It is more relevant to river intake for public water supply.

8. *Shore location.* Usual location of the intake some distance away from the shore to avoid turbid water from surface wash along the banks.

12.5 TYPES OF INTAKES

Depending on the function served and the range in reservoir head under which it is to operate, the discharging capacity and frequency of reservoir drawdown, intakes for hydroelectric projects may be relatively simple submerged intakes or more elaborate structure raised as a tower above maximum reservoir level. Intakes may also be classified as low pressure and high pressure intakes, though there is no clear line of division between the two classes of intakes; the essential requirements of the two types being the same but may differ in details and type of equipment on account of the difference between the depths of water in which they are to be operated. Broadly the intakes are classified as (*i*) Run-of-the river type, (*ii*) Canal intake, and (*iii*) Reservoir type.

1. Run-of River Intake. In run-of-river power plants, intake is appurtenant to power house and draws water from the river without any appreciable storage upstream of the diversion dam, or barrage (Figs. 6.5 & 12.1). Characteristic of river flows, the intake is designed to withstand high peaks and short duration flood flows and high sediment loads. The bellmouth entrance is essentially provided with trashrack.

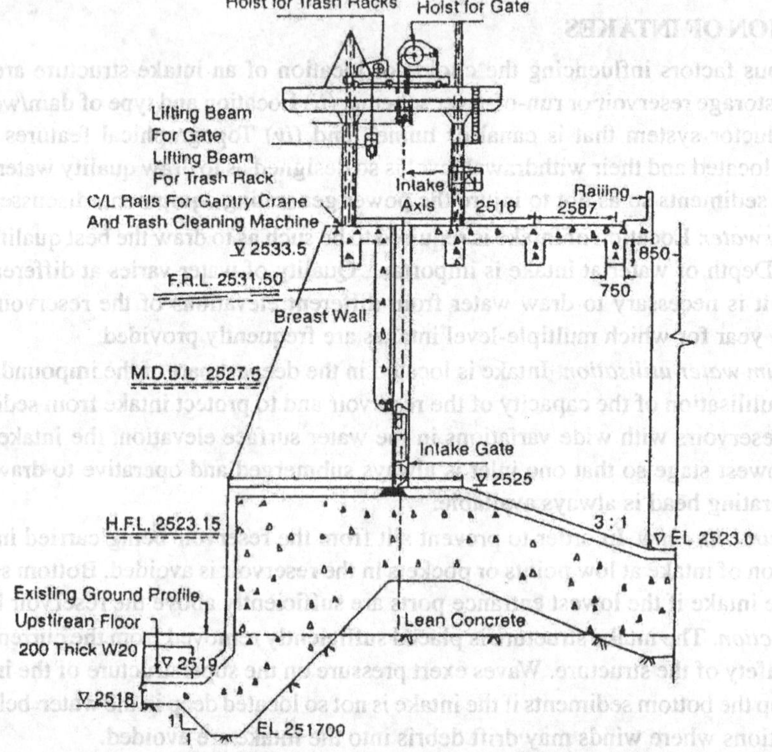

Fig. 12.1 Power plant intake showing gate and operating crane.

Canal Intake. It is also a variant of the run-of-river intake, that is provided adjacent to the diversion weir/barrage to admit water into the canal. It is designed to function under low heads and the topography and geology permit straight reach suitable for it. Sediment excluder is an essential component of the intake. The crest of the intake is generally raised to prevent entry of coarse fraction of bed load into the canal. Trashrack on the upstream and stilling basin and energy dissipation devices on the downstream of the intake are provided. In certain locations desilting tunnels may also have to be provided upstream of the intake.

On large canals, short intake is a part of power house structure forming part of streamlined passage to spiral casing (Fig. 12.2). In another variant, in open canal or free flow conduit intake for penstock is provided in forebay (Fig. 12.2b).

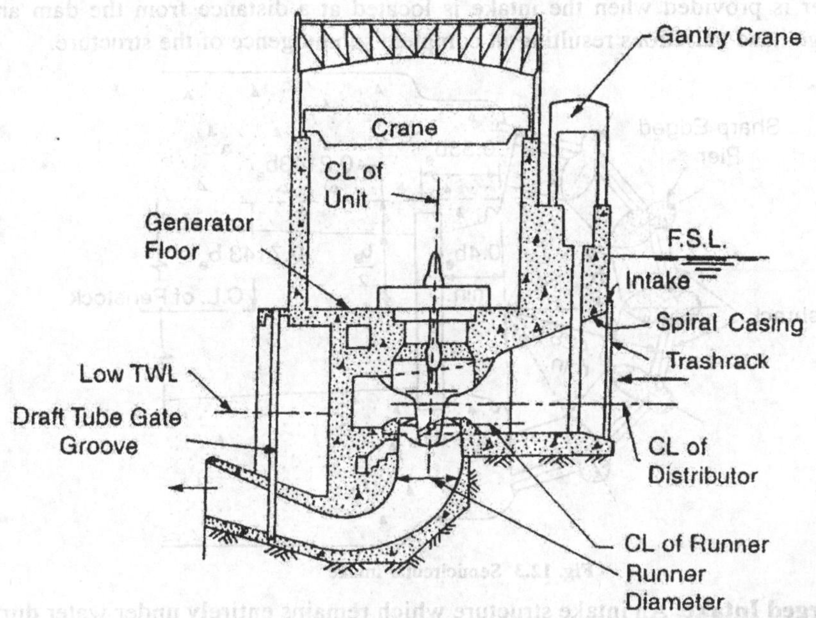

Fig. 12.2(*a*) Canal intake.

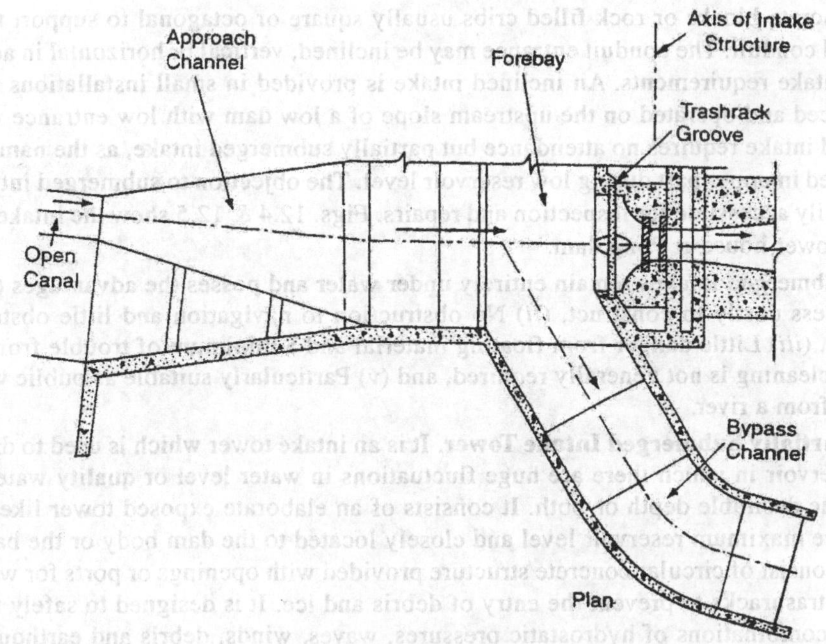

Fig. 12.2(*b*) Forebay intake.

2. Reservoir type Intakes. As the name implies, it is provided on a storage reservoir (Fig. 5.2). It may be high head (above 30m), medium head (15–30m) or low head (upto 15m) depending on the head above the centre line of the penstock. It is semicircular type in power plants located at the toe of concrete or masonry dam (Fig. 12.3). In earthen dam, it is a sloping intake or tower type intake, the latter is provided when the intake is located at a distance from the dam and it is subjected to large head variations resulting in complete submergence of the structure.

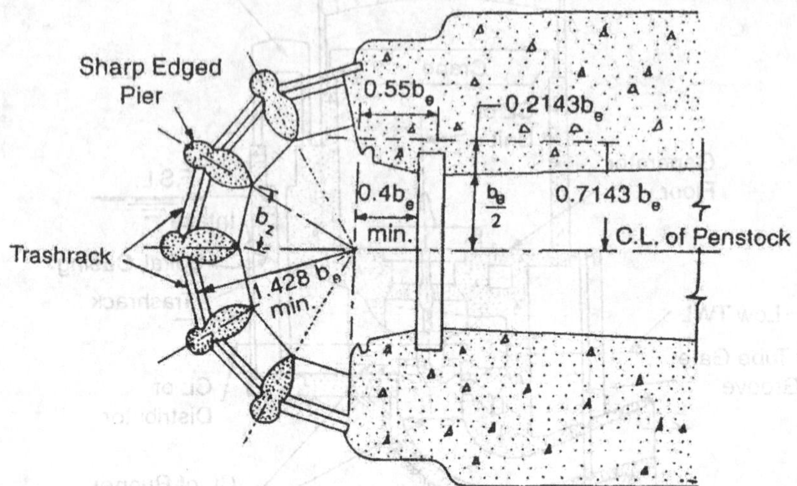

Fig. 12.3 Semicircular intake

(*i*) **Submerged Intake.** An intake structure which remains entirely under water during its operation is termed as submerged intake. It is provided where the structure serves only as an entrance to the conduit and where ordinarily cleaning of trash is not required. It consists of simple concrete blocks or rock filled cribs usually square or octagonal to support the end of withdrawal conduit. The conduit entrance may be inclined, vertical or horizontal in accordance with the intake requirements. An inclined intake is provided in small installations where the gate is placed and operated on the upstream slope of a low dam with low entrance velocities. Submerged intake requires no attendance but partially submerged intake, as the name implies, gets exposed in upper part during low reservoir level. The objection to submerged intake is that it is not easily accessible for inspection and repairs. Figs. 12.4 & 12.5 show the intake provided for pong power house at Pong dam.

The submerged intakes remain entirely under water and posses the advantages (*i*) Simple structure, less costly to construct, (*ii*) No obstruction to navigation and little obstruction to river flows, (*iii*) Little danger from floating material and a minimum of trouble from ice, (*iv*) Trashrack cleaning is not generally required, and (*v*) Particularly suitable as public water supply intake from a river.

(*ii*) **Partially Submerged Intake Tower**. It is an intake tower which is used to draw water from a reservoir in which there are huge fluctuations in water level or quality water is to be drawn at the desirable depth or both. It consists of an elaborate exposed tower like structure rising above maximum reservoir level and closely located to the dam body or the bank of the stream. It consist of circular concrete structure provided with openings or ports for water entry fitted with trashracks to prevent the entry of debris and ice. It is designed to safely withstand the severe combinations of hydrostatic pressures, waves, winds, debris and earthquake pressures. Usually 50 percent of the trashracks are considered clogged to visualise the worst designed conditions for the intake structure. It has the merit that best quality water available at different depths at different seasons of the year can be drawn through ports at different elevations. Large size port openings protected with trashracks are made to minimize the entrance velocity in order to avoid clogging of the port as also to avoid difficulty with entrained air at low velocities. The two types of intake towers are as follows :

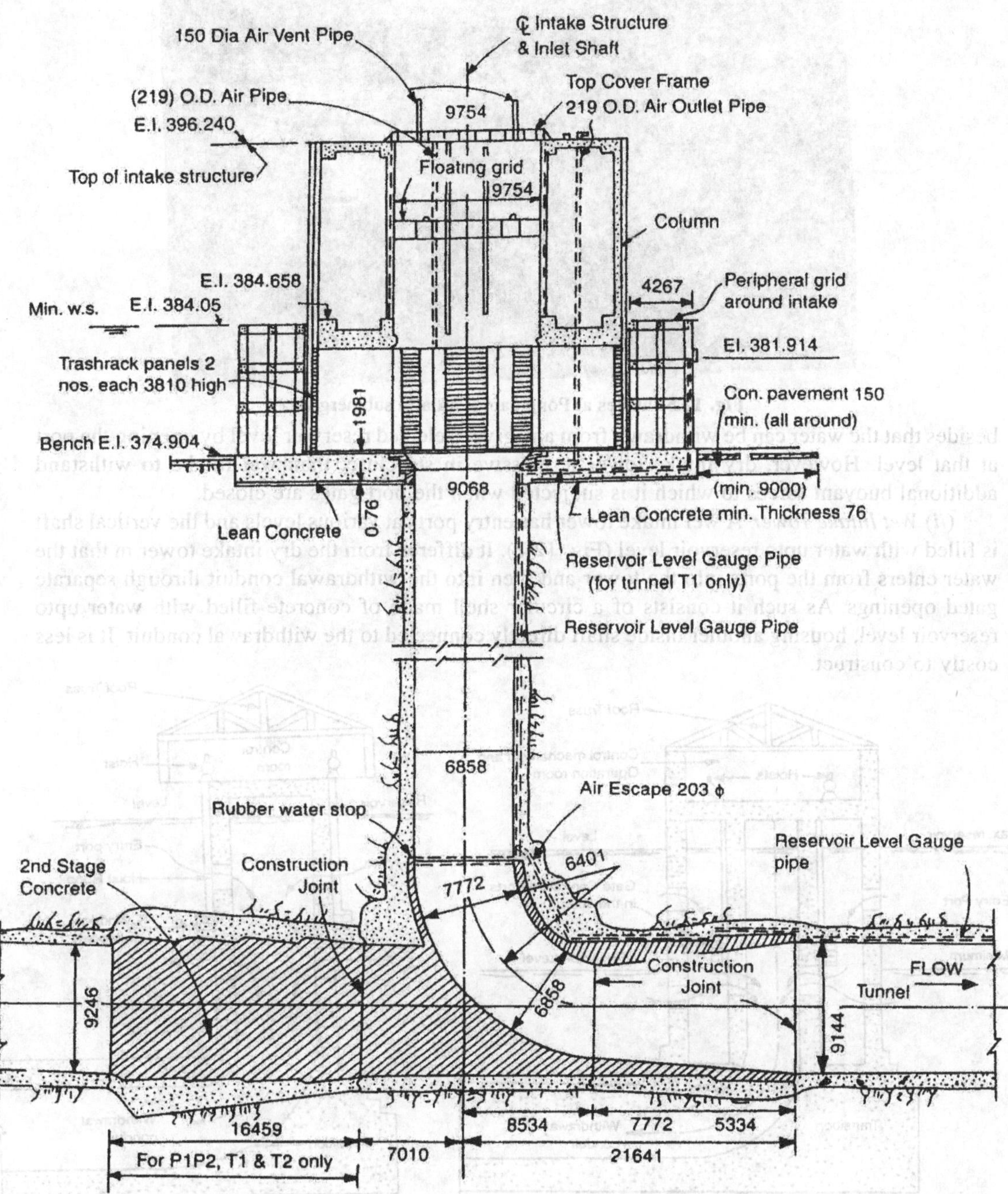

Fig. 12.4 Intake for irrigation and power tunnels at Pong dam.

***(iii)* Dry Intake Tower.** In dry intake tower (Fig. 12.6) the entry ports are directly connected with the withdrawal conduit and water is directly drawn into conduit through the gated entry ports and there is no water inside the tower when gates are in closed position. Dry intake tower has the merit that the interior of the tower being dry is made accessible for inspection and operation

Fig. 12.5 Intakes at Pong dam (Partially submerged).

besides that the water can be withdrawn from any given selected reservoir level by opening the port at that level. However, dry intake tower is massive in structure, than wet intake to withstand additional buoyant forces to which it is subjected when the port gates are closed.

(*ii*) *Wet Intake Tower.* A wet intake tower has entry ports at various levels and the vertical shaft is filled with water upto reservoir level (Fig. 12.7). It differes from the dry intake tower in that the water enters from the ports into the tower and then into the withdrawal conduit through separate gated openings. As such it consists of a circular shell made of concrete filled with water upto reservoir level, housing another inside shaft directly connected to the withdrawal conduit. It is less costly to construct.

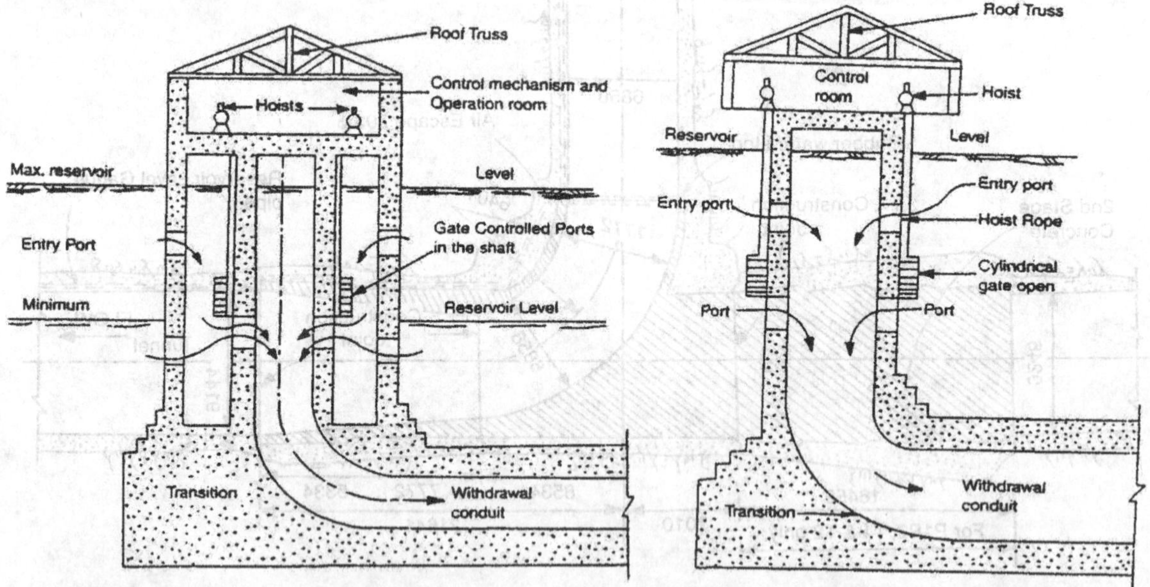

Fig. 12.6 Dry intake tower. Fig. 12.7 Wet intake tower.

3. Low Pressure Intake. It is the usual form of intake used for relatively smaller drawdowns as in the case of hydel power plants of small units catering for daily and weekly water surface variations. Low pressure intake usually performs the functions of a dam and is designed accordingly based on the theory of stability of a dam. The maximum velocity through the gross cross-sectional area of the trashrack is about 0.75 m/s which may be of the order of 1.5 m/s in the case of large units and mechanical racks of wider rack bars spacing. The racks are liable to complete and sudden plugging with ice or debris and are designed considering complete stoppage of flow through the racks.

4. High Pressure Intake. High pressure intake, in general, is used in a storage reservoir and head water for hydel power generation involving considerable drawdown under normal operation conditions. The intake is constructed in a variety of forms depending on the type of dam whether of concrete or earthen dam. In concrete dam, high pressure intake varies little in detail from low pressure intake except in the type of gates installed. In earthen or rockfill dam, the intake is generally a tower built near the foot of the upstream slope of the dam, the structural details of which vary greatly from the low pressure intake. In some cases shafts of outlet tunnels are also used as intake. Considerable high velocities 3 to 4 m/s through gross area of trashracks are adopted.

12.6 TRASHRACKS

Trashrack is defined as a screen or grating provided in front of an intake or penstock to prevent entry of debris, jungle, etc. The shape of the trashrack structure is semicircular in high dams with nearly vertical face or a straight trashrack in low dam and diversion structure. No part of the trashrack structure falls within 80 percent of the entrance height of the intake. Trashracks usually consists of trash sections 1.5 to 2 m wide and not too long for handling, made up of mild steel flats on edge, 5 to 15 cm centre in both directions depending on the maximum size of debris to be excluded from entering the intake. Coarse trashracks are provided near the ports to prevent large drift, such as cakes of ice, roots, tree and timber from being drawn into the intake. In some part of the intake fine trashracks are provided to protect the machine and machine apparatus through which water flows. The spacing of trashracks is kept so that net opening between them is at least 5 mm less than the minimum opening between turbine runner blades to prevent clogging by material which may pass through the trashrack. In cold regions, trashrack is oftenly clogged with frazile ice. Electrical heating for small trashracks is provided to prevent ice formation on the racks. Electrical heating is not economical for large trashracks and instead compressed air is passed along the screens at low level and formed air bubbles remove the ice from the rack and is carried to the water surface from where it is passed away by means of chute provided for the purpose. Another method is the provision of vigorous reversal of flow through the port made when the intake is installed as an expedient for removal of ice from intake ports. The floating debris accumulated, as are denied entry into the intake, are removed with the help of power driven rack-rakes.

Various factors which govern the necessity and type of trashrack and the size of the openings are (*i*) Nature of the trash in the reservoir, (*ii*) Nature of utilisation of water impounded in the reservoir, whether for public water supply, irrigation or hydel power generation, (*iii*) Need for the exclusion of small trash from the outflow, and (*iv*) Accessibility or otherwise for removing accumulated trash.

Design Considerations for Trashracks

Trashracks are closed spaced for excluding small trash when the outlet consists of small conduit with valve controls, and are generally more widely spaced in large sized conduit with large slide gate controls. The trashracks also vary in shape depending on the arrangements provided for their mounting on the intake structure. For a deep inlet intake, trashracks usually of the shape of a cage surmounting the entrance are arranged in an open box placed in front of a vertical entrance or positioned along the front side of a tower structure.

In the case of intakes which are to remain submerged at all operating levels of the reservoir, there is little risk of damage due to clogging of debris or ice, etc. and the trashrack may consist of widely spaced structural members without rack bars such as struts and beams so placed as to exclude only the large trash and other floating debris.

The required area of trashrack is governed by the limiting velocity through the rack and of

course the nature of the trash to be excluded. In trashracks which are inaccessible for cleaning, usually the water velocity through them is limited to 0.75 m/s., while in units with hand racking and mechanical racking velocity of 0.75 and 1.5 m/s respectively is adopted. Where maximum loss value is considered, 50 percent of the rack area is considered as clogged which means twice the velocity through the trashrack.

QUESTIONS

1. **Define water conductor system in a hydro power plant. What role does it play in the power development? Does it vary with the type of power development, discuss?**
2. **What are the considerations which govern the alignment of water conductor system?**
3. **What are the components of water conductor system in a storage and run-of-river developments.**
4. **Define intake. Discuss various types of intakes.**
5. **What do you understand by transhrack in water conductor system? Briefly mention various design considerations for trashracks.**

13

PENSTOCKS AND PRESSURE SHAFTS

13.1 INTRODUCTION

Penstocks are the pressure conduits or shafts, usually of short length, to convey water from reservoir or forebay to scroll case and thence to turbines of a hydroelectric plant. Usually the tunnel(s) earlier used as diversion tunnel(s) is designated as penstock tunnel except the upstream end which is plugged as is the case with earthen dams. In concrete or masonry dams, the penstocks are entirely embedded in the dam body in case the power plant is located at the toe of the dam. In run-of-river plant, when the distance between the forebay and power plant is short, a separate penstock for each unit is provided. In case of high head plants, separate penstock for each unit is provided. For moderate heads and long distance a single large diameter penstock is provided to serve several turbines by a system of multiple bifurcations at the lower end.

The penstocks may be of reinforced cement concrete, prestressed concrete or steel, generally of steel. The design principle is the same as for pressure vessels except that water hammer occasionally occurs due to governor control and turbine gate manipulation and has to be considered. The welded steel penstocks, as are commonly used, are made from boiler quality plates conforming to IS: 2002, 1992, 2041–1982 or 8500–1992, ASTM : A–537 etc.

The fabricated penstocks are tested for 1.5 times the design pressure which includes water hammer pressure rise when the design working stress is 2/3 yield point. Radiographic and ultrasonic testing are done for detecting defects in the welded joints. Stress relieving is done for straight ferrules with plate thickness over 38 mm and for all specials.

Besides thrashrack, gate control is provided at the entrance in forebay or downstream of intake from the reservoir to permit inspection and repairs of penstocks. The gates are closed in case penstocks are to be dewatered for repairs. Downstream from the gate and near to it an air vent or stand pipe which connects the top of penstock with open air and going up above the head water is provided. When the head gates are closed, and water is drawn off through the wheels negative pressures are exerted on the penstocks, the air inlet valve allows air to enter the penstocks to prevent their collapse due to sudden draining out of the water. The head gate is operated under balanced conditions. In case penstock is empty, it is filled through a bypass so that surge of water may not damage the turbines. The air vent pipe facilitates proper escape of air from the penstock.

Sufficient water seal is ensured at the intake of penstock at the reservoir or forebay by keeping intake at low level particularly at low water to avoid formation of vortices and carrying air into the penstocks and thence to the turbine resulting in reduced turbine efficiency. The entrance to the penstocks is essentially smooth and streamlined and sharp bends are avoided along the length to minimise loss of head and the need to provide special anchorages.

Air valves are provided at steep gradients to admit air into penstock at times of flow acceleration. A surge tank is suitably located. Saddles and anchor blocks are provided to support the penstock up to the power plant. A relief valve is provided just before the power house to discharge water when pressure in penstock increases on account of sudden closure.

Manholes, the general size of opening is 50 cm, are provided along the length of the penstock at 120 to 150 m to provide access to the pipe interior for inspection, maintenance and repairs. In underground penstock, the manhole is provided in top position of the pipe. The location is fixed so as to provide natural ventilation to the interior surface. The manhole, in general, consists of a circular nozzle head, or wall, at the opening of the pipe, with a cover plate fitted to it by bolts.

13.2 CLASSIFICATION OF PENSTOCKS

The penstocks are classified, as under:

1. Classification Based on Type of Installation

(i) **Surface Penstocks:** Also called exposed penstocks. The surface penstocks, as the name implies, are laid exposed and are supported above ground by saddle supports or ring girder supports. A variant of the surface type is wherein a combination of surface penstock and penstock in tunnel, as dictated by the topography, is adopted as at Beas sutlej link project in Himachal Pradesh (Fig. 13.1). Expansion joint provided is shown in Fig. 13.1(*a*).

The problem areas in surface penstocks are *(i)* Expansion joints particularly for high head and large diameter pipes which entail drain on foreign exchange, and *(ii)* Depth of overburden at anchor blocks location where measures such as piles, well foundation etc. are to be carefully designed and executed.

Supports: The exposed penstocks laid on surface are generally anchored at bends and sometimes at intermediate points. These are provided at 100 to 125m spacing to ensure that the penstocks do not shift during installation, to prevent sliding down of hill slopes at intermediate support and to resist the forces causing displacement at bends. In between two anchors, the penstocks are supported over a number of piers on which these are free to move. The types of supports generally provided are briefly as under:

Saddle Support: This type of supports are generally provided for penstocks of less than 2.5 m diameter with a restricted spacing. The conduit rests on the saddles with contact angle varying from 90° to 120°. The bearing plate on support is provided with suitable lubrication devices to reduce the friction between the pipe line and the supports during contraction and expansion so as to reduce the load coming on anchors.

Ring Girder with Column Supports: The support of this type is provided when the penstock is carried on rocker or roller type of supports or on sliding type of support and pipe is of bigger diameter with large spans. The load is transferred through the ring to two points on column on side.

Rocker and Roller Supports: These are of cast steel or mild steel plates generally used for bigger diameter pipes. These supports provide very little resistance to movements due to temperature changes.

Advantages of Surface Penstocks

(i) Economical to install, *(ii)* More working space available for construction work, and *(iii)* Easily accessible for inspection, repairs and repainting.

Disadvantages

(i) Prone to rusting and corrosion being exposed, *(ii)* Repeated painting of outer surface is needed, *(iii)* Supporting and anchoring on steep hill slope is difficult and costly, *(iv)* Susceptible to damage by landslides, snowslides etc, *(v)* In very cold climate, the water in conduit may freeze against which protection needs to be provided, *(vi)* Numerous expansion joints are required to avoid high temperature stress, and *(vii)* Requires high tensile strength steel.

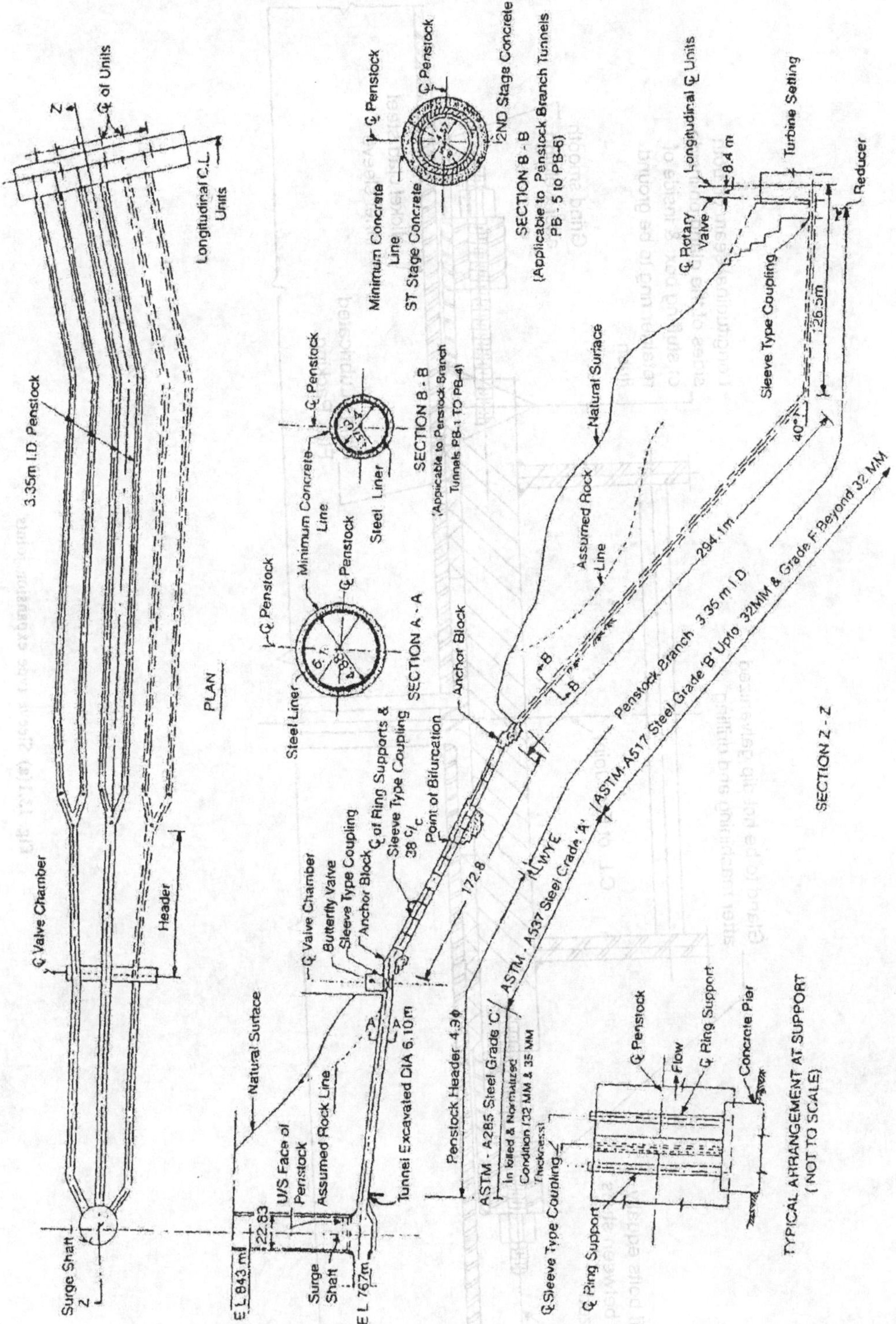

Fig. 13.1 In-tunnel and exposed penstock layout (Beas Sutlej Link Project).

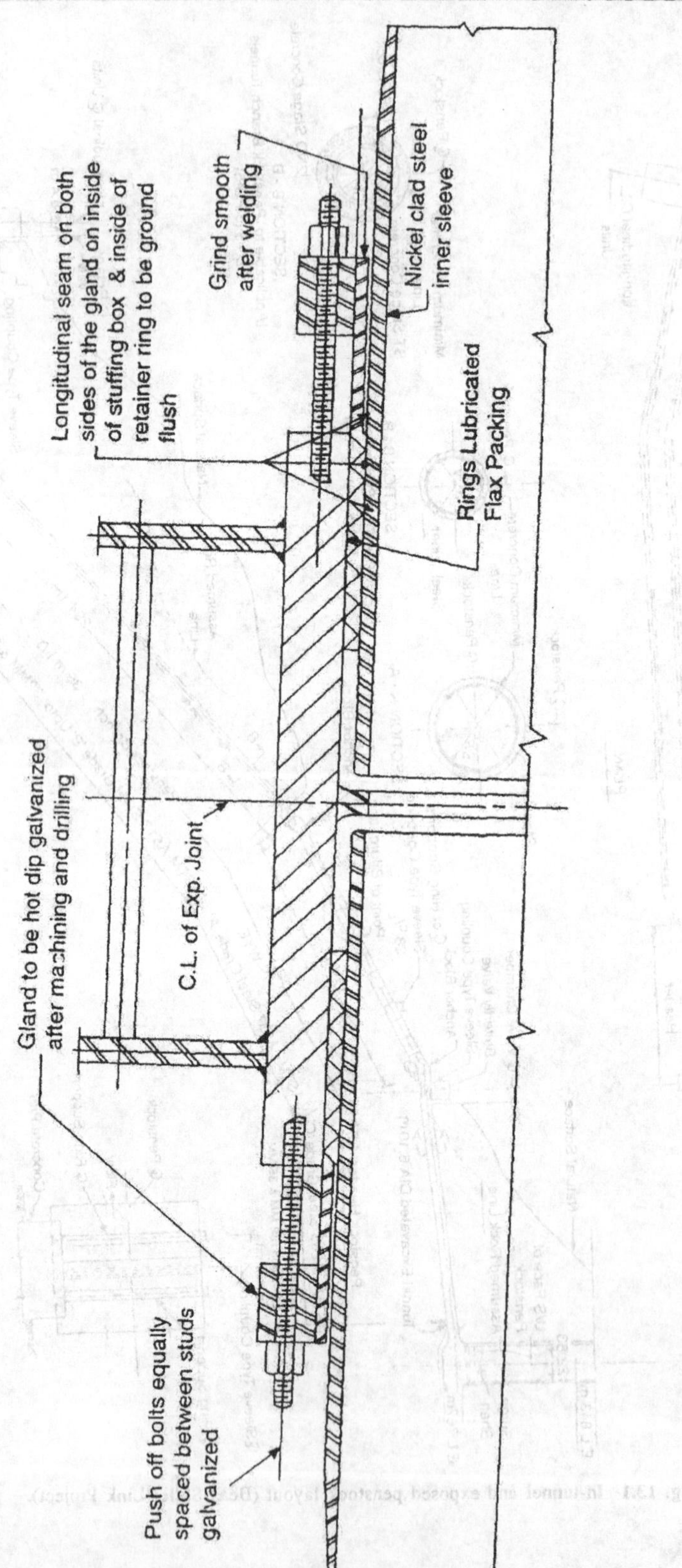

Fig. 13.1(a) Sleeve type expansion joints

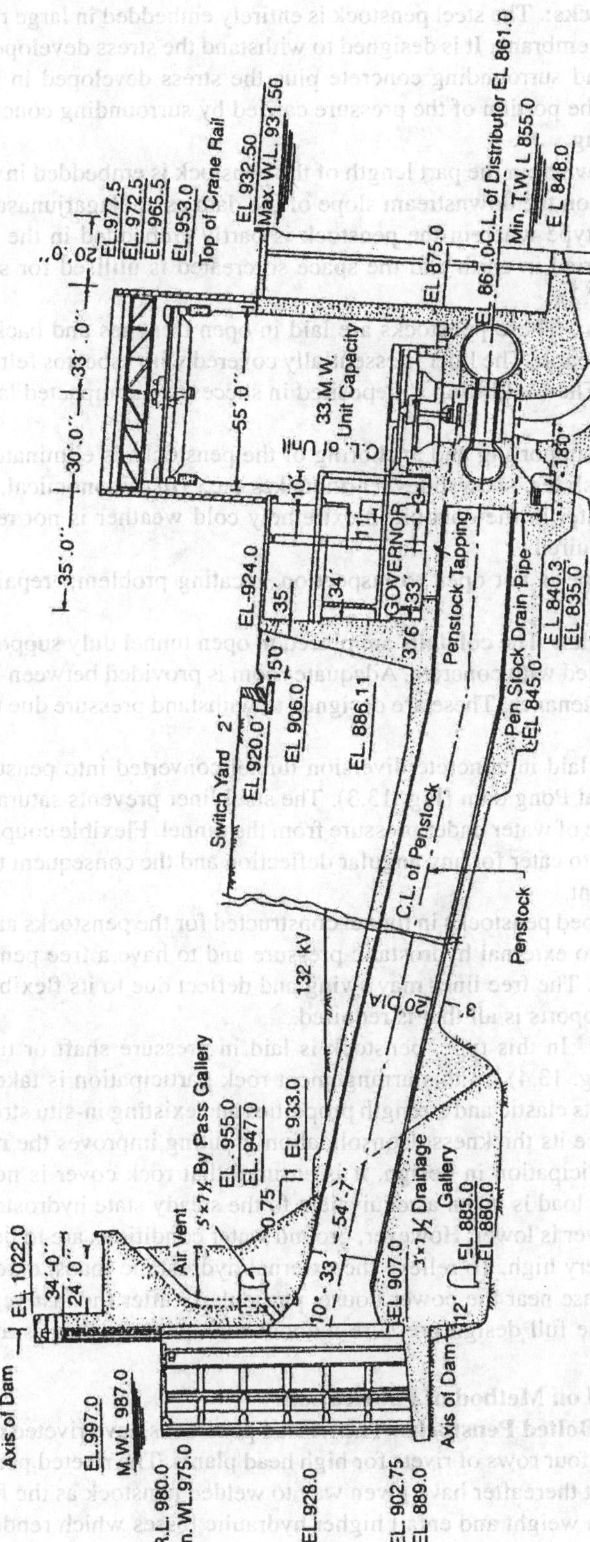

Fig. 13.2 Penstock embedded in concrete tunnel.

(*ii*) **Embedded Penstocks:** The steel penstock is entirely embedded in large mass of concrete dam serving as watertight membrane. It is designed to withstand the stress developed in closing the initial gap between liner and surrounding concrete plus the stress developed in the liner due to remainder of pressure less the portion of the pressure carried by surrounding concrete and rock. It does not require any grouting.

A variant of the type is wherein the part length of the penstock is embedded in the dam and the balance length is supported on the downstream slope of the dam as at Nagarjunasagar dam.

Another variant is the type wherein the penstock is partly embedded in the dam and partly encased in concrete and buried in earth and the space so created is utilised for switchyard as at Jawahar Sagar dam (Fig. 13.2).

(*iii*) **Buried Penstocks:** These penstocks are laid in open trenches and backfilled to give a cover of 1.25 to 1.5 m of the earth. The liner is essentially covered with asbestos felt wrap and white washed before backfilling. The backfilling is deposited in successive compacted layers.

Advantages

(*i*) Difficult and costly supporting and anchoring of the penstocks is eliminated, (*ii*) Not susceptible to damage by landslides, snowslides, earthquakes etc., (*iii*) Economical, (*iv*) Protection needed from freezing of water in the conduit in extremely cold weather is not required, and (*v*) Expansion joints are not required.

However, the buried pipe is not open to inspection, locating problem, repairs difficult and costly.

(*iv*) **In-Tunnel Penstocks:** The conduits are placed in open tunnel duly supported suitably as surface penstocks or backfilled with concrete. Adequate room is provided between liner and tunnel wall for inspection and maintenance. These are designed to withstand pressure due to ground water and grout pressure.

The penstocks are also laid in concrete diversion tunnel converted into penstock tunnel and backfilled with concrete as at Pong dam (Fig. 13.3). The steel liner prevents saturation of the surrounding rock due to seepage of water under pressure from the tunnel. Flexible coupling is provided upstream of the power plant to cater for any angular deflection and the consequent tilt and to guard against differential settlement.

UBBR practice is to embed penstocks in tunnel constructed for the penstocks and the liners are secured from buckling due to external hydrostatic pressure and to have a free penstock liner in a bigger concrete lined tunnel. The free liner may swing and deflect due to its flexibility and provision for movement at the supports is all that is required.

(*v*) **In-shaft Penstock:** In this type, penstock is laid in pressure shaft or tunnel shaft and backfilled with concrete (Fig. 13.4). In this arrangement rock participation is taken into account appropriately depending on its elastic and strength properties and existing in-situ stress levels in the design of steel liner to reduce its thickness. Consolidation grouting improves the rock properties. While considering rock participation in design, it is ensured that rock cover is not less than 0.4 times the head. The external load is taken as equivalent to the steady state hydrostatic head or the depth of overburden whichever is lower. However, ground water conditions are to be established in cases where rock cover is very high. To relieve the external hydrostatic loads, extensive drainage system is installed. In any case near the power house, particularly after the last vertical bend, the steel liner is designed for the full design pressure without rock participation as also where rock cover is less than 40%.

2. Classification Based on Method of Fabrication:

(*i*) **Riveted or Flange Bolted Penstocks:** The riveted penstocks have riveted circumferential joints which involve three to four rows of rivets for high head plants. The riveted penstocks were in vogue in India up to 1950 but thereafter have given way to welded penstock as the former involves more material, are heavier in weight and entail higher hydraulic losses which render them uneconomical.

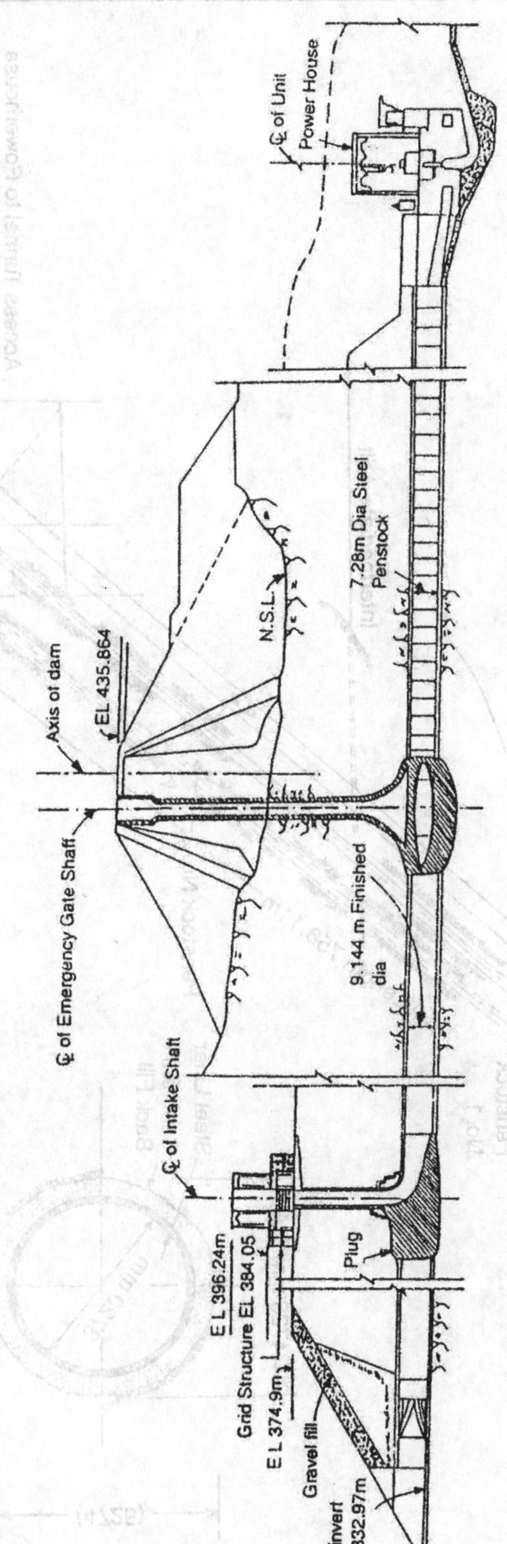

Fig. 13.3 Penstock embedded in concrete in tunnel at Pong dam.

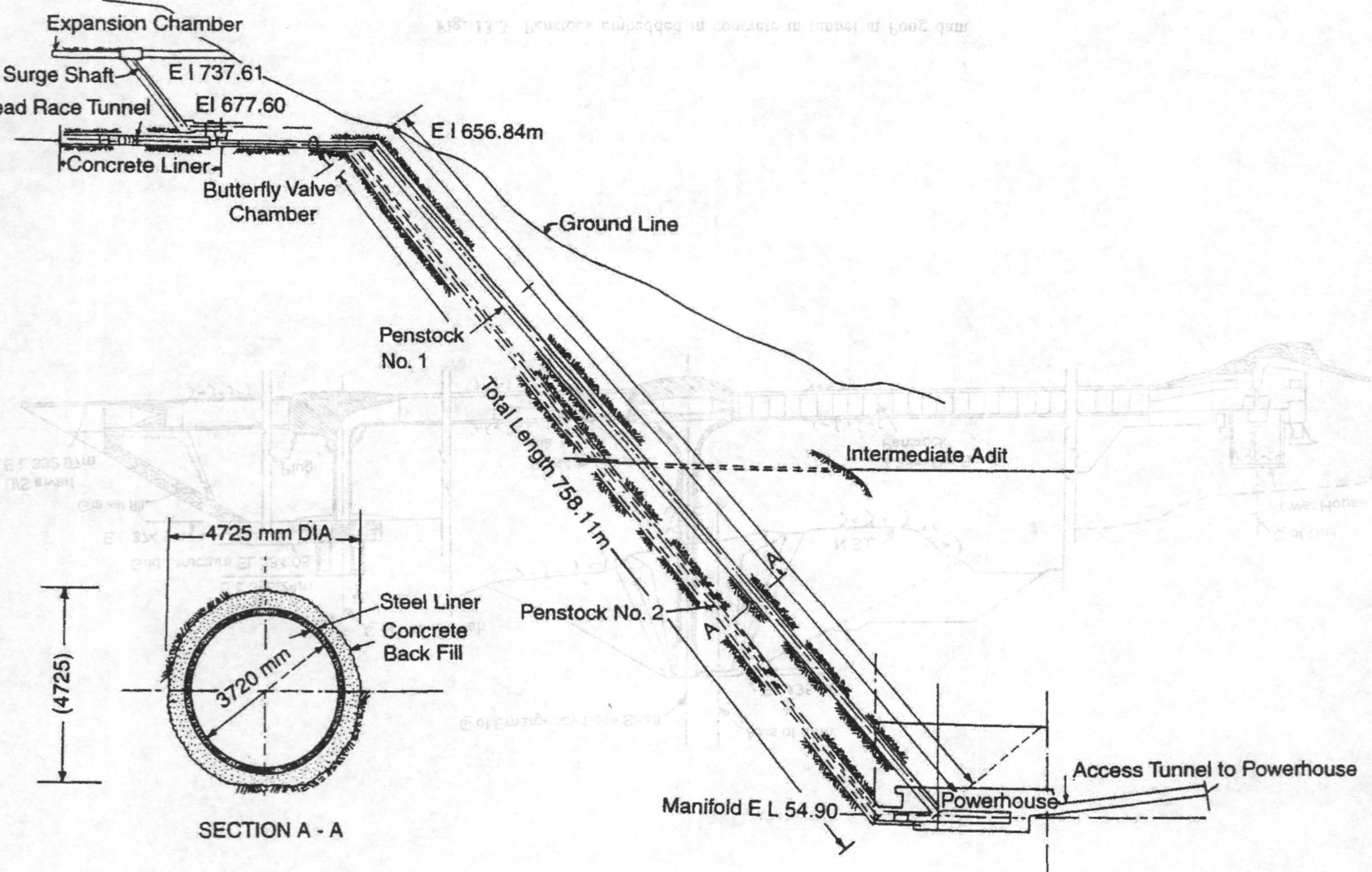

Fig. 13.4 Penstock in shaft tunnel.

(*ii*) **Welded Penstock:** With the advancement in welding techniques and development of higher heads, all welded steel penstocks are mostly used. The pipes are formed by butt welding the longitudinal and circumferential joints. The advantages over riveted penstocks are (*i*) Lesser in weight, (*ii*) Easier to fabricate and erect, and (*iii*) Lesser hydraulic losses.

(*iii*) **Multilayered Penstocks:** It is an improved type of penstock compared to the welded one. It consists of several layers of thin steel plates wrapped around the pre-fabricated central core pipe by a special wrapping machine. The internal pressure is resisted by the interaction of layers.

The advantages over welded penstock are (*i*) Saving in material, and (*ii*) Use of thin plates eliminates stress relieving. However, cost of fabrication and installation is high.

(*iv*) **Banded or Hooped Penstocks:** In this type, bands or hoops are slipped over thin walled penstock pipe (by cold process or hot process) which induce prestress in the pipe as a result of which high operating heads are carried by comparatively thin pipes. The banded penstocks are designed for equal stress in the hoop and in the pipe under maximum operating pressure. Saving in material is the merit but are expensive mainly because of special fabrication process. This type of penstocks have been used in Jogindernagar power house (Himachal Pradesh).

3. Classification Based on Material of Fabrication: The fabrication of penstocks is made from a variety of materials depending upon head, discharge and topography such as steel, reinforced cement concrete and wood. The commonly used materials are steel and reinforced cement concrete.

(*i*) **Steel Penstock:** Steel penstocks are extensively used than that of any other material for heads above 50m. The welded pipe, as is invariably used these days, offers smooth interior surface. The minimum thickness of exposed steel pipe is computed by the USBR formula (D + 50)/400 (cm), where D is diameter of shell (cm). As per IS : 11639 (Part 2) : 1995 minimum handling thickness of steel lining in buried/embedded penstock is:

$$t = (R + 0.25)/200$$

where, t is thickness (m), and R is penstock radius (m). The difference in plate thickness in adjoining ferrules is not to be more than 5 mm.

For large diameter penstocks, over 5 m, stiffening rings are provided to prevent distortion and opening of seams. Allowance of about 1.5 mm is also added to allow for corrosion.

Steel Penstocks at Pong Power House. Two penstock headers of 7.28 m diameter and 305 m long, each bifurcating into two branch penstocks of 5.026 m diameter and further reduced to 4.5 m diameter through combined reducing bends (Fig. 13.5) upstream of the junction with the scroll casing and butterfly valve are provided at Pong power plant. Bifurcation (wye), the largest of its kind in Asia weighing about 190 tonne, was fabricated out of ASTM- A-537 Grade A and ASTM-A-285 grade *C* steel. Each branch penstock feeds 86200 mhp turbine coupled to 60 MW generator. The header and wye were fabricated out of 48 mm thick plates.

The test pressure for the penstock assembly was computed from the formula:

$$P = \frac{1.5(2 \times s \times e)t}{d} \text{ (kg/cm}^2\text{)}$$

where, P is test pressure (kg/cm²), t is thickness (mm) of the thinnest plate in the penstock assembly, d is maximum internal diameter (mm) of penstock assembly, s is design stress of 1052-1640 kg/cm² for the plates used and e is joint efficiency 95%.

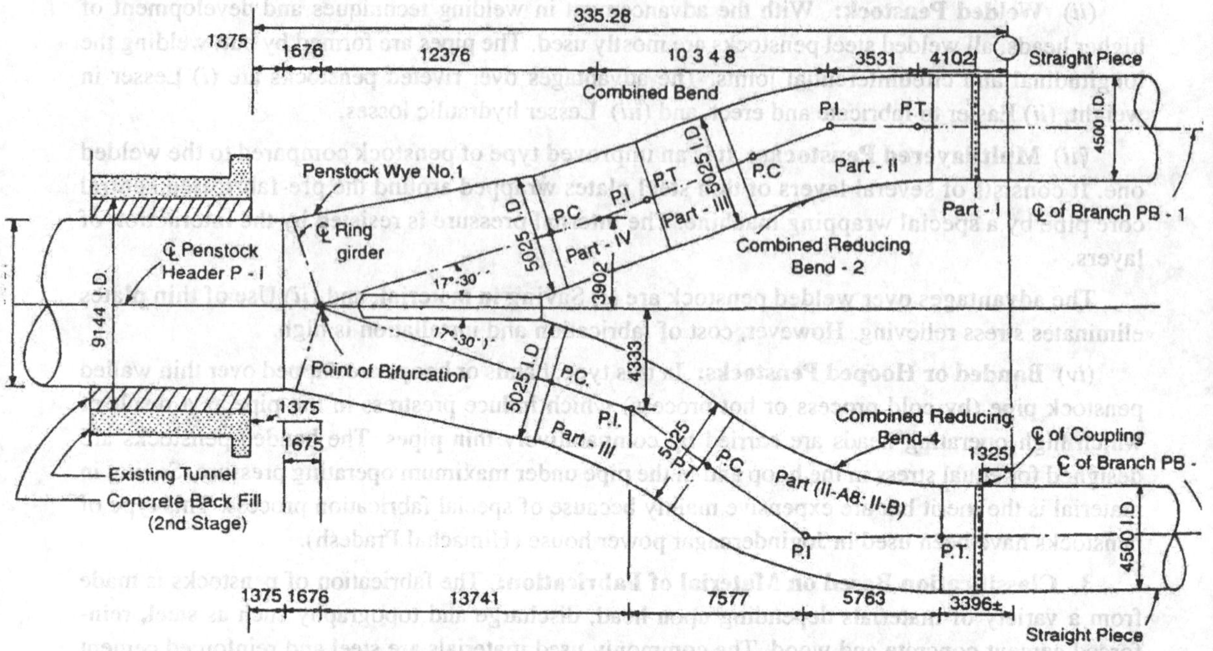

Fig. 13.5 Profile and general arrangement of penstocks at pong power house.

On each penstock branch, one sleeve type expansion coupling supplied by M/s Dresser Manufacturing Co. USA (Fig. 13.5a) was installed at its downstream end to allow for any longitudinal and angular movements or any differential tilt in the power house with respect to concrete block encasing penstock branches.

(ii) **Reinforced concrete Penstock:** R.C.C penstocks are used for heads up to 20 m. R.C.C pipe is more economical than steel for moderate pressures. The thickness of lining is such so as to get a relatively impervious shell. Usually the invert is slightly thicker than the shell. The penstocks are generally buried under 1 m of fill to minimise in expansion and contraction joints and tendency to leakage is omitted. The working stress in the reinforcement is taken as 60 percent of the normally adopted stress to take care of the effects of water hammer and safety margin. The concrete of good quality is used to develop full stress in the concrete. The limitation of R.C.C. penstocks is that the concrete deteriorate rapidly in colder climates under alternative freezing and thawing.

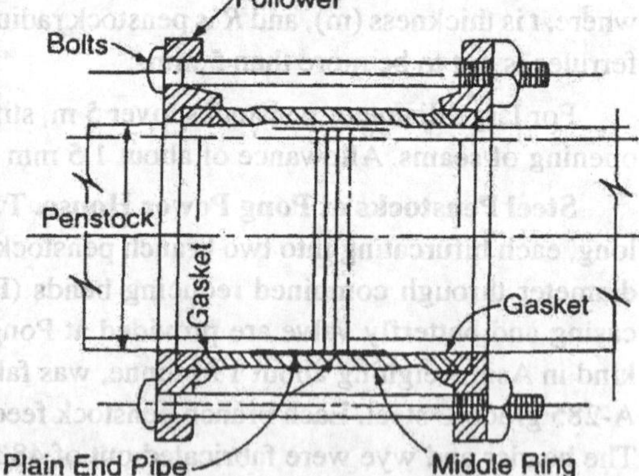

Fig. 13.5(a) Dresser coupling

(iii) **Glass Fibre Reinforced Penstock:** Penstocks of high density polyethylene (HDPE), unplasticised poly vinyl chloride and glass fibre rainforced pipe are being used worldwide and in India. Though these are not cost-effective compared to steel penstocks but have the merit of better hydraulic efficiency, ease in installation, transportation and maintenance.

13.3 LAYOUT OF PENSTOCKS

The layout and arrangement of penstocks depends upon the type of development, site conditions, topography and relative location of dam and power plant. The topographic features of the sloping terrain mainly determine the ideal alignment of the penstocks. The direction of the alignment may have to be changed in order to obtain the most economical profile, avoiding excess of excavation and poor foundation strata as also to keep the aesthetic beauty of the natural scenery. The changes in the direction of alignment may be either in profile or in plan or in both directions.

1. Alignment: The penstock is to follow the shortest route for economy and to minimise head loss. Deviation from straight route is not infrequent in which case suitable horizontal curves are provided. Sharp bends are avoided.

2. Grade: The penstocks are pressure conduits and their hydraulic gradient may or may not be in the direction of flow. The grade of a penstock is that it is always sloping towards the power plant; the grade varies to suit the topography but below its maximum hydraulic gradient, prevailing when peak capacity is reached at low head water level, so as to avoid development of negative pressures and air troubles. In general as flat a grade as is possible, but kept below the hydraulic gradient for maximum length, to minimise pressures and cost of pipe and a sharp dip in a short length near to the power house is adopted for reasons of economy.

3. Branches and Wyes: The wye and branches are given special care to ensure smooth, stream-lined flow from header to the branch and to minimise hydraulic losses by introducing conical reducers and keeping angle of deflection small; between 30° to 75° but normally not less than 45°. However, if this angle is too small, it is difficult during fabrication to have access to all parts for making joints of the assembly and to provide for adequate reinforcement for the assembly.

4. Fall: In case of concentrated fall development, the power plant is located close to dam and the penstocks are generally short and embedded in concrete or masonry of the dam.

In divided fall, the power plant is located a little farther away from the dam and depending upon the utility of space between the dam and power house, the penstocks are encased in concrete before burying in earth as for Jawahar Sagar dam where the space so created is utilised as switch yard (Fig. 13.2).

5. Diversion tunnel conversion: After diversion stage, the diversion tunnel earmarked for the purpose is converted into penstock tunnel by installing penstock liners inside the tunnel and plugging the intake (Fig. 13.3).

6. Tunnel shaft: For a very high head development, where underground power house is located just below the reservoir, the liners are placed in tunnel shafts and backfilled with concrete as at Koyna stage III project. Water may be led through pressure shafts or pressure tunnels with surge tanks as in the case of Tehri dam. The layout should be such that the length of the pressure shaft/ pressure penstock is the minimum consistent with economy.

7. Surface penstock: In high and medium head penstocks, water from dam or diversion structure is conducted by open channel or tunnel up to forebay or surge shaft at a suitable point and from there carried by surface penstocks as in the case of Upper Sileru project and Balimela project.

A combination of surface penstock and penstock in tunnel is resorted to, depending on topography as in Beas Sutlej link project (Fig. 13.1).

13.4 HYDRAULIC DESIGN OF PENSTOCKS

The hydraulic design of a penstock involves determination of hydraulic losses, pressure rise or pressure drop due to turbine or pump operation and ascertaining of most economic diameter of penstock. The provision of surge tank or pressure relief device in the penstock is also considered.

1. Hydraulic Losses

A penstock has rarely straight, uniform alignment and it involves a variety of deviations like partial obstacles, change in section, branches, bends etc., which impose additional losses other than the friction losses. The various losses in the penstock from intake to power house comprise (*i*) head loss at trashrack, (*ii*) head loss at intake entrance, (*iii*) friction losses, and (*iv*) other losses as at bends, bifurcations, transitions, valves etc.

Head losses in penstocks

At trashrack	*At intake entrance*	*Friction loss*	*Bend loss*	*Expansion*	*Contraction*	*In reducer*
$h_t = k_t\left(\frac{V^2}{2g}\right)$	$h_e = k_e\left(\frac{V^2}{2g}\right)$	$h_f = \left(\frac{fLv^2}{2gD}\right)$	$h_b = k_b\left(\frac{V^2}{2g}\right)$	$h_{ex} = k_{ex}\left(\frac{V_1^2}{2g} - \frac{V_2^2}{2g}\right)$	$h_c = k_c\left(\frac{V_2^2}{2g} - \frac{V_1^2}{2g}\right)$	$h_r = k_r\left(\frac{V_c^2}{2g} - \frac{V_n^2}{2g}\right)$

where, k_t is loss coefficient, V is velocity (m/s), g is acceleration due to gravity, k_e is loss coefficient; 0.5 for sharp cornered, 0.23 for slightly rounded and 0.05–0.1 for bell-mouthed entrance, f is loss coefficient depending on type, pipe condition and Reynold number, L is pipe length (m), D is pipe diameter (m), R is hydraulic mean depth (m), S is slope of energy gradient, n is roughness coefficient; 0.012–0.014 for concrete pipe and 0.08–0.012 for steel pipe, k_b is bend loss coefficient, k_{ex} is loss coefficient, depending upon cone angle, V_1 and V_2 velocities at u/s and d/s ends respectively, k_c is loss coefficient for contraction, v_n and v_c are velocities in normal and contraction sections respectively. Losses also occur at control valve and inlet valve in the penstock at the upstream and downstream ends immediately ahead of turbine respectively. Losses also occur in penstock wyes and branches, details for which are given in IS: 11625–1986.

2. Sizing of Penstocks

The number and size of penstocks to be installed in a power plant are determined from water power studies taking into consideration hydrological data and head available for power generation. Large diameter penstock may sometimes be more economical to use, as many penstock lines as the number of units, but, in general, single penstock is used to feed more than one units by proper bifurcations and manifolds.

1. Number of Penstocks: The number of penstocks to be installed depends upon (*i*) Spacing and size of units, (*ii*) Location of take-off of penstocks, (*iii*) Size easy for handling, fabrication, transportation and erection, (*iv*) Head losses occurring at manifolds vs losses in single line, (*v*) Cost of civil works like number of piers, anchor blocks etc., (*vi*) Flexibility in operation, and (*vii*) Thickness of liner not to exceed 50 mm.

The number of penstocks is determined by economical analysis keeping in view the above mentioned factors. A comparison of single penstock to multi-penstock for identical velocity or identical head loss condition is summarised in Table 13.1.

Table 13.1 Single vs multi-penstock

Item	*Single penstock*	*'n' number of penstocks*	
		Identical velocity	*Identical head loss*
Discharge	Q	Q/n	Q/n
Diameter	d	d/n	$d/n^{2/5}$
Velocity	v	v	$v/n^{1/5}$
Head loss	h	$hn^{1/2}$	h
Wall thickness	S	$S/n^{1/2}$	$s/n^{2/5}$
Total weight	G	G	$G/n^{1/5}$

3. Size of Penstock

The diameter to be adopted for a single line or a number of lines is determined by economic diameter studies.

Having determined the number of penstocks and discharging capacity of each, the next step is to determine the size of the penstock. The optimum size of penstock is based upon the loss of head at average load and the corresponding velocities are average velocities. The average loss of head is the average of the losses at the different operating turbine gates rather than the loss corresponding to average velocity. The average wheel-unit load for a 50 percent capacity factor may be about 0.7 of the wheel capacity and the average velocity head about 0.5 that at full gate capacity. As such the penstock size determined on average wheel discharge or wheel capacity × capacity factor gives approximately correct results. However, in long penstocks the losses at full capacity are considered and bigger size penstock is required so that plant capacity is not reduced due to higher loss of head with the plant fully loaded.

The maximum penstock velocity for full wheel discharge is determined as under:

Turbine	*Head (m)*	*Velocity (m/s)*
Reaction	150–175	$0.05–0.07\sqrt{2gh}$
Impulse	above 150	$0.02–0.03\sqrt{2gh}$
h is effective head on plant.		

The next variable to the velocity in the penstock is the determination of its diameter. The various factors to be taken into consideration in the determination of economic diameter of the penstock are value of power lost in friction, maximum permissible velocity, total annual cost, number of penstocks, profile of penstock and cost of anchor blocks, if any, thickness of penstock plate, variation of flow, load factor etc. The friction loss too depends on velocity. Thus larger the diameter for a given discharge, smaller is the loss of head and higher is the net head available to the turbine and hence higher power production. The big size penstock means lesser velocity and higher capital outlay whereas the requirement is evidently for minimum annual costs. Economic evaluation of the various alternatives has to be made to work out optimum diameter accompanied by maximum economy.

13.5 ECONOMIC DIAMETER OF PENSTOCKS

1. I.S.I Method: Precisely the economic diameter of penstock is the diameter for which the annual cost, which includes the cost of power lost due to friction and charges for amortization of construction cost, maintenance, operation etc. is the minimum. The economic diameter is determined by evaluating annual power loss and annual cost for maintenance and equating first derivative with respect to D to zero and is given by the equation based on Manning's formula (as per IS: 11625-1986)

$$D^{22/3} = \frac{2.36\times10^6\times Q^3\times n^2\times ep_f\times C_p}{\left[1.39\,C_e+0.6\,C_c+\dfrac{121\,HC_s(1+i)}{\sigma_a e_j}\right]\times p}$$

Where, C_c is unit cost of concrete lining (Rs/m³), C_e is unit cost of excavation (Rs/m³), C_p is cost of 1 kWh of energy (Rs), C_s is cost of steel (Rs/kg), D is diameter of penstock, e = overall efficiency of plant, e_j is joint efficiency of penstock, H is head on penstock including water hammer (m), i is percentage by which steel in penstock is overweight due to provision of stiffeners, corrosion allowance etc., n is rugosity coefficient in Manning's formula, p is ratio of annual fixed operation and maintenance charges to construction cost of penstock, p_f is annual load factor, and Q is discharge through penstock (cumecs). The derivation of the formula is as under:

The economic diameter of a penstock is the diameter for which the annual cost which includes the cost of power lost due to friction and charges for amortization of construction cost, maintenance cost etc. is minimum.

(*i*) Cost of power lost

Head loss in penstock/m length, (By Manning's formula)

$$h_f = \frac{v^2n^2}{R^{4/3}} = \frac{10.29\,Q^2n^2}{D^{16/3}}$$

Annual cost of power lost,

$$E_f = 9.804\times Q\times h_f\times ep_f\times 8760\,C_p$$

$$E_f = \frac{0.88\times10^6\,Q^3n^2\,ep_f\,C_p}{D^{16/3}}$$

(*ii*) Annual charges on capital cost

Cost of excavation, considering tunnel dia is 0.33 D in excess of penstock dia, total cost/unit length of penstock.
$$= \frac{\pi}{4}(D+0.33\,D)^2\,C_e$$
$$= 1.39\,D2\,Ce$$

(*iii*) Cost of concrete lining taking thickness of lining as 0.165 D
$$= \pi\,(D+0.165\,D)\times 0.165\,D\times C_c$$
$$= 0.6\,D^2\,C_c$$

(*iv*) Cost of steel in penstock

Steel lining thickness,
$$t = \frac{pd}{2\sigma_a e_j} = \frac{0.1\,HD}{2\sigma_a e_j}$$

Penstock cost
$$= \frac{\pi\,D_t\,(1+2)\,7850\times C_s}{2\sigma_a\times 9.81\times e_j}$$
$$= \frac{121\,HD^2\times C_s\,(1+i)}{\sigma_a\times e_j}$$

(*v*) Annual charges on capital cost

Annual cost of penstock
$$E_p = P[D_t\,(1.39\,C_e + 0.6\,C_c + \frac{121\,Hc_s\,(1+i)}{\sigma_a\,e_j}$$

1. Economic Diameter

Total annual cost,
$$E = E_p + E_f$$

Economic diameter, D
$$= \frac{\delta(E_p+E_f)}{\delta D} = 0$$

$$D^{22/3} = \frac{2.36\times 10^6\times Q^3\times n^2\times ep_f\times C_s}{\left[1.39\,C_e + 0.6\,C_c + \frac{121\,H_{cs}\,(1+i)}{\sigma_a\times e_j}\right]p}$$

2. Economical Velocity As A Proportion of the Spouting Velocity
$$V = 0.125\sqrt{2gh}$$
$$D = \sqrt{\frac{Q}{V\times 0.7854}}$$

3. Empirical Formula

(*i*)
$$D = (P/H)^{0.466}$$

(*ii*)
$$D = 4.44\,\frac{P^{0.43}}{H^{0.65}}$$

4. Rational Method: Penstocks are designed to carry water to the turbines with the least possible loss of head consistent with the overall economy of the installation. The most economical diameter of the penstock is the one which gives the minimum value of annual loss of revenue due to loss in energy resulting from friction in the pipe plus the annual operating charges, which include interest on investment, charges for maintenance and depreciation etc.

The economical diameter is dependent upon the cost of penstock, the value of power lost in friction and more or less arbitrary allowances for depreciation or replacement service. These elements vary considerably during the life of a plant.

The computation for economical diameter is done on the basis of the equivalent discharge passing through the penstocks. Equivalent discharge is that constant discharge which, when running through the penstock for a particular period, will cause the same power loss as due to actual varying discharges in day to day variation of power load during the same period. As loss of power due to frictional resistance is proportional to nearly the cube of the discharge passing through the penstock, the equivalent discharge through a penstock during any particular period is:

$$\text{Equivalent discharge} = \sqrt[3]{\frac{\Sigma Q^3 \times \Delta t}{t}}$$

where, Q is the discharge passing through the penstock at any time, Δt is duration of discharge Q, and t is the total time for which the equivalent discharge is required.

The discharge at any time is also dependent upon the load variation factor and the efficiency of the generating units, etc. Ignoring small variation in efficiency in a particular week, the losses are dependent upon the cube of the load variation factor. Consideration is also given to the number of units working to cater for the total discharge through the power plant, as each header feeds two units.

After computing the mean equivalent discharge, loss of power is calculated by finding out the head loss due to friction by Scobey's formula:

$$h_f = \frac{K_s \cdot V^{1.9}}{D^{1.1}} \times \frac{L}{1000}$$

where, K_s is coefficient, value 0.37 for 10 year old pipe, V is velocity (ft/s), D is diameter (ft) and L is length (ft).

4. Sarkaria Formula $\quad D = 0.62\left(\dfrac{P^{0.43}}{H^{0.65}}\right),\ D = 3.55\left(\dfrac{Q^2}{2gH}\right)^{0.25} \quad V = 0.125\sqrt{2gH}$

where, D is penstock diameter (m), P is rated H.P. (metric) of turbine, H is rated head of turbine (m), Q is discharge in penstock (cumecs) and V is velocity in penstock (m/s).

5. Graphical Method: Graphically too the economic size of penstock can be determined by choosing five values of diameters and corresponding cost. Let total annual cost of penstocks corresponding to any diameter is plotted as curve T. Let cost on account of price of energy lost in friction (H.P.) at the prevalent rate of its sale is plotted E. The plot of total cost T_E is the sum of ordinates of T and E corresponding to any diameter. The economical diameter corresponds to point A on the curve T_E (Fig. 13.6).

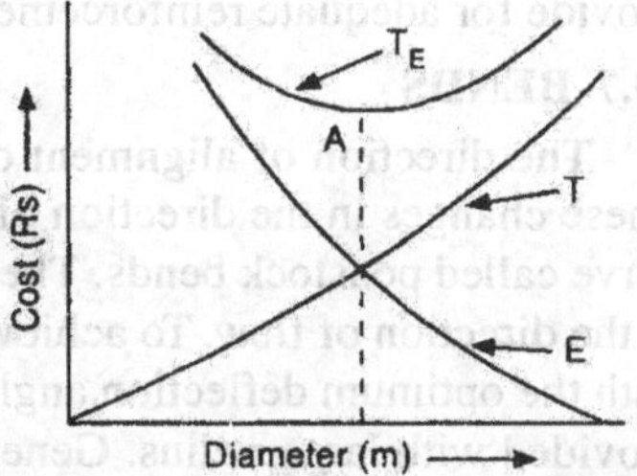

Fig. 13.6 Economical diameter

3. Design Loads

(*i*) **Internal pressure:** The penstock is designed for maximum internal pressure caused due to maximum water level in the reservoir or in surge tank, as the case may be, plus anticipated increase in pressure due to water hammer effect development when arresting or releasing the flow of water

(*ii*) **Pressure wave velocity:** It is computed by the formula:

$$a = \frac{1425}{1 + (d/100t)}$$

where, a is pressure wave velocity (m/s), d is penstock diameter (m), t is penstock shell thickness (*m*).

(*iii*) **External pressure:** The external pressure head is either the difference between ground level vertically above the penstock and the penstock invert level or the maximum level from which the water is likely to find its way around the liner whichever is less.

(*iv*) **Longitudinal stresses caused by radial strain:** Radial expansion of steel caused by internal pressure tends to cause longitudinal contraction with corresponding tensile stress equal to 0.303 times the hoop tension in the circular liner.

13.6 BRANCHES AND WYES

Each unit of power plant is not provided with individual penstocks. Generally a header penstock with branches to feed each unit is provided but it involves loss of head at the junction and is avoided. Branch pipes of bifurcating pipe are generally known as "wye" pieces which may be symmetrical or assymetrical. Depending upon the number of units a single penstock feeds, the penstock branching is defined as bifurcation when feeding two units, trifurcation if feeding three units and manifold when feeding more than three units.

The wye is an intricate hydraulic structure continuously varying in size and shape and is fabricated with utmost care. The joints and welds for imperfections are tested ultrasonically and radiographically. The wye parts and shell thicker than 38 mm are stress relieved. They are then hydrostatically tested at pressure 150 percent of the design pressure of the penstock pipe.

Bifucation Trifurcation Mainfold

Fig. 13.7 Wye branching of penstock

The basic criteria for designing these wyes is structural safety combined with hydraulic efficiency. The branch pipes are proportioned to ensure smooth, streamline flow from the header to the branch and to minimise hydraulic losses, by introducing conical reducers and keeping angle of deflection small. Generally the branches are provided with deflection angles ranging from 30° to 75°; normally not less than 45°. If the angle is too small it becomes difficult during fabrication to have access to all parts for making joints of the assembly and to provide for adequate reinforcements for the assembly.

13.7 BENDS

The direction of alignment of the penstocks have to be changed due to a variety of reasons. These changes in the direction of alignment are negotiated by providing successive segments in a curve called penstock bends. The bends are designed to minimise the hydraulic loss due to change in the direction of flow. To achieve this the successive segments of the curved portion are designed with the optimum deflection angles to avoid sharp changes in the direction of flow and the bend is provided with large radius. Generally the radius of bend is 3 to 5 times the diameter of pipe. The deflection angle between each successive segments is limited to 5° to 10°; the most recommended value is 4° to 6°.

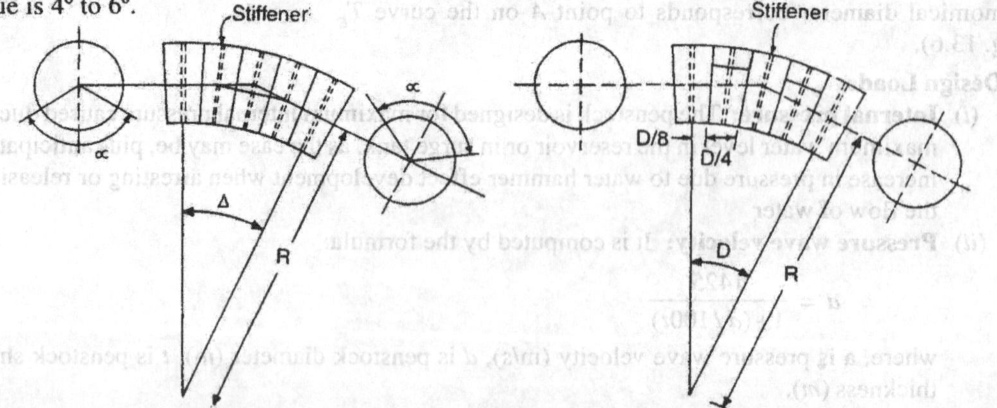

Fig. 13.8(*a*) Simple bend. **Fig. 13.8(*b*)** Compound bend

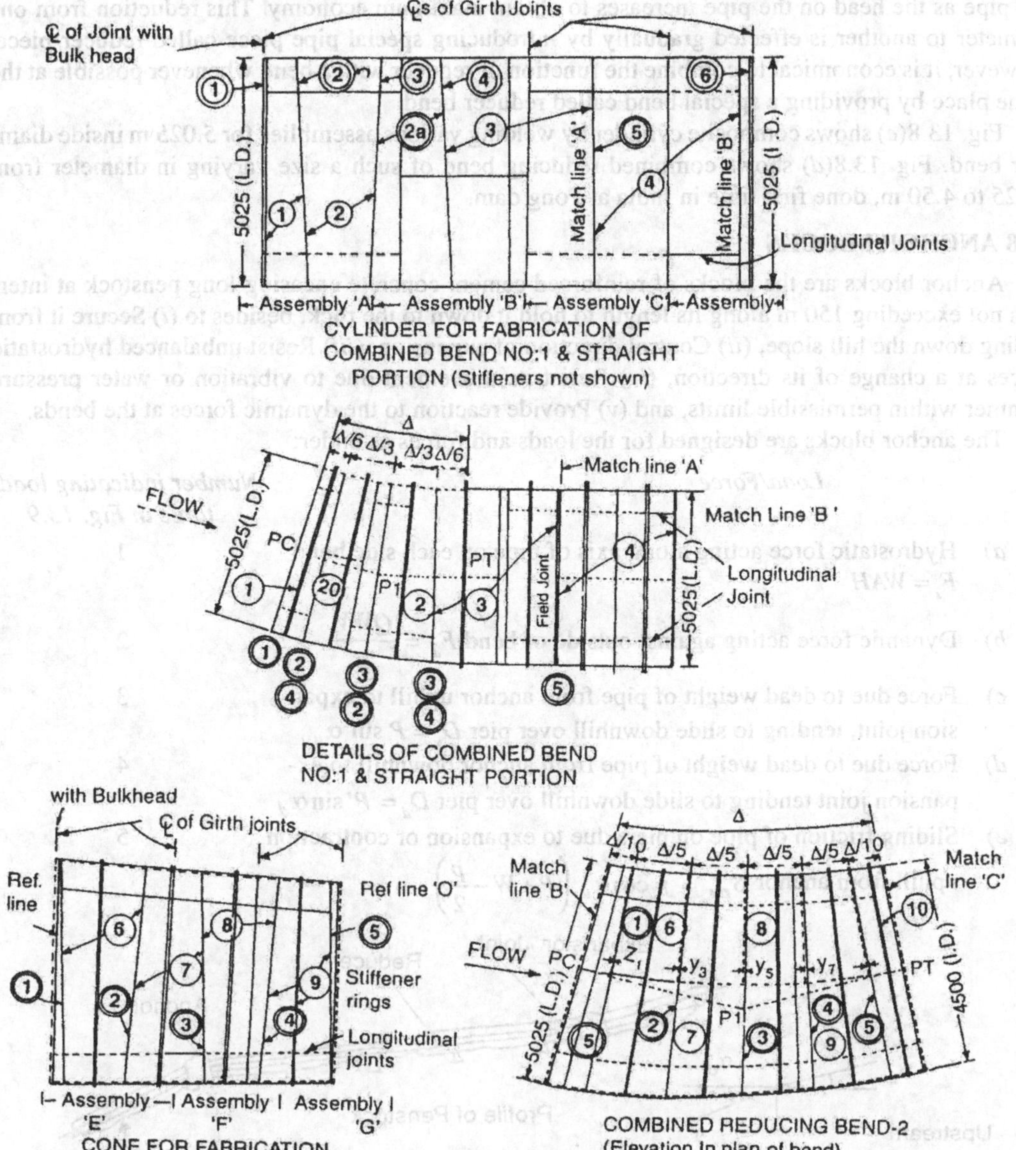

Fig. 13.8(*c*) & (*d*) Combined bend and combined reducing bend.

Types of Bends

Penstock bends may be classified as under:

Simple Bend: When the change in alignment is only in plane that is either vertical or horizontal the bend is called simple bend and the deflection angle is the deviation in the direction of alignment (Fig. 13.8*a*).

Compound Bend: The penstock alignment may be deflected both in horizontal plane and vertical plane to conform to the contour and alignment. The construction is greatly simplified if the pipe is bent to one angle which will accommodate the deflections in both planes. This is called a compound bend (Fig. 13.8*b*).

Reducer Bend: In long and very high head penstocks, it is necessary to reduce the diameter of the pipe as the head on the pipe increases to obtain maximum economy. This reduction from one diameter to another is effected gradually by introducing special pipe piece called reducer piece. However, it is economical to combine the function of reducer with a bend whenever possible at the same place by providing a special bend called reducer bend.

Fig. 13.8(*c*) shows composite cylinder by welding various assemblies for 5.025 m inside diameter bend. Fig. 13.8(*d*) shows combined reducing bend of such a size varying in diameter from 5.025 to 4.50 m, done first time in India at Pong dam.

13.8 ANCHOR BLOCKS

Anchor blocks are the blocks of reinforced cement concrete encasing long penstock at intervals not exceeding 150 m along its length to hold it down to the rock, besides to (*i*) Secure it from sliding down the hill slope, (*ii*) Control direction of expansion, (*iii*) Resist unbalanced hydrostatic forces at a change of its direction, (*iv*) Resist its movement due to vibration or water pressure hammer within permissible limits, and (*v*) Provide reaction to the dynamic forces at the bends.

The anchor blocks are designed for the loads and forces as under:

Load/Force	*Number indicating load/ force in Fig. 13.9*
a) Hydrostatic force acting along axis of pipe on each side bend $F_s = WAH$	1
b) Dynamic force acting against outside of bend $F_d = \dfrac{QWV}{g}$.	2
c) Force due to dead weight of pipe from anchor uphill to expansion joint, tending to slide downhill over pier $D_u = P \sin \alpha_u$	3
d) Force due to dead weight of pipe from anchor downhill to expansion joint tending to slide downhill over pier $D_d = P' \sin \alpha_d$	4
e) Sliding friction of pipe on piers due to expansion or contraction uphill from anchor $S_{pu} = f \cos \alpha_u \left(P + W - \dfrac{P}{2}\right)$	5

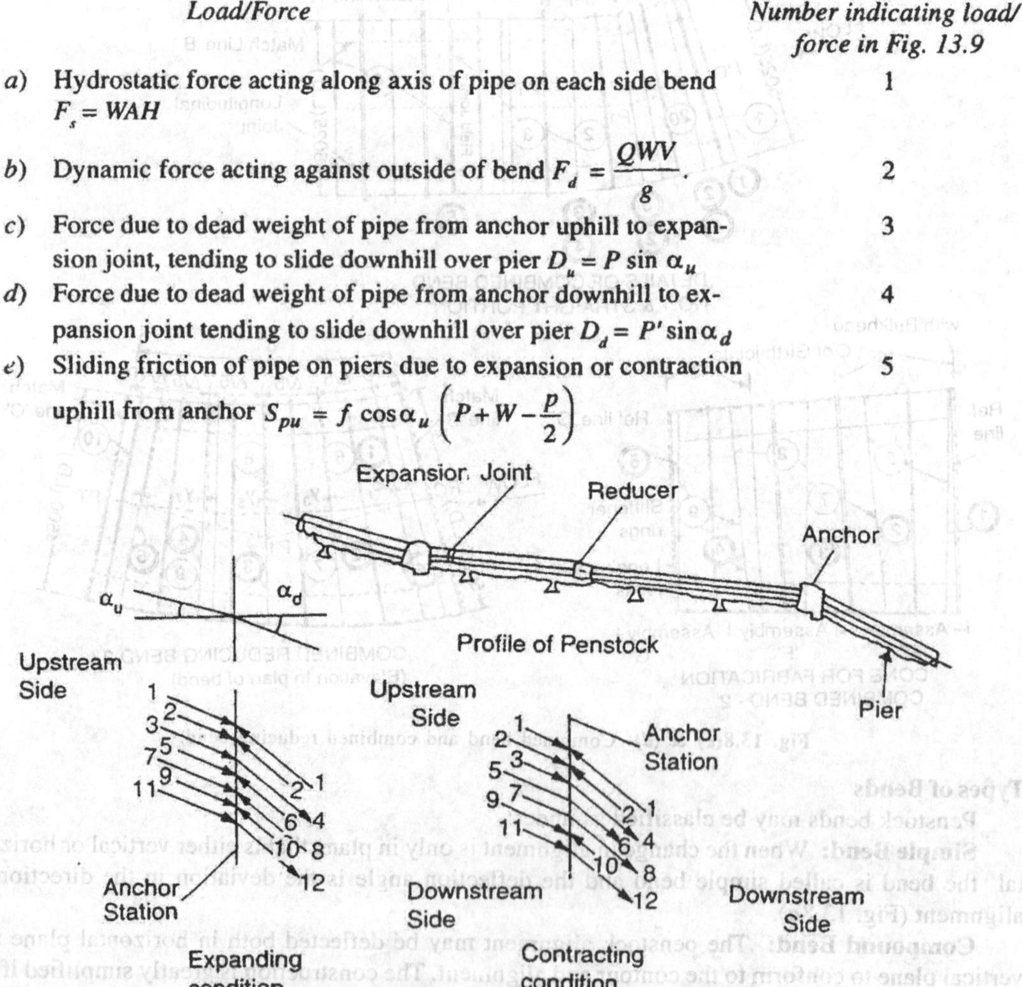

Note — The expansion joint may be provided in the end of a reach as indicated in the figure or in the middle of the reach.

Fig. 13.9 Action of loads/forces on anchor block

	Load/Force	*Number indicating load/ force in Fig. 13.9*
f)	Sliding friction of pipe on piers due to expansion or contraction downhill from anchor $S_{pd} = f \cos\alpha_d \left(P' + W' - \frac{p'}{2}\right)$	6
g)	Sliding friction of uphill expansion joint $S_{eu} = \frac{f'\pi(d+2t)}{1000}$	7
h)	Sliding friction of downhill expansion joint $S_{ed} = \frac{f'\pi(d+2t)}{1000}$	8
j)	Hydrostatic pressure on exposed end of pipe in uphill expansion joint, $F_u = \frac{wH\pi t(d+t)}{10^6} = wah$	9
k)	Hydrostatic pressure on exposed end of pipe in downhill expansion joint $F_d = wa'h$	10
m)	Longitudinal force due to reducer above anchor $L_u = wH(A' - A)$	11
n)	Longitudinal force due to reducer below anchor $L_d = wH(A - A'')$	12

Inertial forces due to earthquake and water hammer pressure at any point is computed. The loads and forces specified above are assumed to act in the directions shown in Fig. 13.9 for the conditions of expansion and contraction and for penstock full and empty conditions.

The anchors are provided at all bends and at intermediate points in long tangents and where the distance between any two bends requiring anchors exceeds 150 m for freely supported penstocks above ground surface or in open tunnel over intermediate supports. For buried penstocks, these are provided at horizontal bends with large deflection angles which produces forces not exceeding the frictional and compressive resistance of the soil, at vertical bends, at summits and at bends adjacent to power house or pumping plant. Fig. 13.9 shows anchor blocks provided on a penstock.

Foundation. The anchor block is better founded on rock base. In case rock is at a greater depth, the stability of overburden material as excavated for the block foundation is checked against sliding as an earthen slope with anchor block on it and in its natural condition. It is essentially secured from erosion.

The foundation of the block is designed so that its maximum pressure on the foundation is within allowable bearing capacity of the soil. The bearing capacity is increased for seismic condition.

The anchor block has to be safe against sliding on foundation. The sliding friction factor computed by dividing the total horizontal forces by total vertical forces is essentially less than 0.50 for concrete on rock, 0.40 for concrete on gravel, 0.33 for concrete on sand, 0.25 for concrete on clayey soil and 0.75 for concrete on solid rock without any weak planes capable of sliding. The resultant of all the forces should fall within the kern of the base. For anchors with stepped bottom, the resultant should fall within the kern of the projection of anchor base on a plane perpendicular to the resultant (Fig. 13.10). Reinforcement is provided for the various forces indicated in (Fig. 13.9).

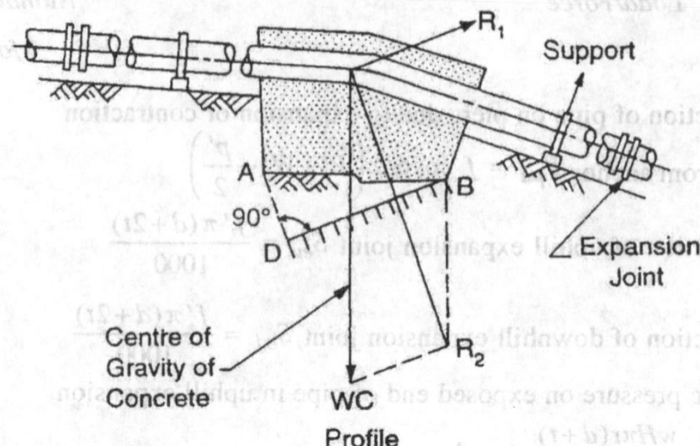

Fig. 13.10 Stability of anchor block with stepped base.

Types of Anchors

1. **Open type: In this type of anchor blocks, the penstock is anchored to the concrete by rings (Fig. 13.11).**

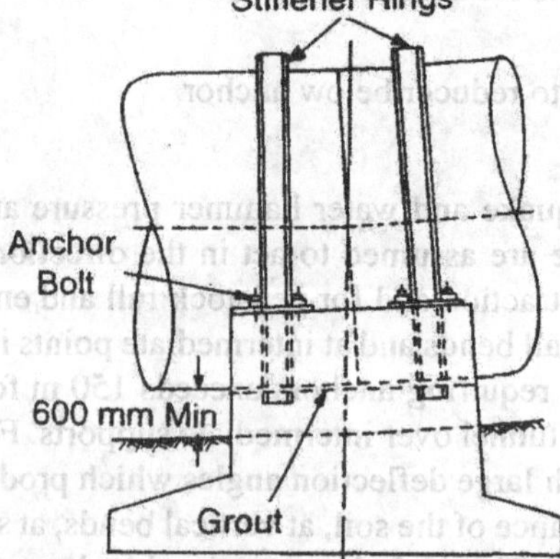

Fig. 13.11 Typical open type anchor block.

2. **Enclosed Type: As the name suggests, the anchor is embedded in concrete (Fig. 13.12 & 13.13).**

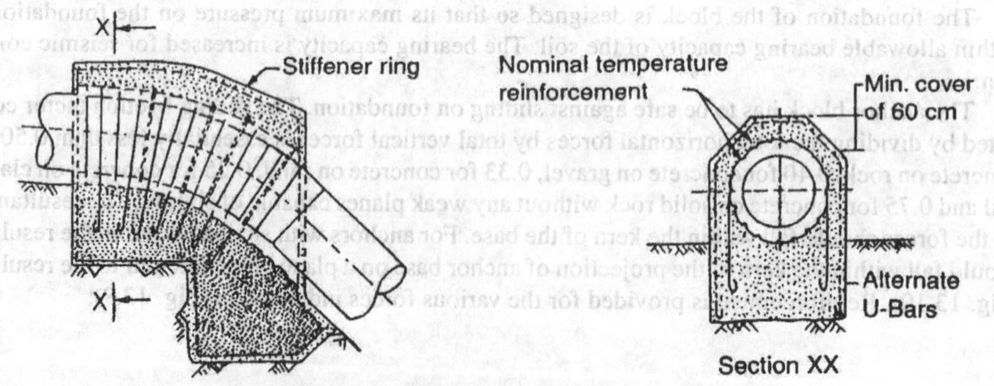

Ordinary Block

Fig. 13.12 Typical enclosed type anchor blocks (IS : 5330 - 1984)

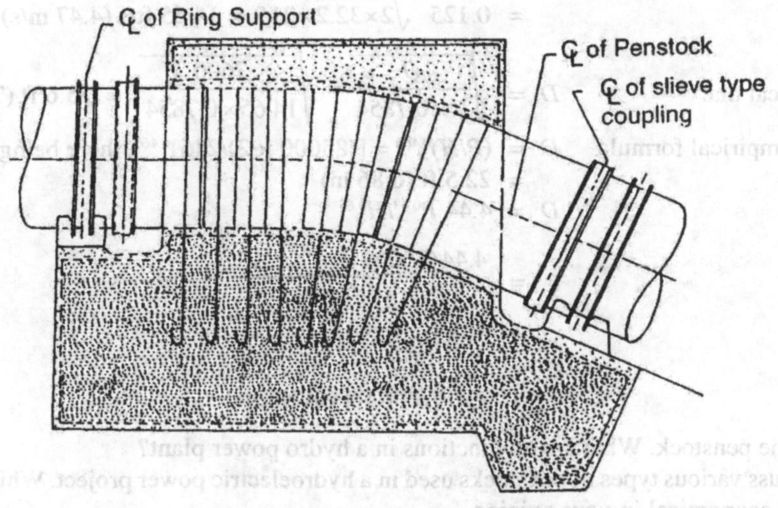

Fig. 13.13 Typical enclosed type anchor blocks (IS: 5330–1984)

3. **Partially Enclosed Type:** In this type of anchor, the block is constructed upto centre line of the penstock.

SOLVED EXAMPLES

Examples 13.1 *Determine economical diameter of penstock pipe 500 m long to carry water to a turbine having output of 1000 kW at 90% efficiency against a head of 150m. Assume f = 0.023.*

Solution.

Transmission loss for minimum diameter $= H/3 = 150/3 = 50\text{m}$

Head available at turbine $= 150 - 50 = 100$ m

Now $Q = \dfrac{P}{g H_\eta} = \dfrac{1000}{9.81 \times 100 \times 0.9} = 1.13$ cumecs

$$V = \frac{4Q}{\pi D^3}$$

$$h_f = \frac{flV^2}{2g D} = \frac{16 flQ^2}{2g\pi^2 D^5}$$

$$\therefore \quad D^5 = \frac{16 flQ^2}{2g\pi^2 h_f}$$

$$= \frac{16 \times 0.023 \times 500 \times 1.13^2}{2 \times 9.81 \times \pi^2 \times 50}$$

$$= 0.024$$

$$D = 0.475 \text{ m}$$

Adopt 0.5 m.

Examples 13.2 *Determine economical diameter for a power plant having a rated head of 215 ft, discharge 7600 cusecs, rated H.P is 85,000 of Francis tubine, there are two identical units fed by the penstock.*

Solution.

(*i*) By economical velocity as a proportion of spouting velocity method:

$$V = 0.125 \sqrt{2gh}$$

$$= 0.125 \sqrt{2\times32.2\times215} = 14.65 \text{ ft/s } (4.47 \text{ m/s})$$

Economical dia. $D = \sqrt{\dfrac{Q}{V\times0.7854}} = \sqrt{\dfrac{7600}{14.65\times0.7854}} = 25.6 \text{ ft } (7.8 \text{ m})$

(*ii*) By empirical formula $D = (P/H)^{0.466} = [(85000 \times 2)/215]^{0.466}$, there being 2 units

$= 22.5 \text{ ft } (6.86 \text{ m})$

Also $D = 4.44\, P^{0.43}/H^{0.65}$

$$= \frac{4.44(85000\times2)^{0.43}}{(215)^{0.65}}$$

$= 23.8 \text{ ft } (7.25 \text{ m}).$

QUESTION BANK

1. Define penstock. What are its functions in a hydro power plant?
2. Discuss various types of penstocks used in a hydroelectric power project. Which one is the most economical in your opinion.
3. Narrate the various factors which govern the layout of penstocks of a storage power development scheme.
4. What is economic diameter of penstocks? Discuss the various factors which govern the determination of economic diameter of a penstock.
5. Define branches and wyes in a power penstock arrangement. Why and where manifold have to be employed?
6. Obtain expression for economic diameter of penstock.
7. Draw sketch showing the forces acting on an anchor block for a penstock in a curve. How do we compute the base width required for anchor block? (*P.U.*)
8. The plant capacity of a hydro power plant is increased. Explain with the help of a load curve. (AMIE)
9. Explain the use of weekly load curve to predict the power requirement for a future. (AMIE)
10. Give expression for economical diameter of penstock. (*P.U. 1999*)

14

HYDRAULIC TRANSIENTS AND SURGE TANKS

14.1 INTRODUCTION

In a closed conduit such as penstock or tunnel feeding water to a power plant, rapid pressure changes occur inside it under unsteady flow conditions i.e. discharge at each section is varying rapidly from one instant to the next. Such pressure changes caused by rapid flow changes are termed as hydraulic transient.

In a turbine-penstock installation, transient conditions are produced under the following operations:

1. By resonance due to the vibration of runner blade and guide vanes, leakage from valve seal and governor hunting.
2. In isolated unit (*i*) at unit start up, (*ii*) load acceptance, and (*iii*) load reduction or total load rejection.
3. In a unit synchronised to a large system at (*i*) load acceptance, and (*ii*) load reduction or total load rejection.

14.2 HYDRAULIC TRANSIENT

For the calculation of hydraulic transient certain basic data as also some constraints set by the operational requirements are required, as under:

1. Data

(*i*) Details of water conductor system i.e. profile length, shape, size and material of penstock, (*ii*) Type, location and characteristics of surge tank and pressure relief valve, (*iii*) Type of turbine and its characteristics, (*iv*) Generator characteristics such as torque, inertia and efficiency, (*v*) Governor, its type and characteristics, and (*vi*) Design discharge and head on turbine.

2. Operational Requirements

(*i*) Maximum reduction in electrical power output that can result from circuit breaker operation, (*ii*) Maximum allowable speed rise in the turbine fed by any one penstock, (*iii*) Maximum rates of load change that generating plant is required to accommodate to satisfy system requirements, (*iv*) Limitations set over the range of variation of transient pressures.

3. Operating Conditions

The pressure rise due to hydraulic transient conditions in power tunnels and penstocks is considered for the normal operating conditions and emergency condition of operations discussed in section 14.6.

4. Factor of Safety

The minimum factor of safety adopted for the design of turbine-penstock, surge tank and other pressure control devices under normal and emergency conditions of operation, as per IS: 12967-Part I-1990 are given in Table 14.1:

Table 14.1 Factor of safety

Operating Condition	*Normal-1 unit*	*Normal-more than 1 unit*	*Emergency*
Factor of safety	2.0	1.6	1.25
The factor of safety for specials like bifurcations is increased by 125 percent.			

14.3 CONTROL OF HYDRAULIC TRANSIENTS

The undesirable transients such as excessive pressure, column separation and turbine overspeed as a result of power failure or load rejection can be reduced or eliminated by the following methods:

1. Reduction of Severity

The severity of the undesirable transients can be reduced by (*i*) Changing profile of penstock, (*ii*) Increasing diameter of the conduit, (*iii*) Reducing water hammer wave velocity, (*iv*) Mechanically coupling rotation inertia of turbine and of generator to turbine, and (*v*) Dual closing of guidance servomotors.

2. Controlling Devices

The surge tank and pressure relief valve are the commonly used devices to control pressure rise due to hydraulic transient. The choice of the type of the device is dictated by an acceptable system response as defined by operation requirement and overall economics.

14.4 WATER HAMMER

Water hammer is defined as the change in pressure, above or below normal pressure, caused by sudden changes in the rate of flow of water. It is experienced at all points in penstocks or long closed hydraulic tunnels between reservoir / forebay or surge tank and the turbines. The water hammer effects are designated as transient phenomena. These disturbances mark the transition stage between two successive steady states of flow. These are initiated by the sudden closing or opening of the turbine gate and are dissipated down to the level of the second steady wave by the operation of damping. This wave movement is propagated with the velocity of sound. With sudden closure of the turbine gate the water hammer pressure provides the necessary force to retard the flow in the closed conduit when load is rejected by the turbine. For very long closed conduits, the water hammer corresponding to normal operation of the turbine may be very great and require extraordinary strength of the conduit or tunnel to withstand it and the violent fluctuations of pressure in it may seriously interfere with proper turbine regulation. In power channel a surging water wave results. Similarly a sudden opening of the turbine gate to take on load due to increased power demand generates water hummer to create vacuum in the conduit or reduction of pressure provides the necessary force to accelerate the water and is correspondingly objectionable.

Hydraulic Gradient Change with Load Variation

The discharge to the turbine from penstock end is controlled by a gate or valve which in turn is controlled by governor to regulate the flow. The sudden change caused in the rate of flow results in a hammering like sound. The two types of water hammer effects in long closed conduits are illustrated in Fig. 14.1. AS′ is the static or no flow gradient. With full load on, the hydraulic gradient shifts to AF; S′F being the loss of head. With sudden decrease in load on the plant, the turbine gates operate to close causing the hydraulic gradient to move from AS′ to AP which position is termed as positive water hammer gradient. The sudden increase in load activates the turbine gates to open and the hydraulic gradient shifts from AS′ to AN which is called negative water hammer gradient. As the gate movement ceases, the hydraulic gradient rises to the steady full load gradient AF.

It follows that every point on the penstock is subjected to positive or negative water hammer pressures. Accordingly the penstocks are designed to withstand an internal pressure corresponding to the maximum positive water hammer pressure gradient AP. The negative pressure water hammer

gradient AN should not be below the top of the penstock at any point along its length as at *K*, otherwise a partial vacuum occurs within the conduit tending to collapse it.

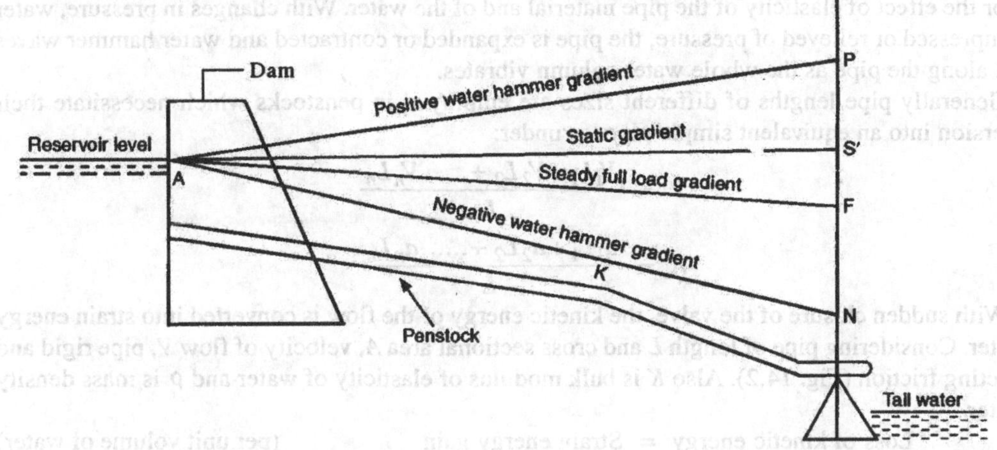

Fig. 14.1 Typical variation in hydraulic gradient with change in load.

14.5 WATER HAMMER THEORY

Water hammer theory was developed by L. Allievi and also by N. Joukowsky. The two basic approaches are Rigid water column theory wherein compressibility effects of water are neglected, and Elastic water column theory which takes into consideration the elastic effects of water. The magnitude of pressure rise due to sudden closure of gate valve depends on (*i*) speed of the valve closure, (*ii*) velocity of flow, (*iii*) penstock length, and (*iv*) elastic properties of the material the penstock pipe is made of and that of water flowing through it.

1. Sudden Closure of Valve

Consider water is flowing in penstock pipe of length *L* with velocity *V*, under a head *H* when suddenly the turbine valve at the pipe end is closed. The elements of water just upstream of the valve come to stop, their kinetic energy gets converted into pressure energy and as a consequence it is compressed by the water flowing against it. It results in a pressure rise which causes the pipe surrounding the element to stretch at the point. In next instant, the forward motion of the successive element of water too is stopped, compressed as water continues to flow in from the reservoir. The process is repeated on the successive elements of water until in a relatively short time the positive pressure wave travels to the reservoir and all the water in the pipe is at rest with the pipe expanded throughout its length. As the positive pressure wave travelling at velocity of sound *C* reaches the reservoir/forebay, there is high hammer pressure in the pipe and hydrostatic pressure head *H* in the reservoir. Since both can not be maintained, the flow starts into the reservoir reducing the pressure in the pipe to that of normal. Now negative pressure wave travels back down the pipe towards the valve at the velocity of sound till it reaches the valve. The time of travel of wave in one round trip is called time of one interval or the critical time of the pipe. In this time all the pipe shrinks to its normal size and water expands until normal pressures exist throughout the length of the pipe.

Due to inertia of water some more water, than that originally stored in the pipe, flows out resulting in fall in pressure at the valve to below the normal. This negative pressure wave now travels back up the pipe towards the reservoir. As the pressure less than hydrostatic can not exist at the reservoir/forebay, a wave of normal pressure now again travels back down the pipe with the flow of water from the reservoir into the pipe. Now the condition has reached which initiated water hammer. The process is repeated until dampened by friction.

During a closure or an opening of the valve, kinetic energy is converted into potential energy or vice versa, and the pressure existing at any instant can be determined by Newton's law of motion but for the effect of elasticity of the pipe material and of the water. With changes in pressure, water is compressed or relieved of pressure, the pipe is expanded or contracted and water hammer waves travel along the pipe as the whole water column vibrates.

Generally pipe lengths of different sizes are employed in penstocks which necessitate their conversion into an equivalent simple pipe, as under:

$$V = \frac{V_1 L_1 + V_2 L_2 + \ldots\ldots V_n L_n}{L}$$

$$A = \frac{a_1 L_1 + a_2 L_2 + \ldots\ldots a_n L_n}{L}$$

With sudden closure of the valve, the kinetic energy of the flow is converted into strain energy of water. Considering pipe of length L and cross sectional area A, velocity of flow V, pipe rigid and neglecting friction (Fig. 14.2). Also K is bulk modulus of elasticity of water and ρ is mass density of water.

Loss of kinetic energy = Strain energy gain (per unit volume of water)

$$\frac{1}{2}\rho ALV^2 = \frac{1}{2}\frac{p^2}{K} \cdot AL$$

or $$p^2 = \rho KV^2$$

or $$p = V\sqrt{K\rho} = V\sqrt{\frac{\rho^2 K}{\rho}} = \rho VC$$

Penstock pipe

Turbine valve

V

L

Fig. 14.2 Water hammer.

$\because$ Velocity of pressure wave, $C = \sqrt{\frac{K}{\rho}}$

$\therefore$ Intensity of pressure wave, $p = V\rho C$

Water hammer pressure head, $h = \frac{p}{w} = \frac{V\rho C}{\rho . g}$

$$= \frac{V}{g}C, \text{ neglecting pipe expansion.}$$

Time taken by pressure wave to travel from valve to reservoir (L) and back to valve,

$$T_c = \frac{2L}{C}$$

Joukowsky's Formula

Consider a section at wave front, ahead of front velocity is V and pressure is normal, but behind it velocity is zero and pressure p above the normal pressure. (Fig. 14.3)

As per Newton's law

Force = Rate of change of momentum

= Mass experiencing change of momentum/s × change in velocity

$$p.A = \rho AC.V$$

or $$p = \rho VC$$

$\therefore$ $$h = \frac{V}{g}C.$$

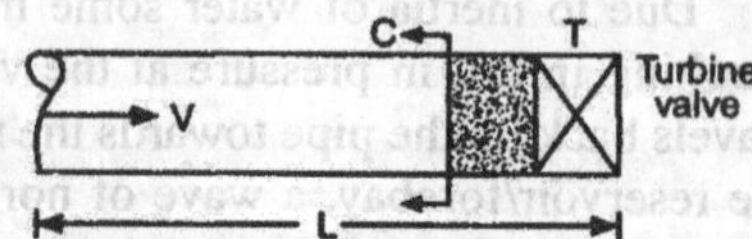

Fig. 14.3 Joukowsky's illustration.

2. Gradual closure of Valve

Assume that valve closure is gradual, $T > \dfrac{2L}{C}$, that is so devised that the water column in pipe flowing with velocity V is brought to rest with a uniform retardation $\frac{V}{T}$ i.e. $\left(\frac{V-0}{T}\right)$ in time T.

∴ Axial force producing retardation = Mass × retardation

$$= \frac{\rho ALV}{T} \qquad (i)$$

Force due to pressure wave = p.A. (ii)

Equating (i) and (ii)

$$p.A = \frac{\rho AVL}{T}$$

or

$$p = \frac{\rho VL}{T}$$

Water hammer pressure head,

$$h = \frac{p}{w}$$

$$= \frac{\rho VL}{wT} = \frac{\rho LV}{\rho.g.T}$$

$$= \frac{VL}{gT}.$$

Fig. 14.4. shows the pressure rise due to sudden valve closure. The pressure moves upstream with a velocity C thereby reducing the incoming velocity to zero, reaches the reservoir (or forebay) at time $T = L/C$. The area of reservoir is large compared to cross sectional area of the penstock pipe. Since velocity in the reservoir is approximately zero, there is no change in head H. Just at the end of $T = L/C$, the pressure in pipe AB is equal to P/γ throughout and velocity is zero. Since this pressure is higher than the head in reservoir, waters starts to flow back to the reservoir, an expanion wave of equal magnitude travels with a velocity C towards the valve. Thus at time $T = 2L/C$, pressure in the pipe in its entire length returns to the normal as it was before the closure of the valve and water has a velocity V in the backward direction. With the valve closed, water is not available to maintain backward flow and as such pressure at B falls below the datum line by p (Fig. 14.5).

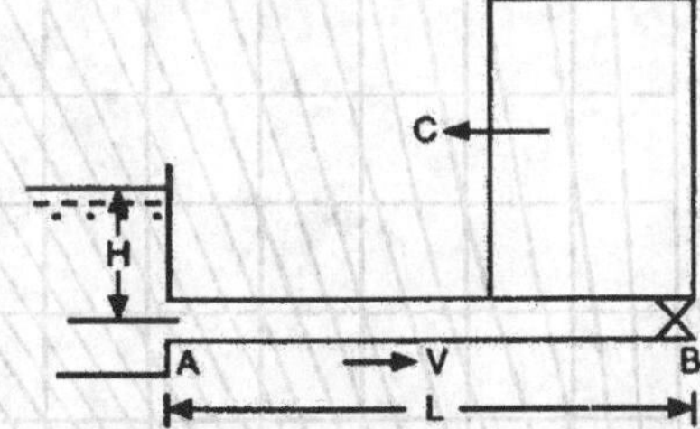

Fig. 14.4 Pressure rise due to sudden valve closure

Negative pressure developed at B travels towards A with velocity C and reaches A at $T = 3L/C$. The pressure in the entire length of the pipe is reduced to $-p$ at the end of $T = 3L/C$. Since reservoir pressure is higher, water flows into the pipe attaining a velocity V towards the valve in the normal condition as the wave proceeds towards B and reaches at $T = 4L/C$. The process is repeated (Fig. 14.5). The pressure is dampened with friction against pipe wall and elasticity of water and ultimately water comes to rest. The actual pressure variation with respect to time is shown by dotted line.

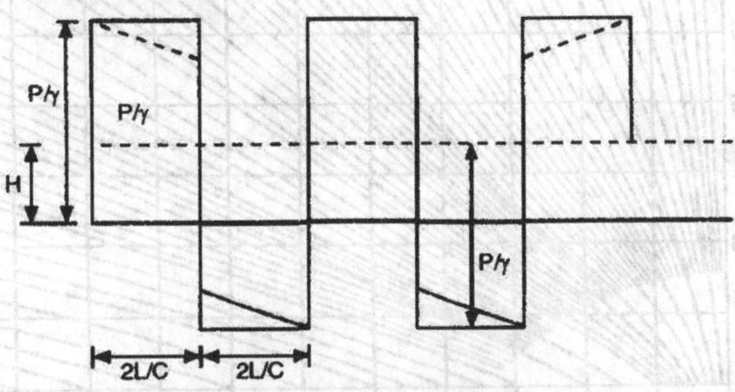

Fig. 14.5 Sequence of pressure variation with time.

Method

For preliminary designs, the maximum pressure rise or drop due to water hammer are estimated from the Allievi's charts shown in Fig. 14.6. It is the widely used of all other methods, despite limitations. The chart was prepared by L. Allievi by combining interlocked series equations, based on two fundamental assumptions, as under:

1. Nozzle area at the control valve is decreased or increased linearly with the time, and

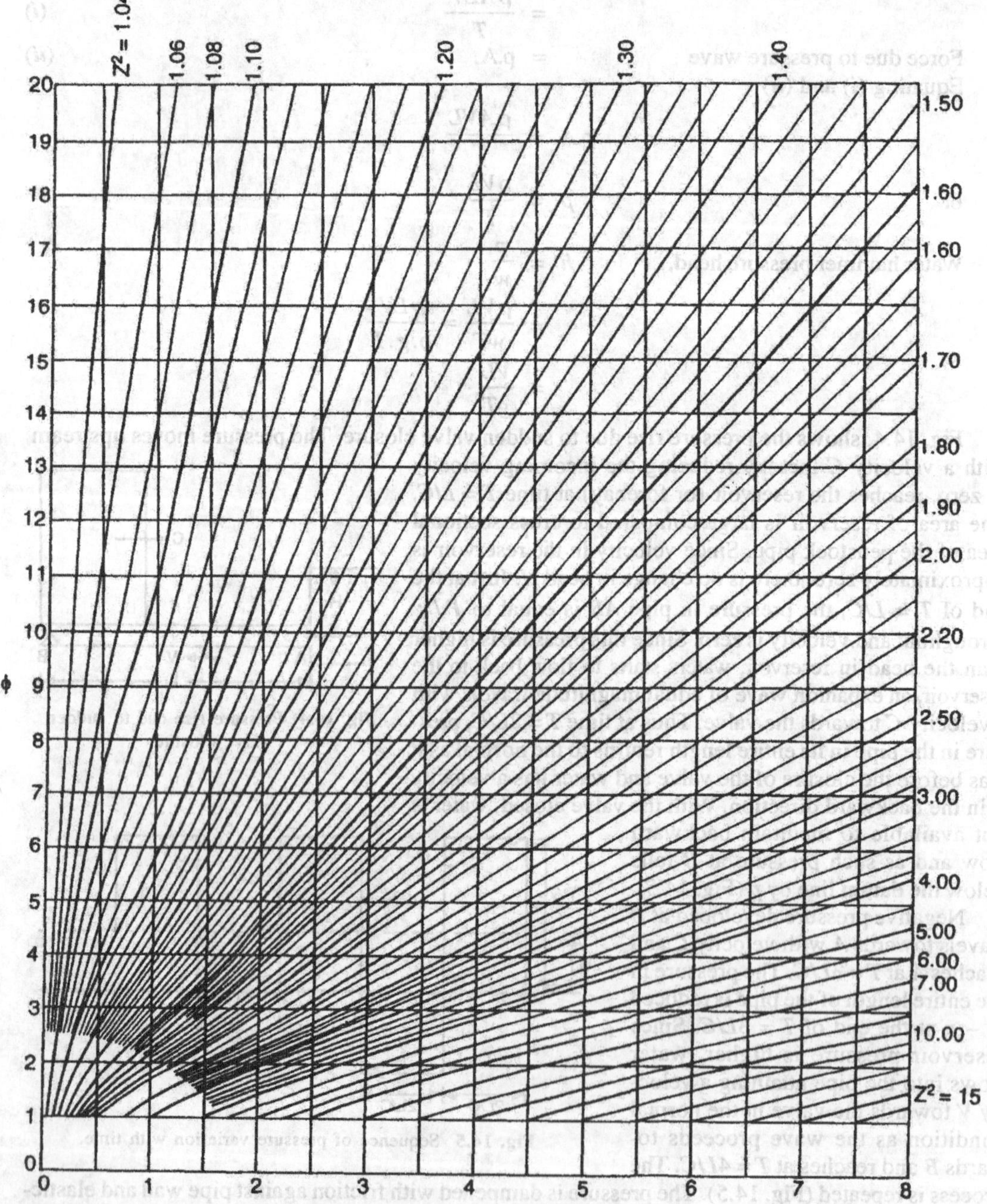

Fig. 14.6 Maximum pressure rise for uniform gate motion and simple conduits (for small values of ρ and θ).

2. The efflux velocity at the nozzle at any instant is proportional to the square root of the sum of the steady state plus water hammer heads.

The penstock pipe is assumed to be of uniform thickness and diameter to eliminate the consideration of wave reflection from points of change in thickness and change in diameter and as such applies to simple pipe or an equivalent simple pipe having constant value of velocity of pressure wave (C) and a uniform initial velocity V.

The coordinates of the chart are dimensionless parameters designated as ρ and θ. The parameter ρ is termed as the characteristic of the pipe and is given by the equation:

$$\rho = \frac{V.C}{2g.H}$$

Parameter θ is called valve operation or time constant. It is the ratio of the total time of gate operation to the time of one interval given by the equation:

$$\theta = \frac{CT}{2L}$$

The symbol Z is a measure of water hammer. Z^2 is defined as the ratio of the maximum total head to the initial head i.e.,

$$Z^2 = \frac{H+h}{H}$$

or

$$h = H(Z^2-1)\,m$$

where, H is static head (m) at lower end of penstock reckoned from water level in forebay or from elevation of water in surge tank at the beginning of gate movement to tailrace level, and h is pressure rise or head due to water hammer (m).

The maximum rise or fall, as the case may be, is taken from a single reading of the applicable chart, the dotted curved lines on the chart for closure indicating the interval in which maximum point pressure occurs.

The method has the merit of saving in time and labour in determining maximum or minimum point values.

Since in an hydraulic turbine operating through a transient range of speeds, the efflux velocity follows the law of machine performance curve for the individual mechanism, the results obtained from the chart vary from true values of transient pressure and velocity.

14.6 SURGES IN POWER TUNNELS AND PENSTOCKS

In tunnels and penstocks conveying water to a power plant, there may be sudden reduction in discharge due to load rejection by turbines. The basic conditions for including the effect of water hammer in the design of tunnels or turbine penstock installations are divided into normal and emergency conditions with a suitable factor of safety assigned to each type of operation.

1. Normal Conditions of Operations

The basic conditions to be considered are (*i*) Turbine penstock installation is operated at any head between the maximum and the minimum values of forebay water surface elevation, (*ii*) Turbine gates are moved at any rate of speed by action of the governor head up to a determined rate, or at a slower rate by manual control through auxiliary relay valve, (*iii*) The turbine is operated at any gate position and be required to add or drop any or all of its load, (*iv*) If the turbine penstock installation is equipped with surge tank, relief valves, governor control apparatus, cushioning stroke device, and any other pressure control device, it is assumed that these devices are properly adjusted and function in all manner for which the equipment is designed.

(*v*) Unless actual turbine characteristics are known the effective area through the turbine gates during the maximum rate of gate movement is taken as a linear relation with reference to time, (*vi*) The water hammer effects are computed on the basis of governor head action for the governor rate which is actually set on the turbine for speed regulation. If the relay valve stops are adjusted to give a slower governor setting than that for which the governor is designed this is determined prior to proceeding with the design of turbine penstock installation and later adhered to at the power plant so that an economical basis for designing the penstock scroll case, etc, under normal operating conditions are established, (*vii*) In those instances, where due to higher reservoir elevation it is necessary to set the stops on the main relay valve for a slower rate of gate movement, water hammer effects are computed for this slower rate of gate movement also, (*viii*) The reduction in head at various points along the penstock is computed for rate of gate opening which is actually set in the governor in those cases where it appears that the profile of the penstock is unfavourable. This minimum pressure is then used as a basis for normal design of the penstock to ensure that sub-atmospheric pressures will not cause a penstock failure due to collapse, (*ix*) If a surge is present in the penstock system, the upsurge in the surge tank is computed for the maximum reservoir level condition for the rejection of the turbine flow which corresponds to the rated output of the governor during the gate traversing time which is actually set on the governor, (*x*) The downsurge in the surge tank is computed for minimum reservoir level condition for a load addition from speed-no-load to the full gate position during the gate traversing time which is actually set on the governor.

2. Emergency Conditions

The basic conditions to be considered as an emergency operations are (*i*) The turbine gates are closed at any time by the action of the governor head, manual control knob with the main relay valve or the emergency solenoid device, (*ii*) The cushioning stroke is assumed to be inoperative, (*iii*) If a relief valve is present, it is assumed inoperative, (*iv*) The gate traversing time is taken as the minimum time for which the governor is designed, (*v*) The maximum head including water hammer at the turbine and along the length of the penstock is computed for the maximum reservoir head condition for final part gate closure to the zero gate position at the maximum governor rate in $2L/C$ seconds, where, L is length of penstock and C is a wave velocity, (*vi*) If a surge tank is present in the penstock system, the upsurge in the tank is computed for the maximum reservoir head condition for the rejection of full gate turbine flow at the maximum rate for which the governor is designed. The downsurge in the surge tank is computed for the minimum reservoir head condition for full gate opening from the speed-no-load position at the maximum rate for which the governor is designed. In determining the top and bottom elevations of the surge tank nothing is added to the upsurge and downsurge for this emergency condition of operation.

3. Emergency Conditions Not to be Considered as a Basis for Design

The other possible emergency conditions of operation are those during which devices of control are assumed to malfunction in the most unfavourable manner. The most severe emergency head rise in a turbine penstock installation occurs from either of the two conditions of operation viz: (*i*) Rapid closure of turbine gates in less than $2L/C$ seconds, when the flow of water in the penstock is maximum, (*ii*) Rhythmic opening and closing of the turbine gates when a complete cycle of gate operation is performed in $4L/C$ seconds.

Since these conditions of operation require a complete malfunctioning of the governor control apparatus at the most unfavourable moment, the probability of obtaining this type of operation is exceedingly remote. Hence, the conditions shall not be used as a basis for design. However, after the design has been established from other considerations, it is desirable that the stresses in the turbine scroll case, penstock and pressure control devices are not in excess of the ultimate bursting strength or twisting strength of structures for these emergency conditions of operation.

14.7 SURGE IN OPEN CHANNEL

In long power channels conveying water to a power plant, there may be sudden reduction in discharge due to load rejection by the turbine. A water wave or surge is produced as a result in the channel which moves upstream. Likewise, a surge is produced in tailrace channel due to regulation of flow by the turbine. This wave moves downwards. The surge may be positive or negative depending upon the depth of water after the surge and the normal depth. Increased surge depth means positive surge and vice versa.

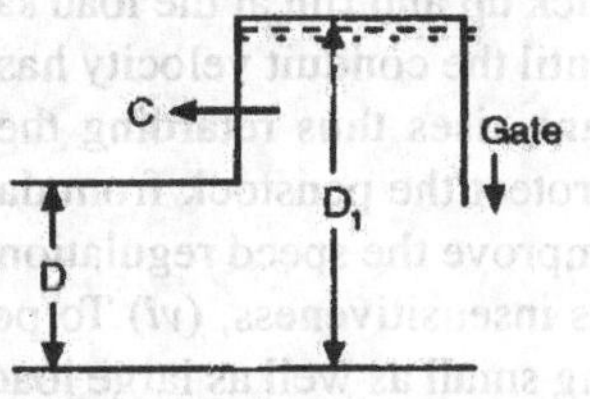

Fig. 14.7 Upstream surge.

A positive surge is produced when turbine gate is closed, whereas negative surge results when the turbine opening increases. The reverse is the case in the case of tailrace channel.

Assuming channel to be rectangular with bed width B and depth D. Referring to Fig. 14.7 when turbine gate is lowered due to load rejection, the discharge intake by the turbine is suddenly reduced which generate an upstream surge moving with a constant velocity V_c (celerity of wave). It is the case of unsteady flow but the wave front can be made stationary by superimposing a velocity + V_c in opposite direction (Fig. 14.8). The continuity equation and momentum equation can now be applied to this hypothetical steady state condition, as under:

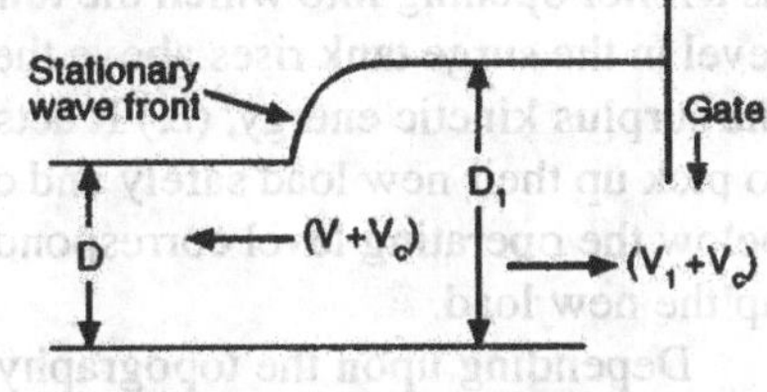

Fig. 14.8 Positive surge.

By continuity equation, $$BD(V+V_c) = BD_1(V_1+V_c)$$

or $$D(V+V_c) = D_1(V_1+V_c)$$

or $$VD = V_1D_1 + V_c(D_1-D)$$

By momentum equation, $$\frac{WB}{2}\left(D_1^2-D^2\right) = \frac{WB}{g}D(V+V_c)(V-V_1) \quad (i)$$

or $$\left(D_1^2-D^2\right) = \frac{2D}{g}(V+V_c)(V-V_1) \quad (ii)$$

$(V+V_c)$ in equation (*i*) is termed as celerity of wave C.

The equations (*i*) and (*ii*) with D_1 and V_c unknown, are solved simultaneously to get D_1. (D_1-D) is the surge to be taken care of in power channel.

The surge wave determination for tailrace channel too is done on the above lines.

14.8 SURGE TANK

Also termed as surge chamber. It is an artificial reservoir introduced along the pressure conduit system at a suitable point upstream and/or downstream of a hydropower station. The tank with a free water surface is usually provided at the transition from tunnel or pipe line to penstock to protect the tunnel or pipe line from excessive changes in pressure, rise in pressure in particular, and which expeditiously maintains equilibrium in the system by storing or furnishing water required for the above purpose.

Fig. 14.9 Surge Tank.

Functions of Surge Tanks:

The surge chamber thus serve the following primary functions: (*i*) To absorb the water hammer or elastic shock waves coming from the penstock or tunnel on closure or opening of the valve or turbine gate, (*ii*) To provide a free reservoir surface close to the turbine as a quick source of compensating water hammer reflections to limit the pressures in the penstock and materially reduce them in the main conduit, (*iii*) To furnish the additional water required by the turbine during increased load demand until the velocity has accelerated to the new steady state value. The quick supply of water is sufficient for the turbine to pick up and run at the load safely and quickly, (*iv*) To temporarily store water during load rejection until the conduit velocity has been decelerated to the new steady state value. The water level in the tank rises thus retarding the flow in the penstock and absorbing surplus kinetic energy, (*v*) To protect the penstock from dangerous pressure rises in case the pressure regulation fails to act and improve the speed regulation for small load decreases when pressure regulation does not act due to its insensitiveness, (*vi*) To positively and quickly dampen the pandulation of water levels following small as well as large load variations, (*vii*) It shortens the distance between the turbine inlet and the nearest free water surface and thereby reduces the intensity of water hammer waves, (*viii*) It acts as a relief opening into which the tunnel flow is diverted at the time of rejection of load. The water level in the surge tank rises above the operating level, thus retarding the tunnel flow and absorbing the surplus kinetic energy, (*ix*) It acts as a reservoir which provides sufficient water to the turbines to pick up their new load safely and quickly at the time of acceptance of load. The water level falls below the operating level corresponding to new load and thus accelerating the tunnel flow to pick up the new load.

Depending upon the topography, the surge tank can be a chamber excavated in the mountain totally underground or with top open to the atmosphere through a ventilation tunnel. It can also be a tower rising high above the surrounding ground (Fig. 14.9). However, rising tower is avoided as far as possible and its location on high ground is preferred to reduce the height of the tower.

14.9 NECESSITY OF SURGE TANK

The provision of surge tank is not necessitated when the power house is located adjacent to the dam in which case the reservoir acts as a surge tank. Likewise, it is not required in the canal fall development wherein the forebay serves as a surge tank.

The adverse water hammer effects may be reduced in the power developments wherein (*i*) The penstocks are of short length, or (*ii*) The valve or turbine gate can be closed very slowly but it is generally not compatible with practical economical speed regulation, and (*iii*) It is possible to provide a bypass to take care of the rejected flow by the installation of a pressure release valve at the turbine., The release valve has the limitation in that it is effective for gate closure only by bypassing the surplus water which means wastage of precious water but not for gate opening since it is not capable to make supply of water to the turbine to meet the demand for increased load and as such has no effect on negative water hammer.

Introduction of an orifice restricts the swing of the transient water levels in the surge tank. Maximum and minimum surge levels can be controlled by the provision of appropriate expansion chambers. This helps in avoiding air entry into the tunnel at low surge, occurrence of negative pressures anywhere in the water conductor and undue head variations on the operating machinery above and below the design/rated head.

The other factors that influence the surge levels are (*i*) Flow velocity adopted in the tunnel, (*ii*) Friction coefficient, (*iii*) Machine operating conditions such as load acceptance, rejection and load combinations like acceptance followed by rejection etc. Adoption of extreme values of friction coefficient coupled with extreme load conditions such as 100–0–100–0, 0–100–0–100 etc. result in very uneconomical design of the surge tank. The limiting operating condition may be 0–50, 50–100–0–50, 10–0–50 for economy in civil works.

14.10 LOCATION OF SURGE TANK

The surge tank is located as close to the power house as possible with a view to have minimal length of penstock. The shorter the length of penstock, the lesser is the intensity of water hammer and vice versa. The longer the length of penstock the higher is its cost due to its strengthening and requirement of speed regulation for the higher intensity of water hammer because surge tank reduces the pressures only in the penstock from the tank to the reservoir. The ideal location is at the turbine inlet but it is not possible in medium or high head developments as it has to be made very high. Generally it is located at the junction of tunnel and penstock to limit its height. The location may be the best at the point where the ground drops rapidly to the power house.

14.11 TYPES OF SURGE TANKS

The selection of location, size, shape and type of surge tank involves rather complex economic considerations as it largely depends on the hydraulics, topography and geology of the site. The various factors govering the selection of most suitable type are (*i*) Economy, (*ii*) Efficiency i.e. relief from water hammer effects, ease of governing, safety against large surges, damping of surges etc, (*iii*) Combining of bypass with surge tank or keeping bypass as separate. A surge tank is not economical on a very high head development due to great height of tank required. Evidently, in order to secure decrease in delivery of water to the turbine a bypass or deflection nozzles offers the alternative in such a situations as they do not require an immediate decrease in flow through the penstock itself.

The surge tanks are usually partially or wholly underground. The exposed surge tanks are of reinforced cement concrete or steel. They have to be protected from freezing in very cold climates by electric heating units. The surge chambers for tunnels are excavated in rock above the tunnel under favourable geological conditions.

Simple Surge Tank

The simplest type of surge tank consists of a cylindrical shaft open to the atmosphere and connected to pressure tunnel directly or by a short connection of cross-sectional area not less than the area of the headrace tunnel (Fig. 14.10) to minimise head loss.

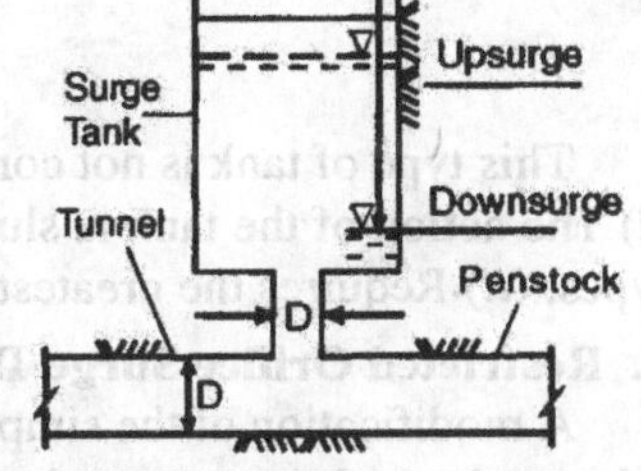

Fig. 14.10 Simple surge tank.

When the load on the turbine is reduced, the flow in the penstocks is retarded and accordingly water level in the surge chamber rises and flow upstream of the chamber is decelerated. When additional load comes on turbines, the surge chamber furnishes the immediate demand which increases flow gradient and as a consequence flow from the reservoir is accelerated. Under steady power load conditions, water level in the chamber remains constant at the level of the normal hydraulic gradient line at the point (Fig. 14.11). The surge chamber shall be high enough to cater for absorbing flow under full load rejection without overflowing. It thus dampens the water hammer pressure.

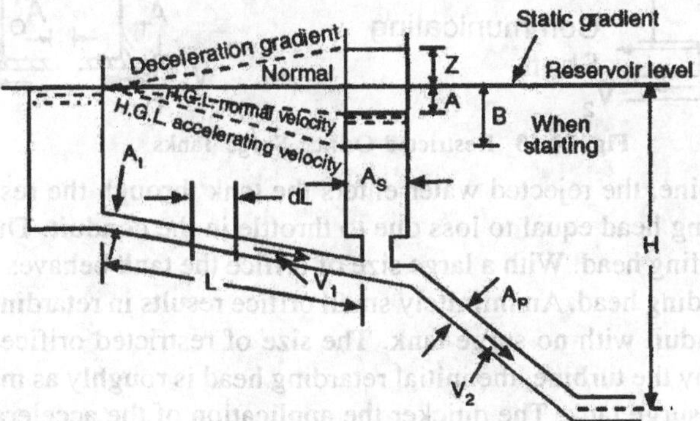

Fig. 14.11 Simple Surge Tank.

A_t is cross sectional area of head race tunnel, A_s is net cross sectional area of surge tank, A_p is cross sectional area of penstock, L is length of head race tunnel, Z is water level in surge tank measured positive above reservoir level, H is gross head on turbine.

The simple surge tank is ideal for ease of governing. The head changes are so gradual that even a very slow acting governor can follow the change of pressure. It is usually adequate for low head development. Under high heads, modifications to improve the damping action and minimise the cost have to be made. Fig. 14.12 shows the action of simple surge tank after a sudden closure of the turbine gates.

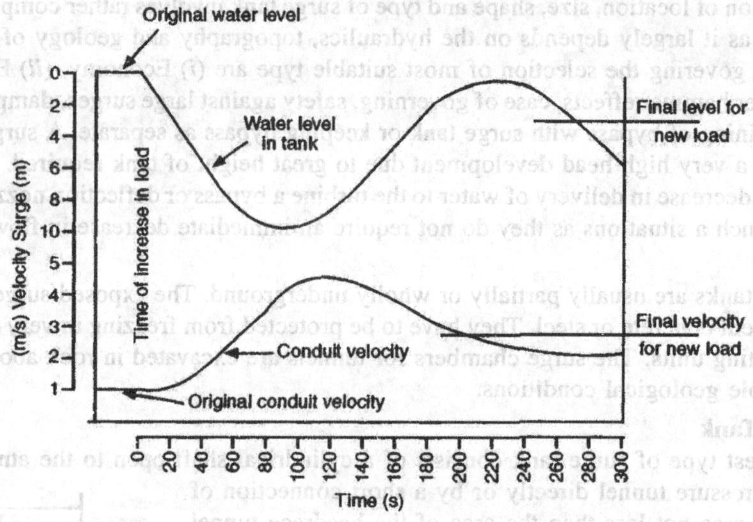

Fig. 14.12 Action of simple surge tank.

This type of tank is not commonly used, except in special cases, due to the limitations such as (*i*) The action of the tank is sluggish, the rise or fall of water is slow as compared to that of other types, (*ii*) Requires the greatest volume, and (*iii*) Expensive.

2. Restricted Orifice Surge Tank

A modification of the simple surge tank is the restricted orifice surge tank in which the area of inlet to the tank is narrowed or throttled to improve damping of oscillations by offering greater resistance (by speeding up pressure change due to greater head loss at the inlet) and is connected to the headrace tunnel with or without a connecting/communicating shaft (Fig. 14.13). With the

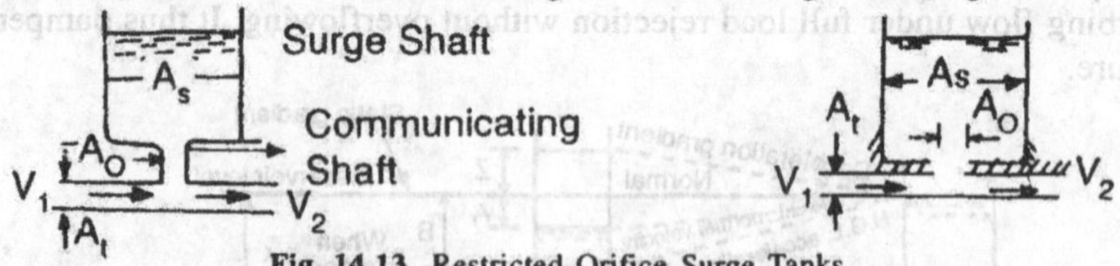

Fig. 14.13 Restricted Orifice Surge Tanks

reduction in load on turbine, the rejected water enters the tank through the restricted orifice, and quickly built up a retarding head equal to loss due to throttle in the conduit. The throttle size may be designed for any retarding head. With a large size of orifice the tank behaves like a simple surge tank with negligible retarding head. An infinitely small orifice results in retarding head equal to the water hammer in the conduit with no surge tank. The size of restricted orifice is so adopted that for full rejection of load by the turbine, the initial retarding head is roughly as much as the ultimate rise of water level in the surge tank. The quicker the application of the accelerating and retarding heads, the more effective is the surge tank in adjusting the conduit discharge, hence less water is stored in or delivered from the tank and as a consequence the tank is smaller in size.

It is economical and more efficient than simple tank but the drawback is that the desirable sudden creation of accelerating and retarding head in the conduit also induces corresponding sudden fluctuations of head on the turbine to which the governor may not easily accommodate. Chibro surge tank is of this type.

3. Differential Surge Tank

It is a compromise between the simple tank and restricted orifice tank. Differential surge tank is a throttled surge tank with an addition of a riser pipe, may be inside the main shaft, connected to main shaft by orifice or ports Fig. 14.14(*i*). The vertical riser has the same diameter as the pipe but of smaller diameter than the connection to the tunnel. The area of port is proportioned to suit the conditions under which the tank is designed to operate and the characteristics of the surge depend on it. The port holes are generally at the bottom of the riser at the sides. The port arrangement has marked damping effect or throttling action and as a result the tank acquires excellent capability to limit and suppress the surges due to synchronous load pulsations.

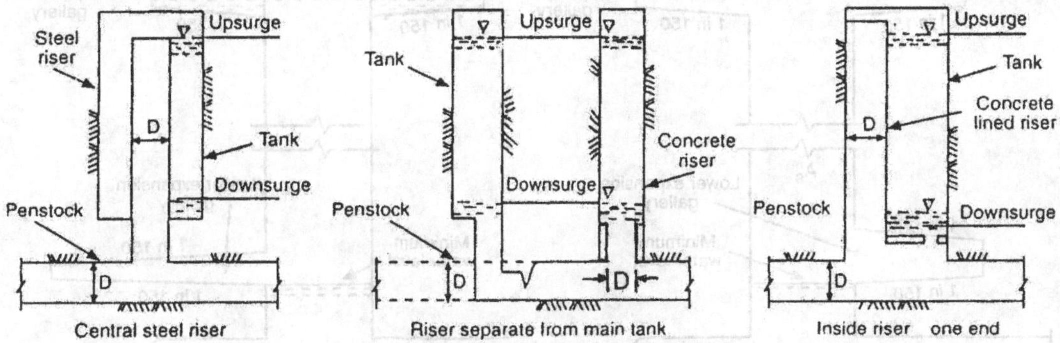

Fig. 14.14 Typical differential surge tanks

Variants of the type are that in which the riser is on one side of the throttled shaft as shown in Fig. 14.14(*ii*) and 14.14(*iii*)

The differential surge tank has the advantages of (*i*) Producing retarding/accelerating head on sudden changes of load due to immediate rise/fall of water level in the riser, (*ii*) Reduces amplitude of surge considerably, (*iii*) The initial pressure changes and the head on turbine, instead of occurring very slowly as in simple tank, occurs quickly enough for good efficiency of tank and is generally still spread over a period long enough to enable the governor to adjust the turbine gates to compensate for the change in load. The water storage function is separated from the conduit acceleration or deceleration functions resulting in more prompt and efficient development of the retarding and accelerating heads and for the same stability effect its capacity may be less than that of simple surge tank and as such economical.

4. Variable Section Surge Tank

It consists of a vertical shaft with an upper enlargement at its top which may be in the form of a reservoir and a lower enlargement at its bottom consisting of one or more horizontal galleries (Fig. 14.15). The upper reservoir absorbs the rising surge while the lower one provides storage of water for starting of turbine. This type of surge tank, unlike the differential tank, is not rapid in its action and is more expensive. Its use is limited in practice to the sites where already excavated chambers or galleries are available to serve enlargements.

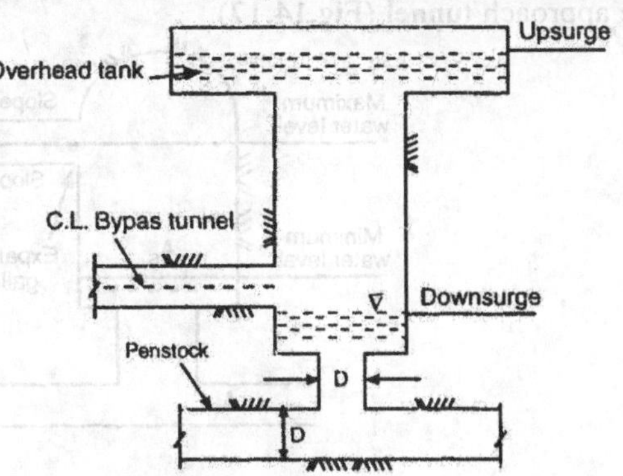

Fig. 14.15 Variable section surge tank.

Combining surge tank with bypass channel embodies the advantage of becoming available some additional volume in the bypass tunnel which, with proper arrangement, can help in reducing the downsurge. The differential type surge tank in bypass can be taken off only from the riser because of the restricted port openings in the main tank being insufficient to pass the full supply discharge of the by-pass. The bypass can also be combined both with simple surge tank and variable section surge tank.

5. Special Surge Tanks

Special surge tanks are the simple or restricted orifice type surge tanks which have expansion galleries (chambers), spilling arrangements, etc. as additional features (Fig. 14.16). Air cushion surge chambers belong to this category. The galleries are preferably D-shaped.

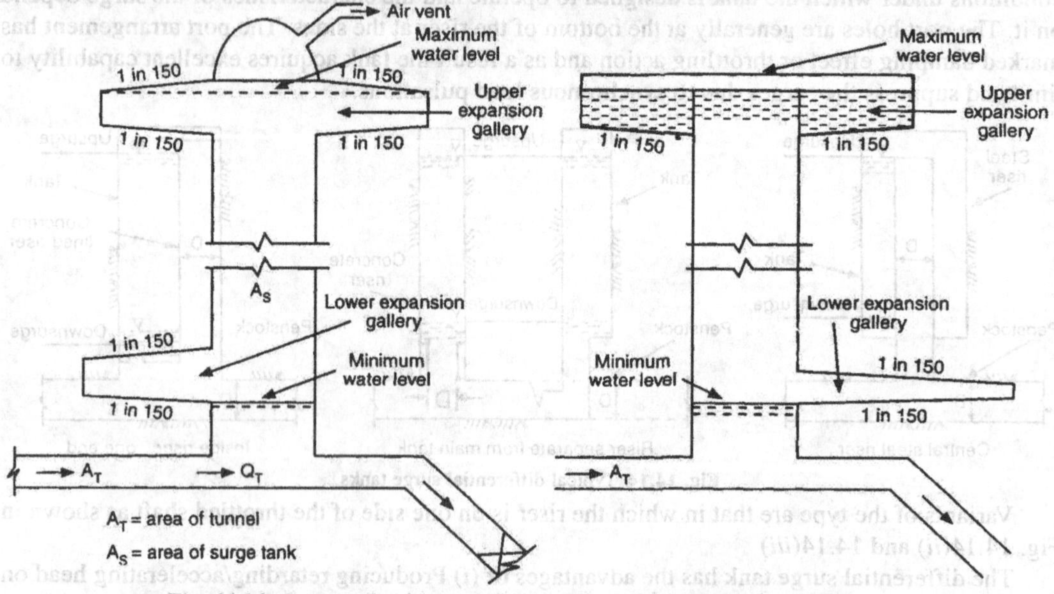

Fig. 14.16 Surge tank with part of approach tunnel as expansion gallery.

The expansion galleries (chambers) are provided when (*i*) The computed maximum and minimum surge levels are higher/lower than the maximum/minimum levels of the surges allowed in the simple, restricted orifice and differential surge tanks discussed heretofore, and (*ii*) In cases where in conventional surge tanks, long approach tunnels are required to be provided or where for reasons of economy a portion of the approach tunnel is utilised as an expansion gallery and the upper portion of the gallery serves as an air vent, if the maximum upsurge remains below the soffit at the mouth of the approach tunnel (Fig.14.17).

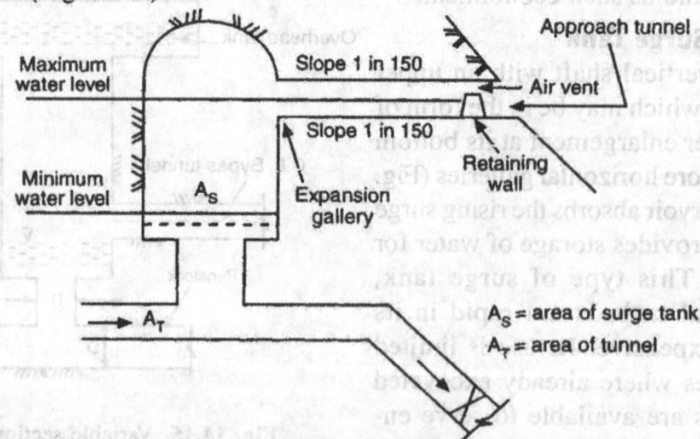

Fig. 14.17 Surge tank with part of approach tunnel as expansion gallery.

14.12 DESIGN OF SURGE TANK

1. Data Required

The data required for the hydraulic design of surge tanks are (*i*) Length, diameter and maximum discharge of head race tunnel, (*ii*) Length, diameter and number of penstocks and discharge of each, (*iii*) Gross head acting on generating units, (*iv*) Operating water levels of the reservoir, (*v*) Geometrical details of entry control structures, bends, transitions etc, which contribute to hydraulic losses, (*vi*) Power generation capacity of the station, (*vii*) Total power in the grid, (*viii*) Normal variation of load in the grid and the percentage shared by the station, (*ix*) Number of machines along with their characteristics, including efficiency, time-closure/opening of gate or deflector, (*x*) Critical load changes, (*xi*) Boundary friction factor and other coefficients for hydraulic losses for maximum, minimum and average conditions, (*xii*) Provision of pressure relief valves, if any, and their characteristics, (*xiii*) Side rock cover (contour plans), (*xiv*) Exploratory geological logs at site for analysis of physical conditions of rock, (*xv*) In-situ value of modulus of elasticity of the rock at different representative depths, (*xvi*) Water table, and (*xvii*) Groundability of rock encountered at different depths.

2. Design Conditions

The surge tank is designed to accommodate full load rejection by the turbine, which might occur after a short circuit on the transmission line. The choice of load increase for which the tank is designed depends upon number of turbines in the power plant, nature of the power market and other considerations. It is designed to accommodate the maximum and minimum water levels anticipated under worst conditions. The maximum upsurge level in the tank is worked out corresponding to (*i*) Full load rejection at the highest reservoir level, and (*ii*) Where considered necessary specified load acceptance followed by full load rejection at the instant of maximum velocity in the headrace tunnel and higher of the two is adopted. The minimum downsurge level is obtained as the worst of the two conditions (*i*) Full load rejection at minimum reservoir level followed by specified load acceptance at the instant of maximum negative velocity in headrace tunnel, and (*ii*) Specified load acceptance at load or speed-no-load condition at the minimum reservoir level. It is assumed that the load acceptance by different turbines in the station shall be such so as not to create superposition of surges.

The extreme water levels in the surge tank for the conditions mentioned above are determined by integrating the equations of mass oscillations for the respective tanks. These equations are non-linear. As such they are integrated either graphically by schoklitsh's method or numerically by Pressel's method.

(*i*) Area of Surge Tank

To ensure the hydraulic stability of the tank, its area is govered by Thoma criteria, as under:

1. Station in isolation:

$$A_{th} = \frac{L A_t}{\beta V_1^2 H_0} \cdot \frac{V_1^2}{2g}$$

where, A_{th} is Thoma area of the tank, L is length of headrace tunnel, A_t is cross-sectional area of headrace tunnel, V_1 is velocity of flow in tunnel at any instant, upstream of surge tank, to be taken as positive when the flow is downstream and negative when flow is upstream, H_0 is net head on turbines, β is coefficient of hydraulic losses such that $h_f = \beta V_1^2$, h_f being the total head loss in headrace tunnel system. β is determined from the formula:

$\beta v^2 = \dfrac{v^2 N^2 L}{R^{4/3}}$ + other losses in tunnel system, v is velocity of flow in the tunnel corresponding to change in discharge upstream of surge tank = $v_i \sim v_f$; v is taken as positive when flow is downstream and negative when flow is upstream.

2. Station to operate in a grid: In case it is to be ensured that the power station is to always operate in a grid, the stabilising effect of the grid is taken into account and the area of the tank is determined from the formula:

$$A_s = A_{th}\,[1 - 1.5\,(1 - k)]$$

Where, A_s is net cross-sectional area of the tank, k is ratio of total power generated by the station to that of grid.

A factor of safety of 2.0 and 1.6 on net area determined by the formulae above is assumed in the case of simple and restricted orifice or differential tanks respectively in order to secure rapid rate of damping required in practical operation.

***(ii)* Height of Surge Tank**

The top of the tank is kept 1.5 m above the highest water level while the bottom is kept 2.0 m below the lowest surge level determined by calculations or by model experiments for the worst conditions. The extra provision at the top may be in the form of a solid retaining wall wherever the tank is brought to the surface and is covered all around by such a wall. It is ensured that there is a sufficient water cover over the tunnel to avoid vortex formation even at the lowest water level in the tank otherwise vortex-suppression devices are considered.

Topography near the shaft permitting, suitable arrangement for spillover is considered so as to restrict the surge height.

The minimum height of the tank is governed by the restriction that it always contains water and in no case the water level in it drops so low that air is drawn into the penstock.

An illustrative example of design of Asia's highest surge tank is given at appendix-III.

14.13 TAILRACE SURGE TANKS

Tailrace surge tank is provided downstream of turbines in an underground power plant to protect long tailrace tunnel under pressure from hammer effect due to fluctuation in load. The surge tank is not required in case the tunnel has free-flow conditions with constant tail water level but the tunnel is designed to accommodate surges, if any, and it becomes uneconomical in the long tunnel than providing a surge tank. The distance between the tailrace surge tank and draft tube should be as short as possible to reduce the negative water hammer waves developing in the draft tube.

Types of Surge Tanks

The surge tank may be a single surge tank downstream of turbine or may form a multiple surge tank system with headrace surge tank depending on the

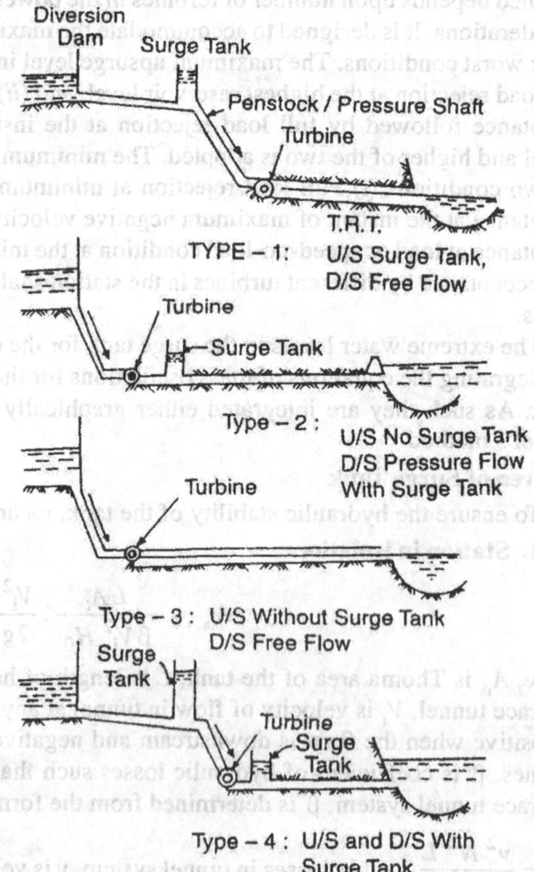

Fig. 14.18 Types of tailrace surge tanks.

type of power development. These surge tanks are classified, from the point of view of hydrodynamics of tailrace surge tank, as under:

Type 1 : Free flow tailrace tunnel without tailrace surge tank. It is the usual arrangement with almost constant tail water level. The tunnel is designed to accommodate surges, if any. In case of long tunnels this may be uneconomical (Fig. 14.18).

Type 2 : Tailrace pressure tunnel with tailrace surge tank. (Fig. 14.18)

Type 3 : Tailrace tunnel flowing either free or under pressure depending on tail water condition. (Fig. 14.18)

Type 4 : System of headrace and tailrace surge tanks hydraulically connected (Fig. 14.18).

The data required in addition to that indicated for upstream surge tanks are (*i*) Length, cross section and maximum discharge through tailrace tunnel, and (*ii*) Maximum and minimum tail water levels at the exit of tailrace tunnel.

The surge tanks are designed to accommodate the maximum and minimum water levels anticipated under worst condition. The maximum upsurge level in the tank is worked out corresponding to (*i*) Full load acceptance at the highest downstream tail water level, and (*ii*) Where considered necessary, load rejection followed by specified load acceptance at the instant of maximum negative velocity in the tailrace tunnel, the downstream tail water level being at its highest and higher of the two is adopted. The minimum downsurge level in the tank is determined corresponding to (*i*) Full load rejection at the lowest downstream tail water level, and (ii) Where considered necessary, specified load acceptance followed by full load rejection at the instant or maximum positive velocity in the tailrace tunnel, the downstream tail water level being at its lowest and the lower of the two is adopted.

SOLVED EXAMPLES

Example 14.1 *A rectangular power channel 20 m wide and 3m deep supplies 120 cumecs discharge to a power plant with four turbines. Two turbines are suddenly closed. Determine magnitude and speed of the surge developed.*

Solution.

$$V = Q/A = 120\,(20 \times 3) = 2 \text{ m/s.}$$

$$Q_1 = (B \times D_1)\,V_1 = B \times V_1 D_1$$

$\therefore$ $$V_1 D_1 = Q_1/B = 60/20 = 3 \text{ m}^2\text{/s per m width}$$

or $$V_1 = 3/D_1$$

By continuity equation $$VD = V_1 D_1 + V_c\,(D_1 - D)$$

$$2 \times 3 = 3 + V_c\,(D_1 - 3)$$

or $$V_c = \frac{3}{D_1 - 3}$$

By momentum equation $$D_1^2 - D^2 = \frac{2D}{g}\,(V + V_c)\,(V - V_1)$$

$$D_1^2 - 3^2 = \frac{2 \times 3}{9.81}\left(3 + \frac{3}{D_1 - 3}\right)(3 - V_1)$$

$$D_1^2 - 9 = 0.61\left(3 + \frac{3}{D_1 - 3}\right)\left(3 - \frac{3}{D_1}\right)$$

By trial $D_1 = 4.1$ m and $V_1 = 3/D_1 = 3/4.1 = 0.73$ m/s.

$\therefore$ Height of surge $= 4.1 - 3.0 = 1.1$ m.

Velocity of surge, $V_c = \dfrac{3}{D_1 - 3} = \dfrac{3}{4.1 - 3.0} = 2.73$ m/s.

Examples 14.2 *Determine pressure rise due to sudden closure of valve at the end of a steel penstock pipe 500 m long carrying water at a velocity of 5 m/s. Assume σ = 102.*

(b) *What shall be the pressure rise if the valve is closed in 2.5 seconds.*

Solution.

(*a*) Sudden closure of valve,

$$p = \rho\, Vc = 102 \times 5 \times 1460$$

$$= 744600 \text{ kg/m}^2$$

(*b*) Time interval of pipe, $\dfrac{2L}{C} = \dfrac{2 \times 500}{1460} = < 2.5$ seconds.

It is thus a case of gradual closure of valve.

∴ Pressure rise, $p = (\rho\, VL)/T$

$$= \frac{102 \times 5 \times 500}{2.5}$$

$$= 102000 \text{ kg/m}^2.$$

Examples 14.3 *A penstock 2000 m long and 4 m in diameter has a surge tank 20 m in diameter for a discharge of 30 cumecs. Friction factor f is 0.018. Normal reservoir level is 500 m. Determine maximum and minimum water levels in the tank.*

Solution. Maximum water hammer pressure,

Velocity in tunnel, $V = \dfrac{30}{\dfrac{22}{7}\dfrac{(4)^2}{4}} = 2.39$ m/s

Frictional head loss in penstock,

$$h_f = \frac{fLV^2}{2gD}$$

$$\therefore \quad h_f = \frac{0.018 \times 2000 (2.39)^2}{2 \times 9.81 \times 4.0} = 2.62 \text{ m.}$$

Maximum upsurge (neglecting friction loss) $Z_{max} = V\sqrt{\dfrac{L.A_T}{g.A_s}}$

$$= 2.39\sqrt{\frac{2000\left(\dfrac{22}{7} \times \dfrac{4^2}{4}\right)}{9.81\dfrac{(20)^2}{4} \times \dfrac{22}{7}}} = 6.83 \text{ m}$$

∴ Maximum water level in surge tank $= 500 + 6.83 - 2.62$

$= 504.21$ m

Maximum downsurge $= \sqrt{Z^2 + h_f^2} = \sqrt{(6.83)^2 + (2.62)^2} = 7.31$ m.

∴ Minimum water level in tank $= 500 - 7.31 = 492.69$ m.

Examples 14.4 *(a) Determine pressure rise through a water flowing elastic pipe 5 km long of 2 m diameter, velocity of water 3 m/s is suddenly stopped by valve at turbine end. Assume* $\rho = 102$ *kg/m*^{3}*,* $E = 2.1 \times 10^6$ *kg/cm*^{2} *and* $K = 21{,}000$ *kg/cm*^{2}*. Pipe thickness is 10 mm.*

(b) What is the length of pipe subjected to peak pressure.

Solution.

Pressure wave velocity, $$C = \sqrt{\frac{\frac{K}{\rho}}{1+\frac{KD}{Et}}}, \quad \text{Here } \frac{K}{E} = \frac{21{,}000\times10^4}{2.1\times10^{10}} = \frac{1}{10^2}$$

$$\therefore \quad C = \sqrt{\frac{\frac{(21000\times10^4)}{102}}{1+\frac{1}{100}\times\frac{2.0}{0.01}}}$$

$$= 828.42 \text{ m/s}$$

Critical time of pipe $= 2\,L/C$

$$= \frac{(2\times5000)}{828.42}$$

$$= 12.07 \text{ seconds.}$$

Since the valve is closed in 5 seconds, it is a case of sudden valve closure.

∴ Maximum water hammer pressure head

$$= \frac{V}{g}\times C$$

$$= \frac{3.0}{9.81}\times 828.42$$

$$= 253 \text{ m.}$$

Length of pipe subjected to water hammer

$$= L - \frac{CT}{2}$$

$$= 5000 - \frac{828.42\times5}{2}$$

$$= 2929 \text{ m.}$$

Examples 14.5 *(a) In a hydel power plant, a penstock 1000 m long carries 10 cumecs of water at a velocity of 5 m/s to the turbine; the head acting being 100 m. The turbine gate is closed in 5 seconds. Determine water hammer pressure.*

(b) The pipe is compound, first half length has an area of 5 m^2 *and the rest 4 m*^{2}*. Allievi* Z^2 *is 2.8.*

(c) Velocity in 200 m length is 5 m/s, next 400 m is 4 m/s and in rest 3 m/s.

Parts	*1*	*2*	*3*
Length	*200*	*300*	*500*
V	*5*	*4*	*3*
V_c *(celerity of pressure waves)*	*900*	*950*	*1000*

Solution.

$$\rho = \frac{V.V_c}{2gH} = \frac{5\times1400}{2\times9.81\times100} = 3.57$$

$$\theta = \frac{V_c\,T}{2L} = \frac{1400\times5}{2\times1000} = 3.5$$

$$Z^2 = 2.8 \text{ (given)}$$

or $$\frac{H+h}{H} = Z$$

or $$h = H(Z^2 - 1)$$
$$= 100\,(2.8 - 1) = 180 \text{ m.}$$

(*b*) Equivalent pipe area, $$\frac{L}{A} = \frac{L_1}{A_1} + \frac{L_2}{A_2}$$

$$\frac{1000}{A} = \frac{500}{5} + \frac{500}{4}$$

or $$A = 4.44 \text{ m}^2.$$

$$V_E = \frac{Q}{A} = \frac{10}{4.44} = 2.25 \text{ m/s}$$

$$\rho = \frac{V_E \,.\, V_C}{2g\,H} = \frac{2.25 \times 1400}{2 \times 9.81 \times 100} = 1.61$$

$$\theta = \frac{V_C \,.\, T}{2L} = \frac{1400 \times 5}{2 \times 1000} = 3.5$$

$$Z^2 \text{ (read from Fig. 14.6)} = 1.58$$
$$h = H(Z^2 - 1) = 100\,(1.58 - 1) = 58 \text{ m}$$

(*c*) $$V_E = \frac{200 \times 5 + 300 \times 4 + 500 \times 3}{1000} = 3.7 \text{ m/s}$$

$$V_c = \frac{200 \times 900 + 300 \times 950 + 500 \times 1000}{1000} = 965 \text{ m/s}$$

$$\rho = \frac{V_E \,.\, V_C}{2g\,H} = \frac{3.7 \times 965}{2 \times 9.81 \times 100} = 1.82$$

$$\theta = \frac{V_C \,.\, T}{2L} = \frac{965 \times 5}{2 \times 1000} = 2.41$$

$$Z^2 = 2.1$$
$$h = H(Z^2 - 1)$$
$$= 100\,(2.1 - 1) = 110 \text{ m.}$$

Examples 14.6 *Determine discharge through a cast iron pipe 20 cm diameter having 2 cm wall thickness if valve at the end is suddenly closed. The stress in the pipe is not to exceed 200 × 10^4 kg f/m². Assume E_s = 130 × 10^8 kg f/m² and k = 200 × 10^6 kg f/m².*

Solution.

$$V = \frac{p\sqrt{1 + \dfrac{KD}{E_t}}}{\sqrt{K\rho}},$$

$$= \frac{\left(\dfrac{2ft}{D}\right)\sqrt{1 + \dfrac{KD}{Et}}}{\sqrt{K\rho}}, \text{ putting } p = \frac{2ft}{D}$$

$$= \frac{2 \times 200 \times 10^4 \times 0.020 \sqrt{1 + \dfrac{200 \times 10^6 \times 0.20}{130 \times 10^8 \times 0.02}}}{0.20 \sqrt{200 \times 10^6 \times \dfrac{1000}{9.81}}} = 3.01 \text{ m/s}$$

$$\text{Discharge} = \text{A.V} \quad = \frac{\pi (0.20)^2}{4} \times 3.01 = 0.095 \text{ cumec.}$$

$$\text{Pressure rise,} \quad P = \frac{V\sqrt{K\rho}}{1 + \dfrac{KD}{Et}} = \frac{3.01 \sqrt{200 \times 10^6 \times 1000 / 9.81}}{\sqrt{1 + \dfrac{200 \times 10^6 \times 0.20}{130 \times 10^8 \times 0.02}}}$$

$$= 4.08.$$

QUESTIONS

1. **What is meant by hydraulic transient? What are the operations in which transient conditions are produced in a turbine-penstock installation?**
2. **Mention briefly how the hydraulic transients as a result of power failure or load rejection are controlled.**
3. **What is meant by water hammer? Does it affect long penstocks? If so why?**
4. **Illustrate with the help of a sketch variation in hydraulic gradient in a penstock with the change of load.**
5. **Derive expressions for water hammer in a rigid pipe and an elastic pipe. Why these two vary?**
6. **Derive an expression for the celerity of wave in power channels.**
7. **Why do you understand by a surge tank? Illustrate with sketches its different types.**
8. **Do we need surge tanks in tailrace tunnels? Illustrate various types with suitable sketches.**
9. **What is surge tank? What are its functions?**
10. **(*a*) What is a surge tank? What are its functions? What are the different loading conditions to determine the height of surge tank?**

    (*b*) A steel pipe line of 0.6 m diameter, 8 mm thick and 3200 m long conveying water at the rate of 0.5 cumecs is closed completely by a quick acting valve in 1 second. Find the pressure rise due to water hammer. ***(P.U. 1999)***
11. **(*a*) What is a surge tank? What are its functions? What are the different loading conditions to determine the height of surge tank.**

    (*b*) A steel pipe line of 0.6 m diameter, 8m thick and 3200 m long conveying water at the rate of 0.5 cumes is closed completely by a quick acting valve in 1 sec. Find the pressure rise due to water hammer. ***(P.U. 1999)***

15

TRANSMISSION LINES

15.1 INTRODUCTION

The electricity generated at a power plant is essentially transmitted to grid sub-stations or from one load centre to another. The generated power is evacuated from the power plant to switch yard where it is stepped up to extra high voltage for transmission to the grid. The distribution, on the other hand, is the conveyance of the electricity from the sub-stations to the consumer at high and medium voltage.

The hydro power plants are usually located in mountainous areas and on this account involve long transmission lines for interconnection to the grid, traversing very difficult terrain and at a very high cost. It thus substantially adds to the cost of hydro power generation besides considerable time in its erection. Nevertheless, it is cheaper to transmit power than to transport coal from coal heads to distant located thermal power plants. It is pertinent to mention that the principal parts of a transmission system, apart from transmission lines, consists of insulators, conductors, transformers, protective and control devices. However, the discussion is limited to transmission lines.

15.2 FUNCTIONS OF TRANSMISSION LINES

Various power plants, located as they are in different places as per site suitability, power demand and economic considerations, essentially feed power into the common grid system, except isolated small hydel plants, for integrated, centrally controlled, optimum and economical operation of the system as a whole. The power can thereby be transferred from one part of the system to another in the region. The northern and southern regions have more hydro power than thermal. In the eastern and western regions thermal predominates over hydro power. The load patterns too are different . The interconnection of all the regions, to create All-India power grid, will essentially facilitate controlled power interchange like import of off-peak surplus from western to southern regions.

The other functions of the interconnection through transmission lines briefly are (*i*) It can limit system fault levels, (*ii*) It can optimise hydrothermal mix of the integrated grid, (*iii*) It enables division of load between stations for economical operation, (*iv*) It increases firmness of the hydel power, (*v*) It improves overall economy of the power system, by delivering low capacity addition run-of-river plants with substantial pondage in hydro-thermal mix, by using on the peak of load and transmit at a low load factor a capacity higher than that capable of carrying as an isolated plant on a commercial load factor, (*vi*) It enhances the load factor of the hydel power system by making available very cheap, surplus hydel power output, and (*vii*) It ensures greater economy in the construction and operation of hydro and thermal plants.

The transmission of power in highest voltage economises in conductor material but at the same time it results in increased cost of the insulating conductors, transformers, switchgear and other terminal apparatus. In view of the above, the relative economics of the limit to the higher voltage for transmission vis-a-vis increased cost of the insulating conductors, transformers etc. is considered in deciding the voltage for transmission.

15.3 TYPES OF TRANSMISSION LINES

The two types of transmission lines are briefly as under:

1. *Overhead Lines*: These are the commonly adopted type of transmission lines wherein the electric current is carried on wires held on poles or towers. It is generally preferred on account of low initial cost but is susceptible to disturbances due to storms, lighting strikes and inclement weather conditions.
2. *Underground Lines*: As the name suggests, the power is transmitted through cables laid underground. These are laid on the ground in pipes or ducts. These are suitable for dense urban areas, factories, power houses and sub-stations. The insulated cables are manufactured from copper and aluminium but mostly of the latter metal due to high cost and shortage of the former. The cables are essentially provided with protective covering, depending on their intended application, to ensure that these are resistant to decay, moisture, abrasion, chemical reaction with the soil.

   The underground cables have the advantage of economy in the use of overground space, safety from damage due to storms, lightening, rains etc but the initial cost is very high, repairs difficult.

15.4 COMPONENT PARTS OF TRANSMISSION LINES

The component parts of a transmission line depend on the type of line, overhead or underground.

The essential components of an overhead transmission line are (*i*) Supporting structure for the line, (*ii*) Conductor, (*iii*) Insulators, (*iv*) Cross arms, and fittings like clamps, mountings etc.

1. Supporting structures

The type of supporting structure depends on the load due to span of the wire.

(*i*) **Poles:** The lines are generally run on wooden, concrete and steel poles.

(*a*) *Wooden Poles*: These poles are wooden with wooden or steel cross arms (Fig. 15.1). These find use in rural electrification being cheap with usual span of 30-60m. The conductors are carried on porcelain pin insulators, the rigidly attached conductors, as they are over short span, themselves act as additional support. The portion of the pole anchored into the ground is suitably treated with bitumen coatings. Likewise, at the top a zinc or aluminium cap is provided (Fig. 15.1). They have good insulating quality. They are suitable for carrying one circuit only.

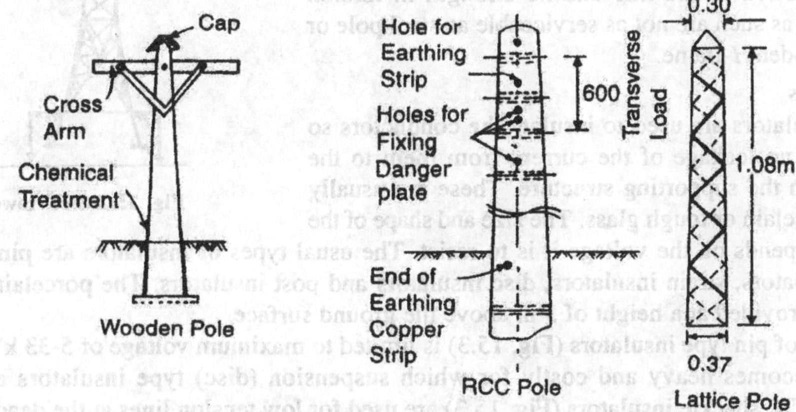

Fig. 15.1 Typical Poles.

(*b*) *Concrete Poles*: The reinforced cement concrete poles are usually square in cross section, tapering towards the top (Fig. 15.1). They find use in distribution system in urban areas for reasons of decorative considerations, as also in remote areas because these can be manufactured locally. They are strong and durable but very heavy in weight posing problem in transportation. They shatter if hit by a vehicle, otherwise have a long life and little maintenance. However, prestressed poles are comparatively lighter in weight but still difficult to handling in transportation and erection. Being single base poles, they are less strong against storms.

(c) *Steel Poles*: The steel poles have conductors supported from steel cross arms by strings of pin type insulators. One or two circuits, preferably one with pin type insulator are usually carried. They are used for light low-voltage lines and span 60-150 m. These are single based poles and as such the problem of right of way, as in double base poles, in not involved. The three types of steel poles are briefly as under:

(i) *Stepped Poles*: These are manufactured from a single tube with diameter reduced in parallel steps in the manufacturing process. These are suitable for lines up to 66 KV. These are light in weight, strong and have long life.

(ii) *Swaged Poles*: These are manufactured from tubes of different diameters swaged together when still hot.

(iii) *Lattice Poles*: These are narrow-base steel poles (Fig. 15.1). They can be of welded as well as of bolted fabrication. They are economical, light in weight and the bolted Assembly can be done at site. Galvanised steel fabrication is preferred.

Another variant is old and used rails and R.S. Joists employed as supports for overhead lines.

(ii) **Towers**

Towers constitute the supporting structure for high-voltage lines and heavily loaded moderate-voltage lines. These may be fabricated out of wood or steel.

(a) *Wooden Towers*: The wooden *A* or *H* frames are two legged towers used for reasons of economy in cost. They are self supporting under load in a direction transverse to the line only. For stability in the direction of the line, they depend on dead-end towers interposed at intervals of 5 to 10 spans.

(a) *Flexible Steel Towers*: There are two-legged flexible towers and are economical in cost (Fig. 15.2). These are used for 60–150m spans with pin type insulations. They require support in the direction of the line and no strength in torsion and as such are not as serviceable as steel pole or wooden *H*-frame.

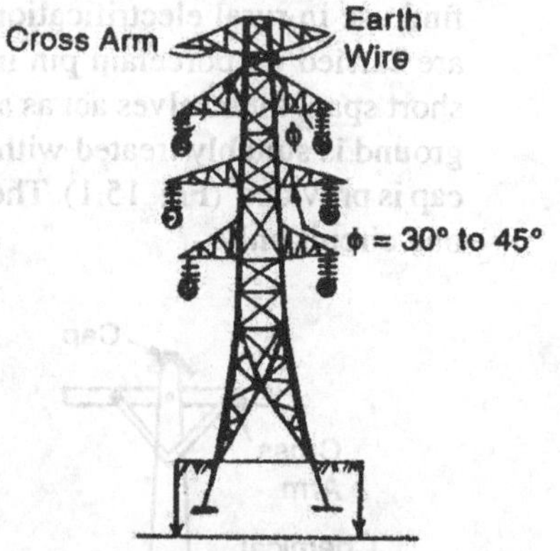

Fig. 15.2 Steel tower

3. Insulators

The insulators are used to insulate the conductors so that there is no leakage of the current from them to the earth through the supporting structure. These are usually made of porcelain or tough glass. The size and shape of the insulators depends on the voltage it is to resist. The usual types of insulators are pin insulators, shackle insulators, strain insulators, disc insulators and post insulators. The porcelain insulators are usually provided at a height of 3 m above the ground surface.

The use of pin type insulators (Fig. 15.3) is limited to maximum voltage of 5-33 kV; at higher voltage it becomes heavy and costly for which suspension (disc) type insulators are suitable (Fig. 15.3). The shackle insulators (Fig. 15.3) are used for low tension lines at the dead ends.

4. Cross Arms

The cross arms 90 × 120 mm on wooden poles are generally of wooden (shisham or fir) for lines up to 66 kV and need replacement after 5 years or so depending on weather conditions.

Fig. 15.2a 220 kV double circuit transmission tower.

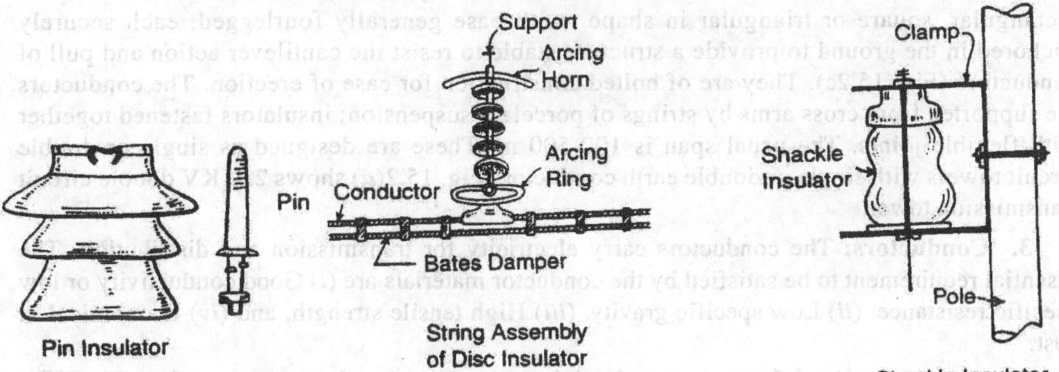

Fig. 15.3 Typical insulators

The steel cross arms usually consisting of angle iron 60 × 60 × 6 mm and 75 × 75 × 6 m and channel sizes 75 × 37 mm and 100 × 50 mm are generally used. The general shape of the cross arms is as depicted in Fig. 15.4.

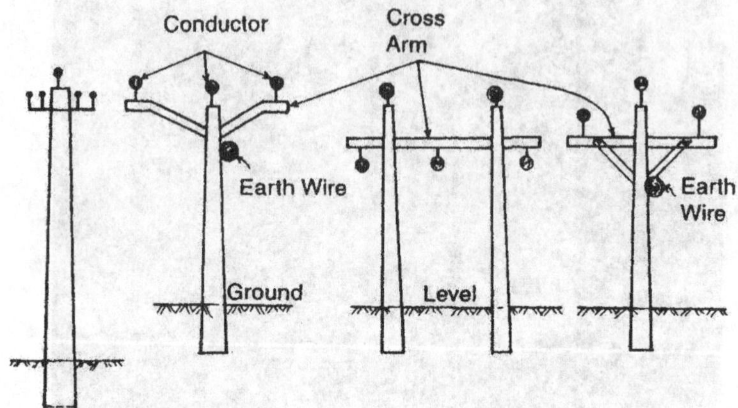

Fig. 15.4 Typical cross arms

15.5 DESIGN OF OVERHEAD TRANSMISSION LINES

The design of overhead transmission lines from electrical and mechanical considerations are briefly as under:

1. **Voltage Selection:** For efficient transmission the voltage of electric power has essentially to be high. The choice of voltage for transmission depends on the distance and the amount of power to be transmitted. The greater the distance and the amount of power, the higher is the economic voltage. The optimum overall economy, however, depends on certain other factors also, briefly as under:

(*i*) Although the cost per kilowatt transmitted is less while transmitting large amount of power, over long distances at higher voltage but the number of terminal stations impair its economy due to high cost of high voltage transformers and switching equipment of such stations.

(*ii*) The number of circuits too affect the overall cost; the minimum the number of circuits, the lower is the cost.

The economical voltage is determined by the following formula :

$$V = 5.5\sqrt{\frac{L}{1.6} + \frac{P}{100}}$$

where, V transmission voltage (kV), L is span (km) and P is load (kW).

2. **Steel Towers**: The steel towers are substantionally built of structural steel. These are rectangular, square or triangular in shape with base generally fourlegged; each securely anchored in the ground to provide a structure stable to resist the cantilever action and pull of conductors (Fig. 15.2a). They are of bolted construction for ease of erection. The conductors are supported from cross arms by strings of porcelain suspension; insulators fastened together with flexible joints. The usual span is 100-500 m. These are designed as single or double circuit towers with single or double earth conductors Fig. 15.2(*a*) shows 220 KV double circuit transmission tower.

3. **Conductors:** The conductors carry electricity for transmission and distribution. The essential requirement to be satisfied by the conductor materials are (*i*) Good conductivity or low specific resistance, (*ii*) Low specific gravity, (*iii*) High tensile strength, and (*iv*) Economical in cost.

These are made of hard-drawn copper, aluminium, aluminium conductor-steel reinforced (ACSR), cadmium copper and steel. The choice of the material the conductor is made of depends on topography, climate, serviceability, span length, quantity of power, losses and economics. Copper, aluminium and ACSR conductors are commonly used. An ACSR conductor is usually composed of central core of one or more galvanised steel wires surrounded by one to three layers of aluminium wires (Fig. 15.5). Size of wires is the same. These are used in long span high voltage transmission lines where a large diameter is required for reasons of strength.

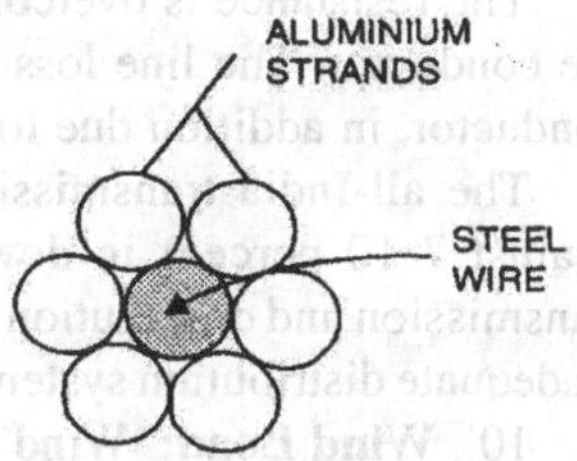

Fig. 15.5 ACSR conductor.

A vident glow around the conductor accompanied by a hissing noise, known as corona, is experienced in high voltage transmission lines on account of partial breakdown of air in the vicinity of the energised line. The power loss on this account is termed as corona loss. The cable used in underground transmission lines consists of one or more insulated conductors laid together and accordingly known as single core, twin core or three core cables. A wire, used in overhead lines, on the other hand, is an insulated or uninsulated conductor uniform in diameter and circular in cross section.

The size of the conductor is selected from the considerations of (*i*) Current carrying capacity of the conductor, (*ii*) Allowable voltage drop or line regulation and (*iii*) breaking strength of the conductor.

The spacing of the conductors depends on voltage and span. It is determined by the following formula:

$$S = \sqrt{d + \frac{v}{150}}$$

where, S is spacing of conductor (*m*), d is sag (*m*), and V is voltage (kV).

4. **Sag:** The span is as large as the reasonable sag in the conductor permits. The sag increases as the square of the span.

5. **Clearances:** The minimum ground clearances are given in Table 15.1.

Table 15.1 Clearances

Ground clearances along and across streets		*Horizontal clearances*	
Class of line	*Min. Clearance (m)*	*Class of line*	*Min. Clearance (m)*
Low, medium, H.V. up to II kV, if bare	4.5	H.V. up to and including 11 kV	1.25
Low, medium and H.V. insulated	4.0	H.V. above and up to 33 kV	1.75
H.V. above 11 kV	5.0	H.V. line up to 12 kV	1.75 plus 0.30 m for every 33 kV or part thereof.

6. **Poles:** The height of poles depends on (*i*) minimum permissible ground clearance for the working voltage, (*ii*) maximum sag, (*iii*) embedded length in ground (1/6th of total length), and (*iv*) arrangement of conductors/cross arms configuration. Poles of 8 m length are suitable for L.T. lines and 9m and above for 11, 33 and 66 kV lines.

The size of poles is determined from the formula:

$$M = fZ$$

where, M is sum of all bending moments caused by wind at ground level, Z is modulus of section, and f is allowable stress (30 kg/sq. mm) and working stress (10 kg/sq. mm).

7. **Economic:** The ratio of base to height of tower may vary between 1 to 3 for single-circuit and 1 to 4 for double circuit type; the height is taken at the intersection of the slope of the legs. The span length of conductor too is dictated by economic considerations. Lighter towers mean more number of towers and insulators than higher and fewer towers and a small number of insulators.

8. **Voltage Drop or Regulation:** The percentage regulation of a line is given by the relation:

$$\frac{E_s - E_r}{E_r} \times 100$$

where, E_s and E_r are sending and receiving end voltages.

9. **Line Losses:** Part of power is expended in transmission termed as line loss. The line loss occurs on account of resistance of the conductor to the electricity flow.

The resistance is overcome by voltage drop which is proportional to electricity flow through the conductor. The line loss is the product of the square of the current and the resistance of the conductor, in addition due to the effect of inductance capacity, leakage, corona and skin effect.

The all-India transmission and distribution losses are of the order of about 23 percent as against 7-10 percent in developed countries. The higher losses are attributable to long sub-transmission and distribution lines, low power factor loads, multiplicity of stages of transformation, inadequate distribution system in urban area, pilferage of electricity and improper load management.

10. **Wind Load:** Wind load on a flat surface is given by the relation :

$$W_l = 0.006V^2$$

where, W_l is wind load (kg/m²) and V is wind velocity (km/hr.).

QUESTIONS

1. What do you understand by transmission and distribution lines.
2. Transmission line is a costly component of the hydro power development, comment.
3. Briefly mention the various functions served by the transmission line in its role in interconnection with the grid.
4. Discuss briefly the various components of a transmission line.
5. How will you differentiate between an electric pole and a elecric tower? What is the difference between a flexible tower and a tower (rigid) in a transmission line ?
6. Why is voltage stepped up for transmission? What are its advantages?

16

HYDRAULIC TUNNELS

16.1 INTRODUCTION

Water conductor system occasionally necessitate an underground water passage, without removing the overlying rock or soil, through high ground, mountain, in rugged terain, urban area or where it is preferred on account of convenience and economy over a surface pipe line or an open channel. It provides a direct route, geology permitting, rather than long detour or deep cutting through the topography thereby effecting economy in the project cost.

Tunnelling came to the aid of water power projects in Sweden in 1914. In India almost all major hydroelectric projects involve extensive and intricate tunnelling. Tunnelling is a costly, time consuming and challenging component of hydropower development especially in underground and pumped storage power houses as also in surface power houses in Himalayas. Tunnel at Nathpa Jhakri project has a static head as high as 200 m rising up to 300 m during transient conditions. Of late, it is becoming simpler, safer and faster with the advent of latest equipment and techniques of tunnel driving and its lining.

The hydraulic tunnels may be free flowing, pressure tunnel or flowing partly full. They may also be lined or unlined but power tunnels are essentially R.C.C. lined or steel lined to discharge silt laden water under high velocities. In pressure tunnels internal loads are considered. The lining is usualy reinforced if the depth of rock cover is less than 1.5 times the internal pressure head.

16.2 NECESSITY FOR TUNNELS

Tunnelling in waterpower development projects is essentially necessitated on account of the following: (*i*) It provides a direct route for the water conductor system thereby resulting in saving in head which is very important in hydropower development as also results in saving in the project cost. (*ii*) It permits tunnelling work to be started at various heads, through adits, along its alignment, (*iii*) Tunnelling is getting easier and faster with the development of latest tunnel boring techniques, (*iv*) The landscape of the area is not disturbed compared to an open channel excavation, (*v*) The concepts of rock mechanics and the latest tool of stress analysis by using finite element method, has established structural stability of tunnels.

16.3 CLASSIFICATION OF TUNNELS

The tunnels appurtenant to hydropower plants are classified as under:

1. Classification on the Basis of Function Served

(*i*) Water Carrying Tunnels

Diversion Tunnels: These are the tunnels which are amongst the first construction works to be started at a hydropower project with the aim of diverting the river flow away from the construction area. They have their intake at the river bed and of the capacity to carry both normal as well as high monsoon flows. On completion of the dam these are converted into spillway tunnel but usually penstock tunnels by plugging their intakes from the river bed. These are usually of the short length.

Headrace Tunnels: These are the tunnels which convey water from dam, barrage or forebay to the power plant. These are pressure tunnels and accordingly termed as power tunnels or pressure tunnels (Fig. 6.11).

Tailrace Tunnels: These are the tunnels which carry away the water from power plant. These are usually free flow tunnels. In underground power plants these are pressure tunnels as also when the turbine setting is below tail water level (Fig. 6.11).

Spillway Tunnels: These are the tunnels which act as spillway to dispose of flood water, usually called shaft spillway.

(*ii*) Non-Water Conveying Tunnels

Dry Tunnels: Also called service tunnels. These tunnels serve diverse purposes, other than conveying water, and are classified as under:

Cable Tunnels: These are meant to convey cables from the power house to switch yard in underground power plants. They also find their use in surface power plants. Fig. 16.1 shows the cable tunnel provided at Pong power plant.

Ventilation Tunnels: These are the tunnels provided to convey fresh air to the underground tunnels.

Adit Tunnels: These are provided to facilitate starting tunnelling at various heads along main tunnel (Fig. 6.11).

Access Tunnels: Also termed as approach tunnels. They serve as passage from surface to underground installations of underground power plant (Fig. 6.11).

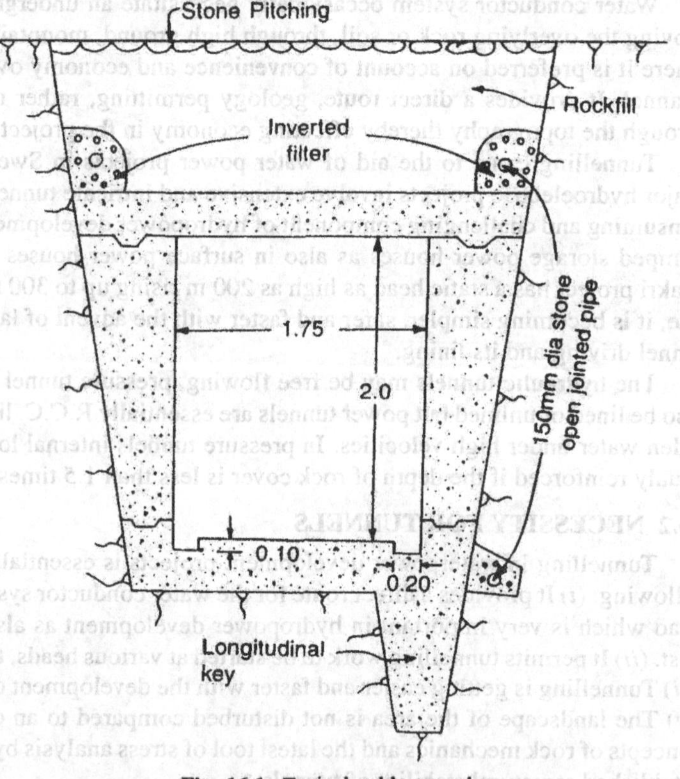

Fig. 16.1 Typical cable tunnel.

2. Classification Based on Shape

The tunnels shape may be circular, rectangular or any combination of these forms such as horse shape, D-shaped, egg shaped and egglipse shaped and are accordingly called circular tunnel, horse shoe tunnel etc.

3. Classification Based on Protective Surface

Lined Tunnel: The power tunnels are usually lined to economise on loss of head, smaller cross section and hydraulically efficient. The lining may be of concrete or steel.

Unlined Tunnel : In good sound rock, the tunnel may be unlined.

4. Classification Based on Alignment

Tunnel: The term tunnel implies here the one with a slight bed slope or virtually a horizontal tunnel.

Shaft: It is a tunnel with alignment vertical. Inclined shaft is one which is steeply inclined to the horizontal.

5. Classification Based on Characteristic of Flow

Low Pressure Tunnel: It is a tunnel with low internal water pressure i.e., water head less than 10 m. Diversion tunnels are usually low pressure tunnels.

Medium Pressure Tunnel: It is a tunnel with water head of 10–100 m.

High Pressure Tunnel: The high pressure tunnel has high internal pressure corresponding to a head of over 100m.

Drainage Tunnel: It is a tunnel provided in the abutments of an earthen dam to collect seepage water from rainfall, reservoir for safe disposal lest it may create sloughing of the abutment hill sides.

16.4 INVESTIGATIONS AND SURVEYS

1. Preliminary Investigation

Preliminary investigation for aligning the tunnel is carried out on 1:50,000 survey of India topographical sheets to establish general feasibility of the tunnel. Detailed strip topographic maps along the tunnel alignment are then prepared to a scale 1:15,000 with 10 m contour interval. At portal faces the contour interval may be 2m. Wherever possible aerial (photographic) survey is carried out and the stereoptic covering extending for at least 3 km on either side of the foreseen limits of the tunnel alignment.

2. Geological Investigations

The geological investigations are carried out to determine (*i*) Origin and type of rock along the alignment and study of regional geological maps of the area, (*ii*) Geological section along the tunnel alignment giving rock types and their disposition, location and attitude of all structural features of rock such as faults, thrusts, joints, dips, strikes and other geological features including pattern, extent and contents of fissures; pressure of water in small or large quantities and their probable pressure at tunnel grade etc., (*iii*) Any geological feature which may affect the magnitude of rock pressure to be anticipated along the proposed alignment, (*iv*) Cover on tunnel, position of subsurface rock and overburden contacts, (*v*) Physical, mechanical and strength properties of rock to determine supporting arrangements, and also resistance to driving tunnel through rock, and (*vi*) Hydrological data and information regarding location, type and volume of water and injurious or troublesome gases contained in subsurface strat around tunnel grade.

The core samples collected from bore holes are classified and specimen from each group is tested to determine physical properties such as specific gravity, modulus of elasticity, poisson's ratio, tensile strength, compressive strength, triaxial shear strength, hardness of rock, swelling index and porosity, grain size and cementing material for sand stones and similar rocks to ward off geological surprises generally encountered in tunnelling. Investigations for selecting the alignment of tunnels are done by means of selective drilling and by making drifts and test tunnels to obtain as much information as possible of the geological conditions prevailing at the site.

Certain construction difficulties may also affect the tunnel alignment such as (*i*) presence of methane gas, (*ii*) presence of jointed rock, aquifers confined by impervious strata which might lead to heavy inflow of water, and (*iii*) squeezing rock conditions and information on underground gases and soil temperatures. The explorations are done to determine top cover, quality of subsurface rock, surface drainage conditions, position, type and volume of water and gases contained in subsurface rock and physical properties and resistance to tunnelling offered by the rock encountered. The best strata for tunnel construction is solid rock with little ground water.

3. Cover

Cover on a tunnel in any direction is the distance from the tunnel soffit to the rock surface in that direction. However, where the thickness of overburden is sizable its equivalent weight may also be reckoned provided that rock cover is more than three times the diameter of the tunnel.

For tunnels located near mountain slopes, the lateral cover rather than the vertical cover is often the governing criteria. In such cases the effective vertical cover equivalent to the actual lateral cover is determined by drawing a profile of the ground surface, perpendicular to the contour lines, and fitting the curve shown in Fig. 16.2 in such a way that it touches the ground surface. The vertical distance 'C_v' is the effective vertical cover and shall be greater than the internal pressure head in the tunnel.

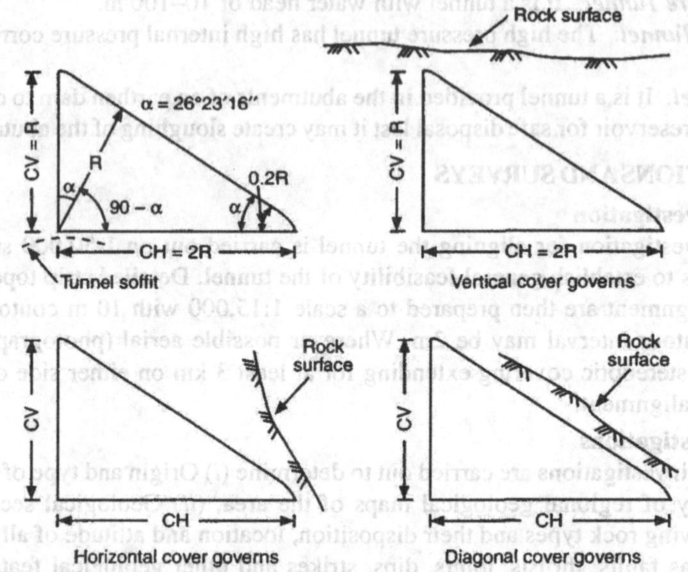

Fig. 16.2 Estimation of effective cover.

16.5 TUNNEL ALIGNMENT

The alignment of tunnel is guided by the nature of soft or hard rock strata and its formation, general topography, geology of the area etc. The various points considered while selecting alignment of a hydraulic tunnel are briefly as under:

1. Shortest Route: The tunnel length should be as short as possible between the intake and surge shaft as the connecting points. It is to ensure minimum losses and economy in cost.

2. Safe route: The alignment should avoid dangerous folds, faults, joint planes, etc. Sound, isotropic and solid rock formations are ideal for tunnelling.

3. Avoid Bends: Bends in the alignment involve losses at all such bends with increase in cost.

4. Easily Accessible: An easy access near the entrance and exit, besides adit points along the length of the tunnel are essential for expeditious construction.

5. Exit and Entry Portals: Careful selection of the entry and exit locations with minimum length and depth of approach cuttings and no weathered, loose fractured layers slope towards portals.

6. Rock Cover: All along the alignment the tunnel should have sufficient rock cover.

However, it may not be possible to follow a straight alignment because of the following parameters affecting the design of hydraulic tunnels:

1. Topography: There may not be sufficient vertical and/or lateral cover.

2. Geology: The geological section along the tunnel alignment may show certain difficult strata through which the process of tunnelling may be difficult and costly.

3. Water loads: Ground and/or rock water loads along the tunnel alignment may be excessive thus increasing the overall cost of the tunnel.

4. Rock mechanics properties: The in-situ stresses.

16.6 TUNNEL CROSS SECTION

The various factors governing the selection of cross section of the tunnel are (*i*) Geological conditions prevailing along tunnel alignment, (*ii*) Maximum required flow at available slope, (*iii*) Specified clearance for vehicles and materials transported in the tunnel, (*iv*) Hydraulic requirements that is the cross section is sufficient to carry the maximum discharge on the head available, (*v*) Structural considerations, (*vi*) Functional requirements, (*vii*) Methods of driving the tunnel, (*viii*) Material and strength of the tunnel lining, (*ix*) Economic studies to obtain economic section, considering velocity, loss due to friction, lined or unlined, cost of gates and hoists, capital and maintenance cost etc. The basic requirement is adequate size to carry the requisite discharge and at the same time conforming to construction requirements, minimum size is 2.5 × 2.5 m.

Shape of Tunnel: The tunnel shape is governed by structural efficiency for sustaining external loads as also the hydraulic efficiency. The tunnel cross section may be circular, rectangular or any combination of these forms. In hard rock, tunnel section excavated with an arched roof may do. In loose and weak rocks, a circular section or horse shoe section may suit. The conventional tunnelling methods are suitable for driving horse shoe section and flat arched section and less economical for circular section. Rectangular section is easy to excavate. A circular section is best hydraulically and also where the ground is unstable and safety reduces with other shapes. However, from construction point of view, D-shape for smaller tunnels and horse shoe for larger tunnels is adopted.

The general requirements and design parameters for the various shapes of tunnels generally used are briefly as under:

Circular Section: It is the most suitable and stable shape (Fig. 16.3) from structural considerations where the tunnel is subjected to high inward or outward radial pressures. Also where the tunnel is subjected to high internal pressure but does not have good quality rock and/or adequate rock cover around it but difficult for excavation, particularly where cross sectional area is small. The maximum flow occurs when the ratio of depth to diameter is 0.93, using Manning's formula. For a depth of 0.80, flow is the same as for full circle and is used to compute the minimum free board.

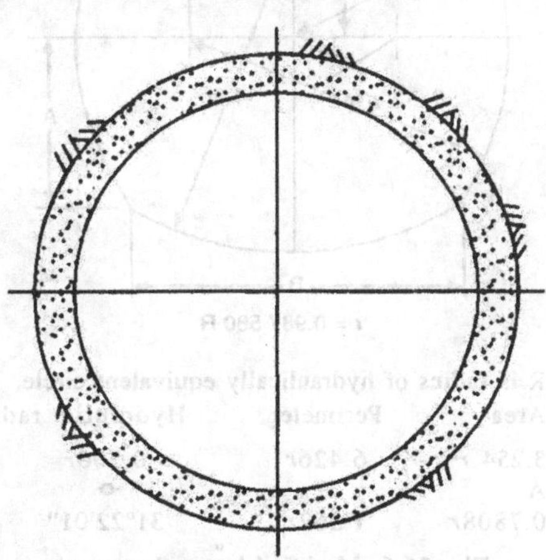

Fig. 16.3 Circular section.

D-Shaped Section: It is suitable in good quality, intact sedimentary rocks and massive external igneous hard, compacted, metamorphic rocks where the external pressure due to rock and water are not very large and also where the lining is not designed against internal pressure. Its merit over horse-shoe section is the added width of the invert which gives more working floor space in the tunnel during driving and flatter invert which helps to eliminate the tendency of wet concrete to slump and draw away from the tunnel sides as also permits use of concurrent lining of the tunnel (Fig. 16.4).

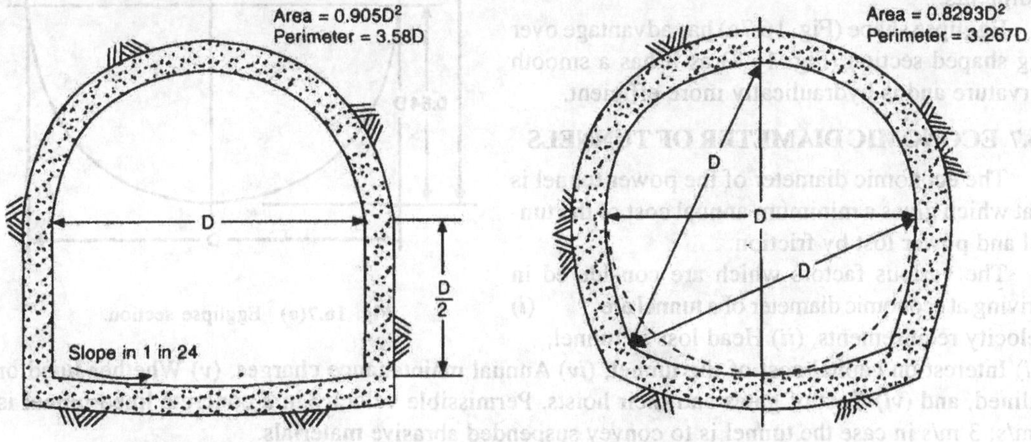

Fig. 16.4 D-section **Fig. 16.5.** Horse-shoe section.

Horse-Shoe Section: It is a compromise between circular and D-shaped sections (Fig. 16.5). It is structurally strong to withstand external rock and water pressures. It is suitable in a moderately good rock as the tunnel has to resist internal pressure also. The maximum flow occurs when the ratio of depth to diameter is 0.94 and at depth of 0.82 the flow is the same as for full section.

The modified horse-shoe section (Fig. 16.6) provides the advantage of a flatter invert for constructional ease. It also affords easy change over to circular section with minimum additional cost in reaches where rock quality is poor or rook cover is inadequate.

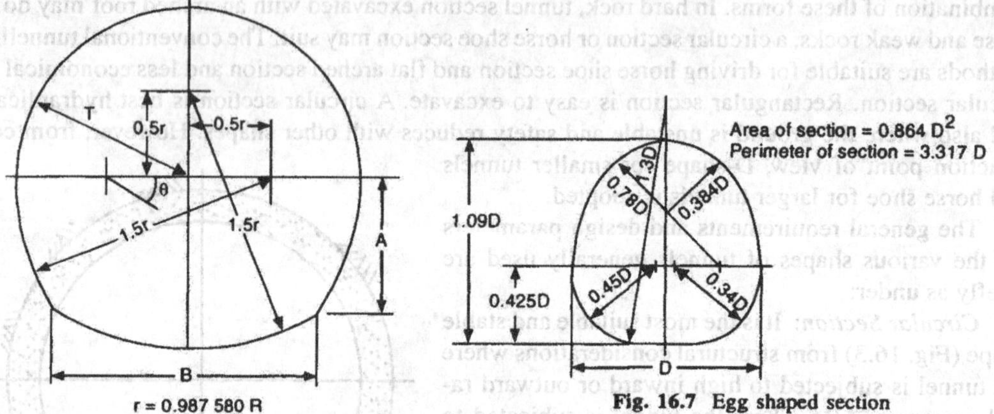

R is radius of hydraulically equivalent circle.

Area	Perimeter	Hydraulic radius
$3.254\ r^2$	$6.426r$	$0.506r$
A	B	θ
$0.7808r$	$1.562r$	31°22'01"

Fig. 16.6 Modified horse-shoe section.

Fig. 16.7 Egg shaped section

Egg-shaped and Egglipse Section: These sections are considered where the rock is stratified, soft and very closely laminated (as laminated sand stone, slates, micaceous schists etc.) and where the external pressures and tensile forces in the crown are likely to be high so as to cause serious rock falls. These are suitable for tunnels carrying sediments.

Egglipse shape (Fig. 16.7a) has advantage over egg shaped section (Fig. 16.7) as it has a smooth curvature and is hydraulically more efficient.

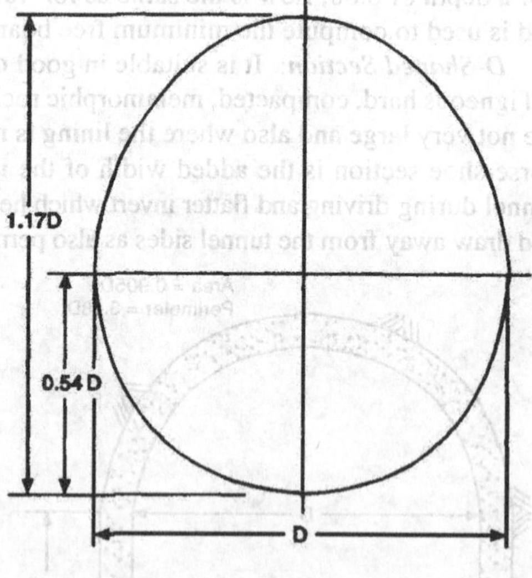

Fig. 16.7(a) Egglipse section.

16.7 ECONOMIC DIAMETER OF TUNNELS

The economic diameter of the power tunnel is that which gives a minimum annual cost of the tunnel and power lost by friction.

The various factors which are considered in arriving at economic diameter of a tunnel are (*i*) Velocity requirements, (*ii*) Head loss in tunnel, (*iii*) Interest on capital cost of the tunnel, (*iv*) Annual maintenance charges, (*v*) Whether lined or unlined, and (*vi*) Cost of gates and their hoists. Permissible velocity in a concrete lined tunnel is 6 m/s; 3 m/s in case the tunnel is to convey suspended abrasive materials.

From practical considerations of space requirements for excavating equipment, the minimum height is 2 m. For mechanised handling of excavated material a minimum section of 2.5 × 2.5 m is required.

In sound rock the unit cost of excavation decreases as the diameter increases to a point that permits the use of full sized shovel equipment up to 10m. In weak rock the unit cost may increase as the size increases owing to extra cost of supports.

For the computation of frictional head losses, Manning's formula is widely used with coefficient of friction 0.011 to 0.018 and in practice an average value is adopted. The coefficient is greatly influenced by the tunnel diameter and it inversely varies with the size. However, in the analysis of transient flow conditions minimum value is adopted for maximum upsurge and maximum value for low surge levels.

16.8 HYDRAULIC DESIGN

Discharge Characteristics: The discharge characteristics of a tunnel i.e. nature of flow are determined by pressure head, slope, size, shape, length, surface roughness, inlet and outlet shapes and maximum and minimum levels at the head and tail ends. In certain locations in addition to hydraulic gradient which has to be positive in the direction of flow, energy gradient, principles of momentum, transient conditions like water hammer, surges etc. are considered. In free flow tunnels proper aeration is ensured. The pulsating conditions are minimised. The expected variation in the friction factor is considered.

In pressure tunnels, the depth of intake is such that no air is sucked in. The slope is fixed to meet the requirement of dewatering. The location of outlet of the tunnel is such that the entry of air does not adversely affect tunnel operation and safety provided to ensure preventing of air locks. Negative pressures are avoided. The residual positive pressure is not to be less than 3 m of water in concrete lined tunnels.

Velocity: Manning's formula is used to determine velocity. The value of rugosity n varies depending on nature of rocks values in very rough surface (0.04 – 0.06), in trimmed surface (0.025 – 0.035) and surface trimmed and invert concreted (0.020 – 0.030). In concrete lined tunnels, value of n varies from 0.012 to 0.018. The velocity in unlined tunnel in rock is limited to 2.5 m/s. In concrete lined, the velocity may be 6 m/s and 9 m/s in steel lined tunnels. The pressure shafts are usually steel lined with velocity up to 8 m/s. Where water carries abrasive material in suspension and as bed load, the recommended velocity as per IS: 4880 (Part III)–1976, is 2.5 m/s.

Overburden: The overburden of rock in pressure tunnels serves to balance the effect of upward force due to internal water pressure. The minimum height of overburden as recommended by Mesonyi is 0.4 to 0.8 hi, where hi is internal water pressure head. Reinforced cement concrete lining is needed where depth of overburden is less than that indicated above.

The diversion tunnels at Pong dam, 5 No. of 9.14 m finished diameter, three of which were later converted into penstock tunnels and two as irrigation (outlet tunnels), were designed for a maximum discharge of 1260-1340 cumecs at a velocity as high as 20.4 m/s. The tunnels were provided with reinforced cement concrete lining 75 cm thick, the amount of reinforcement being 2,753 kg/m (Fig. 16.8.).

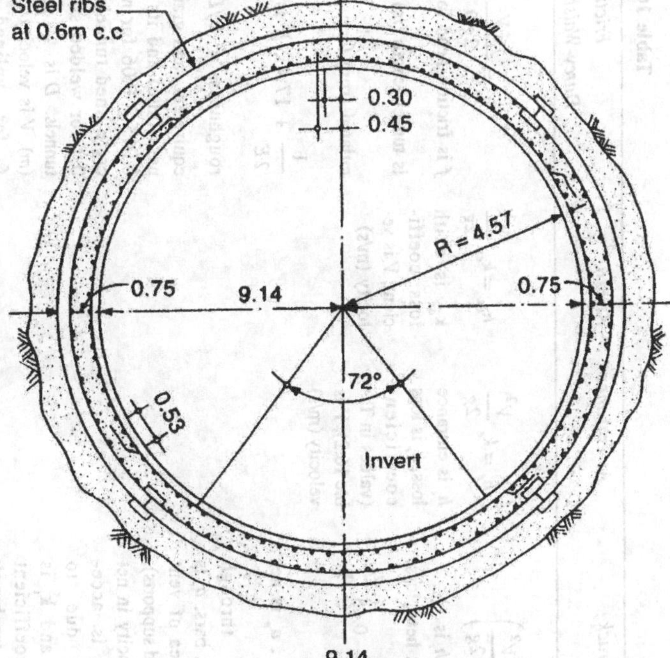

Fig. 16.8 Typical cross-section of reinforced concrete lining.

Table 16.1. Hydraulic losses

At trashrack	*At entrance*	*At exit*	*Friction losses (formulae)* *Darcy-Waisbach*	*Friction losses (formulae)* *Manning's*	*Expansion*	*Contraction*	*Gate loss*
$h_t = k_t\left(\frac{V^2}{2g}\right)$ where, h_t is trashrack head loss, k_t is $1.45 - 0.45\frac{a_n}{a_t} - \left(\frac{a_n}{a_t}\right)^2$, a_n is net area through trashrack bars, a_t is gross area of vent (racks and supports), V is velocity in net area, g is acceleration due to gravity and k_t is loss coefficient through trashrack bars	$h_e = k_e\frac{V^2}{2g}$ h_e is entrance loss, k_e is loss coefficient (values in Table 16.2), V is velocity (m/s)	$h_{ex} = k_{ex}\frac{V^2}{2g}$ k_{ox} is exith loss coefficient, V is velocity (m/s)	$h_f = \frac{fLV^2}{2gD}$ f is friction head loss (m), L is tunnel length (m), f is calculated from $(\frac{1}{\sqrt{f}} = 2\log_{10}\frac{1}{2E} + 17.4)$, E is relative roughness $= k_s/D$, k_s is equivalent sand grain roughness (mm) and its value is 0.015 – 0.06 for new concrete lined tunnels, 0.05 – 0.1 for welded steel lined tunnels. D is tunnel dia. (m), V is velocity (m/s). f for unlined tunnels $= 0.00257\delta$ where, δ is $\frac{A_{99} - A_1}{A_1} \times 100$, A_{99} is area corresponding to 99% frequency and A_1 is area corresponding to 1% frequency. For non-circular section, $h_f = \frac{fLV^2}{8gR}$, R is hydraulic mean depth.	$h_f = \frac{V^2n^2L}{R^{4/3}}$ n is roughness coefficient, value 0.012 – 0.018 for concrete lined tunnel. For unlined, n = 0.04 – 0.06 (rough surface), 0.025 – 0.035 trimmed and 0.020 – 0.030 trimmed and invert concreted), L is tunnel length.	$h_f = k_e\left(\frac{V_1^2}{2g} - \frac{V_2^2}{2g}\right)$ k_e is loss coefficient, $3.5\left(\tan\frac{\alpha}{2}\right)^{1.22}$; α is angle of tunnel wall surface w.r.t C.L., V_1 is velocity at beginning of transition, V_2 is average velocity (m/s) at the end of transition.	$h_c = k_c\left[\frac{V_2^2}{2g} - \frac{V_1^2}{2g}\right]$ k_c is loss coefficient for contraction, value 0.1 for gradual and 0.5 for abrupt contractions. V_2 is velocity in contracted section, V_1 is velocity in normal section.	$h_g = k_g\frac{V^2}{2g}$ k_g is loss coefficient for gate, average value 0.10 for partly open gate and 0.19 for wide open gate.

Pressue Flow Losses: The hydraulic losses in tunnels flowing full are determined as given in Table 16.1.

Table 16.2. Entrance loss

Type of entrance	*Loss coefficient k_e*		
	Maximum	*Minimum*	*Average*
Gate in thin wall-unsuppressed contraction	1.80	1.00	1.50
Gate in thin wall-bottom and sides suppressed	1.20	0.50	1.00
Gate in thin wall-corners rounded	1.00	0.10	0.50
Square-cornered entrances	0.70	0.40	0.50
Slightly rounded entrances	0.60	0.18	0.25
Fully rounded entrances $\frac{\gamma}{D} \geq 0.15$	0.27	0.08	0.10
Circular bellmouth entrances	0.10	0.04	0.05
Square bellmouth entrances	0.20	0.07	0.16
Inward projecting entrances	0.93	0.56	0.80

16.9 TUNNEL EXCAVATION

In tunnelling at least two working faces, one on each end of the tunnel are available. Depending upon the length of tunnel and the schedule of its completion, the tunnel is provided with one or more intermediates; each such adit offering two more working faces through which tunnelling can progress. The intermediate adits may be in the horizontal plane for mucking, in certain cases only vertical intermediate adit is possible. The horizontal adits have no hydraulic function and have to be plugged after the tunnel is completed but the vertical adits may act as small surge shafts and hence not plugged.

1. Tunnelling through Firm Ground

The method of tunnel excavation depends upon the size and shape of tunnel, rock characteristics, the equipment available, support system envisaged for the tunnel and actual stand up time and the overall economics. In firm ground comprising soft and stratified rocks such as sand stone, shale, cemented sand and gravel, hard clays etc, which have a bridge action period just enough to install supports within the mucking period tunnels can be excavated either by conventional drilling and blasting method or by the use of tunnelling machines.

Conventional Methods

Full Face Method: In this method, the entire cross sectional area of the tunnel to be excavated is attacked simultaneously. It is generally adopted for small size tunnels in good rock conditions where major rockfalls are not anticipated.

Top Heading and Benching Method: In the method, a top heading is excavated first, either to full length or part length of the tunnel, and is supported simultaneously. The benching is then removed slowly (Fig. 16.9.). The method is used where the tunnel has large cross sectional area or where the rock is not of good quality.

Bottom Heading and Stoping Method: In this method, a bottom heading is made first and the overhead stope is removed later as shown in Fig. 16.9. It is used where the rock is consistent and sound and the tunnel section is very large.

Drift Method: The method is used where in a large tunnel it is economical to drive a small tunnel, called drift or a pilot tunnel, prior to excavating the full face. The drift may be excavated in the centre, side, bottom or top as shown in Fig. 16.9.

The special methods, as under, are used as the tunnel bore increases and as the quality of rock deteriorates.

Side Drift Method: The method is adapted to large size tunnels through bad rock which require support before mucking. The side drifts (Fig. 16.9) are driven ahead of the main excavation for some convenient distance. The support of these drifts includes the main tunnel supports and wall plates. Just prior to shooting the main bore, the drift support is removed leaving the main posts and wall plates in position which project above the muck pile thus permitting erection of the main arch ribs before mucking out.

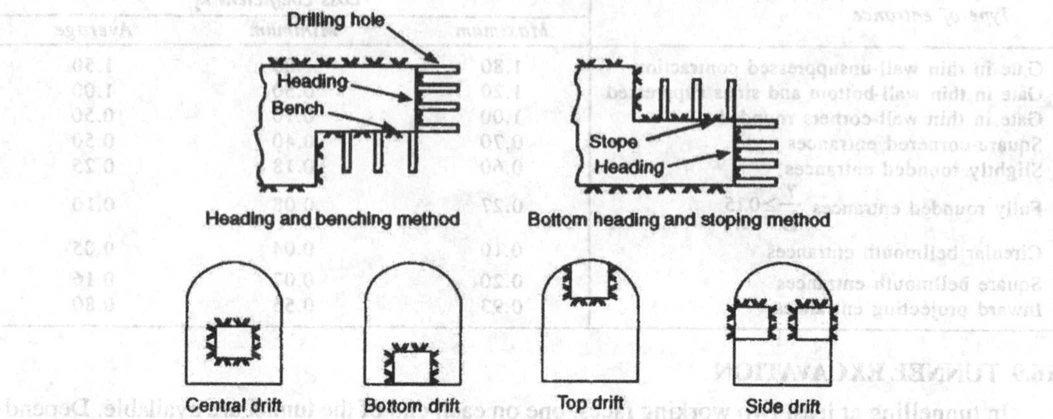

Fig. 16.9 Tunnel excavation methods.

Multiple Drift Method: It is a combination of the side drifts and top drift. It is suitable for large openings in crushed rock formations in fault zones which may behave like earth, even though the rock is compacted enough to require light blasting. In the method, a side drift is driven first on each side.

2. Tunnelling through Soft Ground

A tunnel may suddenly pass from good rock to the material with no cohesion such as clean sand and gravel, termed as soft or running ground. It requires use of forepoling method in combination with steel supports. The boards which are driven ahead to support the ground ahead of the last rib are known as spiles. They act as cantilevers which carry the weight of ground until their forward ends are supported by installing the next rib. The spiles are installed as far down around the sides of the tunnel as is necessary. Then the top breast boards are removed, the exposed ground is excavated and the breast boards are reinstalled ahead. This process is continued until the excavation arrives at floor level whereupon the next rib can be installed. At sometimes afterwards the tails of the spiles are cut-off.

3. Tunnelling through Soft, Water Bearing Strata

The seepage towards tunnels located below water table shortens or even eliminates the bridge-action period, reduces the bearing capacity of the tunnel bottom which might cause its heaving up. In soft, porous ground it may create the danger of loss of ground due to soil being washed into the tunnel. Free draining ground such as clean sand or gravel cannot be tunnelled through unless it is previously drained or grouted.

New Austrian Tunnelling Method

The new Austrian tunnelling method (NATM) is based on the principle to take maximum advantage of rock's capacity to support itself by careful measures and deliberate guidance of the force during the readjustment process from primary to the secondary state of stress around the excavations. By employing careful excavation and timely installation of support elements, the local progressive loosening can be limited. The support elements are shortcrete, welded wiremesh, rock bolts, steel ribs, forepoling, steel lagging. The adequacy of the support elements can be assessed by deformation monitoring by the installation of extensometer or tape extensometer.

The NATM constitutes a method by which the surrounding rock or soil formation of a tunnel are integrated into an overall ring like support structure and as such these formations themselves become a part of the primary support.

TBM Method

The tunnel boring machine (TBM) used for tunnelling saves overbreak costs/concreting costs which alone can pay for the cost of tunnel boring machine just in about 6–8 km drivage. Advance rates of TBM are around 50 m a day *i.e.*, 10 to 15 times more than drill and blast method. The other advantages of TBM are (*i*) Less support is required as the ground around and adjacent to a tunnel is not disturbed, (*ii*) Thinner concrete lining is adequate, (*iii*) Lesser labour cost; 40-60% of drill and blast excavation, (*iv*) Lesser mucking cost, (*v*) Lesser access roads cost, (*vi*) Automatic excavation and muck removal, (*vii*) Clear air environment as TBM is equipped with dust collector and is operated by electricity, (*viii*) Expeditious tunnelling and thereby bringing power on line early.

16.10 LOADS AND REACTIONS

The lining is considered as an independent structural element assuming that it deflects under the active external loads and its deflection is restricted by the passive resistance developed in the surrounding rock mass. Any point on the periphery of the lining thus may be stated by the following equation:

$$\Delta p = \Delta r + \Delta e + \Delta w - \Delta y$$

where, Δp is deflection due to passive resistance, Δr is deflection due to rock load, Δe is deflection due to self weight of lining and the water contained in it, Δw is deflection due to external water pressure, if any, and Δy is yield of surrounding rock mass due to abutting of the lining against it.

The loads and reactions involved in the design of lining are rock load, external water pressure if any, grout pressure, if any, self weight of lining, weight of water contained in the tunnel, reaction due to active vertical loads and lateral passive pressure due to the deformation of lining.

(*a*) *Rock Load*: Rock load is assumed as an equivalent uniformly distributed load over the tunnel soffit over a span equal to the width or diameter of the tunnel as the case may be (Fig. 16.10).

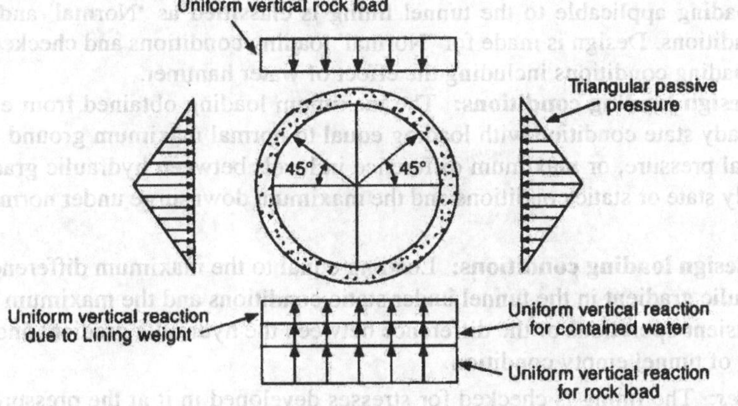

Fig. 16.10 External loads on lining.

Rock load is estimated by the following methods:

1. Rock load at depths less than or equal to 1.5 $(B + H_t)$ is taken as equal to depth of actual rock cover, where B is the width and H_t is the height of tunnel opening. In the case of circular tunnels B and H_t both are equal to the diameter of tunnel D.

Rock load (H_p) on the roof of support in tunnel, with width B and height H_t at depth of more than 1·5 ($B + H_t$) is assumed to be as given in Table 16.3.

Table 16.3 Rock load (IS: 4880 (Part IV–1971)

Rock condition	*Rock load H_p*	*Remarks*
Hard and intact	Zero	Light lining required only if spalling or popping occurs.
Hard stratified or schistose	0 to 0.50 B	Light support
Massive, moderately jointed	0 to 0.25 B	Load may change erratically from point to point
Moderately blocky and seamy	0.25 B to 0.35 $(B + H_t)$	No side pressure
Very blocky and seamy	(0.35 to 1.10) $(B + H_t)$	Little or no side pressure
		Considerable side pressure
Completely crushed but chemically intact	1.10 $(B + H_t)$	Requires continuous support for lower ends of ribs or circular ribs.
Squeezing rock, moderate depth	(1.10 to 2.10) $(B + H_t)$	Invert struts required. Circular ribs recommended
Squeezing rock, great depth	(2.10 to 4.50) $(B + H_t)$	
Swelling rock	Up to 80 m irrespective of value $(B + H_t)$	Circular ribs required. In extreme cases use yielding support.

2. Rock load is also determined by using Fenner's ellipse (Fig. 16.11) by the equation:

$$a = \frac{b}{2}(m-2)$$

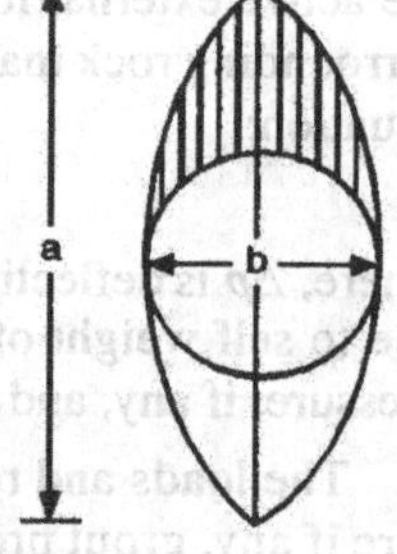

Fig. 16.11 Fenner's ellipse

where, a is main axis of ellipse of rock load, b is tunnel diameter (excavated) and m is inverse of poisson's ratio for rock; usually 2 to 7.

The weight of rock in the shaded portion is taken as rock load and is considered as uniformly distributed on the diameter of tunnel.

(*ii*) *External Water Pressure*: Lining is designed for external water pressure, if any. However, in areas where drainage holes are provided, lining is not designed for external water pressure. Where effective pressure grouting has been done, the external water pressure is assumed to act on the whole grouted cylinder which may be taken as a structural element for computing the stresses and deformation of lining.

The design loading applicable to the tunnel lining is classified as 'Normal' and 'Extreme' design loading conditions. Design is made for 'Normal' loading conditions and checked for safety under 'Extreme' loading conditions including the effect of water hammer.

1. **Normal design loading conditions:** The maximum loading obtained from either maximum steady or steady state condition with loading equal to normal maximum ground water pressure and no internal pressure, or maximum difference in levels between hydraulic gradient in the tunnel, under steady state or static conditions and the maximum downsurge under normal transient operation.

2. **Extreme design loading conditions:** Loading equal to the maximum difference in levels between the hydraulic gradient in the tunnel under static conditions and the maximum downsurge under extreme transient operations or the difference between the hydraulic gradient and the tunnel invert level in case of tunnel empty condition.

Grout Pressures: The lining is checked for stresses developed in it at the pressure on which grouting is done and it is ensured that the stresses are within the limits depending on the strength attained by concrete by then.

Vertical Reactions: Reactions due to the vertical loads are assumed reasonably to be vertical and uniformly distributed on the invert of the lining (Fig. 16.10).

Lateral Passive Pressure: It is estimated either by considering the lining as a ring restrained by elastic medium with a suitable modulus of deformation or by restricting the maximum deflection of the lining at the horizontal diameter to an assumed value, depending upon yield of the surrounding rock mass, by the lateral passive pressure. The latter method is used for the design of circular lining. For design of non-circular linings, the former method is used. The stress pattern is determined by photo-elastic studies.

Internal Water Pressure: Lining is considered as a part of composite thick cylinder consisting of peripheral concrete and surrounding rock mass subjected to specified boundary conditions.

1. Normal design loading conditions: Maximum static conditions corresponding to maximum water level in the head pond, or loading equal to the difference in levels between the maximum upsurge occurring under normal transient operation and the tunnel invert.

2. Extreme design loading conditions: Loading equal to the difference between the highest level of hydraulic gradient in the tunnel under emergency transient operation and the invert of the tunnel.

16.11 TUNNEL LINING

The power tunnels which connect directly to the penstock are essentially lined with cement concrete or reinforced concrete or steel lined. The functions served by a concrete lining are (*i*) To provide smooth surface thereby reducing head losses in the system, (*ii*) To protect steel ribs provided in the tunnel from deteriorating, (*iii*) To prevent leakage of water, and (*iv*) To protect the turbines from the loose rock particles falling into the water and being carried to the turbines.

The structural design of the tunnel lining is based on thorough study of rockmass, its effective cover, in-situ modulus of elasticity, poisson's ratio, state of stress, crushing strength and presence or absence of water in the rock. The concrete lining is designed for external loads considering it as an independent structural member. For internal water pressure lining is considered as a part of composite thick cylinder consisting of peripheral concrete and surrounding rock mass subjected to specific boundary conditions. Pressure grouting of rock mass surrounding the tunnel is done.

Cement Concrete Lining

Generally plain cement concrete lining is provided in power tunnels. In pressure power tunnels with high hydrostatic loads, reinforcement is provided to withstand bursting where cover is inadequate i.e. less than the internal water pressure head to minimise leakage and to ensure structural stability. Circular lining is generally adopted where the effective cover in good rock is less than the internal water pressure head and in poor rock where the effective cover is less than 1.25 times the internal load. The thickness of lining is designed such that stresses in it are within permissible limits when the most adverse conditions occur. The minimum thickness of lining, as per IS: 4880 (Part IV)–1971 is 15 cm, roughly it is taken as 6 cm per metre of finished diameter of the tunnel in reasonably stable rock. For reinforced concrete lining, the minimum thickness is 30 cm, the reinforcement, however, is arranged in the crown to allow for proper placement of slick line. The portion of the tunnel which is to be reinforced, amount of reinforcement required depends on the physical features of the tunnel, geological factors and internal water pressure. The steel supports, where used, are considered as reinforcement only if these are of high tensile bolts at the joints or by welding the joints. A minimum cover of 15 cm is provided over the inner flange of steel supports and 8 cm minimum over the reinforcement bars. Near the portals of both pressure and free-flow tunnels, adequate amount of reinforcement both longitudual and circumferential reinforcement, in addition to steel supports, is provided, if required, to resist loads resulting from loosened rock headings or from sloughing of the portal cuts.

For free-flow tunnels, precast cement concrete lining is provided but normally reinforcement is not needed to resist external loads. Such loads are resisted and taken care of by steel supports and/ or precast concrete rings. However, reinforcement is provided, where required, to resist external loads due to unstable ground or grout or water pressures. Drainage holes are provided in the crown above full supply level where the external water pressure is substantially higher than the internal water pressure. At successive sections, one vertical hole is drilled in the crown alternating with two drilled horizontal holes, one in each side wall, extending to a depth of at least 15 cm beyond the back of lining.

Steel Lining

Steel lining is essentially required where (*i*) The tunnel has to withstand high pressures, (*ii*) The rock cover is inadequate, cavitation of lining is expected due to high velocity of water, (*iii*) Erosion can occur as in silt flushing tunnels, (*iv*) Seepage of water through the lining is likely to involve heavy loss of water and the structural stability of the rock mass around the tunnel may be affected adversely or may cause damage to the tunnel or the adjoining structures, (*v*) Heavy seepage of water is anticipated, provision is made for grouting with cement and/or chemicals or extra drainage holes and also feasibility of steel lining is considered, (*vi*) Pressure shaft is to be lined because of sharp inclination, (*vii*) Portion of the water conductor system between emergency gate and power plant is to be lined.

The main function of steel lining is to protect the concrete and to stop leakage of water from the tunnel. The other purposes served by the steel lining are (*i*) To prevent water loss from the tunnel, (*ii*) To resist the bursting pressure of water carried by the tunnel, (*iii*) To provide protection from seepage of water from the surrounding mass like rocks, concrete etc, and (*iv*) To provide a smooth surface of flow of water.

The steel lining has to be strong enough to withstand the initial pressure which is not taken by the rock surrounding the lining and must be capable of taking the full external pressure. The steel lining consists of a steel plate of adequate thickness conforming to boiler and pressure vessel steel provided to the inner surface of the tunnel. The minimum thickness required to provide rigidity during fabrication and erection is given by the equation:

$$t = (d + 50)/400 \quad \text{cm}$$

where, t is minimum thickness of plate (cm), and P is internal diameter of the tunnel (cm). The thickness of steel plate is also determined from hoop stress and longitudinal stress by the formula:

$$t = (PD)/2f_s e$$

where, t is steel plate thickness (cm), p is maximum internal pressure after rock participation (kg/cm^2), D is internal diameter of finished surface of tunnel (cm), f_s is design stress in steel liner (kg/cm^2) and e is efficiency of longitudinal joint.

Maximum permissible velocity is 9 m/s. Generally a circular section is best adapted for steel lining. *D*-shaped and rectangular section too are used.

Shotcrete Lining

Shotcrete lining or pneumatically applied concrete in rock tunnelling is provided to protect and support zones of fractured, crushed, disintegrated or spalling rock, and to preserve and prevent further deterioration caused by the action of water or the atmosphere or aging. The need for temporary supports is eliminated. Shotcrete lining offers an effective, economical and safe alternative where conventionally placed concrete is not possible. It is very effective in areas where ground water is discharging from the freshly exposed surface and in filling up and effectively controlling wide seams. Shotcrete in conjunction with wire mesh and rock bolts can be used to built up structural members which can permanently support even very large size tunnels under adverse geological conditions at very low costs.

The design parameters for the lining depending upon the rock conditions and the effective span to be bridged are (*i*) Thickness of layer, (*ii*) Size and grading of aggregates, (*iii*) Amount of cement to be used, (*iv*) Amount of sand and coarse aggregates to be used, (*v*) Water cement ratio, and (*vi*) Use of accelerators.

Improvements in ductility and toughness by adding steel fiber to the shotcrete mix are sufficient to replace the wire mesh reinforcement. The steel fiber reinforced shotcrete (SFRS) can be applied with the conventional equipment with dry as well as wet shotcreting methods. When using state of the art technology, the technical performance of fiber reinforcement is generally equally good or better than traditional mesh reinforcement. Additionally, it gives a number of other advantage such as: (*i*) Overall productivity when applying on drill and blast rock surfaces is often more than doubled, (*ii*) Substantially improved safety, shotcrete with reinforcement can be placed by remote controlled manipulator, (*iii*) No areas of poor compaction behind overlap areas of 3 to 4 layers of mesh, causing very poor concrete quality and high risk of subsequent mesh corrosion and concrete cover spalling, (*iv*) The intended thickness and overall quantity of shotcrete can be achieved quite accurately and the problem of excess quantity to cover the mesh on rough substrata is avoided, (*v*) One layer of mesh will be placed at varying depth in the shotcrete layer and cannot be placed in the cross section to target tension zones. The fibers will be present in the whole cross section, irrespective of where the tension will occur, (*vi*) Logistic advantage of avoiding handling and storage of reinforcement mesh under ground, (*vii*) It can follow the exact contours of the rock thereby achieving uniform thickness, better bond between the rock and shotcrete and less rebound.

To be able to achieve the above advantages the wet mix application method has to be used. The reason for this is the lack of control and the large amount of fiber rebound in the dry process. Typically the steel fiber rebound in the dry process is above 50 per cent and using synthetic fibers it is likely to be even higher. In comparison, steel fiber rebound is typically 10 to 15 per cent in the wet mix process.

The thickness of lining varies from 30 to 100 mm, depending upon the quality of rock and the span to be bridged. For 100 mm thickness, the shotcrete is spread in two layers each of 50 mm thickness.

16.12 TUNNEL GROUTING

The space between the supporting system and the excavated surface has necessarily to be backfilled to arrest the further displacements and movement of the strata around an excavated and supported tunnel by means of placing tightly packed concrete behind the supports. The treatment of rock around the tunnel bore as also the plane of contact between the rock and concrete lining and that between the concrete and steel lining in the tunnel is done by pressure grouting. Even the design of lining is based on the assumption of properly executed grouting of the rock and contact zones, grouting is done to fill them and to provide a proper bond between the concrete and the rock surface. In steel lined tunnel, grouting is done to fill up the gap formed between the steel liner and backfill concrete because of shrinkage of concrete.

Types of Grouting

The two types of tunnel grouting is briefly as under:

1. Backfill or Contact Grouting: The contact grouting is done to fill up all voids and cavities between the concrete lining and the rock by sand-cement grout. The contact grouting is also known as low pressure grouting or backfill grouting as it is normally done at pressures not exceeding 5 kg/cm^2 and is considered as a part of concreting. It is done throughout the length of concrete lining not earlier than 21 to 28 days after placement of the lining i.e. the concrete has cooled off. The grouting is limited to the arch portion of a *D*-shaped tunnel or cavity whereas in vertical shafts and steel lined tunnels it is done all along the full periphery. The depth of grout holes at each location is such that the holes extend 30 cm beyond the concrete lining into the rock. The spacing of the sections may normally be 3 m centres.

2. Consolidation Grouting: Also called pressure grouting. The consolidation grouting is done to consolidate the rock shattered around the cavity and fills any gaps caused by shrinkage of concrete, by filling up the joints and discontinuities in the rock. The rock quality thus gets improved thereby increasing the resistance of the rock to carry the internal pressure. The consolidation grouting is done after the backfill grouting is completed in a length of atleast 60 m ahead of the point of grouting. The grouting is done all around the cavity and depth of grout holes is at least equal to the diameter of the tunnel from the finished concrete. For small tunnels, rings of grout holes may be spaced at 3 m centres, depending upon the nature of rock. Each ring may consist of 4 grout holes distributed at about 90° around the periphery with alternate rings placed at 45° axis. The grouting is done at a maximum practicable pressure consistent with the strength of lining and safety against uplift of overburden with increasing pressures. As a thumb rule, the grout pressure of 1.5 times the water pressure in the tunnel may be used. The normal range of pressure grouting is 7 to 10 kg/cm^2.

The process of grouting consists of (*i*) drilling holes, (*ii*) cleaning and washing holes, (*iii*) testing holes, (*iv*) grouting holes, and (*v*) testing of grouted zone for efficacy of grouting.

Rock Bolts: With the advancement made in the state-of-art of rock mechanics, it is now considered expedient to reinforce the rock by means of rock bolts and/or shotcrete in conjunction with wire mesh etc. rather than support it with steel supports. The rock bolt consists of a steel bar inserted in a hole drilled into the rock. The end away from the rock surface has a device which permits it to be firmly anchored into the hole. The projecting end is fitted with a plate which bears against the rock surface. The bolt is placed in tension between the anchor and the plate, thereby exerting a compressive force on the rock between the ends of the bolt. The commonly used rock bolts are classified in three groups by the type of anchorage used i.e. (*i*) Dive-set or slot and wedge type, (*ii*) Torque-set or Expansion type, and (*iii*) Grouted type. The most commonly used. bolts are the expansion anchorage type. Epoxy type resins for anchorage of the bolts are also used in water bearing strata.

16.13 SURGES IN TUNNELS

In long closed hydraulic tunnels water hammer is created by the sudden closure of the turbine gates. The water hammer provides the necessary force to retard the flow in the tunnel when load is rejected by the turbine. For very long tunnels, the water hammer corresponding to normal operation of the turbine may be very great and may require extraordinary strength of the tunnel to withstand it and the violent fluctuations of pressure in the tunnel may seriously interfere with proper turbine regulation. Likewise, for sudden opening of the gates, the resulting negative water hammer, or reduction of pressure, provides the necessary force to accelerate the water and is correspondingly objectionable for very long tunnels.

These positive and negative water hammer pressures can be eliminated by providing a surge tank at the lower end of the tunnel. The steady state water level in the tank fluctuates up and down as the turbine rejects or accepts the load. The maximum and minimum pressures occurring all along the length of the tunnel can be calculated conveniently with the help of digital computers. The tunnel has to be designed to withstand the maximum excess pressure that is likely to occur. Similarly it is very essential to work out the sub-normal pressures in the surge tank for sudden acceptance of the load. It is to be ensured that the pressure in the tunnel never becomes negative which may cause the tunnel to collapse.

For headrace tunnels flowing full, maximum pressures occur at the time of load rejection and minimum pressures at the time of load acceptance. However, for tailrace tunnels flowing full, minimum pressures exist at the time of sudden load rejection whereas maximum pressures may occur at the time of load acceptance. As such a surge tank may be provided downstream of the power house also for the safety of the tailrace tunnel.

16.14 AIR LOCKING

Air may enter and accumulate in a tunnel (*i*) During filling air may be trapped along the crown at high points or at changes in cross-sectional size or shape, (*ii*) Air may be entrained at intake either by vortex action or by means of hydraulic jump associated with gate opening and (*iii*) Air dissolved in the flowing water may come out of solution as a result of decrease in pressure along the tunnel.

The presence of air in a pressure tunnel is not desirable because of (*i*) The localisation of an air pocket at the high point in a tunnel or at a change in slope which occasions a marked loss of head and diminution of discharge, (*ii*) The slipping of a pocket of air in a tunnel and its rapid elimination by an air vent can provoke a water hammer by reason of the impact between two columns, and (*iii*) The supply of a mixture of air and water to a turbine affects its operation by a drop in output and efficiency thus adversely affecting the operation of generator. The presence of air in a pelton nozzle can be the cause of water hammer shocks. Admission of air to a pump may occasion loss of priming, (*iv*) If the velocity exceeds a certain limit air would be entrained causing bulking.

The entry of air in a tunnel can be prevented. Intakes may be designed properly. A shallow intake is likely to cause air being sucked in. Throughout the length of tunnel, the velocity should either remain constant or increase towards the tail end. It should be checked that at no point on the tunnel section negative pressures are developed. Vortices that threaten to supply air to a tunnel should be avoided. However, if inevitable, they should be suppressed by floating baffles, hoods or similar devices, (*vi*) Partial gate openings that result in hydraulic jumps should be avoided, and (*vii*) Traps or pockets along the crown should be avoided.

16.15 INSTRUMENTATION

Various instruments are installed in a tunnel to measure rock stresses and deformities briefly as under. These instruments are provided at more than three sections or at the typical representative reaches met with during excavation. The range of instruments to be installed depends upon rock cover, internal pressure and geological features and properties of rock mass and are fixed after due analysis.

Bore Hole Extensometer: A multiple position bore hole extensometer is used to evaluate the deformations and stresses from there and zone of plastic flow. A minimum of three per section are used.

Strain Meters: These are embedded in concrete lining for the measurement of stress in lining and reinforcements.

Tunnel Load Cells: These are mounted on steel supports for measurement of rock load coming on support. A minimum of three per section are used.

Contact Pressure Cells: These are placed between lining and rock to measure rock loads on lining and transfer of internal pressure to surrounding rock.

QUESTIONS

1. Hydraulic tunnels cause time and cost over run in hydro electric projects construction, why?
2. Why do we need hydraulic tunnels?
3. Discuss briefly the classification of tunnels in a hydroelectric project.
4. What are the salient factors to be considered in deciding the alignment of a tunnel in a hydel project.
5. What are the various types of lining used in hydraulic tunnels of a hydro power project? Under what conditions each type is suitable.

PART–II
DAMS ENGINEERING

- Reservoirs and Dams
- Earth and Rockfill Dams
- Gravity Dams
- Arch and Buttress Dams

17

RESERVOIRS AND DAMS

17.1 INTRODUCTION

A reservoir is created with the impounding of runoff from the catchment upstream by the construction of a dam across a river or stream. Storage is done during the period when the flow is in excess of the demand for release during the lean supply period so as to maintain continuous hydel power generation, besides meeting requirements of irrigation, water supply, etc. Besides storage, the other types of reservoirs appurtenant to hydro power projects are as under:

Single Purpose Reservoir: It is the reservoir to serve only one purpose which may be power generation, flood control or irrigation or municipal water supply, recreation etc.

Multipurpose Reservoir: Also termed as multiuse reservoir. In this reservoir, the storage and release cater for a combination of two or more purposes such as irrigation, hydel power generation, flood control, public water supply, navigation, recreation, fisheries, etc. A reservoir which is designed for one of the purposes, but provides other incidental benefits because every reservoir serves or should serve more than one purpose on account of indirect benefits that may result from its operation, is also referred to as multipurpose reservoir. Keeping with the diverse purposes served by a multipurpose reservoir, it is usually of large capacity so as to (*i*) Generate maximum firm power, (*ii*) Reserve a certain minimum storage capacity at all times for flood control, and (*iii*) Provide steady supply for kharif and rabi crops . Usually, these enormous benefits accruing from a multipurpose project repay the cost of building the dam within a couple of years.

Balancing Reservoir: It is a reservoir, usually of limited capacity, located downstream of (or subsidiary to) a main reservoir to (*i*) Store the water let down from the reservoir in excess of that required for irrigation or additional power generation, (*ii*) Provide flexibility of operation to the distribution system, and (*iii*) At certain locations primarily for permitting regulated supply to the power penstocks with a view to cater for the fluctuations in requirement of water supply to the turbines depending on power load.

System of Reservoirs: These consist of a group of single or multipurpose reservoirs which may be operated in an integrated manner for optimum utilisation of water and power resources of the river system.

17.2 STORAGE ZONES OF A RESERVOIR

Various zones of storage in a reservoir, illustrated in Fig. 17.1, are briefly as under:

Dead Storage: Dead storage in a reservoir, about 10-25% of the gross storage capacity, is provided to cater for sediment deposition by the impounded sediment-laden waters as very small portion of the sediment passes down through dam outlets and for the minimum drawdown in the case of power projects. It is equivalent to the volume of sediment expected to be deposited in the reservoir during the designed life of the reservoir, usually taken as 100 years. The level attained by the reservoir corresponding to the dead storage is termed as dead storage level. It is the level below which a reservoir is not susceptible to release water by the in-built outlet means. In power projects, releases are allowed up to minimum drawdown level only instead of dead storage level so as to maintain the minimum head required for power generation.

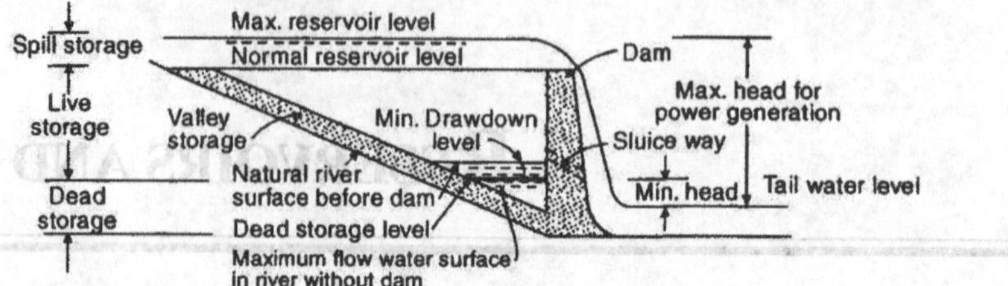

Fig. 17.1 Storage zones of reservoir.

Live Storage: It is the storage capacity of the reservoir above dead storage level which constitutes useable portion of the total storage. It is thus the difference of gross storage capacity and the sum of dead storage capacity and inactive capacity for hydro power generation. Inactive capacity is the storage capacity exclusive of dead storage below which evacuation is not contemplated because of minimum irrigation and power load requirement. However, in practice dead storage capacity and inactive capacity are considered the same. Live storage has to be sufficient so that the project is successful for (*i*) 75% of its life period in an irrigation project, (*ii*) 90% of its life period in an hydel power project and (*iii*) 100% of its life period in water supply project.

Flood Storage: Flood storage is the storage contained between the maximum reservoir level and the full reservoir level. It is the storage space provided in a reservoir for storing flood water temporarily to moderate the releases downstream. Flood storage varies with the spillway capacity for a given design flood. When the spillway capacity is equal to the inflow design flood, there is no flood storage and the full reservoir level and maximum reservoir level are the same. Effective flood storage is the difference between the amount of water stored in the reservoir during the period of flood peak reduction and the amount of natural valley storage during the same period.

Valley Storage: Rivers in dry season flowing within its banks have small water flow compared to that when in floods. Valley storage is the storage in the river or stream in floods after it has overflowed its banks. It is an important element in the design of large size flood control reservoirs where the valley storage at the time of flood is of significant proportions in regard to the reservoir storage volume.

17.3 STORAGE CAPACITY DETERMINATION

The storage capacity of a reservoir to meet the demand of continuous supply is determined with the help of the observed discharge data of a stream or a river on which the dam is to be built. The flow values for the driest years in as long a period as is available, say 25 to 30 years; minimum 10 years is considered. The storage capacity is determined by the following methods:

1. **Analytical Method:** The inflow and demand values in each month are determined. The demand includes prior rights, if any, evaporation losses, rainfall etc. The deficit and surplus of water which is the departure of inflow volume from the demand volume is determined. The maximum value of cumulative excess of demand over the inflow represents the minimum storage necessary to meet the demand. The cumulative excess inflow volume starting from each demand withdrawal from storage is also determined which indicates the filling up of the reservoir and the volumes in excess of storage to be spilled over as also whether the reservoir fills up after a demand and when.

2. **Mass Curve Method:** In this method the storage capacity is determined from mass curves of demand and inflow for the years of minimum flow (Fig. 17.2). An examination of the demand curve and inflow curve drawn to the same scale would indicate when storage is required. When inflow rate is low, *i.e.,* the inflow curve is flat but at that time if the slope of demand curve is more,

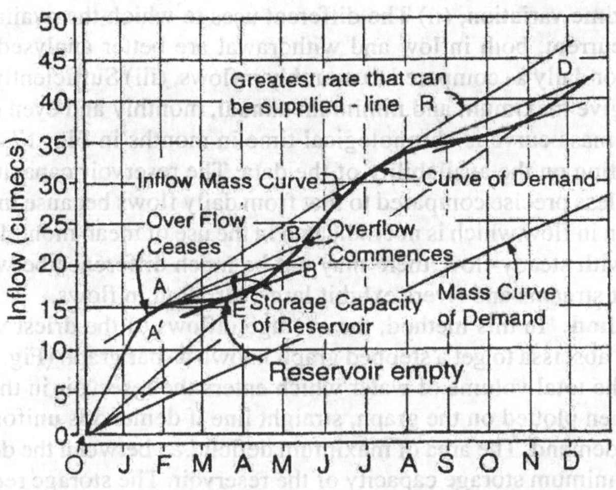

Fig. 17.2. Mass curve.

it indicates that storage is required to meet the demand. The degree of accuracy of results is dependent upon the accuracy and extent of the available observed data; the shorter the period, the greater the probability of error. Error is also caused by the available ground storage, but since it is indeterminate and offers a factor of safety in design, it is seldom allowed for. The salient characteristics of the capacity determination by the method are as under:

1. The reservoir is assumed full at the beginning of the dry period i.e., when the inflow rate is less than the demand (withdrawal) rate.
2. In case demand curve is a straight line the required storage capacity is the maximum vertical distance (s) between the line drawn straight parallel to the demand line and tangential to hump at the beginning of the mass curve of inflow; line AB in Fig 17.2 and a line drawn parallel to this line and tangential to the mass curve at the trough portion; line A'B' to meet the demand. If flow data is available for a large period, the demand lines are drawn tangentially at various other crest points and the largest of the storages obtained is taken as the minimum storage capacity of the reservoir. Where demand curve is variable, the maximum intercept between the two curves gives the minimum storage capacity required to meet the demand.
3. The period between the point of tangency and the point where the tangent intersects the mass curve of runoff represents the period during which water in the reservoir is lower than the spillway; overflow stops at the starting point (O) and commences at the intersection point.
4. Reservoir is empty at the bottom of the trough of inflow (E) where the vertical line indicates the storage capacity.
5. Slope of line OD, joining the first and the last points of a mass curve, represents the average discharge over the whole period of record.
6. Greatest rate of demand as can be supplied by the stream, provided reservoir surface area is not increased, is represented by slope of the line (R) above mass curve of inflow, tangent to it at two or more points but not intersecting the inflow mass curve at any point.

7. Demand so great as reservoir is never filled, a line G, drawn tangent to and below the mass curve of inflow and parallel to the demand curve will not intersect the inflow mass curve to the left of the point of tangency.

Limitations

(i) The method assumes a definite sequence of events but rainfall, the prime source of inflow, is subject to considerable time variation, (ii) The different uses to which the available storage is put are not necessarily concurrent, both inflow and withdrawal are better analysed in time intervals sufficient small, weekly or daily as compared to monthly inflows, (iii) Sufficiently long term record of inflow is required to give maximum and minimum annual, monthly and even daily inflows, and (iv) The abscissa of the mass curve is chronological time in months in Fig. 17.2. It may be days, weeks or months depending on the availability of the data. The reservoir capacity computed from mean monthly flows are less precise compared to that from daily flows because in each month there is a considerable variation in flow which is not indicated in the use of mean monthly flows. However, in the case of a stream with steady flow, there may not be much difference between the daily and monthly mass curves but streams and rivers exhibit large variation in flows.

3. **Bar Graph Method:** In this method, the average inflows of the driest year are plotted as ordinates against time as abscissa to get a stepped graph known as bar graph (Fig. 17.7). The area in the diagram represents the total volume of water which enters the reservoir in that year. Likewise, the average demand is then plotted on the graph, straight line if demand is uniform and stepped in the case of non-uniform demand. The area of maximum deficit *i.e.*, between the demand and inflow represents to scale the minimum storage capacity of the reservoir. The storage required to meet the maximum uniform demand that can be met with without loss of water over spillway is given by the area of surplus over the demand.

Determination of reservoir capacity: (*a*) Area of water spread or reservoir area is computed in million cubic metre or hectare metre by measuring the area at definite contours by a planimeter. A curve showing the water spread at various elevations, called area-capacity curve (Fig. 17.3) is prepared for reference, (*b*) *Capacity*: Volume of storage (*V*) or capacity of the reservoir is computed by taking contour areas at equal intervals and totalling up by the formula, as follows:

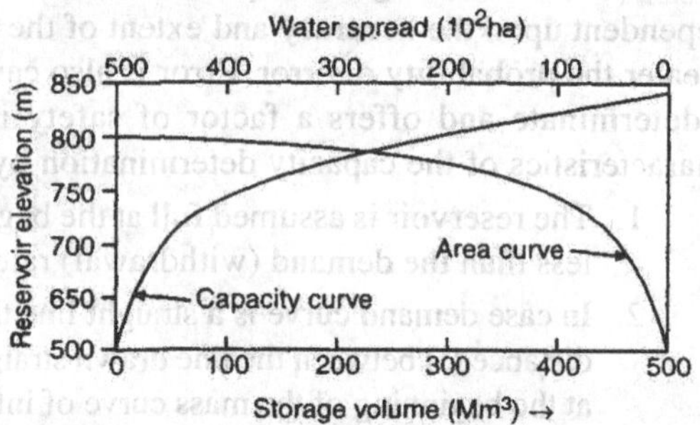

Fig. 17.3 Area-capacity curves.

1. *Contour Area Interval Method*: The capacity between successive unit elevations is worked out arithmetically by taking the average submergence area and multiplying by the unit height.

2. *Cone formula*: $V = \frac{H}{3}(A_1 + A_2 + \sqrt{A_1 A_2})$, where A_1 is area at one contour, A_2 is area at another contour (sq km) and H is contour interval (m).

3. *Modified Prismoidal Formula*: $V_s = 2H/6\ (A + 4B + C) - V_y$ Where, V_s is capacity between middle and top contours, H is contour interval between A and B contour, A is area of the bottom contour, B is area of middle contour, C is area of top contour and V_y is capacity between middle and bottom contours previously determined. The capacity below the lowest contour is computed by contour area interval method which provides V_y. After finding the volume below the lowest contour, the formula is used progressively for each succeeding higher contour.

4. *Average End Area Length Method*: $V = (L/2)\ (A_1 + A_2)$ where, V is capacity at elevation 'X' between two successive range lines, A_1, A_2 are areas of successive cross sections at elevation 'X', and L is distance between range lines.

17.4 TRAP EFFICIENCY

Reservoir sedimentation is measured in terms of trap efficiency. Trap efficiency is defined as the ratio of sediment retained in the reservoir to the sediment brought by the water into the reservoir. The trap efficiency primarily depends on the sediment load characteristics, and the detention time of the inflow, method of reservoir operation and age of reservoir.

The two methods for determining trap efficiency of a reservoir are Brune's method and Churchil method. In Brune's method, trap efficiency is related to inflow of water into the reservoir and its storage capacity. Fig. 17.4 gives the generalized trap efficiency based on actual observations from existing reservoirs. From the curves the amount of sediment which a new reservoir will trap can be read against known value of the capacity-inflow ratio. It indicates that trap efficiency is low for reservoirs having small capacity-inflow ratio and it gradually increases as capacity-inflow ratio increases.

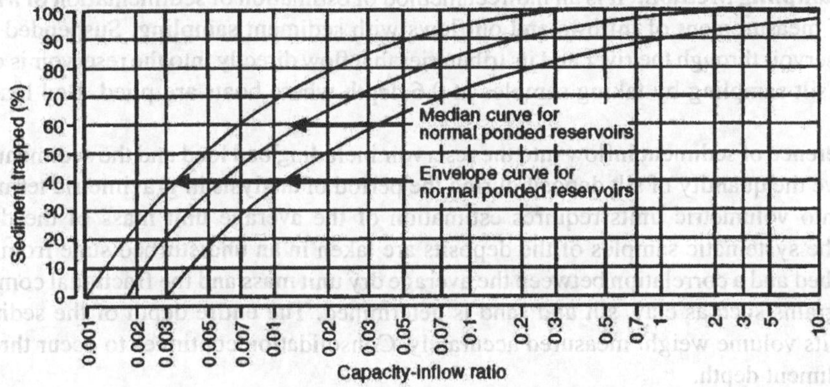

Fig. 17.4 Trap efficiency (After Brune)

17.5 RESERVOIR SEDIMENTATION

The river water carries silt. The flow impounded in the reservoir during monsoons is heavily silt laden. The fine sediment in the river water is carried in suspension, as suspended load, while the larger solids move on or near the bed as bed load. When the river water reaches a reservoir, its velocity and turbulence are reduced and as a consequence the bed load generally gets deposited along the thawleg or pre-reservoir course, but the suspended sediments which take a long time to settle down spread more evenly all over the reservoir bed, and frequently the thickest deposit of such sediment are below the deepest standing water nearer the dam. A small amount of sediment deposit is passed down along with the supply released through the outlets. The bed load and a part of suspended sediment is deposited in the reservoir bed if the reservoir discharges the surplus freely for long time which is the case when runoff of the river is large in comparison to the reservoir capacity, *i.e.*, small capacity-inflow ratio. However, if the capacity-inflow ratio is large, *i.e.* the reservoir is large in comparison to the annual runoff, all the water-borne sediment brought by runoff is deposited in the reservoir bed.

The principal adverse effects of sedimentation of a reservoir are (*i*) Reduction of active storage capacity which may reduce the capability of a reservoir to deliver the benefits which could have been delivered by the reservoir but for sedimentation. The progressive reduction of the storage capacity affect the output, (*i*) Reduction in the availability of firm water and power supply, (*ii*) May block the outlet intakes hampering operation of gates especially those which are not frequently operated, (*iii*) Increased passage of sediments in irrigation canals, power houses etc. which may become serious with the passage of time, (*iv*) Increase in flood levels upstream may cause additional submergence, formation of marshy lands, (*v*) River regime at entry to the reservoir is affected. Delta formation and braided river pattern may result and this may be unsightly, (*vi*) With large sediment deposits, humps are formed in reservoir bed. The deleterious effect of hump formation is early

reduction of live storage capacity, (*vii*) Turbidity of the reservoir water is increased resulting in deterioration of water quality detriment to fish survival, and generating equipment.

Results of sedimentation surveys of Indian reservoirs have revealed that the rate of sedimentation for major catchments vary between about 4.75 to 14.25 ha m annually per 100 sq. km. and the reservoirs are losing a capacity of half a per cent to one per cent annually. IS: 6518–1972 has recommended provision of sedimentation at the rate of 0.1 to 0.2 ha m/year/sq km of drainage area for the entire economic life of the reservoir.

Estimating Silt Load into a Reservoir

The estimate of sediment flow into a reservoir is based upon stream flow sampling and by silt survey method.

1. Silt Sampling Method: It is an indirect method of estimation of sedimentation of a reservoir consisting of measurement of inflows and outflows with sediment sampling. Suspended silt load reaching a reservoir through the river and its tributaries that flow directly into the reservoir is observed by means of silt sampling by taking samples at 0.6 depth where boats are plied. Bed load is also measured.

The difference of sediment inflow into the reservoir including bed load and the sediment outflow therefrom give the quantity of silt deposit during the period of analysis in gravimetric terms and its conversion info volumetric units requires estimation of the average unit mass of the deposited sediments. The systematic samples of the deposits are taken in an undisturbed state from all over the reservoir bed and a correlation between the average dry unit mass and the fractional composition of different grains such as clay, silt and sand is determined. The entire depth of the sediments is sampled and its volume weight measured accurately. Consolidation continues to occur throughout the entire sediment depth.

2. Capacity Survey Method: It is a direct measurement method carried out periodically to assess the volume of deposit as also its location in a reservoir, usually when a major reservoir losses 3 per cent of its capacity. The two methods are range line method and contour method. In range line method, the reservoir area is surveyed and range monuments, bench mark pillars etc. are fixed. The reservoir area is divided into suitable sections called silt ranges, along the main river and its tributaries that flow directly into the reservoir. The ranges may be generally parallel to each other fixed at 0.75 to 2 km interval depending on the length of the reservoir. The ranges are kept closer in upper reaches where movement of deposits is more frequent due to drawdown and filling operations of the reservoir. Range monuments and bench mark pillars are fixed on both the banks of the main river and its tributaries for observing cross-sections. Cross-sections of the river bed and its tributaries are observed along the range lines by sounding by means of echo-sounding equipment. The echo-sounding equipment, operated from a motor launch which runs along the predetermined lines or cross-sections defined by monuments, determine water depths by utilising supersonic sound pressure waves and precision timing. It gives a permanent and continuous record on a calibrated chart of depths of water through which the sound waves are sent. To determine the value of silt deposit the observed cross-sections are superimposed on the corresponding previous cross-sections to indicate the extent of silt deposit at these cross-sections. The total quantity of silt deposited in the reservoir is computed from the value of silt deposited at each cross-section. The method does not indicate origin of the sediment yield sub-catchmentwise.

Control of Sediment Deposition:

The deposition of sediment in a reservoir can be controlled to a certain extent by operating gates or other outlets in the dam in such a manner as to permit selective withdrawal of water having sediment content higher than the average. If the water could be escaped at those levels where the concentration of suspended sediment is the highest, considerable amount of sediment is passed down without settling into the reservoir.

Watershed erosion control measures are effective in preventing or delaying the movement of silt from the point of origin in the watershed to the reservoir; watershed management and afforestation in the catchment area are very effective to reduce the silt flow into the reservoir besides other erosion control and soil conservation measures such as regrassing and grass land cultivation processes, gully control, wild life protection, reservoir shore line protection, highway and railway line erosion control, etc.

Removal of silt deposits: Sediment deposits can be removed by excavation, dredging and sluicing with hydraulic and mechanical agitation. Dredging is expensive and is not economically feasible for most water storage projects. Sluicing with hydraulic or mechanical agitators for stirring up the silt deposits along with the flood flushing may be possible in small reservoirs.

Design Criteria

The design criteria followed in India to cope with the sedimentation are briefly as under:

(*i*) Feasibility service time and full service time should not be less than 100 and 50 years respectively for irrigation projects and power projects supplying power to a community and 70 and 25 years for power projects supplying power to grid.

(*ii*) The silt level of permanent outlets should be above zero elavation after feasible service time.

(*iii*) Revised area-capacity curves expected after full service time should be used in simulation studies.

(*iv*) If annual sediment load is less than 0.1 percent of gross reservoir capacity sediment distribution studies are not necessary.

The steps involved in reservoir sedimentation are:

(*i*) Assessment of sediment yield and trap efficiency.

(*ii*) Assessment of new zero elevation after feasible service time.

(*iii*) Distribution of sediment in the reservoir to obtain revised area-capacity curves after full service time.

17.6 LIFE OF A RESERVOIR

The term life of a reservoir denotes the period during which or a specified fraction of its total or active capacity is lost.

Every reservoir is ultimately to silt up. In earlier practice, the life of a reservoir was considered as number of years when the storage allocated as dead storage is filled up. It was based on the assumption that all the sediment load first deposits into dead storage and then encroaches on live storage. This concept is not in line with the actual process of sedimentation of reservoirs as heavier material begins to deposit on impinging the reservoir backwater, a portion of which in live storage. Thus capacities reserved for the irrigation, hydel power and flood control are encroached upon simultaneously with the deposition of finer material in dead storage zone.

Central Water Commission (India) practice is to fix minimum drawdown level a little above the zero elevation which is likely to develop after 100 years of operation. Generally, the life of a reservoir in India is adopted as 100 years and to provide for silt storage space (dead storage) equal to the estimated total volume of silt deposition during its entire life. In this context, there are certain relevant terms, as under.

Full Service Time: For a specific purpose, the period or notional period for which a reservoir provided or is /was expected to provide, a part of the full planned benefits inspite of sedimentation. Full service time of an irrigation project shall not be less than 50 years after the start of operation as per IS : 12182-1987.

Economic Life: If at any point of time, the benefits likely to accrue in further operation of a reservoir compare unfavourably under the relevant economic criteria with the further costs involved in the operation and maintenance of the system but excluding any element to cover the past costs incurred, the reservoir is said to have reached the end of economic life.

17.7 RESERVOIR EVAPORATION LOSSES

The stored water in a reservoir suffers loss due to evaporation from its water spread.

In case observed data is not available from the neighbouring reservoirs for application with due correction, the evaporation losses given in the Table 17.1 may be assumed. Evaporation in hot months of May and June is two to five times than in winter months of December and January. For the Indian subcontinent loss of water by evaporation generally varies from 150 to 200 cm depth of stored water annually; a maximum of 350 cm in parts of Rajasthan and Saurashtra and a minimum of 50 cm in parts of Jammu and Kashmir.

Table 17.1 Reservoir evaporation loss

Month	*Evaporation Loss (cm)*		
	South and Central India	*North India*	*Design assumption at pong dam (including seepage losses)*
January	10	7	6.5
February	10	9	7.4
March	18	13	11.9
April	23	16	17.7
May	25	27	20.5
June	18	24	15.4
July	15	18	9.7
August	15	14	8.4
September	15	14	7.7
October	13	13	6.1
November	10	9	5.2
December	10	8	5.5
Total	182	172	122

17.8 WATERSHED MANAGEMENT

Water storage reservoirs created at a very high cost are national assets of great importance. One of the important problems in their operation and maintenance is the loss of storage capacity due to continuous deposition of water borne sediments. The soil erosion originates in the catchment, is transported through the streams and rivers and deposited in the reservoirs. The reservoirs provide space for the accumulation of such sediments but they do not confine themselves in the allocated spaced, called "dead storage" and simultaneously encroach upon the live storage at a rate higher than the design assumption which means not only their inability to provide sustained water supply and power generation, but most importantly reduction in their operational life.

Soil erosion and consequent sedimentation of reservoirs is a complex process dependent on a number of factors such as geology, climate, landscape features, soil characteristics which are not readily influenced by human resources but has effect on land disturbance which includes land management practices like tillage, overgrazing, mining, urban development, logging etc. Construction activities and tourism are the major factors under human control. The river ecosystems are increasingly subjected to greater stress due to these human activities. Their adverse effect is more pronounced in the catchment area. It requires soil and land use survey, catchment stabilisation and watershed management which includes soil conservation, afforestation and engineering measures in moderating floods and reducing sediment discharge, maintenance and enhancement of the ecology and land productivity. Effective watershed management requires three level approach *i.e.*, state level, regional level and individual farm level.

The benefits of watershed management are (*i*) Reduction in runoff and damaging floods, (*ii*) Quality management of land resources of soil, water and vegetation for sustained availability and development, (*iii*) Maintenance and improvement of water quality, (*iv*) Soil fertility conservation and improvement in agricultural production, (*v*) Prolonging life of storage reservoirs by reduction in sediment inflow, (*vi*) Wild life and forest protection, (*vii*) Improvement in environment, and (*viii*) Restoration and revegetation of seriously eroded areas.

The integrated watershed management was launched in the country during Fifth Plan (1974-78) and soil conservation measures in the catchment area of a few river valley projects were also taken up as a centrally sponsored scheme. The results obtained from earlier treatments have shown that soil conservation measures result in considerable reduction in silt load inflow into the reservoir.

17.9 DAMS

A dam is a hydraulic structure constructed across a river or a stream to impound part of runoff from the catchment upstream of the dam. Streamflow is variable within very wide limits from season to season largely conforming to the pattern of rainfall and snow melting in snow fed rivers. The streamflow also varies from year to year. The rate at which water is required to meet the demands of hydropower, irrigation, water supply, flood control rarely conform to the pattern of water availability. Hence the need to conserve or impound water during wet periods for use during lean periods. This is achieved through construction of a dam.

Selection of dam site

The various considerations which govern the final selection of a dam site, the general location having been chosen as a result of investigations are (*i*) Good catchment upstream of the dam site as shall contribute adequate supply of water to meet the intended water demand, (*ii*) Minimum length of dam for maximum storage for a given height. Narrow gorge opening out upstream to admit large storage in a deep reservoir less susceptible to weed growth and minimum evaporation losses is preferable, (*iii*) Minimum height of dam for the maximum intended storage, cost of dam is proportional to square of the dam height, (*iv*) Suitable foundations as for high concrete or masonry dams solid rock foundation is essential, (*v*) Major portion of the dam is on high ground compared to river bed for economy in cost and better drainage, (*vi*) Availability of suitable location for spillway. A deep gorge for the dam and a flank for the spillway separated by a high hillock is the best site for earthen dam, (*vii*) Availability, locally or within economical leads of suitable construction materials such as concrete ingredients, pervious and impervious soils and stones (for rip rap) in the case of concrete and earthen dams respectively, (*viii*) Ensure adequate storage capacity of the reservoir, (*ix*) Rim of dam watertight. Reservoir basin should also be reasonably watertight. Cup shaped basin with fairly watertight bed and sides is preferred, (*x*) Stable banks and adjoining hill slopes, (*xi*) Minimum construction and maintenance cost, (*xii*) Availability of suitable site for economical construction of residential colony, office, construction equipment and workshop complex, (*xiii*) Requirements of coffer dam, diversion and dewatering operations. The site should be suitable for diversion of river during construction period, (*xiv*) Minimum value of land and property compensation. The land to be submerged should be of low value, and not densely wooded as far as possible, (*xv*) Location of dam and appurtenant works in one State, if possible, to avoid interstate disputes, (*xvi*) Good quality water for intended uses and essentially carrying minimum sediment load, high silt contributing tributaries are avoided, as far as possible, and (*xvii*) Healthy locality. Very cold climate and heavy rainfall impede construction activities.

17.10 SELECTION OF TYPE OF DAM

Various considerations in the selection of type of dam are topography, geology and nature of foundations, spillway location, seismicity of the area, economic life of dam etc.

A summary of the character of site conditions vis-a-vis type of dam suitable for the situation is as follows:

Character of site	Type of dam
(i) Foundation character	
(a) Impervious solid rock such as schist and gneiss.	Any type
(b) Gravel and coarse sands	Earth-rockfill dam
(c) Grit, sand and silt clay	Earthen dam
(d) Very fine and uniform sands	Earthen dam
(e) Clay, plastic clay	Earthen dam with very flat slopes
(f) Silt and fine sand	Low gravity dam or earthen dam
(g) Heavy over-burden on rock layer	Earthen or buttress dam
(ii) Character of gorge	
(a) V-shaped narrow gorge	Arch dam
(b) U-shaped narrow gorge	Overflow concrete dam
(c) Low-wide gorge	Earth/rockfill dam
(iii) Character of construction materials	
(a) Ingredients of concrete, *i.e.*, sand, shingle and cement-locally available	Concrete dam
(b) Sand, shingle and cement involve long distance carriage	Buttress dam
(c) Pervious and impervious soils and stone (for riprap) locally available	Earth/rockfill dam
(d) Stone of required specific gravity and size available	Stone masonry dam
(iv) Spillway requirement	
(a) Spillway location separate from main dam	Earthen dam
(b) Adequate width for dam only	Gravity dam with overflow spillway
(c) Large spillway requirement	Overflow concrete dam
(v) Character of the region	
(a) Seismic zone	Earth dam and concrete gravity dam
(b) Excessive uplift pressure exerted	Arch dam

17.11 APPORTIONMENT OF COST OF A MULTIPURPOSE RESERVOIR PROJECT

River valley projects are essentially multipurpose for hydel power generation irrigation besides indirect benefits which accrue from them such as drinking water, flood protection, pisciculture, navigation, tourism etc. The cost of construction of the project is apportioned on the basis of the uses for which the reservoir is planned. The various methods of cost apportionment are as under:

Common Allocation Procedure: The different methods for cost allocation generally have a common four step procedure *i.e.*, (*i*) To determine specific costs which serve each of the purposes, (*ii*) To assign such specific costs to various purposes from the capital cost and to find out the remaining joint cost, (*iii*) To distribute joint cost equitably to various purposes, and (*iv*) To sum up, for each purpose, the specific cost and the share of joint cost chargeable to that purpose.

1. Alternative Cost Method: In this method, the alternative cost of a new single purpose project is split up into two parts, namely, the specific cost equal to that in the multipurpose project,' and the balance. The joint cost of the multipurpose project is then distributed in the ratio of the balance. The allocated joint cost is then added to the respective specific cost to arrive at the total cost allocated to each function.

2. Alternative Justifiable Cost Method: This method is similar to the alternative cost method, the only difference being that the specific cost is subtracted from the justifiable cost. The joint cost is then distributed in the ratio of the balance as in the alternative cost method and the cost allocated to each purpose/function is worked out by adding the respective specific cost to the allocated joint cost.

3. Bearability Concept: This concept suggests that the net revenue from the minor advantage (gross revenue minus the operation, maintenance and depreciation on both the specific cost and share of the joint cost of the minor advantage) shall be equal to the interest charges on the capital investment on the minor advantage, both specific cost and share of joint cost chargeable to minor advantage.

4. Benefits Methods: In this method, the cost is allocated to each function by its specific cost plus a share of the joint cost in direct proportion to the estimated net benefits accruing to that function.

5. Ceiling Allocation (or Priority of Use) Method: The priority of use method assigns the specific cost entirely to the individual purpose and assigns the remaining joint costs to project purposes in a descending order of priority. Thus, the top priority purpose is charged with joint costs equal to the lesser of two amounts, namely (*i*) the benefits, minus specific costs assigned to that purpose, or (*ii*) the cost of the most economical alternative minus specific costs assigned to that purpose. The remaining joint costs are then allocated in the same manner to the other functions in order of priority. In this way, the higher priority functions are assigned the maximum allocation represented by alternative joint costs for these purposes, thereby relieving the lower priorities from sharing the joint costs.

6. Equal Apportionment Method: In this method, the joint cost is distributed equally among the different purposes served by multipurpose project. It is based on an arbitrary assumption that an equal portion of the joint cost is reasonably chargeable to each of the purposes.

7. Separable Cost: Also termed as remaining benefits method. It is a refined form of alternative justifiable cost method method in so far as the concept of specific cost for each purpose is replaced by the concept of separable costs. Separable cost for each purpose is subtracted from the justifiable cost and the remainder which is really the 'remaining alternative costs' is termed in this method, as the 'remaining benefits' for the purpose. The joint cost, which in this case is the difference between entire project cost and the sum of separable cost, is, then, distributed in the ratio of remaining benefits for each purpose/function. When separable cost for all purposes/functions consists of specific cost only the separable cost-remaining benefits method is identical to the alternative justifiable cost method in computation and results. It is rational and sufficiently accurate method but it is difficult to work out the benefits or its alternative costs.

8. Use of Facilities Method: This method is based on the premise that the joint costs are proportional among the various functions according to their amount of use of the joint facilities. The usual two approaches for assessing the use for each purpose are the 'capacity approach' and 'water released approach'. In the 'capacity approach', the joint cost is distributed in the ratio of storage assigned to different purposes, whether or not those storages are actually used. Two or more purposes may make use of the same reservoir capacity alternately or simultaneously. In the 'water released approach', the joint cost is distributed in the ratio of total quantity of water released for each purpose. It is a simple but not an accurate method.

9. Vendibility Method: The vendibility theory assumes that the products of a multipurpose project are sold in an entirely free competitive market. The method assumes that the revenues from all benefits of a multipurpose project shall at least be equal to the total cost of the project. The revenues of each benefits should be sufficient to cover the specific costs and in addition should contribute towards the joint cost in such a manner as to render the entire project remunerative with revenues realized from each of the benefits at competitive prices. In this method, a particular purpose, the revenues from whose benefits are not sufficient to meet its costs are subsidized by other purposes the benefits from which are produced at more competitive price.

The method of dividing joint cost is (*a*) The specific cost is subtracted from the estimated revenues for that purpose, (*b*) The remainder is capitalized at the prevailing rate of interest, and (*c*) The remainder is the share of the joint cost or is the upper limit of the share of the joint cost for this particular purpose.

10. Incremental Method: In this method the separable costs are allocated to their respective functions and the total joint cost to one basic function considered the basic or principal function.

SOLVED EXAMPLES

Examples 17.1 *The yield of water from a catchment area during each successive month is (1.4, 2.1, 2.8, 8.4, 11.9, 11.9, 7.7. 2.8, 2.52, 2.24, 1.96, 1.68) × 10^6 m^3. Determine the minimum capacity of a reservoir required to allow the above volume of water to be drawn off at a uniform rate assuming that there is no loss of water over the spillway.*

Solution. 1. By Analytical Method

Total inflow: = (1.4 + 2.1 + 2.8 + 8.4 + 11.9 + 11.9 + 7.7 + 2.8 + 2.52 + 2.24 + 1.96 + 1.68) × 10^6 M^3 = 57.4 × 10^6 m^3

Monthly demand = (57.4 × 10^6)/12 = 4.78 × 10^6 m^3

Storage required = [(8.4 + 11.9 + 11.9 + 7.7) × 10^6] – [(4.78 × 4) × 10^6]
= 20.78 × 10^6 m^3

2. *By Mass Curve Method*: The step by step procedure is as follows:

(*i*) The cumulative inflow works out as (1.4, 3.5, 6.3, 14.7, 26.6, 38.5, 46.2, 49.0, 51.52, 53.76, 55.72 and 57.40) × 10^6 m^3.

The mass inflow curve (dotted line in Fig. 17.5) is drawn by plotting the cumulative inflow as ordinate against time (months) as abscissa.

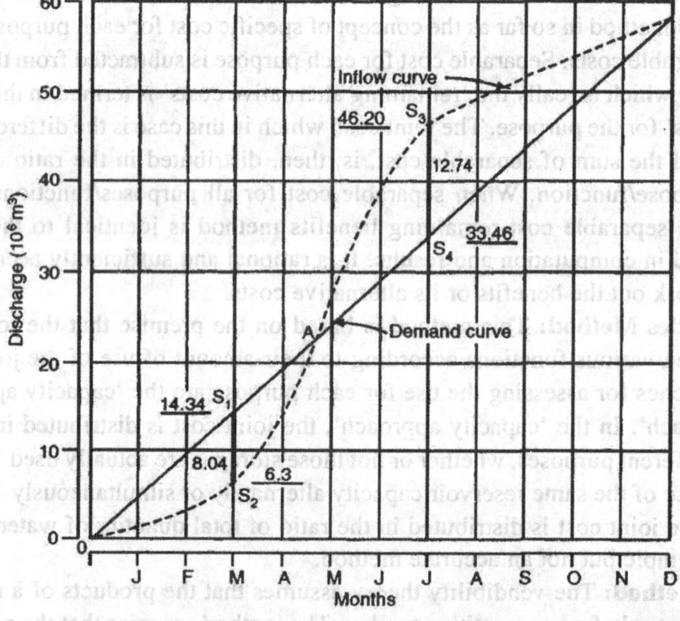

Fig. 17.5. Mass curve.

(*ii*) The monthly demand = (57.40 × 10^6)/12 = 4.78 × 10^6 m^3. The demand curve is drawn as straight line for the uniform monthly demand.

(*iii*) It is evident from the curve that (*a*) S_2 is the lowest point on the depression or trough of the mass inflow curve indicating that the inflow is just equal to the outflow, (*b*) S_3 is the highest point on the hump of the curve indicating maximum reservoir level, (*c*) from 0 to S_2 the slope of the mass inflow curve is less than that of the demand curve indicating that inflow into the reservoir is less than outflow from it and the reservoir is emptying. S_2 represents the minimum reservoir level, (*d*) from S_2 to *A* the slope of mass inflow curve is more than that of the demand curve indicating that inflow is more than outflow and the reservoir water level is rising during the period S_2 to *A*, (*e*) at *A* the reservoir water level is the same as it is in beginning *i.e.*, corresponding to point 0, (*f*) $S_1 - S_2$, *i.e.*, the vertical through S_2 cutting demand curve at S_1 represents, to scale, the minimum initial storage required to cater for deficit during the period 0 to S_2. At S_2 reservoir has fallen to the minimum water level (*g*) from *A* to S_3 the slope of the mass inflow curve is more than that of the demand curve indicating that the water level in the reservoir is rising. S_3 represents full reservoir level, *(h)* $S_3 - S_4$, *i.e.*, the vertical through S_3 cutting the demand curve at S_4 represents the storage, between the initial water level, corresponding to point 0, and full reservoir level, and *(i)* from S_3 upwards, the slope of the mass inflow curve is less than demand curve indicating that the inflow is less than the outflow and the reservoir is again falling.

The minimum storage capacity required to meet the demand

$$= (S_1 - S_2) + (S_3 - S_4)$$

$$= (8.04 + 12.74) \times 10^6 \text{ m}^3 = 20.78 \times 10^6 \text{ m}^3.$$

Initial storage = 8.04 Mm³.

3. *By Bar Graph Method*: Bar graph (Fig. 17.6) area shows that total volume of 57.4 × 10⁶ m³ has entered the reservoir in the year and required minimum storage is 20.83 × 10⁶ m³ which is the area hatched between the bar graph above the average demand graph of 4.78 × 10⁶ m³ drawn as a horizontal line. It also represents, maximum cumulative excess of outflow over inflow during the year which is also the maximum cumulative deficit of water. The value of storage 20.83 × 10⁶ m³ slightly differs from that obtained by mass curve method. It is because of the curvilinear variation of inflow volumes in mass curve as against linearly from one month to the other in this method.

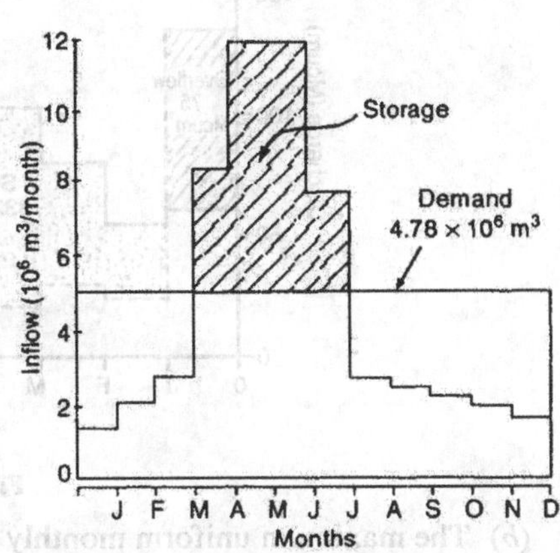

Fig. 17.6 Bar graph.

Example 17.2 *(a) The runoff data for a river along with the probable demands (M cu m) is given below. Can the flow in the river cater to the demands? If so, how?*

(b) What is the maximum uniform demand that can be met and what is the storage capacity required to meet the demand.

Month	*J*	*F*	*M*	*A*	*M*	*J*	*J*	*A*	*S*	*O*	*N*	*D*
Inflow	*135*	*23*	*27*	*21*	*15*	*40*	*120*	*185*	*112*	*87*	*63*	*42*
Demand	*60*	*55*	*80*	*102*	*100*	*121*	*38*	*30*	*25*	*59*	*85*	*75*

Solution: *(i) By Analytical Method*: Evaporation losses and prior water rights of the downstream, if any, are not given and as such ignored. Solution in tabular form is as under:

Month	*Inflow (M m³)*		*Demand (M m³)*	*Deficit (M m³)*	*Cumulative deficit (M m³)*	*Cumulative excess flow (M m³)*
January	135		60	+ 75		75
February	23		55	– 32	32	—
March	27	Total	80	– 53	85	—
April	21		102	– 81	166	—
May	15	=	100	– 85	251	—
June	40		121	– 81	332	—
July	120	870	38	+ 82		82
August	185		30	+ 155		237
September	112		25	+ 87		324
October	87		59	+ 28		352
November	63		85	– 22		
December	42		75	– 33	55	427

The maximum cumulative deficit of 332 M m³ as in the month of June is the storage required to meet the indicated demand. The column of cumulative excess flow shows that the reservoir shall be filled again in the month of October.

2. *By Bar Graph Method*: Fig. 17.7 shows the bar graph for the inflow and demand. The minimum storage is indicated by the portion of the bar graph between demand and inflow (shown by dots) which works out to 332 M cum.

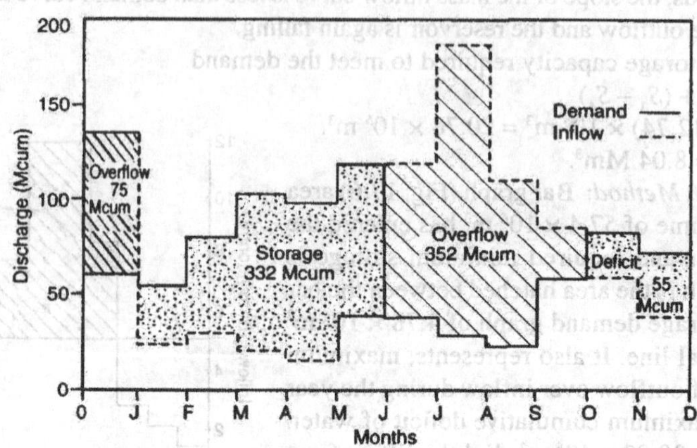

Fig. 17.7 Bar graph.

(*b*) The maximum uniform monthly demand that can be met with from cumulative inflow of 870 M. cum is 72.5 M cum (*i.e.*, 870/12). The storage required to meet this demand is 276.5 Mm^3

i.e., $[(135 + 120 + 185 + 112 + 87) - (72.5 \times 5)]$

Example 17.3 *Monthly flow rate in M m^3 of a river are given below. What would be the maximum storage capacity when 85 cumecs of water is drawn at a uniform rate for power generation? If the reservoir capacity is fixed at 710 M cu m, what uniform rate of withdrawal is possible? Estimate also the percentage change of the generation capacity. Assume a month is equal to 30.4 days, if needed.*

Jan.	*Feb.*	*Mar.*	*Apr.*	*May*	*June*	*July*	*Aug.*	*Sept.*	*Oct.*	*Nov.*	*Dec.*
290	*237*	*164*	*132*	*79*	*66*	*171*	*580*	*790*	*497*	*303*	*290*

Solution:

Month	*Inflow ($M\ m^3$)*	*Demand ($M\ m^3$)*	*Deficit ($M\ m^3$)*	*Cumulative deficit ($M\ m^3$)*	*Excess ($M\ m^3$)*	*Cumulative excess ($M\ m^3$)*
Jan.	290	223*			67	
Feb.	237	223			14	81
Mar.	164	223	59	59		
April	132	223	91	150		
May	79	223	144	294		
June	66	223	157	451		
July	171	223	52	503		
Aug.	580	223			357	357
Sept.	790	223			567	924
Oct.	497	223			274	1198
Nov.	303	223			80	1278
Dec.	290	223			67	1345

* Demand = $85 \times 30.4 \times 24 \times 60 \times 60 = 223\ Mm^3$.

Reservoir capacity required = 503 M m^3. It is also obtained from Fig. 17.8 by drawing a line parallel to demand line with slope of 85 cumecs. Uniform rate of demand met with if storage capacity is 710 M m^3, as read from Fig. 17.8, is 101 cumecs.

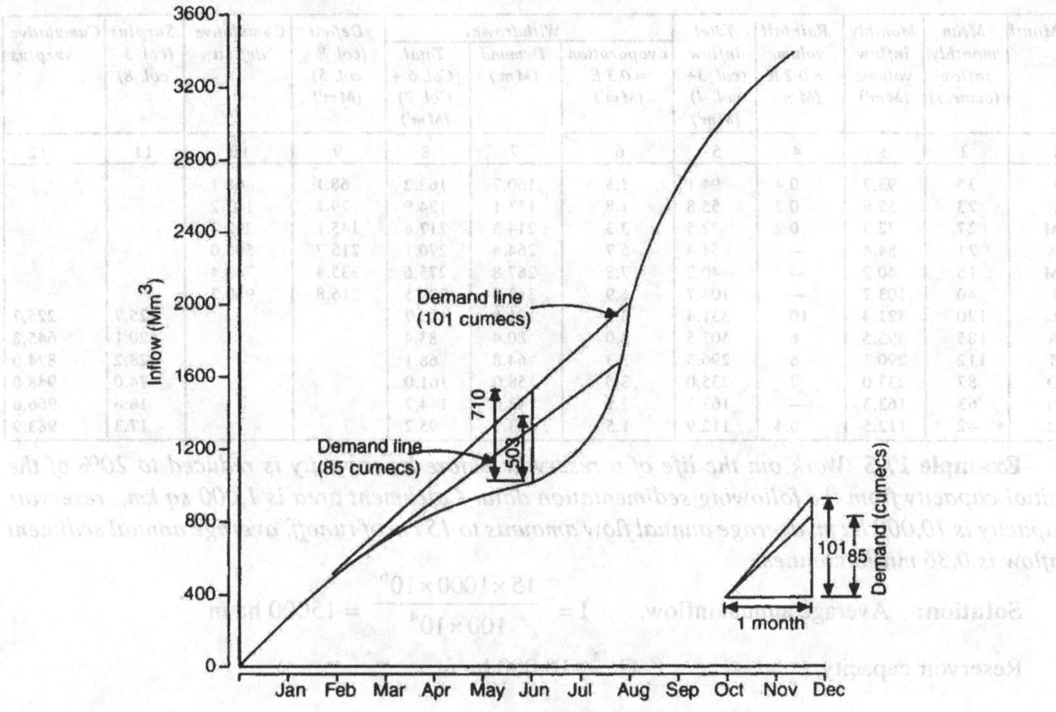

Fig. 17.8 Storage capacity

$$\text{Percentage increase in generation capacity} = \frac{101-85}{85} \times 100 = 19\%.$$

Cumulative excess shows that the reservoir shall be full in September.

Example 17.4 *The following table gives the mean monthly inflow and demand from a reservoir in cumecs:*

Month	*J*	*F*	*M*	*A*	*M*	*J*	*J*	*A*	*S*	*O*	*N*	*D*
Inflow	*35*	*23*	*27*	*21*	*15*	*40*	*120*	*185*	*112*	*87*	*63*	*42*
Demand	*60*	*55*	*80*	*102*	*100*	*121*	*38*	*30*	*25*	*59*	*55*	*35*

The mean monthly evaporation and rainfall (cm) are as under:

Evaporation	*5*	*6*	*11*	*19*	*26*	*23*	*13*	*10*	*11*	*10*	*7*	*5*
Rainfall	*2*	*1*	*1*	*—*	*—*	*—*	*50*	*40*	*30*	*10*	*—*	*2*

Determine storage required to meet the demand. Assume runoff coefficient of reservoir submerged area as 0.33. Average reservoir area is 30 km².

Solution:

Inflow volume (I) = I × No. of days in a month × 8.64 × 10^4 m³

(in January) = 35 × 31 × 8.64 × 10^4 = 93.7 M m³ (*i.e.*, col. 3 of solution in Table (Page 287) and so on

Volume of demand (D) = 60 × 31 × 8.64 × 10^4 = 160.7 M m³ (col. 7) and so on

Volume of evaporation loss = (E/100) × reservoir area (m³)

= (E/100) × 30 × 10^6 = 0.3 E M m³

Rainfall volume = (R/100) × (1 – 0.33) 30 × 10^6 = 0.2 R Mm³

Table (Page 287) shows the computation of storage capacity.

The minimum capacity of the reservoir required is 960.2 M m³ which is the cumulative deficit. The reservoir shall be full again in November and flow in excess of 960.2 M m³ spills over till end December.

Month	Mean monthly inflow (cumecs)	Monthly inflow volume (Mm^3)	Rainfall volume = 0.2 R (Mm^3)	Total inflow (col. 3+ col. 4) (Mm^3)	Withdrawal			Deficit (col. 8 - col. 5) (Mm^3)	Cumulative deficit	Surplus (col. 5 - col. 8)	Cumulative surplus
					Evaporation = 0.3 E (Mm^3)	Demand (Mm^3)	Total (Col. 6 + Col. 7) (Mm^3)				
1	2	3	4	5	6	7	8	9	10	11	12
J	35	93.7	0.4	94.1	1.5	160.7	162.2	68.1	68.1		
F	23	55.6	0.2	55.8	1.8	133.1	134.9	79.1	147.2		
M	27	72.3	0.2	72.5	3.3	214.3	217.6	145.1	292.3		
A	21	54.4	—	54.4	5.7	264.4	270.1	215.7	508.0		
M	15	40.2	—	40.2	7.8	267.8	275.6	235.4	743.4		
J	40	103.7	—	103.7	6.9	313.6	320.5	216.8	960.2		
J	120	321.4	10	331.4	3.9	101.8	105.7			225.7	225.7
A	185	495.5	8	503.5	3.0	80.4	83.4			420.1	645.8
S	112	290.3	6	296.3	3.3	64.8	68.1			228.2	874.0
O	87	233.0	2	235.0	3.0	158.0	161.0			74.0	948.0
N	63	163.3	—	163.3	2.1	142.6	144.7			18.6	966.6
D	42	112.5	0.4	112.9	1.5	93.7	95.2			17.3	983.9

Example 17.5 *Work out the life of a reservoir before its capacity is reduced to 20% of the initial capacity from the following sedimentation data. Catchment area is 1,000 sq km,. reservoir capacity is 10,000 ha m, average annual flow amounts to 15 cm of runoff, average annual sediment inflow is 0.36 million tonnes.*

Solution: Average annual inflow, $I = \dfrac{15 \times 1000 \times 10^6}{100 \times 10^4} = 15000$ ha m

Reservoir capacity, $C = 10{,}000$ ha m

Capacity inflow ratio, $\dfrac{C}{I} = \dfrac{10000}{15000} = 0.67$

Average annual sediment inflow (assuming density of sediment deposits as 1.2 gm/cc) $= \dfrac{0.36 \times 10^6 \times 10^3}{1.2 \times 10^3 \times 10^4} = 30$ ha m

The capacity inflow ratio decreases as the sedimentation increases and trap efficiency also decreases.

Capacity (ha m)		Capacity inflow ratio, C/I	Trap efficiency (percent)		Annual sedimentation (ha m) (30 × col 5/100)	Loss of capacity (ha m) (10,000 × 20/100)	No. of years in which capacity is lost (col. 7/col. 6)
%	Volume		for C/I (read from Brune's curve)	Av. for increment			
1	2	3	4	5	6	7	8
100	10,000	0.67	98	97			
80	8,000	0.53	96	95	29.1	2000	69
60	6,000	0.40	94	92	28.5	2000	70
40	4,000	0.27	90		27.6	2000	72
20	2,000	0.13	84	87	26.1	2000	77
						Total	288 yrs.

Example 17.6 *Calculate the life of a reservoir from the following data:*

Dead storage capacity $= 1.0 \times 10^9\ m^3$

Average annual suspended load in dead storage $= 0.02 \times 10^9\ m^3$

Assume coarse silt 1%, medium silt 15% and fine silt 84%.

Solution. (*i*) Suspended sediment load deposition

Average annual suspended load in dead storage	$= 0.02 \times 10^9$ m^3
Coarse silt at 1%	$= 0.0002 \times 10^9$ m^3
Medium silt @ 15%	$= 0.003 \times 10^9$ m^3
Fine silt @ 84%	$= 0.0168 \times 10^9$ m^3
Total	$= 0.02 \times 10^9$ m^3

Assume that all coarse silt, 80% of medium silt and 20% of fine silt is deposited in the reservoir. Further assume that fine silt is deposited in a semi-fluid state with 45% silt and 55% water.

Silt deposited in the reservoir:

Coarse silt	$= 0.0002 \times 10^9$ m^3
Medium silt $= 0.8 \times 0.003 \times 10^9$	$= 0.0024 \times 10^9$ m^3
Fine silt $= 0.2 \times 0.0168 \times 10^9 \times 100/45$	$= 0.0075 \times 10^9$ m^3
Total	$= 0.0101 \times 10^9$ m^3

(*ii*) Bed load deposition, assume 10% of total load $= 0.002 \times 10^9$ m^3

Total sediment load deposited $= (0.0101 + 0.002) \times 10^9$ m^3 $= 0.0121 \times 10^9$ m^3

Life of reservoir $= \dfrac{1.0 \times 10^9}{0.0121 \times 10^9} = 83$ years

Example 17.7 *Determine trap efficiency of a reservoir of 100 × 10⁶ m³ and 85 × 10⁶ m³ capacity at FRL 600 m and mean operating pool elevation 580 m, average inflow 2000 × 10⁶ m³/yr, reservoir length at MRL 25 km.*

Solution: 1. By Brune's Method

Capacity-inflow ratio, C/I $= 100 \times 10^6/2000 \times 10^6 = 0.05$

Trap efficiency for C/I $= 0.05$ $=$ 76 per cent, read from Fig. 17.4 median curve

Example 17.8 *Apportion the cost among the irrigation and power components of a multipurpose river valley project. Total cost of project is Rs. 1000 crores, Cost of dam portion is Rs. 700 crore, Cost of irrigation system is Rs. 300 crore, Cost of power specialities like penstock, trashrack structures, etc. is Rs. 100 crore, Cost of irrigation specialities is Rs. 50 crores, Height of dam is 150 m, Live storage is 2000 M cu m, Storage utilized for power is 1200 M cu m, and Storage utilized for irrigation is 800 M cu m.*

Single purpose dam:

(*i*) *for power generation, (1200 M cu m. live storage)* = *Dam height 135 m, cost Rs. 700 crore*

(*ii*) *for irrigation (800 M cu m. live storage)* = *Dam height 105 m, cost Rs. 500 crore*

Solution:

(*i*) By Equitable Justifiable Method

Cost of common civil works (700-100-50) = Rs. 550 crore

According to this method, the amount is to be apportioned among power and irrigation facilities in the ratio of cost of dam estimated, assuming it was constructed for providing irrigation and power benefits separately.

Hence, share cost debitable to power $=$ Rs. $\dfrac{550 \times 700}{700 + 500} = 320$ crore

Share of cost debitable to irrigation $=$ Rs. $\dfrac{550 \times 500}{700 + 500} = 230$ crore

Total cost to be borne by irrigation $=$ Rs. $230 + 50 + 300 = 580$ crore

Total cost to be borne by power $=$ Rs. $320 + 100 = 420$ crore

Total cost of transmission lines and the power house are not included in the above analysis as it is to be borne by power.

(*ii*) By Equal Apportionment Method

	Irrigation	*Power*	*Total*
Specific cost	50	100	= Rs. 150 crore
Share of joint cost $\frac{(1000-100-50)}{2}$	425	425	= Rs. 850 crore
Total allocation	475	525	= Rs. 1000 crore

Example 17.9 *Apportion the cost of a multipurpose project, total cost is Rs. 1000 crore.*

Estimated use — storage capacity

(a) For flood control	*= 5000 M cu m.*
(b) For power	*= 2500 M cu m*
(c) For irrigation	*= 2500 M cu m*
Separable cost, flood control	*= Rs. 100 crore*
Irrigation	*= Rs. 500 crore*
Power	*= Rs. 300 crore*
Total	*= Rs. 900 crore*

Solution.

Joint cost (total — separable cost)

= 1000 – 900 = Rs. 100 crore

For flood control, share of joint cost

(in ratio of use) = (100 × 5000)/1000 = Rs. 50 crore

For irrigation, share of joint cost

(in ratio of use) = (100 × 2500)/10000 = Rs. 25 crore

For power, share of joint cost,

(in ratio of use) = (100 × 2500)/10000 = Rs. 25 crore

Therefore, the total cost chargeable to:

Flood control	= 100 + 50	= Rs. 150 crore
Irrigation	= 500 + 25	= Rs. 525 crore
Power	= 300 + 25	= Rs. 325 crore
	Total	= Rs. 1000 crore

QUESTIONS

1. Define reservoir. Discuss various types of reservoirs. Distinguish between single purpose and multipurpose reservoirs.
2. What are single purpose and multipurpose projects? Under what conditions do you go for multipurpose projects?
3. What do you understand by storage zones of a reservoir? Discuss live storage and dead storage. Does dead storage constitute a loss in a reservoir feeding a hydel power plant?
4. Define mass inflow curves and demand curves. Explain the procedure for finding storage capacity of a reservoir in order to meet a particular rate of demand for power generation.
5. Where are storage power schemes necessary? Describe different methods used for arriving at the storage capacity of a reservoir.
6. What is a storage elevation curve for a reservoir? How is it obtained from the contour plan of a reservoir site?
7. Discuss the factors which are considered for the selection of site for a proposed dam. It is assumed that the type of dam has already been selected for the project.
8. Explain causes of silting of reservoirs and method for its control. Discuss various methods of reservoir sediment control.

9. Discuss the geological and topographical features which effect the selection of type of dam.
10. The following table gives the mean monthly runoff of a river. Draw a mass curve and find the capacity of the reservoir for 100% utilization of water at constant regulated flow for power generation.

Month	*Runoff* (m^3/s)	*Month*	*Runoff* (m^3/s)
April	80	October	30
May	40	November	20
June	9	December	40
July	10	January	50
August	20	February	60
September	40	March	120

11. Define the term "dam". What are single purpose and multipurpose projects? Under what condition is each type provided?
12. (*a*) What preliminary investigations should be conducted for the selection of a dam site?
    (*b*) Under what conditions would you adopt a gravity dam or earthen dam?
13. What are the preliminary investigations involved in the planning of an hydel project?

18

EARTH AND ROCKFILL DAMS

18.1 INTRODUCTION

Dams form important adjunct of hydro power development. The dams may be of various types, earth and rockfill dam, concrete and masonry dams. The highest earthen dam for hydel power generation is 260.5 m high Tehri dam under construction in Uttar Pradesh. To keep pace with the ever-increasing demands of water for irrigation, drinking water supply and hydel power generation etc., more and more earthen dams, in preference to other types of dams, are expected to be constructed in the country in times to come, due to the reasons (*i*) The number of dam sites favourable for masonry or concrete structures are constantly decreasing, (*ii*) The construction materials like cement and steel required in huge quantities for building concrete or masonry dams are getting costlier day by day, (*iii*) Earthen dams with recent advancement in the technology of design and construction and the resultant stability, adaptability and economy are becoming more commonly adopted type of dam, (*iv*) Earth dams are made of locally available materials, like clay, gravel, sand and silt. The two basic requirements to be satisfied by an earthen dam are (*i*) Imperviousness, and (*ii*) Stability under all conditions of operation. Evidently these two conditions can not be satisfied by sandy soils and clay soils alone, necessitating their suitable blending for use in embankment construction of an earthen dam.

18.2 TYPES OF EARTHEN DAMS

The type of embankment to suit various site conditions depends on (*i*) Quality and quantity of available materials, (*ii*) Foundation conditions and cutoff requirements, (*iii*) Types of available earthmoving equipment, (*iv*) Diversion considerations, (*v*) Construction schedule, (*iv*) Climatic conditions, and (*vii*) Safety considerations. Various types of earthen dams are as under:

1. Homogeneous Dam

Homogeneous dam, as the name implies, is constructed almost entirely of a single soil material even though it may have an internal drainage system and upstream slope protection. It may be composed of relatively pervious sands and sand-gravel mixtures but is usually made of impervious or semi-pervious soils. It is used when (*a*) Only one type of soil is locally available in required quantity, (*b*) The dam is of very low height so that it is economical to construct it homogeneous rather than zoned, or (*c*) It is uneconomical to construct a zoned embankment type dam due to paucity of free-draining material. Fig. 18.1 indicates homogeneous earthfill dam (Kuttiyadi dam) in Kerala state.

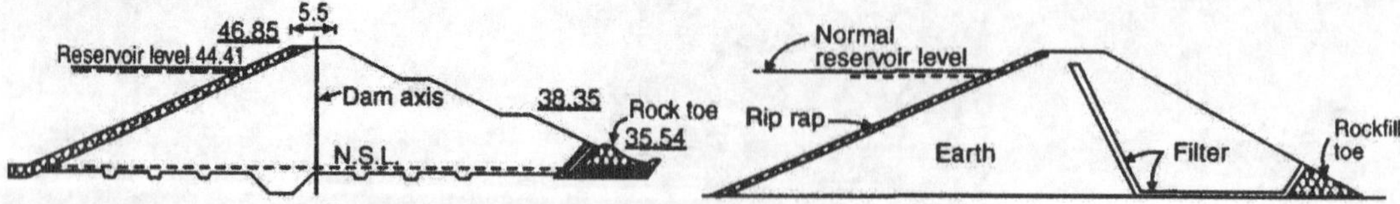

Fig. 18.1 Homogeneous earthfill, Kuttiyadi dam (Kerala)

Fig. 18.2 Modified homogeneous

A purely homogeneous dam is essentially of low height and of relatively bigger section with flatter slopes so as to be safe against piping. The finer the soil, the higher the dam and flatter are the slopes. The essential requirements of a homogeneous dam are (*a*) The material comprising the dam is sufficiently impervious so as to be effective in providing barrier to the passage of water and develop required shear strength on compaction, maintain most of it on saturation with reservoir filling, (*b*) The side slopes are sufficiently flat (2 : 1 to 4 : 1) for stability. Upstream slope is safe against sudden drawdown of the reservoir and the downstream slope to resist sloughing when saturated to a high level, and (*c*) Seepage that emerges at the downstream due to long term storage in the reservoir, regardless of the flatness of the slope and impermeability of the soil, is taken care of by providing rock toe. It is on this account that instead of purely homogeneous section a modified homogeneous section wherein an internal drainage system is provided to keep the phreatic line well within the embankment is preferred (Fig. 18.2).

2. Rolled Fill Dam

Rolled fill dam embankment is constructed by controlled placement of earth material in thin layers and compacted by means of sheep foot rollers and other compaction machinery. The earth is spread in layers and water is sprinkled over the layer. Only additional moisture needed on the fill to replace moisture loss by evaporation, in order to maintain the material at desired moisture content, is added by light sprinkling of water. The compaction is done by sheep foot rollers hauled by crawler tractors properly weighted so as to ensure the requisite pressure. The compaction is also carried out by heavy pneumatic rubber tired rollers. A number of coverages are given to ensure compaction of the material. The placement and compaction is carried out at the optimum moisture content. But method of rolling and optimum moisture content varies with the types of soils placed; better compact a soil at somewhat below optimum water content than above.

3. Zoned Dam

It is the most common type of rolled dam in which use is made of more than one types of the soil material. It essentially consists of a central impervious core zone followed by transition zones of comparatively pervious material and finally covered by free draining pervious outer zones on both sides. The impervious core is protected on both sides by filter, draining and transition zones. Two shells of sufficiently thick material, rockfill, ensure a high degree of permeability. Thus the permeability increases from the centre towards each slope. The entire dam, except the impervious core, forms part of drainage system. Filters allow the water to pass through them and retain the fine materials. Transition zones do the same thing with respect to the filters. The shells allow passage of filtered water through them. However, drainage system in the form of a horizontal drain or rock toe is provided. Fig. 18.3 shows a typical zoned earth dam.

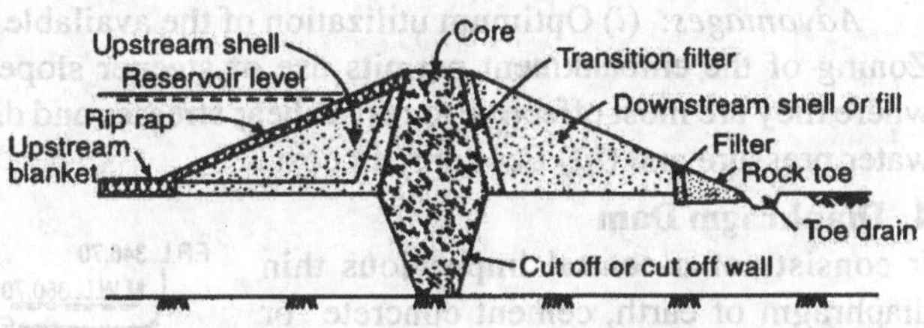

Fig. 18.3 Typical zoned earth dam

A dam is termed as of zoned embankment type if the horizontal width of the impervious zone at any elevation equals or exceeds the height of the embankment above that elevation in the dam and is not less than 3 m. Fig. 18.4 (*a*) depicts zoned earthen Pong dam on Beas river. Fig. 18.4 (*b*) shows section of Ramganga dam.

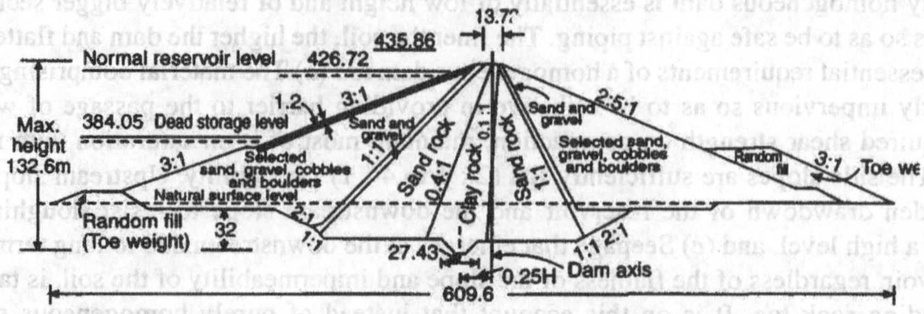

Fig. 18.4(*a*). Earthen dam at Pong (Himachal Pradesh)

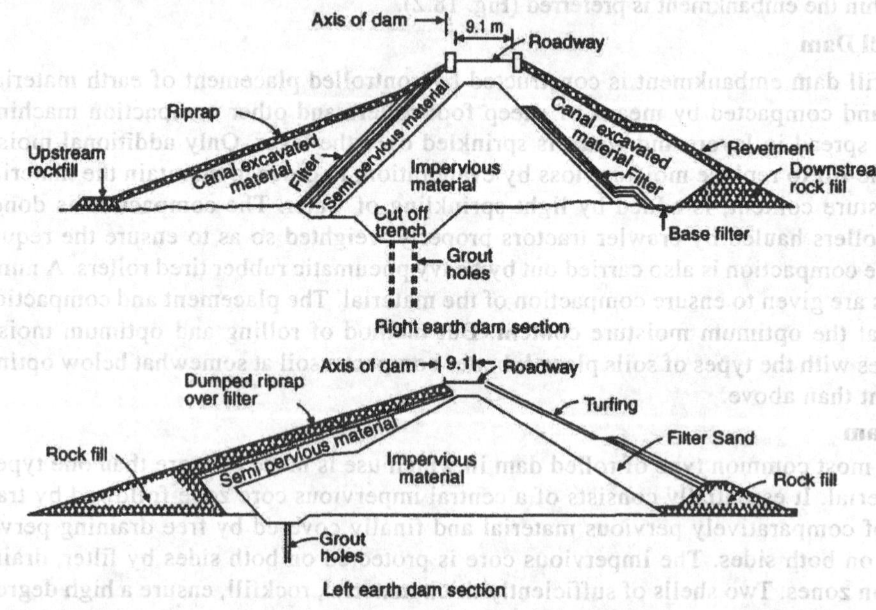

Fig. 18.4(*b*). Ramganga dam section

Advantages: (*i*) Optimum utilization of the available materials of different permeabilities, (*ii*) Zoning of the embankment permits use of steeper slopes because stronger materials are placed where they are most effective to resist shear stresses and drainage system controls and reduces pore water pressure and (*iii*) Economy in cost.

4. Diaphragm Dam

It consists of a central impervious thin diaphragm of earth, cement concrete or bituminous concrete surrounded by earth or rockfill. The diaphragm extends through the cutoff to the impervious foundation to check seepage through the dam. It may be placed either centrally or at the upstream face as a blanket. The inclined diaphragm provides slightly better stability against earthquake. Earthen blanket on the upstream slope is not preferred, except for small dams, because it has to be buried to protect against erosion by wave action and as such it is not available for inspection or repairs. When the diaphragm is of earth, by definition, it has thickness of the impervious core at any elevation less than 3 m or less than

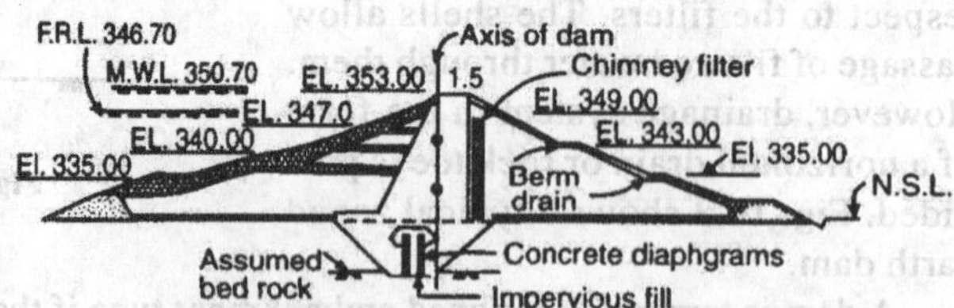

Fig. 18.5. Earth dam with concrete diaphragm

the height of the embankment above the corresponding elevation. However, if the diaphragm thickness equals or exceeds these limits, it is considered to be of zoned embankment type. Earthen core diaphragm is preferred. The core has base width of 0.3 to 0.5 times the water head. Concrete diaphragm is susceptible to rupture due to settlement of the dam or its foundation.

5. Hydraulic Fill Dam

In hydraulic fill dam, as the name implies, the soil material is excavated, transported and placed by using water (Fig. 18.6). Along the outer edges of the embankment, pipes–called flumes are laid at a suitable falling gradient. The material at the borrow pits is mixed with water to form slush and is pumped into these pipe flumes. The slush is discharged through the outlets in the flumes at suitable intervals along their length. The coarser materials of the slush settle soon after the discharge at the outer edge while the finer materials settle at the centre forming an impervious central core. Compaction is not done. However, high pore pressures develop in the core material. The drainage from the core is slow and as such settlement of embankment occurs over long periods. The hydraulic fill is suitable for sandy and gravelly materials. Hydraulic fill dams are not constructed now-a-days.

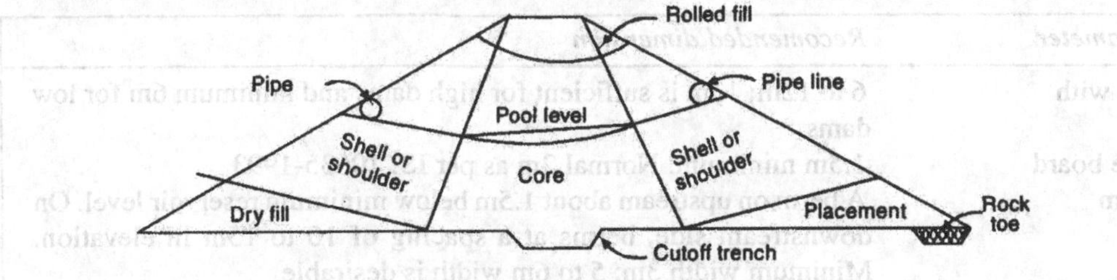

Fig. 18.6 Hydraulic fill dam

18.3 DESIGN CRITERIA FOR EARTH DAMS

The essential requirements of design of an earth dam are safe and stable structure at a minimum construction and maintenance cost. The essential design criteria are (*i*) Safe against overtopping during design flood by providing adequate spillway and outlet capacity, (*ii*) Spillway be of sufficient capacity to pass peak flood, (*iii*) Safe against overtopping by wave action by providing adequate free board, (*iv*) Side slopes stable during construction and under all conditions of reservoir operation. Upstream slope is safe against rapid drawdown conditions while the downstream slope is safe against sloughing, (*v*) Side slopes upstream and downstream are flat enough so that the shear stress induced in the foundation is within the shear strength of the material comprising the foundation with a suitable factor of safety, (*vi*) Upstream slope is protected against erosion by wave action while the crest and downstream slope is protected against erosion due to wind and rain. Horizontal berms at suitable intervals in upstream and downstream faces may be provided for this purpose, (*vii*) Downstream slope is safe during steady seepage under full reservoir condition, (*viii*) Portion of the dam downstream of the impervious core is properly drained by the provision of suitable drainage system, (*ix*) Seepage flow through the dam and foundation is so controlled that there is no danger of fine particles getting washed out from the downstream by the efflux of seepage. Moreover, quantity of seepage loss is restricted to the minimum, (*x*) There is no possibility of free flow of water from upstream through either the dam or the foundation, (*xi*) Adequate impervious core to act as water barrier. Top of impervious core is maintained higher than the maximum reservoir level, (*xii*) Seepage line, *i.e.*, phreatic line is well within the downstream face so that no sloughing of the slope takes place, (*xiii*) Seepage flow through the embankment, foundation and abutments is controlled so that no internal erosion takes place. The amount of water lost through seepage must be controlled so that it does not interfere with planned project functions, (*xiv*) Transition filters are incorporated to satisfy the two essential conditions of filters, viz., to prevent choking of the filter by soil and have minimum of head loss in the filter, (*xv*) Dam as a whole is earthquake resistant. The seismic conditions of the region are investigated with reference to geological map of vicinity. India has been divided into five seismic zones. Horizontal seismic coefficient for static design in different zones, is taken as under:

Zone	Horizontal seismic coefficient h
I	0.01
II	0.02
III	0.04
IV	0.05
V	0.08

Vertical seismic coefficient, where applicable, is taken as half of horizontal seismic coefficient as indicated above.

18.4 DAM SECTION

Salient parameters of dam section are briefly as given in Table 18.1.

Table 18.1. Dimensions of dam section.

Parameter	*Recomended dimension*
Top with	6 to 12m; 13m is sufficient for high dams and minimum 6m for low dams.
Free board	1.5m minimum. Normal 2m as per IS: 10635-1993
Berm	A berm on upstream about 1.5m below minimum reservoir level. On downstream side, berms at a spacing of 10 to 15m in elevation. Minimum width 3m; 5 to 6m width is desirable.
Side slopes	Usually 3:1 and 2.5:1 on upstream and downstream sides respectively.

18.5 SLOPE PROTECTION OF DAM

1. Upstream Slope Protection

Upstream slope protection is provided for the protection to the embankment material against erosion due to wave action, velocity of flow, rain wash, wind action, etc. by placing a protective layer, called riprap (Fig.18.7 and 18.8). The riprap may be hand placed or dumped riprap.

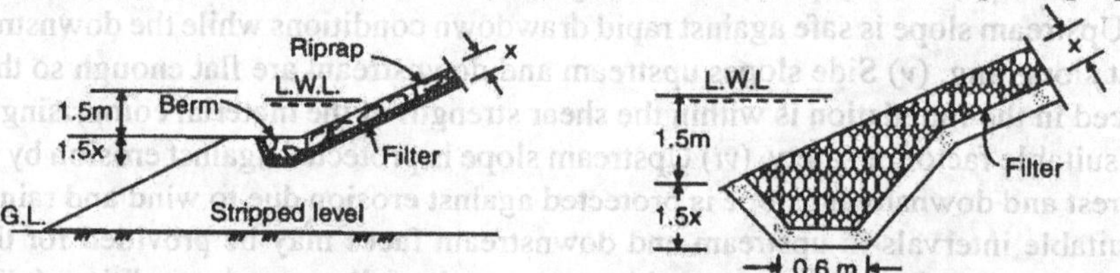

Fig. 18.7 Pitching with berm below low water level. **Fig. 18.8** Pitching with no berm below dead storage level.

1. ***Hand Placed Riprap***: It consists of stones, boulders, cement concrete blocks or soil cement blocks of suitable size placed by hand in more or less definite pattern with a minimum amount of voids and with reasonably even and uniform top surface or alternatively panelwise concrete slabs or precast concrete interlocking type blocks. It has the drawback that the wave run-up height on the slope for the same wave height and embankment slope is about 1.3 to 1.5 times more than that for dumped riprap. The riprap consists of one man stone laid on edge. It is preferably laid in one course such that the layer thickness is the same as the stone size. However, at least 80 per cent of the area of riprap should have stones weighing more than 45 kg, spread uniformly in area.

Where stones of required size are not available to cover the entire thickness of riprap, the same may be laid in two layers. Alternatively, as adopted at Pong dam, boulders pitched on end are grouted over a portion of the area so that stones are not dislodged by wave action (Fig. 18.9) or alternatively riprap is laid in two layers and grouted by pouring fluid cement mortar worked into the riprap.

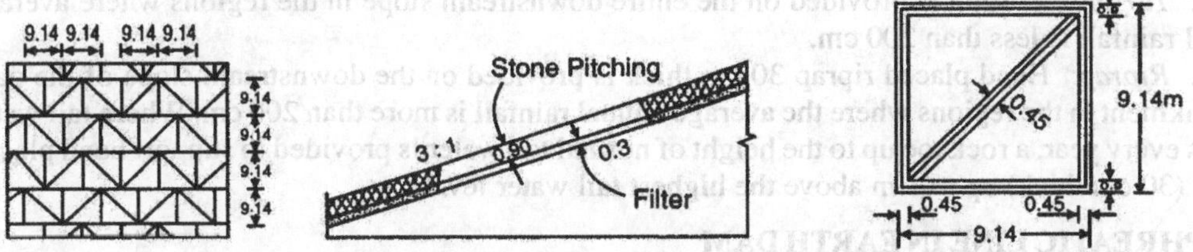

Fig. 18.9 Slope protection at Pong dam.

The minimum thickness of riprap is 30 cm. IS: 8237-1985 has specified minimum thickness of riprap as per Table 18.2.

Table 18.2 Minimum thickness of riprap

Wave height	*Hand placed riprap*	*Dumped riprap*	
(m)	*Minimum thickness (cm)*	*Min. average rock size D50 (cm)*	*Minimum riprap thickness (cm)*
Less than 1.0	30	30	60
1.0 – 2.0	45	40	75
Larger than 2.0	60	70	100

The advantages of hand placed riprap are (*i*) Thickness of hand placed riprap is one-half of the thickness required for dumped riprap, and (*ii*) Better appearance of the finished surface.

The *disadvantages* of this type of riprap are (*i*) Not so flexible as dumped riprap as it cannot adjust itself as well to local settlements and is not suitable where considerable settlement is expected, and (*ii*) Stone must be of excellent quality.

2. *Dumped Riprap*: Dumped riprap consists of boulders or blasted rock free from quarry fines and dumped in place by mechanical means over a properly graded filter, two layers, coarse and fine each of 20 cm thickness. It may be dumped over the upstream face from the embankment level as the embankment is being raised up or after the embankment has been completed.

The minimum thickness of dumped riprap is 45 cm. IS: 8237-1985 has specified the minimum thickness and average rock size as per Table 18.3.

Table 18.3 Thickness and gradation of riprap on 3 : 1 slope.

Reservoir fetch	*Nominal thickness*	*Gradation, % age of stones of various weights (kg)*			
(km)	*(cm)*	*Maximum size*	*At least 25% greater than*	*45 to 75 per cent between*	*Not more than 25 per cent less than*
1.5 and less	45	45	15	0.5 – 1.5	0.5
4	60	65	30	1.5 – 30	1.5
8	75	115	45	2.5 – 45	2.5
16	90	230	90	5 – 90	5.0

The *advantages* of dumped riprap are (*i*) It provides excellent wave protection, (*ii*) It is very durable, (*iii*) Usually less costly, (*iv*) Low maintenance cost, (*v*) Mechanized laying requires a minimum of expensive hand work, (*vi*) Failure has been found in 5% cases due to improper size of stones against 30% in the case of hand placed riprap, and (*vii*) Preferable for high dams.

The *disadvantages* of dumped riprap are (*i*) Individual stone must be of sufficient weight to resist displacement of wave action, and (*ii*) Final surface is not of pleasing appearance as that of hand placed riprap.

2. Downstream Slope Protection

It is provided for protection to the dam embankment material against erosion due to rain wash, wind and tail water. Usual type of protection provided is as under:

1. *Turfing*: Turfing is provided on the entire downstream slope in the regions where average annual rainfall is less than 200 cm.

2. *Riprap*: Hand placed riprap 30 cm thick is provided on the downstream slope of the dam embankment in the regions where the average annual rainfall is more than 200 cm. Where tail water occurs every year, a rock toe up to the height of normal tail water is provided to support hand placed riprap (30 cm thick) up to 1 m above the highest tail water level.

18.6 PHREATIC LINE IN EARTH DAM

Phreatic or seepage line is defined as the line within a dam embankment below which there are positive hydrostatic pressures in the embankment. The hydrostatic pressure on the line itself is atmospheric or zero. Above the line, there is a zone of saturation due to capillary action which is neglected. Phreatic line separates the saturated region of flow from dry or moist zone of soil mass. Above the phreatic line soil is dry (moist) and below the line it is submerged. The determination of its locus in an earth dam is necessary to assess the soil properties while testing the stability of the earth dam embankment.

1. Dam with a Filter

In Fig. 18.10 *AB* is the upstream face and *BL* (equal to *L*) is its horizontal projection. Locate a point *C*, on water surface, at a distance equal to 0.3 *L* from *B*, *i.e.*, *BC* = 0.3*L*. *F*, the starting point of the filter is the focus of basic parabola. The filter length usually kept is 25% of the distance of toe of dam to the centre of dam crest. With point *C* as centre and *CF* as radius draw an arc cutting the line *LB* extended in *D*. Draw a vertical tangent to the curve *FD* at *D* such that *DE* is the directrix, *CD* being equal to *CF*. Point *G* is located midway between *F* and *E*. Thus a base parabola *CC'G* is drawn with focus at *F*. The base parabola is now corrected by eye, a short transition curve (*BB'*) being drawn to connect point *B* with base parabola. The transition curve should start normal to the upstream face at the point of intersection. *BB'C'G* now represents the phreatic line. Equation of base parabola—with *F* as focus and any point *P* (*x*, *y*) on the parabola (such that the distance of the point *P* from focus is the same as that from the directrix), is as under:

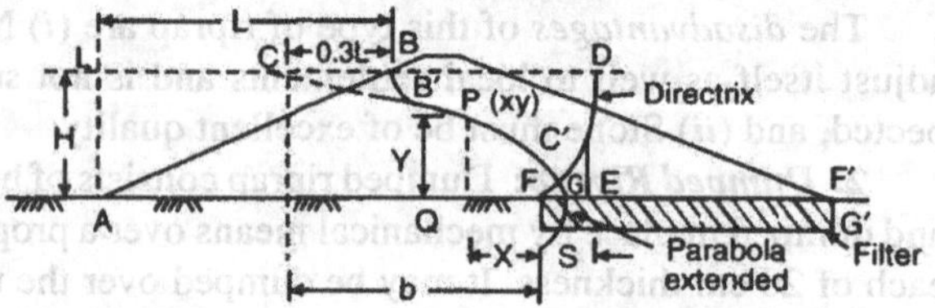

Fig. 18.10 Determining phreatic line in a dam with horizontal drainage filter (Casagrande's method)

The discharge can now be calculated from known coefficient of permeability *K* and focal distance *S* by the formula $q = ks$.

For determining *q* in terms of distance *b* and height *H*,

$$S = \sqrt{x^2 + y^2} - x$$

at $\quad C, x = b \text{ and } y = H$

Therefore $\quad S = \sqrt{b^2 + H^2} - b$

Hence $\quad q = KS$

or $\quad q = K(\sqrt{b^2 + H^2} - b)$

2. Dam without Filter

In this case, the focus of parabola is at the toe of the dam. The phreatic line is determined in the same manner as that for a dam with a filter. The base parabola *BJG* cuts the downstream slope at *J* and extends beyond the downstream slope of the dam (Fig. 18.11) up to point *G*, vertex of the parabola. The phreatic line will, however, emerge out at *K* meeting the downstream slope tangentially there. The portion *KF* is termed as discharge face and always remains saturated. The correction $JK = (\Delta a)$ by which the parabola is to be shifted downward is found by the value of $\dfrac{\Delta a}{a+\Delta a}$ as given by Casagrande (Table 18.4 for various slopes α of discharge face with horizontal). Also $(a+\Delta a) = S/(1-\cos\alpha)$ and $\Delta a = [(a+\Delta a)(180-\alpha)]/4$.

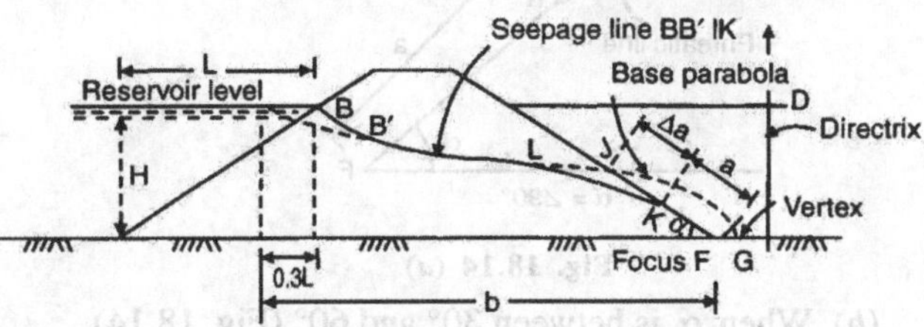

Fig. 18.11 Determining phreatic line in a dam without drainage filter

Intermediate values of α are read from Fig. 18.12. Value of α varies, < 90° where no drainage is provided, 90° or more in rockfill toe, and 180° where horizontal filter is provided.

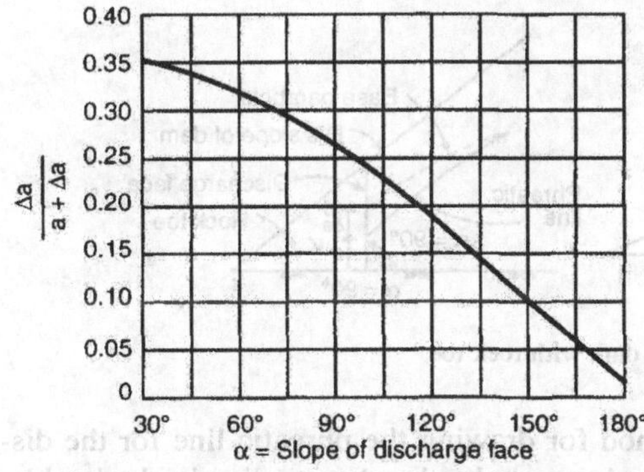

Fig. 18.12 α vs(Δa / a+Δa)

Table 18.4 Casagrande values of Δ *a*/(*a* + Δ *a*)

α (*degrees*)	$\dfrac{\Delta a}{a+\Delta a}$
30	0.36
60	0.32
90	0.26
120	0.18
135	0.14
150	0.10
180	0

Value of a position of point *K*, *i.e.*, where the phreatic line intersects the downstream slope, is determined by Schaffernak and Van Iterson relation, as under:

(*a*) When α < 30° (Fig. 18.13)

$$a = \frac{d}{\cos\alpha} - \sqrt{\frac{d^2}{\cos^2\alpha} - \frac{H^2}{\sin^2\alpha}}$$

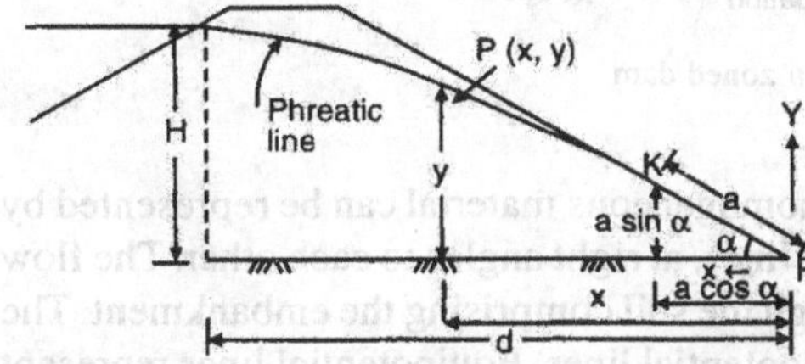

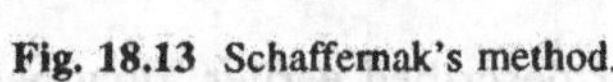

Fig. 18.13 Schaffernak's method.

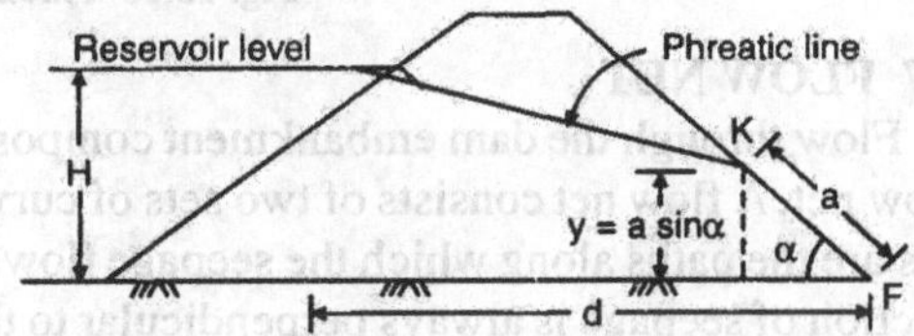

Fig. 18.14 Slope α = 30° to 60°.

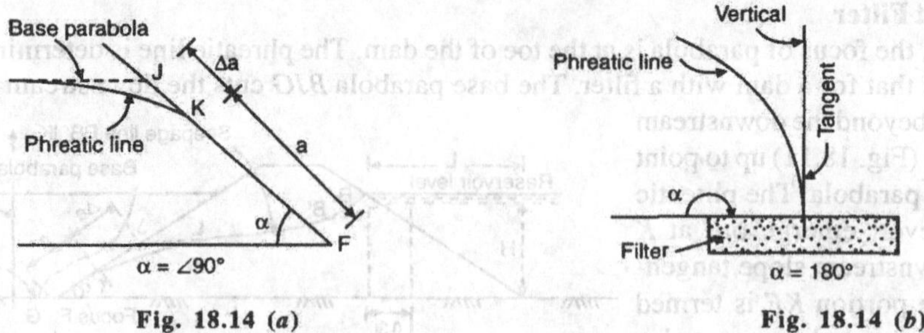

Fig. 18.14 (a) Fig. 18.14 (b)

(b) When α is between 30° and 60° (Fig. 18.14).

Casagrande relation, $a = \sqrt{H^2 + d^2} - \sqrt{d^2 - H^2 \cot^2 \alpha}$. Figs. 18.14(a) & (b) indicate when α = < 90° and α = 180° respectively.

3. Dam with Rock Toe

Focus of base parabola is located at the heel (the inner toe—Fig. 18.15) and basic parabola is drawn in the usual manner. The value of α is measured clockwise from the horizontal base. Correction for $\dfrac{\Delta a}{a + \Delta a}$ is read from Fig. 18.12 and applied.

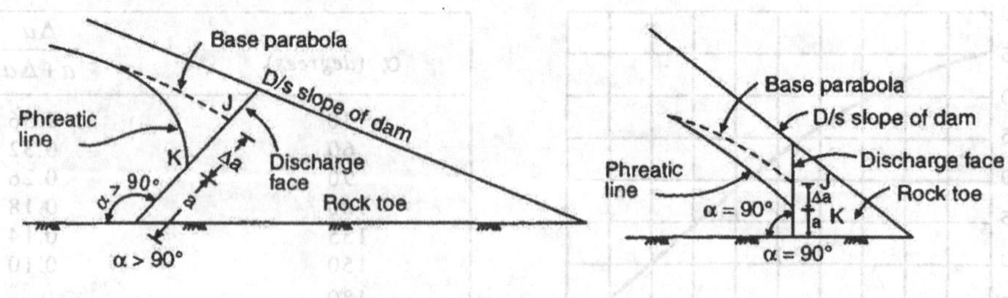

Fig. 18.15 Phreatic line in dam with rock toe

4. Zoned Dam

IS: 7894-1975 recommended approximate method for drawing the phreatic line for the discharge face of the core is that instead of drawing a basic parabola, the phreatic line is obtained by joining the entrance point and breakout point on the discharge face. The breakout point is located on discharge face at a height 0.5*h* from base of the dam, *h* being the head on upstream above the impervious base. At the entrance a short transition curve is introduced.

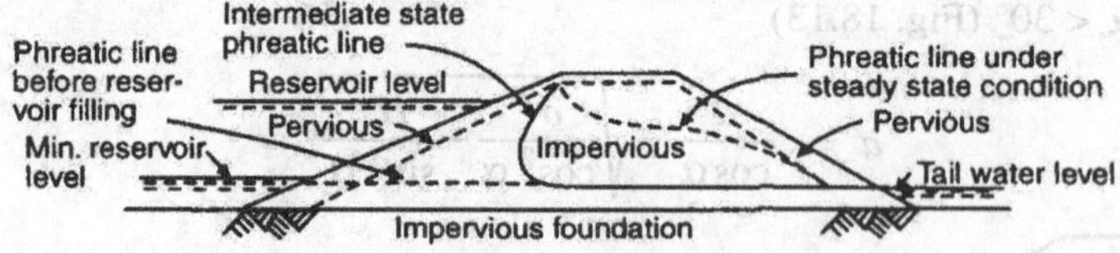

Fig. 18.16 Phreatic line in zoned dam

18.7 FLOW NET

Flow through the dam embankment composed of homogeneous material can be represented by a flow net. A flow net consists of two sets of curvilinear lines, at right angles to each other. The flow lines are the paths along which the seepage flows through the soil comprising the embankment. The direction of seepage is always perpendicular to the equipotential lines. Equipotential lines represent contours of equal head (potential).

The phreatic line having been determined in the manner described earlier, flow net can be drawn by plotting (*i*) Dam section to a suitable scale — same scale for horizontal and vertical directions, (*ii*) Drawing flow net by free hand sketching and making suitable adjustments and corrections by observing the properties of flow net *i.e.*, (*a*) The flow lines and equipotential lines intersect at right angles to each other, (*b*) The distance between the flow lines is made equal to the distance between the potential lines, so as to form a series of squares, (*c*) The rate of flow between each pair of flow lines, called *flow channels*, is the same and the energy drop between any two successive potential lines is the same, (*d*) Smaller the dimensions of the field, greater will be the hydraulic gradient and velocity of flow through them, (*e*) In a homogeneous soil, every transition in the shape of curves is smooth, being either elliptical or parabolic in shape.

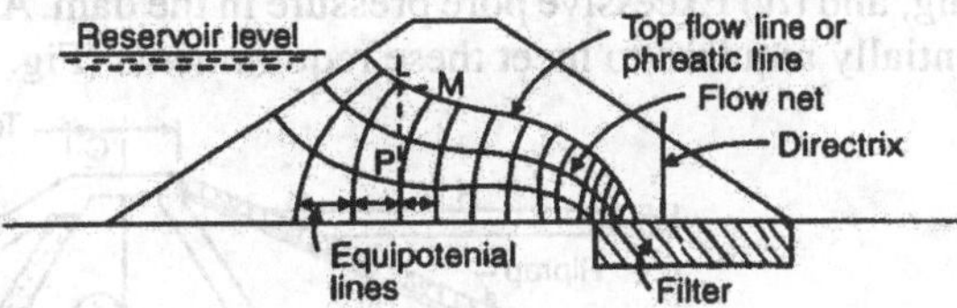

Fig. 18.17 Flow net by graphical method.

A limited number of flow lines and equipotential lines are drawn for convenience. The seepage discharge and pore pressures can be determined from flow net.

1. *Seepage Discharge*: The seepage rate (q) can be determined from the flow net using Darcy's law. The relation for discharge passing through a flow net are as follows:

(*i*) *Isotropic soils*, $q = K_H N_f/N_d = (K_H/N_d)$ (number of flow channels)

($K_H = K_V$, *i.e.*, horizontal permeability = vertical permeability)

(*ii*) *Non-isotropic soils*, $q = \sqrt{K_H K_V} \dfrac{HN_f}{N_d}$

Where, q = rate of seepage, K = coefficient of permeability, H = total pressure head causing flow, N_f = number of flow channels in flow net, and N_d = number of equal potential drops in flow net.

In non-isotropic soils, wherein horizontal and vertical permeabilities differ in the dam section which is generally the case, the flow net is drawn in the manner as for isotropic soils. The dam section is drawn to the same vertical scale but to transformed horizontal scale. The transformed horizontal scale is actual horizontal dimensions reduced by multiplying with $\sqrt{K_V / K_H}$. Where K_V is permeability in vertical direction and K_H is permeability in the horizontal direction. For example, if horizontal permeability is four times the vertical permeability, the horizontal dimensions of the dam shall be reduced by half for plotting.

2. *Pore Pressures*: Under steady state seepage, pore pressures at any point in the dam embankment is equal to the hydrostatic head due to reservoir minus the head loss in seepage through the embankment up to that point. The pore pressure at the point in the body of the dam can be found from flow net.

Pore pressure at point *P* (Fig. 18.17) is equal to difference in elevation between *M* and *P*, where *M* is the point of intersection of phreatic line with the equipotential line through *P*. Pore pressure at *P* is equal to *LP*, where *LM* is a horizontal line through *M*.

18.8 SEEPAGE IN EARTH DAMS

In earth dam, the entire dam, except for the core, forms part of the drainage system. Filters, made of select materials, allow the water to pass through and in turn the transition zones do the same thing with respect to the filters. Finally the shells are draining elements and the filtered water is channelled through them downstream as far as the river or towards the reservoir. Thus in fill dams, drainage flows are generally expelled through their own filters or through the permeable downstream shell. Those from the abutments are channelled to the river bed in order to avoid surface erosion.

Seepage through the earth dams is unavoidable. The reservoir water seeps through the body of the dam or through its contact with foundation and abutments. The adverse effects of seepage in earth dam are (*i*) Loss of reservoir storage capacity, (*ii*) Sloughing of downstream toe due to saturation and progressive development of local shear failures, (*iii*) Weakening of soil strength as finer soil particles are moved through coarse particles resulting in internal erosion or piping failure, (*iv*) Foundation blow out due to excessive uplift, and (*v*) Reduction in effective strength of soil as seepage causes pore pressures.

Seepage Control Measures

The purpose of seepage control measures is to (*i*) Contribute to the stability of the dams and their foundations, (*ii*) Effect reduction of the loss of water, (*iii*) Avert failure of the dam structure by piping, and (*iv*) Excessive pore pressure in the dam. A suitably designed internal drainage system is essentially required to meet these requirements (Fig. 18.18).

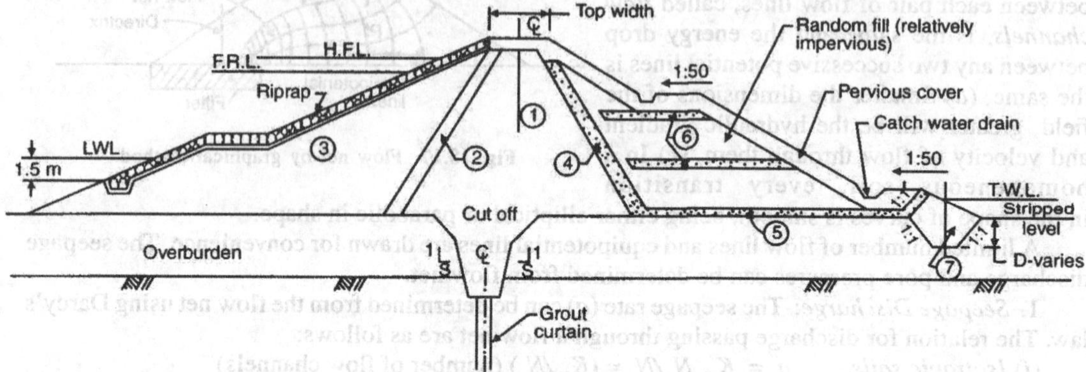

1. Central impervious core, 2. Transition zone, 3. Upstream pervious shell, 4. Inclined filter, 5. Horizontal filter, 6. Horizontal filters at intermediate level, 7. Rock toe and toe drain. (The horizontal filters at intermediate levels are also provided in upstream shell, if the material is of impervious nature).

Fig. 18.18 Section of earth dam showing seepage control measures.

18.9 ROCKFILL DAM

A dam with the diaphragm comprising mostly of rock materials is termed as rockfill dam. Term rockfill dam also refers to the dams constructed entirely of rock. Alternatively, a rockfill dam is defined as an embankment which uses rock of various sizes to provide stability and an impervious membrane to impart watertightness. The mass stability is developed by the friction and interaction of one particle on another rather than by any cementing agent binding the particles together. The water pressure is resisted only by the weight of the rock.

Rockfill dams are less flexible than earthen dams but more flexible than gravity dams. The foundation requirements are more rigid than earthen dams but not so strict and rigid as for gravity dams. The foundation requirements for rockfill dams are (*i*) Essentially the foundation consists of hard durable rock which is not softened or eroded appreciably by water percolating from the reservoir, (*ii*) Foundation free from faults, shear zones or other structural weakness, and (*iii*) Minimum of foundation settlement.

Rockfill dams are suitable for areas where there is scarcity of earthfill materials but durable, hard and sound rock such as unweathered igneous and metamorphic rocks are available for the construction of the dam. Sedimentary rocks and shales which slake in the presence of air are not suitable. A rockfill dam is not suitable when the normal operation of the reservoir does not permit periodic inspection of the impervious membrane constructed on the upstream slope. It is also essential that the dam is not overtopped in time of flood and as such spillway, usually of side channel type, of adequate capacity is provided.

18.10 CLASSIFICATION ROCKFILL OF DAMS

Rockfill dams may be classified on the basis of location of impervious membrane, as (*i*) Rockfill dam with upstream impervious membrane, and (*ii*) Earth core type rockfill dam as illustrated in Fig. 18.19. The components of the dams are also depicted therein.

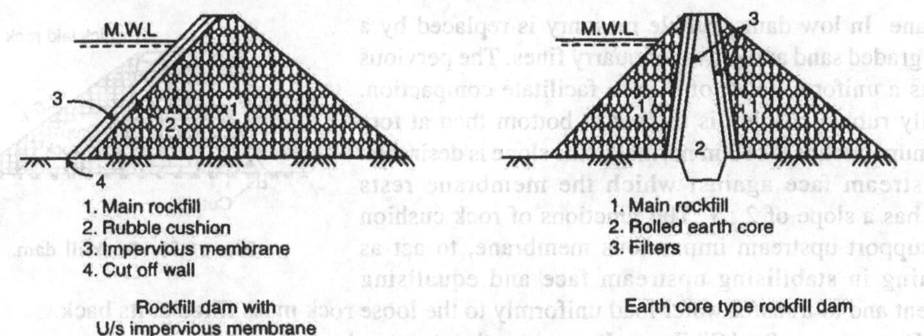

Fig. 18.19 Types of rockfill dams

18.11 DESIGN PARAMETERS

Top width: The top width is normally fixed according to working space required at top, the minimum being 6 m.

Side slope: The upstream slope usually has hand placed stones and as such slope of ½ : 1 to 1 : 1 is adopted. For dams over 50 m height, the slope adopted is 1.3 : 1. The downstream slope made of dumped rock usually corresponds to the natural slope of the rock varying from 1.3 : 1 to 1.5 : 1. For high dams, berms are provided.

Base width: The overall base width usually ranges between 2.5 to 3 times the dam height.

Settlement: The settlement of compacted rockfill after placing the rubble cushion and the impervious facing is generally 1% of the height. It takes place during construction and up to the period the reservoir is filled. The horizontal movement of rockfill also takes place on filling of reservoir and like vertical settlement it continues over a long time.

18.12 DAM SECTION

A rockfill dam section comprises upstream impervious membrane, upstream rock cushion, downstream rockfill and upstream cutoff. The basic requirement is to ensure (*i*) Safety against overtopping, (*ii*) Stability, and (*iii*) Safety against internal erosion.

1. *Upstream Impervious Membrane*: This watertight membrane is constructed on the upstream slope where it is available for inspection and repairs during the drawdown condition of the reservoir. The other advantages of upstream membrane are (*i*) It has greater margin of safety against shear failure because of low pressures in rockfill, larger rockfill mass to resist water pressure and water pressure having larger downward component, (*ii*) Dam can be raised by dumping rock on the downstream side and then extending the membrane upward on the sloping surface, (*iii*) Permits grouting of the rock below cutoff wall at the upstream toe while embankment is being constructed, and (*iv*) Dam has comparatively lesser volume of fill.

The membrane usually consists of asphaltic concrete or cement concrete slabs with suitable joints duly filled with bitumen filler. Reinforced cement concrete membrane is the most common type because of its durability and ease of construction. The joints between the slabs are 25 to 50 mm wide in horizontal direction and 50 to 75 mm wide in vertical direction. Horizontal joints are avoided as far as possible. The joints are filled with a plastic or compressible material. Thickness of concrete slab is about 1% of water head or 30 cm minimum reinforced with steel equal to 0.5 per cent of concrete area, in both directions. Impervious membrane if of earth core and placed centrally is designed as in case of earth dams. Asphaltic concrete membrane is economical and more flexible than concrete slab.

2. *Upstream Rock Cushion*: It may consist of stone masonry or rubble masonry or pervious zone (Fig. 18.20). The masonry provides a smooth, compact bedding for the impervious

membrane. In low dams, rubble masonry is replaced by a zone of graded sand and gravel or quarry fines. The pervious zone has a uniform width of 5 m to facilitate compaction. Generally rubble cushion is thicker at bottom than at top. A minimum thickness of 3 m normal to the slope is desirable. The upstream face against which the membrane rests usually has a slope of 2 : 1. The functions of rock cushion are to support upstream impervious membrane, to act as cushioning in stabilising upstream face and equalising settlement and to transfer water load uniformly to the loose rock mass filled at its back.

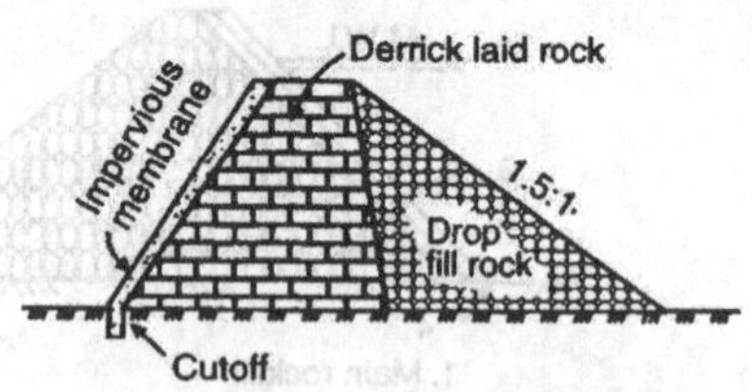

Fig. 18.20 Rockfill dam.

3. *Downstream Rockfill Zone*: It provides the structural support to the impervious membrane or core of the dam by its weight and internal stability against water pressure fromt the reservoir. The downstream rockfill zone (Fig. 18.21) serves to minimize total settlement and possibility of damage to the impervious membrane. The rockfill is dumped on the embankment and spread in layers with a maximum thickness of 1 m. The rockfill is to have minimum number of large voids and is compact. The downstream slope usually adopted is natural angle of the rockfill, *i.e.*, 1.4 : 1. In seismic regions, the slopes of rockfill are flattened beyond the angle of repose of the material to include the ratio of probable earthquake acceleration to the gravity acceleration. The requirements of crest width and free board are the same as indicated for earth dams.

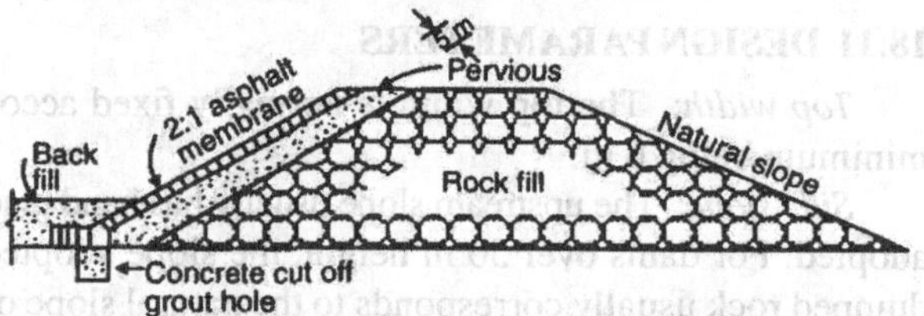

Fig. 18.21 Rockfill dam (d/s rockfill).

4. *Earth core*: Earth core, either vertical or inclined upstream, forms an impervious barrier in an earth core type rockfill dam (Fig. 18.20).

5. *Cutoff*: A watertight cutoff is provided at the upstream toe of the dam along with the contact of the impervious membrane with the foundation and abutments to prevent seepage under the dam and provide support for the weight and thrust of the impervious membrane. The contact length on foundation should be adequate to provide a minimum seepage path, 1/20 to 1/10th of head of water but not less than 3 m depending on rock quality. It should provide at least 1 m rockfill under membrane to permit the membrane to deflect normal to its face. Grouting is done regardless of the apparent quality of the foundation rock. The dam section is checked against sliding caused by water pressure on the upstream. The resisting force comprises vertical component of water and reduced weight of the rockfill due to buoyant effect. The coefficient of sliding varies between 0.6 and 1.0. Fig. 18.22 shows rockfill dam, Hidkal dam in Karnataka.

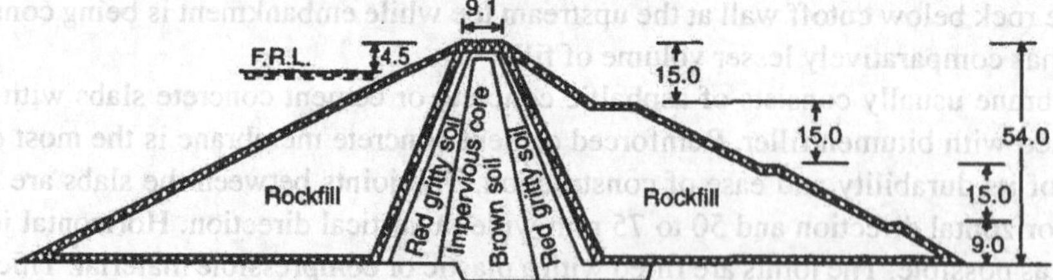

Fig. 18.22 Rockfill dam—Hidkal dam (Karnataka).

18.13 ADVANTAGES AND DISADVANTAGES OF ROCKFILL DAMS

Advantages of rockfill dam are (*i*) Cheaper where suitable hardrock is available, (*ii*) Suitable where suitable materials for earth dam are not available, (*iii*) Economical in remote locations where cost of cement for concrete dam is high, (*iv*) Suitable where foundation is not suitable for concrete dam, (*v*) Can be constructed with relatively unskilled labour, (*vi*) Can be raised subsequently, if so required, without much difficulty.

Disadvantages: (*i*) Time taken in construction is usually more than that required for a concrete dam. This is a crucial factor, (*ii*) More construction equipment is required, (*iii*) Foundation requirements are more rigid than earthen dam, and (*vi*) High maintenance cost.

SOLVED EXAMPLES

Example 18.1 *Determine the seepage line through the dam section (Fig. 18.23). Coefficient of permeability of soil material comprising the dam is 5×10^{-4} cm/sec.*

Solution.

$$LB = 20 \times 3 = 60 \text{ m}$$
$$AB = 0.3 \times 60 = 18 \text{ m}$$

Fig. 18.23 Seepage line through earthen dam.

Coordinates of any point (x, y) on a parabola of equation:

$$\sqrt{x^2 + y^2} = x + s$$

Coordinates of point F are 64.5 *i.e.*, (136.50 – 30 – 60 + 18) and 20, *i.e.*,

$$\sqrt{(64.5)^2 + (20)^2} = 64.5 + S$$

or
$$S = 3.03$$

Vertex C of the parabola is 1.52 m *i.e.*, $S/2$ beyond F and directrix D is at 3.03 from F

Now, equation of parabola $y = \sqrt{2Sx + S^2}$

$$= \sqrt{2 \times 3.03 \times x + 3.03 \times 3.03}$$
$$= \sqrt{6.06\,x + 9.18}$$

Coordinates of parabola for various values of x are as under:

x	0	10	20	30	40	50	60	64.5
y	3.03	8.35	11.42	13.82	15.86	17.67	19.31	20.0

Fig. 18.23 shows the plotting of phreatic line.

Seepage discharge $q = KS$

$$= (5 \times 10^{-4} \times 10^{-2}) \times 3.03$$
$$= 15.15 \times 10^{-6} \text{ cumec per m length of dam.}$$

Example 18.2 *Determine seepage through foundation of the dam section shown in Fig. 18.24.*

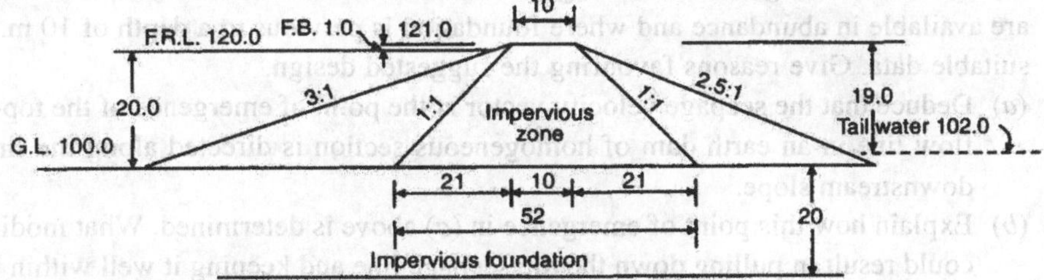

Fig. 18.24 Seepage through dam foundation.

Solution. Darcy's formula, $Q = KiA$

where, Q = discharge, k = coefficient of permeability of foundation

$= 0.2 \times 10^{-6}$ m/sec

i = hydraulic gradient = h / l

$i = (120.0 - 102.0)/52 = 0.346$

A = gross area of foundation through which flow takes place,

$= 1 \times 20$ sqm for a dam foundation width of 1 m

Now, $Q = 0.2 \times 10^{-6} \times 0.346 \times 20$

$= 1.38 \times 10^{-6}$ cumec

Example 18.3 *A flow net is plotted for a homogeneous earthen dam of height 25 m and length 2000 m with free board 2 m. The results obtained indicate number of potential drops as 10 and number of flow channels as 4. The dam has a horizontal filter of 30 m at the downstream end and the coefficient of permeability of the dam material is 5 × 10⁻⁴ cm/sec. Calculate the discharge through the dam.*

Solution.

$$q = KH\frac{N_f}{N_d} = \frac{5\times10^{-6}\times23\times4}{10}$$

$= 46 \times 10^{-6}$ cumec/m length of dam

$Q = 46 \times 10^{-6} \times 2000 = 0.092$ cumec.

QUESTIONS

1. Explain how the following parameters effect design of an earthen dam: (*i*) optimum moisture content, (*ii*) C and ϕ value of soil, (*iii*) permeability of soil, (*iv*) sudden drawdown of reservoir.
2. Illustrate with neat sketch the following parts of an earthen dam and state their functions briefly:
   (*i*) rock toe, (*ii*) horizontal drainage blanket, (*iii*) cutoff, (*iv*) riprap.
   (*b*) Give a few typical cross-sections of earthen dams to suit locally available material and foundations.
3. Show with the help of simple outline sketches suitable broad designs of earth dam for different available materials and the governing geology of the site.
4. Briefly discuss the checks that are required to be made to investigate the stability of an earthen dam.
5. Discuss control of seepage in earthen dams.
6. Explain the various steps that can be taken in the design and in the construction of an earthen dam to reduce seepage from it.
7. Discuss hydraulic fill dam.
8. Discuss cutoff trench.
9. Give a suitable design for a 20 m high dam for a site where both clay silt and sand gravel are available in abundance and where foundation is pervious to a depth of 10 m. Assume suitable data. Give reasons favouring the suggested design.
10. (*a*) Deduce that the seepage velocity vector at the point of emergence of the top-seepage flow line in an earth dam of homogeneous section is directed along the line of the downstream slope.
    (*b*) Explain how this point of emergence in (*a*) above is determined. What modifications could result in pulling down the top seepage line and keeping it well within the body of the dam?

11. (*a*) What is a flow net? What are its salient properties?

(*b*) What points are to be noted while sketching a flow net?

(*c*) A homogeneous earthen dam is 21.5 m high and has a free board of 1.5 m. A flow net was constructed and the following results were noted:

Number of potential drops = 12

Number of flow channels = 3

The dam has a horizontal filter of 15 m length at its drawdown end. Calculate the discharge per metre length of the dam if the coefficient of permeability of the dam material is 2.7×10^{-3} cm/s.

Solution:

$$q = 2.7 \times 10^{-3} \times 10^{-2} \times (21.5 - 1.5) \times \frac{3}{12}$$

$$= 1.35 \times 10^{-6} \text{ cumec.}$$

19

GRAVITY DAMS

19.1 INTRODUCTION

Gravity dam is defined as a structure so proportioned that it derives its stability from its own weight to resist the external forces. A gravity dam is most durable and requires little maintenance. It may be constructed of masonry or concrete. Concrete gravity dams are greatly preferred and their construction is common. A gravity dam is generally straight in plan and termed as straight gravity dam. However, it may also be slightly curved in plan. Nevertheless the external forces are resisted by the weight of the dam and not by arch action. In a gravity dam no bending stress is introduced at any point of the dam and as such it is called solid gravity dam. Gravity dams transfer the load to foundations by cantilever action and require strong foundation rock to resist the weight of the dam. They can be built to any height. They are most suited across narrow gorges with steep side slopes where earthen dam might slip. Bhakra Dam over river Sutlej is a straight concrete gravity dam.

Cement concrete dams have earned a second place in the construction of gravty dams in the country on account of good speed of construction possible due to the deployment of machinery.

Various terms relating to dam parameters are discussed as under:

1. *Dam axis*: The line of upstream face of the dam or the line of crown of the dam, if upstream face is sloping, is taken as reference line for the purpose of layout. It is called *base line* of the dam or *dam axis.*

2. *Structural height*: It is the height between the top of dam and the lowest foundation level, exclusive of narrow fault zones, if any.

3. *Hydraulic height.* It is the height between the highest controlled water surface and the lowest point in the original river bed level at the axis of the dam.

4. *Dam length.* It is the length of the dam measured along the axis of the dam at the level of top of dam, *i.e.* from one abutment to the other.

19.2 SELECTION OF DAM SITE

Various considerations in the selection of a good dam site have already been discussed in chapter 17. Those specific to gravity dams are (*i*) A narrow gorge, opening upstream, at the dam site, (*ii*) Strong rock foundation to safely withstand static and dynamic forces including earthquakes, (*iii*) Strong and water tight abutments, (*iv*) Stable side slopes of abutments, (*v*) Suitable location for power house and spillway, (*vi*) Availability of construction materials nearby, (*vii*) Accessibility by rail and road, (*viii*) Availability of electric supply for construction, and (*ix*) Dam height is reasonable for the required storage capacity.

19.3 FORCES ACTING ON A GRAVITY DAM

Various forces acting on a gravity dam as considered in design are as under:

Dead Load: It comprises the major resisting force. It includes weight of the concrete or the dam is considered. The cross-section of the dam is divided into several triangles and rectangles. Weight of each triangle and rectangle and their point of application at respective centre of gravity are computed. The resultant of all these downward forces is thus found out by taking moments of the component forces which constitute the total weight of the dam acting at its centre of gravity. Unit weight of concrete and masonry is usually taken as 2400 kg/m^3 and 2300 kg/m^3 respectively.

Reservoir and Tail Water Loads: Water pressure corresponding to full reservoir level constitutes the major external force acting on the dam. The water pressure acts horizontally and its intensity varies triangularly with zero intensity at the water surface and equal to $WH^2/2$ at the base, where W is the unit weight of water (1000 kg/m^3) and H is the water depth upstream. The resultant force $WH^2/2$ acts at $H/3$ from the base of the dam (Fig. 19.1). When the upstream face is partly vertical and partly inclined (Fig. 19.2), the resultant water pressures comprise (*i*) horizontal component $P = 1/2\ \gamma_w h^2/2$, where γ_w is unit weight of water in S.I. units. ($P = WH^2/2$ in MKS units) acting at $H/3$ from the base of the dam, and (*ii*) vertical component P_1 which is the weight of the column of water *ABCDA* acting at *CG* of the area. In low overflow dams, the dynamic effect of the velocity of approach is also considered.

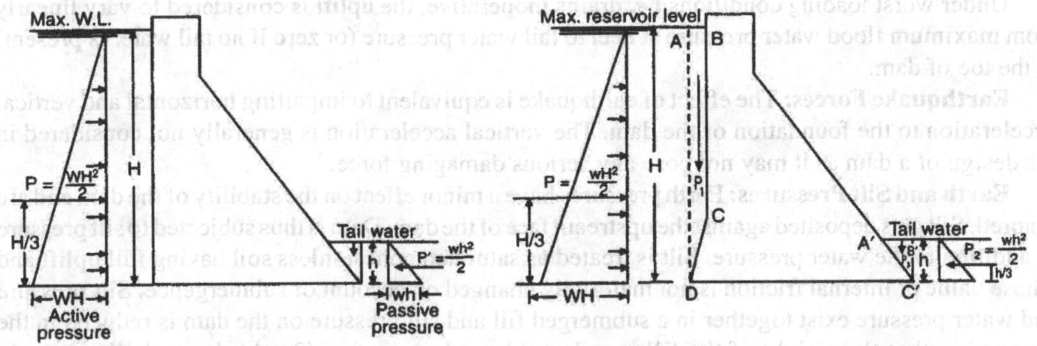

Fig. 19.1 Water pressure acting on dam with u/s face vertical and tail water.

Fig. 19.2 Water pressure acting on dam with u/s face partly inclined and tail water.

Tail Water Pressure. Tail water pressure, adjusted for any retrogression, is taken at full value for non-overflow sections and at a reduced value for overflow sections depending on the type of energy dissipation arrangement provided and the anticipated water surface profile downstream. The full value of corresponding tail water is, however, considered in the case of uplift. The tail water of height h on the downstream side exerts both horizontal and vertical pressures; horizontal pressure, $P_2 = 1/2\ \gamma_w h^2/2$ in S.I. units ($P_2 = wh^2/2$ in MKS units) acting at $h/3$ from the base and vertical pressure, P_3 = weight of column of water *A'B'C'* (refer Fig. 19.2).

3. Uplift Pressure: It is the second major external force acting upward due to seeping water. The uplift forces occur as internal pressures in pores, cracks and seams within the body of the dam, at the contact surface between the dam and its foundation and within the foundation. The pressure intensities at the heel and toe of the dam are taken equal to the hydrostatic pressures at the heel and toe joined by a straight line in between (Fig. 19.3). Drainage galleries are sometimes provided in the dam body to relieve uplift pressure. Accordingly, the uplift pressure distribution in the body of the dam is assumed to have an intensity which at the line of the formed drain/drainage holes exceeds the tail water head by one-third the differential between reservoir level and tail water level. The pressure gradient is then extended linearly to heads corresponding to reservoir level and tail water level (Fig. 19.4). The pressure is assumed to act over 100 percent of the area. The uplift criteria in case of dams founded on compact and unfissured rock is as specified above.

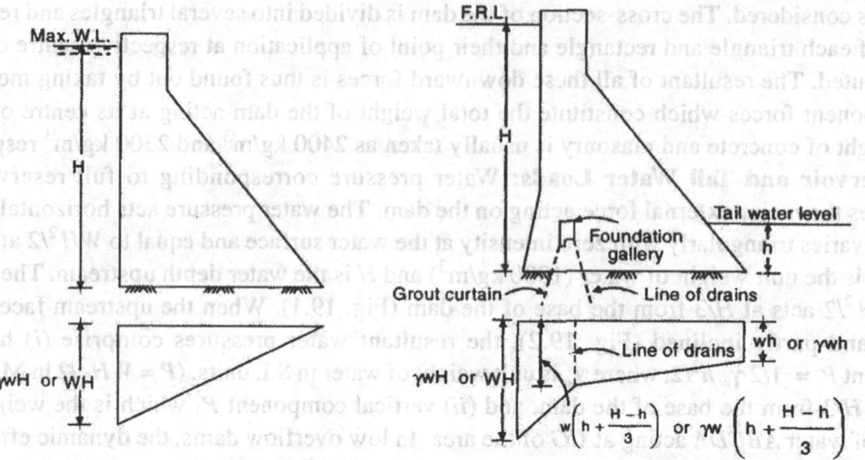

Fig. 19.3 Uplift pressure diagram without tail water.

Fig. 19.4 Uplift pressure diagram with drainage gallery and tail water.

Under worst loading conditions *i.e.* drains inoperative, the uplift is considered to vary linearly from maximum flood water pressure at heel to tail water pressure (or zero if no tail water is present) at the toe of dam.

Earthquake Forces: The effect of earthquake is equivalent to imparting horizontal and vertical acceleration to the foundation of the dam. The vertical acceleration is generally not considered in the design of a dam as it may not pose any serious damaging force.

Earth and Silt Pressures: Earth pressures have a minor effect on the stability of the dam and are ignored. Silt gets deposited against the upstream face of the dam. Dam is thus subjected to silt pressure in addition to the water pressure. Silt is treated as saturated cohesionless soil having full uplift and whose value of internal friction is not materially changed on account of submergence. Silt pressure and water pressure exist together in a submerged fill and silt pressure on the dam is reduced in the proportion that the weight of the fill is reduced by submergence. The horizontal silt pressure $P_s = W_s \frac{h^2}{2}\left(\frac{1-\sin\beta}{1+\sin\beta}\right)$ is assumed to be equivalent to that of a fluid weighing 1360 kg/m^3. Vertical silt and water pressure is determined as if silt and water together have a density of 1925 kg/m^3. β is the angle of internal friction taken as 30° for sand, gravel, clay and silt, W_s is unit weight of silt or earth submerged in water, and h is the height of silt submerged in water above the base of dam. In a dam with inclined upstream face, the vertical weight of silt resting against the slope also acts as a vertical force.

Wind Pressure: Wind loads on dams are ignored.

Ice Pressure: Ice pressure is not encountered in the design of dams in India. Ice if formed on the water surface of the reservoir expands and contracts with change in temperature. The dam face is subjected to force due to expansion of ice. Ice pressure is taken as 250 kN kg/m^2 applied to the face of dam over the anticipated area of contact of ice with the face of the dam as per IS:6512-1984.

Wave Pressure: Upper portions of the dam are subjected to impact of waves. Wind pressure depends on the height of wave. The wave pressure diagram is approximated by triangle 1-2-3 in Fig.19.5. The total wave force P_w is given by the area of triangle or $P_w = 2000\ hw^2$. The centre of application is at a height of 0.375 hw above the still water level *T.* Saville method recommended by IS : 6512–1984 for the computation of wave height is discussed inFig. 19.5.

(*i*) Maximum unit pressure, $P_w = 2400\ h_w$ kg/m^2 acting at 0.125 h_w above the still water level.

(*ii*) Total wave force, $P_w = 2400 H_w \times 1/2 \times 5/3\ H_w, = 2000\ H^2{}_w$, $(P_w = 2\gamma_w H_w^2 = 2\times 9.81\ H_w^2$ kg N/m in S.I. units) acting at 0.375 h_w above the still water level assuming pressure distribution is triangular, and height equal to $\frac{5}{3} H_w$.

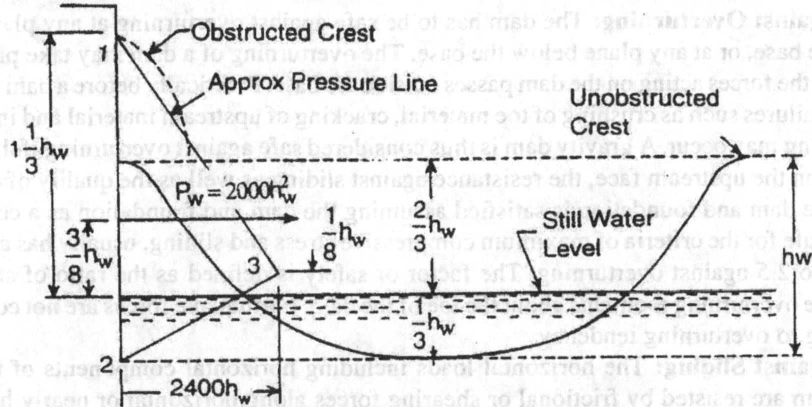

Fig. 19.5 Wind pressure and wave diagram.

(*iii*) When a maximum wind velocity is known, the same is used for FRL condtion and 2/3 times for MWL condition.

Thermal Loads: Stresses in the dams are affected due to temperature variation in the dam body. The cyclic variation of air temperature and solar radiation on the downstream side and reservoir temperature on the upstream side also affect the stresses in the dam. Even deflection of the dam is maximum in morning and goes on reducing to a minimum value in the evening. Magnitude of deflection is generally less when spillway is working than when it is not working. While considering the thermal load, temperature gradients are assumed depending on location, orientation, surrounding topography etc.

Reaction of Foundation: Some water under pressure percolates between the dam and foundation, through horizontally bedded joints under the foundation and/or through pour joints in the dam. Part weight of the dam is supported by water and the direct foundation reaction is correspondingly reduced which effect is termed as uplift. The resultant of all horizontal and vertical forces including uplift must be balanced by an equal and opposite reaction at the foundation consisting of the total vertical reaction and the total horizontal shear and friction. Further, for the dam to be in static equilibrium, the force is so located that summation of moments is zero.

The greatest load out of the various loads mentioned above is the weight of the gravity dam as it is the only active force that assists failure of the dam. The safety of the dam is roughly proportional to the density of concrete or masonry it is made of.

19.4 COMBINATION OF FORCES FOR DESIGN

A dam is designed for the most adverse combination of load conditions as have reasonable probability of simultaneous occurrence. Such load combinations are : (*i*) Load Combination A-Construction condition, dam completed but reservoir empty and no tail water, (*ii*) Load Combination B-Normal operating condition with reservoir full (elevation) and normal dry weather tail water, normal uplift, ice and silt (if applicable), (*iii*) Load Combination C-Flood discharge condition. Reservoir at maximum flood level, all gates open, tail water at flood elevation, normal uplift and silt (if applicable), (*iv*) Load Combination D-Combination A, with earthquake, (*v*) Load Combination E-Combination B, with earthquake but no ice, (*vi*) Load Combination F-Combination C, but with extreme uplift (drains inoperative), (*vii*) Load Combination G-Combination E, but with extreme uplift (drains inoperative).

19.5 MODES OF FAILURE — SAFETY CRITERIA

A dam has to be stable against overturning, sliding, compression or crushing, and tension. Criteria for structural stability of a gravity dam in this context is as follows:

Safety against Overturning: The dam has to be safe against overturning at any plane within the dam, at the base, or at any plane below the base. The overturning of a dam may take place if the resultant of all the forces acting on the dam passes outside the base. Practically before a dam overturns bodily, other failures such as crushing of toe material, cracking of upstream material and increase in uplift and sliding may occur. A gravity dam is thus considered safe against overturning if the criteria of no tension on the upstream face, the resistance against sliding as well as the quality of concrete/ masonry of the dam and foundation is satisfied assuming the dam and foundation as a continuous body. A dam safe for the criteria of maximum compressive stress and sliding, usually has a factor of safety of 1.5 to 2.5 against overturning. The factor of safety is defined as the ratio of stabilizing moments to the overturning moments about the toe of the dam. Earthquake forces are not considered as contributing to overturning tendency.

Safety against Sliding: The horizontal loads including horizontal components of the loads acting on a dam are resisted by frictional or shearing forces along horizontal or nearly horizontal planes in the body of a dam, on the foundation or on horizontal or nearly horizontal seams in the foundation. It follows that the total magnitude of the forces tending to induce sliding shall be less than the minimum total available resistance along the critical path of sliding, *i.e.*, along that plane or combination of planes which mobilizes the least resistance to sliding. The sliding resistance is a function of the cohesion inherent in the materials and at their contact and the angle of internal friction of the material at the surface of sliding. The stability of a dam against sliding is evaluated by comparing the minimum total available resistance along the critical path of sliding (that is along that plane or combination of planes which mobilises the least resistance to sliding) to the total magnitude of the forces tending to induce sliding.

The factor of safety against sliding (F) is calculated, as under, on the basis of partial factor of safety in respect of friction (Fϕ) and partial factor of safety in respect of cohesion (F_c) as per Table. 19.2.

Table 19·2 Partial factors of safety against sliding

Sl. No.	*Loading condition*	*Fϕ*	*F_c* For dams and the contact plane with foundation	*F_c* For foundation: Thoroughly investigated	*F_c* For foundation: Others
(i)	A, B, C	1.5	3.6	4.0	4.5
(ii)	D, E	1.2	2.4	2.7	3.0
(iii)	F, G	1.0	1.2	1.35	1.5

$$F = \frac{1}{P}\left[\frac{(w-u)\tan\phi}{F\phi} + \frac{CA}{F_c}\right]$$

where, F = factor of safety against sliding, minimum 1.0, w = total mass of dam, u = total uplift force, tan ϕ = coefficient of internal friction of the material, c = cohesion of the material at the plane considered, A = area under consideration for cohesion, Fϕ = partial factor of safety in respect of friction, F_c = partial factor of safety in respect of cohesion, and P = total horizontal force.

Safety against Crushing: Safety against crushing is ensured if the compressive stresses produced are within the allowable stresses.

$$\text{Normal stress at toe} = \frac{W}{B}\left(1+\frac{6e}{B}\right)$$

$$\text{Normal stress at heel} = \frac{W}{B}\left(1-\frac{6e}{B}\right)$$

When pressures at toe and heel are both compressive, (Fig. 19.8) the maximum compressive stress occurs at the toe (when eccentricity) $e = B/6 = \frac{2W}{B}$ *i.e.* $\frac{W}{B}\left(1+\frac{6}{B}\times\frac{B}{6}\right)$, and the corresponding compressive stress at heel will be zero. This maximum compressive stress shall be less than the allowable compressive stress. The excessive stresses at toe and heel can be brought within permissible limits by providing fillets sloping at 1 : 1 at the upstream and 2 : 1 at the downstream. The height of fillets is given by the relation:

$$h_{heel} = 6H - 0.658\,H^2 - 1.07$$

$$h_{toe} = 6.5H - 1.1H^2 - 0.9$$

where h = fillet height (m), and H = dam height (in 100 m)

The strength of concrete is to satisfy early load and construction requirements and at the age of one year it should be 4 times the maximum computed stress in the dam or 140 kg/cm^2, whichever is more. The allowable working compressive stress in any part of the dam is usually taken as 70 kg/cm^2 or 700 kgN/cm^2 for concrete.

Safety against Tension: A gravity dam is usually designed so that no tension is developed in the dam body and for that the resultant must lie within the middle third. IS: 6512-1984 recommends that no tensile stress occurs at the upstream face of the dam for load combination B. Normal tensile stresses, however, may be permitted in other load conditions but not exceeding values given in Table 19.3. In the case of high dams, tensile stress up to a maximum of 5 kg/cm^2 or 500 kg, N/cm^2 is sometimes permitted for reasons of economy and keeping in view that such worst loading condition may occur rarely and for a very short duration.

Table 19.3 Permissible tensile stress

Load combination	*Permissible tensile stress*	
	Concrete	*Masonry*
C	0.01 *fc*	0.005 *fc*
E	0.02 *fc*	0.01 *fc*
F	0.02 *fc*	0.01 *fc*
G	0.04 *fc*	0.02 *fc*
fc = cube compressive strength of concrete/mortar for masonry		

19.6 PRINCIPAL AND SHEAR STRESSES

The principal and shear stresses are determined by considering an elementary triangular section at either the heel or toe of the dam section such that stress intensities may be assumed to be uniform on its face. With reservoir full, the stresses at the toe are greater and as such stress analysis is done for the downstream face. Consider a triangular section *ABC*. (Fig. 19.6), *AB* and *AC* are the principal planes, mutually at right angles, plane *AB* will also have only a normal stress on it, and will be the other principal plane. Let the small lengths *AB*, *BC* and *AC* be d_r, d_y and d_s respectively. Let p = intensity of water pressure acting on face *AC*, σ = principal stress on plane *AB*, τ = shear stress along face *BC*, P_n = normal stress on face *BC*.

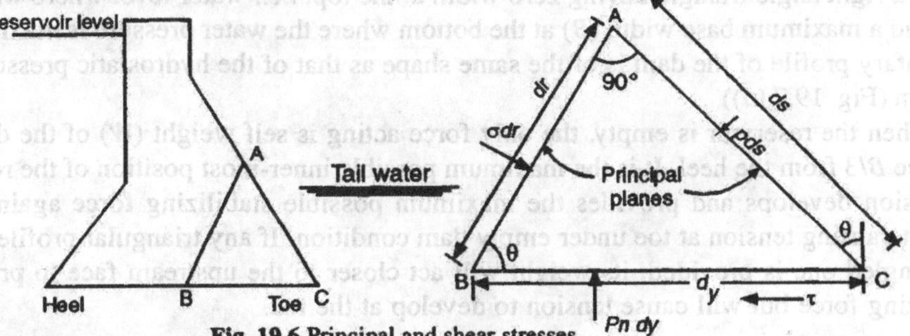

Fig. 19.6 Principal and shear stresses.

Consider unit length of the dam, the normal forces on the planes *AB*, *BC* and *CA* are respectively, σd_r, $p_n\, d_y$ and $p.ds$. Resolving all the forces in the vertical direction, we get :

$$p_n\, d_y = p.ds.\sin\theta + \sigma\, dr \cos\theta$$

But $$d_r = d_y \cos\theta \text{ and } d_s = d_y \sin\theta$$

$\therefore$ $$p_n\, d_y = p.d_y.\sin^2\theta + \sigma\, d_y \cos^2\theta$$

or $$p_n = p\sin^2\theta + \sigma\cos^2\theta$$

Hence $$\sigma = (p_n - p\sin^2\theta)/\cos^2\theta$$

or $$\sigma = p_n \sec^2\theta - p\tan^2\theta \quad ...(i)$$

The principal stress relationship given by equation (*i*) is applicable to both upstream and downstream faces of the dam. However, the worst condition for the downstream face will be when there is no tail water, and hence *p* will be zero. In such case, the principal stress σ_1 is given by

$$\sigma_1 = p_n \sec^2\theta$$

Considering hydrodynamic pressure intensity P_e of tail water due to an earthquake, the principal stress at downstream face is given by

$$\sigma = P_n \sec^2\theta - (P - P_e)\tan^2\theta$$

Similarly considering hydrodynamic force to u/s, principal stress at d/s face is given by

$$\sigma = P_n \sec^2\theta - (P + P_e)\tan^2\theta$$

However, for u/s face σ_1 is always less than *P*. Hence σ is minor principal stress and *P* is major principal stress for the upstream side.

Shear Stress: Resolving all the forces in horizontal direction, we get:

$$\tau d_y = \sigma\, d_r \sin\theta - p\, ds \cos\theta$$

$$\tau = \sigma\sin\theta\, d_r/d_y - p\, d_s/d_y \cos\theta$$

Substituting $\sin\theta$ for d_s/d_y and $\cos\theta$ for d_r/d_y

$$\tau = (\sigma - p)\sin\theta\cos\theta$$

Substituting σ from Eqn. (*i*), we get

$$\tau = (P_n \sec^2\theta - p\sec^2\theta)\cos\theta\sin\theta$$

or $$\tau = (P_n - p)\tan\theta. \quad (ii)$$

The equation (*ii*) gives shear stress at the *d/s* face only. For u/s side, the magnitude of τ remains the same but the direction is reversed. Hence for upstream, $\tau = -(P_n - p)\tan\theta$ ...(*iii*)

The shear stress at downstream face, if tailwater is neglected is given by

$$\tau = P_n \tan\theta$$

Considering the hydrodynamic pressure due to earthquake, the shear stress of *d/s* is given by

$$\tau = [P_n - (P - p'_e)\tan\theta]$$

Similarly, shear stress for u/s face $= \tau = -[P_n - (P + P_e)]\tan\theta$.

19.7 ELEMENTARY PROFILE OF A GRAVITY DAM

The elementary profile of a gravity dam subjected only to the water pressure on the upstream face is a right angle triangle having zero width at the top, *i.e.*, water level where water pressure is zero and a maximum base width (*B*) at the bottom where the water pressure is maximum. Thus the elementary profile of the dam is of the same shape as that of the hydrostatic pressure distribution diagram (Fig. 19.7 (*a*)).

When the reservoir is empty, the only force acting is self weight (*W*) of the dam acting at a distance *B*/3 from the heel. It is the maximum possible inner-most position of the resultant so that no tension develops and provides the maximum possible stabilizing force against overturning without causing tension at toe under empty dam condition. If any triangular profile other than the right angled one is provided, its weight will act closer to the upstream face to provide a higher stabilizing force but will cause tension to develop at the toe.

Vertical Stress Distribution when Reservoir is Empty (Fig. 19.7)

$$P_{max} = \frac{W}{B}\left(1+\frac{6e}{B}\right)$$

$$P_{min.} = \frac{W}{B}\left(1-\frac{6e}{B}\right)$$

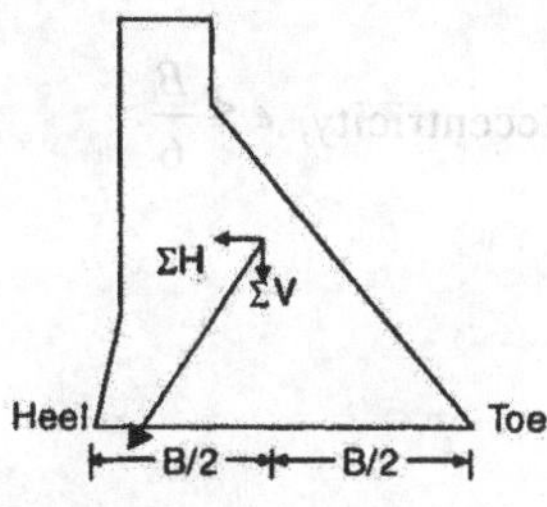

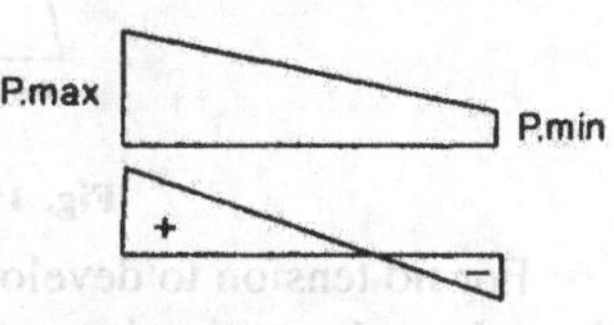

Fig. 19.7. Vertical stress distribution for reservoir empty.

Forces acting on the dam are:

(*i*) Weight of dam, $W = \frac{1}{2} BH \rho w$

where, ρ = unit weight of dam material (2400 kg/m^3 of concrete), and

w = unit weight of water (1000 kg/m^3)

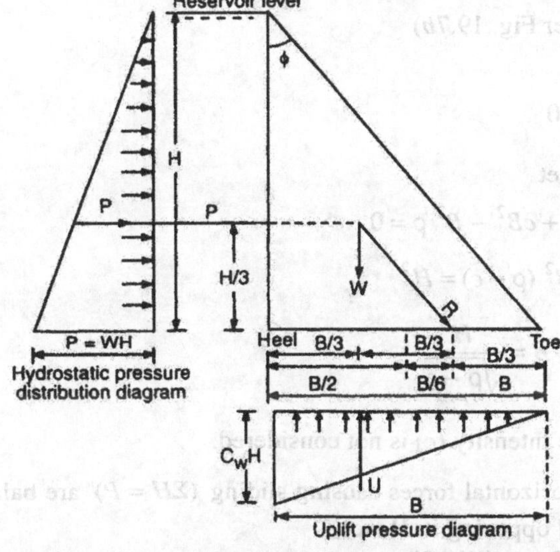

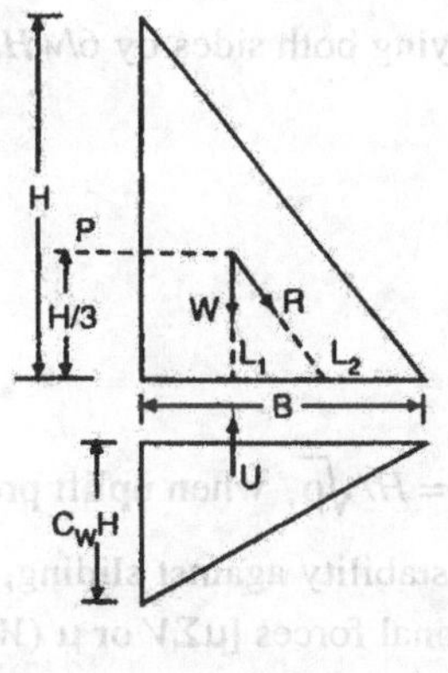

Fig. 19.7. (*a*). Hydrostatic pressure distribution diagram. (*b*). Elementary profile.

(*ii*) Water pressure, $P = \frac{1}{2} wH^2$, acting at *H*/3 from the base

(*iii*) Uplift pressure, $U = \frac{1}{2} cwBH$

where, c = coefficient of uplift pressure intensity, its value is *1*.

Normal stress at toe $P_{max} = \frac{W}{B}\left(1+\frac{6e}{B}\right)$, without considering uplift pressure. (Fig. 19.8)

$P_{max} = \frac{(W - u)}{B}\left(1+\frac{6e}{B}\right)$, with uplift pressure duly considered

Normal stress at heel $P_{min} = \frac{W}{B}\left(1-\frac{6e}{B}\right)$, without uplift pressure

$P_{min} = \frac{(W - u)}{B}\left(1-\frac{6e}{B}\right)$, with uplift pressure duly considered.

Eccentricity, $e \leq \frac{B}{6}$.

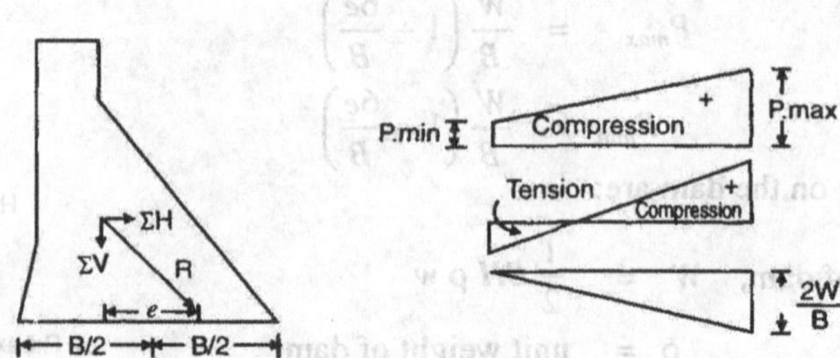

Fig. 19.8 Vertical stress distribution when reservoir is full.

For no tension to develop (*a*) The resultant should pass at outer third point. Taking moments about L_2 and equating it to zero (because moment of R about L_2 is zero), we get

$$P\frac{H}{3} + U\frac{B}{3} - W\frac{B}{3} = R \times 0 \text{ (refer Fig. 19.7}b\text{)}$$

or $$\frac{wH^2}{2}\frac{H}{3} + \frac{1}{2}cwBH\frac{B}{3} - \frac{1}{2}BH\rho w\frac{B}{2} = 0$$

Multiplying both sides by $6/wH$, we get

$$H^2 + cB^2 - B^2\rho = 0$$

$$B^2(\rho - c) = H^2$$

$$B = \frac{H}{\sqrt{\rho - c}}$$

Also $B = H/\sqrt{\rho}$, when uplift pressure intensity (c) is not considered.

(*b*) For stability against sliding, the horizontal forces causing sliding ($\Sigma H = P$) are balanced by the frictional forces [$\mu\Sigma V$ or $\mu(W - \mu)$] opposing it. Hence

$$P = \mu(W - U)$$

or $$\frac{1}{2}wH^2 = \mu\left(\frac{1}{2}BH\rho w - \frac{1}{2}cBwH\right)$$

or $$B = \frac{H}{\mu(\rho - c)}$$

or $$B \geq \frac{H}{\mu(\rho - c)}$$

$$\geq \frac{H}{\mu(\rho - 1)}, \text{ if } C = 1$$

Also, $B = \frac{H}{\mu\rho}$ when uplift is not considered.

Base width B should be maximum of the two values given by the above two conditions.

Normal stress, $P_n = \frac{(W - u)}{B}\left(1 + \frac{6e}{B}\right)$, at the toe

Under limiting condition $e = \frac{B}{6}$, therefore

$$P_n = \frac{W - u}{B}(1 + 1)$$

$$= 2\frac{W - u}{B}$$

or
$$P_n = \frac{2}{B}\left(\frac{1}{2}BH\rho w - \frac{1}{2}cBwH\right)$$

or
$$P_n = wH(\rho - c)$$

$$P_n = wH(1 - 1), \text{ at heel}$$

$$= 0$$

Principal stress near the toe (σ) is the maximum normal stress in the dam, that is,

$$\sigma_1 = P_n \sec^2\theta - P' \tan^2\theta$$

or
$$\sigma_1 = P_n \sec^2\theta, \text{ when there is no tail water } (P' = 0)$$

$$\sigma_1 = wH(\rho - c)\left[\left(\frac{B}{H}\right)^2 + 1\right]$$

But,
$$B = \frac{H}{\sqrt{\rho - c}} \text{ or } B^2 = \frac{H^2}{\rho - c} \text{ or } \frac{B^2}{H^2} = \frac{1}{\rho - c}$$

or
$$\sigma_1 = wH(\rho - c)\left[\frac{1}{\rho - c} + 1\right]$$

or
$$\sigma_1 = wH(\rho - c + 1)$$

Also
$$\sigma_1 = \gamma_w H(\rho - c + 1) \text{ (in S.I. units)}$$

Shear stress (τ) at a horizontal plane near the toe,

$$\tau = P_n \tan\theta$$

$$\tau = wH(\rho - c)\tan\theta$$

$$\tau = wH(\rho - c)\left(\frac{B}{H}\right)$$

But,
$$B = \frac{H}{\sqrt{\rho - c}} \text{ or } \frac{B}{H} = \frac{1}{\sqrt{\rho - c}}$$

$\therefore$
$$\tau = wH(\rho - c)\frac{1}{\sqrt{\rho - c}}$$

$$\tau = wH\sqrt{\rho - c}$$

The principal stress and shear stress at the heel are zero since normal stress at the heel is zero.

19.8 LOW AND HIGH GRAVITY DAMS

Low Dam. A low dam is of limiting height such that the resultant of all forces passes through the middle third and the maximum compressive stress at the toe does not exceed the permissible limit, *i.e.*,

$$\sigma_1 = wH\left(\rho - c + 1\right) = f$$

or limiting height,
$$H = \frac{f}{w(\rho - c + 1)}$$

or
$$H = \frac{f}{w(\rho + 1)},$$ ignoring uplift to be on the safe side.

Also
$$H = \frac{f}{\gamma_w(\rho + 1)}$$ (in S.I. units)

The limiting height, for the usual stress of dam material, viz. $w = 1000 \text{ kg/m}^3$ $\rho = 2400 \text{ kg/m}^3$ and $f = 30 \text{ kg/cm}^2$ (or 300 tonne/m^2), gives

$$H = \frac{300}{1.0(2.4 + 1)} \approx 88 \text{ m}$$

Also
$$H = \frac{300 \; kN/m^3}{9.81(2.4+1)} \simeq 90 \text{ m}$$

High Dam: A dam the height of which exceeds the limiting height of low dam is termed as high dam. In a high dam the maximum permissible compressive stress is exceeded if the resultant of all forces were to pass through the middle third; to avoid excessive stresses the resultant is maintained still near the centre of the base for which purpose the downstream slope is flattened and the upstream slope is also provided with a batter. In this case, the resultant of water pressure and self weight of the dam passes out of the middle third at certain sections in the dam body indicative of development of tension. The dam profile has, therefore, to be adjusted by adding extra slope or batter on the upstream and downstream sides to limit the compressive stress within the permissible limit (Fig. 19.10).

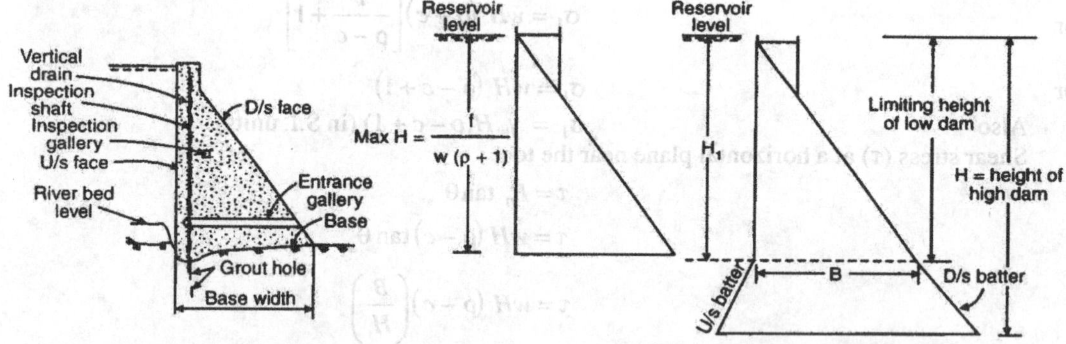

Fig. 19.9. Gravity dam of triangular section.

Fig. 19.10. Low dam and high dam.

Practical Profile of a Dam

The elementary profile of a dam composed of a triangle with apex at the maximum water surface is only a theoretical one and has to be modified from practical considerations; *(a)* Dam has to be raised above the maximum reservoir level to provide suitable free board to guard against overtopping by wave action, *(b)* Top of dam has to be of a certain width, instead of zero width of the elementary triangular profile, to provide suitable roadway.

The modifications due to *(a)* and *(b)* above cause the resultant to shift towards the heel crossing the inner middle third (under empty dam condition) indicative of development of tension at the toe.

Batter has, therefore, to be provided on the upstream face to avoid development of tension. Figs. 19.9 and 19.10 show the profile of low dam and high dam respectively modified from practical considerations. Various considerations in fixing the parameters of the dam section are as under:

1. *Top Width.* The top width of the dam is generally dictated by the requirement of roadway to be provided. The most economical top width is 14% of the dam height. It is also taken equal to 0.55 $H^{1/2}$, where H is the maximum water depth. Usually the width provided varies from 6 to 10 m.

2. *Free Board.* The upstream parapet of the roadway above the top of dam is usually a solid wall but does not form part of free board. The free board provided is maximum of (*i*) $1.33h_w$, (refer Fig. 19.5), (*ii*) 3 to 4%, usually 5% of the dam height, whichever is more.

3. *Base Width.* The base width of the dam shall be safe against overturning, sliding, and no tension in dam body. The base width (B) is determined from the equation, $B = H/\sqrt{(\rho - c)}$. The upstream face is kept vertical up to a height H_1 to be determined by trial. Approximately H_1 is equal $2a\sqrt{(\rho - c)}$, where a is top width of the dam. Below height H_1 batter is provided to upstream and downstream faces to increase the base width with a view to ensure that no tension develops in the body of the dam. For this, the resultant force shall be located within the inner middle third and outer middle third for reservoir in empty and full conditions respectively. Fig. 19.11 shows dimensioned practical profile of a gravity dam. Usually upstream slope starts from MWL or FRL with 0.15:1, varying between 0.05 to 0.15, slope. Likewise, the downstream slope starts at about 5 to 7m below top of dam with only one slope if the dam height is less than 30 m. Between 30 to 50m height, two slopes can be provided for economy without affectin ease of construction.

SOLVED EXAMPLES

Example 19.1 *What should be the maximum height of elementary profile of a dam, if the safe limit of stress on the masonry should not exceed 350 tonnes/m²? Assume weight of masonry 2.4 tonnes/cum. (a) Determine base width also. (b) Determine H and B, if uplift intensity factor is 0.67 and factor of safety is 2.*

Solution. The limiting height of elementary profile of a masonry dam,

$$H = \frac{f}{w(\rho + 1)}$$

Putting given values

$$H = \frac{350}{1(2.4 + 1)} = 103\,\text{m}$$

Base width

$$B = \frac{103}{\sqrt{e}} = \frac{103}{\sqrt{2.4}} = 66\ m$$

(*b*)

$$H = \frac{350}{2 \times (2.4 - 0.67 + 1)} \approx 64\text{m}$$

$$B = \frac{64}{\sqrt{2.4 - 0.67}} \approx 49\ \text{m}$$

Example 19.2 *(a) Draw uplift pressure diagram for a dam holding 50 m water depth at upstream vertical face, top and bottom width 10 m and 30 m respectively. Uplift may be considered to be acting on 60% of the area of section. Tail water depth is 5 m.*

(b) Also if there is a drainage gallery at 6 m from face.

Solution. (*a*) Uplift at heel = 50 × 0.6 = 30 m

Uplift at toe = 5 × 0.6 = 3.0 m

Pressure diagram is shown in Fig. 19.11.

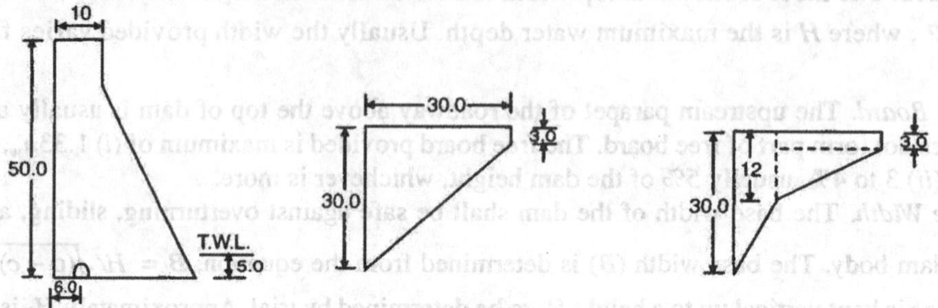

Fig. 19.11. Uplift pressure diagram. **Fig. 19.12 Uplift pressure diagram with drainage gallery.**

(*b*) Uplift at drainage gallery line = 3.0 + 1/3 (30.0 – 3.0) = 12.0 m.

Pressure diagram is indicated in Fig. 19.12.

Example 19.3 *A concrete gravity dam has maximum water level 305.0 m, bed level 225.0 m, top R.L. of dam 309.0 m, d/s face slope starts at R.L. 300.0 m, d/s slope 2 : 3, tail water is nil, upstream face of dam is vertical, centre line of drainage gallery is 8 m d/s of u/s face, uplift pressure is 100% at heel, 50% at line of gallery and zero at toe, weight of concrete is 2.4 t/m³. Considering only weight, water pressure and uplift, determine (i) Maximum vertical stresses at the toe and heel of dam, (ii) Major principal stresses at toe of dam, and (iii) Intensity of shear stress on a horizontal plane near the toe.*

Solution.

Height of dam = 309.0 – 225.0 = 84 m

Depth of water = 305.0 – 225.0 = 80 m

1. Top width of dam:

14% of the height = 0.14 × 84 ≈ 12 m or $0.55\ H^{1/2} = 0.55\ (80)^{1/2} \approx 5$ m

Adopt top width of dam as 12 m

2. Bottom width of dam = 12.0 + (300.0 – 225.0) × 2/3 = 62 m

(*a*) Consider 1 m length of dam and $w = 1$ t/m³. (Referring to Fig. 19.13)

Designation	*Force*	*Lever arm (m)*	*Moment about toe (t.m)*
Weight of dam			
$W_1 = 12 \times 84 \times 1 \times 2.4$	= 2419 t	50 + 12/2 = 56	135464
$W_2 = 50 \times 75/2 \times 1 \times 2.4$	= 4500 t	50 × 2/3 = 33.33	149985
$W = W_1 + W_2$	= 6919 t		ΣM_1 = (+) 285449
Uplift $U_1 = 40 \times 8/2$	= 160 t	54 + 8 × 2/3 = 59.33	9493
$U_2 = 40 \times 8$	= 320 t	54 + 8/2 = 58.0	18560
$U_3 = 54 \times 40/2$	= 1080 t	54 × 2/3 = 36.0	38880
ΣU	= 1560 t		ΣM_2 = (–) 66933 t.m

Water Pressure

$$P = \frac{Wh^2}{2} = \frac{1}{2} \times 1 \times 80 \times 80 = 3200 \qquad \frac{80}{3} = 26.67 \qquad \Sigma M_3 = (-)\ 85344 \text{ t.m}$$

$$V = W - u = 6919 - 1560 = 5359 \text{ t} \qquad M = M_1 - M_2 - M_3 = 285449 - 66933 - 85344 = (+)\ 133172 \text{ t.m}$$

Position of resultant from toe, $\bar{x} = \frac{M}{V} = \frac{133172}{5359} = 24.85$ m

Eccentricity, $e = \frac{62}{2} - 24.85 = 6.15$ m

Normal compressive stress at toe, $P_n = \frac{V}{B}\left(1 + \frac{6e}{B}\right)$

$$= \frac{5359}{62}\left(1 + \frac{(6 \times 6.15)}{62}\right) = 138\text{ t/m}^2$$

Normal compressive stress at heel, $P'_n = \frac{V}{B}\left(1 - \frac{6e}{B}\right)$

$$= \frac{5359}{62}\left(1 - \frac{(6 \times 6.15)}{62}\right) = 35\text{ t/m}^2$$

(*b*) Principal stress at toe, $\sigma = P_n \sec^2 \alpha - P' \tan^2 \alpha$

= Here, P' is zero, there being no tail water,

$\tan \alpha = 2/3$

$$\sec^2 \alpha = 1 + \tan^2 \alpha = 1 + \left(\frac{2}{3}\right)^2 = \frac{13}{9}$$

Therefore, $\sigma = P_n \sec^2 \alpha = 138 \times 13/9 = 199\text{ t/m}^2$.

(*c*) Intensity of shear stress on a horizontal plane near toe, $\tau_0 = (P_n - P') \tan \alpha$

$= (P_n - 0)\ 2/3$, P' is zero there being no tail water

$= 138 \times 2/3 = 92\text{ t/m}^2$.

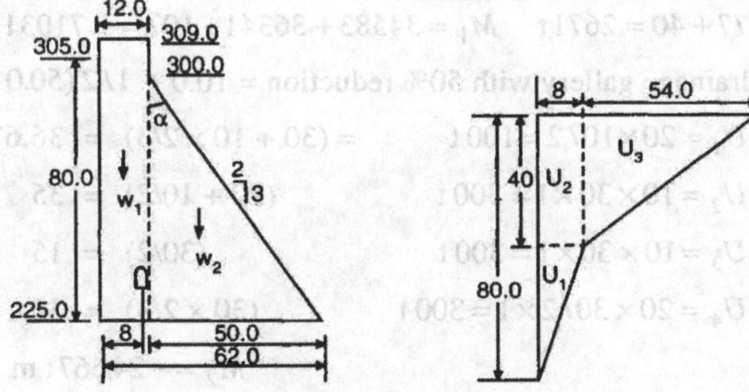

Fig. 19.13 Uplift pressure diagram.

EXAMPLE 19.4 *A concrete gravity dam has maximum reservoir level 150.0 m, base level of dam 100.0 m, tail water elevation 110.0 m, base width of dam 40 m, location of drainage gallery 10 m from u/s face which may be assumed as vertical. Compute hydrostatic thrust and uplift force per metre length of dam at its base level. Assume 50% reduction in net seepage head at the location of the drainage gallery.* (*AMIE*)

Solution. Consider 1 m length of dam (Fig. 19.18).

Free board = 5% of dam height = 0.05 × 50 = 2.5 say 3 m

Dam height = 50 + 3 = 53 m

Top width $= 0.14H$ or $0.55H^{1/2}$, where H is height of the dam

$= 0.14 \times 53 = 7.5$ or $0.55\ \sqrt{53}\ 1/2 = 4$ m, adopt 7.5 m

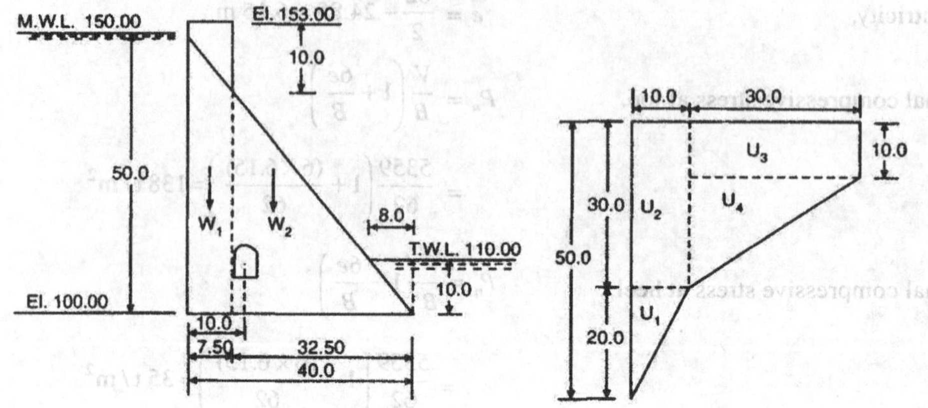

Fig. 19.14 Uplift pressure diagram.

Designation	*Force (t)*	*Lever arm (m)*	*Moment about toe (t.m)*
Weight of masonry			
$W_1 = 7.5 \times 53 \times 1 \times 2.4$	= 954	$\left(32.5 + \frac{7.5}{2}\right) = 36.25$	34583
$W_2 = 32.5 \times \frac{43}{2} \times 1 \times 2.4$	= 1677	$\left(32.5 \times \frac{2}{3}\right) = 21.67$	36341
Weight of tail water			
$W_3 = 8 \times \frac{10}{2} \times 1 \times 1$	= 40	$\left(8 \times \frac{1}{3}\right) = 2.67$	107

$W = 954 + 1677 + 40 = 2671$ t $\quad M_1 = 34583 + 36341 + 107 = +\ 71031$ t.m

Uplift pressure at drainage gallery with 50% reduction = 10.0 + 1/2 (50.0 – 10.0) = 30 t

Uplift pressure $\quad U_1 = 20 \times 10/2 = 100$ t $\quad = (30 + 10 \times 2/3) = 36.67 \quad$ 3667 t.m.

$U_2 = 10 \times 30 \times 1 = 300$ t $\quad (30 + 10/2) = 35 \quad$ 10500 t.m.

$U_3 = 10 \times 30 \times 1 = 300$ t $\quad (30/2) = 15 \quad$ 4500 t.m.

$U_4 = 20 \times 30/2 \times 1 = 300$ t $\quad (30 \times 2/3) = 20 \quad$ 6000 t.m.

$U = 1000$ t $\quad M_2 = -24{,}667$ t.m.

$V = W - U = 2671 - 1000 = 1671$t, $M = M - M_2 = 71{,}031 - 24{,}667 = 46{,}364$t.m.

Water pressure, u/s face $\quad P = \frac{wh^2}{2} = \frac{(1 \times 50 \times 50)}{2} = 1250 \quad \frac{50}{3} \quad (-)\ 20{,}833$

d/s face, $P' = \frac{wh'^2}{2} = 1 \times 10 \times \frac{10}{2} = 50 \quad \frac{10}{3} \quad +\ 167$

Pressure acting downstream = 1250 – 50 = 1200, $M_3 = (-)\ 20833 + 167$

$= (-)\ 20{,}666$

Net moment = 46,364 – 20,666 = 25,698 t.m.

Position of resultant from toe, $\bar{x} = \dfrac{M}{V} = \dfrac{25698}{1671} = 15.38 \text{ m}$

Eccentricity, $e = \dfrac{B}{2} - \bar{x} = \dfrac{40}{2} - 15.38 = 4.62 \text{ m}$

Normal compressive stress at toe, $P_n = \dfrac{V}{B}\left(1 + \dfrac{6e}{B}\right)$

$$= \frac{1671}{40}\left(1 + 6 \times \frac{4.62}{40}\right) = 70.73 \text{ t/m}^2$$

Normal compressive stress at heel $= \dfrac{V}{B}\left(1 - \dfrac{6e}{B}\right)$

$$= \frac{1671}{40}\left(1 - \frac{6 \times 4.62}{40}\right) = 12.82 \text{ t/m}^2$$

Maximum principal stress at toe, $\sigma = P_n \sec^2 \alpha - P' \tan^2 \alpha$, here $P' = 10$

$\tan \alpha = 40/50$ and $\sec^2 \alpha = 1.64$

Therefore, $\sigma = 70.73 \times 1.64 - 10(40/50)^2 = 110 \text{ t/m}^2$

Intensity of shear stress on a horizontal plane near toe,

$\tau_0 = (P_n - P') \tan \alpha$

$\tau_0 = (70.73 - 10)40/50 = 49 \text{ t/m}^2$

QUESTIONS

1. (*a*) What preliminary investigations should be conducted for the selection of a dam site?
   (*b*) Under what conditions would you adopt a gravity dam and an earthen dam?
2. (*a*) Prove that the maximum shear stress intensity in an elementary profile is twice that the average.
   (*b*) Determine the safe height of a dam having elementary profile for the maximum compressive stress of 140 t/m^2 with an uplift intensity factor of 0.67.
3. How will you check the stability of the gravity dam section for reservoir full condition.
4. Define gravity dams. Enumerate the points considered while selecting site for the construction of a gravity dam.
5. Give the list of the forces affecting the stability of a gravity dam.
6. State the design criteria of a concrete gravity dam and explain briefly how you will determine the various factors of safety against possible failures.
7. Discuss the main principles of design of a masonry gravity dam. Explain how the design profile is evolved.
8. Explain how uplift considerations affect the design of a gravity dam. What measures can be adopted to reduce the undesirable effects due to uplift in such cases?
9. Write a brief note on the necessity and methods of foundation treatment of dams.
10. Discuss the evolution of the final profile of a gravity dam from its elementary triangular profile and explain the main principles of its design.
11. How are pore pressures and uplift pressures controlled in rigid dams?
12. What is meant by elementary profile of a gravity dam and how is it determined? What should be the maximum height of elementary profile of a dam, if the safe limit of stress on the masonry were not to exceed 150 t/m^2? Assume weight of masonry 2.4 t/m^3.

20

ARCH AND BUTTRESS DAMS

20.1 ARCH DAM

An arch dam is just a curved beam, the ends of which are restrained and the way in which the loads are resisted is termed as *arch action*. Arch dam may thus be defined as a curved masonry or concrete dam, convex upstream, which depends principally on arch action for its stability; itself resisting part of pressure of the reservoir by its own weight. The arch mostly transfers horizontally the water load to the canyon walls or abutments depending on the amount of curvature and as a result considerable economy is effected in the quantity of material used compared to gravity dams. Since the arch can be made to abut against the walls of the canyon under the hydrostatic loads acting on its extrados face, arches are efficiently employed as structural members in dams where the valley is narrow, the usual recommended ratio of crest length to dam height being 5 or less. For the arch action to be more effective, the rock constituting ends of arch should not yield. It follows that the abutments should be of sound rock. Uplift on the base of dam is not of importance in arch dams. Fig. 20.1 shows section of Idikki arch dam.

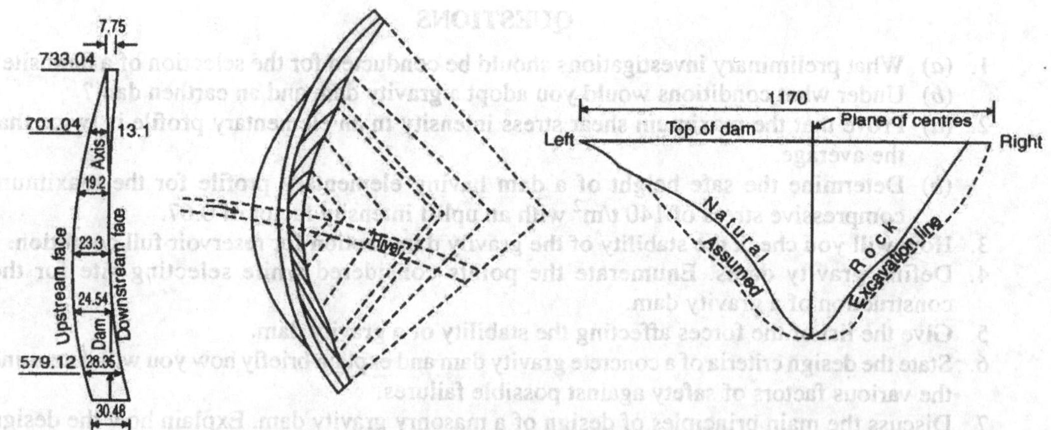

Fig. 20.1 Idikki arch dam.

20.2 TYPES OF ARCH DAMS

Arch dams are classified on the basis of thickness, symmetry with respect to crown section or characteristics of extrados and intrados curves but mainly as constant radius, variable radius or constant angle types.

1. Constant Radius Arch Dam: A constant radius arch dam is an arch dam, the arch centre of which for the upstream face, the downstream face and the centre line are all coincident with the axis centre at all elevations (Fig. 20.2). In simple types the upstream face is vertical so that its radius at all elevations is constant but thickness is proportional to the height. As a result the central angles of the arch rings of the intrados vary at various elevations, *i.e.*, goes on reducing from top to base, the maximum being at the top of dam and a certain minimum at the bottom of the dam (Fig. 20.2). This type of dam is triangular in cross-section with increased thickness at the crest. It is particularly adapted to U-shaped canyon where cantilever action carries a relatively large proportion of water load at lower elevations. Form work required for construction is simple. However, it is less economical compared to constant angle arch dam.

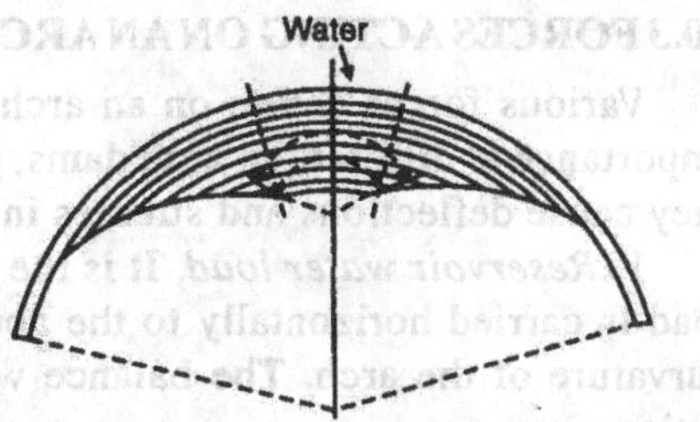

Fig. 20.2. Constant radius arch dam.

2. Variable Radius Arch Dam. Also termed as variable centre arch dam. In this type of dam, both the radius and central angle are not constant (Fig. 20.3). It is a compromise between the constant radius and the constant angle arch dam. The characteristic of this type of dam is variable radius, *i.e.*, radii of extrados surface or arch rings corresponding to upstream face and intrados surface or arch rings corresponding to downstream face, vary at various elevations; maximum at the top and minimum at the base. The radii gradually decrease as the depth below the crest increases. The central angle is as large and as nearly constant as possible to obtain maximum arch efficiency at all elevations. Central angle is different for different arches and usually varies from 180° to 150°. The upstream face is generally vertical or even overhanging at the upstream side near abutments and that at the downstream side near the crown. The centres of the horizontal arch rings at various elevation do not lie on a single vertical line, rather the centres of different radii of extrados or intrados fall on a smooth curve which has lent it the name of variable centre arch dam. This type of dam is suitable for narrow V or U-shaped valleys as arch action can be depended at lower levels.

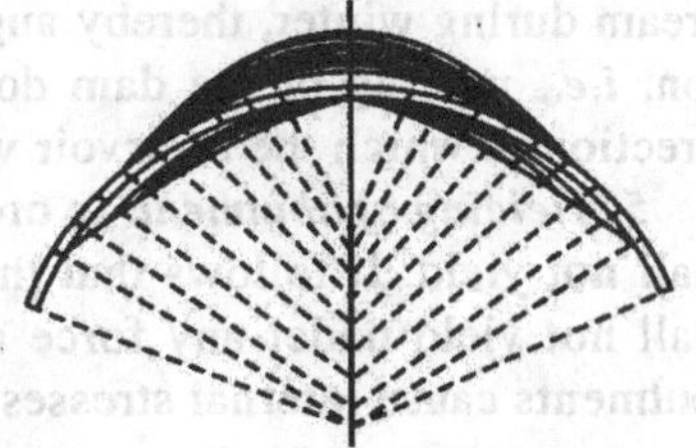

Fig. 20.3 Variable centre arch dam.

Advantages. (*i*) Greater arch efficiency, (*ii*) Saving in concrete; over 50% saving in concrete compared to constant radius arch dam, (*iii*) Useful to meet foundation requirements, and (*iv*) Preferred than constant radius arch dam.

3. Constant Angle Arch Dam. It is a type of arch dam in which all the arch rings have a constant angle (Fig. 20.4). It is thus an improvement on variable radius type arch dam in that the central angle of the horizontal arch rings at all elevations is constant. The central angle of the arch for most economical arch section is usually 133°. In a U-shaped valley, central angle of about 133° can be adopted but a V-shaped valley gives very low angle at the bottom. The constant angle arch dam is not easy to adapt to existing contour lines. However, it is most economical, 30 to 40 per cent saving in concrete compared to constant radius arch dam.

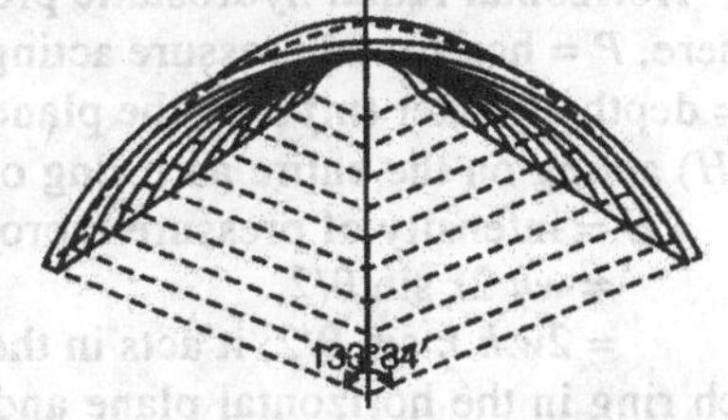

Fig. 20.4 Constant angle arch dam.

Modern dams are often neither constant angle, nor constant radius dams, but thin shells adapted to the shape of the valley.

20.3 FORCES ACTING ON AN ARCH DAM

Various forces acting on an arch dam are the same as for gravity dam but their relative importance is different in arch dams, as under. The temperature changes are very important as they cause deflections and stresses in arch dams.

1. *Reservoir water load.* It is the principal live load acting on the dam. Most of the water load is carried horizontally to the abutments by arch action depending on the amount of the curvature of the arch. The balance water load is transferred to the foundation by cantilever action.

2. *Uplift pressure.* Unlike gravity dams, uplift pressure in arch dams is not so important because arch dams are very thin section at the base. However, if cracks occur, uplift in cracks results in increase in downstream deflection, changes in load distribution accompained by increases in maximum compressive stresses in the arch and cantiliver elements.

3. *Ice pressure.* Ice pressure causes heavy concentrated loading on top arch, more precisely along the arch element at the level of the ice. The internal stresses caused by ice pressure are important in arch dams.

4. *Temperature changes.* Internal stresses caused by temperature changes are important, their effect being to (*i*) move the dam upstream during summer and (*ii*) to move the dam downstream during winter, thereby augmenting reservoir loads. In stress analysis, the latter condition, *i.e.*, moving of the dam downstream during winter is very important as it acts in the direction in which the reservoir water load acts.

5. *Yielding of abutment.* In order that the arch section is more effective, the ends of the arch shall not yield. It follows that the rock of the canyon constituting the abutments of the arch shall not yield under any force and movement developed at that section. Slight yielding of abutments cause internal stresses in the arch.

20.4 DESIGN OF ARCH DAM

The various methods commonly used for the design of arch dam are as under:

1. Thin Cylinder Theory. In thin cylinder theory of arch design it is assumed that (*i*) The arch is cylindrical, the restraint being provided in its radial movement, (*ii*) Stresses in the arch are approximately the same as in a thin cylinder of equal outside radius, (*iii*) Pressure on the ring is equal to the full hydrostatic pressure at the elevation, (*iv*) Weight of concrete and water is directly carried to the foundation, (*v*) Horizontal water thrust is taken up entirely by the arch action, and (*vi*) Stress distribution across arch thickness is uniform. Fig. 20.5 shows the section of arch as part of a cylinder of mean radius *r*, thickness *t* and central angle θ.

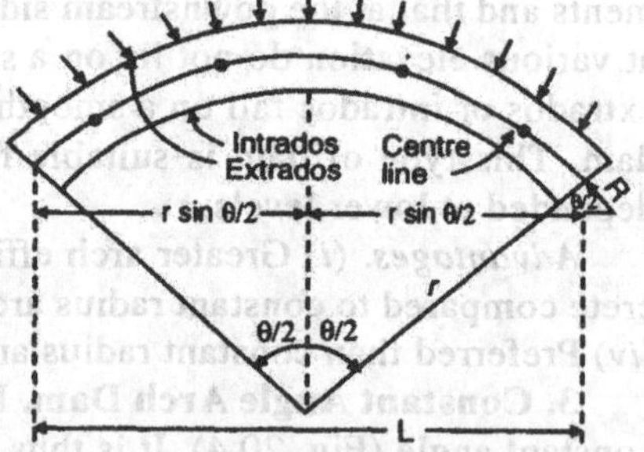

Fig. 20.5 Thin cylinder theory.

Horizontal radial hydrostatic pressure, $P = wh$ (t/m²)

where, P = horizontal pressure acting in radial direction, w = unit weight of water (1 t/m³), and h = depth of water (m) over the plane in consideration. The total hydrostatic by pressure force (PH) acting on the entire arch ring of unit height is given :

PH = intensity of pressure × projected area

$= wh\, 2r \sin \theta/2,$

$= 2w.h.r. \sin \theta/2$; it acts in the downstream direction along the axis of symmetry of the arch ring in the horizontal plane and is balanced by the components of reactions at the abutments in the upstream direction.

If R is the reaction at each abutment, its component acting upward is $2R \sin \theta/2$.

For equilibrium, $2R \sin \theta/2 = 2.w.h.r. \sin \theta/2$

or $$R = whr$$

This reaction at the abutments is equal to the maximum permissible compressive stress σ introduced in the arch. Therefore,

$$\sigma = R/t = \frac{whr}{t}$$

or
$$t = \frac{whr}{\sigma}$$

Also
$$t = \frac{w.h.r_i}{\sigma - wh}, \text{ where } r_i \text{ is intrados radius}$$

$$t = \frac{w.h.r_m}{\sigma - 0.5\,wh}, \text{ where } r_m \text{ is mean radius}$$

Central angle. The central angle of the top arch influences the curvature of the entire dam. The central angle is selected from the consideration of economy. For a minimum volume of concrete in the dam, the surface of each individual arch element is made small consistent with low stresses:

$$V = \theta rt, = \frac{whr^2\theta}{\sigma}, t \text{ being} = \frac{whr}{\sigma}$$

$$l = 2r \sin \theta/2$$

$$r = l/2 \sin \theta/2$$

Therefore,
$$V = \frac{\theta l.w.h.l}{2 \sin \theta/2 \; \sigma 2 \sin \theta/2}$$

$$V = \frac{\theta l^2\, wh}{4\sigma \sin^2 \theta/2}$$

$$V = \frac{whl^2\, \theta}{4\sigma \sin^2 \theta/2}$$

For minimum volume,
$$\frac{dV}{d\theta} = 0$$

or
$$\frac{dV}{d\theta} = \frac{\sin^2(\theta/2) - \theta \sin(\theta/2)\cos(\theta/2)}{\sin^4(\theta/2)} = 0$$

or
$$\tan \theta/2 = \theta$$

or
$$\theta = 133° - 34'$$

Thus the most economical angle is 133° 34′. For practical reasons of economy in cost of shuttering, the angle is usually less than most economical theoretical value and may vary between 110° to 125°. A large central angle has the merit that reaction at the support is correctly oriented with respect to the sides of the gorge so that the mass of rock supporting the reaction is sufficient for transmission of the force.

Limitations. The cylinder theory has the limitations that (*i*) The arch sections are not thin cylinders, (*ii*) The arch ring is not free to move at the abutments and the fixing reduces movements and shears, (*iii*) Shearing and bending stresses, temperature stress and ice thrust besides stress due to rib-shortening and that due to yielding of the abutments are not taken into consideration, and (*iv*) The formula for thin cylinders yields unreliable results and is used only as a rough approximation.

2. Thick Cylinder Theory. Contrary to the thin cylinder theory wherein the stress distribution at a cross-section at any level is considered constant due to small thickness of the arch as compared to the radius, in thick cylinder theory of design of arch dam the actual variation of stress is considered. All the radial cross-sections are assumed to be subjected to similar stress-whereas the formula gives higher stress at the intrados at all sections, yet in bottom arches this is true only near the abutments, while at the crown the stresses are reversed. For top arches the conditions are quite the opposite.

3. Elastic Arch Theory. In this theory it is assumed that the entire external load acting on the dam is transferred horizontally to the abutments through arch action only and that the entire dam consists of a number of horizontal arch rings placed one over the other, their ends fixed and each ring is independent of the other. The cantiliver action is neglected. The thin cylinder theory has certain defects as it takes into consideration only uniformly radial pressure, but the elastic arch theory takes into account variable loads along the length of the arch as also ice thrust, etc. In elastic theory of design of arches, additional stresses due to rib shortening, temperature changes, shrinkage and yielding of abutments are also considered. The ends of the arch dam are fixed and when water pressure acts radially from upstream, side rib shortening effect takes place which causes additional stresses. Similarly, variation in temperature, shrinkage and yielding of abutment cause additional stresses.

4. Trial Load Analysis. The assumption in thin cylinder theory and elastic theory that any horizontal arch ring is independent of the arch ring below and above it is not true. The trial load method developed by USBR gives reasonably satisfactory approximation. The trial load analysis hinges on the assumptions that (*i*) Foundation deformations are independent of the shape of the foundation surface, (*ii*) Rock formations in foundation and concrete in dam are homogeneous and isotropic. Modulus of elasticity of concrete and rock foundation is the same for compression and tension, (*iii*) Water pressure on reservoir walls do not cause differential movements at dam site, (*iv*) Direct stresses vary linearly from upstream to downstream faces, (*v*) Plane sections before bending remain plane after bending, (*vi*) Hooke's law is applicable, and (*vii*) Temperature strains are proportional to temperatures.

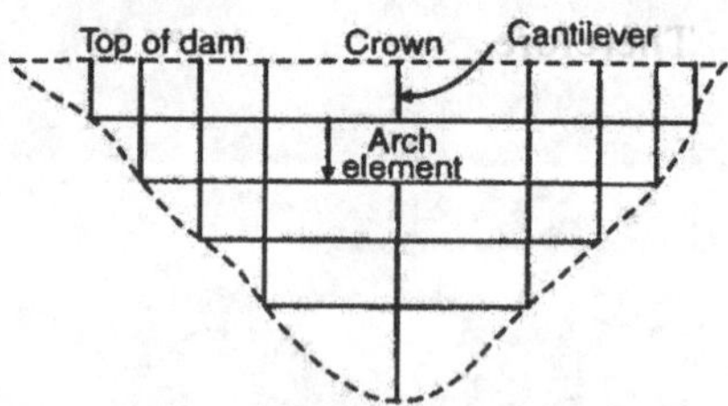

Fig. 20.6 Trial load analysis

The dam is assumed to be made up of two systems of elements, viz. horizontal arches transmitting thrust to abutments and vertical cantilevers fixed at the foundation, both the horizontal and vertical elements are considered a unit distance apart and representative units are selected for analysis (Fig. 20.6). Each system occupies the whole dam structure and the loading is assumed to be divided between them in such a manner that equal movements result in all directions at points of intersection of these elements. The horizontal component of water load is resisted jointly by arch action and cantilever action. The arches are considered to carry part of water pressure and total load due to other forces, while the cantilever takes the remaining water pressure and total load due to other forces. The distribution of load between the arch system and cantilever system is determined by trial and error. The stresses in the dam are computed on the basis of final load distribution. The dam dimensions fixed by the formentioned methods are analysed by this method and modified to arrive at final design. The trial load method is widely used in USA for arch dam designs; the computations are programmed for high speed electronic computers.

5. Finite Element Method. The method has the underlying principle to represent a continuous 2-dimensional system as an assemblage of finite 2-dimensional flat elements interconnected at a finite number of nodal points. The elements may be rectangular as in cylindrical shells or triangular as in curved shells. In the analysis, the stiffness of individual elements are first determined and then superimposed to obtain total stiffness.

Preliminary Design of Arch Dam

Salient dimensions of double curvature arch dam as per USBR practice are as under:

(*i*) **Thickness at crest (T_C)** $= 0.01\,(H + 1.2\,L_1)$ m

(*ii*) **Thickness at base (T_B)** $= 3\sqrt{0.0012\,H\,L_1\,L_2\left(\dfrac{H}{122}\right)^{\frac{H}{122}}}$ m

(*iii*) **Thickness at 0.45 H, $T_{0.45H}$** $= 0.95\ T_B$

(*iv*) **Crown cantilever projections**

(*a*) **At crest**

Maximum u/s projection, (USp) $= 0.0$

Maximum d/s projection (DSp) $= T_c$

(*b*) **At base**

USp $= 0.67\ T_B$

DSp $= 0.33\ T_B$

(*c*) **at 0.45 H**

USp $= 0.95\ T_B$

DSp $= 0.0$

Concrete volume (v) $= V_1 + V_2$ m³

where,

$$V_1 = 0.000177\ H^2.L_2\left[\frac{(H + 0.8L_1)^2}{L_1 - L_2}\right]$$

$$V_2 = 0.0108\ H.L_1\ (H + 1.1\ L_1)$$

The upper end limits of above equations are :

$$30 \le H \le 370$$
$$30 \le L_1 \le 1825$$
$$4.5 \le L_2 \le 370$$

Fig. 20.7 illustrates the dimensions of a double curvature arch dam as per USBR practice. L_1 is straight line distance at crest level between abutments excavated to sound rock and L_2 is the straight line distance at 0.15 H above base, between abutments excavated to sound rock.

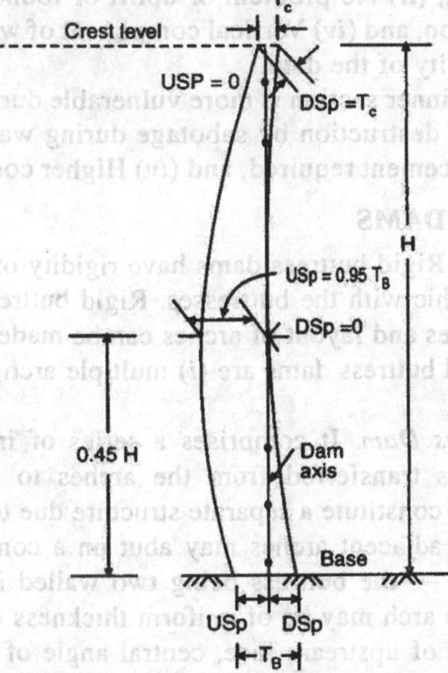

Fig. 20.7 Typical double curvature arch dam.

20.5 BUTTRESS DAMS

In a gravity dam the quantity of concrete is determined by the dead weight required for the stability of the dam, rather than by the strength of the concrete. Thus in gravity dams there are lower stresses in the dam body in central part of the dam and the high strength concrete there is not stressed as it could be, and hence lesser efficient use of concrete strength, large uplift force on dam base and hence increased dam section for stability. A buttress dam envisages to reduce (*i*) The total volume of concrete by concentrating the material where the stresses are higher and (*ii*) Other undesirable features of gravity dam but has higher cement content per cubic metre of concrete than a gravity dam. On steep hill slopes on both sides the buttresses are subjected to severe torsion effects which present formidable design problems requiring heavy anchorages and reinforcement for torsion moment and shear. Spillway arrangement would be difficult and costly.

A buttress dam is defined as a dam consisting of a relatively thin water supporting facing or deck supported by buttresses generally in the form of equally spaced triangular walls or counterforts that transmit the water load and deck weight to the foundations. The upstream face being inclined, usually at an angle of about 45°, water pressure exerted by it add to the stability of the dam. Buttress dams are adaptable to overflow and non-overflow conditions. In overflow dam, another deck at the downstream face is provided to guide the falling flow. The essential components of a buttress dam are:

(*i*) *Deck.* The upstream sloping membrane or deck which supports the water load and transmit it to a series of buttresses or counterforts. The water tight deck is continuous at the foundation, where cutoff is provided to reduce or prevent seepage.

(*ii*) *Buttresses.* The buttresses or counterforts are provided at certain intervals at right angles to the axis of the dam to support the deck and for transference of the load to the foundation. Corbels help transmit load from the upstream deck to buttresses.

(*iii*) *Lateral braces.* A system of cross bracing between buttresses is provided to effect lateral stiffness and resistance to buckling of counterforts.

(*iv*) *Foundation.* Mat foundation provides distribution of the pressure onto the soil.

Advantages. (*i*) Thin section with on- third to one-half economy in concrete compare to gravity dam of similar height, (*ii*) No problem of uplift or foundation drainage, (*iii*) Can be constructed on weak foundation, and (*iv*) Vertical component of water resting against upstream sloping deck add to the stability of the dam.

Disadvantages. (*i*) The thinner section is more vulnerable due to deterioration of concrete, (*ii*) Susceptible to damage or destruction by sabotage during war, (*iii*) Saving in concrete is largely offset by more reinforcement required, and (*iv*) Higher cost of form work.

20.6 TYPES OF BUTTRESS DAMS

1. Rigid Buttress Dams. Rigid buttress dams have rigidity of construction. The upstream sloping deck is made monolithic with the buttresses. Rigid buttress dams are suitable as high dams. The spacing of buttresses and layout of arches can be made to get best advantage of the foundation. Two types of rigid buttress dams are (*i*) multiple arch dam, and (*ii*) multiple dome buttress dam.

(*i*) *Multiple Arch Buttress Dam.* It comprises a series of inclined arches supported by buttresses. The water load is transferred from the arches to the foundation through the buttresses. Each arch does not constitute a separate structure due to the thrust at arch springing which are balanced. The two adjacent arches may abut on a common buttress (Fig. 20.8 *a*), or abut on independent walls — the buttress being two walled interconnected by horizontal connections (Fig. 20.8 *b*). The arch may be of uniform thickness or of variable thickness. The main considerations are slope of upstream face, central angle of each arch (between 150° to 180°) and buttress spacing.

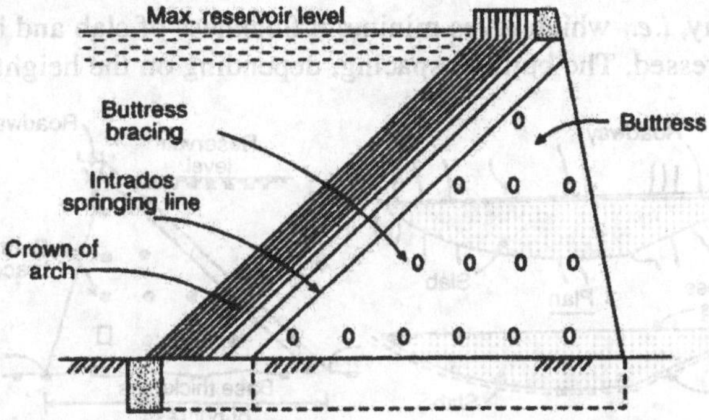

Fig. 20.8 (*a*) Multiple arch dam with adjacent arches abutting on a common buttress.

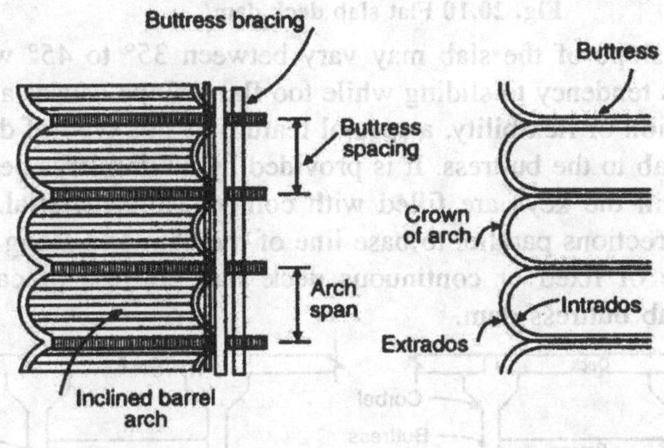

Fig. 20.8 (*b*) Multiple arch dam with buttress interconnected horizontally.

Advantages. (*i*) Economical in cost, (*ii*) Relatively stronger than flat slabs. Thinner deck permits a wide economical spacing of buttresses, (*iii*) Most suited for high dams, (*iv*) Suited for all types of foundations and better adapted to temperature changes.

Disadvantages. (*i*) Buttresses are thicker and more substantial than flat slab buttresses, (*ii*) Thin arches respond rather rapidly to climatic changes, (*iii*) Variation in pressures due to alternate filling and emptying of the reservoir causes frequent reversal of movements tending to reduce the life of the dam, and (*iv*) Unequal settlement of foundation is objectionable. Stability of any unit depends upon the adjacent unit so that any settlement in one may effect other units.

(*ii*) *Multiple Dome Buttress Dam.* It is buttress dam is which massive buttresses spaced far apart support a set or series of domes on which the water rests (Fig. 20.9). The arches act as water retaining membranes stretching across the length of the dam. The arches are under compression.

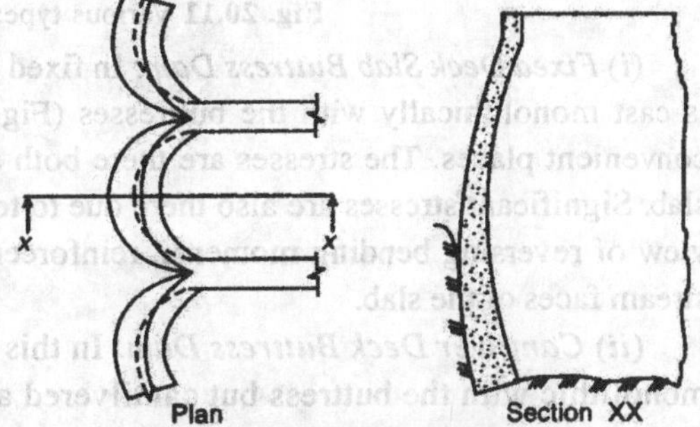

Fig. 20.9 Multiple dome dam.

2. Deck Slab Buttress Dams. Deck slab buttress dam (Fig. 20.10) is made of a flat reinforced concrete slab spanning across a series of parallel buttresses which rest directly on the foundation or upon a concrete slab resting on the foundation material. The water pressure is transmitted by the slab to the buttresses. The main layout considerations are spacing of the buttresses and shape of buttress head to support the slab. The buttress spacing is governed by the consideration of economy, *i.e.*, which gives minimum thickness of slab and buttress and concrete at all sections is fully stressed. The buttress spacing, depending on the height of dam, varies from

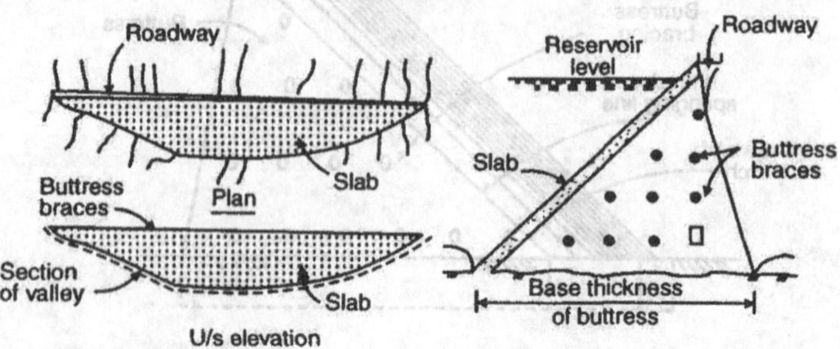

Fig. 20.10 Flat slab deck dam.

5 to 15 m. The upstream slope of the slab may vary between 35° to 45° with the horizontal. Too steep a slope increases tendency to sliding while too flat a slope causes a very high inclined stress at the heel. Articulation of flexibility, a special feature of this type of dam, is provided by not rigidly attaching the slab to the buttress. It is provided by having keys between the slab and the support. The recesses in the keys are filled with compressible material. It is to allow for expansion of the slab in directions parallel to base line of the dam and along the slope. The flat slab buttress dam may be of fixed or continuous deck slab buttress or cantilever deck slab buttress or simple deck slab buttress dam.

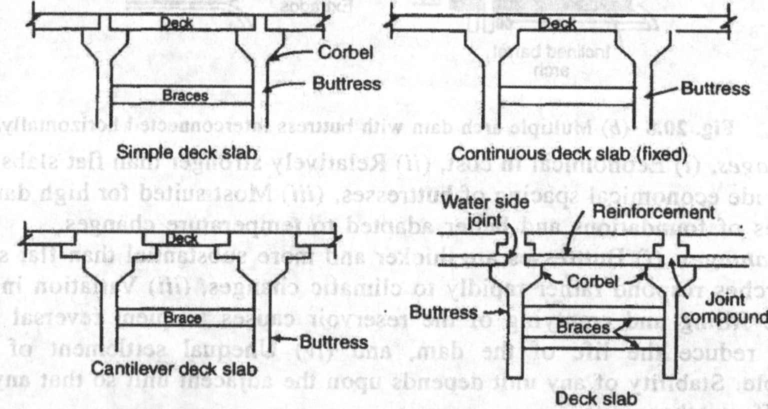

Fig. 20.11 Various types of deck slab buttress dam.

(*i*) *Fixed Deck Slab Buttress Dam:* In fixed or continuous deck slab buttress dam, the deck slab is cast monolithically with the buttresses (Fig. 20.11) with the provision of expansion joints at convenient places. The stresses are there both at the upstream and downstream faces of the deck slab. Significant stresses are also there due to temperature variation, foundation settlement, etc. In view of reversing bending moments, reinforcement is provided both on the upstream and downstream faces of the slab.

(*ii*) *Cantilver Deck Buttress Dam:* In this type of dam, as the name implies, the slab is laid monolithic with the buttress but cantilvered at both ends, with proper joints at both ends or at the centre (Fig. 20.11). The water load is transferred to the supporting buttress by cantilever action of the deck and reinforcement is provided at upstream face in view of cantilever action. It is subjected to high stresses due to temperature, shrinkage and foundation settlement. It is suitable for low dams.

(*iii*) *Simple Deck Slab Buttress Dam:* Also termed as free deck type or Amburson dam (Fig. 20.12). The reinforcement is placed at the downstream face of the deck slab and as such is not susceptible to rusting. The influence of temperature, shrinkage and foundation settlement is less. This type of dam is suitable for a wide valley with poor foundation. It requires thicker slab.

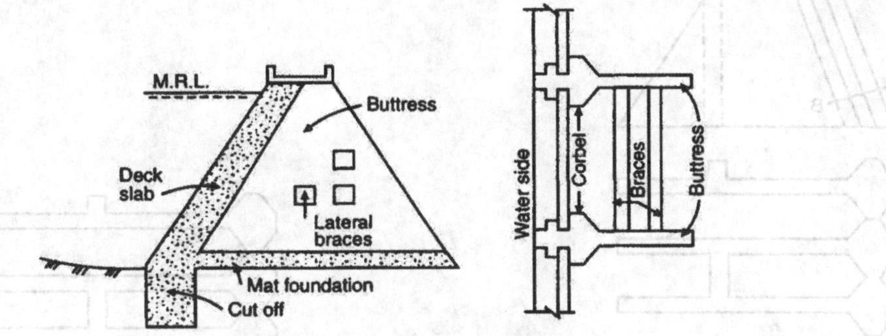

Fig. 20.12 Deck slab buttress dam.

3. Bulk Head Buttress Dam. It is a type of buttress dam in which the face slab is replaced by flaring the upstream edge of the buttresses to span the distance between buttress walls. It consists of a series of buttresses with massive head forming the water supporting member. The massive heads and buttresses, though independent structural units, behave as one monolithic unit for stability of the dam section. The flaring portion may be of different geometrical shapes.

Advantages: (*i*) Simple in construction, (*ii*) Water pressure is compressive and no reinforcement is required in the deck, (*iii*) Free from appreciable uplift, (*iv*) Stability against overturning is achieved with less concrete than in a gravity dam, (*v*) Minor relative settlement is tolerable as each unit is structurally independent of others, and (*vi*) Buttress can be spaced to suit the availability of good rock material in the foundations.

Disadvantages: (*i*) More form work is required, and (*ii*) Excavation is more if foundation overburden is large.

Types of Bulk Head Buttress Dam

1. Massive Head Buttress Dam. It consists of a series of buttresses with upstream massive heads flat as in deck slab abutting each other. The head and buttress together resemble letter T (Fig. 20.13).

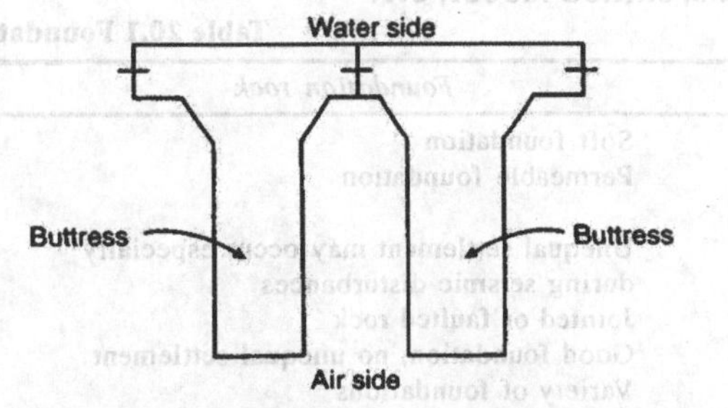

Fig. 19.13 Massive head buttress dam.

2. Round Head Buttress Dam. In this type of dam, the massive heads forming the water supporting member are circular (Fig. 20.14); the shape of round head is such that water pressure acts radially and is compressive. The purpose of round head is to eliminate reinforcement but buttresses are made massive and unreinforced. At the two adjacent massive heads water tightness is achieved by providing two layers of copper seals with an asphalt seal in between.

3. Diamond Head Buttress Dam. It is the type of dam in which the massive heads are sharp and like a polygon (Fig. 20.15). The pressures act perpendicularly to the surface of the diamond head and are compressive.

4. Miscellaneous Types. (*i*) *Columnar Buttress Dam.* It consists of a continuous flat slab supported on a number of inclined columns rather than buttresses. The columns are stiffened by a two-directional system of struts.

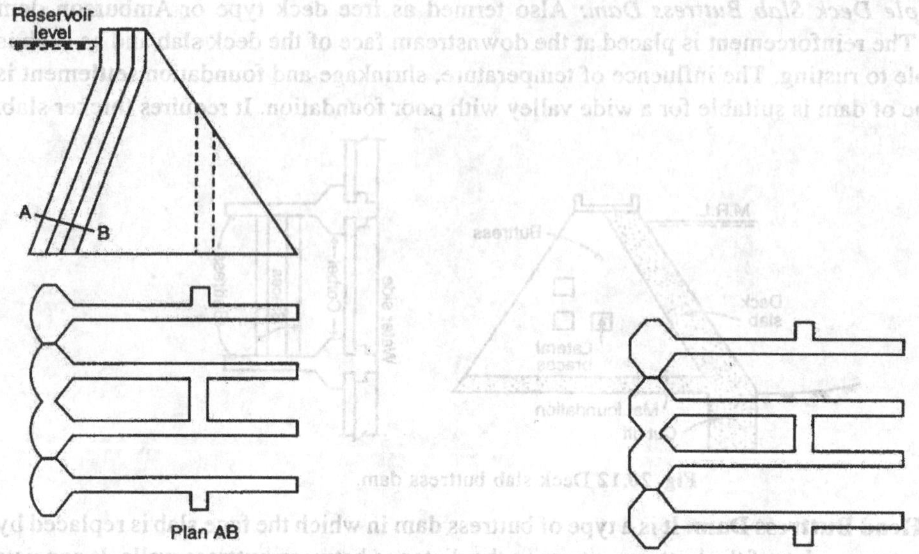

Fig. 20.14 Round head buttress dam. **Fig. 20.15** Diamond head buttress dam.

(*ii*) *Truss Buttress Dam.* It is similar to columnar buttress dam with the exception that the deck slab is supported on reinforced concrete trusses.

20.7 SELECTION OF TYPE OF BUTTRESS DAM

The selection of various types of buttress dam depends on:

1. *Foundation conditions.* For good foundation any type of buttress dam is suitable. The type of dam suitable for various foundation rocks is roughly as stated in Table 20.1.

2. *Requirement of spillway, irrigation or power outlets.* Spillway can be conveniently provided at the top and outlets at the base of the buttress dam. The control equipment can be housed in the chamber constructed inside the buttresses.

3. Other considerations are economy governed by the availability of construction materials, skilled labour, etc.

Table 20.1 Foundation rock and dam type

Foundation rock	*Suitable type of dam*
Soft foundation	Any of buttress types dam with spread footing.
Permeable foundation	Slab and buttress dam with bottom slab under buttresses. Percolation and uplift are considered.
Unequal settlement may occur especially during seismic disturbances	Slab and buttress dam with buttresses strutted.
Jointed or faulted rock	Slab and buttress dam.
Good foundation, no unequal settlement	Multiple arch dam.
Variety of foundations	Massive head buttress dams.

20.8 FORCES ACTING ON BUTTRESS DAM

Forces acting on buttress dam are similar to those acting on gravity dam and arch dams except that (*i*) Downward component of water pressure is greater, (*ii*) Uplift is less due to lateral escape of the uplift pressure under the buttresses. Uplift from water head on rock foundation is usually neglected. However, the foundation is drilled for drainage where the rock is liable to uplift pressure on horizontal seams. Full uplift from tail water is considered, (*iii*) Ice pressure is not important as ice tends to slide up the sloping deck, (*iv*) Wind pressure is important if a diagonal wind of high velocity can reach the downstream side. It is, however, of little importance in the case of well braced or double walled buttresses, and (*v*) Because of small

mass relative to rigidity of individual units, slab and round head, buttress dams are efficient in resisting earthquake loadings.

The buttress dams are analysed for stability in a manner similar to gravity dams. The design element is taken as a full panel, rather than a slice of unit thickness. The essential design procedure envisages fixing the economical spacing and upstream slope of buttress dam, design of deck and buttresses and overall stability check.

QUESTIONS

1. Define arch dam. Discuss various types of arch dams and their merits and demerits.
2. For a fixed combination of span, loading and permissible cylinder theory stress, prove that the economic arch dam is one which subtends a constant angle of 133° 34′ at centre.
3. What are the situations to which the various types of dams are most suited, viz. arch dam, earth dam, and rockfill dam? What are the basic principles of their design?
4. What are arch and rockfill dams? Illustrate with sketches and mention site conditions favourable for construction of such dams.
5. What are arch and buttress dams? Illustrate with sketches and mention site conditions favourable for construction of such dams.
6. Discuss the various forces which act on an arch dam.
7. What is buttress dam? Discuss the functions served by its component parts.
8. (*a*) Explain the theory of cylinder action in the case of an arch dam.

   (*b*) Design a 100 m high constant angle arch dam by thin cylinder theory for a valley 20 m wide at the bottom and 220 m wide at 100 m height. The permissible compressive stress is 500 t/m².

   Solution :

Central angle = 133° 34′ (assumed)

Chord length, l of top arch = 220 m

Radius of intrados, r_i = 120.68 m

($0.544\, l = 0.544 \times 220$)

$$\text{Thickness of arch, } t = \frac{Pr_i}{\sigma - P} = \text{cal.5} \div \text{cal.6}$$

COMPUTATION

Depth of arch (m)	l (m)	r_i (m)	$P = wh$ (t/sqm)	Pr_i (t/m)	$\sigma - p$ (t/sqm)	t (m) cal.5÷cal.6	$r_e = r_i + t$ (m)	r_m (m) cal.8–t/2
1	2	3	4	5	6	7	8	9
0	220	120.68	0	0	500	0	120.68	120.68
10	200	108.8	10	1088	490	2.22	111.02	109.91
20	180	97.92	20	1958.40	480	4.08	102.0	99.96
30	160	87.04	30	2611.20	470	5.56	92.60	89.82
40	140	76.16	40	3046.40	460	6.62	82.78	79.47
50	120	65.28	50	3264.00	450	7.25	72.53	68.91
60	100	54.4	60	3264.00	440	7.42	61.82	58.11
70	80	43.52	70	3046.40	430	7.08	50.60	47.06
80	60	32.64	80	2611.20	420	6.22	38.86	35.75
90	40	21.76	90	1958.40	410	4.78	26.54	24.15
100	20	10.88	100	1088.00	400	2.72	13.60	12.24

9. (*a*) What are the advantages of buttress dams over other types? [P.U.]

(*b*) Show with the aid of a neat sketch the general dimensions for buttress corbel and flat slab deck.

10. Discuss cylinder theory for arch and obtain an expression for the thickness or arch ring in terms of external radius, height of dam and allowable stress.

11. What are arch dams? Illustrate with examples and mention site conditions favourable for construction of such dams.

PART–III

HYDROLOGY

- Precipitation Measurement and Analysis
- Runoff and Flood Estimation

21

PRECIPITATION MEASUREMENT AND ANALYSIS

21.1 HYDROLOGY

Hydrology is the science which deals with the water of the earth in all its states, *i.e.* occurrence, distribution and circulation, through the unending hydrologic cycle of precipitation, consequent runoff, stream flow, infiltration and storage, eventual evaporation and reprecipitation. It is concerned with the atmosphere, the medium through which water is brought to and removed from the earth's surface. The study of hydrology concerns itself with three different forms of water *i.e.* (*i*) above land as atmospheric water or precipitation, (*ii*) on land surface as surface water or runoff, and (*iii*) below the land surface as subsurface water or groundwater.

Engineering hydrology includes the science which deals with the estimation of runoff and its transmission from one place to another, besides the quality of water. For a power engineer, surface water is of direct importance but owing to lack of data of stream flow, it is usually necessary to study precipitation, evaporation and percolation to indirectly determine the availability of surface and groundwaters. For planning and development of power projects, an engineer has to determine, various hydrology related aspects viz. (*i*) Water power potential of a river or stream, (*ii*) Dependable yield for hydel power generation,and may be for irrigation because river valley projects are usually of multipurpose concept with hydropower development as the essential component, (*iii*) Probability of floods, their pattern and magnitude, (*iv*) Sediment yield and transportation, (*v*) Hydro-meteorological data, (*vi*) Streamflow forecasting by the use of precipitation data, (*vii*) Water quality, erosion and pollution control of rivers and streams, and (*viii*) Operation and maintenance of river valley projects.

21.2. HYDROLOGY AND WATER POWER DEVELOPMENT

The hydrology plays a vital role in the planning of water power projects for sustained development compatible with quality and environmental requirements. India has a great diversity of climatic conditions from continental to oceanic and a wide range of temperatures and precipitation levels. Most of the rainfall occurs under the influence of the south-west monsoon, during three to four months (June to September), but precisely for a period of about 100 days, in most parts of the country. The balance precipitation occurs during the North-East monsoon of November and December. Rainfall is unevenly distributed, generally exceeding 100cm/year in areas to the east of longitude 78°E, increasing to 250 cm along the coast, the western Ghats and over most of Assam and Sub-Himalayan West Bengal; Cherapunji records a maximum rainfall of 1142 cm. It diminishes from 50 cm to less than 15 cm in the extreme west; Rajasthan, Kutch and high Ladakh plateau receive rainfall as low as 10 to 50 cm in a year. The average annual rainfall in the country is computed as 120 cm.

The two basic elements of hydro power development are (*i*) River/stream flow and (*ii*) Fall or available head for the generation of electricity. The pre-requisite for planning a large and a small hydel project is the determination of these two parameters by carrying out investigations and studies. The assessment is essentially made of precipitation, volume of water in the river/stream to be available as run-of-the river power plant or quantum of storage to be made, its variation and flood pattern, infiltration and evaporation to compute runoff from rainfall, probability of availability, sediment transport, water quality, sediment deposition in reservoir vis-a-vis probable life of the reservoir etc. Analysis thereof leads to finally selection of (*i*) The site with minimum environmental impacts, (*ii*), Power potential, and (*iii*) Units of power plants. The water availability for thermal and nuclear energy too is determined by the use of hydrology and hydrological data.

To be more precise water availability has essentially to be adequate during the various months of the year. The availability of minimum flow in dry months determines whether or not pondage/storage is needed. The availability of flows in wet season dictates the quantum to be stored for optimum utilisation vis-a-vis power development. There can be two possibilities in the assessment of availability of water, that is runoff data is available for adequate number of years; weekly, daily or monthly. The available discharge data is used to determine the water availability of 90% dependability by statistical approach. The other alternative is that adequate rainfall data is available but runoff data only for a few years. The water availability in this case is determined by correlating rainfall and runoff of the site. The runoff at the desired dependability is computed accordingly. Herein lies the role of essential elements of hydrology.

21.3 PRECIPITATION

It is the general term for all forms of moisture emanating from clouds and falling to the surface of the earth. It is thus the total supply of water derived from the atmosphere in the form of rain, snow, mist, frost, hail, sleet, etc. It is usually expressed as depth of liquid water on a horizontal surface in a day, month or year and designated as daily, monthly or annual precipitation. Precipitation is thus the sum of the amounts of liquid precipitation *i.e.* rainfall and liquid equivalent of frozen precipitation, *i.e.* snow, hail, sleet, freezing rain.

Occurrence of Precipitation

For precipitation to occur, the three essential requirements are (*i*) Some mechanism is required to cool the air sufficiently to cause condensation and droplets growth, (*ii*) Condensation nuclei are also necessary, but they are usually present in the atmosphere in adequate quantities, and (*iii*) Large scale cooling is needed for significant amounts of precipitation. It is achieved by lifting of the air. In order for precipitation to take place, it is first necessary for a portion of the atmosphere to be cooled until the air is saturated with water vapour. Condensation thereupon occurs, resulting in the formation of minute droplets which are visible collectively in the form of clouds or fog. These droplets will not in themselves form into rain drops and may remain indefinitely as droplets in a condition of colloidal suspension. Rain drops, or other forms of precipitation, result when the minute droplets of moisture collect on particles in the air termed hygroscopic nuclei. The commonest hygroscopic nuclei are salt particles from oceans and products of combustion in the atmosphere over cities. The exact form of precipitation, whether as rain drops, snowflakes or the ice particles depends upon the temperature of the air in which the precipitation is formed and the temperature of lower air layers through which it falls to the earth. It is necessary for moist air to be cooled to the saturation point (dew point) before condensation can occur.

Types of Precipitation

1. *Convective Precipitation*: It is the precipitation resulting from the upward movement of air that is warmer than its surrounding. The effect may be caused by the temperature difference between hot, bare fields and cool woods, between land and lakes than the surrounding countryside. Convective precipitation is generally of a showery nature with rapid changes of intensities.

2. *Orographic Precipitation*: It is the precipitation caused by dynamic cooling of air as an air current rises over a mountain barrier. When moisture bearing winds usually blowing from ocean to land surface are forced up by the influence of coastal mountain ranges, the cooling and condensation processes take place and precipitation occurs along the windward slope of mountains. The precipitation in Himalayan region is due to orographic ascent of air mass laden with moisture content because of long travel over ocean. Orographic precipitation is ordinarily of low intensity.

3. *Cyclonic Precipitation*: It is the precipitation associated with the passage of depressions or cyclones. It results from the lifting of air converging into a low pressure area. Cyclonic precipitation may be heavy and prolonged. Most of the winter rains in Punjab and Haryana are due to cyclones. Cyclonic precipitation may be divided into Frontal precipitation, and Non-frontal precipitation.

Frontal precipitation: Cyclonic storms and their resulting precipitation are associated with the planes of contact (frontal surfaces) between the air masses of different temperatures and moisture contents. The boundary of discontinuity between the two air masses is termed as front. When a moving warm moist air-overrides a cold air mass or is forced up by a cold air wedge, it cools down at high altitudes resulting in precipitation. A front may be of two types (*i*) warm front, or (*ii*) cold front depending whether there is active or passive ascent of warm air mass over the cold air mass. Warm front preciptiation, generally light to moderate, is formed in warm air moving gradually upward over a wedge of cold air. Cold front precipitation, generally of showery nature, is formed in warm air which is forced upward by an advancing wedge of cold air.

Non-frontal precipitation: It results when a moving cold air mass meets a rather stationary moist warm air mass. The warm air mass, being lighter, is lifted up over cold air which cools down at higher altitudes resulting in precipitation.

4. *Precipitation Due to Turbulent Ascent*: Air mass is forced up due to increased friction of earth surface after its travel over the ocean and friction on condensation results in precipitation. Winter rainfall in Tamil Nadu is mainly due to this process.

21.4 RAINFALL

Rainfall is the total liquid product of precipitation or condensation from the atmosphere as received and measured in a raingauge.

21.5 RAINFALL MEASUREMENT

Rainfall is measured on the basis of the vertical depth of water accumulated on a level surface during an interval of time, if all rainfall remained where it fell. It is measured in an instrument called raingauge. Raingauge measures the quantity of rain that falls at a given place and time usually 8.30 A.M. Rainfall is measured in mm. An amount less than 0.1 mm is recorded as a trace. The sum total of all the measurements during the previous 24 hours is entered as the rainfall of the day.

Selection of Site

The raingauge is installed at a site selected with the considerations that (*i*) It is true representative of the area for which the raingauge station is to give rainfall, (*ii*) It is in an open place, (*iii*) It is in a level ground, not upon a slope or terrace, (*iv*) Away from trees, buildings or other obstructions by a minimum distance of four times but never less than twice the height of obstruction, (*v*) Site is shielded from strong wind. The ground should not fall away steeply on the side of the prevailing wind, and (*vi*) Easily accessible at all times. In hills minimum level space required is 6 × 6 m shielded from high winds.

Network of Gauges

The rainfall varies from place to place and as such a number of raingauges are set up at various locations. A network of raingauges is the simplest and most common means of measurement of rainfall. The number of stations are required to be adequate to represent reliable picture of the intensity and duration of rainfall in the area. The network of raingauges design is considered as saturated when in a network any addition of gauge to it will not appreciably alter the rainfall assessed. It is in regard to important project wherein maximum accuracy is desired. The optimum network design is that in which, by interpolation between values of different stations, it is possible to determine with sufficient accuracy the characteristics of the basic hydrometeorological elements at any point of interest. In planning network it has to be ensured that it may not contribute towards faulty assessment of water resources and power producing potential. As per IS : 4987-1994 for optimum network in plains one raingauge station up to 500 km² is sufficient while in hilly areas, with general average elevation one km above sea level, one raingauge in 250 to 400 km² is sufficient. In predominently hilly areas with heavy rainfall, one raingauge in not more than 150 km²; the denser stations being preferable. In areas where optimum network may not be feasible, it may be designed wherein at least one raingauge may represent 750 km² in plain areas, 100-250 km² in hilly areas and 1,500–10,000 km² in arid areas.

21.6. OPTIMUM NUMBER OF RAINGAUGE STATIONS

Optimum number of raingauge stations required to compute average rainfall in the basin within 10% error is determined by the procedure as under:

Total rainfall from *n* number existing gauges (R)	Mean rainfall (m)	Sum of squares of all gauges (SS)	T	Square of standard deviation (S^2)	Coefficient of variation (Cv)	Optimum number of gauges within 10% error	Additional raingauges required
$P_1 + P_2 .... + P_n$	$\frac{R}{n}$	$P_1^2 + P_2^2 + ... P_n^2$	$\frac{R^2}{n}$	$\frac{SS - T}{n - 1}$	$\frac{100\sqrt{S^2}}{m}$	$\left(\frac{C_V}{10}\right)^2$	$(N - n)$

21.7 TYPES OF RAINGAUGES

The two principal types of raingauges are non-recording and recording raingauges, as under :

1. Non-recording Raingauge: It is also termed as non-automatic raingauge. Symon's raingauge extensively used earlier in India consists of a receiving funnel 127 mm in diameter, the base, the body and the receiver with a brass rim 1.25 cm high (Fig. 21.1). The rain falling in the receiver is funnelled into a 175 cm capacity container and measured. The funnel shank is inserted into the neck of bottle which is 75 to 100 mm in diameter. The rainfall is measured by a measuring glass which can be read to the nearest 0.1 mm. Recently India meteorological department has introduced fibreglass reinforced polyester raingauge which is an improvement over the Symon's gauge. Non-recording raingauge gives only the total depth of rainfall for 24 hours and not its intensity and duration during different time intervals of the day. In remote areas, storage gauge is used to measure total seasonal precipitation. It consists of a vertical 60 cm diameter steel pipe of sufficient length to place its 20 cm catch ring above maximum accumulated snow. To convert the snow which falls into the gauge to liquid state an antifreeze solution is placed in the receiver.

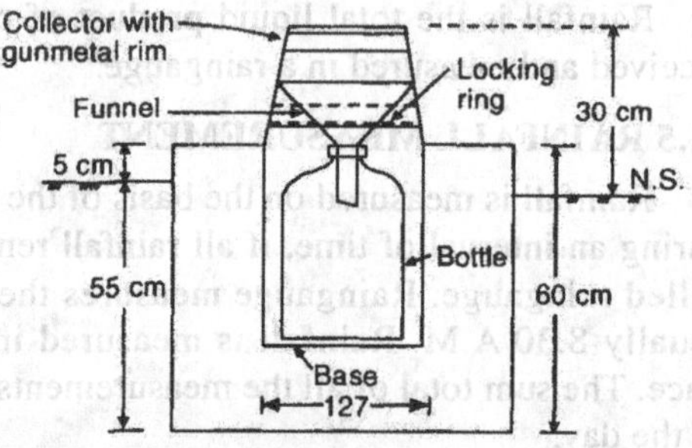

Fig. 21.1 Standard raingauge

2. Recording Raingauges: Recording raingauge, an improvement over the non-recording raingauge, is automatic which gives record of rainfall. Since the raingauge represents cumulative rainfall, it is also termed as integrated raingauge or continuous raingauge. It is specially adapted to inaccessible places in hilly areas where it is not possible to visit the station daily. The mechanical arrangement in the recording raingauge records rainfall for extended periods in the form of a pen trace on a clock-driven chart. The graph of the cumulative rainfall is termed as mass curve of the rainfall. The chart indicates (*i*) total amount of rainfall since the record is started at the station, (*ii*) duration of rainfall, *i.e.* times of starting and ending of the rainfall, and (*iii*) intensity of the rainfall for any given period. It is desirable to equip at least 20% of the raingauge stations with recording gauges to know intensities of rainfall. Self-recording raingauge may also be installed in conjunction with a non-recording gauge.

The raingauge is installed on a concrete or masonry platform 60 × 60 × 35 cm, sunk to a depth of 27.5 cm, truly horizontal and at a height of exactly 7.5 cm above the ground surface.

Advantages: (*i*) Only means to get rainfall data from inaccessible hilly areas, (*ii*) Automatic recording of rainfall, free from error due to human judgement, (*iii*) Records intensity of rainfall at any time besides total rainfall, (*iv*) Duration of rainfall is also indicated, and (*v*) Large capacity of the gauge.

Limitations: (*i*) High initial cost, (*ii*) Recording may be interrupted due to development of any fault in mechanical as well as electrical recording mechanism or mechanical defects.

The various types of recording gauges are as under:

1. Weighing Bucket Raingauge: Weighing bucket gauge is used for recording rainfall as well as snow. It consists of a receiver bucket supported by a spring or lever balance (Fig. 21.2). The catch from funnel empties into a bucket mounted on a weighing scale. The weight of the rain falling into the bucket depresses the entire mechanism. This movement is transmitted to a recording pen through levers and links recording the increased weight of the bucket and its content on a chart held by a clock-driven drum giving a trace of accumulated amounts of rainfall with time. It gives mass curve of rainfall for as long as one week.

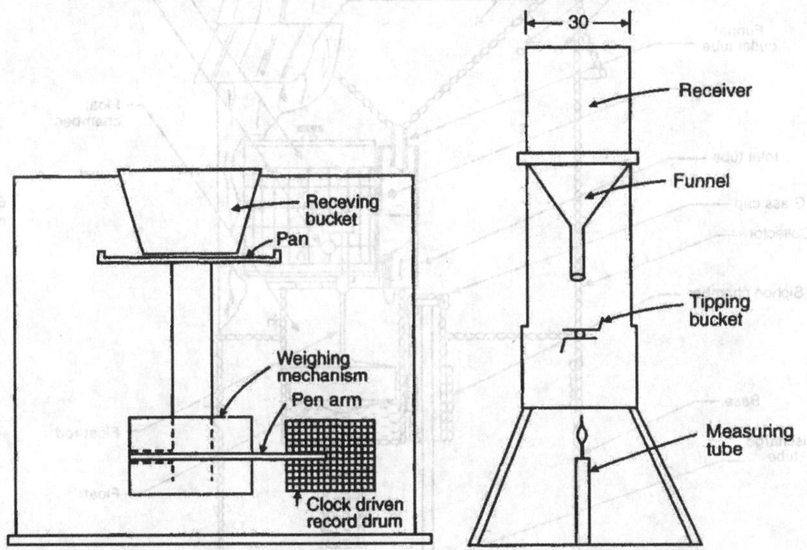

Fig. 21.2 Weighing bucket raingauge. **Fig. 21.3** Tipping bucket raingauge.

Advantages: (*i*) Simple mechanism, and (*ii*) Suitable for digitalising of output signal for the transmission of telemetry.

Disadvantage: The main disadvantage is that rainfall of less than the capacity of bucket compartments is not recorded.

2. Tipping Bucket Raingauge: It consists of a sharp edge receiver, 30 cm in diameter, and a funnel. The rain water collected in the receiver is funnelled into a twin compartment tipping bucket (Fig. 21.3). When 0.25 mm of rain collects to fill one compartment, it tips and empties the water into a storage reservoir and simultaneously second compartment of the bucket moves under the funnel to receive rain. Thus the bucket is tipped by each 0.25 mm of rain. As the bucket is tipped, it actuates an electrical circuit causing a pen to mark on a revolving drum. The movement of the tipping of bucket can be transmitted electronically over distances to a control room to record rainfall measurement directly and in this case graph paper or drum is not installed in the gauge. It is ideally suited for digitalising of the output signal.

3. Float Recording Gauge: It consists of a rectangular float chamber, carrying a funnel and a drum on either side on its top and a light hollow float at its bottom provided with a self-starting siphoning arrangement (Fig. 21.4). The funnel receives the rain and conducts it into the chamber. The float gets raised as the rain water level in the chamber rises. This movement is transmitted through a suitable mechanism to the pen which moves on a revolving chart actuated by a clockwork. When the chamber gets filled, thereby causing the float to touch the top, siphonic action starts and the accumulated water is drained off. The clock mechanism revolves the drum once in 24 hours. Rewinding is needed once in a week when the chart fixed around the drum is also replaced with a new one.

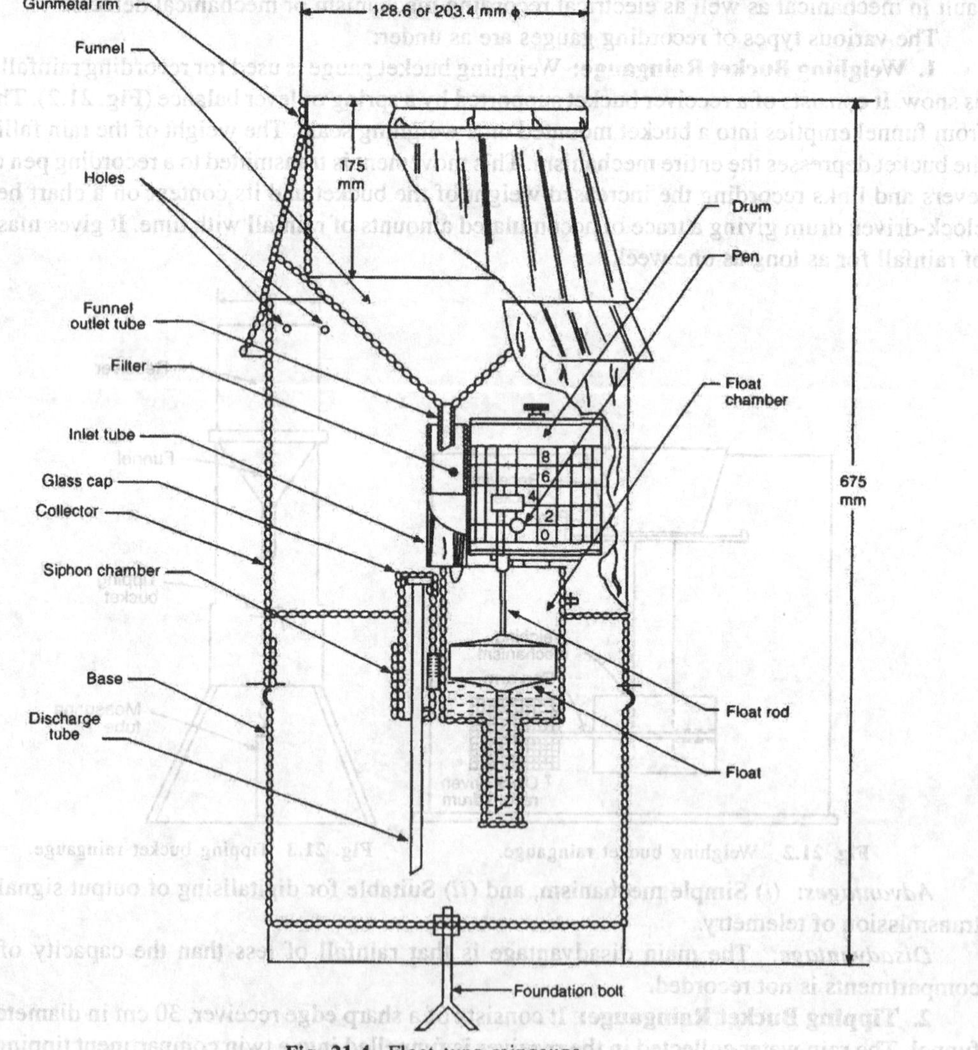

Fig. 21.4 Float type raingauge.

21.8 ESTIMATION OF MISSING RAINFALL RECORD

Rainfall records of certain stations suffer a break because of non-observation of the rainfall by the observer or failure of the instrument installed. The missing rainfall record is approximated as follows:

1. From the record of Three Stations

(i) Arithmetic Mean Method: This method is applicable if the normal annual rainfall at the missing station X is within 10 per cent of the normal annual rainfall at the adjoining three stations A, B and C. Arithmetic mean of the rainfall of the three stations is taken as the rainfall at the station with missing record *i.e.* $P_x = \frac{1}{3}(P_A + P_B + P_C)$.

(ii) Normal Ratio Method: The method is applicable if the normal annual rainfall at any of the nearby operative stations A, B, C differ from that at the station 'x' with missing data, value P_x, by more than 10 per cent. In this method the rainfall values at three stations are weighted by the ratios of the normal long term (annual) rainfall value (N).

$$P_x = \frac{1}{N}\left[\left(\frac{N_X}{N_A}\right)P_A + \left(\frac{N_X}{N_B}\right)P_B + \left(\frac{N_X}{N_C}\right)P_C + \ldots + \left(\frac{N_X}{N_n}\right)P_n\right]$$

N_A, N_B or N_C differs from N_X by more than 10 per cent. N is number of index (nearby) stations.

2. From the Record of Two Stations

Rainfall values of two or even one adjoining station, where three adjacent stations are not there, may be used to estimate the missing rainfall record of a station.

3. Ratio Method

In this method the missing record of rainfall is computed from the ratio of the rainfall for the particular year, value in respect of which is missing at one station, to that of the rainfall of complete years at both stations.

21.9 AVERAGE RAINFALL OVER A BASIN

Where there is only one raingauge station, the rainfall recorded there is considered applicable over the entire area. Where more than one raingauge stations are established in a basin, the computation of the average rainfall which will represent the whole basin is done by the following methods:

1. Arithmetic Unweighted Mean Method

It is the simplest method of determining the average depth of rainfall over a basin by taking simple arithmetic mean of the rainfall gauges recorded at various stations. The method gives good results in flat regions if the gauges are uniformly distributed and the rainfall at various stations do not vary widely from the mean.

$$\text{Average rainfall over a basin, } (P_{av}) = \frac{P_1 + P_2 \ldots + P_n}{n} = \frac{\Sigma P}{n}$$

Where, $P_1, P_2 \ldots, P_n$ = rainfall recorded at the stations, and n = number of raingauge stations.

2. Thiessen Polygon Method

This method is used for non-uniform raingauge network by assigning weights to station data in proportion to the space between the raingauge stations. Thus the method simply assumes linear variation of precipitation between the raingauge stations and assigns equal segment of the area to the nearest station. It is also called weighted mean method. It is more accurate than arithmetic mean method and is commonly used for medium size catchment areas.

The average rainfall over the basin is determined from the equations.

$$P_{av} = \frac{A_1P_1 + A_2P_2 + \ldots + A_nP_n}{A_1 + A_2 + \ldots + A_n} = \frac{\Sigma(A \times P)}{\Sigma A}$$

Where, P_{av} = average depth of rainfall in the basin, $P_1, P_2, \ldots, P_n$ = rainfall depths at various stations, and $A_1, A_2, \ldots, A_n$ = areas surrounding the raingauge stations.

Limitations: The method has the limitation of (*i*) Inflexibility, as a new Thiessen diagram is required every time there is a change in the gauge network, (*ii*) Orographic influence is not considered, (*iii*) Assumption that raingauge represents the area which is close to it may not be valid when the rainfall is influenced by topography or results from intense convection.

3. Isohyetal Method

It is probably the most accurate method for the determination of average depth of rainfall over a large basin and provides a means of considering the orographic (mountain) effect. The method takes into account the actual special relationship of the raingauges and that of the nearby gauges and makes corrections accordingly. In this method, the stations and rainfall values are plotted on a map to a suitable scale. Then the contours of equal precipitation *i.e.* isohyetes* are drawn. In constructing the isohyetal maps, full use of the orographic effects of elevation on rainfall and storm movements can be made so that the final map may represent a more realistic pattern of rainfall. The procedure of isohyetel method is to (*i*) Plot rainfall at each station, (*ii*) Prepare isohyetal map of the area enclosed, (*iii*) Areas between successive isohyetes are measured by means of a planimeter, (*iv*) Each of these areas is multiplied by the average rainfall intensity between the successive isohyetes, and (*v*) Average rainfall is determined from the relation:

$$P_{av} = \frac{A_1 \dfrac{(P_1+P_2)}{2} + A_2 \dfrac{(P_3+P_4)}{2} + \ldots + A_{n-1} \dfrac{(P_{n-1}+P_n)}{2}}{A_1 + A_2 + \ldots \ldots A_n}$$

or $$P_{av} = \frac{\Sigma\,[A_1(P_1+P_2)/2]}{\Sigma A}$$

Advantages: (*i*) Most accurate method when the stations are large in number and is used by an experienced analyst, and (*ii*) Enables interpretation of all available data.

21.10 MEAN ANNUAL RAINFALL

The term 'mean annual rainfall' is synonymous with 'rainfall' at a place. Mean annual rainfall is determined by taking simple average of the total rainfall of several consecutive years at the place. Since the annual rainfall varies at the same station, rainfall records for a considerable number of years are required. The greater the number of years considered, the more accurate is the mean value of rainfall so obtained. In India, a rainfall cycle of 35 years is considered. The value is recomputed every decade. According to Alexander Binnie, mean rainfall worked out from 35 years of rainfall record is within 1.8% of the true value. If the rainfall record is for a period shorter than 35 years, the error in the computed mean value will be as given in Table 21.1.

Table 21.1 Error in average rainfall value computed from record of period less than 35 years.

Period (years)	5	10	15	20	25	30	35
Error in mean rainfall (percentage)	14.9	8.2	4.8	3.3	2.8	2.3	1.8

U.S. Weather Bureau has considered somewhat longer periods of rainfall record. According to it, about 40 years observations are required to give results within 1.5% of the true mean.

IS 4987:1994 has recommended the following number of years of rainfall record:

Catchment layout	Island	Shore	Plain	Mountains
No. of years	30	40	40	50

Studies by Central Water and Power Commission, India show that 33 to 40 years of rainfall record is necessary to predict a 100-year flood and 80 to 90 years record is required to predict a 250-years flood.

21.11 ADJUSTMENT OF RAINFALL RECORD

The inconsistency of the record of a station arises due to (*i*) Change in location of raingauge, (*ii*) Change in exposure, and (*iii*) Changes in instrument or observational procedure. The consistency is tested by comparing its accumulated annual or seasonal rainfall with the concurrent values of mean rainfall of a group of adjoining stations termed as base stations. A group of minimum 10 base stations in the immediate surrounding with similar environment of a station, the record of which is being tested (say *X*) is selected. The data of rainfall (annual or monthly) of the station *X* and the concurrent accumulated values of several surrounding stations; better the mean of the surrounding stations is used. In case of consistent data, the plot is a straight line. The inconsistent data are exhibited by a change in slope or break at a point of the occurrence of inconsistency, shown by the year 1980 in Fig. 21.5. The slope of the line AB is 'a' and of BC is '*b*'. The adjustment of the inconsistent data is made by multiplying each value of rainfall before the break point by the recent conditions following the break. The average long-term rainfall of the base stations are arranged in reverse order, starting with the latest record and ending with the entry of the oldest record. The accumulated rainfall of the station *X* and the accumulated values of the average of group of base stations are determined starting from the latest record and plotted for various consecutive time periods (Fig. 21.5). A change in the slope of the double mass curve (if it persists for more than 5 years) indicates change in the rainfall regime of the station *X*. The rainfall values of station *X* beyond the point of change of slope is corrected to make the record compatible with the new regime of the station by applying the following correction factor:

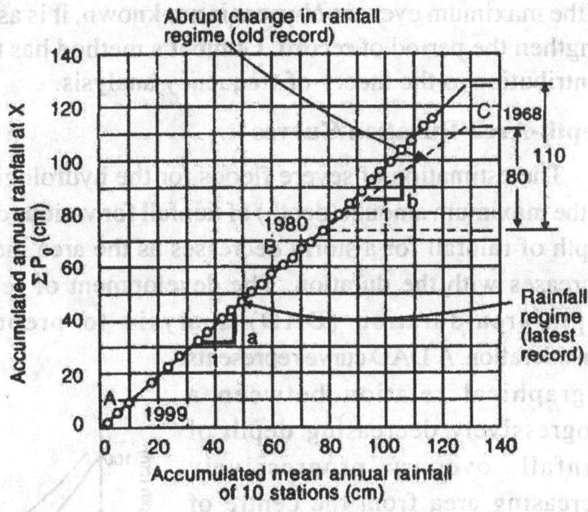

Fig. 21.5 Double mass curve

$$P_C = P_o(S_c/S_o)$$

Where, P_C is corrected rainfall at station *X* at any time *t*, P_o is original recorded rainfall at station *X* at time period *t*, S_c is corrected slope of double mass curve, and S_o is original slope of the double mass curve.

Fig. 21.5 shows change in regime in the year 1980. Factor 80/110 = 0.7273 is to be applied to records prior to 1980.

21.12 RAINFALL FREQUENCY ANALYSIS

Rainfall frequency analysis is defined as the rainfall which can be expected to be equalled or exceeded on the average once every *N*-years, the *N*-year rainfall. *Exeedence interval* is the average number of years between the occurrence of an event and a greater event. The *recurrence interval* is the average number of years within which a given rainfall event will be equalled or exceeded. Frequency of occurrence of maximum and minimum rainfall is determined to predict the future rainfall trends for the design of river protection works, drainage works, etc.

Various methods for determining recurrence interval are:

1. California method (1923): $T = N/m$
2. Hazen's method (1930): $T = 2N/2m–1 = N/m – 1/2$
3. Weibull method, $T = (N + 1)/m$
4. Gumbel's method, $T = \dfrac{N}{m+C-1}$

where, T = recurrence interval, N = number of years of record, m = serial number of a particular event arranged in descending order so that maximum rainfall is at top to find out maximum rainfall value and in ascending order to determine minimum rainfall value, and C = Gumbel's correction, values are as under:

m/N	1.0	0.9	0.8	0.7	0.6	0.5	0.4	0.3	0.2	0.1	0.08	0.04
C	1	0.95	0.88	0.845	0.78	0.73	0.66	0.59	0.52	0.40	0.38	0.28

California method, also expressed as $a = y/b$, where a = recurrence interval, y = years of rainfall record, and b is ranking of rainfall, has the merit of simplicity. It is a conservative method from the point of view of safety, but the method is not applicable if it is required to determine rainfall having recurrence interval of more than the total number of years of which the records are available. Hazen's method is widely used for giving most probable recurrence interval. Hazen's formula assigns a recurrence interval of $2N$ to the maximum event on record. Since the actual frequency of occurrence of the maximum event in N-years is not known, it is assumed to be $2N$. But no statistical method can lengthen the period of record. Gumbel's method has the best theoretical basis and is the latest best contribution to the theory of frequency analysis.

Depth-Area-Duration Curves

The estimation of severe floods for the hydrological design of dams require the determination of the maximum amount (depth) of rainfall for various durations over areas of various sizes. Maximum depth of rainfall for a storm decreases as the area increases; for a given area the maximum depth increases with the duration. The development of relationship for the three parameters is called depth-area-duration (DAD) analysis to prepare DAD curves for the region under consideration. A DAD curve represents a graphical relation between a progressively decreasing depth of rainfall over a progressively increasing area from the centre of storm outward to its edges. The DAD curves are prepared for total duration of the storm or selected durations of 6, 12, 18, 24 and 36 hours incremental; corresponding isohyetel for these time intervals being available. Maximised DAD curves are used for the estimation of probable maximum precipitation (PMP). Fig. 21.6 shows typical DAD curves.

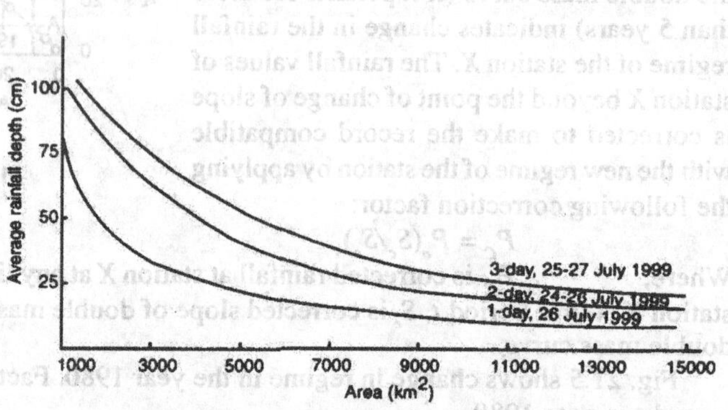

Fig. 21.6 Typical depth-duration-area curves.

21.13 INTENSITY-DURATION-FREQUENCY ANALYSIS

The total rainfall, its intensity and duration are important for the determination of peak flow. The rainfall intensities of different durations and different return periods are also required as may cause severe floods in a river during the life of a hydraulic engeneering structure e.g., the intensity-duration-frequency data are used in the rational method for the computation of design flood from a small catchment.

From the record of rainfall which indicates the accumulated amount of rainfall with respect to time, the maximum rainfall of say 5 minutes duration in each year is determined. The rainfall values are arranged in the descending order and the return period for each value is computed. The frequency of rainfall intensity for a given duration is expressed in terms of recurrence interval i.e., the value which will be equalled or exceeded once in T-years on an average. It is given by the law of probability, the respresentative type is:

$$T = (n + 1)/m$$

where, T is recurrence interval, n is number of years of recorded rainfall and m is ordinal number of each event, arranged in descending order of magnitude.

Similar analysis is done for the other selected durations say 10, 15 minutes. For each frequency level determined, the rainfall depths are plotted for different durations. These are the depth-duration-frequency curves (Fig. 21.8). The rainfall depths can be converted into intensities and plotted as intensity-duration-frequency curves (Fig. 21.7). For higher or lower return periods, interpolation between the return periods curves can be made.

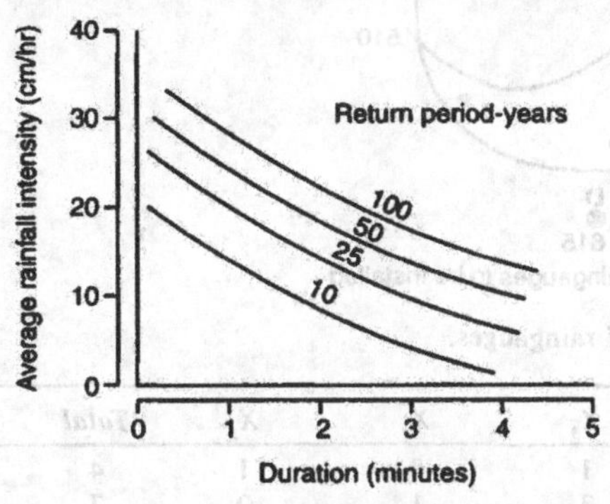

Fig. 21.7. Intensity-duration-frequency curves.

Rainfall depth (cm)
40
30
20
10
0
Return period-years
100
50
25
10
0
1
2
3
4
5
Duration (minutes)

Fig. 21.8. Depth-duration-frequency curves

SOLVED EXAMPLES

Example 21.1 *Fig. 21.9 shows a typical river basin with 4 raingauge stations and 4 outside as per IS: 4987-1994. Determine adequacy of the gauges and allocate additional gauges so that percentage error is less than 10%.*

Solution:

(*i*) *Adequacy of raingauges*

Total rainfall, $R = 800 + 540 + 445 + 410 = 2195$ mm

Mean rainfall, $m = R/4 = 2195/4 = 548.75$ mm

Sum of squares, $SS = (800)^2 + (540)^2 + (445)^2 + (410)^2 = 1297725$

$T = R^2/4 = (2195)^2/4 = 1204506$

Square of standard deviation, $S^2 = (SS–T^2)/4–1 = (1297725–1204506)/3$

$= 31073$

Coefficient of variation, $C_v = [100\,(S^2)^{1/2}]/m = [100 \times (31073)^{1/2}]/548.75$

$= 32.12$

Optimum number of gauges, $N = (C_v/P)^2 = (32.12/10)^2 = 10.32 \simeq 11$

Additional gauges required $= 11 - 4 = 7.$

(*ii*) *Allocation of additional gauges*: (*i*) Draw isohyets at equal intervals (380, 510, 640, 770 and 900 mm) taking into consideration rainfall outside the catchment (Fig. 21.9), (*ii*) Measure the area bounded by two consecutive isohyets within the catchment with a planimeter. The proportional areas of the different isohyetal zones marked are $X_1 = 0.1229$, $X_2 = 0.3760$, $X_3 = 0.3392$, $X_4 = 0.1021$, and $X_5 = 0.0598$, (*iii*) 11 gauges are to be allocated to different zones in proportion of their areas. Thus the No. of raingauges in different isohyetal zones are $X_1 = 1.4$, $X_2 = 4.1$, $X_3 = 3.7$, $X_4 = 1.1$ and $X_5 = 0.7$, (*iv*) Additional No. in each zone are:

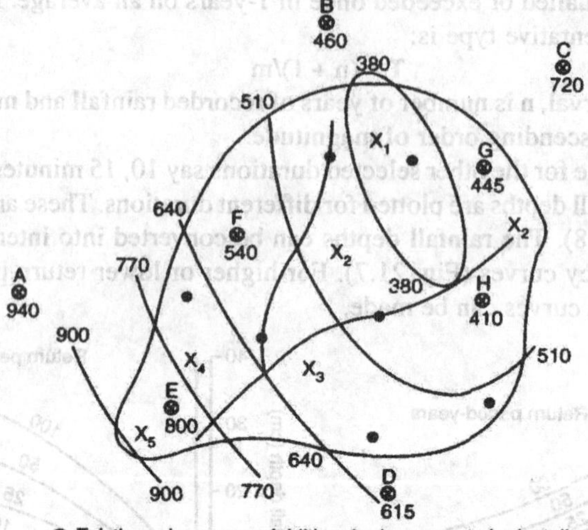

Fig. 21.9 Distribution of raingauges.

Zone	X_1	X_2	X_3	X_4	X_5	*Total*
Existing No.	0	2	1	0	1	4
Additional No. of gauges	1	2	3	1	0	7
Total No.	1	4	4	1	1	11

The additional raingauges are indicated by circles in Fig. 21.9.

Example 21.2 *Compute the average precipitation by the arithmetic average method and Thiessen polygon method from the following data:*

Station No.	*Precipitation (mm)*	*Area (sq km)*	*Station No.*	*Precipitation*	*Area*
1	*30.8*	*45*	*3*	*32.0*	*30*
2	*34.6*	*40*	*4*	*24.6*	*38*

Solution:

By arithmetic average method, $P = (30.8 + 34.6 + 32.0 + 24.6)/4 = 30.50$ mm

By Theissen polygon method, $$P = \frac{P_1A_1 + P_2A_2 + P_3A_3 + P_4A_4}{A_1 + A_2 + A_3 + A_4}$$

$$= \frac{30.8\times45+34.6\times40+32.0\times30+24.6\times38}{45+40+30+38}$$

$$= 30.49 \text{ mm.}$$

Example 21.3 *Estimate the total volume of rainfall water received in m³ over a basin of 4444 ha for the following observations:*

Station	*A*	*B*	*C*	*D*	*E*
Polygon area (ha)	*518*	*777*	*906*	*1495*	*748*
Observed rainfall (mm)	*267*	*198*	*142*	*114*	*81*

Solution: Average depth of rainfall over catchment.

$$= \frac{P_aA_a + P_bA_b + P_cA_c + P_dP_d + P_eA_e}{A_a + A_b + A_c + A_d + A_e}$$

$$= \frac{267\times518+198\times777+142\times906+114\times1495+81\times748}{518+777+906+1495+748} = 146.67 \text{ mm}$$

Total volume of rainfall water over the basin = ($4444 \times 10^4 \times 146.67/10^3$) = 6.52×10^6 m³.

Example 21.4 With the information given below, calculate the mean rainfall over the area:

Steps between isohyets (cm)	*Area (km²)*	*Steps between isohyets (cm)*	*Area (km²)*
127.0 – 139.7	181	165.1 – 177.8	311
139.7 – 152.4	388	177.8 – 190.5	104
152.4 – 165.1	337	190.5 – 203.2	26

Solution:

Average precipitation (P)	*Area (A)*	*Product (P.A.)*	*Average precipitation (P)*	*Area (A)*	*Product (P.A.)*
(127.0 + 139.7)/2 = 133.35	181	24,136	(165.1 + 177.8)/2 = 171.45	311	53,321
(139.7 + 152.4)/2 = 146.05	388	56,667	(177.8 + 190.5)/2 = 184.15	104	19,152
(152.4 + 165.1)/2 = 158.75	337	53,499	(190.5 + 203.2)/2 = 196.85	26	5,118

Total A = 1347 P.A.= 211,893

Average precipitation, P_{av} = (2,11,893)/1347 = 157.31 cm

Example 21.5 *Determine mean rainfall for the catchment from the rainfall data as under:*

Station	*Station weight*	*Rainfall (mm)*	*Station*	*Station weight*	*Rainfall (mm)*
A	*0.04*	*400*	*D*	*0.15*	*500*
B	*0.06*	*300*	*E*	*0.27*	*240*
C	*0.08*	*200*	*F*	*0.40*	*175*

Solution:

Weighted mean rainfall = ($0.04 \times 400 + 0.06 \times 300 + 0.08 \times 200 + 0.15 \times 500 + 0.27 \times 240 + 0.40 \times 175$) = 259.8 mm.

Example 21.6 *The following table gives the data of an isohyetal map. The total area of the basin is 300 sq km. Compute the average depth of rainfall over the basin.*

Isohyet (cm)	*40*	*35*	*30*	*25*	*20*	*15*	*10*
Basin area covered by isohyet (sq km)	—	*17*	*45*	*75*	*165*	*215*	*300*

Solution: *(P.U.)*

Mean isohyets (P)	*Area (A)*	*P × A*	*Mean isohyets (P)*	*Area (A)*	*P × A*
40 – 35 = 37.5	17	637.5	25 – 20 = 21.5	90	2025.0
35 – 30 = 32.5	28	910.0	20 – 15 = 17.5	50	875.0
30 – 25 = 27.5	30	825.0	15 – 10 = 12.5	85	1062.5

Average rainfall, $P_{av} = \frac{\Sigma P \times A}{\Sigma A} = \frac{6335}{300} = 21.12 \text{cm}.$

Example 21.7 *The annual average rainfall of four stations A, B, C and D in an area is 105, 122, 95 and 102 cm respectively. In a particular year station C did not function and the rainfall figures of A, B and D were 90, 97 and 82 cm respectively. Compute the missing precipitation at station C.*

Solution:

	A	*B*	*C*	*D*
Annual average rainfall (cm)	105	122	95	102
Annual rainfall (cm)	90	97	—	82

Since annual rainfall at any of the stations (122 cm at station *B* in this case) differs from that at the station with missing data (95 cm at station *C*) by more than 10 per cent, normal ratio method is applicable.

$$\therefore \quad P_c = N_x/3\,[P_A/N_A + P_B/N_B + P_C/N_C]$$

$$= 95/3\,(90/105 + 97/122 + 82/102] = 77.78 \text{ cm}$$

Example 21.8 *Maximum values of 24 hour precipitation (mm) at a raingauge station are 140, 113, 132, 115, 130, 118, 127, 127, 123, 121. Estimate maximum and minimum precipitation having a recurrence interval of 5 and 15 years. Explain limitations of the method used, if any.*

Solution:

Maximum rainfall			*Minimum rainfall*		
Rainfall (mm)	*m*	*(T = 2n/2m-1)*	*Rainfall (mm)*	*m*	*T = 2n/2m -1*
140	1	20	113	1	20
132	2	6.67	115	2	6.67
130	3	4.0	118	3	4.0
127	—		121	4	1.86
127	5*	2.22	123	5	2.22
123	6	1.82	127	—	—
121	7	1.54	127	7*	1.54
118	8	1.33	130	8	1.33
115	9	1.18	132	9	1.18
113	10	1.05	140	10	1.05

*** Using higher value of *m*, when two rainfall events of the same value.**

Fig. 21.10 indicate the maximum value of precipitation with 5 and 15 years frequency as 131 mm and 137 mm respectively. The minimum values with these recurrence interval are 117 mm and 113.75 mm (Fig. 21.18).

Since the actual frequency of occurrence of maximum event in 21 years is not known and no statistical device can lengthen the period of record, the assumption of recurrence interval of 2*n* to the highest event on the record is the limitation of the Hazen's method used.

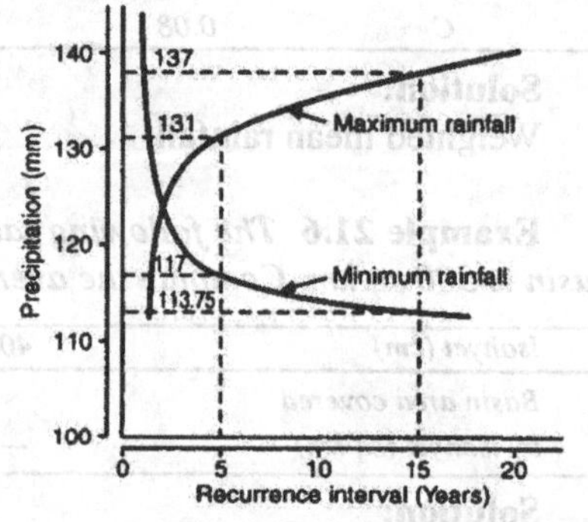

Fig. 21.10 Maximum & minimum precipitation.

Example 21.9 *Estimate the magnitude of rainfall the probability of whose exceedence would be 1 in 10,000 years from the observed values of annual rainfall (in cm) from 1961 to 1985. What shall be rainfall whose return period is 1 in 1,000 years?*

95	*87*	*113*	*98*	*78*	*125*	*140*	*105*	*88*	*67*	*115*	*103*	*127*
99	*102*	*65*	*91*	*130*	*132*	*68*	*60*	*58*	*50*	*82*	*100*	

Solution:

S.No. (m)	*Rainfall (cm)*	*Exceedence probability F = (m/N + 1) × 100*	*S.No. (m)*	*Rainfall (cm)*	*Exceedence probability F = (m/N + 1) × 100*	*S.No. (m)*	*Rainfall (cm)*	*Exceedence probability F = (m/N + 1) × 100*
1	140	(1/25 + 1) × 100 = 3.85 and so on				18	82	69.23
2	132	7.69	10	102	38.46	19	78	73.08
3	130	11.54	11	100	42.31	20	68	76.92
4	127	15.38	12	99	46.15	21	67	80.77
5	125	19.23	13	98	50.00	22	65	84.62
6	115	23.08	14	95	53.85	23	60	88.46
7	113	26.92	15	91	57.69	24	58	92.31
8	105	30.77	16	88	61.54	25	50	96.15
9	103	34.62	17	87	65.38			

The exceedence probability values, as above, are plotted on probability paper (Fig. 21.11).

The rainfall values for return periods of 1 in 10,000 years (exceedence 0.01 per cent) read from the figure comes to 1850 mm and that for 1 in 1,000 years (exceedence 0.1 per cent) as 1700 mm.

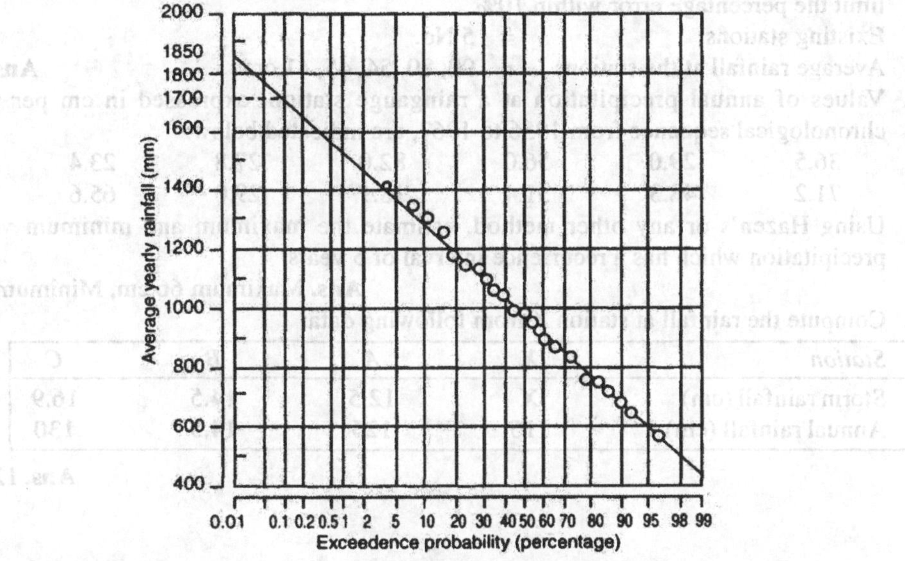

Fig. 21.11 Exceedence probability.

QUESTIONS

1. Define hydrology. What role does hydrology play in the water power planning?
2. Discuss water power engineering application of hydrology.
3. Explain the terms precipitation, stream flow and evaporation. How are these measured? List different methods of finding average depth of rainfall over an area.
4. (*a*) What are the various factors that effect the rainfall characteristics ?
   (*b*) What is the intensity of rainfall and annual rainfall average?
5. (*a*) How would you obtain the average rainfall over a catchment area?
   (*b*) Discuss the various methods of obtaining peak flows.
   (*c*) How long range runoff from a catchment area is estimated?
6. How is the rainfall at a place measured? What are the factors that effect rainfall at a place?
   (*a*) What points will you keep in view while selecting a raingauge station?
   (*b*) Give details of recording and non-recording types of raingauges.
7. What is the practical usefulness of rainfall data analysis?
8. Define raingauge. Name various types of raingauges and discuss with the help of sketch non-recording type rainguage.
9. Discuss the merits and demerits of recording and non-recording raingauges. Explain weighing bucket raingauge.
10. Name the principal sources of errors in rainfall measurements.
11. How will you compute average rainfall over a basin? Which method is superior?
12. Explain how rainfall is measured at a place.
13. Describe different types of recording and non-recording rainguages. What is the difference between continuous and discontinuous type gauges? Draw neat sketches to illustrate any one of them. Explain how will you compute rainfall in catchment from gauge readings.
14. Discuss the various methods of measurements of rainfall.
15. Describe the various methods of calculating mean rainfall from point rainfall records in a given area.

16. (*a*) Give optimum number of raingauge stations required as per I.S.I recommendation.
(*a*) Determine optimum number of raingauges in a basin from the following data so as to limit the percentage error within 10%

Existing stations	=	5 No.
Average rainfall at the stations	=	90, 80, 54, 45, 41 cm.

Ans. 8 No.

17. Values of annual precipitation at a raingauge station, expressed in cm per year in chronological sequence from 1956 to 1967, are indicated below:

36.5	29.0	56.0	82.0	27.8	23.4
71.2	48.3	31.4	18.1	29.0	65.6

Using Hazen's or any other method, estimate the maximum and minimum value of precipitation which has a recurrence interval of 5 years.

Ans. Maximum 66 cm, Minimum 27 cm

18. Compute the rainfall at station *X* from following data:

Station	*X*	*A*	*B*	*C*
Storm rainfall (cm)	X	12.5	14.5	16.9
Annual rainfall (cm)	115	125	14.5	130

Ans. 12.65 cm

22

RUNOFF AND FLOOD ESTIMATION

22.1 RUNOFF AND FACTORS EFFECTING IT

When rain falls, before runoff or surface flow can occur, the precipitation must satisfy the demands of evaporation, interception, infiltration, surface storage, surface detention and channel detention. Runoff is thus the flow collected from the drainage basin and appearing at an outlet of the basin. It includes surface runoff received into the streams after rainfall as perennial or intermittent flow, delayed runoff that enters the streams after passing through portion of the earth, and other delayed runoff that has been temporarily detained as snow cover or stored in natural lakes or swamps. The various factors effecting the runoff are discussed as under:

1. Precipitation Characteristics

(*i*) *Type of precipitation*: Precipitation may be rainfall, snowfall or both. Rainfall immediately produces runoff. Snowfall is a delayed runoff temporarily detained as snow cover.

(*ii*) *Rainfall intensity*: Rainfall intensity, *i.e.* the amount of rain occurring in an unit interval of time, generally expressed in mm per hour, has pronounced effect on the runoff. Rainfall intensity influences both the rate and volume of runoff. Higher the rainfall intensity, greater is the runoff.

(*iii*) *Frequency of intense rainfall*: More frequent the intense rainfall, higher is the runoff.

(*iv*) *Duration of rainfall*: Storm of short duration may produce no runoff, whereas a storm of the same intensity but of long duration results in runoff. More the duration of rainfall, more prolonged is the runoff, as the soil gets saturated and consequently reduced infiltration loss and humid atmosphere slows evaporation loss. The storm duration corresponding to the maximum rate of runoff is termed as time of concentration. It may be defined as the longest time taken by water to travel by overland surface flow from any point of the watershed to the outlet. It depends on various factors such as topography, size and shape of watershed.

(*v*) *Distribution of rainfall*: Distribution of rainfall over the basin has great effect on runoff. When rainfall is concentrated over the basin and the entire watershed contributes, the runoff is maximum. An intense storm on one portion of the catchment results in higher runoff than a moderate storm over the entire watershed.

(*vi*) *Direction of storm movement*: A storm moving in the direction of stream flow results in shorter period of runoff and higher peak than a storm moving upstream.

(*vii*) *Snow*: Snow retards runoff from rainfall and also tends to store a part of such rainfall. Snow accumulates on land surface till rise in temperature melts it to produce runoff.

2. Drainage Basin Characteristics

(*i*) *Size of catchment*: A small area tends to have higher discharge per square kilometre than a large area. Peak flow decreases as the area increases because intense storm is usually over a small area and period of surface runoff increases with area. The square, broad and short watersheds (Fig. 22.1) have almost symmetrical drainage pattern but shorter distance to the outlet.

(*ii*) *Shape of catchment*: Runoff is influenced by the shape of catchment area (Fig. 22.1). Long narrow watersheds are likely to have lower runoff rates than compact watersheds of the same size because runoff from the former does not concentrate as quickly as it does from the latter and long catchments are less likely to be covered by intense storm. Fan-shaped catchments (Fig. 22.1) are indicative of higher runoff because all the tributaries are approximately of the same size and peak flow from the tributaries tends to reach the main stream roughly at the same time.

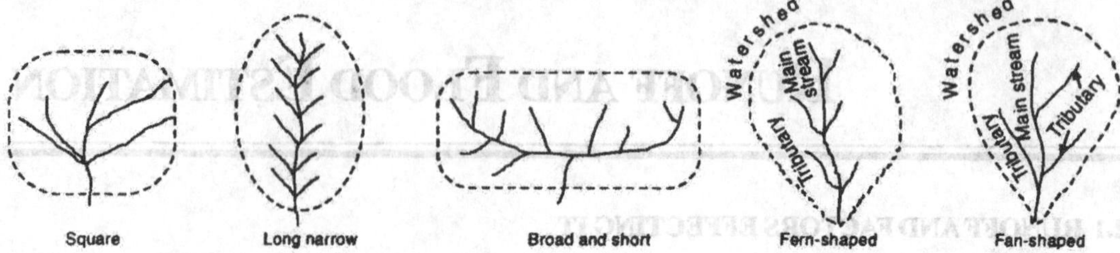

Fig. 22.1 Shapes of catchment.

Fern-leaf catchment gives lesser runoff because the tributaries are generally of different lengths and meet main stream at regular intervals thereby the discharges are likely to be distributed over a large period of time.

(*iii*) *Slope of catchment*: Catchment areas with flat slope and without surface outlets have lower runoff than areas with steep slope and well defined drainage pattern. Runoff from a long slope is generally slower but lasts longer after the rainfall ceases. Effect of slope is less pronounced during the period of prolonged and intense rainfall when runoff usually becomes constant and reverse is the case when rainfall is of short duration.

(*iv*) *Orientation of catchment*: Orientation, *i.e.* general direction of slope of a catchment, with respect to prevailing winds, storm movement and sun's position has considerable effect on the runoff.

3. Geological Characteristics

Type of soil and its permeability greatly influence runoff. Runoff is less if soil and subsoil are pervious as seepage is more. Runoff is more from rocky or impervious soil because absorption is least. Fault zones, fissures and cracks result in rain water finding its way through these openings and surface runoff is less. Seepage water may flow to the stream during low flow if it has outlet within the stream catchment.

4. Meteorological Characteristics

Temperature, humidity and wind effect runoff. High temperature, low humidity and strong wind result in higher evaporation loss and reduced runoff.

5. Geographical Characteristics

(*i*) *Altitude*: Higher catchments usually have steep slope and are less pervious and as such produce greater total runoff though dry season runoff is much lower due to less ground storage. Snow at higher altitudes tends to retard runoff from rainfall and store a portion of such rainfall.

(*ii*) *Drainage net*: Drainage net has an important effect on runoff. More dense and efficient the drainage network, the more flashy is the stream flow and vice versa. Catchments without surface outlets have lower runoff than areas with steep, well defined drainage patterns.

(*iii*) *Nature of catchment surface*: The vegetation and cultivation greatly influence runoff. Thick vegetation and well maintained soil reduce runoff to some extent. Effect of vegetation on low water flow is relatively small compared to that of geological and topographical characteristics of the catchment. Cultivation results in decreased surface runoff.

6. Storage Characteristics

Lakes, reservoirs, check dams, swamps, etc. act as flood moderators and reduce peak flow. They also tend to increase water losses due to evaporation.

7. Hydrological Characteristics:

The volume of runoff depends on the soil and its hydrological condition i.e., moisture control at the time of storm, its organic content, humus and temperature.

8. Other Factors

These include urbanisation. Urbanisation on individual catchment basis leads to an increase in the area of impervious surfaces and alteration of natural drainage patterns resulting in increased volumes and rates of water runoff, increased peak flood over a shorter period of time and consequent high potential for flooding.

22.2 ESTIMATION OF RUNOFF

The estimation of runoff involves the processes of computation of rate of rainfall and how much of it becomes runoff to arrive at volume or peak rate of runoff from a catchment to determine its safe yield and/or discharge expected relative to a design storm over the given watershed. Various methods of estimation of runoff from a catchment are as under:

1. Rainfall-Runoff Coefficient

Runoff coefficient is the ratio of volume of stream flow above base flow to the volume of rainfall. It is expressed as $R = KP$.

Where, R is runoff (cm), P is rainfall (cm), K is runoff coefficient; usual values of K are as given in table 22.1:

Table 22.1 : Runoff coefficient (IS: 8835–1978)

Type of soil	*Runoff coefficient*	*Type of soil*	*Runoff coefficient*
Plateau lightly covered	0.70	Loam, largely cultivated and suburbs with gardens, lawns, macademised roads	0.30
Clayey soils stiff and bare and clayey soils lightly covered	0.55	Sandy soils, light growth	0.20
Loam, lightly cultivated or covered	0.40	Parks, lawns, meadows, gardens, cultivated area	0.05-0.20

2. Runoff Percentage

(*i*) *Binnie's percentage*: Alexander Binnie's runoff percentages, based on data of two rivers in Madhya Pradesh, are as given in Table 22.2 :

Table 22.2 : Runoff Percentage

Annual rainfall (cm)	*Runoff (%)*	*Rainfall (cm)*	*Runoff (%)*
50	15	90	35
60	21	100	38
70	25	110	40
80	29		

(*ii*) *Barlow's percentage*: T.G. Barlow's runoff percentages, based on studies of catchments mostly under 130 km² in Uttar Pradesh, are as under:

Class	*Description*	*Runoff (%)*
A	Flat, cultivated and black cotton soils	10
B	Flat, partly cultivated various soils	15
C	Average	20
D	Hills and plains with little cultivation	35
E	Very hilly and steep, with hardly any cultivation	45

These percentages are for average type of monsoon and are modified by the application of following coefficients according to the nature of season:

Nature of season	Class of catchment				
	A	B	C	D	E
Light rain, no heavy downpour	0.70	0.80	0.80	0.80	0.80
Average or varying rainfall, no continuous downpour	1.0	1.0	1.0	1.0	1.0
Continuous downpour	1.50	1.50	1.60	1.70	1.80

3. Tables and Curves

1. *Strange's Tables and Curves*: Strange gave tables and curves, based on data of Maharashtra, between rainfall and runoff. Strange's table and curves give runoff for daily rainfall for good, average and bad catchments and surface conditions, viz., dry, damp and wet prior to rain (Table 22.3).

Table 22.3 Daily runoff and yield according to Strange runoff percentage and yield

Daily rainfall (mm)	Percentage	Original stage of ground surface				
		Dry	Damp		Wet	
		Yield (mm)	Percentage	Yield (mm)	Percentage	Yield (mm)
5	—	—	4	0.2	7	0.35
10	1	0.10	5	0.5	10	1.0
20	2	0.40	9	1.8	15	3.0
25	3	0.75	11	2.75	18	4.5
30	4	1.20	13	3.9	20	6.0
40	7	2.8	18	7.2	28	11.20
50	10	5.0	22	11.0	34	17.0
60	14	8.46	28	16.8	41	24.6
75	20	15.0	37	27.75	55	41.25
100	30	30.0	50	50.0	70	70.0

For good or bad catchments, add or deduct up to 25 per cent of yield. Fig. 22.2 depicts the Strange's curve.

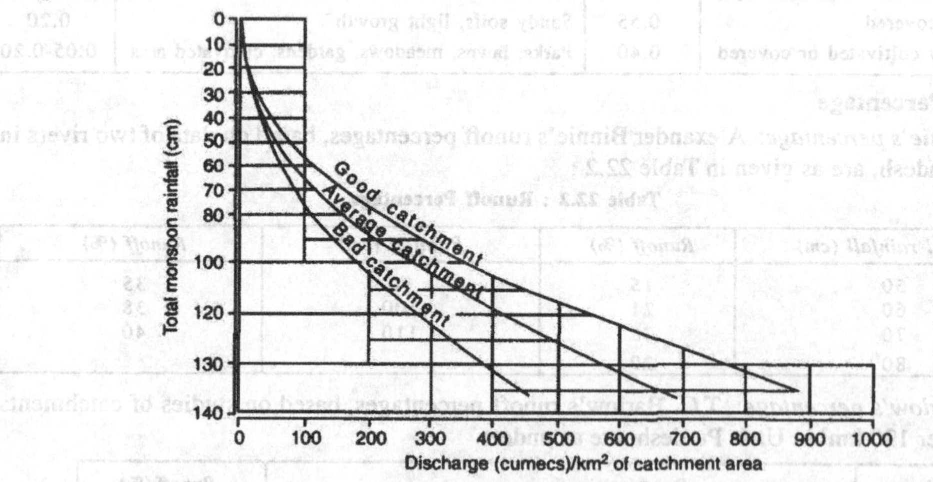

Fig. 22.2 Strange's curve

4. Empirical Formule

1. *Lacey's Formula*

$$R = \frac{P}{1 + \frac{304.8}{P}\left(\frac{F}{S}\right)}$$

Where, R is runoff, P is rainfall, F is monsoon duration factor varying between 0.5 – 1.5 and S is catchment factor depending upon slope; value varies from 0.25 for flat cultivated areas to 3.45 for very hilly areas.

2. *Khosla's formula (1960)*

$$R = (P - T/3.74)$$

where, T is temperature (°C), P is rainfall (cm) and R is runoff (cm).

The formula gives fairly good results for the annual yield for preliminary studies.

3. *Inglis and Desouza's formula (1929)*

$$R = 0.85\,P - 30.5 \text{ for ghat areas of western India}$$

$$R = \frac{(P-17.8)\,P}{254} \text{ for Deccan plateau.}$$

5. Infiltration Method

It is the most reliable method of estimation of runoff. Infiltration is defined as the entrance of water into the soil through the interstices or pores of the soil. Infiltration occurs both prior to and during the occurrence of surface runoff but decreases with time until a minimum rate of infiltration is reached. In rainfall, infiltration begins at high rate and decreases to a minimum rate as rain continues. At any instant, infiltration capacity of a soil is the maximum rate at which water enters the specified soil in given condition. The rate at which infiltration actually takes place during a rainfall, expressed in depth of water per unit of time, usually in mm per hour, is known as *infiltration rate*. It is equal to the infiltration capacity or the rainfall rate whichever is less. The limiting value of infiltration capacity is controlled by the soil permeability and can be determined experimentally by subjecting an experimental plot to rainfall rate in excess of infiltration capacity and by measuring surface runoff (Fig. 22.3). Infiltration capacity curve (Fig. 22.4) is the curve showing the infiltration rate at any period during a specified storm if the rainfall intensity were to equal or exceed the capacity at that instant.

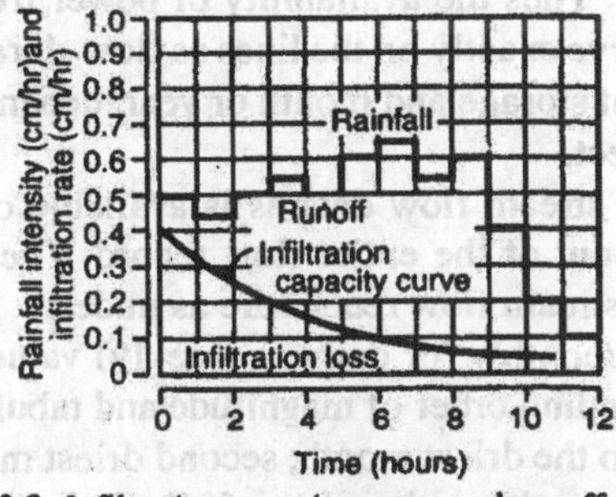

Fig. 22.3 Infiltration capacity curve and runoff

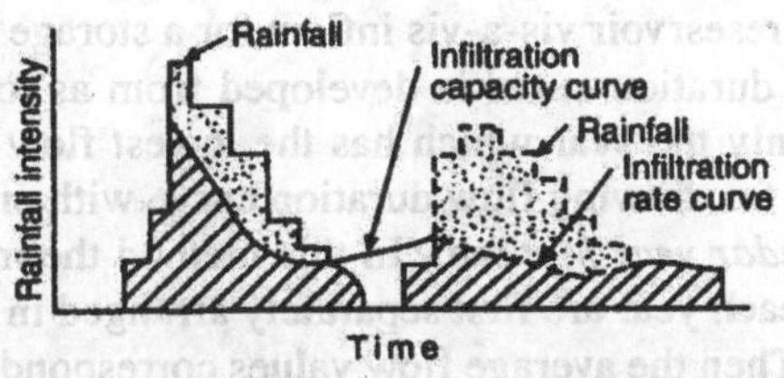

Fig. 22.4 Typical infiltration rate and infiltration capacity curve.

Infiltration index is the average rate of loss such that the volume of rainfall in excess of that rate will be equal to the direct runoff. Runoff from large area is determined by using infiltration indices. Three common types of infiltration indices are:

1. ϕ*-index*: The average rainfall intensity above which the volume of rainfall equals the volume of observed runoff and equals total basin recharge divided by duration of rainfall provided rainfall intensity is continuously in excess of basin recharge. The index (Fig. 22.5) is termed as ϕ -index:

Fig. 22.5 Typical ϕ -index and runoff.

$$\phi = \frac{\text{Basin recharge}}{\text{Rainfall duration}}$$

2. *W-index*: *W*-index is the average rate of infiltration during the time the rainfall intensity exceeds the infiltration capacity, that is

$$W = (P - R)/t_r \text{ cm/hr}$$

where, t_r is duration of rainfall (hr), *P* is precipitation (cm), *R* is runoff (cm). Essentially, *W*-index is equivalent to ϕ-index minus the average rate of retention or depression storage.

3. W_{min}*-index*: With very wet conditions, when the retention rate is at minimum, the values of *W*-index and ϕ-index are almost identical and *W*-index is termed as W_{min}-index.

The infiltration index can be used to estimate runoff coefficient *K* from the following relation:

$$K = (P - W)/P$$

Where, *P* = rainfall intensity (cm/hr).

The use of infiltration indices does not constitute a rational application of infiltration theory and results obtained are considered empirical.

22.3. FLOW-DURATION CURVE

It is a curve which represents the relation between flow and time. The curve is a plot of stream flow (Q) in ascending order of magnitude as ordinate against the percent of time as abscissa, the curve thereby shows the flow equalled or exceeded for any desired percentage of time covered by the record. The stream flow data may be daily, ten daily or monthly values; the curve of average daily discharges gives more accurate results than those obtained from the monthly average values. The curve differs from the hydrograph in that the flow values are plotted irrespective of the time of occurrence whereas in the hydrograph these are in the calendar order or actual order of occurrence. The area under the curve gives the total amount of flow during the period.

The flow-duration curve is a useful tool for studying the stream flow variability for water power development. The curve representing a long period data can be used for the estimation of the amount of power that can be produced under average conditions by the conversion of discharge scale into power units corresponding to the available head. Thus the availability of power from a hydro plant can also be represented by a power-duration curve exactly on the lines as flow-duration curve. The time period may be a day for a diversion without storage and month or year, depending on the size of resesrvoir vis-a-vis inflow for a storage project.

The flow duration curve is developed from as long a stream flow data as is available or by considering only the year which has the lowest flow rate out of the entire flow record. The two methods used for drawing flow duration curve with entire stream flow record are as under:

(*i*) *Calendar year Method*: In this method the mean monthly (or daily or weekly) values of flow rates of each year are first separately arranged in ascending order of magnitude and tabulated in that order. Then the average flow values corresponding to the driest month, second driest month and so on up to the wettest month are determined by taking arithmetic mean of all values of the same rank. These average values are then plotted to give flow duration curve. The curve has 12 points or months of the average year based on the number of years for which the record is available. The number of points on the curve are more in case day or week is adopted as a time unit with percent of time scale as abscissa.

(*ii*) *Total period Method*: In this method the mean monthly (daily or weekly) flow rates for the entire period, irrespective of calendar year in which they occurred, are arranged in ascending order of magnitude. With these values the flow duration curve is prepared. For 10 years of monthly record; there are 120 points to draw the curve. Alternatively, continuous ranges of flow varying by convenient amount are considered for each year and separately the number of days during which the daily stream flow is recorded in a particular range are noted. The number of days, for all the years, are considered and those having the daily flow in the same range are added and percent of time the daily flow occurred in a particular range is determined. The lower value of the range of flow rate is then plotted against the corresponding percent of time to give flow duration curve. The total period method gives more correct results than calendar year method which averages out extreme values.

Effect Of Time Unit on Curves

The flow duration curve based on daily flow data is more precise as it gives more flow details than the curves based on weekly or monthly flow data because the latter averages out values and on this account tend to be smoother. A plot of the weekly and monthly flows reveal that the former is above the latter in the wet period of the year and below in the dry period (Fig. 22.6).

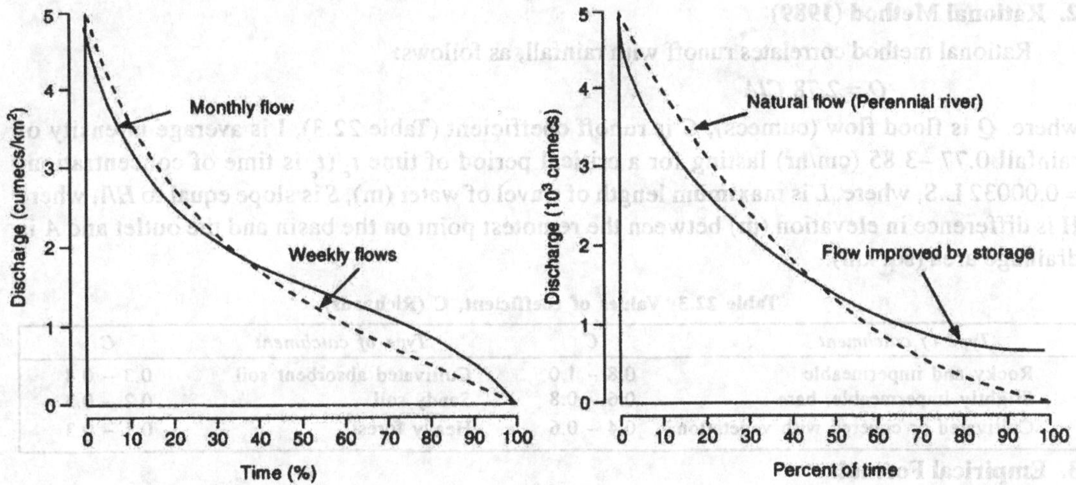

Fig. 22.6. Flow duration curves.

Effect of Storage on Curve

The effect of storage on flow duration curve is to raise the curve above the values below the mean and to lower it at high flow portion. Since usually in a typical flow duration curve, the flow is equalled or exceeded 30 to 40 percent of time, the effect of storage is to tend to raise the lower two-thirds of the curve and to lower the upper one-third.

Limitation

(*i*) The considerable variation in the monthly flows is not reflected in the curve as it is drawn from mean monthly flows, (*ii*) It does not represent the flows in the natural sequence of occurrence, and (*iii*) It is not possible to indicate whether the lowest flows occurred in consecutive periods or scattered throughout the period of data.

22.4 FLOOD ESTIMATION

Flood is defined as the flow pattern in a stream constituting a distinct progressive rise culminating in a peak or summit together with the recession that follows crest. Flood is the maximum rate of runoff that is unusually high stage of the river. The highest peak discharge in a water year is termed as annual flood.

Flood estimation is essential for the design of irrigation, hydel power and flood control projects. On proper selection of the design flood depends the safety and cost of such structures. An excessive flood implies unnecessarily costly structures and too small a design flood, on the other hand, involves a high risk not only of the total failure of the structure and services rendered by it, but also safety of persons and property located downstream.

Methods of Estimation of Design Flood:

Various methods of estimation of flood discharge are as under:

1. Factor of Safety to Maximum Historical Flood

It is the earliest method of estimating flood discharge, obtained by application of a suitable factor of safety, usually 1.5, to the observed maximum flood at site or at nearby site on the stream. An extension of the method is that from high flood marks the cross sectional area of the river is computed, water surface slope determined and flood discharge calculated by slope area method. Coefficient of rugosity is assumed or computed by substituting data from a few measured floods.

2. Rational Method (1989)

Rational method correlates runoff with rainfall, as follows:

$$Q = 2.78\ CIA$$

where, Q is flood flow (cumecs), C is runoff coefficient (Table 22.3), I is average intensity of rainfall 0.77 –3.85 (cm/hr) lasting for a critical period of time t_c (t_c is time of concentration) = 0.00032 L.S, where, L is maximum length of travel of water (m), S is slope equal to H/l, where H is difference in elevation (m) between the remotest point on the basin and the outlet and A is drainage area (sq. km).

Table 22.3 Values of coefficient, C (Richards)

Type of catchment	*C*	*Type of catchment*	*C*
Rocky and impermeable	0.8 – 1.0	Cultivated absorbent soil	0.3 – 0.4
Slightly impermeable, bare	0.6 – 0.8	Sandy soil	0.2 – 0.3
Cultivated or covered with vegetation	0.4 – 0.6	Heavy forest	0.1 – 0.3

3. Empirical Formulae

Dicken's formula (1865), $Q = CA^{3/4}$

Where, Q = discharge (cumecs), A = catchment area (km²), C = coefficient; values are northern India plains = 6, northern India hilly = 11-14, central India = 14 – 28, western India = 22 – 25, Maximum value = 35, Actual observed value in Northern India = 22. Dicken's formula is used in central and northern India.

Ryve's formula (1884), $Q = CA^{2/3}$

Where, C = 6.8 within 80 km of east coast, 8.5 within 80-160 km of east coast, 10.2 limited area near hills, and 40 actual observed value.

The formula is used in Tamil Nadu and parts of Andhra Pradesh and Karnataka states.

Inglis's formula, $Q = \dfrac{124A}{\sqrt{A+10.4}} \simeq 124\,A^{1/2}$

The formula is used in Maharashtra state for small catchments.

Ali Nawaz Jang Bahadur's formula, $Q = C(0.3906A)^{[0.925 - (1/14)\log 0.3906\,A]}$

Where, C = coefficient, values, north India = 60, South India = 48 and Maximum value = 86.

The formula gives better results than Dicken's and Ryve's formulae for areas below 25,000 sq km.

Fuller's formula: $Q_{max} = CA^{0.8}\,(1 + 0.8 \log T)\,(1 + 2.67\,A^{-0.3})$

Where, Q_{max} = maximum flood (cumecs) having recurrence interval T years, T = number of years in which maximum flood is considered, A = area of drainage basin (sq km), and C = constant, value varying from 0.026 to 2.77.

Creager's method: Creager's flood frequency method is used in case sufficient data is not available

$$Q = CA^{0.5}\left[2 - 1.374\left(e^{-0.04A^{-0.3}}\right)\left\{1 - \frac{\log_e^{0.1T}}{3}\right\} + \frac{\log_e^{0.1T}}{3}\right]$$

Where, Q = discharge (cumecs) for a period of T years, A = area (sq km), and C = constant of value 1405 for areas most favourable to large floods.

Percentage increase method: Central Water and Power Commission formulae for obtaining percentage increase (P) to the maximum known flood in a record of N-years, to evaluate 100 years flood are:

For North and Central Indian rivers $P = 50 - 0.4\,N$

For South Indian rivers $P = 50 - 0.45\,N$

Where, P = percentage increase to be made to obtain design flood, and N = number of years for which known flood is the maximum (minimum 25 to 30 years).

The empirical formulae have regional application and do not give clue to the probable frequency at the computed flood figure.

4. Envelope Curve

Creager Envelope Equation: Justin, Creager and Hinds obtained following equation, for the envelope curve of maximum floods:

$$Q = (A/2.95)^{0.894\,(A/2.95)^{-0.048}}$$

Kanwar Sain envelope curve: Fig. 22.7 indicates envelope curves for South, North and Central India rivers.

The method is better than empirical formulae in that selection of coefficient based on judgement is eliminated. The order of frequency of floods derived is somewhat greater than the longest period of record.

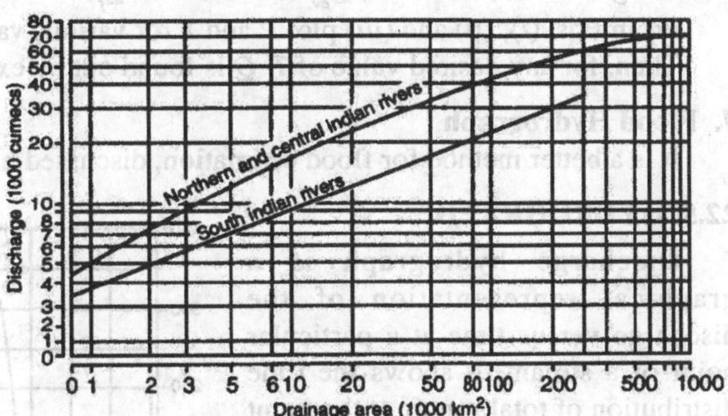

Fig. 22.7 Envelope curve

5. Probability or Flood Frequency Method

The probability method requires data for fairly long period; minimum of 30 years. In this method, record of peak flood discharges are interpreted in terms of future probabilities of occurrence to determine the recurrence interval of floods of given magnitude. The recurrence interval, also called return period, is the average interval of time within which the magnitude X of a flood event will be equalled or exceeded once. The recurrence interval can be determined, if a particular flood discharge has a serial number m, by the following methods:

1. *California method (1923)*: $T = N/m$
2. *Hazen's method (1930)*: $T = 2N/2m-1$
3. *Weibull method (1839)*: $T = (N+1)/m$
4. *Gumbel's method*: $T = \dfrac{N}{m+C-1}$

Where, C = Gumbel's correction. It depends on m/N ratio, given in Table 22.4, T = recurrence interval (years), n = number of events *i.e.*, years of record, and m = order of rank of event, highest flood (m = 1) and $m = n$ for the lowest flood events.

Table 22.4 Gumbel's C

m/N	1.0	0.9	0.8	0.7	0.6	0.5	0.4	0.3	0.2	0.1	0.08	0.04
c	1.0	0.95	0.88	0.845	0.78	0.73	0.66	0.59	0.52	0.40	0.38	0.28

In order to find any design flood of desired frequency a plot of Q vs T (or F) is prepared on a probability graph. Unknown flood for any value of T is determined from the plot.

Let $Q_1, Q_2, \ldots\, Q_n$ comprise a series of extreme values of floods, the probability of occurrence of a value equal to or less than Q is given by the relation,

$$P = e^{-z} \qquad \ldots(i)$$

Where,

$$Z = e^{-y} \qquad \ldots(ii)$$

such that $Y = a(Q - Q_f)$ ...(*iii*)

where, Q_f = value of flood discharge having highest frequency

$= Q_{av} - 0.45\ \sigma x$

Q_{av} = average value of discharge

$$\sigma x = \sqrt{\frac{(Q_1 - Q_{av})^2 + \ldots + (Q_n - Q_{av})^2}{N}}$$

a = constant = $1.28/\sigma x$

Recurrence interval of Q in years, $T = 1/(I - P)$

For given data, find σx, Q_{av} and calculate Q_f.

From Eqs. (*i*), (*ii*) and (*iii*) plot P and T for various values of y.

Then, for any desired value of T, Q is found out by extrapolation.

7. Flood Hydrograph

It is a better method for flood estimation, discussed hereinafter:

22.5 HYDROGRAPH

Discharge hydrograph is a graphical representation of the discharge versus time at a particular point of a stream. It shows the time distribution of total runoff at the point of measurement. Fig. 22.9 shows a typical single peaked simple hydrograph resulting from an isolated storm. A hydrograph consists of:

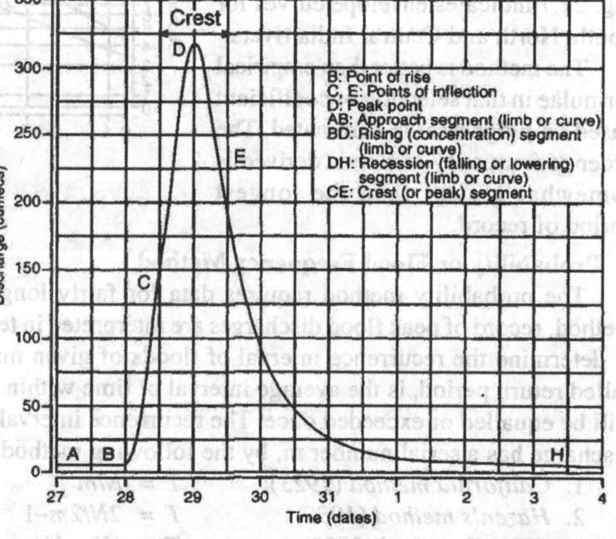

Fig. 22.8 Typical single peaked hydrograph.

(*a*) A rising limb or concentration curve:

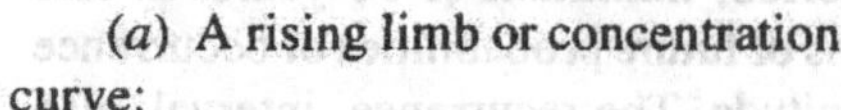

(*b*) A peak or crest segment:

(*c*) Falling limb or recession curve:

Base Flow Separation

Three procedures for the separation of base flow are as under:

1. A simple procedure consists of drawing a straight line from the point of rise to an arbitrary point on the lower portion of falling limb of the hydrograph line *bd* (Fig. 22.9). An arbitrary point is roughly located at a time $N = 0.84A^{0.2}$ days after the peak of hydrograph, A is the drainage area in sq km.

2. Another simple procedure is to extend the recession prior to storm to a point under a little beyond the peak of the hydrograph, line *bc* (Fig. 22.9). From this point, a straight line is drawn to the arbitrary point on the hydrograph determined as above.

3. The third procedure makes use of composite base flow recession curve. Curves *aa′* and *bb′* (Fig. 22.10) corresponding to recessions of lower end and approach segment of the hydrograph are first drawn to match the composite curve. The connecting curve *cc′* is then drawn arbitrarily to join the two portions *aa′* and *bb′* smoothly in such a way that its crest point lies below the point of inflection.

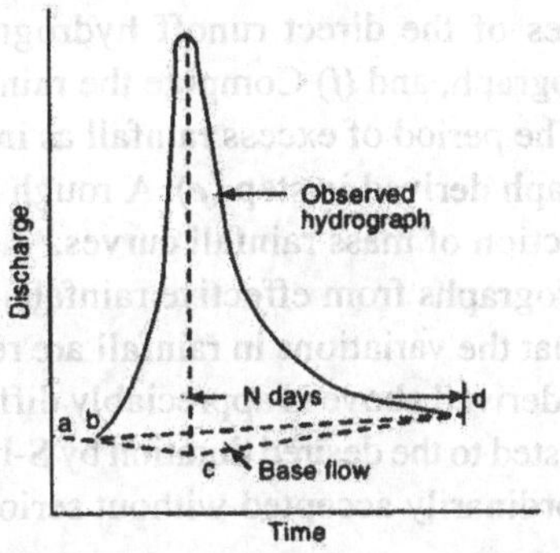

Fig. 22.9 Base flow separation by simple methods.

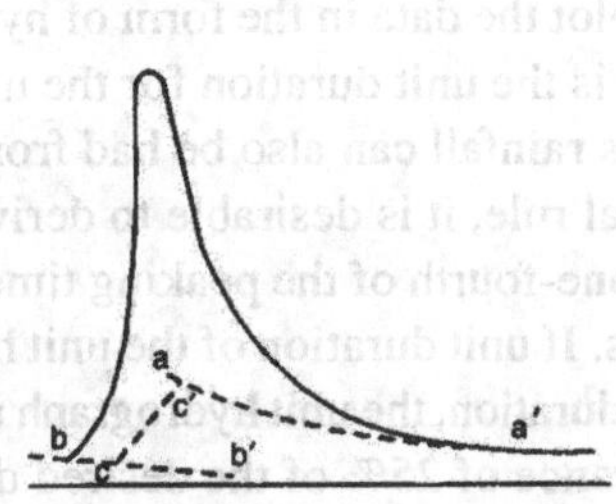

Fig. 22.10 Base flow separation by composite base flow recession curve.

22.6 UNIT HYDROGRAPH METHOD

Unit hydrograph method is widely used for flood estimation and flood prediction. Unit hydrograph of a drainage is a hydrograph of direct runoff resulting from 1 cm of effective rainfall of a unit duration generated uniformly over the basin area at a uniform rate in a unit of time. It is assumed that the hydrograph resulting from the storms of same duration have similar shapes having ordinates proportional to the total direct runoff. The unit rainfall duration refers to the smallest period of time or duration applicable to and commensurate with the catchment in which the rainfall can be assumed to be uniform. It should not exceed the period during which the design storm rainfall is uniform in intensity in various portions of the basin under study. The best unit duration is about one-fourth of the basin lag time, *i.e.*, the time from the middle of the block of excess rainfall to the peak of the observed hydrograph. It may vary from one day in the case of large catchments to a few hours for small catchments, but usually taken as 6, 12 or 24 hours. For example, a 6-hour hydrograph results from 6-hour unit rainfall duration, meaning thereby a discharge hydrograph produced by surface runoff from a storm of 6-hour duration and yielding a rainfall excess of 1 cm.

The basic premise of unit hydrograph principle is that *R* cm effective rainfall over the basin in the same unit duration will produce a direct runoff hydrograph of the same time base as the unit hydrograph but having ordinates *R*-times those of the unit hydrograph. Thus a rainfall excess of 2 cm within the unit duration will produce a runoff hydrograph having ordinates twice as great as those of unit hydrograph.

1. Unit Hydrograph from Simple Hydrograph

The most direct method of deriving a unit hydrograph involves the analysis of a simple hydrograph resulting from an isolated unit storm that produces reasonably uniform effective rainfall rates within its duration. The procedure of unit hydrograph construction is (*i*) From the past discharge records, select an isolated, well defined and single peaked flood rises of short duration, (*ii*) Runoff volume for all such floods are computed roughly from the daily average flows and those having considerable volume are selected, (*iii*) The instantaneous hydrograph for each selected floods is plotted using the gauge records and suitable rating curve of the site, (*iv*) Each hydrograph is then analysed; various steps involved are (*a*) Analyse the rainfall data related to the hydrograph and prepare mass curves and the hyetograph for the storm period under consideration, (*b*) Separate the base flow from the stream discharge hydrograph to get hydrograph of the direct runoff. The observed hydrograph may have to be modified to eliminate the runoff from preceding or subsequent rainfall, (*c*) Find the ordmates of direct runoff subtracting the ordinates of base flow from the total ordinates, (*d*) Calculate the volume under hydrograph of direct runoff (in cm of depth over the basin) by planimeter or by counting squares (1 cm runoff from 1 sq km = 2.78 cumec hours), (*e*) Calculate the ordinates of unit hydrograph by dividing the ordinates of the direct runoff hydrograph by its volume as computed in step (*d*) to obtain the unit hydrograph, and (*f*) Compute the rainfall excess quantities and plot the data in the form of hyetograph. The period of excess rainfall as indicated by the hydrograph is the unit duration for the unit hydrograph derived in step (*e*). A rough idea of the period of excess rainfall can also be had from the inspection of mass rainfall curves.

As a general rule, it is desirable to derive unit hydrographs from effective rainfalls of periods equal to about one-fourth of the peaking time in order that the variations in rainfall are reflected by the hydrographs. If unit duration of the unit hydrograph derived above is appreciably different from the desired unit duration, the unit hydrograph may be adjusted to the desired duration by S-hydrograph method. A tolerance of 25% of the desired duration is ordinarily accepted without serious error.

2. Unit Hydrograph from Complex Hydrograph

Where simple hydrographs from isolated unit storm are not available for deriving unit hydrograph, a complex hydrograph (hydrographs) resulting from a storm of varying intensity is studied. In case of two peaked complex hydrograph resulting from sufficiently separated storms, the two storms are separated by the use of a direct runoff depletion curve and base flow separated. The unit hydrographs are then developed from each storm. However, in the case of extended period of rainfall wherein hydrograph cannot be separated into portions contributed by each rain burst, the complex hydrograph is analysed by the method of successive approximation or Collin's method*.

3. S-Hydrograph (S-Curve)

When a unit hydrograph of a desired unit duration cannot be developed by analysis of available data, it is derived by conversion of the available unit hydrograph. In general, if a unit hydrograph of a given unit duration is available, unit hydrograph of other durations can be derived by the application of the principle of superposition. If other durations are integral multiples of the given duration, the computations are relatively easy. For example, to derive a 6-hour unit hydrograph from a 2-hour unit hydrograph, three 2-hour unit hydrographs with their initial points separated by 2-hour are summed up to yield a hydrograph of 3 cm of runoff during a 6-hour period. If the ordinates of this summation graph are divided by 3, a 6-hour unit hydrograph is obtained. A general method based on the principle of superposition is the S-hydrograph method.

The S-hydrograph is a graph showing the summation of the ordinates of a series of unit hydrographs spaced at unit rainfall duration intervals. It represents the hydrographs of average rate of rainfall excess of the unit duration continued indefinitely. This method is quite suitable when uniform spells of rainfall have been recorded at greater time interval than the unit duration applicable to the catchment. However, it is not so suitable for smaller catchment areas where rainfall is recorded at shorter intervals not exceeding the unit duration.

The S-hydrograph is essentially a hydrograph produced by a continuous effective rainfall at a constant rate for an indefinite period. It can be shown that the ordinates of the S-hydrograph ultimately approach the rate of effective rainfall at a time of equilibrium. If I cm per hour is the constant rate of continuous effective rainfall and A is the drainage area in sq km, the discharge at the time of equilibrium is equal to $2.78\,AI$ cumecs. The S-hydrograph can be constructed graphically by summing up a series of identical unit hydrographs spaced at intervals equal to the duration of effective rainfall from which they were developed.

4. Synthetic Unit Hydrograph

Synthetic unit hydrograph is a unit graph developed on the basis of estimation of coefficients expressing various physical features of a catchment. Synthetic unit hydrograph is developed, based on known physical characteristics of the basin, where adequate rainfall, runoff data is lacking. Among several known methods of synthetic unit hydrograph, the one developed by Synder is most

* For details, see authors *Hydrology and Water* Resources Engineering.

commonly used having the following relations:

Lag time, $t_p = C_t (L.\ L_c)^{0.3}$ (*i*)

Standard duration of net rain, $t_r = t_p/5.5$ (*ii*)

Peak flow, $Q_{PR} = (C_p.\ A)/t_p$ (cumecs) (*iii*)

Time base, $T = 3 + 3\ (t_p/24)$ (days) (*iv*)

Peak flow/sq km, $q_p = Q_p/A$ (*v*)

For any duration (t_r') other than standard duration (t_r), the basin lag time

$$(t_p') = t_p + \frac{(t_r' - t_r)}{4}$$

$$= \frac{21}{24} t_p + \frac{t_r'}{4}$$

Time to the beginning of rising limb $= \frac{t_r'}{2} + t_p'$

$$\therefore \quad Q_p = C_p A / t_p'$$

Where, t_p = lag time from mid-point of effective rainfall duration t_r to peak of a unit hydrograph (hours), t_r = standard duration of effective rainfall (hours), Q_p = peak discharge of unit hydrograph for standard duration t_r, C_p = coefficient; value ranging from 4 to 5, C_t = a coefficient value ranging from 0.2 to 2.2, L = basin length (km) measured along the water course from the basin upstream boundary to the gauging station, L_c = distance (km) along the main water course from the gauging station to a point opposite the centroid of basin, and T = base length of unit hydrograph for t_r in days.

For sketching the unit hydrograph, U.S. Army Corps of Engineers relations for widths of unit hydrograph at 50% and 75% of the peak discharge are:

$W_{50} = 5.6/q_p^{1.08}$ and $W_{75} = 3.21/q_p^{1.08}$, q is peak discharge (cumecs/km²).

22.7 STREAM GAUGING

Discharge, also termed as rate of flow, is defined as the volume of water flowing through a cross-section in a unit time such as cubic metres per second (cumecs). The objects of discharge measurement are accurate assessment of the available and dependable natural water resources of a river, their tributaries and streams for the development of irrigation and power potential, flood control, water supply, navigation, etc., and issue of flood warnings, regulation and equitable distribution of supplies in a canal system, determination of seasonal and annual variations in runoff and estimation of water losses and regeneration in different reaches of the river and planning for inland navigation.

The need for extensive collection of stream flow data on a scientific basis is well recognised. The World Meteorological Organisation has laid down norms as given in Table 22.5, for the minimum network of stream gauging stations:

Table 22.5. WMO norms

Terrain	*WMO norms (1 station for)*
Flat region	3000 – 10,000 km²
Hilly region	1,000 – 5,000 km²
Arid region	No norms

22.8 SELECTION OF GAUGE AND DISCHARGE SITES

The stage or gauge is the elevation of water surface of the stream measured relative to a datum which can be related to mean sea-level or any arbitrary datum connected to the former.

The water depth is read at gauge site and the rate of flow is determined at discharge site. The essential requirements of a gauge site are, (*i*) It offers a place where, once fixed, the gauge is permanent, *i.e.* immune from shifting due to silting or scouring, and (*ii*) Registers always the same elevation of water surface corresponding to the given discharge. The various types of gauges are

Vertical staff gauge, Inclined gauge, Hook gauge, Temporary gauge, and Crest stage gauge.

Discharge Site:

Gauge reading and discharge observation are complementary operations and as such the site selected for the installation of the gauge shall preferably be suitable for discharge observations as well. If site conditions are such that the two sites have to be distinct, these shall be as close as possible. Besides, the two essential requirements indicated for gauge site, the discharge measurement site in plains shall meet the additional requirements such as (*i*) At or as close to the gauge site as possible in the case of velocity measurement by current meter or near the middle of a straight reach where velocity measurement is done by a float, (*ii*) Located in a straight reach of such length as is the least of 3 to 4 times the river width or 0.75 km during high floods, both upstream and downstream of the normal cross section where velocities are to be measured, (*iii*) The flow in the straight reach is streamlined, steady, free from cross currents, reverse slope and eddies and showing less departure from normal flow at all stages of the river, (*iv*) The water flow is preferably in a single channel. Two straight channels are, however, preferable to one defective channel, (*v*) Site is free from aggradation and degradation, (*vi*) Located where the greatest range of fluctuations in river stage occurs, (*vii*) Reasonably uniform cross section of the river channel in the selected reach, (*viii*) Site, as far as possible, is free from trees and obstruction as is likely to interfere with flow and clear vision during observation, (*ix*) Section line be as nearly as possible at a right angle to the direction of the flow, (*x*) Site is free from the tendency of formation of vortex or development of backward flow, (*xi*) Located preferably away from bridges or other hydraulic structures which are likely to affect the flow of water, (*xii*) Located neither on the upstream of a confluence nor in its close proximity, (*xiii*) Reasonably stable bed and banks for affording a regular gauge-discharge relationship, (*xiv*) The banks are high enough to contain HFL and have more or less regular side slope to permit fixation of sloping gauges to record both low supplies and high floods at the same location, (*xv*) Easily accessible at all times of the year.

Besides the above, the discharge measurement sites in the hills shall meet the requirements (*i*) Straight reach of minimum length of 150 m above section line, (*ii*) Free from projections in the bed and sides, (*iii*) Minimum 150 m removed from incoming torrent upstream and downstream from the section line, (*iv*) A bridge without piers or rope bridge, and (*v*) Nearest to a proposed dam site.

22.9 SEGMENTATION

The river cross section is divided into suitable compartments or vertical sections and total discharge is computed as the sum of the discharges in various sections on the basis of the depths and velocities observed at these verticals. Fixing the numbers and positions of verticals is termed as segmentation (Fig. 22.11). In general, the interval between two verticals is not more than 1/20 of total width and discharge between two verticals is not more than 10% of the total discharge. Table 22.6 indicates the intervals at which the depths of water are measured along the cross section on rivers of different widths as specified in IS: 1192-1981.

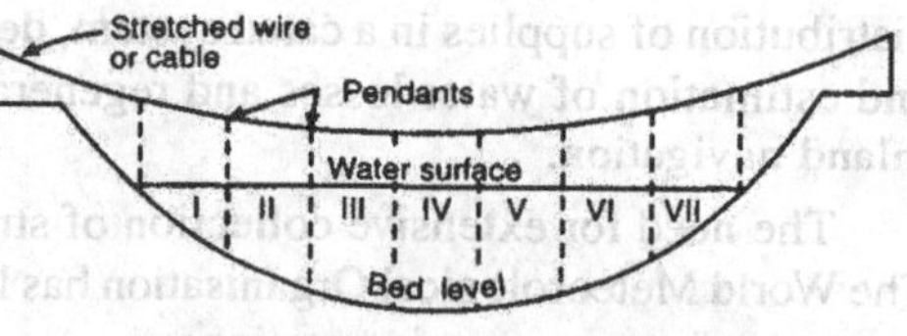

Fig. 22.11 Segmentation.

Table 22.6 Width of interval for measuring depths

Rivers channel width (m)	*Observation points (No)*	*Segment width (m)*
Less than 15 m or in channels where river bed changes abruptly	15	1.5
Between 15 to 90	15	6.0
Between 90 to 180	15	15.0
Over 180	25	Variable

* For details refer "Hydrology and Water Resources Engineering" by the authors.

The verticals for depth and velocity measurements are kept the same according to Table 22.6. In velocity measurement the maximum spacing between adjacent verticals is so maintained that the mean velocity does not differ by more than 20% with respect to the lower value of the two velocity measurements.

22.10 DISCHARGE MEASUREMENT

The methods for discharge measurement are direct and indirect. The important direct methods are velocity-area method using current meter, floats and chemical dilution method. The indirect methods include slope-area method and using flow-measuring structures such as gated structures, flumes, weirs etc.

1. Velocity Measurement by Current Meter

The current meter is the instrument used for velocity measurement. The velocity varies approximately as a parabola, from zero at the bottom of the channel to a maximum at or near the water surface at top. The maximum velocity is observed to occur at 0.15 or more precisely between 0.10 to 0.20 of the depth below water surface at the vertical. The ratio of maximum velocity to the mean velocity is approximately 1.2. The mean velocity on any vertical section occurs at a depth 0.6 of the total depth.

Velocity is also measured at 0.2 and 0.8 of depth of water and the average of two is taken as the mean velocity in the vertical. For more accuracy the velocity is measured at 0.2, 0.6 and 0.8 of the depth of water at each vertical and mean of the three values is taken as the mean velcoity in the verticals. Alternatively 0.6 depth measurement is weighted and Vmean = 0.25 ($V_{0.2} + 2V_{0.6} + V_{0.8}$). When the river is in flood the current meter is submerged either to a constant depth below the water surface or to a certain percentage of depth surface velocity is observed with a float. The mean velocity is usually taken as 0.89 of the observed surface velocity.

2. Velocity Measurement by Floats

Velocity measurement by floats (Fig. 22.12) is resorted to when it is not possible to use current meter due to unsuitable velocities or depths or quick observation is to be made. It is a very simple method. The surface velocity of flow is ascertained from the time noted during which a float traverses a known distance called float run. Three cross sections are selected along a straight reach of the channel spaced far apart so as to give a travel time of 20 seconds or less in high velocities. Velocity measurements by floats are less accurate (+ 10%) than current meter observations. The surface velocity is reduced to mean velocity by the application of a suitable reduction coefficient, usually taken as 0.89, 0.85 as per IS: 3911-1994.

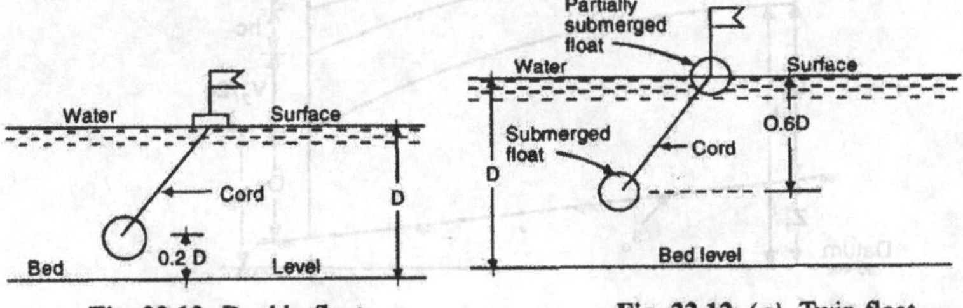

Fig. 22.12 Double float. **Fig. 22.12** *(a)* Twin float.

22.11 DISCHARGE COMPUTATION

The various methods of discharge computation are briefly as under :

1. Velocity area Method: It is the most accurate method, making use of observed depth and velocity measured by current meter or float. The two methods of discharge computation which is essentially a product of velocity and area are as under:

(*i*) *Mid-section Method*: The cross section is considered to be made up of a number of compartments each bounded by two adjacent verticals. The width is taken as the sum of half the distances to adjacent verticals. The first and last verticals are sited near to the banks as flow at the edges is omitted. Mean velocity is observed at 0.6 depth at verticals. Referring to Fig. 22.16, $q_1 = a_1 V_1$, $a_1 = d_1 [(w/2 + (w_i + 1)/2])]$, $i = 1, 2, 3, \ldots N\text{-}2$, d_i is depth at ith segment and V_1 is average velocity at ith vertical. For first segment $i = 1$ and last segment $i = N - 1$. V_d in two half widths next to banks is estimated.

Total $Q = V_1 d_1 = \left[\frac{b_1 + b_2}{2}\right] = V_2 d_2 = \left[\frac{b_2 + b_3}{2}\right] + \ldots\ldots\ldots$

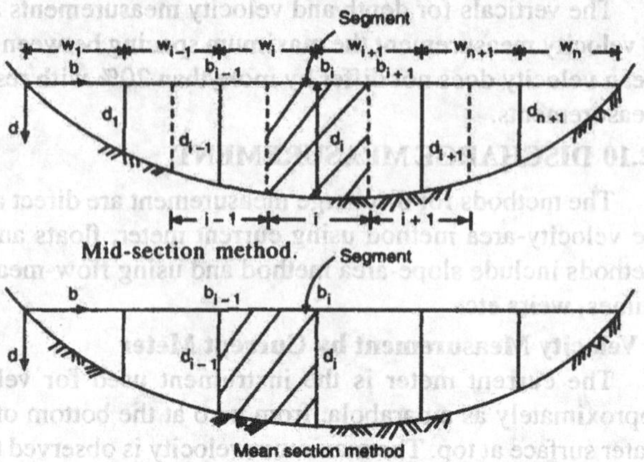

Fig. 22.13 Discharge Computation

(*ii*) *Mean section method*: Width of strip is the distance between two verticals. Mean depth is the average of the depths in two verticals.. Velocity is the average value of mean velocities in two adjacent verticals (Fig. 22.13). Discharge in each strip $q = b\left[\left(\frac{d_1 + d_2}{2}\right)\left(\frac{V_1 + V_2}{2}\right)\right]$ and $\Sigma q = Q$.

The first and last verticals are sited as near to the banks as possible, since discharge in end segments is omitted.

2. Slope area Method: This method provides a rough estimate of the discharge and is used when actual measurements by velocity area method are not possible. In this method the average cross sectional area is multiplied with the velocity derived from observed water surface slope (Fig. 22.15). The cross sectional area is measured at three locations, two at the ends and one in the centre of a straight river reach of length at least 10 times the river width or 300 m whichever is more.

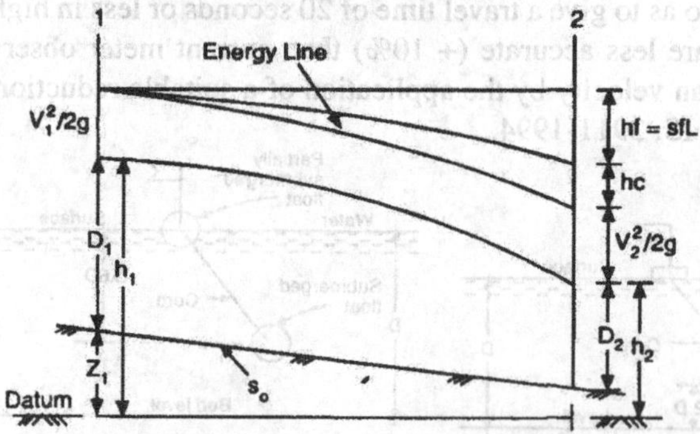

Fig. 22.14. Slope-area method.

Mean cross sectional area, $A = (A_1 + 2A_2 + \ldots + (2A_{n-1}) + A_n)/2(n-1)$.

Where, $A_1, A_2, \ldots, A_n$ are the cross sectional areas measured at various locations.

Mean wetted perimeter, $P = (P_1 + 2P_2 + \ldots + 2P_{n-1} + P_n)/2\,(n-1)$.

Water surface slope is computed from gauges installed at least in three cross sections, one on either river bank, making a total of six gauges. The water surface drop or slope in the reach shall not be less than 150 mm, the minimum being 30 mm. Mean velocity is determined from the observed water surface slope by Chezy formula $V = C\sqrt{RS}$ or Manning's formula, $V = \frac{1}{n} R^{2/3} S^{1/2}$.

2. Float Method

Each float is released far enough the cross section to attain a constant velocity before it reaches the first cross section. The time at which the float crosses each of the three cross sections are noted by stop watch. The floats are uniformly distributed over the channel width. Repeated observations are made. The surface velocity determined by dividing the travel time by the distance between cross sections is converted into mean velocity. The water depth is measured. The discharge is the sum of the flow in each panel ascertained by multiplying the average area of cross section with mean velocity.

22.12 GAUGE-DISCHARGE CURVE

The estimation of discharge from gauge record for periods for which no discharge observations are available is best done by using a proper stage-discharge curve. The curve is obtained by plotting an adequate number of observations of discharges along the abscissa against gauges along the ordinate, suitably distributed throughout the whole range of stage (Fig. 22.15).

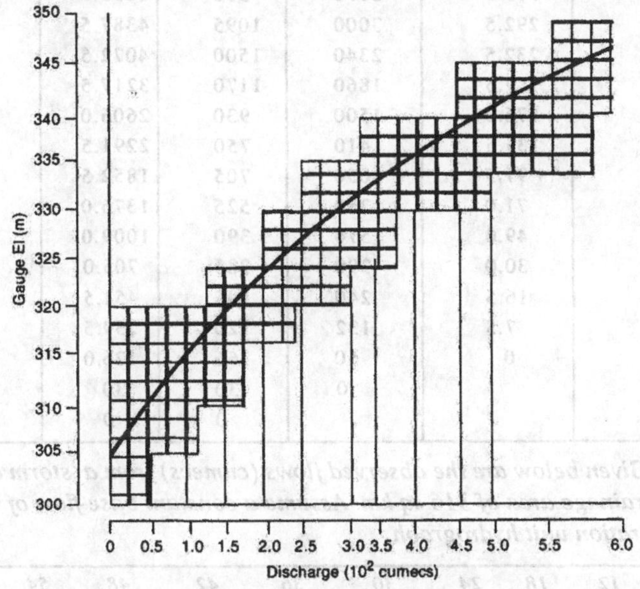

Fig. 22.15. Gauge-Discharge Curve

SOLVED EXAMPLES

Example 22.1 *Find out the ordinates of a storm hydrograph resulting from a 3 hour storm with rainfall of 2.0, 6.75 and 3.75 cm during subsequent 3 hours intervals. The ordinates of unit hydrograph are:*

Hr	*3*	*6*	*9*	*12*	*15*	*18*	*21*	*24*	*3*	*6*	*9*	*12*	*15*	*18*	*21*	*24*
OUH (cumecs)	*0*	*110*	*365*	*500*	*390*	*310*	*250*	*235*	*175*	*130*	*95*	*65*	*40*	*22*	*10*	*0*

Assume an initial loss of 5 mm, infiltration index of 2.5 mm/hr and base flow of 10 cumecs.

Solution:

1. Rainfall excess
   (*a*) During 1st three hours = 2.0 – (0.25 × 3) – 0.5 = 0.75 cm
   (*b*) During 2nd three hours = 6.75 – (0.25 × 3) = 6.0 cm
   (*c*) During last three hours = 3.75 – (0.25 × 3) = 3.0 cm
2. Rainfall excess as ratio of unit rainfall of 1 cm during subsequent 3-hour intervals is 0.75 cm, 6 cm and 3 cm.

3. Runoff due to 0.75 cm rainfall starts from 3 hour, because of 6 cm rainfall from 6 hour, and 3 cm rainfall from 9 hour. The computation of ordinates of storm hydrograph is tabulated in Table 22.7.

Table 22.7 : Storm hydrograph

Time (hr)	*Ordinate of (3 hr) U.H. (cumecs)*	*Rainfall excess (cm)*	*Surface runoff from rainfall excess during successive unit periods (cumecs)*				*Base flow (cumecs)*	*Ordinate of storm hydro-graph (cumecs)*
			0.75 (col. 2 × 0.75)	*6.0 (col. 2 × 6.0)*	*3.0 (col. 2 × 3.0)*	*Total*		
1	2	3	4	5	6	7	8	9
3	0	0.75	0				10	10.0
6	110	6.0	82.5	0		82.5	10	92.5
9	365	3.0	274.0	660	0	934.0	10	944.0
12	500		375.0	2190	330	2895.0	10	2905.0
15	390		292.5	3000	1095	4387.5	10	4397.5
18	310		232.5	2340	1500	4072.5	10	4082.5
21	250		187.5	1860	1170	3217.5	10	3227.5
24	235		176.0	1500	930	2606.0	10	2616.0
3	175		131.5	1410	750	2291.5	10	2301.5
6	130		97.5	1050	705	1852.5	10	1862.5
9	95		71.0	780	525	1376.0	10	1386.0
12	65		49.0	570	390	1009.0	10	1019.0
15	40		30.0	390	285	705.0	10	715.0
18	22		16.5	240	195	451.5	10	461.5
21	10		7.5	132	120	259.5	10	269.5
24	0		0	60	66	126.0	10	136.0
3				0	30	30	10	40.0
6					0	0	10	10.0

Example 22.2 *Given below are the observed flows (cumecs) from a storm of 6 hour duration on a stream with a drainage area of 316 sq km. Assume a constant base flow of 17 cumecs, derive and plot a 6-hour duration unit hydrograph.*

Time (hr)	*0*	*6*	*12*	*18*	*24*	*30*	*36*	*42*	*48*	*54*	*60*	*66*	*72*
Flow	*17*	*113.2*	*254.5*	*198*	*150*	*113.2*	*87.7*	*67.9*	*53.8*	*42.5*	*31.1*	*22.64*	*17*

Solution: Separating base flow of 17 cumecs, surface runoff is as under:

Time (hr)	0	6	12	18	24	30	36	42	48	54	60	66	72
Surface run-off (cumecs)	0	96.2	237.5	181	133	96.2	70.7	50.9	36.8	25.5	14.10	5.64	0

$$\text{Runoff depth} = \frac{6\times 60\times 60}{3}[0+0+2(96.2+181+96.2+50.9+25.5+5.64) + 4(237.5+131+70.7+36.8+14.1)]$$

$$= \frac{20.7308\times 10^6}{316\times 10^6}\times 100 = 6.56 \text{ cm.}$$

Ordinate of unit hydrograph $= 96.2/6.56 = 14.66$ and so on.

Time (hr)	0	6	12	18	24	30	36	42	48	54	60	66	72
OUH	0	14.66	36.20	27.59	20.27	14.66	10.78	7.76	5.61	3.89	2.15	0.86	0

Fig. 22.16 shows the unit hydrograph.

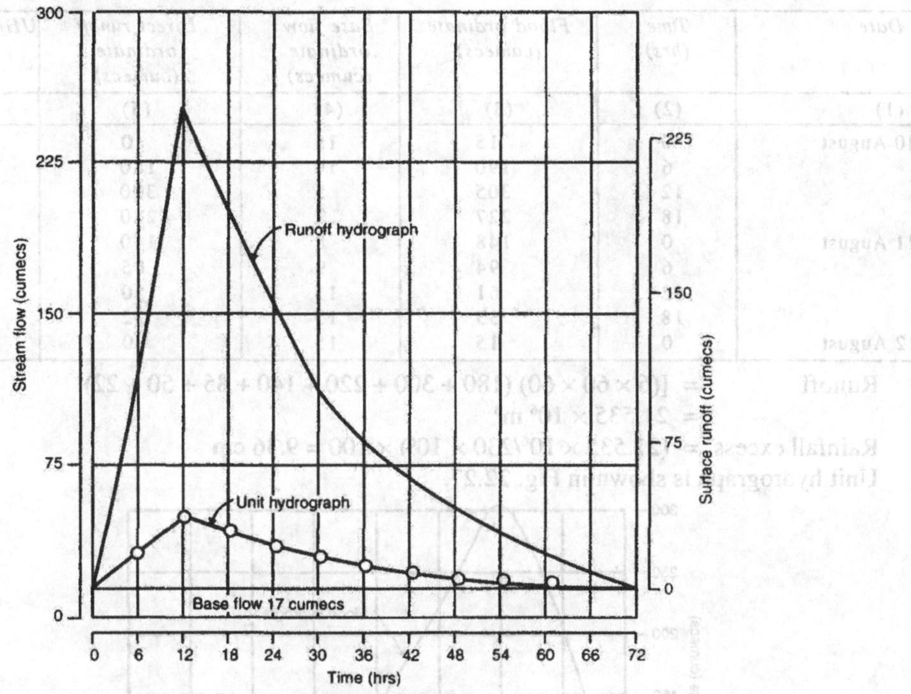

Fig. 22.16. Unit hydrograph.

Example 22.3 *Given below are the ordinates of a unit hydrograph for a storm of 4-hour duration. Find ordinates of flood hydrograph when the maximum flood observed was 4000 m³/s and base flow was 250 m³/s.*

Time (hours)	*0*	*4*	*8*	*12*	*16*	*20*	*24*
Surface runoff (cumecs)	*0*	*1500*	*1200*	*600*	*220*	*80*	*0*

Solution: Maximum surface runoff = 4000 – 250 = 3750 cumecs.

Rainfall excess = 3750/1500 = 2.5 cm

Time	0	4	8	12	16	20	24
O.U.H. of surface runoff	0	1500	1200	600	220	80	0
O.S.R. = 2.5 × O.U.H. (cumecs)	0	3750	3000	1500	550	200	0
Base flow	250	250	250	250	250	250	250
O.F.H. (cumecs)	250	4000	3250	1750	800	450	250

Example 22.4 *From a small catchment (230 km²) flows for a 6-hour storm are known. Determine rainfall excess and draw unit hydrograph.*

Date	*Time*	*Flow (cumecs)*	*Base flow (cumecs)*
10 August	*0000*	*15*	*15*
	0600	*190*	*10*
	1200	*305*	*5*
	1800	*227*	*7*
11 August	*0000*	*148*	*8*
	0600	*94*	*9*
	1200	*61*	*11*
	1800	*35*	*13*
12 August	*0000*	*15*	*15*

Solution:

Date	*Time (hrs)*	*Flood ordinate (cumecs)*	*Base flow ordinate (cumecs)*	*Direct runoff ordinate (cumecs)*	*UHO (col. 5/9.36) (cumecs)*
(1)	(2)	(3)	(4)	(5)	(6)
10 August	0	15	15	0	0
	6	190	10	180	19.23
	12	305	5	300	32.05
	18	227	7	220	23.50
11 August	0	148	8	140	14.96
	6	94	9	85	9.08
	12	61	11	50	5.34
	18	35	13	22	2.35
12 August	0	15	15	0	0

Runoff $= [(6 \times 60 \times 60)(180 + 300 + 220 + 140 + 85 + 50 + 22)$

$= 21.535 \times 10^6 \text{ m}^3$

Rainfall excess $= (21.535 \times 10^6/230 \times 10^6) \times 100 = 9.36$ cm

Unit hydrograph is shown in Fig. 22.27.

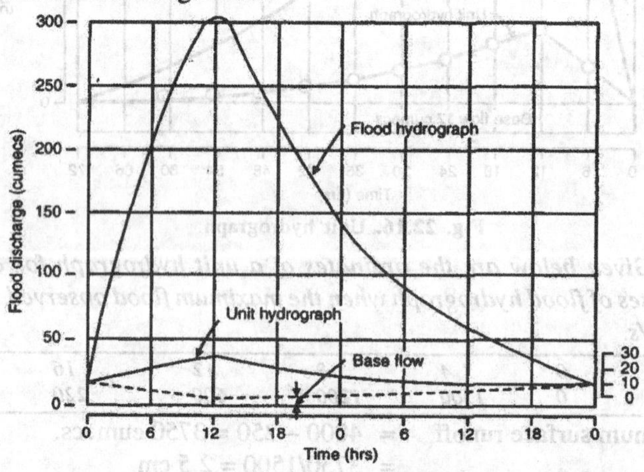

Fig. 22.17 Unit hydrograph

Example 22.5 *The following are the ordinates of a 12 hour unit hydrograph.*

Time (hr)	*0*	*12*	*24*	*36*	*48*	*60*	*72*	*84*
Flow (10^2 cumecs)	*0*	*16*	*29*	*26*	*14*	*7*	*1.5*	*0*

If the successive 12 hour rainfall excesses are 1.5 cm, 3.0 cm and 0.75 cm for the catchment, plot the resulting flood hydrograph on a graph sheet and mark the lag time and flow expected for the river.

Solution:

Time (hrs)	*O.U.H (10^2 cumecs)*	*Rainfall excess*	*Surface runoff from rainfall excess in successive periods (10^2 cumecs)*			*O.F.H. (10^2 cumecs)*
			0.5	*3.0*	*0.75*	
0	0	1.5	0			0
		3.0				
12	16	0.75	24	0		24
24	29		43.5	48	0	91.5
36	26		39	87	12	138
48	14		21	78	21.75	120.75
60	7		10.5	42	19.5	72
72	1.5		2.25	21	10.5	33.75
84	0		0	4.5	5.25	9.75
96				0	1.125	1.125
					0	0

The lag time from the hydrogaph is 21.6 hours (from C.G. of rainfall to C.G. of peak) and peak flow for the river is 13,800 cumecs (Fig. 22.18).

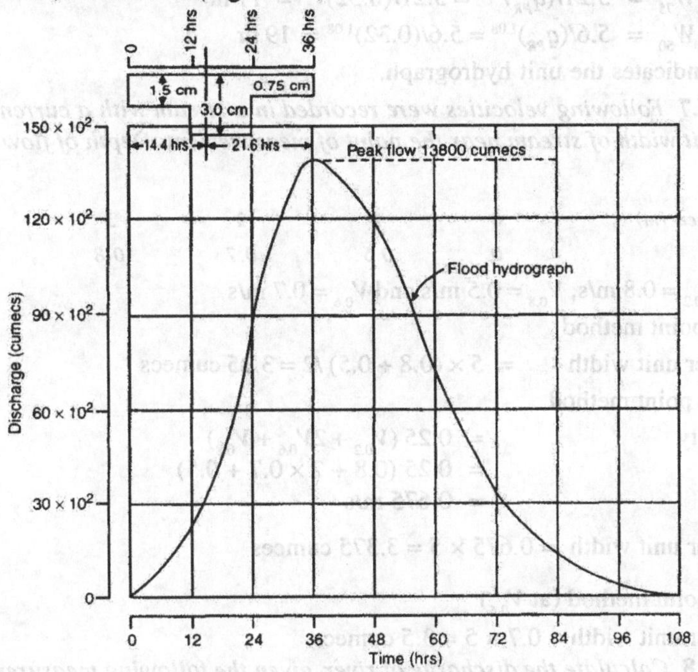

Fig. 22.18 Flood hydrograph.

Example 22.6. *Determine a 6 hr synthetic unit hydrograph for a catchment of 100 sq km, length of main stream is 10 km, distance of outlet to centroid of basin is 4 km C_t = 1.5 and C_p = 0.69.*

Solution:

Basin lag, $t_P = C_t\,(L_c\,.\,L)^{0.3} = 1.5\,(4 \times 10)^{0.3} = 4.54$ hr.

$t_r = t_p/5.5 = 4.54/5.5$

$= 0.83$ hr

$t_{PR} = t_p + (t_R - t_r)/4$

$= 4.54 + (6 - 0.83)/4$

$= 5.83$ hr

Using t_R (given) $= 6$ hr

$q_{PR} = C_P / t_{PR}$

or $Q_P = 2.78\, C_p A/t_{PR}$

$= 2.78\, C_p A/t_{PR}$

$= (2.78 \times 0.69 \times 50)/5.83$

$= 16$ cumecs

Base period length,

$T = 3 + 3\, t_{PR}/24$

$= 3 + (3 \times 5.83)/24$

$= 3.73$ days $= 90$ hr

Time from beginning of rising limb to peak $= t_{PR} + t_R/2$

$= 5.83 + 6/2 = 8.83$ hr.

$q_{PR} = 16/50$

Fig. 22.19 Synthetic unit hydrograph

= 0.32 cumec/sq km

$W_{75} = 3.21/(q_{PR})^{1.08} = 3.21/(0.32)^{1.08} = 11$ hr

$W_{50} = 5.6/(q_{PR})^{1.08} = 5.6/(0.32)^{1.08} = 19$ hr

Fig. 22.19 indicates the unit hydrograph.

Example 22.7 *Following velocities were recorded in a stream with a current meter. Find the discharge per unit width of stream near the point of measurement. Depth of flow at the point was 5 m.*

Depth above bed (m)	*0*	*1*	*2*	*3*	*4*
Velocity (m/s)	*0*	*0.5*	*0.7*	*0.8*	*0.8*

Solution: $V_{0.2} = 0.8$ m/s, $V_{0.8} = 0.5$ m/s and $V_{0.6} = 0.7$ m/s

(*a*) By two point method

Discharge per unit width $= 5 \times (0.8 + 0.5)/2 = 3.25$ cumecs

(*b*) By three point method

Mean velocity
$$= 0.25\,(V_{0.2} + 2V_{0.6} + V_{0.8})$$
$$= 0.25\,(0.8 + 2 \times 0.7 + 0.5)$$
$$= 0.675 \text{ m/s}$$

Discharge per unit width $= 0.675 \times 5 = 3.375$ cumces

(*c*) By one point method (at $V_{0.6}$)

Discharge per unit width $= 0.7 \times 5 = 3.5$ cumecs.

Example 22.8 *Calculate the discharge of river, given the following measurements made with a flow meter.*

Distance from one bank (m)	*0*	*10*	*20*	*30*	*40*	*50*	*60*	*70*
Depth of bed (m)	*0*	*0.7*	*1.2*	*1.5*	*1.8*	*1.5*	*0.9*	*0*
Average velocity V, (m/s)	*0*	*0.15*	*0.24*	*0.30*	*0.36*	*0.33*	*0.24*	*0*

Solution:

Discharge (*i.e.*, $d \times v$)	0	0.11	0.29	0.45	0.65	0.50	0.22	0

Total discharge
$$= 10\,[(0 + 0)/2 + 0.11 + 0.29 + 0.45 + 0.65 + 0.50 + 0.22)]$$
$$= 22.2 \text{ cumecs.}$$

Example 22.9 *Compute the stream flow for the measurement data below:*

Distance (m)	*0*	*0.6*	*1.2*	*1.8*	*2.4*	*3.0*	*3.6*	*4.2*	*4.8*	*5.4*	*6.0*	*6.6*
Depth, d (m)	*0*	*0.3*	*1.29*	*2.16*	*2.55*	*2.22*	*1.68*	*1.41*	*1.05*	*0.63*	*0.42*	*0*
Velocity, $V_{0.2}$ *(m/s)*	*0*	*0.42*	*0.57*	*0.78*	*0.87*	*0.81*	*0.75*	*0.69*	*0.63*	*0.54*	*0.45*	*0*
$V_{0.8}$	*0*	*0.21*	*0.36*	*0.54*	*0.60*	*0.30*	*0.51*	*0.45*	*0.39*	*0.33*	*0.30*	*0*

Solution:

$(V_{0.2} + V_{0.8})/2$	0	0.32	0.46	0.66	0.74	0.55	0.63	0.57	0.51	0.44	0.37	0
Discharge ($d \times V_{av}$)	0	0.10	0.59	0.43	1.89	1.22	1.06	0.80	0.54	0.28	0.16	0

(*a*) Total discharge by trapezoidal rule
$$= 0.6\,[(0 + 0)/2 + 0.10 + 0.59 + 1.43 + 1.89 + 1.22 + 1.06 + 0.80 + 0.54 + 0.28 + 0.16] = 4.84 \text{ cumecs}$$

(*b*) By simpson rule
$$= 0.6/3\,[0 + 4\,(0.10 + 1.43 + 1.22 + 0.80 + 0.28) + 2\,(0.59) + 1.89 + 1.06 + 0.54) + 0.16] + 0.6\,(0.16 + 0)/2 = 4.75 \text{ cumecs.}$$

Example 22.10 *A river cross section measurements by surface float method are given as under. Estimate the discharge passed through the section at the gauging site, assuming velocity factor of 0.89.*

S.No.	Width of section (m)	Mean depth in each section (m)	Surface velocity, observed (m/sec)	S.No.	Width of section (m)	Mean depth in each section	Surface velocity observed (m/sec)
1	24	1.60	0.80	7	30	1.70	0.98
2	30	2.27	1.00	8	30	2.40	0.68
3	30	2.43	1.07	9	30	1.40	0.68
4	30	2.30	0.98	10	30	0.30	0.46
5	30	1.20	0.94	11	30	0.40	0.40
6	30	1.20	0.75	12	30	1.00	1.00

Solution in tabular form is as under :

S. No.	Section width (b)	Mean depth (d)	Area (b.d.)	Surface velocity (V_s)	Mean velocity $V_s \times 0.89$	Discharge (area × V_m)
1	24	1.60	24 × 1.6 = 38.4	0.80	0.71	27.26
2	30	2.27	30 × 2.27 = 68.1	1.00	0.89	60.61
3	30	2.43	30 × 2.43 = 7.29	1.07	0.95	69.26
4	30	2.30	30 × 2.30 = 69.0	0.98	0.87	60.03
5	30	1.20	30 × 1.20 = 36.0	0.94	0.84	30.24
6	30	1.20	30 × 1.20 = 36.0	0.75	0.67	24.12
7	30	1.70	30 × 1.70 = 51.0	0.98	0.87	44.37
8	30	2.40	30 × 2.40 = 72.0	0.68	0.61	43.92
9	30	1.40	30 × 1.40 = 42.0	0.68	0.61	25.62
10	30	0.30	30 × 0.30 = 9.0	0.46	0.41	3.69
11	30	0.40	30 × 0.40 = 12.0	0.40	0.36	4.32
12	30	1.00	30 × 1.0 = 30.0	1.00	0.89	26.70
						Total discharge = 420.14 cumecs

Example 22.11 *The following mean monthly discharge data are available for a river.*

Month	*J*	*F*	*M*	*A*	*M*	*J*	*J*	*A*	*S*	*O*	*N*	*D*
Discharge (Mm^3)	50	20	30	70	100	300	1000	900	600	200	90	60

Draw flow duration curve and determine (*i*) percentage of time the average flow is available, (*ii*) flow with 35% dependability, and (*iii*) flow with 70% dependability.

Solution: **Mean monthly flow = (50 + 20 + ... + 60)/12 = 285 Mm^3.**

Monthly average discharge is arranged in ascending order of magnitude assigning a rankof 1.0 to minimum flow. The record is available for 12 months i.e. $n = 12$.

Discharge (Q)	No. of months during which available (m)	Percentage time (m/n)	Discharge (Q)	m	m/n
20	12	100	100	6	50
30	11	91.6	200	5	41.5
50	10	83.3	300	4	33.3
60	9	75	600	3	25
70	8	66.6	900	2	16.6
90	7	58.3	1000	1	1

Flood duration curve is obtained by plotting the discharge against percentage time (Fig. 21.12) and the flow with 35% and 70% dependability are 285 Mm^3 and 47 Mm^3 respectively.

QUESTIONS

1. What is runoff? What are the factors that effect the runoff from a catchment area.
2. List the factors effecting runoff.
3. List the methods of working out runoff from a drainage area and describe any two methods in detail.
4. Discuss the various methods of determining runoff from a drainage area? State three important formulae.
5. Explain the effect over runoff of (*a*) thick vegetal cover, (*b*) snow cover, (*c*) grass, (*d*) permeable soil and (*e*) uneven land.
6. (*a*) What do you understand by catchment area and runoff?
   (*b*) Explain the various factors that influence the runoff from a catchment area.
   (*c*) How will you measure rainfall?
7. (*a*) What are the different factors effecting the runoff from a catchment area?
   (*b*) How will you estimate the flood from a basin where the stream runoff data are scarce?
8. Define fan-shaped and fern-shaped catchments. Discuss their chief characteristics.
9. What is infiltration? Define (*a*) infiltration capacity, and (*b*) infiltration rate. How would you use infiltration capacity curve to calculate runoff from a small catchment?
   What is the difference between *W*-index and ϕ-index.
   Explain the different methods of measuring infiltration.
10. Explain with sketches, if necessary, the following methods for calculating net storm rain: (*a*) Infiltration capacity method, (*b*) *W*-index method.
    A three-hour storm occurs over a 62 sq km area. From the following data, estimate the net storm rain for the whole area and its hourly distribution.

Sub-area (km^2)	ϕ-index (cm/hr)	Hourly rain		
		1st	2nd	3rd
14.0	2.0	1.5	5.0	0.8
20.0	3.5	1.5	5.0	0.8
28.0	1.0	1.5	5.5	1.5

11. Describe the terms as used in hydrologic practices (*i*) Infiltration index, (*ii*) Time of concentration, (*iii*) Runoff, (*iv*) Isohyetals.
12. Discuss the Rational method for the qualitative determination of urban storm water runoff. What are the limitations of the method?
13. What is flood? Describe the various methods of flood estimation.
14. What are the various factors which effect the peak flood.
15. List various formulae to obtain the peak flow and discuss their limitations.
16. Distinguish between maximum flood and spillway design flood.
17. What is the importance of calculating the maximum flood discharge, total runoff from a river for the design of a water power project?
18. What are the various purposes for which calculation of maximum probable flood are necessary? What methods are used for the calculation of flood discharge? Describe briefly.
19. Explain briefly how peak flood discharge can be estimated from frequency analysis of stream flow data.
20. Find the values of *C* and *n* in the empirical formula, $Q = CA^n$ for flood discharge, *Q* in m^3/sec from a catchment for area *A* km^2, from the following data:

A (km^2)	4	10	50	100	200
Q (cumec)	100	158	354	500	706

$(C = 50, n = 0.5)$ Ans.

21. Describe briefly how you will determine the maximum flood by using (*i*) rational formula, and (*ii*) Dicken's formula.

22. Determine the net runoff and total rainfall and *W*-index for the following data when ϕ index = 3 cm/hr.

Time of rainfall (minutes)	30	30	30	30	30
Rates of rainfall (cm/hr)	2.0	2.5	12.0	8.0	2.0

Net rainfall = 7.0 cm, total rainfall = 13.25 cm, Windex = 2.5 cm/hr. Ans.

23. Describe the methods of river gauging in different situations.

24. What are the requirements of a site for area velocity method of measurement of discharge in a river?

25. (*a*) Write a note on the velocity measurement by current meter.

    (*b*) How will you measure the discharge of a stream with the help of float?

PART–IV

FLUID POWER ENGINEERING

- Centrifugal Pumps
- Reciprocating Pumps
- Miscellaneous Hydraulic Machines

23

CENTRIFUGAL PUMPS

23.1 INTRODUCTION

In 17th century B.C. in Cairo pumps with an endless chain were operated to raise water from a depth of 91.5m. In 2nd half of the 19th century startling progress was made in the development of centrifugal pumps. In modern practice, machines for delivering liquids are known as pumps.

Pump is a machine designed to move liquids and impart energy to them thereby lifting them from lower level to a higher level, itself driven by an external source. Alternatively, the pump may be defined as a machine that converts mechanical energy into hydraulic energy which is a process reverse of a prime mover.

A pump differs from a turbine in that the flow in it is decelerated that is it takes place from low pressure towards the higher pressure which is the reverse of a turbine in which the flow takes place from higher pressure side towards low pressure side and is accelerated.

23.2 CLASSIFICATION OF PUMPS

The pumps are variously classified, as under:

1. Classification Based on Mechanical Principles Involved:

(*i*) Displacement — Reciprocating and rotary pumps
(*ii*) Velocity — Centrifugal, deep well turbine and jet pump
(*iii*) Buoyancy — Air lift pump
(*iv*) Impulse — Hydraulic ram.

However, they fall under two broad categories — centrifugal and reciprocating pumps.

2. Classification Based on Type of Casing.

These fall under centrifugal type of pumps.

(*i*) *Volute Pump*: In this type of centrifugal pump, the impeller discharges into a spiral casing of such proportions as to produce same velocity of flow around the circumference of the impeller. The velocity of water is reduced at entry into the discharge pipe thereby creating the needed pressure head.

(*ii*) *Turbine Pump*: In this type of centrifugal pump, the impeller is surrounded by stationary guide vanes, circular and concentric with the impeller, which reduce the velocity of water before it enters the casing thereby converting the velocity head into pressure head in the casing itself.

3. Classification Based on Number of Impellers

Single Stage and Multistage Pumps : The centrifugal pumps wherein only one impeller is used are known as single stage pumps and those using two or more impellers are called multistage pumps. These pumps are used for high lifts.

4. Classification Based on Guide Housing of Vane

(*i*) Open, (*ii*) Semi-Open, (*iii*) Enclosed.

5. Classification Based on Pump Positioning

The centrifugal pumps have other variants called bore hole pumps or deep well turbine pumps and submersible pumps. These may be described as suction lift pumps and positive-submergence pumps. Suction lift pumps are located above the well and draw water by suction lift.

(*i*) *Positive-submergence Pumps*: The positive-submergence pumps are used where the pumping level is within the suction capability of the pump and are installed within the casing with the pumping inlet usually submerged below the pumping level and as such under positive pressure head.

(*ii*) *Deepwell Turbine Pumps*: The deep well turbine centrifugal pumps have several impellers installed on a vertical shaft which is suspended and rotated by motor placed on ground surface.

(*iii*) *Submersible Pumps*: In submersible pump, the motor is placed beneath the pump bowels below water level in the well.

23.3 TYPES OF PUMPS

The two types of pumps, on the basis of transfer of mechanical energy, are roto-dynamic pumps and displacement pumps.

1. Roto-dynamic Pumps: The roto-dynamic pumps have a wheel or impeller which rotates the water in a casing thereby imparting energy to the latter. The increase in energy level is effected by a combination of centrifugal energy, pressure energy and kinetic energy. These may be radial flow (centrifugal) pump, axial flow pump and mixed flow pump. Centrifugal pumps are widely used. The centrifugal pumps are of two types viz. (*i*) Volute type centrifugal pump, and (*ii*) Diffuser or turbine type centrifugal pump.

2. Displacement Pumps: These pumps are constant displacement and variable displacement pumps. Theoretically speaking constant displacement pump discharges the same volume of water regardless of the head against which it works and is used where discharge is low. The variable displacement pump has inverse relationship between the rate at which it delivers water and the head against which pumping takes place. The pumping rate decreases as the pumping head increases. These pumps work on the principle of mechanically induced vacuum in a chamber and thus drawing in water which is mechanically displaced and forced out of the chamber.

Reciprocating pumps and rotary pumps are its two types. These pumps depend for their operation on the intake and displacement of liquid by pistons, gear teeth moving in working chambers.

3. Miscellaneous Types of Pumps: These include jet pumps, air lift pumps, hydraulic ram (impulse pump).

23.4 SELECTION OF PUMPS

The selection of a pump is made on the basis of adaptation to the particular conditions of operation. The various considerations in the selection of a pump unit are as under:

(*i*) *Capacity and Lift*: The pump is capable to lift the required quantity of water against the given total head. It caters to the range of variation in the total head due to fluctuations in pumping level. Forthermore, suction conditions are important in the selection and installation of a pump. High suction lift causes serious reduction in capacity and efficiency. The vertical turbine pumps are commonly used for pumping from deep wells.

(*ii*) *Speed*: The speed of a suction pump has a considerable effect on its performance, the quantity varies directly to the speed and the head varying as the square of the speed. However, increase in speed tend to increase maintenance. However, high speed pumps and motors generally cost less than low speed pumps.

(*iii*) *Water Quality*: The quality of water whether clear water or sewage water or water containing solids determine the type of pump to be used. For clear water, pump with single or double suction closed type impellers and for muddy water pumps with open impellers are suitable.

(*iv*) *Standardisation and number of Units*: Standardised pump units, where a number of pumps are involved, are desirable for easy interchangeability of spare parts and reduce diversity of spare parts to be stocked for operation and maintenance. Determination of proper number of units and their hydraulic characteristics are important.

(*v*) *Efficiency*: The pump should have high efficiency. The efficiency tends to improve as the speed increases which in turn depends on the design of pump. It is advisable to select a pump that gives satisfactory efficiency over the operating range. In the case of borehole turbine and centrifugal pumps, the efficiency and power consumed vary with the operating heads.

(*vi*) *Cost*: It is cheap in initial cost, operation and maintenance.

(*vii*) *Duration and rate of pumping*: The duration for which the pump is to run in a day and rate of pumping and ease of maintenance is considered in selecting the pumping set. Pump in which moving parts are easily accessible are preferable for best service from maintenance point of view.

(*viii*) *Location of Pump*: The location of pump with respect to pumping level in the source dictates the type of pump.

(*ix*) *Space Requirement*: Pump with minimum space (floor area) requirement for pump location is preferred.

(*x*) *Reliability*: It provides dependable and trouble free service.

(*xi*) *Diameter of Well*: Vertical turbine pumps require smaller diameter of pump compared to centrifugal pump.

(*xii*) *Overloading*: The pump may not get overloaded due to wear or misalignment of components. It is, therefore, necessary that driving unit may be provided with some margin of power.

(*xiii*) *Wattage*: The wattage of the pump is suitably selected taking into consideration the worst pumping conditions that may be encountered.

(*xiv*) *Life*: Long life pump is preferred; its depreciation cost is low. Low speed pumps have longer life than high speed pumps.

The manufacturers of the pumps can make recommendations, on the basis of the above data, with regard to the suitable type of pump and selection can be made depending on the price, water quality, efficiency, space requirement and overall economics. The criteria in this regard broadly is as under:

1. Centrifugal Pumps: Reciprocating pumps are outdated and generally centrifugal pumps are widely used as they are economical and provide satisfactory service. They can also be employed for pumping waste water and water containing solids. The centrifugal pumps offer horizontal shaft pump, vertical spindle bore hole pump and submersible pump. The horizontal shaft centrifugal pumps are cheap and are generally used for pumping under a wide range of pumping conditions except for pumping large quantity of water against low heads and pumping from wells and bore holes. The vertical spindle pumps and submersible pumps are used for pumping large amount of water under low heads. The centrifugal pumps need priming if placed above the water. These may be placed under water for pumping large quantities of water.

2. Reciprocating Pumps: Reciprocating pumps are suitable for pumping clear water against very high but variable head with a higher suction lift. Sufficient floor area is required for its installation. Its cost is high.

3. Rotary Pumps: For small discharges rotary pumps are used because they are as good as centrifugal pumps.

4. Air Lift Pumps: Air lift pumps are cheap for pumping simultaneously from a number of wells as one compressor unit can operate all the pumps but they have low efficiency. These are suitable for wells where discharge is not large and water contains suspended impurities.

5. Jet Pumps: These are used for pumping water from small wells.

6. Impulse Pumps: Impulse pumps (hydraulic rams) are used if large amount of water is available in abundance at moderate head to lift small quantity out of it to a higher head without the aid of any prime mover.

23.5 CENTRIFUGAL PUMPS

Centrifugal pumps are so called because they operate under the action of centrifugal force and lift effects. These pumps correspond to Francis turbines with reversed flow direction.

Centrifugal pumps convert mechanical power, in the form of shaft torque and speed, into hydraulic power in the form of flow rate raised in pressure level. The pump receives a flow of liquid at a certain inlet pressure, raises this pressure to a higher value and discharges the liquid through the outlet. The pump depends on rotational speed to generate a head, which appears as a difference of pressure between inlet and outlet branches. The discharge pumped and the power required depends upon the dimensions of the system to which the pump is attached.

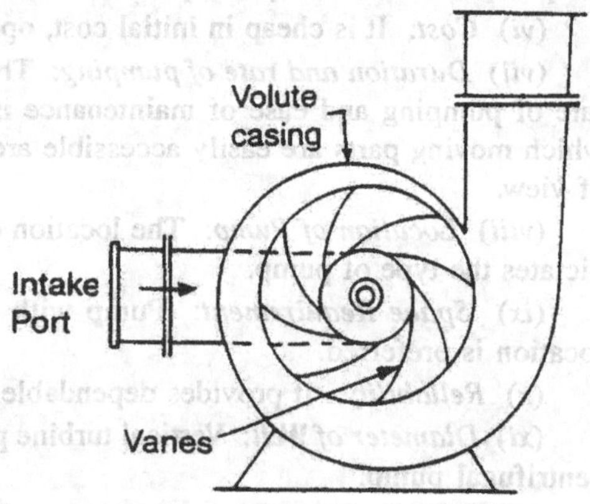

Fig. 23.1 Contrifugal pump arrangement.

Basically, a centrifugal pump comprises an impeller and a casing together with driving shaft. (Fig. 23.1) The pump impeller is provided with vanes and enclosed in casing. As the impeller rotates, the liquid is moved by centrifugal force from the centre towards the periphery and thrown into a volute chamber to enter the discharge pipe. The liquid is sucked in through intake port. The impellers may be single entry type or the double entry type. Impellers may be radial flow, mixed flow or axial flow according to the direction of water within the impellers.

The radial flow pump has impeller shaped to force water outward in a direction at right angles to its axis, mixed flow pump has axial as well as radial velocity and force water in axial direction only in axial flow pump. The radial flow and mixed flow pumps are generally called centrifugal pumps. Axial flow machines are called axial flow pumps Fg. 23.2).

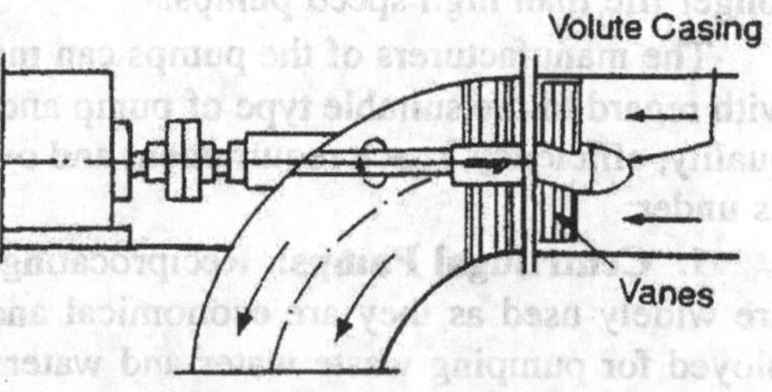

Fig. 23.2 Axial-flow pump schematic.

In practice the pump can work under a suction head of 6 m. A sump well is usually provided with bottom plugged and of sufficient depth to place the pump so that the suction head plus fluctuations in water table is limited to 6-8m. Multistage pumps comprise several impellers on one shaft in order to generate a greater head than can efficiently be obtained from one stage. The discharge pressure is directly proportional to the number of stages, other things being equal. Centrifugal pumps have five distinct designs namely volute type, turbine type, mixed type, axial type and regenerative type. The regenerative type pump has unique feature in that the liquid is carried to and from the periphery of the impeller.

Centrifugal pumps are built on horizontal as well as vertical shafts. Electrical driven pumps are generally connected directly to the shaft of the motor. Horizontal-shaft centrifugal pumps, as these are usually set at a minimum safe height above the surface of water to avoid submergence during high water, are dependent upon the atmospheric pressure to force water up to the pump and are relatively free from trouble, have high efficiency and stand high speeds.

It is desirable that suction line to the pump is short and straight and at least one size large than the pump suction nozzle and submerged to a minimum depth of about four diameters with low entry velocity.

Power required to drive the pump is given by the relation:

$$P = \frac{wQH_{ma}}{\eta_0} \text{ kW} = \frac{w_1 QH}{75} \text{ H.P.}$$

Where, w is 9.81, Q is discharge (cumec), H_{ma} is monometric head of water (m), w_1 is unit weight of water (1000 kg/m^3) and η_0 is overall efficiency of the pump.

A centrifugal pump installation requires a large size deep well sump with bottom plugged with concrete to house it at or near the ground water table. Because of its initial cost it has given place to vertical turbine bore hole pumps.

The discharge of a centrifugal pump is given by the relation:

$$Q = \pi D b V_f = \pi D_1 b_1 V_{f1}$$

Where, Q = discharge, D and D_1 = diameter of impeller at inlet and outlet respectively, b and b_1 = width of impeller at inlet and outlet respectively, and V_f and V_{f1} = velocity of flow at inlet and outlet respectively.

Advantages

(*i*) Most widely used pump for irrigation on account of high dependability and discharge, (*ii*) Low initial cost, (*iii*) Low maintenance cost, (*iv*) Simple mechanism; few wearing parts, (*v*) Easy to install and operate, vibration free running, (*vi*) Can be operated at very high speeds, (*vii*) Give constant steady discharge, (*viii*) Light and compact; requires lesser floor space, (*ix*) Superior performance characteristics, (*x*) Can pump sediment containing water, and (*xi*) Long life.

Disadvantages

(*i*) Not suitable for high pressure and low duties, (*ii*) Needs priming, (*iii*) Low efficiency at high heads and discharges, (*iv*) Limited suction lift, (*v*) Efficiency is high only for low discharge and head, and (*vi*) Rate of flow cannot be regulated.

23.6 VERTICAL TURBINE PUMPS

Also called borehole or deepwell turbine pump. These are multistage diffuser-type pumps capable to discharge up to 2,000 litres/s and develop heads upto 500m. The number of stages depends upon the head to be developed. The high head pumps may have 20 or more stages. These pumps may be oil or water lubricated. In the oil-lubricated pumps, the oil-lubricated shaft is enclosed in a tube, whereas in water-lubricated pumps the water being pumped serves as lubricant. Practice tends towards oil rather than water lubrication.

A vertical turbine pump comprises impellers, mounted on a shaft and driven by a vertical spindle motor at the ground level which rests on a stool containing Michel bearing housing for the vertical drive shaft, the final discharge bend and the gland. The pump being of small diameter is lowered in the casing which is kept wider upto that depth from the top than the normal bore to accommodate pump bowl. A sump well necessary for a centrifugal installation is not required and as such economical. A length of water flow pipe and drive shaft sufficient to reach the lowest water level is suspended from the stool. The pump is attached to the lower end of the water flow pipe below the water level. It is specially indicated for tubewells where pumping water level is too low to install volute centrifugal pump. Impellers are usually closed or semi-open. The diffusers extend upward in the pump bowl. For average water conditions, the material used for the impeller includes bronze, close-grained gray iron, high nickel iron and porcelain-enamelled iron.

Advantages

(*i*) Smaller diameter of pump compared to centrifugal pump, (*ii*) Adapted to high lifts, (*iii*) High efficiency, (*iv*) Cheap as sump is not required to be constructed, (*v*) Better alternative to horizontal shaft centrifugal pump, and (*vi*) Wider casing pipe in top portion to accommodate pump bowl.

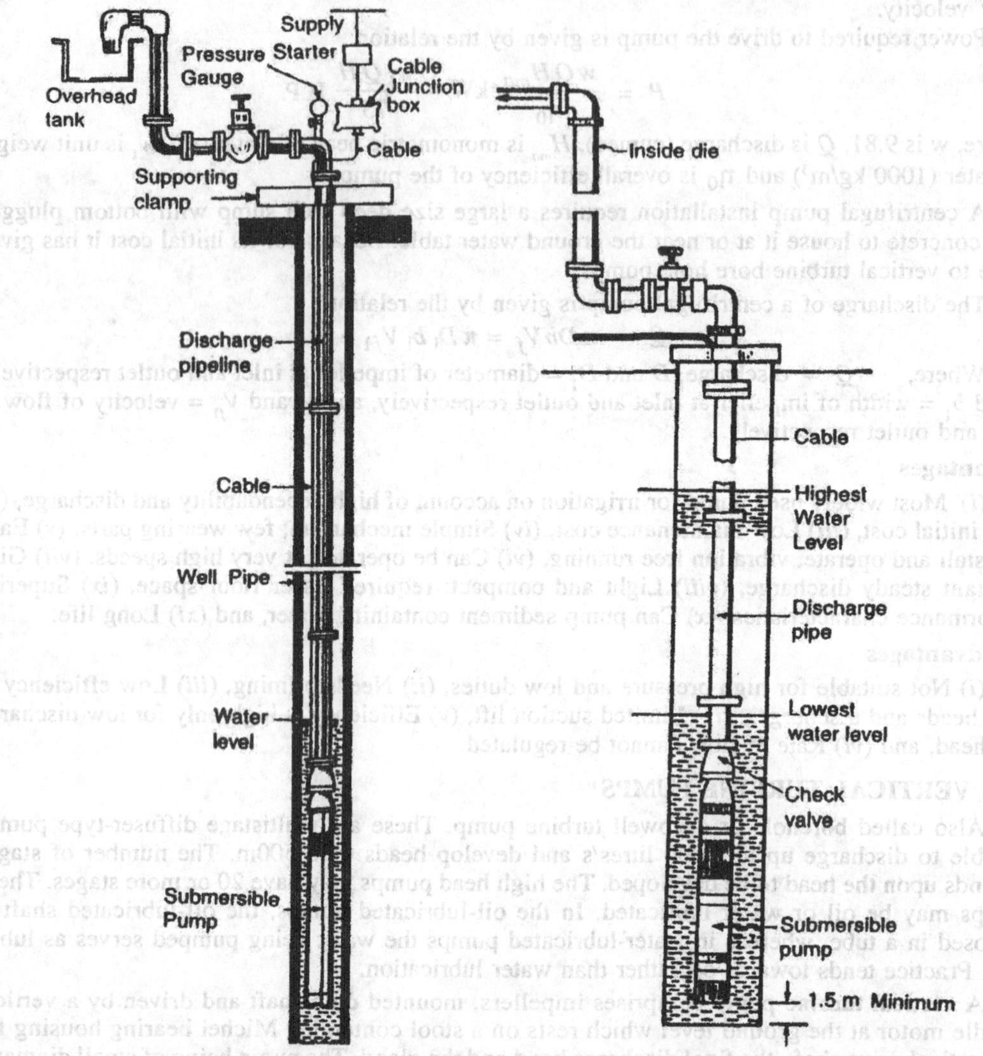

Fig. 23.3 Submersible pump (courtesy M/s Calama Industries Pvt. Ltd. New Delhi).

Disadvantages

(*i*) High initial cost, and (*ii*) Difficult to install and repair, pump being lower down in the casing.

23.7 SUBMERSIBLE PUMPS

Submersible pumps (Fig. 23.3), the entire assembly of which including the electric motor and the pump are submerged under the water in the well, are recent developments in pumping water from deep wells. These pumps can be installed in small wells of 10 cm diameter. These can be set at any pumping depth within the limits of horse power and pipe size. The pump inlet has to be completely submerged to provide necessary cooling. A check valve is installed to outlet of the pump to prevent the delivery pipe from emptying each time the pump stops. A level control switch is installed in case water level in the well has large fluctuation. There is no limitation on the depth from which water can be pumped if the well is large enough in diameter to accommodate the discharge pipe, the motor and the pump. Use of current sensing type of overload relays is indicated as the motor installed in the pumped water may not be subjected to the same ambient temperature as the overload relays in the starter.

The pump is placed above the motor with the water inlet between the units. Delivery of water is through a riser pipe on which the assembly is suspended; the driving shaft is thus dispensed with. Regardless of the downward conditions of the well, the motor is never out of the water. The pump is generally a multistage centrifugal type, the impellers and diffusers so arranged in series that the pump can deliver water against pressure head sufficient to force it up out of the well, discharging it against adequate pressure needs at the surface. The suction and cover contains a cylindrical strainer arranged between the pump motor so that water can be drawn radially inwards from the well.

The electric motor is connected to the control system on the ground level by means of a special submarine type electrical cable. Motors for submersible pumps are generally of two classes, 'dry motor' and 'wet motor'. Dry motors are hermetically sealed to completely exclude the water of the well. In wet motors, the water of the well has access to the inside of the motors; and the rotor and bearing actually revolve in the well water. The windings of the stator are completely sealed off from the rotor by means of an inner-liner preventing the well water from getting to the windings.

Advantages

(*i*) Easy to install; the installation requires no sump. The unit with pipe is held by a supporting clamp which rests on the upper edge of the casing pipe, (*ii*) Economical; in place of a normal pump house there is a small well head covered with a lid which prevents impurities contaminating the water, (*iii*) Pumping unit is safe from surface hazards as fire, floods or vandalism, (*iv*) No suction trouble; as the pump works below water level and water flows automatically and no suction difficulties arise. There are no glands hence no leakage in the pump, (*v*) Noiseless operation; as the pump is of centrifugal design, it is nearly noiseless in operation and works deep down in the well, (*vi*) Simple mechanism; there are no belts, gears, packing or rods that wear out and need replacement, (*vii*) High efficiency; the overall efficiency of a submersible pump is higher than an ordinary shaft type pump as the farmer does not always operate under a constant load but over a range of loads, the motor performance varies within reasonable defined limits. Therefore, rated output of submersible pumps is high enough to avoid over-loading during the period of maximum load, (*viii*) The pump can be controlled manually or automatically depending on the type of installation, (*ix*) Its slender construction enables the submersible pump to operate in small and crooked bore wells, (*x*) No priming or lubrication is ever required, (*xi*) Random lengths of pipe can be used between the pump and suction service, (xii) They meet wide fluctuations of water surface without necessitating a resetting of the pump, (*xiii*) No maintenance; as the bearings are water-lubricated, no periodical oiling, greasing is needed. Besides, greasing contaminates the water, (*xiii*) Useful for deep as well as shallow wells, and (*xiv*) Reduced bearing friction, there being no long drive shaft from ground suface to the pump.

Disadvantages

(*i*) Difficulty of inspection as operating parts are inaccessible, (*ii*) Lack of inspection entails running after bearings are worn out or until the pump fails to operate, before repairs are provided, (*iii*) Entire pump has to be lifted from deep well, in case motor breaks down, for repairs.

23.8 MISCELLANEOUS TYPE OF PUMPS

1. Jet Pumps

Jet pumps are best suited for pumping from wells in which pumping level is more than 7.5 m vertical distance below the pump location and up to 60 m with capacities up to 12,000 litres/hr.

net discharge. Jet pumps combine a single stage centrifugal pump at the top of the well with a jet or ejector nozzle located at the suction screen in the well. A portion of the water discharged by the pump flows down and through the ejector, where it helps to improve flow into the pump and up the discharge tube. For wells with lift up to 7.5 m, the jet is often located on the surface in the pump casing, instead of in the well. This permits easier maintenance. For greater depths the jet is in the well, and the pump, which may be either horizontal or vertical, is on the surface. A foot valve is installed on the bottom of jet or lifting unit. Jet pump schematic is illustrated in Fig.23.4. The motive fluid flows through a nozzle which decreases with length. The flow velocity builds up gradually. The fluid kinetic energy thus grows and peaks as the flow issues from the nozzle into the chamber, the increased kinetic energy of the flow results in a pressure drop in the chamber; the difference in pressure between the atmosphere and the chamber causes the suction liquid to rise to the chamber where it is entrained by the jet to motive fluid and the liquid being moved enter the divergent chamber and thence carried through delivery pipe into the tank at height.

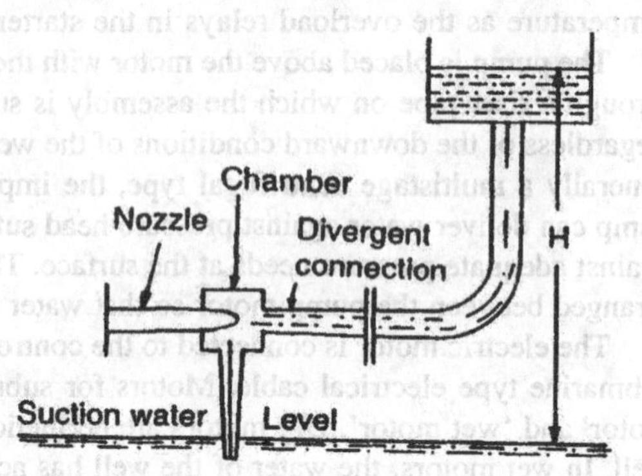

Fig. 23.4. Jet Pump-schematic.

Fig. 23.5 illustrates the principle upon which the ejector works. Water under pressure discharges through a nozzle which smoothly, but rather abruptly, reduces the area through which the flow must pass. The increased velocity flow thus created reduces pressure of water. If the discharge velocity at the nozzle is great enough, the pressure at point 'A' will be lowered sufficiently to draw water into the venturi assembly through an opening at this point, this being added to the total volume of water flowing beyond point 'A'. Most of the pressure in the pipe is recovered at point 'B' through gradual enlargement in the venturi tube. A centrifugal pump acts as a prime mover which produces flow to the nozzle and continues the combined flow beyond this point.

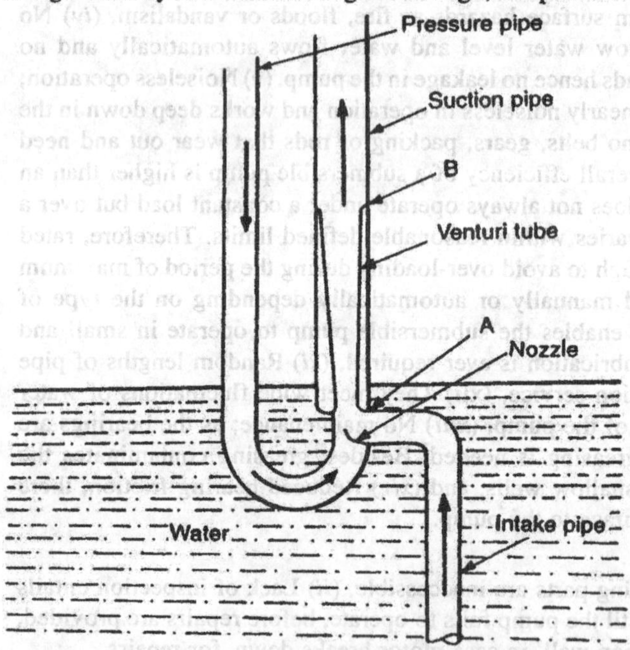

Fig. 23.5. Operating principle of jet pump

The efficiency of jet pumps is very low; up to 25%. However, they find use for lifting sediment laden (muddy) water as also in mines for pumping oil and dewatering excavation trenches.

Advantages

(*i*) Adaptability to small wells down to 5 cm inside diameter in deep lift installations, (*ii*) Portable, light and compact, (*iii*) Easier maintenance, all moving parts are at ground surface and accessible, (*iv*) Simple, economical in first cost and maintenance, (*v*) Adaptability to being installed with the moving parts offset from the well, (*vi*) Can handle muddy water, (*vii*) Simple in design and construction, (*viii*) Free from moving parts, and (*ix*) Can handle liquids and gases.

Disadvantage : Low efficiency.

2. Air Lift Pumps

Air lift pumps are rarely used as a permanent installation due to low efficiency. These are extensively used for the development and preliminary testing of wells and removal of highly corrosive or abrasive water which is likely to damage parts of the pump submerged in it. They also find use in pumping water from deep wells with high lifts up to 150m.

Fig. 23.6 shows the general arrangement of an air lift pump. The operating principle of the pump is shown in Fig. 23.6*a*. Air lift pump when in operation pumps out most of the air (or other gas) from an enclosed space resulting in a partial vacuum. Compressed air is released through holes evenly to form a uniform mixture. Air bubbles mix with the water and reduce the specific gravity of the column of water sufficient to lift it to the surface. With continued supply of air, the water column in the discharge pipe is forced upwards to discharge from outlet at the top.

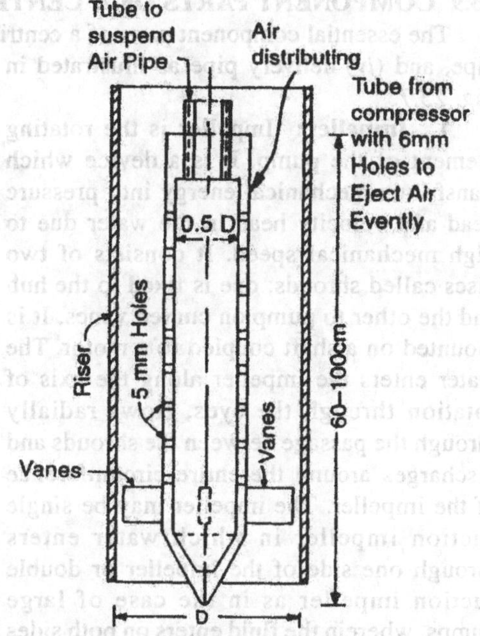

Fig. 23.6. Air lift pump.

For efficient operation, the lower end of the discharge pipe must be submerged in the well water. The extent of submergence, before the air is admitted, varies from 70 per cent for 6 m lift to 40 per cent for 200 m. When the air is admitted into the well, the water recedes from static level to the bottom of the discharge pipe. This displaced column of the water rises up the discharge pipe, and as air flow continues, it enters the pipe, aerating the water and lowering the specific gravity of the mixture. Pressure in the well is momentarily decreased and then increased as the bottom end of the pipe is uncovered and covered. The cycle repeats rapidly thereby giving a nearly constant flow from the top of the discharge pipe. The riser is of large diameter to deliver liquid to a considerable height.

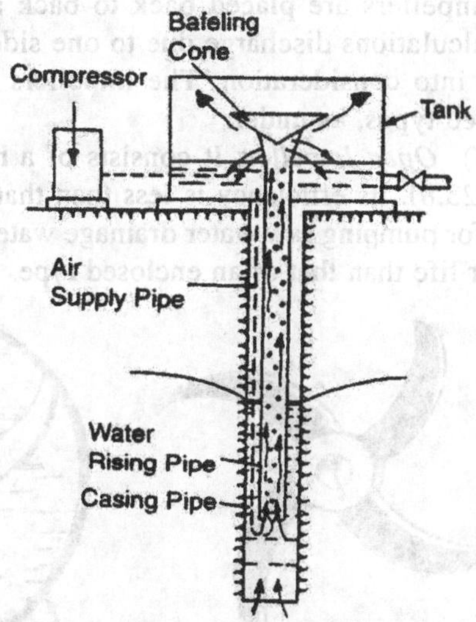

Fig. 23.6*a* Operating principle of air lift pump

Advantages

(*i*) Adaptability to boreholes of any practicable (small) size, where reciprocating and centrifugal pumps cannot be employed, (*ii*) Adaptability to crooked holes not suited to any type of pump, (*iii*) Capable to draw more water from a small diameter well, with sufficient capacity to deliver it than any other type of pump, (*iv*) Capable to operate a group of wells from a centrally located air compressor station, (*v*) Simple and trouble-free operation, there being no remote or submerged moving parts, (*vi*) Can draw alkaline and acidic water to a small hgiht as in chemical and food industries, (*vii*) Cheap, (*viii*) Reliable, (*ix*) No moving parts in contact with water, (*x*) Discharge can be increased by increasing more amount of compressed air, and (*x*) Simplicity of design.

Disadvantages

(*i*) Low efficiency, (*ii*) Large submergence; in order to obtain adequate submergence of the air outlet well is sometimes made deeper than is required for other types of pumps, (*iii*) Intermittent nature of flow from the well, (*iv*) Low efficiency, 25 to 50%, (*v*) Contamination of supplied liquid by compressed air, (*vi*) Greatly increased oxygen content of supplied liquid, and (*vii*) Inferior to deep well centrifugal pumps.

23.9 COMPONENT PARTS OF A CENTRIFUGAL PUMP

The essential component parts of a centrifugal pump are (*i*) impeller, (*ii*) casing, (*iii*) suction pipe, and (*iv*) delivery pipe as illustrated in Fig. 23.7.

1. Impeller: Impeller is the rotating element of the pump. It is a device which transforms mechanical energy into pressure head and velocity head in the water due to high mechanical speed. It consists of two discs called shrouds; one is fixed to the hub and the other to pump on curved vanes. It is mounted on a shaft coupled to a motor. The water enters the impeller along the axis of rotation through the eyes, flows radially through the passage between the shrouds and discharges around the entire circumference of the impeller. The impeller may be single suction impeller in which water enters through one side of the impeller or double suction impeller as in the case of large pumps, wherein the fluid enters on both sides of the impellers to avoid axial thrust. In it two impellers are placed back to back and for calculations discharge due to one side is taken into consideration. The impellers are of three types, as under:

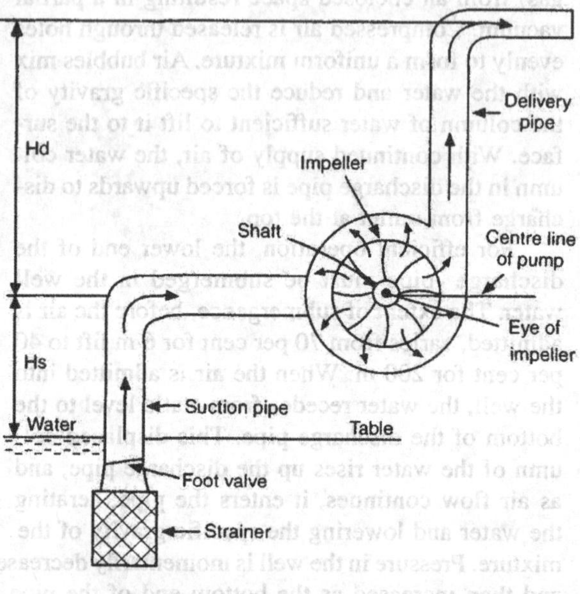

Fig. 23.7 Component parts of a typical centrifugal pump

(*i*) *Open Impeller*: It consists of a hub with attached vanes which are open on both sides (Fig. 23.8). Its efficiency is less than that of a closed impeller centrifugal pump. It is generally used for pumping raw water drainage water with limited sand, sewage water, paper pulp etc. It has longer life than that of an enclosed type.

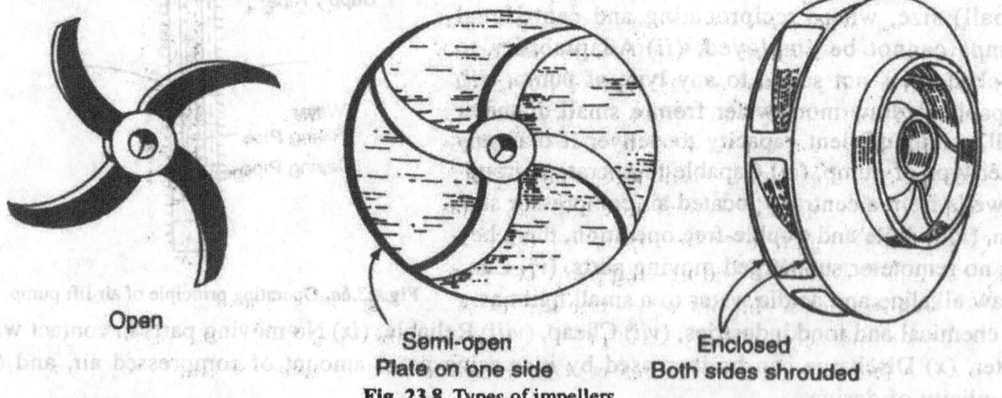

Fig. 23.8 Types of impellers.

(*ii*) *Semi-Open Impeller*: It is the impeller in which the vanes have only base plate, devoid of crown plate (Fig. 23.8). It finds use for pumping high viscous liquids having sediments and other foreign matter held in suspension.

(*iii*) *Closed Impeller*: Also called shrounded impeller. It has metal cover plates or shrouded on both sides of the vanes (Fig. 23.8). It is adopted for pumping clear liquids of low viscosity and has high efficiency but has slightly high initial cost.

(*iv*) *Propeller impeller*: For handling large quantities of water at no lift and low heads propeller type impellers are used to obtain a flow in the direction of the axis of rotation.

The impellers may be of steel, cast iron, gun metal etc. Steel impellers with special packings are used for lifting hot water. Cast iron impellers are used for pumping clear water, fresh water and are cheap. Gun metal or stainless steel impellers are used where corrosion is apprehended due to salty water or chemicals.

2. Casing: The impeller of a centrifugal pump is enclosed in a water-tight chamber called casing. The casing directs the fluid to and from the impeller and leads to the delivering pipe under high pressure by partially converting the kinetic energy of water into pressure energy thereby increasing the pump efficiency. It contains suction and discharge arrangements, bearing support facilities to house the rotor assembly and stuff box. The three types of casings on the basis of energy conversion, are briefly as under:

1. *Volute Casing*: In this type of casing, the impeller is surrounded by a spiral casing (Fig. 23.9*a*) so proportioned as to provide a gradual increase in area of flow from the impeller outlet to the delivery pipe to produce equal velocity flow around the circumference of the impeller and to reduce gradually the velocity of water as it flows from the impeller to the discharge pipe to change velocity head to pressure head. The volute casing only slightly increases the pump efficiency, considerable loss takes place in eddies.

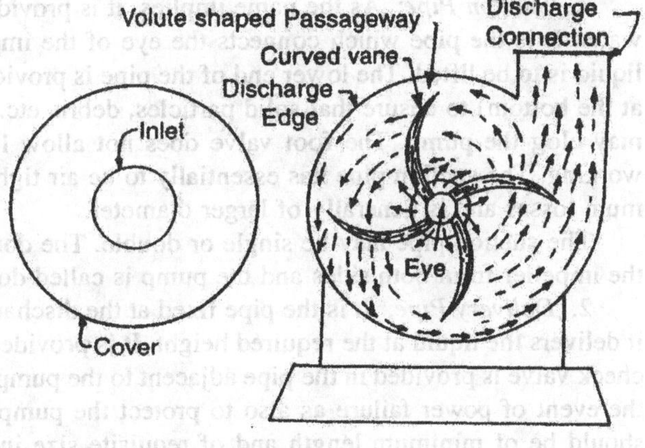

Fig. 23.9*a* Volute casing pump

2. *Vortex Casing*: It is an improved version of the volute casing wherein the spiral casing is combined with a circular chamber (Fig. 23.9*b*) resulting in reduced formation of eddies and hence higher efficiency.

3. *Guide Blades*: It is another variant of volute casing (Fig. 23.9*c*) wherein the water is caused to flow through passages of increasing area formed by guide vanes. The impeller is surrounded by a series of vanes or guide blades mounted on a ring which is called diffuser. The guide

Fig. 23.9*b* Vortex casing

Fig. 23.9*c* guide Blade casing

blades are provided at such an angle that the water enters without shock and is surrounded by a volute chamber by which the water reaches the delivery pipe that has decreased velocity of flow and the kinetic energy is converted into pressure energy resulting in higher efficiency than that in a volute casing pump. The pump with a diffuser assembly arrangement resembles a pressure turbine and as such called turbine pump.

The casings can also be classified, on the basis of construction, as vertically split casing or horizontally split casing. The vertical-split casing is used for small capacity pumps being cheaper. The horizontal-split casing is used for large capacity pumps as it permits removal of the cover to provide access to the impeller.

3. Piping System: The piping system consists of two limbs viz., suction pipe and delivery pipe. It is essentially of a proper design with regard to its size and layout to ensure efficient working of the pump.

1. *Suction Pipe*: As the name implies, it is provided at the suction end of the pump to draw water. It is the pipe which connects the eye of the impeller to the well or sump wherefrom the liquid is to be lifted. The lower end of the pipe is provided with a strainer (fitted with a foot-valve at the bottom) to ensure that solid particles, debris etc. are not drawn into the piping system and may clog the pump. The foot valve does not allow liquid to drain out when the pump is not working. The suction pipe has essentially to be air tight with minimum of bends to ensure minimum losses and is generally of larger diameter.

The suction pipe may be single or double. The double suction pipe envisages water entry to the impeller from both sides and the pump is called double suction or double acting pump.

2. *Delivery Pipe*: It is the pipe fixed at the discharge end of the pump. As the name implies, it delivers the liquid at the required height. It is provided with a check valve and a gate valve. The check valve is provided in the pipe adjacent to the pump to prevent back flow through the pump in the event of power failure as also to protect the pump from water hammer. The discharge pipe should be of minimum length and of requisite size in accordance with the requirements at the discharge end.

23.10 PRIMING AND WORKING OF A CENTRIFUGAL PUMP

Priming of a centrifugal pump is defined as the evacuation of any air, gas or vapours from the suction pipe, casing and a portion of the delivery pipe by completely filling them with the water before starting the operation of the pump to lift the water. In case priming of the pump has not been done before starting it, air pockets, inside the impeller may result in formation of vortices causing discontinuity of the flow, rubbing of the wearing rings and damaging the pump.

In small capacity pumps, the priming is done by opening the air vent valve of the casing and pouring water into the funnel provided on the pump till all air is escaped from the suction pipe, impeller and casing through the valve. In large pumps, priming is done by evacuating the casing and suction pipe by a vacuum sump or ejector and water is thus drawn from the pump or well into the suction pipe and filling the pump. A water glass is placed near the top to the pump casing to indicate complete priming.

In the self priming pumps, the priming is automatic from the arrangement (containing the supply of water in the suction pipe) provided in the pump.

The pump is generally started against a closed discharge valve, priming done and motor started. The speed of impeller is gradually increased with the attainment of normal speed by the impeller. The pump is started with discharge valve closed because in this condition the pump operates at only 35 to 50 percent of full load. After a pump is stopped and it remains idle for some time, it gradually loses its priming by partly draining out through the glands.

As regard pump working, water is led into the centre of the impeller where it is set in rotation by the revoling blades of the impeller. The rotation of the water generates centrifugal force resulting in a pressure at the outer diameter of the impeller and when flow takes place the water passes out of the impeller with considerable velocity and pressure and is connected in the gradually expanding passageway of the casing and carried up through discharge pipe. The valve is then opened to commence the flow.

23.11 WORK DONE BY A CENTRIFUGAL PUMP

The work done by a centrifugal pump and its blade angles are determined by drawing inlet and outlet triangles of velocities (Fig. 23.10) in the same way as for a turbine with the exception that inlet triangle here is outlet and outlet triangle is the inlet triangle. It is assumed that (*i*) The water enters the eye of impeller radially, (*ii*) Whirl component V_{w_1} of the inlet absolute velocity V_1 is zero, (*iii*) Flow component, V_{f_1} = absolute velocity V, α = is 90°, (*iv*) No loss due to shock at entry, (*v*) No energy loss, due to friction and eddy formation, in the impeller, and (*vi*) Velocity distribution is uniform through narrow passages formed between two adjacent vanes.

Fig. 23.10 Velocity triangles

Here, D is diameter of impeller at inlet, R is $D/2$, V is absolute velocity of water at inlet, v is tangential velocity of impeller at inlet, also termed as

$$\text{peripheral velocity at inlet} = \frac{\pi DN}{60} = \frac{2\pi RN}{60} = wR.$$

V_r is relative velocity of water at inelt, V_f is velocity of flow at inelt, α is angle made by the absolute velocity, V, at the inlet with the direction of motion, ϕ is the angle made by relative velocity, V_r, at inlet with the direction of motion of the vane, also called vane angle at inlet, N is speed of impeller (r.p.m.) and μ_1, V_{r_1}, V_1, β and ϕ are the corresponding values at the outlet.

$$\text{Work done per kN of water} = \frac{V_{w1}\, v_1}{g} \text{ k N.m.}$$

23.12 HEAD, EFFICIENCY, POWER AND SPEED OF CENTRIFUGAL PUMPS

The main parameters of a pump are capacity and head it develops. Fig. 23.11 is a logarithmic curve of these parameters for reciprocating, centrifugal and axial flow pumps.

The total head or lift against which a pump works includes suction head, delivery head and loss of head due to friction, entrance and exit.

The usual practice is to consider lift separately for it is necessary to compare the lift with the permissible suction lift of the pump, the total head of the pump being the sum of static vertical lift (straight vertical distance between the well water level and the water of offtake point) plus hydraulic friction losses in the entire pipe line, suction as well as delivery plus velocity head loss.

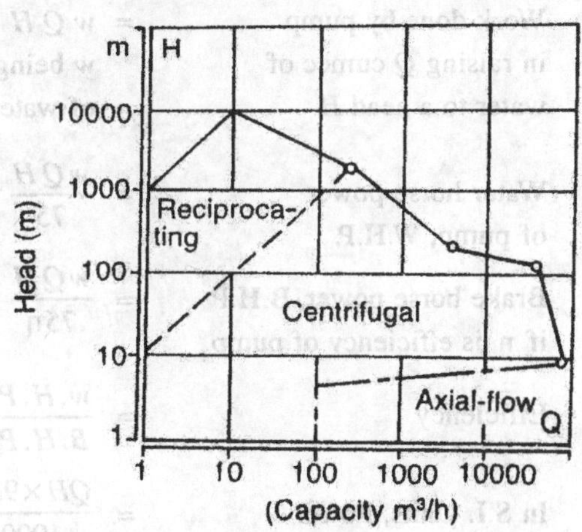

Fig. 23.11 Capacity-head curve for pumps

(*i*) *Suction Head*: It is the difference in level between the axis of the pump and the level of the suction liquid, usually denoted by *Hs*. It is related to the pump speed, discharge and total head and as such is a function of the specific speed.

(*ii*) *Delivery Head*: It is the height through which the liquid is lifted above the pump axis, usually denoted by H_d.

(*iii*) *Static Head*: It is the sum of suction head and delivery head *i.e.*, $H_s + H_d$.

(*iv*) *Manometric Head*: It is the head against which a centrifugal pump has to work. It is indicated by a differential manometer connected at the inlet and outlet of the pump. In addition to the static head, the pump has to work against loss of head in suction side, h_{fs} and delivery side, h_{ds} besides velocity head at exit of delivery pipe, V_d.

∴ Manometric head, (*i*) $H_m = H_s + H_d + h_{fs} + h_{fd} + V_d^2/2g$

(*ii*) H_m = Total head at pump outlet-total head at pump inlet.

(*v*) *Total Head*: The total or gross head is the sum of static head and all the head losses in flow before, through and after the impeller.

2. Efficiencies

The efficiency of the pump depends on the form of the impeller. It is the ratio of power output to the power input. The maximum efficiency of a pump varies with the specific speed and the discharge capacity.

A centrifugal pump has the following types of efficiencies:

(*i*) *Manometric Efficiency*: It is defined as the ratio of manometric head developed by the pump to the head imparted by the impeller to the liquid.

$$\eta_{ma} = \frac{H_{ma}}{\frac{V_{w_1} V_1}{g}}$$

(*ii*) *Mechanical Efficiency*: It is the ratio of energy delivered by the impeller to the liquid to the energy input to the pump shaft.

(*iii*) *Overall Efficiency*: It is the ratio of power output of the pump to the power supplied by the prime mover.

$$\eta_0 = \frac{wQH_{ma}}{P}$$

3. Horse Power

The horse power of a pump is determined by computing the work done by it in lifting water to the requisite height *H* (m).

Work done by pump in raising *Q* cumec of water to a head *H* $= wQH$ kg m/s, *w* being specific weight of water (kg/m³)

Water horse power of pump, W.H.P. $= \dfrac{wQH}{75}$

Brake horse power, B.H.P if η is efficiency of pump $= \dfrac{wQH}{75\eta}$

Efficiency $= \dfrac{w.H.P.}{B.H.P.}$

In S.I. Units, Power $= \dfrac{QH \times 9.81}{1000}$ kW.

4. Speeds of Pumps

(i) Minimum Starting Speed

On starting a centrifugal pump, there is no flow of liquid until the pressure difference in the impeller is enough to overcome the manometric head. The pressure head caused by the centrifugal force on rotating water, when the impeller is rotating, is $\frac{v_1^2}{2g} - \frac{v^2}{2g}$.

Flow is not to commence until this is greater than H_m, V_d being zero when flow starts

$$H_{ma} = \frac{V_{w1} v_1}{g}$$

The minimum speed to commence flow is

$$\frac{v_1^2}{2g} - \frac{v^2}{2g} = \frac{V_{w1} v_1}{g}$$

The flow actually starts when

$$\frac{u_1^2}{2g} - \frac{u^2}{2g} = \eta_m \frac{V_{w1} v_1}{g}$$

Where, η_m is manometric efficiency.

(ii) Specific Speed

The specific speed (N_s) of a centrifugal pump is defined as the speed at which the pump would deliver one cubic metre of liquid per second against a head of one metre. The specific speed is a characteristic of pumps to compare the performance of different pumps. In determining specific speed it is assumed that (*i*) all the pumps are similar, so that all linear dimensions are in proportion to the diameter of the impeller, (*ii*) velocity diagrams for all pumps are similar, and (*iii*) all velocities are proportional to the square root of the manometric head.

Using notation, Q is discharge (cumec), D and B are diameter and width of impeller, N_s is specific speed, N is speed (r.p.m.), H_{ma} is manometric head (m).

Discharge, $Q \propto$ area of flow × velocity.

or $\quad Q \alpha \pi Db \times V_f \quad$ (*i*)

$V_f \alpha \sqrt{H_m} \quad$ (*ii*)

Also $\quad Q \alpha \pi D^2 V_f$, b being $\alpha D \quad$ (*iii*)

Further, tangential velocity $v \alpha DN \quad \left(\because v = \frac{\pi DN}{60}\right) \quad$ (*iv*)

Also $\quad v \alpha V_f \alpha \sqrt{H_m} \quad$ (*v*)

Also $\quad Q \alpha \pi Db \times V_f$

$\alpha D^2 \sqrt{H_m} \quad \left(\because b \alpha D, \text{ and } V_f \alpha \sqrt{H_m}\right) \quad$ (*vi*)

Also $\quad D \alpha \frac{\sqrt{H_m}}{N} \quad (v = wD/2,\ w \alpha N \text{ or } v \alpha ND \text{ or } D \alpha \frac{v}{N} \quad$ (*vii*)

and $\quad v \alpha \sqrt{H_m}\,)$

From (*vi*) and (*vii*)

$$Q \alpha \frac{H_m}{N^2} \sqrt{H_m}$$

$$Q \alpha \frac{H_m^{3/2}}{N^2}$$

or $$Q = k\frac{H_m^{3/2}}{N^2} \quad \text{(K is constant of proportionality)}$$

Specific speed, when Q is 1 cumec and H_{ma} is 1 m

$$1 = \frac{K}{N_s^2}$$

or $$K = N_s^2$$

$$\therefore \quad Q = \frac{N_s^2 H_{ma}^{3/2}}{N^2}$$

or $$N_s = \frac{N\sqrt{Q}}{H_{ma}^{3/4}}$$

H_{ma} in a multi-stage pump is total head developed divided by the number of stages. Q in a double suction pump is taken as half of the actual discharge delivered by the pump.

It is evident from the equation that increasing the speed and capacity (discharges) of a pump increases its specific speed but increase in head decreases the specific speed.

The specific speeds of various types of pumps are given in Table. 23.1. In general, high head pumps have low specific speed and low head pumps have high specific speeds.

Table 23.1 Specific speeds of pumps

Type of Pump	*Slow speed (radial flow)*	*Medium speed (radial flow)*	*High speed (radial flow)*	*High speed (mixed flow)*	*High speed (axial flow)*	*Very High speed*
Specific speed	10–30	30–50	50–80	80–160	160–500	Above 500

23.13 CHARACTERISTICS OF CENTRIFUGAL PUMPS

The centrifugal pumps are designed and manufactured for performance characterised by a relationship between discharge of the pump, head of water against which it operates, speed, power requirement and efficiency of operation. These characteristics depending on diameter, shape and speed of impeller within the casing are different for different models of the pumps. Ordinarily, it operates under maximum efficiency condition. However, its performance differs when it is run at conditions different from the designed. The efficiency of a pump is low when pumping lesser discharge than the quantity for which it is designed or pumping higher discharge than designed at a head lower than the normal. The manufacturers of the pumps indicate the performance characteristics of the pumps representing graphically the inter-relations between discharge, head, speed and horse power which are termed as characteristic curves (Fig. 23.12). These curves are useful in the selection of pumps and enable to make adoption of the pump to the operating conditions to obtain high efficiency and low operating cost. These curves are obtained by plotting at constant speed the variable head, efficiency and brake horse power against the corresponding discharge. The head curve shows that the pump can operate over a wide range of discharge and total head.

It transpires from the curves that for a given speed:

1. The efficiency depends on discharge and total head. Maximum efficiency is obtained at a particular discharge known as normal or rated discharge or rated capacity of the pump at a given speed. The variable quantity of discharges at maximum efficiency are obtained by running several pumps of smaller capacities in parallel so that a variable number are run at capacity as per requirement at near maximum efficiency.

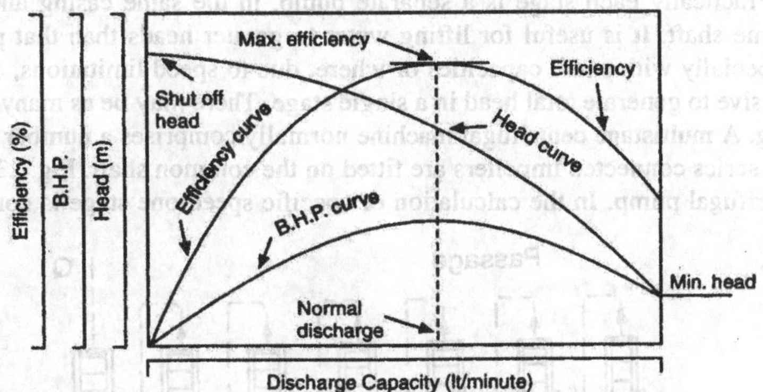

Fig. 23.12 Typical characteristic curves.

2. The maximum efficiency of a pump varies with the specific speed and discharge capacity.
3. The discharge increases and correspondingly the head decreases and vice versa. The maximum discharge corresponds to minimum head. At zero discharge pump operates with discharge valve closed, the head developed is maximum termed as shut off head.
4. The break horse power required from motor to run the pump increases as the discharge increases.

23.14 SINGLE AND MULTI-STAGE PUMPS

A centrifugal pump in which only one impeller is used is called single stage pump. It is used for pumping against low or moderate heads. The single admission type pumps (Fig. 23.13) are generally used for the small sizes. Single admission pumps are made in one or more stages. The double admission pump shown in Fig. 23.14 may be either single or multi-stage. Double stage or

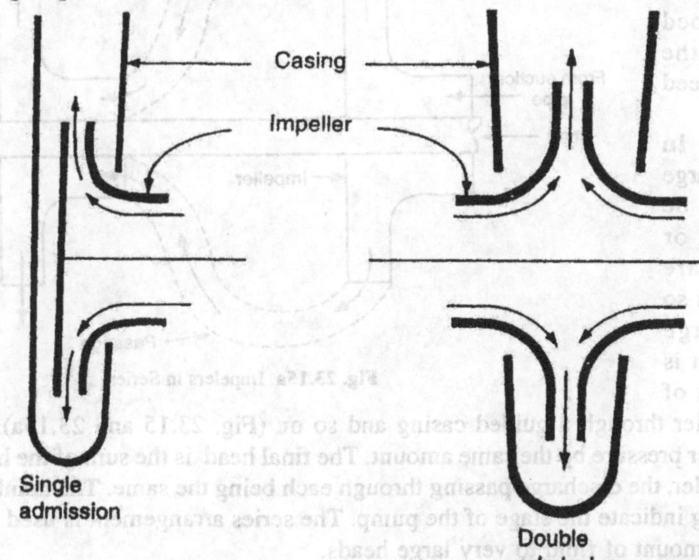

Fig. 23.13 Single admission impeller

Fig. 23.14 Double admission impeller

multi-stage pump is that which has two or more identical impellers so arranged that discharge from one impeller enters the eye of the next impeller. Such a pump is essentially a high head or high pressure pump and consists of two or more stages depending on head against which liquid is to be pumped. Practically each stage is a separate pump, in the same casing and impellers attached to the same shaft. It is useful for lifting water to greater heads than that possible with a single pump, especially with small capacities or where, due to speed limitations, diameter of the impeller is excessive to generate total head in a single stage. There may be as many as eight stages in a single casing. A multistage centrifugal machine normally comprises a number of single-stage machines whose series connected impellers are fitted on the common shaft. Fig. 23.15 indicates a seven-stage centrifugal pump. In the calculation of specific speed one stage is considered.

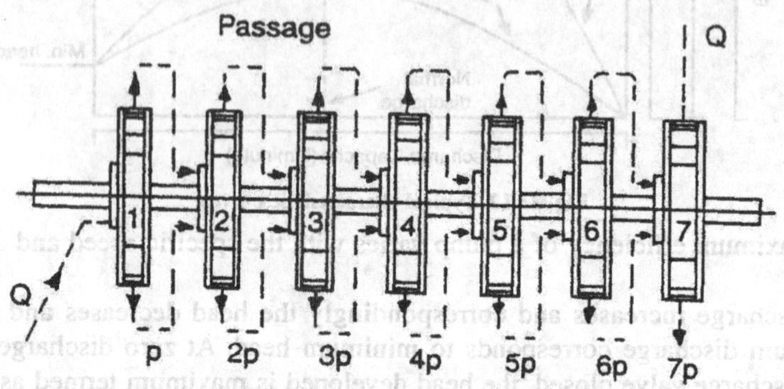

Fig. 23.15 Functional diagram of multistage pump

The centrifugal pumps can be arranged, depending on the application, type of drive and other requirements, with a horizontal shaft or vertical shaft.

23.15 PUMPS ARRANGEMENT

The pumps may be arranged in series or in parallel. The head developed depends on the diameter and speed of the impeller.

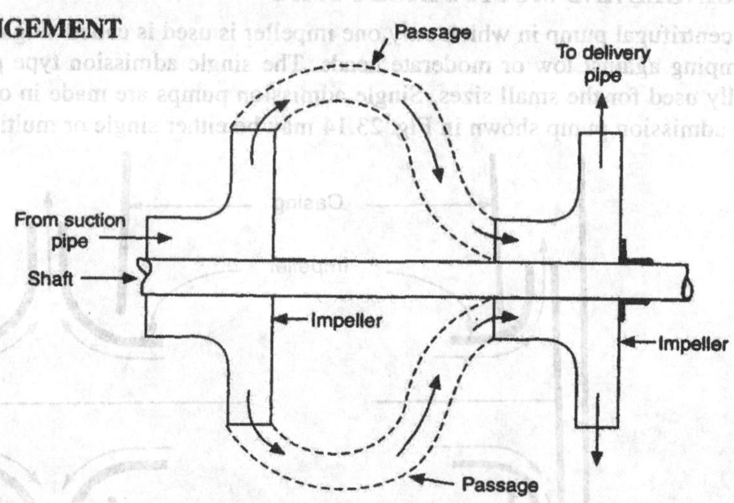

Fig. 23.15a Impelers in Series

1. Pumps in Series: When large heads are to be developed two or more pumps are arranged in series so that the discharge from first impeller is led into the inelt of the second impeller through a guided casing and so on (Fig. 23.15 and 23.15a). Each impeller increases the water pressure by the same amount. The final head is the sum of the heads developed by the each impeller, the discharge passing through each being the same. The number of impellers fitted in the casing indicate the stage of the pump. The series arrangement is used for delivering a relatively small amount of fluid to very large heads.

Advantages

(*i*) Higher suction lift is achieved due to lower specific speed of individual impellers, (*ii*) Flexibility in choosing number of stages by fitting required number of impellers in the casing, (*iii*) Low friction losses, (*iv*) Number of stages may be chosen to suit the pump speed vis-a-vis motor speed.

2. Pumps in Parallel: Wherever large quantity of liquid is to be pumped than that possible by one impeller against a relatively small head, two or more pumps are arranged in parallel (Fig. 23.16). In the arrangement each pump lifts liquid indipendently against the same height and delivers into a common pipe for carrying to the required height. The total discharge pumped is the sum of individual pump discharge times the number of pumps employed. The pumps used should essentially have identical characteristics and their shutoff heads should be about the same.

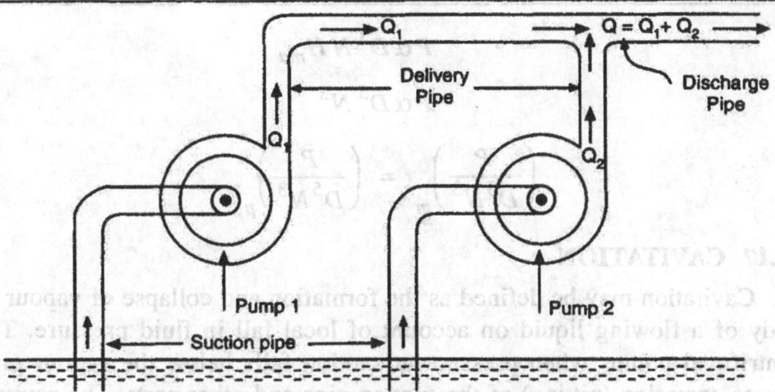

Fig. 23.16 Centrifugal pumps in parallel

Advantage

(*i*) Large amount of water can be pumped to a relatively small height.

23.16 MODEL TESTING

The performance of prototype can be ascertained by testing of model of geometrically similar pumps. The important parameters to be satisfied are speed, head, discharge and power.

1. Specific speed

Specific speed of model = Specific speed of prototype

$$(N_s)_m = (N_s)_p$$

$$\left(\frac{N\sqrt{Q}}{H_{ma}^{3/4}}\right)_m = \left[\frac{N\sqrt{Q}}{H_m^{3/4}}\right]_p$$

2. Head

$$\left(v_1 \,\alpha \sqrt{H_{ma}}\right)_m = \left(v_1 \,\alpha \sqrt{H_{ma}}\right)_p$$

Also $$v_1 = \left(\frac{\pi DN}{60}\right)_m = \left(\frac{\pi DN}{60}\right)_p$$

$$\therefore \quad \frac{(DN)_m}{(DN)_p} = \frac{\left(\sqrt{H_{ma}}\right)_m}{\left(\sqrt{H_{ma}}\right)_p}$$

3. Discharge

$$Q \,\alpha\, D^2 \times V_f$$

Also $$V_f \,\alpha\, v \,\alpha\, DN$$

$$\therefore \quad Q \,\alpha\, D^2 . DN = D^3 N$$

or $$\left(\frac{Q}{D^3 N}\right)_m = \left(\frac{Q}{D^3 N}\right)_p$$

4. Power

$$P = w\,Q\,H_{ma}$$ **Ignoring losses**

or $$P \,\alpha\, Q\,H_{ma}$$

$$P \alpha D^3 N H_{ma}$$

$$P \alpha D^5 N^3 \qquad (\because \sqrt{H_{ma}} \alpha DN)$$

$$\therefore \quad \left(\frac{P}{D^5 N^3}\right)_m = \left(\frac{P}{D^5 N^3}\right)_p$$

23.17 CAVITATION

Cavitation may be defined as the formation and collapse of vapour filled bubbles within the body of a flowing liquid on account of local fall in fluid pressure. The cavitation appears in centrifugal pumps when pressure at suction falls below the vapour pressure of the liquid and causes corrosion (pitting) of the suction pipe and other parts. The cavitation increases as the net positive suction head (NPSH) decreases.

Thoma cavitation factor σ (sigma), as is used in turbines, is applicable which indicates that cavitation begins when value of σ is less than critical value σ_c

$$\sigma_c = \frac{H_a - H_v - H_s}{H_{ma}} = \frac{H_{sv}}{H_{ma}}$$

Where, H_a is atmospheric pressure head (m of water head), H_v is vapour pressure (*m* of water head), H_s is total suction head $\left(H_s + H_{fs} + \frac{V_s^2}{2g}\right)$, H_{sv} is net positive suction head and H_{ma} is manometric head.

The cavitation occurs due to limited suction, very high specific speed, very high temperature of the liquid and high runner speed.

Vapour cavities (bubbles of vapour) appear whenever the pressure at any point in a flow falls to the vapour pressure of the liquid at that temperature. The cavities so formed due to motion of liquid are carried to zones of high pressure where the vapours condense and collapse. The space so created is occupied by the adjoining liquid which rushes with a great velocity thereby creating spherical shock waves and noise.

As the cavities collapse at the surface of the body at collapsing pressure of the order of 100 times the atmospheric pressure, the metal particles give way due to repeated hammering action. The erosion of the metal is termed as pitting. The cavitation affects the hydraulic machinery by (*i*) effecting reduction in discharge, power output and efficiency, (*ii*) Vibration of various parts and noise, and (*iii*) Loss of material causing roughening of surface. It limits the head against which a pump can operate satisfactorily. The axial flow pump is more prone to cavitation and as such used for low heads.

SOLVED EXAMPLES

Example 23.1 *Determine horse power of a centrifugal pump to lift 2 cumecs of water through a head of 10m, the head loss in suction and delivery pipes is 1m. The pump efficiency is 75%.*

Solution.

$$\text{Water H.P.} = \frac{wQH}{75} = \frac{1000 \times 2.0 \times (10+1)}{75}$$

$$= 293 \text{ H.P.}$$

$$\text{Brake H.P.} = 293/0.75 = 391 \text{ H.P.}$$

Example 23.2 *A centrifugal pump has impeller of external diameter of 500 mm and internal diameter of 200 mm, water enters the impeller at 3 m/s. The vane angles at the inlet and outlet are 25° and 45° respectively. Determine (i) impeller speed, and (ii) work done.*

Solution.

(*i*) Tangential velocity at inlet,

$$v = \frac{V}{\tan 25^\circ} = \frac{3}{0.4663}$$
$$= 6.43 \text{ m/s}$$

Also
$$v = \frac{\pi DN}{60}$$
$$6.43 = \frac{\pi \times 0.2\, N}{60}$$

or
$$N = 613.77 \text{ r.p.m.}$$

(*ii*) Work done

Tangential velocity at outlet,

$$v_1 = v \times D_1/D$$
$$= 6.43 \times 0.50/0.20 = 16.08 \text{ m/s}$$

Fig. 23.17

Velocity of whirl at outlet,

$$V_{w1} = V_1 - \frac{V_{f1}}{\tan 45^\circ} = 16.08 - \frac{3}{1} = 13.08 \text{ m/s} \quad (\text{Since } V = V_f = V_{f1})$$

Work done per kN of water,

$$W = \frac{V_{w1} \cdot V_1}{g} = \frac{13.08 \times 16.08}{9.81} = 21.44 \text{ kN.m.}$$

Example 23.3 *A centrifugal pump has external and internal diameters as 500 mm and 200 mm respectively. The vanes are set back at an angle of 30° to the outer rim. The pump runs at 1200 r.p.m. Velocity of flow (radial) is constant at 3 m/s. Determine (i) vane angle at inlet,* β *angle, (ii) Velocity of whirl at outlet, and (iii)* Work done per kN of water.

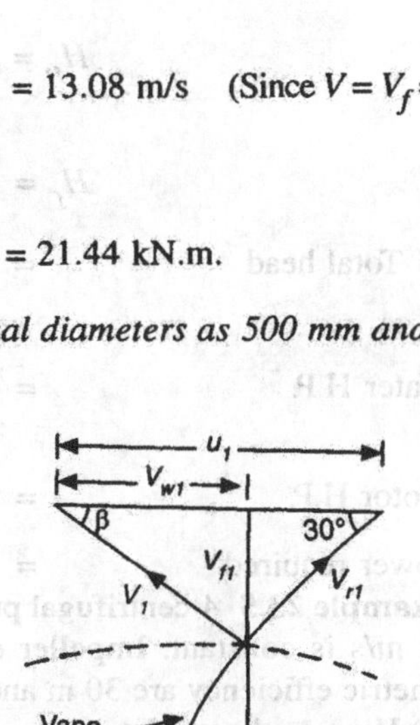

Fig. 23.18

Solution.

(*i*) Vane angle at inlet, θ

$$v = \frac{\pi DN}{60} = \frac{\pi \times 0.2 \times 1200}{60} = 12.57 \text{ m/s}$$

$$\tan\theta = \frac{V_f}{v} = \frac{3.0}{12.57} = 0.2387$$

$$\therefore \quad \theta = 13.42^\circ. \quad \textbf{Ans.}$$

(*ii*) Angle made by absolute velocity at outlet with direction of motion.

Tangential velocity of impeller at outlet,

$$v_1 = \frac{\pi D_1 N}{60} = \frac{\pi \times 0.5 \times 1200}{60} = 31.43 \text{ m/s}$$

(*iii*) Velocity of whirl at outlet,

$$V_{w1} = v_1 - \frac{V_{f1}}{\tan 30^\circ} = 31.43 - \frac{3}{0.5774} = 26.23 \text{ m/s}$$

$$\therefore \quad V_1 = \sqrt{V_{f1}^2 + V_{w1}^2} = \sqrt{(3)^2 + (26.23)^2} = 26.40 \text{ m/s}$$

$$\tan \beta = \frac{V_{f1}}{V_{w1}} = \frac{3.0}{26.23} = 6.52°$$

(*iv*) Work done per kN of water,

$$W = \frac{V_{w1} \times v_1}{g} = \frac{26.23 \times 31.43}{9.81} = 84.04 \text{ kN.m} \quad \textbf{Ans.}$$

Example 23.4 *A centrifugal pump lifts 2000 lt/minute of water against a head of 50 m, the losses in pipes are 5 m. The delivery pipe is of 20 cm diameter and 200 m long. Assume pump efficiency as 75% and motor efficiency 80%, f = 0.01.*

Solution.

$$Q = 2000/1000 \times 60 = 0.033 \text{ cumec.}$$

$$V = \frac{(0.033)}{\frac{(\pi(0.20)^2)}{4}} = 1.05 \text{ m/s}$$

$$H_f = \frac{flV^2}{2gd}$$

$$\therefore \quad H_f = \frac{0.01 \times 200 \times (1.05)^2}{2 \times 9.81 \times 0.20} = 0.564 \text{ m.}$$

∴ Total head = 50 + 0.56 = 50.56 m, neglecting entrance and exit losses.

Water H.P. $= \frac{wQH}{75} = \frac{1000 \times 0.033 \times 50.56}{75} = 23$ H.P.

Motor H.P. $= \frac{23}{0.75 \times 0.80} = 38$ H.P.

Power required = 38 × 0.735 = 28 kW.

Example 23.5 **A centrifugal pump works at 100 r.p.m, vane angle at outlet is 60°. Velocity of flow 3 m/s is constant. Impeller diameter at exit is twice that at inlet, manometric head and manometric efficiency are 30 m and 70 percent respectively. Find (*i*) diameter of impeller at exit, and (*ii*) Vane angle at inlet.**

Solution.

(*i*) Impeller diameter at exit, D_1

$$\eta_{ma} = (gH_{ma}) / V_{w1}\, v_1$$

$$\therefore \quad 0.70 = (9.81 \times 30) / V_{w1}\, v_1$$

or $$V_{w1}\, v_1 = 420$$

or $$V_{w1} = 420/v_1 \qquad (i)$$

$$\text{Tan } \phi = (V_{f1}) / v_1 - V_{w1}$$

or $$v_1 - V_{w1} = \frac{V_{f1}}{\tan \phi} = \frac{3.0}{\tan 60°} = \frac{3.0}{1.732}$$

$$v_1 - V_{w1} = 1.732 \qquad (ii)$$

From (*i*) and (*ii*)

$$v_1 - 420/v_1 = 1.732$$

or $$v_1^2 - 1.732\, v_1 - 4.20 = 0$$

or $$v_1 = 21.38 \text{ m/s}$$

Fig. 23.19

Tangential velocity at outlet, $v_1 = \pi D_1 N / 60$ m/s

$$21.38 = (\pi D_1 \times 100)/60$$

Impeller dia. at exit, $D_1 = 4.08$ m **Ans.**

(*ii*) Vane angle at inlet, θ

$$\tan\theta = \frac{V_f}{v} = \frac{3.0}{21.38/2} \qquad (\because \; D = D_1/2)$$

$\therefore$ $\theta = 15.68°$ **Ans.**

Example 23.6 *A three stage centrifugal pump has vane angle at exit 45°. The impeller is 500 mm in diameter and 25 mm wide; thickness of vanes is 10% of total area. The pump runs at 1000 r.p.m. to deliver 0.075 cumecs of water. Find out (i) pump power, (ii) manometric head, (iii) specific speed. Mechanical and manometric efficiencies are 85 per cent and 75 per cent respectively. Assume radial velocity at inlet.*

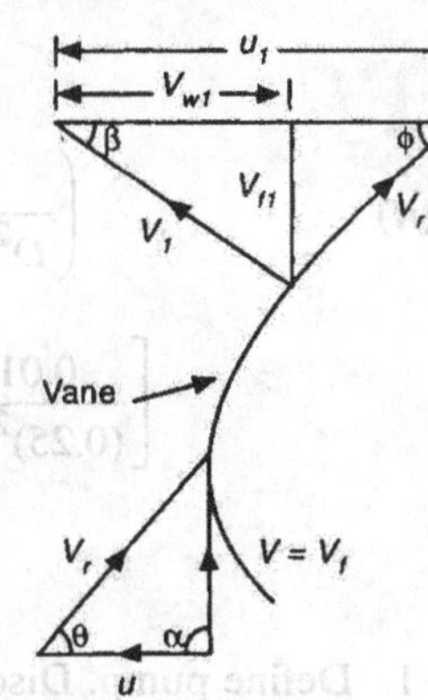

Fig. 23.20

Solution.

(*i*) Manometric head

Tangential velocity at outlet

$$v_1 = \frac{\pi D_1 N}{60} = \frac{\pi \times 0.5 \times 1000}{60}$$

$$= 26.19 \text{ m/s}$$

Net outlet area $= 0.90\,\pi\, D_1 B_1$

$$= 0.90 \times \pi \times 0.5 \times 0.025 = 0.035 \text{ m}^2$$

Velocity of flow at outlet, V_{f1}

$$V_{f1} = Q/\text{net outlet area}$$

$$= 0.075/0.035 = 2.14 \text{ m/s}$$

$$\tan\phi = \frac{V_{f1}}{v_1 - V_{w1}}$$

$$\tan 45° = \frac{2.14}{26.19 - V_{w1}}$$

or $V_{w1} = 24.05$ m/s

$$H_{ma} = \eta_{ma}\,(V_{w1}\,v_1)/g = (0.75 \times 24.05 \times 26.19)/9.81$$

$$= 48.16 \text{ m}$$

$H_{total} = 48.16 \times 3 = 144.48$ m **Ans.** (∵ of 3–stage)

(*ii*) Power of pump, P

$$\eta_0 = \eta_m \times \eta_{ma} = 0.85 \times 0.75 = 0.6375$$

$$P = \frac{wQH_{ma}}{\eta_0} = \frac{9810 \times 0.075 \times 48.16}{1000 \times 0.6375} = 55.58 \text{ kW} \quad \textbf{Ans.}$$

(*iii*) Specific speed, N_s

$$N_s = \frac{N\,Q^{1/2}}{(H_{ma})^{3/4}}$$

$$= \frac{1000(0.075)^{1/2}}{(48.16)^{3/4}} = 15 \quad \textbf{Ans.}$$

Example 23.5 *A centrifugal pump of 0.050 cumec capacity operates at 1500 r.p.m. against a total head of 20 m. If the impeller diameter of the prototype and model is 0.5 m and 0.25 m respectively. Determine discharge and head, if the model is run at 3000 r.p.m. Ignore efficiency.*

Solution.

(*i*)

$$\left(\frac{Q}{D^3 N}\right)_m = \left(\frac{Q}{D^3 N}\right)_P$$

$$\left[\frac{Q}{(0.25)^3 \times 3000}\right]_m = \left[\frac{0.050}{(0.5)^3 (1500)}\right]_P$$

or

$$Q_m = \frac{0.050 (0.25)^3 \times 3000}{(0.5)^3 (1500)}$$

$$= 0.0125 \text{ cumec.}$$

(*ii*)

$$\left(\frac{Q}{D^2 \sqrt{H}}\right)_m = \left(\frac{Q}{D^2 \sqrt{H}}\right)_P$$

$$\left[\frac{0.0125}{(0.25)^2 \sqrt{H}}\right]_m = \left[\frac{0.050}{(0.50)^2 \sqrt{20}}\right]_P$$

$$H_m = 20\text{m.}$$

QUESTIONS

1. Define pump. Discuss various types of pumps.
2. Give an account of the classification of various types of centrifugal pumps.
3. Discuss the working of a single acting centrifugal pump.
4. What do you understand by casing of a centrifugal pump. Discuss its various types with the help of illustrative sketches.
5. Define static head, manometric head and total head of a centrifugal pump.
6. Define efficiency of a centrifugal pump. What do we understand by manometric efficiency, volumetric efficiency and mechanical efficiency?
7. What are single stage and multi-stage centrifugal pumps? Discuss.
8. Define specific speed of a centrifugal pump. Derive an expression for the same. (AMIE)
9. What are the two arrangements for the installation of centrifugal pumps? What are the merits of the each?
10. What do you understand by model testing relative to a centrifugal pump?
11. Discuss characteristic curves of a centrifugal pump.
12. (*a*) Explain with neat sketches the working of a centrifugal pump.

    (*b*) Discuss the various methods adopted to increase the efficiency of a centrifugal pump by altering the shape of the casing of the chamber surrounding the impeller.

    (*c*) Two centrifugal pumps have the head Vs discharge characteristics as follows:

Q (l/s)	0	4	8	12	16	20	24
Pump 1, H_1 (*m*)	50.0	51.8	50.8	48.0	42.5	32.5	18.3
Pump 2, H_2 (*m*)	46.7	45.9	44.2	40.3	34.3	26.0	17.0

    Both pumps are installed together and are required to pump water through a pipe of 150 mm dia having $f = 0.02$. Calculate the heads under which pumps are working and discharge (l/s) pumped by them if the pumps are connected in series (static lift is 65 m and suction and delivery pipes are 800 m long). (*AMIE '01*)

13. A centrifugal pump delivers 60 l/s of water to a height of 20 m through a 90 m long and 100 mm dia. pipe. If the overall efficiency of the pump is 75%, obtain power required to drive the pump. Take friction coefficient in pipe f as 0.012. Also draw velocity triangles for radial entry. *(AMIE w'0)*
14. The impeller of a centrifugal pump is 1.2 m in dia. and rotates at 1200 r.p.m. The blades are curved backwards and make an angle 30° with the tangent at the outer periphery. Obtain power required if flow velocity at the outlet be 20 m/s. Also estimate the head by which water is lifted when a difference casing reduces the outlet velocity by 60%. Draw outlet velocity triangle. *(AMIE s'99)*
15. A centrifugal pump having an impeller 25 cm outside dia. rotates at 1000 r.p.m. The vanes are radial at exit and are 7.5 cm wide. The velocity of radial flow through the impeller is 3m/s. The velocities in suction and delivery pipes are 2.5 m/s and 2 m/s respectively. Determine:
   (*i*) Height through which the pump lifts water.
   (*ii*) The power required by the pump. *(AMIE s' 2000)*
16. (*a*) A centrifugal pump of 30 cm impeller dia. is discharging water at a rate of 200 l/s, while running at 1200 r.p.m. The discharge area at the outlet is 0.1 sq. m and vanes are bent back at 35° to the tangent at outlet. The dia. of suction and delivery pipes are 30 cm and 22.5 cm respectively. The gauges fixed on the suction and delivery pipes close to the pump show heads of 3.5 m below and 18.5 m above atmosphere. If power required be 96 H.P. find:
   (*i*) overall efficiency, and (*ii*) manometric efficiency. Also draw velocity triangles at inlet and outlet.
   (*b*) A propeller pump 20 cm dia. is found to be the most efficient when delivering 5600 l/s of water at 1200 r.p.m. against a head of 6 m. A similar pump is required to deliver 1400 l/s at 700 r.p.m. Calculate the dia. of the second pump and the head developed by it. *(AMIE s'98)*
17. (*a*) A centrifugal pump having an effective blade angle of 45° at exit is required to lift a head of 20 m, the shaft speed being 700 r.p.m. and the flow velocity 2 m/s. Calculate impeller dia. Why is the suction height of a pump limited? On what factors does it depend?
   (*b*) What are the effects of variation of speed of a centrifugal pump on its power consumption, discharge and total head?
   (*c*) A centrifugal pump delivers water at the rate of 1320 l/m, when running at 1450 r.p.m. and consuming 4.8 kW of input power. The impeller is 230 mm in dia. and has a width of 19 mm at its outlet. There is no whirl at impeller inlet and at the outlet of the impeller. The blades are curved backwards making an angle of 30° with the tangent. The cross sectional areas of the suction pipe and delivery pipes are 14.2×10^{-3} m² and 10.3×10^{-3} m² respectively. A vacuum gauge fitted close to the pump casing on the suction pipe reads 3 m (vacuum) and a pressure gauge fitted similarly on the delivery pipe reads 12 m. There two gauges are located at same horizontal level. Neglecting the effects of lade thickness and assuming that actual whirl component of velocity at impeller output is two-third of the theoretical value, find: (*i*) total head developed by the pump, (*ii*) theoretical or Euler head developed by the pump, (*iii*) head loss in impeller and casing, and (*iv*) overall efficiency of the pump. *(AMIE s'96)*
18. Answer the following questions with sketches, if necessary:
   (*i*) Why gear pump is not recommended for high pressure applications?
   (*ii*) Why an axial flow pump with fixed blades is not started with its delivery valve closed?

(*iii*) Why the actual shut-off head of a centrifugal pump is much less than the theoretical 'shut-off' head? (*AMIE w'94*)

19. (*a*) Explain the conditions under which two pumps may be put in series or in parallel operation.
(*b*) A single acting centrifugal pump has an impeller of 0.3 m dia which rotates at 1780 r.p.m. and discharges 4000 l/minute of water to a height of 25 m at an efficiency of 70%. Obtain the number of stages of a similar multistage pump to lift 5000 l/minute of water to a height of 175 m at 1500 r.p.m.
(*c*) A three-stage centrifugal pump has an impeller of 0.4 m dia. and 25 mm width at outlet. The vanes are curved back at an angle of 30° and reduce the circumferential area of flow by 15%. The manometric efficiency of each stage is 85% and overall efficiency is 75%. Determine the head generated by the pump when running at 1200 r.p.m. and dischanging 0.06 cumec of water. Also find power required to drive the pump. (*AMIE w'96*)

20. Explain with sketches, if necessary:
(*i*) Why the input head of an impeller of centrifugal pump with finite number of vanes is lower than the Euler's head?
(*ii*) The functional difference between a gear pump and a vane pump.
(*iii*) Why the efficiency of centrifugal pump is usually less than that of a reaction turbine. (*AMIE n'96*)

21. (*a*) Justify that a foot-valve must be fitted to the suction pipe of a centrifugal pump.
(*b*) A centrifugal pump discharges 7050 l/min. against a total head of 26.5 m when running at 600 r.p.m. The outer diameter of impeller is 60 cm and inner diameter is 30 cm. The area of flow through the wheel is 615 cm^2. The vanes are set back at an angle of 45°. Water enters the wheel radially without shock. Calculate:
(*i*) manometric efficiency, (*ii*) vane angle at inlet, (*iii*) head loss at inlet to impeller if the discharge is reduced by 50%, the speed remaining the same. (*AMIE w'95*)

22. Write note on priming of a centrifugal pump. (*AMIE w'95*)

23. A centrifugal pump discharges 1800 l/s of water against a head of 6 m. Its impeller has 1.22 m dia. 200 r.p.m. speed 0.75 m^2 area at outlet, 25° angle of vane at outlet and a ratio of external to internal dia. as 2:1. Neglecting slip, find (*i*) useful power, (*ii*) hydraulic efficiency and (*iii*) minimum starting speed against a head of 6 m. (*AMIE w'94*)

24. (*a*) Define and explain the manometric efficiency, mechanical efficiency and overall efficiency of a centrifugal pump.
(*b*) What is priming? Why is it necessary?
(*c*) A centrifugal pump impeller has an outer dia of 30 cm and an inner diameter of 15 cm. The pump runs at 1200 r.p.m. The impeller vanes are set at blade angle of 30° at the outlet. If the velocity of flow is constant at 2m/s, Calculate: (*i*) velocity and direction of water outlet, (*ii*) head developed if manometric efficiency is 0.85, and (*iii*) blade angle at the inlet. (*AMIE 'w 93*)

25. (*a*) Discuss various types of impellers of a pump and give their applications.
(*b*) A centrifugal pump running at 1000 r.p.m. discharges 80 l/s against a net head of 16 m. The volumetric efficiency of pump is 85%. If the vane angle at outlet is 35 and flow velocity is 1.5 m/s, find diameter and width at impeller of outlet. (*AMIE s'97*)

26. (*a*) Define and explain the manometric efficiency, mechanical efficiency and overall efficiency of a centrifugal pump.
(*b*) What is priming? Why is it necessary?
(*c*) A centrifugal pump impeller has an outer diameter of 30 cm and an inner dia. of 15 cm. The pump runs at 1400 r.p.m. The impeller vanes are set at a blade angle of 30° at the outlet. If the velocity of flow is constant at 3 m/s. Calculate: (*i*) velocity and

direction of water outlet, (*ii*) head developed if manometric efficiency is 0.85, (*iii*) blade angle at inlet.

27. (*a*) What is the 'pumping theory' of a positive displacement pump?
    (*b*) Describe working of an external gear pump with a neat sketch. (*AMIE w'2000*)
28. Write note an model testing of pumps. (*AMIE s'93*)
29. (*a*) The whole of the energy corresponding to velocity of whirl at exit is wasted, (*ii*) 40% of this energy is converted into useful pressure energy. Neglect friction. Find also impeller width at inlet, if the dischange is 170 l/s. Neglect blade thickness.
    (*b*) Compare and contrast the performance characteristics of a centrifugal pump and an axial flow pump. (*AMIE w'98*)
30. (*a*) Explain the phenomenon of cavitation in centrifugal pumps.
    (*b*) Briefly explain various losses that occur in a pump. (*AMIE w'98*)
31. A centrifugal pump running at 1200 r.p.m. delivers 225 l/s against a head of 114 m. The diameter of the impeller is 30 cm at inlet and 60 cm at exit. The breadth at inlet is 15 cm and at exit 7.5 cm. The vane angle at exit is 30°. The water enters radially at inlet. The delivery pipe is 30 cm in dia. Determine: (*i*) theoretical head developed, (*ii*) manometric efficiency, (*iii*) pressure rise in the impeller assuming losses as 10% of the velocity head at exit, (*iv*) loss of head in volute casing, (*v*) power required to drive the pump assuming an overall efficiency of 60%. (*AMIE w'97*)
32. (*i*) What is the effect of speed variation on discharge, head and power of a centrifugal pump? Give reasons using principle of similarity to account for speed variations.
    (*ii*) Why are piston pumps usually made of odd number of cylinders?
    (*iii*) Explain why pumps are generally less efficienct than turbines.
    (*iv*) Give reasons why hydraulic pumps cannot be used as hydraulic motor?
    (*v*) Can a relief valve be used as a sequence valve? If so, is it necessary to make any modifications? Explain. (*AMIE w'97*)
33. (*a*) Show that pressure rise in the impeller of a centrifugal pump is given by:

$$\frac{P_2 - P_1}{w} = \frac{1}{2g}\left[V_{f_1}^2 + M_r^2 - V_{f_2}^2 \ \text{cosec}^2\ \phi\right], \text{ with usual notations.}$$

    (*b*) A single stage centrifugal pump runs at 500 r.p.m. and delivers 300 m³/mm of water against a head of 120 m. The pump impeller is 2 m in dia. and has a positive suction lift of 3 m. Laboratory tests are conducted on a model with 45 cm impeller and on a reduced head of 95 m. Calculate the speed, discharge and suction lift for the laboratory tests. Assume H_v = 0.34 m of water and H_{atm} = 10.15 m of water. (*AMIE s'93*)
34. (*a*) By means of a neat sketch explain the working of a vane pump.
    (*b*) Explain how is the overall efficiency of a centrifugal pump determined experimentally. Draw a table for such a pump test. (*AMIE s'93*)

24

RECIPROCATING PUMPS

24.1 RECIPROCATING PUMPS

The reciprocating pump is a positive displacement pump as in it the fluid is displaced by a

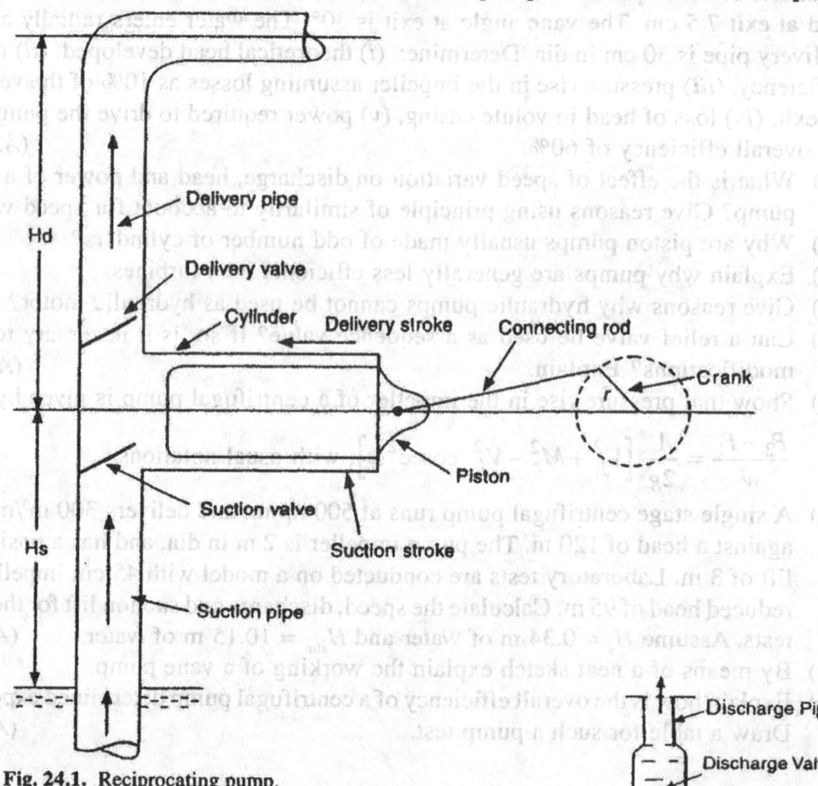

Fig. 24.1. Reciprocating pump.

linear movement of a plunger or piston in a cylinder. It consists of a cylinder in which a piston or plunger, constituting the reciprocating part of the pump, is moved backward and forward thereby creating alternatively a vacuum pressure and positive pressure in the cylinder and raising of the fluid results (Fig. 24.1). Fig. 24.1(*a*) Shows the basic design of the pump. Power from an external source drives the pump. The pump may be single acting or double-acting depending on whether the fluid is displaced during only one direction of the stroke or during both directions. In a

Fig. 24.1*a*. Reciprocating pump schematic.

single acting pump, the fluid is suck into the piston on the outward stroke and forced out during the inward stroke (Fig. 24.2). In double acting, the fluid acts on both sides of the piston and is sucked and delivered during one stroke (Fig. 24.3). There may be one or more pistons/plungers and accordingly the pumps are called single, duplex, triplex etc.

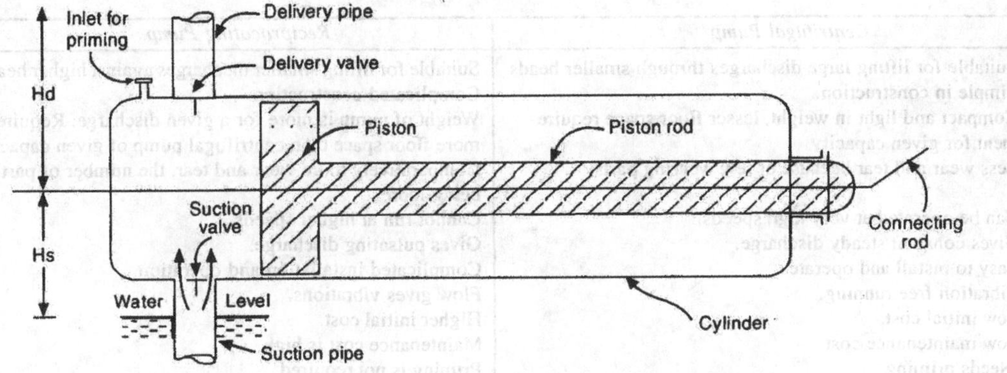

Fig. 24.2. Single-acting pump.

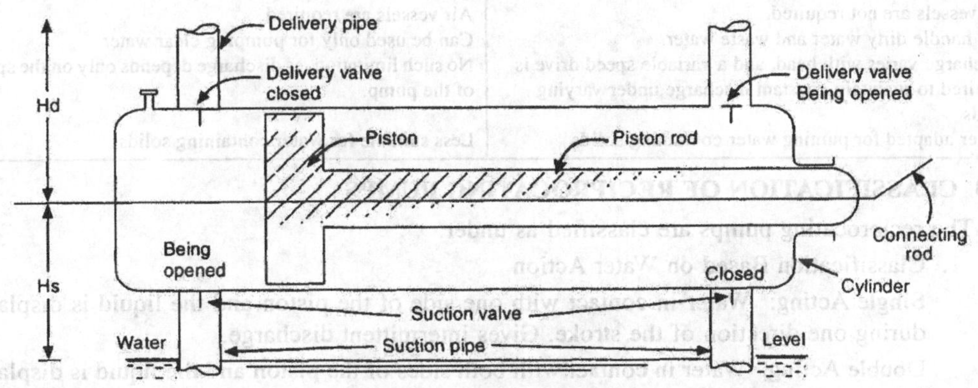

Fig. 24.3. Double-acting pump.

Reciprocating pumps can be used only for pumping clear water as suspended impurities may interfere with the working of the pump.

Plunger: It is a solid cylinder that enters a larger space forcing the fluid out of that space.

Piston: It consists of a cylindrical cap at the end of the driving rod. The rim of cap is in lubricated contact with the cylinder wherein the piston moves backward and forward.

The principal components of a reciprocating pump (Fig. 24.1) are cylinder, plunger/piston, crank and connecting rod, suction valve, delivery valve, suction pipe and delivery pipe. The piston/plunger works in the cylinder and obtains movement by a connecting rod which in turn is connected to rotating crank. The suction valve admits water into the cylinder from the suction pipe and correspondingly the delivery valve discharges water from the cylinder into the delivery pipe.

Advantages

(*i*) Furnishes high pressures, (*ii*) Higher efficiency than centrifugal pump, (*iii*) Best adapted for lifting small discharges over high heads, and (*iv*) Does not require priming.

Disadvantages

(*i*) Complicated construction, installation and operation, (*ii*) Weight of pump is more for a given discharge, (*iii*) Requires more floor space than centrifugal pump of the given capacity, (*iv*) Can be used for pumping clear water only, (*v*) More wear and tear, (*vi*) High initial cost, (*vii*) High maintenance cost, and (*viii*) Pulsating flow accompanied by vibration, shock and loss of energy if single acting.

24.2 COMPARISION OF CENTRIFUGAL AND RECIPROCATING PUMPS

Centrifugal Pump	*Reciprocating Pump*
Suitable for lifting large discharges through smaller heads	Suitable for lifting smaller discharges against higher heads.
Simple in construction.	Complicated construction.
Compact and light in weight, lesser floor space requirement for given capacity.	Weight of pump is more for a given discharge. Requires more floor space than centrifugal pump of given capacity.
Less wear and tear because of few wearing parts.	Comparatively more wear and tear, the number of parts being more.
Can be operated at very high speeds.	Cannot run at higher speeds.
Gives constant steady discharge.	Gives pulsating discharge.
Easy to install and operate.	Complicated installation and operation.
Vibration free running.	Flow gives vibrations.
Low initial cost.	Higher initial cost.
Low maintenance cost.	Maintenance cost is high.
Needs priming	Priming is not required.
Not suitable for high pressures.	Can develop much higher pressure.
Low efficiency.	Comparatively higher efficiency.
Air vessels are not required.	Air vessels are required.
Can handle dirty water and waste water.	Can be used only for pumping clear water.
Discharge varies with head, and a variable speed drive is required to maintain constant discharge under varying heads.	No such limitation, as discharge depends only on the speed of the pump.
Better adapted for puming water containing solids.	Less suitable for water containing solids.

24.3 CLASSIFICATION OF RECIPROCATING PUMPS

The reciprocating pumps are classified as under:

1. Classification Based on Water Action

   Single Acting: Water in contact with one side of the piston and the liquid is displaced during one direction of the stroke. Gives intermittent discharge.

   Double Acting: Water in contact with both sides of the piston and the liquid is displaced during both directions of the stroke. It gives more uniform discharge than single acting pump.
2. Classification Based on Number of Cylinders.
   Single Cylinder Pump: One throw pump.
   Double Cylinder Pump: Two throw pump.
   Triple Cylinder Pump: Three throw pump.
3. Classification Based on Fitting of Air Vessel.
   With Air Vessel Pump — Pulsating discharge
   Without Air Vessel Pump — Intermittent discharge.

24.4 WORKING OF RECIPROCATING PUMPS

Fig. 24.1 shows a single acting reciprocating pump. The piston is connected by a connecting rod to a crank of radius *r* which is rotated by a prime mover. When the crank rotates, the piston moves backward and forward in the cylinder. It moves from one end of the cylinder to the other end for every half revolution of the crank, which is termed as stroke.

During the suction stroke, the piston moves outwards i.e., towards the right and a vacuum is created in the cylinder and water is sucked in from the source through suction valve, which causes the suction valve to close and water enters into the cylinder. During the delivery stroke, the piston moves towards the left and the increase in pressure closes the suction valve and opens the delivery valve thereby forcing the water up into the delivery pipe to the required height. At the end of the delivery stroke, the piston occupies the extreme left position.

In the delivery pipe the velocity of flow varies with the change in the velocity of the piston. The flow as such is not constant but pulsating and gives vibration, shock and loss of energy which can be remedied by using a double-acting pump.

In a double-acting reciprocating pump, there are two working strokes per revolution of the crank that is simultaneous occurrence of the suction and delivery strokes. The pump has two sets of suction and delivery valves (Fig. 24.3). It gives more uniform discharge than single-acting pump, the discharge quantum depends on the number of revolutions per minute of the pump. The piston or plunger may be from one to five and correspondingly the pump is called simplex (single), duplex, triplex, quadruplex and quintuplex.

Theoretically the volume of water lifted per revolution is the same as the stroke volume of the cylinder in a single-acting pump and twice to the volume in a double-acting pump but actual discharge is less due to losses. The difference between theoretical discharge and actual discharge is called 'slip of the pump'. The coefficient of discharge that is the ratio between actual and theoretical discharge is less than unity. However, in case pump is running at high speeds, suction pipe is long and delivery pipe is short, the pressure due to inertia of the column of the suction pipe becomes large compared to that outside the delivery pipe causing the delivery valve to open before completion of the suction stroke and some water is pushed into the delivery pipe before the beginning of the delivery stroke. The actual discharge is more than the theoretical discharge and the slip is called negative and the coefficient of discharge is more than unity.

The delivery rate of the pump depends only on the speed and is independent of head. The efficiency of the pump depends on the slippage of water.

Discharge Formula

Let r is radius of crank and L is stroke length in metres and A is cross sectional area of the piston in sq. metres.

∴ Discharge stroke = L A.

Work done/second = weight of water raised × total head

Work done $= W\left(H_s + H_d + \dfrac{V_d^2}{2g}\right)$ kg m/s

Discharge of pump, $Q = \dfrac{LAN}{60}$ cumec

Power required to drive pump $= \dfrac{wLAN}{60\times1000}(H_s+H_d)$ kW.

As V_d is usually small, it is neglected unless the total lift is very small, w is the density of liquid (N/m³) and N is number of revolutions per second.

In double-acting pump, the work done/second

$$= \frac{2wLAN}{60}(H_s+H_d)$$

Discharge of pump = $2LAN/60$

Power required to drive pump, $p = \dfrac{2wLAN}{60\times1000}(H_s+H_d)$ kW

Where, w is specific weight of water (N/m³ in SI units).

Horse power of pump is determined by computing the work done by a pump in lifting water. Consider a pump lifts W kg of water to height H metres,

Work done by pump = $w.H$ m kg

$$= wQH \text{ m kg/s}$$

Water horse power $= \dfrac{wQH}{75}$

The actual horse power required is greater than this due to frictional resistance of water and mechanical parts besides leakage.

Brake horse power $= \dfrac{wQH}{75\times\eta}$.

24.5 INDICATOR DIAGRAM

Indicator diagram may be defined as the graph of the work done by the reciprocating pump during a complete cycle. Alternatively, it is a graph of the pressure head of the water in the cylinder and stroke length of the piston for one revolution corresponding to the various positions of the piston during the suction and delivery strokes. Such a diagram is automatically obtained in case an indicator is placed on the cylinder and pump is working at a low speed. The indicator diagram in which the loss of head due to friction in suction and delivery strokes is neglected is termed as ideal indicator diagram.

Referring to Fig. 24.4, the diagram shows the pressure head as ordinate and stroke length as abscissa. The line *AP* represents atmospheric pressure. Line *SS'* represents the pressure in the cylinder during suction stroke; *AP* being the atmospheric line. Likewise, the line *DD'*, represents the pressure in the cylinder during the delivery stroke $H_d + V_d^2/2g$ above the atmospheric line. The area *SS' AP* represents the work done on the suction stroke. The total work done per revolution by a single-acting pump is given by the area *SS' DD'* while that by a double-acting pump is twice this area.

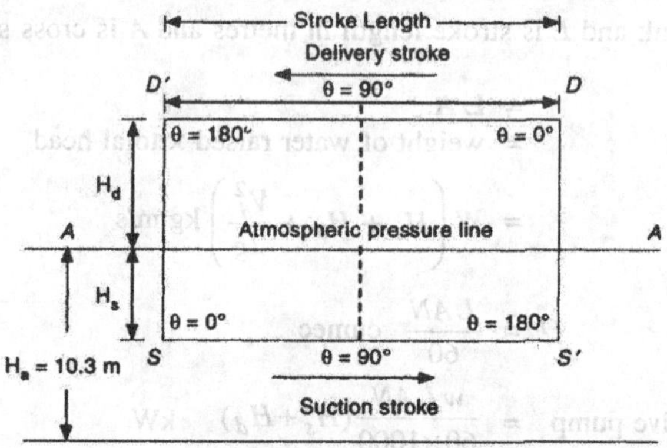

Fig. 24.4 Ideal indicator diagram.

Work done by pump/s $= wQ(H_s + H_d)$

$$= \frac{wLAN}{60}(H_s + H_d)$$

Work done by pump/s $= \dfrac{wAN}{60}$ (area of indicator diagram)

Thus indicator diagram represents to scale work done by the pump.

Variation in Pressure due to Acceleration of Piston

The piston is considered to be moving with simple harmonic motion, assuming connecting rod is very large compared to the crank length. This causes acceleration at the beginning and a retardation at the end of each stroke, as also variation of pressure in the cylinder with transmission of corresponding acceleration and retardation to the water in the suction and delivery pipes respectively.

Fig. 24.5 shows diagrammatic view of crank and connecting rod of a reciprocating pump. Consider the crank is rotating with an angular velocity ω.

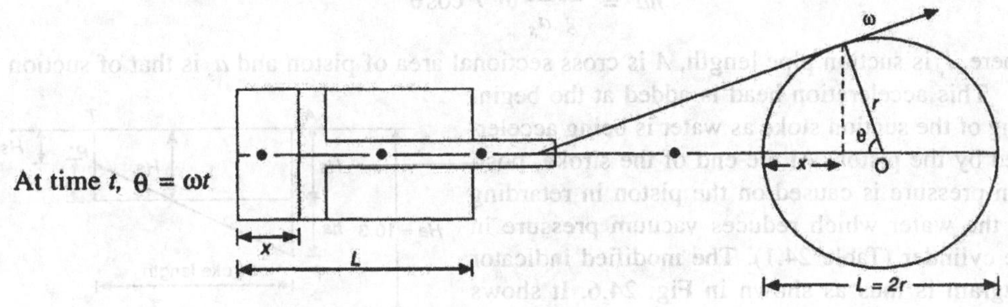

Fig. 24.5 Schematic view of single acting pump.

Displacement of piston from stroke end,

$$x = r - r\cos\omega t.$$

$\therefore$ Piston velocity, $v = \dfrac{dx}{dt} = \omega r \sin \omega t$

Piston acceleration, $f = \dfrac{d^2x}{dt^2} = \omega^2 r \cos\omega t$

Let area of piston is A and that of pipe connected to pump be a. Since volume of outflowing water from pipe per second is equal to the volume of inflowing water into the cylinder,

Water velocity in pipe, $v = \dfrac{A}{a}\omega r \sin\theta$

Water acceleration in pipe, $f = \dfrac{A}{a}\omega^2 r\cos\theta$

The velocity change in water in the pipe is due to corresponding external force causing it. The external force, considering mass of water involved in velocity change, is $\dfrac{\gamma a l}{g}$

$$F = \frac{\gamma a l}{g}\frac{A}{a}\omega^2 r\cos\theta$$

or

$$ha = \frac{F}{a\gamma} = \frac{l}{g}\frac{A}{a}\omega^2 r\cos\theta.$$

The pressure head due to acceleration acting on piston varies with the angle θ. The value of acceleration is maximum at the beginning of the stroke and minimum at the end of the stroke.

ha at beginning of stroke $= \dfrac{l}{g}\dfrac{A}{a}\omega^2 r$ $\quad\because \theta = 0, \ \cos\theta = 1$

ha at mid stroke $= 0$ $\quad\because \theta = 90, \ \cos\theta = 0$

ha at end of stroke, $= -\dfrac{l}{g}\dfrac{A}{a}w^2 r$ $\quad\because \theta = 180, \ \cos\theta = -1$

Effect of Acceleration in Suction Pipe

The indicator diagram of the suction stroke (Fig. 24.6) needs modification to take cognisance of the acceleration head.

$$ha = \frac{l_s}{g}\frac{A}{a_s}\omega^2 r\cos\theta$$

where, l_s is suction pipe length, A is cross sectional area of piston and a_s is that of suction pipe.

This acceleration head is added at the beginning of the suction stoke as water is being accelerated by the piston. At the end of the stroke, position pressure is caused on the piston in retarding by the water which reduces vacuum pressure in the cylinder (Table 24.1). The modified indicator diagram is thus as shown in Fig. 24.6. It shows that the negative head is increased, at the beginning of the stroke ($H_s + h_a$). In case simple harmonic motion is not assumed, the line PP' would be slightly curved. The pertinent points worth noting from the indicator diagram are that: (*i*) There is no change in the work done during one full stroke, (*ii*) The suction lift gets limited, (*iii*) The outflow is not steady, (*iv*) Friction loss is increased, and (*vi*) Speed cannot be increased to enhance the flow ratio. However, the effect of acceleration head gets reduced in case air vessels are fitted near the pump at the delivery and suction sides or employing multi-cylinder pump.

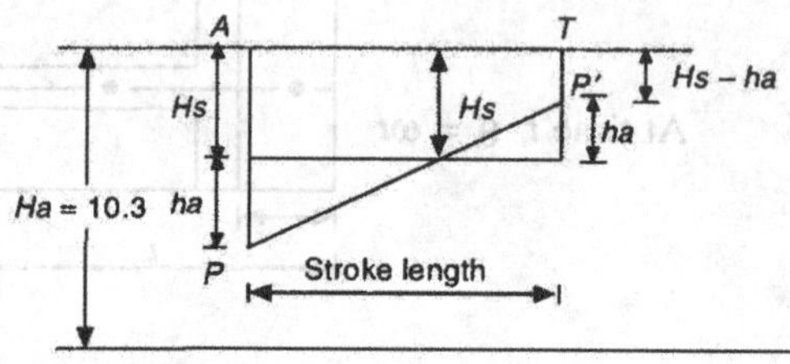

Fig. 24.6. Effect of acceleration in suction stroke on indicator diagram.

Table 24.1 Pressure head in suction stroke

Instant of stroke	*Suction stroke (absolute)*
At beginning	$H_a - (H_s + h_{as})$
At middle	$H_a - H_s$
At end	$H_a - H_s - h_{as}$

Effect of Acceleration in Delivery Pipe

Just as in the suction pipe, the water column in the delivery pipe is accelerated at the beginning of the stroke and retarded at the end of the stroke. Since delivery pipe is usually of longer length compared to suction pipe, the accelerating head is large. Fig. 24.7 shows the modified

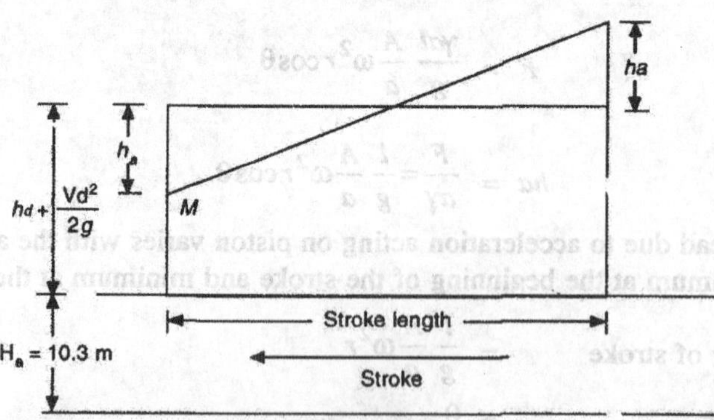

Fig. 24.7. Effect of acceleration in delivery stroke on indicator diagram

indicator diagram. The acceleration head is added compared to the diagram shown in Fig. 24.6. The pressure head on piston in delivery stroke is given in Table 24.2.

Table 24.2

Instant of stroke	*Delivery stroke (absolute)*
At beginning	$H_d + h_{ad}$
At middle	H_d
At end	$H_a + H_d - h_{ad}$

Minimum pressure head in the cylinder is at *M*, equal to $H_d + \frac{V_d^2}{2g} - H_a$ above atmosphere. The limiting condition so that separation does not take place is when:

$$H_a = 8.0m + H_d + \frac{V_d^2}{2g}$$

It is worth noting that in case delivery pipe is vertical, both sides of the equation will increase with the length of pipe and H_d may be more than H_a. A better layout of delivery pipe is that as shown in Fig. 24.8(b). In these two layouts, delivery pipes are of same length and same lift but bent in two different fashions (Fig. 24.8), the delivery pipe *A* is vertical first and horizontal afterwards whereas pipe *B* is horizontal first and then vertical. The conditions at point '*a*' is the same in both the cases, the H_a and H_d being the same. But in pipe *A* at point *T*, H_a is zero but there is considerable pipe length thereafter which affects H_a and as a consequence separation takes place at the point.

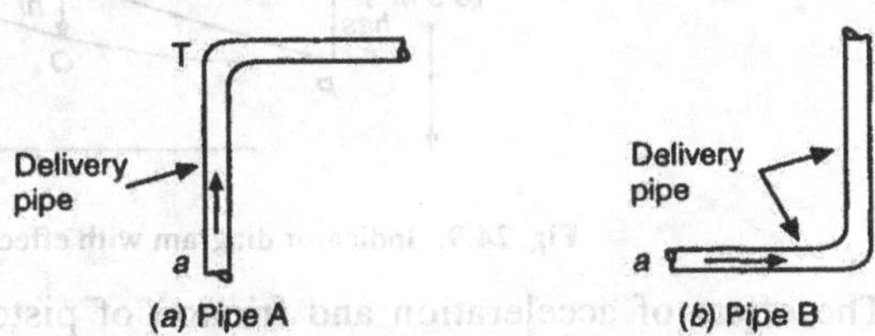

Fig. 24.8 Delivery pipe bent in two fashions

Effect of Friction in Pipes

As in pipes, water flowing through suction and delivery pipes of a reciprocating pump offers resistance to the flow.

Velocity in pipe at any instant of stroke

$$v = \frac{A}{a} wr \sin\theta$$

Head loss due to friction

$$h_f = \frac{fl}{d} \times \frac{v^2}{2g} = \frac{fl}{2gd}\left(\frac{A}{a} wr \sin\theta\right)^2$$

1. Effect on beginning of stroke

(*i*) At the beginning of stroke, $\theta = 0$ and velocity of water in pipe is zero and as such there is no loss of head due to friction.

(*ii*) At the middle of stroke, $\theta = 90°$ or $\sin\theta = 1$.

∴ Loss of head due to friction, $h_f = \frac{fl}{2gd}\left(\frac{A}{a} wr\right)^2$. The equation for h_f is a parabola.

(*iii*) At the end of stroke, $\theta = 180°$ and velocity of water is zero and as such there is no loss of head due to friction.

The combined indicator diagram by adding the frictional head in suction stroke (Fig. 24.6) and delivery stroke (24.7), is shown in Fig. 24.9. The parabola *PQP'* is the work done against friction in suction pipe and the parabola *MRN* represents the work done against friction in delivery pipe.

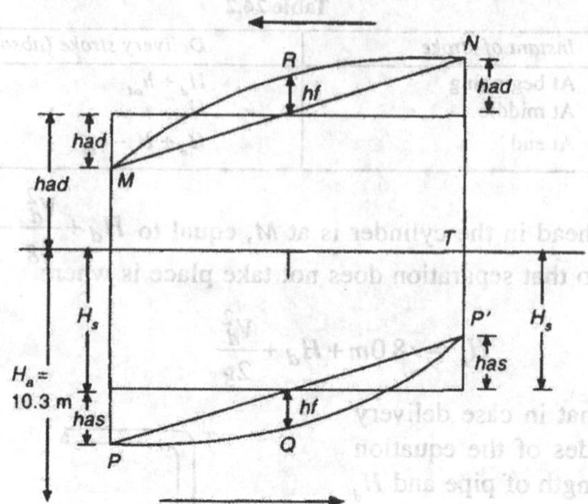

Fig. 24.9. Indicator diagram with effect of acceleration and friction.

The effect of acceleration and friction of piston on pressure head on piston (m of water absolute) on the indicator diagram (Fig. 24.9) reveals that:

Table 24.8. Pressure head (absolute)

Instant of stroke	*Suction stroke*	*Delivery stroke*
At beginning	$H_a - (H_s + H_{as})$	$H_a + (H_d + H_{ad})$, H_f being zero
In middle	$H_a - (H_s + H_{fs})$	$H_a + (H_d + H_{fd})$, h_{ad} being zero
At end	$H_a - (H_s - H_{as})$	$H_a + (H_d - H_{ad})$, h_f being zero

24.6 AIR VESSELS

Air vessel is a closed cast iron chamber with an opening at the base for the water to flow. The inflow and outflow is resulted by the pressure difference between the inside of the vessel and the point of connection in the delivery pipe as well as suction pipe. The upper part of the chamber contains compressed air and water is in the lower part. It is assumed that the air vessel is large compared to the area of the delivery pipe so as to reduce the variation in air pressure in the vessel and as such the air pressure in the vessel is taken as constant. The air vessel is fixed close to the cylinder near the suction valve in the suction pipe as well as near delivery valve on the delivery pipe.

Objectives of Air Vessel

The various objects of fitting the air vessels are (*i*) To so regulate the outflow as to give a uniform discharge, (*ii*) To run the pump without any danger of separation to take place, (*iii*) To effect reduction in acceleration head which limits the maximum speed of the pump, and (*iv*) To economise on power needed to run the pump as a result of reduction in acceleration and friction heads.

Working of Air Vessel

With regard to the working of the air chamber, as the level of water in the chamber rises, the trapped air in the upper portion is compressed which forces the water out as soon as the water pressure falls.

Air vessels are fitted to suction and delivery pipes of a reciprocating pump (Fig. 24.10).

1. Air Vessel in Suction Pipe: During first half of the suction stroke, the water first enters into the air vessel from the suction pipe and compresses the air which in turn forces the water into the cylinder during the second half of the suction stroke when pressure in the cylinder falls. As a result the velocity of water in the suction pipe is practically uniform, the accelerated water column is limited to the small amount between the air vessel and the cylinder, both being located as near as possible.

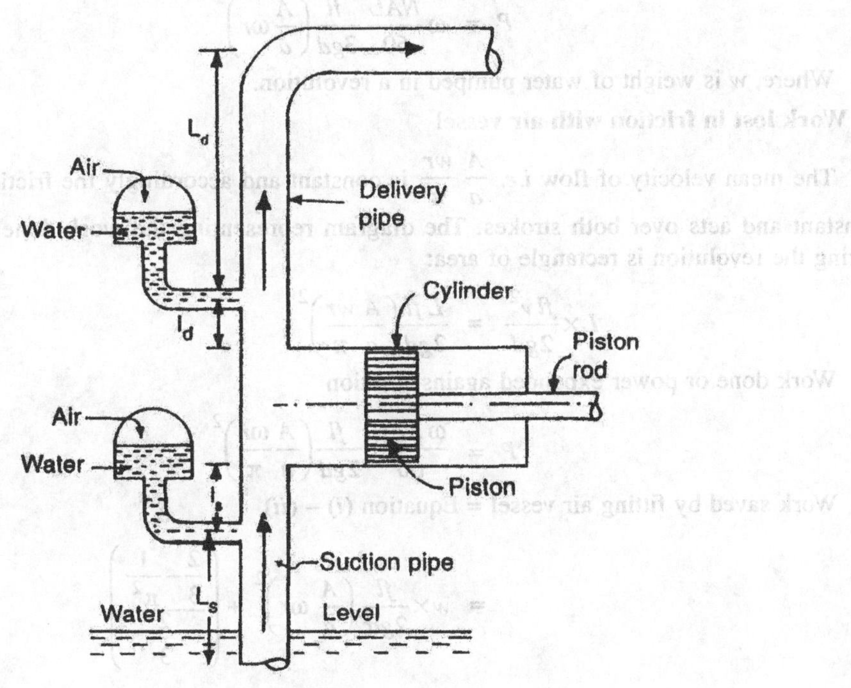

Fig. 24.10. Air vessels.

2. Air Vessel in Delivery Pipe: During the mid of the delivery stroke as the water is forced by the piston into the delivery pipe at a velocity higher than the mean velocity, the additional water flows into the air vessel. At the end of the stroke when the water is forced into the delivery pipe at a velocity less than the mean velocity, the water flows out of the air vessel to make up the deficiency thereby maintaining constant flow in the delivery pipe beyond the air vessel. However, the volume of water which is accelerated in the delivery pipe between the air vessel and the cylinder is very small as the air vessel is fitted close to the cylinder.

Saving in Work with Air Vessel

Consider a single acting reciprocating pump with and without air vessel fitted on suction and delivery pipes nearest to the pump.

1. Work lost in friction without air vessel

The velocity of water, $$v = \frac{A}{a}\omega r \sin\theta$$

Loss of head due to friction, $$h_f = \frac{flv^2}{2gd}$$

The diagram representing the work lost in friction during a revolution is a parabola, the equation being:

$$h_f = \frac{fl}{2gd}\left(\frac{A}{a}\omega r\right)^2$$

The area of the diagram $= L\times\frac{2}{3}\frac{fl}{2gd}\left(\frac{A}{a}\omega r\right)^2$ *(i)*

Work done or power expended agaist friction

$$P = \omega\frac{NAL}{60}\frac{fl}{3gd}\left(\frac{A}{a}\omega r\right)^2$$

Where, *w* is weight of water pumped in a revolution.

2. Work lost in friction with air vessel

The mean velocity of flow *i.e.* $\frac{A}{a}\frac{wr}{\pi}$ is constant and accordingly the frictional loss is also constant and acts over both strokes. The diagram representing the work done against friction during the revolution is rectangle of area:

$$L\times\frac{flv^2}{2gd} = \frac{Lfl}{2gd}\left(\frac{A}{a}\frac{wr}{\pi}\right)^2$$

Work done or power expended against friction

$$P = \frac{\omega NAL}{60}\frac{fl}{2gd}\left(\frac{A}{a}\frac{\omega r}{\pi}\right)^2 \quad (ii)$$

Work saved by fitting air vessel = Equation *(i)* – *(ii)*

$$= w\times\frac{fl}{2gd}\left(\frac{A}{a}\omega r\right)^2 + \left(\frac{\frac{2}{3}-\frac{1}{\pi^2}}{\frac{2}{3}}\right)$$

Percentage of work saved $= \left(\frac{\frac{2}{3}-\frac{1}{\pi^2}}{\frac{2}{3}}\right)\times 100 = 84.8\%$ of frictional work.

Now consider that the pump is double acting,

$$v = Z\frac{A}{a}\frac{wr}{\pi}$$

Percentage of work saved $= \frac{\frac{2}{3}-\frac{4}{\pi^2}}{\frac{2}{3}} = 39.2$ per cent.

24.7 ROTARY PUMPS

Rotary pumps are another variant of positive displacement type of pumps. Rotary pumps which are suitable for small discharges use rotary motion instead of reciprocating motion in reciprocating pump. The rotary motion is obtained by means of two rotating elements called cams, gears or screws fitted closely in casing which mesh together and rotate in opposite directions. Unlike centrifugal pump, the rotary pump is not started against a closed valve. The pump has internal leakage and as a result the actual discharge is less than the displacement, the difference is termed as slip. The amount of slip varies with the viscosity of the liquid.

The water drawn through suction pipe is trapped between the cams or gears and the casing and is forced around and out in each revolution through discharge pipe. These pumps are self priming, require no valves and can be operated at higher speeds giving larger capacity with smaller size.

The four major types of rotary pumps are (*i*) Gear pumps, (*ii*) Rigid rotor vane (Sliding gate) pumps, (*iii*) Rotary piston pumps, (*iv*) Screw pumps.

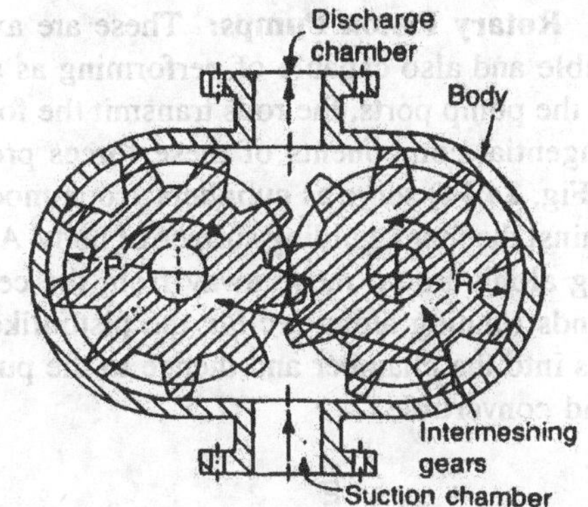

Fig. 24.11. Rotary pumps.

1. Gear Pumps: In the pump (Fig. 24.11) intermeshing gears run in small clearance contact with the inside of body. The drive gear is fitted with a shaft extending through the body via a stuffing box, the other gear (driven one) is idle. As the gears rotate in the direction indicated by arrow, liquid is trapped between the gear teeth and conducted from suction chamber to discharge chamber where the liquid is forced out of the tooth spaces by the intermeshing gears. The capacity of the pump is determined by its geometrical dimensions and the rotative speed of its shaft. These pumps are reversible in that changing the sense of rotation of gears reverses the direction of flow in the pipe lines connected to the pump.

2. Rigid Rotor Vane Pump: The pump has two basic patterns; the external supply of liquid to the pump (Fig. 24.12) and internal supply of liquid to the pump Fig. 24.13). Both the types are reversible and convertible also i.e., capable of developing a shaft torque if the liquid is supplied to the pump under an initial pressure. The volumetric efficiency of the pumps allow for leakage through clearances from the delivery to the suction side; its numerical value is within 0.94 to 0.98.

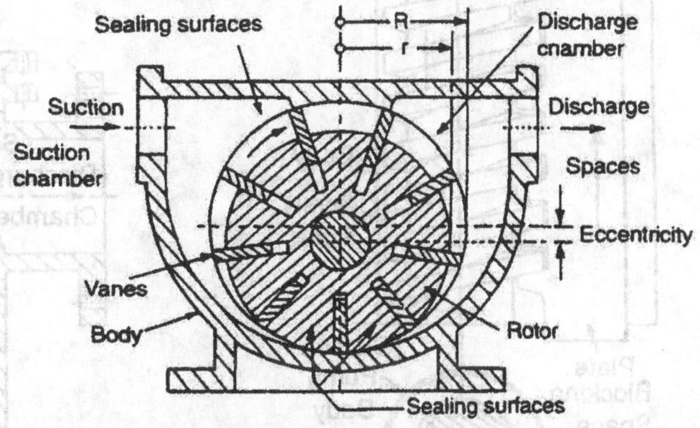

Fig. 24.12. Rigid rotor vane pump with external liquid supply.

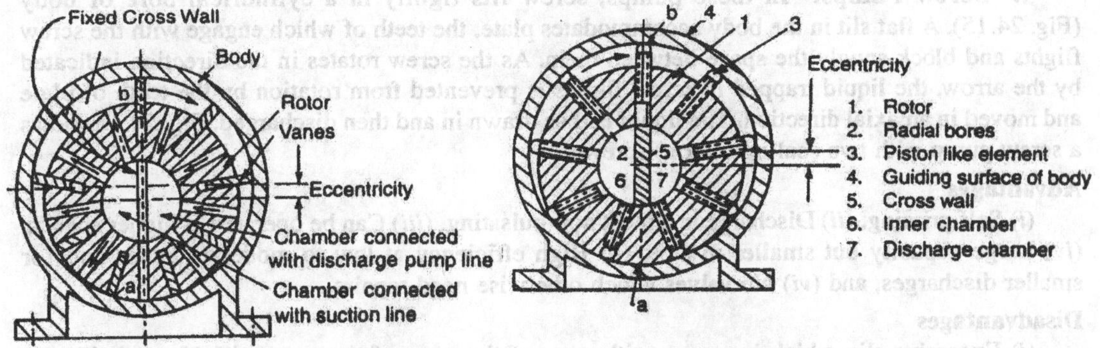

Fig. 24.13. Rigid rotor vane pump with internal liquid supply

Fig. 24.14. Radial rotary piston pump

3. Rotary Piston Pumps: These are axial and radial pumps. The axial piston pumps are reversible and also capable of performing as a driver. When liquid under pressure is supplied to one of the pump ports, the rods transmit the forces acting on the pistons to the face of plate where the tangential components of these forces produce a torque on the plate shaft. Radial bores in rotor (Fig. 24.14) serve as cylinders accommodating piston like elements, the other ends of which are against the inner, guiding surface of body. As the rotor rotates clockwise, the pistonlike elements running along arc *ab* move away from the centre and draw in liquid from inner chamber. With their ends running along arc *ba*, the pistonlike elements move towards the centre and the liquid reaches into the chamber and thence to the pump discharge connection. These pumps are reversible and convertible.

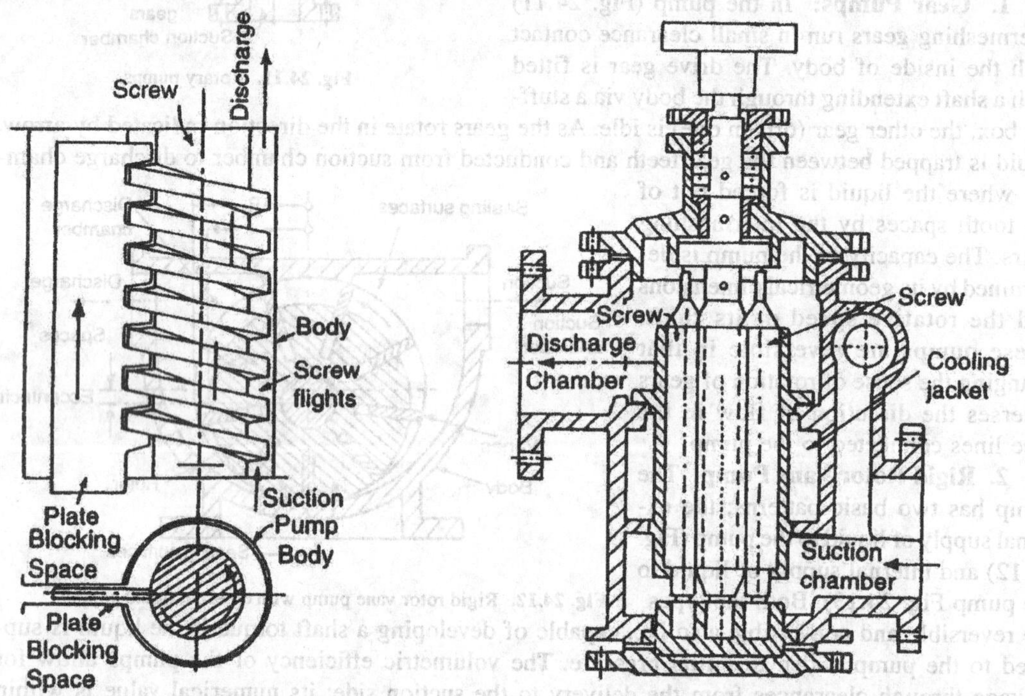

Fig. 24.15. Screw pump-theory of operation.

Fig. 24.16. Screw pump with one driven and two sealing worm screws.

4. Screw Pumps: In these pumps, screw fits tightly in a cylindrical bore of body (Fig. 24.15). A flat slit in the body accommodates plate, the teeth of which engage with the screw flights and block snugly the space between them. As the screw rotates in the direction indicated by the arrow, the liquid trapped in screw flights is prevented from rotation by the teeth of plate and moved in an axial direction. The liquid is thus drawn in and then discharged. Fig. 24.16 shows a screw pump with two sealing (worm) screws.

Advantages

(*i*) Self priming, (*ii*) Discharge is almost non-pulsating, (*iii*) Can be operated at higher speeds, (*iv*) Large capacity but smaller in size, (*v*) High efficiency at low to moderate heads and for smaller discharges, and (*vi*) No valves which otherwise need repairs.

Disadvantages

(*i*) Excessive slip which increases with wear of the parts of the pump, (*ii*) Not suitable for pumping water containing suspended matter, (*iii*) High initial cost, and (*iv*) Not durable, need frequent replacement of cams or gears.

SOLVED EXAMPLES

Example 24.1 *A single acting reciprocating pump supplies 0.02 cumec of water at 40 r.p.m. The suction and delivery heads are 5m and 10m respectively. Piston diameter and stroke length are 300 mm and 500 mm respectively. Determine (i) theoretical discharge, (ii) percentage slip, and (iii) power to run pump.*

Solution.

(*i*) Theoretical discharge, $Q_{th} = \frac{LAN}{60} = \frac{0.5[\pi(0.3)^2/4]40}{60}$

$= 0.0236$ cumec **Ans.**

(*ii*) Coefficient of discharge, $C_d = 0.020/0.0236 = 0.85$ **Ans.**

(*iii*) Percentage of slip $= \frac{0.0236 - 0.020}{0.0236} \times 100 = 11\%$ **Ans.**

(*iv*) Power to run pump $= \frac{wLAN}{60 \times 1000}(H_s + H_d)$

$= \frac{9810 \times 0.0236}{1000}(5 + 10)$

$= 3.47$ kW **Ans.**

Example 24.2 (*a*) *Determine pressure head at the beginning and end of stroke if a single acting reciprocating pump with piston diameter 150 mm and 500 mm stroke length lifts water from 5 m (up to cylinder axis) at 30 r.p.m. The suction pipe is 100 mm in diameter and 8 m long and delivery pipe of the same size as suction is 20 m long, delivery head is 15 m.*

(*b*) *Also determine power required to run the pump.*

Solution.

(*a*) Piston area, $A = \pi(0.15)^2/4 = 0.0176 \text{ m}^2$

Suction pipe area, $a = \pi(0.1)^2/4 = 0.0079 \text{ m}^2$.

Angular velocity of crank,

$\omega = 2\pi N/60 = 2\pi \times 30/60 = 3.14$ rad/s.

Acceleration pressure head, in suction pipe $h_{as} = \frac{A}{a} \cdot \frac{l_s}{g} \cdot \omega^2 r$

$= \frac{0.0176}{0.0079} \times \frac{8}{9.81} \times 3.14^2 \times 0.5/2 = 4.48$ m

Pressure head due to acceleration head in delivery pipe

$$h_{ad} = \frac{l_d}{g} \times \frac{A}{ad} \times \omega_r^2 \cos\theta \ (\cos\theta = 0 \text{ at the beginning})$$

$$= \frac{20}{9.81} \times \frac{0.0176}{0.0079} \times (3.14)^2 \times 0.50/2 = 11.20 \text{ m}$$

Table 24.4 Pressure head on piston (m of water absolute)

At instant of stroke	*Suction stroke*	*Delivery stroke*
Beginning	$10.30 - (H_s + h_{as})$ $= 10.30 - (5.0 + 4.48)$ $= 0.82$	$H_d + H_{ad}$ $= 15 + 11.20 = 26.20$ (gauge) $= 10.30 + 26.20 = 36.50$
Middle	$10.30 - H_s$ $= 10.30 - 5.0 = 5.30$	H_d $= 15$ m (gauge) $= 10.30 + 15.0 = 25.30$
End	$10.30 - (H_s - h_{as})$ $= 10.30 - (5.0 - 4.48)$ $= 9.78$	$H_d - h_{ad}$ $= 15 - 11.20 = 3.80$ (gauge) $= 10.30 + 3.80 = 14.10$

(*b*) Power to run the pump,

Discharge, $Q = \dfrac{ALN}{60} = \dfrac{0.0176 \times 0.5 \times 30}{60} = 0.0044$ cumec

Work done, $w\,Q(H_s + H_d) = 9810 \times 0.0044\,(5.0 + 15.0) = 863.28$ N m/s.

Power to run pump, $P = 863.28$ watts. **Ans.**

Example 24.3 *A single acting reciprocating pump has piston diameter of 150 mm and stroke length of 500 mm. The suction and delivery pipes are 100 mm in diameter and 7 m and 30 m long. The suction depth and delivery height 5 m and 20 m reckoned from the axis of the cylinder. N is 25 r.p.m. Determine the pressure head on the piston at the beginning, middle and end of the suction and delivery strokes. Assume any data not given.*

Solution.

Area of piston, $A = [\pi (0.15)^2]/4 = 0.0177\ m^2$

Area of suction pipe, $a = [\pi (0.10)^2]/4 = 0.0079\ m^2$

Angular velocity of crank, $\omega = 2\pi\ N/60 = 2\pi \times 25/60 = 2.62$ rad/s

Acceleration pressure head, $h_{as} = \dfrac{l_s A}{g\,as}\,\omega^2 r$

$$= \frac{7}{9.81} \times \frac{0.0177}{0.0079}\,(2.62)^2 \times 0.50/2 = 2.74\ m$$

1. Suction stroke

(*i*) Pressure head on piston at the beginning of stroke $\theta = 0$

$= (H_s + h_{as}) = 5.0 + 2.74 = 7.74$ m of water (vacuum)

$= 2.56$ m of water (absolute) **Ans.**

(*ii*) Pressure head at middle of suction stroke

Velocity, $v = \dfrac{A}{a}\,\omega r \sin\theta \qquad \theta = 90°.$

$$= \frac{0.0177}{0.0079} \times 2.62 \times 0.50/2 \times 1 = 1.47\ m/s$$

Head loss due to friction, $h_{fs} = \dfrac{flv^2}{2gd}$

$$= \frac{0.01 \times 7 (1.47)^2}{2 \times 9.81 \times 0.1} = 0.075\ m$$

$\therefore$ Pressure head in middle of stroke

$= (H_s + h_{fs}) = (5.0 + 0.075)$

$= 5.075$ m of water (vacuum)

(*iii*) Pressure head at piston the end of suction stroke $\theta = 180°$

$$H = (H_s - H_{as})$$
$$= (5.0 - 2.74) = 2.56 \text{ m of water (vacuum)}$$

2. Delivery stroke

Acceleration pressure head, $h_{ad} = \dfrac{l_d}{g} \times \dfrac{A}{ad} \omega^2 r$

$$= \frac{30}{9.81} \times \frac{0.0177}{0.0079} (2.62)^2 \times 0.50/2$$
$$= 11.76 \text{ m of water}$$

Pressure at beginning of stroke $= H_d + H_{ad}$ $\quad \theta = 0°$

$$= 20 + 11.76 = 31.76 \text{ m of water above atmosphere}$$

Pressure at middle of stroke $= (H_d + h_{fd}) = 20 + \dfrac{fl_d}{d_d\, 2g} \dfrac{A}{ad} \omega^2 r$ $\quad \theta = 90°$

$$= 20 + \frac{0.01 \times 20 (2.62)^2}{0.1 \times 2 \times 9.81}$$
$$= 20.22 \text{ m of water (above atmospheric)}$$

Pressure at end of stroke $= H_d - h_{ad}$ $\quad \theta = 180°$

$$= 20 - 11.76$$
$$= 8.24 \text{ m of water (above atmosphere)}$$

Power required to run the pumps

Work done $= \dfrac{w\,ALN}{60}\left(h_s + h_d + \dfrac{2}{3} h_{fs} + \dfrac{2}{3} h_{fd}\right)$

$$= \frac{9810 \times 0.0177 \times 0.50 \times 30}{60}\left(5.0 + 20 + \frac{2}{3} \times 0.075 + \frac{2}{3} \times 0.22\right)$$
$$= 1120 \text{ N m/s}$$

Power to run pump $= \dfrac{1120}{0.82} = 1.366$ kW (assume $\eta_{mech.} = 0.82$)

Example 24.4 *Determine friction head saved in friction by fitting air vessel in suction and delivery pipes of a single acting reciprocating pump pumping water. The diameter of pump cylinder (single) is 25 cm, stroke 50 cm, speed 50 r.p.m. The diameter of suction and delivery pipes is 20 cm and 15 cm, their lengths being 10 m and 50 m respectively. Assume f = 0.02.*

Solution.

Volume of water pumped $= \dfrac{LAN}{60} = \dfrac{0.5 \times \dfrac{\pi (0.25)^2}{4} \times 50}{60} = 0.02$ cumec

1. Suction pipe

Angular velocity of crank,

$$\omega = \frac{2\pi N}{60} = \frac{2 \times \pi \times 50}{60} = 5.24 \text{ rad/s}$$

Velocity of water in delivery pipe $= \dfrac{A \omega r}{a \pi}$

$$= \left(\frac{0.25}{0.20}\right)^2 \times \frac{2\pi \times 50}{\pi \times 60} \times \frac{0.50}{2}$$

$$= 0.65 \text{ m/s}$$

Friction head saved $= \frac{fl}{2gd}\left(\frac{A}{a}wr\right)^2\left(\frac{2}{3}-\frac{1}{\pi^2}\right)$

$$= \frac{0.02 \times 10}{2 \times 9.81 \times 0.20}(0.65\pi)^2\left(\frac{2}{3}-\frac{1}{\pi^2}\right)$$

$$= 0.12 \text{ m}$$

2. Delivery pipe

Velocity of water $= \left(\frac{0.25}{0.15}\right)^2 \times \frac{2\pi \times 50}{\pi \times 60} \times \frac{0.50}{2} = 1.16 \text{ m/s}$

Friction head saved $= \frac{0.02 \times 50}{2 \times 9.81 \times 0.15}(1.16\pi)^2\left(\frac{2}{3}-\frac{1}{\pi^2}\right)$

$$= 2.55 \text{ m}$$

Total head saved $= 0.12 + 2.55 = 2.67$ m

QUESTIONS

1. (*a*) Define slip, negative slip and coefficient of discharge of a reciprocating pump.
   (*b*) Explain the advantages and disadvantages of fitting an air vessel on suction and delivery sides of a reciprocating pump.
   (*c*) A single acting reciprocating pump has a stroke length of 400 mm and piston dia. of 400 mm. The pump runs at 20 r.p.m. The suction and delivery heads are 4 m and 20 m respectively. The length of suction and delivery pipes are 6 m and 25 m respectively, and the diameter of the suction and delivery pipes is 200 mm. Estimate the power required to drive the pump if coefficient of friction in pipes, *f*, is 0.02. *(AMIE s'01)*
2. (*a*) Compare the basic features of centrifugal and reciprocating pumps. Define slip. Can slip be negative?
   (*b*) A single-acting reciprocating pump has bore and stroke each of 200 mm and discharges 300 litres of water per minute at 60 r.p.m. Neglecting losses, obtain: (*i*) Theoretical dischange (l/minute), (*ii*) coefficient of discharge, and (*iii*) slip of the pump. *(AMIE w')*
3. Discribe the function of an air vessel and show that an air vessel decreases frictional power lost. How does this compare with priming? *(AMIE w'0)*
4. (*a*) A single acting cylinder pump lifts 60 l/s of water against a net static head of 20 m. The cylinder measures, stroke 400 mm, piston dia 300 mm. Estimate pump speed if slip be 1%. Also estimatic force required to operate pump.
   (*b*) How do acceleration and friction heads modify the indicator diagram of a reciprocating pump? *(AMIE s'99)*
5. Distinguish between:
   (*i*) Reciprocating pumps and rotodynamic pumps,
   (*ii*) Impulse turbines and reaction turbines.
   (*iii*) Hydraulic coupling and hydraulic torque converter. *(AMIE w'99)*

6. **Differentiate between:**
   (*i*) **Reciprocating pump and gear pump.**
   (*ii*) **Shut-off head and delivery head,**
   (*iii*) **Priming and cavitation,**
   (*iv*) **Unit speed and specific speed.** (*AMIE s*'98)
7. **A single acting reciprocating pump has plunger dia. and stroke as 25 cm and 45 cm respectively. The suction pipe is 12.5 cm in dia. and 12 m long with a suction lift of 3 m. An air vessel is fitted to the suction pipe at a distance of 1.5 m from the cylinder and 10.5 m from the sump water level. If the barometer reads 10 m of water and separation takes place at 2.5 m varcuum, find the speed at which the crank can operate without separation. Take friction factor $f = 0.01$.** (*AMIE s*'98)
8. (*a*) **Sketch the typical indicator diagrams for a reciprocating pump, (*i*) working very slowly, and (*ii*) working at high speed. No air vessel is fitted.**
   (*b*) **A double-acting reciprocating pump has a plunger of dia 200 mm and stroke 300 mm. The length of suction and delivery pipes are 8 m and 60 m respectively. The diameter of delivery pipe being 100 mm, suction lift is 4 m and delivery head is 30 m. An air vessel is fitted on the pump on the delivery side. The pump runs at 30 r.p.m. Find the diameter of the suction pipe in order to avoid separation which occurs when the absolute pressure falls to 2.5 m of water on the suction side of the pump. What is the power required to drive the pump? Take $f = 0.025$. Minor losses may be neglected. Assume barometer reads 10 m of water.** (*AMIE w*'97)
9. (*a*) **Show that for single-acting reciprocating pump fitted with air vessel on suction and delivery sides. The ratio of work done against friction with and without air vessel is $3/2\pi^2$.**
   (*b*) **Find the power worked in fitting an air vessel on delivery side to a single-acting reciprocating pump, with plunger diameter 30 cm, stroke 50 m, delivery pipe diameter 10 cm, length of delivery pipe 70 m. Take $N = 60$ r.p.m. $4f = 0.04$ and atmospheric pressure = 10 m of water.** (*AMIE s*'97)
10. (*a*) **Obtain the expressions for the instataneous valves of the discharge going into or coming out of the air vessels fitted to the suction and delivery sides of a reciprocating pump.**
   (*b*) **A single-acting reciprocating pump, running at 40 r.p.m. is to be used for lifting water. The pump cylinder has a bore of 120 mm and the length of stroke of the piston is 300 mm, while the connecting rod is 650 mm long. The suction pipe is 100 mm in diameter and 10 m in length. The pressure required to force the suction valve open is 2.943 kN/m². If no air vessel is fitted on the suction side, what would be the maximum permissible suction lift?**
   The saturation vapour pressure of water at the prevailing temperature is 12.36 kN/m³ and the local atmosphere pressure is 91.23 kN/m². (*AMIE s*'96)
11. (*a*) **Show that for a single-acting reciprocating pump, the work sourced against friction in delivery pipe by fitting an air vessel is 84.8%.**
   (*b*) **A single-acting reciprocating pump is to raise a liquid of density 1200 kg/cm m from 2.5 m below the pump axis to 9 m above it. The plunger is of 125 mm dia. and has a stroke of 225 mm. The suction and delivery pipes are of 75 mm dia. each and 3.5 m and 13.5 m long respectively. There is a large air vessel fitted on the delivery pipe near the pump axis. If separation takes place at 0.9 kq/cm² below atmosphere, find, (*i*) The maximum speed at which the pump can be run without separation and (*ii*) power required to drive the pump. Assume friction factor f as 0.08.** (*AMIE w*'98)

12. Why the size of air vessel used in a double-acting reciprocating pump is less than that used in a single-acting reciprocating pump. (*AMIE w'96*)
13. Write short notes on any two of the follows:
    (*i*) Cavitation in pumps (*ii*) Working of a gear pump
    (*iii*) Hydraulic ram, (*iv*) fluid coupling (*AMIE s'96*)
14. A single acting reciprocating pump 25 cm bore and 50 cm stroke delivers water through 10 cm pipe to a tank 12 m above it and 25 m horizontally from it. If separation occurs at 0.225×10^5 N/m² absolute, find the safe speed of the pump, if delivery pipe is fistly horizontal and than vertical. The crank-connecting rod ratio is 1/5 and atmospheric pressure is 10.3 m of water. (*AMIE s'94*)
15. Show that work sanked in overcoming friction is about 40% for a double acting reciprocating pump. (*AMIE s'94*)
16. (*a*) Show that the maximum accelerating head in a reciprocating pump without air vessel is given by

$$H_a = \frac{L}{q} \times \frac{A}{a} \; w^2 r, \text{ with usual notations.}$$

    (*b*) A single acting reciprocating pump has piston dia. 12.5 cm and stroke length 30 cm. The centre of pump is 4 m above water level in the sump. The dia. and length of suction pipe are 7.5 cm and 7 m respectively. The separation occurs if the absolute pressure head in the cylinder during suction stroke falls below 2.5 m of water. Calculate maximum speed at which the pump can run without separation. Take atmospheric pressure head 10.3 m of water. (*AMIE w'93*)
17. (*a*) How do you classify reciprocating pumps?
    (*b*) A double acting reciprocating pump has a piston of 250 mm dia. and piston rod of 50 mm diameter, which is on one side of the piston only. Length of the piston strokes is 380 mm and the speed of the crank moving the piston is 60 r.p.m. The suction and discharge heads are 5 m and 20 m respectively. Find the force required to work piston during the 'in' and 'out' strokes. Neglect friction. Determine the quantity of water raised by the pump and the power required.
    (*AMIE s'97*)
18. (*a*) Show that the accelerating head in case of reciprocating pump does not affect its power requirement.
    (*b*) A double-acting reciprocating pump has a plunger 15 cm in dia. and a stroke of 45 cm. The suction lift is 3.5 m and length of suction pipe is 6 m, the pipe dia being 10 cm. Find the maximum speeds at which the pump would run without separation if: (*i*) no air vessel is used, (*ii*) air vessel is used on suction side close to the pump, (*iii*) Take atmospheric p_a = 10.3 m and separation pressure as 2.5 m absolute. Assume $4f$ = 0.024.
    (*AMIE s'93*)

25

MISCELLANEOUS HYDRAULIC MACHINES

25.1 INTRODUCTION

Hydraulic machines are the devices in which power is transmitted by means of a fluid (water or oil), making use of the principles of fluid states and fluid kinematics. Various hydraulic machines, discussed hereinafter are:

1. Hydraulic accumulator
2. Hydraulic intensifier
3. Hydraulic crane
4. Hydraulic ram
5. Hydraulic lift
6. Hydraulic press

25.2 HYDRAULIC ACCUMULATOR

It is a device designed to accumulate energy and is used for temporarily storing the energy of water under pressure and making it available, as a secondary source of power, to hydraulic lifts and cranes. The hydraulic lifts and cranes require energy (water under pressure) to be supplied during the upward motion of the load only (i.e., a small interval of working time followed by idle period as no energy is used during the downward motion) which is supplied from the hydraulic accumulator. Since the pump are supplying the energy continuously, it is stored in the accumulator during the idle periods of the machine and supplied at an increased rate during the working period and as such the energy supplied by the pumps may not be that large as is required by the lift or crane for doing maximum work as the same is drawn from the accumulator.

The hydraulic accumulator (Fig. 25.1) consists of a vertical cylinder containing a sliding ram to which is fixed a container filled with a heavy weight or the ram is loaded with weights. The cylinder has an inlet connected to the pump and an outlet to the machine.

When the machine is in idle period and water is no longer required for its working, it is admitted under pressure into the cylinder through the inlet. The water pressure raises the heavy ram until the cylinder is full of water which means that the hydraulic accumulator has stored its maximum amount of energy, called the capacity of the accumulator. The machine during the period of maximum work, when it requires maximum amount of energy, draws it from the accumulator and the ram descends to its initial position.

Capacity of Accumulator

Work done in lifting ram = weight (w) × ram lift (L)

= p. A.L. kg m

It is the equal to the capacity of accumulator

∴ Capacity of accumulator = p. A.L

= p × volume of accumulator

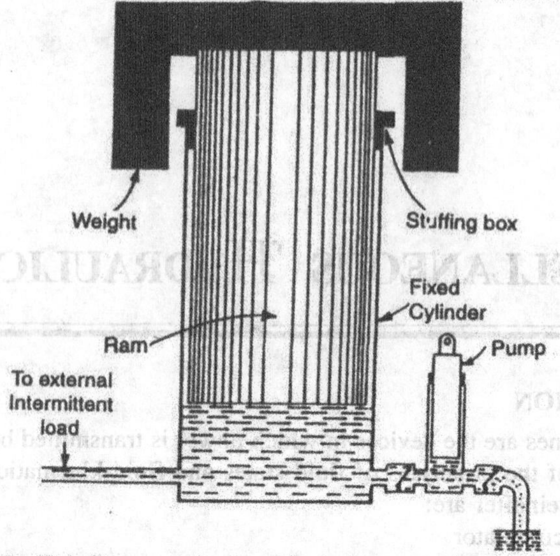

Fig. 25.1 Diagrammatic view of hydraulic accumulator

Where, w is total weight of ram including the load on it, A is area of sliding ram, L is lift of ram and p is intensity of pressure of the water supplied.

25.3 HYDRAULIC INTENSIFIER

Hydraulic intensifier is a device for increasing the intensity of water pressure by means of the energy of a large quantity of water at low pressure. It finds use when the pressure of water supplied to be machine is of insufficient intensity.

A hydraulic intensifier consists of a fixed ram through which water at high pressure flows to the machine. A hollow sliding ram containing high pressure water is mounted externally on the fixed ram which in turn is encased by a fixed cylinder containing low pressure water from the main supply.

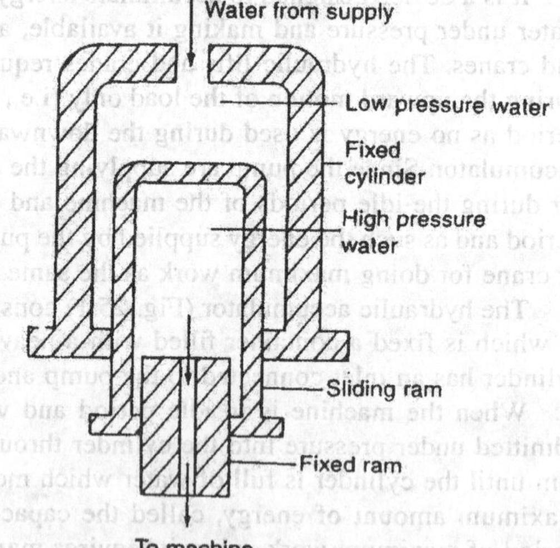

Fig. 25.2. Hydraulic intensifier

The water under low pressure presses on the end of the sliding ram thereby forcing it downwards on to the fixed ram which increases the water pressure in the sliding ram.

Since total upward force = total downward force

$$pa = PA$$

$$\therefore \quad p = \frac{PA}{a}$$

Where, p is intensity of pressure of high pressure water in sliding ram, P is the intensity of pressure of low pressure water in fixed cylinder, A is external area of the end of sliding ram and a is the area of the end of the fixed ram. Considering loss due to friction at each packing:

$$p = \frac{A}{a}P(1-k^2)$$

This intensifier supplies high pressure water during the downward stroke only and as such is a single acting intensifier. A double acting intensifier gives continuous supply of high pressure water.

25.4 HYDRAULIC CRANE

It is a device by means of which heavy load up to 250 tonne (up to 25 MN) are lifted and find extensive use for loading and unloading materials and equipment in workshops, warehouses, docks etc.

A hydraulic crane consists of a crane and a hydraulic jigger. The crane essentially consists of a mast fixed on a revolving pedestal to which a tie and a jib are fixed. The mast is capable of revolving about a vertical axis, the jib swinging with it. The jib can be raised or lowered to reduce or increase the radius within which the crane can be operated for lifting or lowering loads. Load to be lifted is suspended by a wire rope which passes over pulleys fixed at the end of jib and over the tie rod to the hydraulic jigger. The jigger attached to the mast consists of a fixed cylinder and a sliding ram to which pulley block is attached. (Fig. 25.3); a set of pulleys is fixed to the ram and the other to the cylinder with cable mounted over both sets of pulleys. The pulleys increase the velocity ratio between the wire rope and the ram thereby the load moves much faster than the speed of the ram. A five sheave pulley block system has a velocity ratio of 5 : 1. Supply of water under high pressure is made through a pipe connection to the cylinder of the jigger.

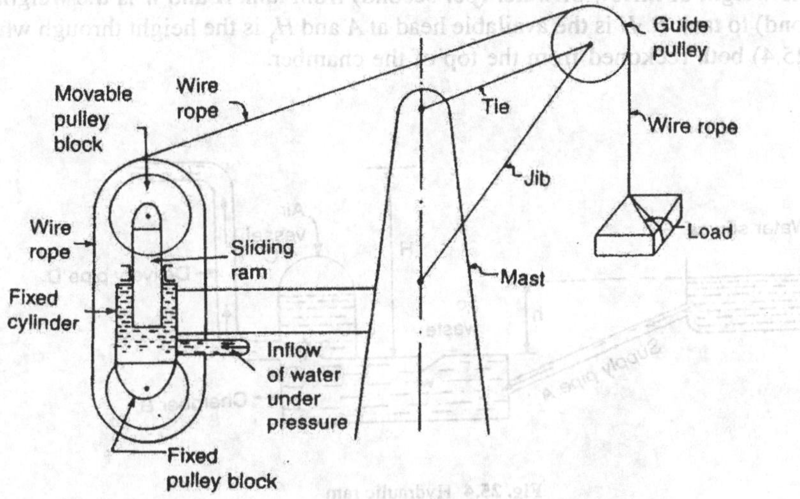

Fig. 25.3 Hydraulic crane

For lifting the load, high pressure water is admitted into the cylinder forcing the ram out which in turn increases the distance between the two set of pulleys thereby winding in the cable. The hydraulic cranes have now given way to electric crane.

25.5 HYDRAULIC RAM

It is an automatic device which does not require any external power as its automatic action is due to the inertia forces of water itself. With the help of a ram, a large quantity of water available at a smaller head is utilised for lifting a small quantity of water to a greater height. Thus higher lift achieved is at the cost of wastage of considerable amount of water. It works on the principle of water hammer; moving mass of water suddenly stopped. The incoming water valve is automatically opened and suddenly closed thereby creating an impulse in the ram which lends it the name impulse pump.

Fig. 25.4 shows a diagrammatic view of the hydraulic ram. It essentially consists of a supply pipe A, waste valve *w* which opens inwards, water chamber *B*, delivery valve *v* which opens outwards, air chamber *C* and delivery pipe *D* attached to the base of the air vessel (chamber). The valve remains open due to its own weight acting downward. Large amount of natural supply of water S is available at a moderate head h. A small quantity of this water is raised to a greater height H by means of the ram as explained hereinafter.

Working: The water flow commences from the source S into the supply pipe A and thence into the chamber C. The waste valve being open, the water runs through it to waste. As the inflow continues and velocity of water in A increases, the dynamic pressure on valve w builds up and ultimately exceeds the weight of the valve lid causing to suddenly close it. The sudden closure of the valve brings the water in A to sudden rest with consequent increase of pressure in C due to the development of positive water hammer. The increased pressure lifts the valve V open and some of the water flows into the air vessel compressing the air in it and consequent increase in the pressure. This increased air pressure in the vessel closes the valve V and forces the water up into the tank B. When the momentum of the water in the chamber is destroyed, the waste valve w opens which causes the flow of water from tank A to run again to waste till the valve is closed and some of its water is forced into the tank B goes via air chamber and delivery pipe. The cycle goes on repeating. The ram may make 200 strokes per minute and can be used to lift water up to about 30m.

Let W is the weight of inflow of water (per second) from tank A and w is the weight of water raised (per second) to tank B. H is the available head at A and H_1 is the height through which water is lifted (Fig. 25.4) both reckoned from the top of the chamber.

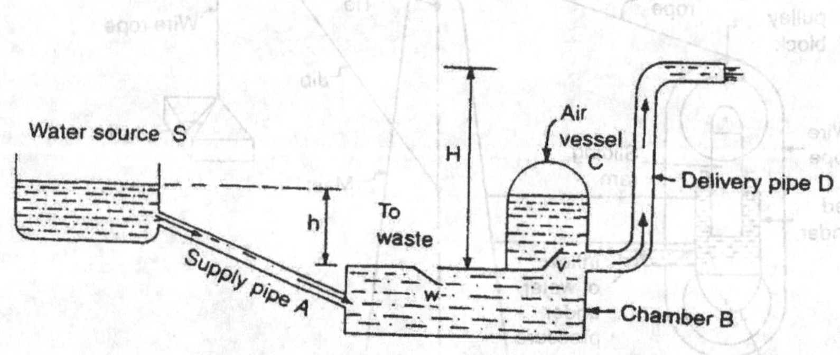

Fig. 25.4 Hydraulic ram

Since energy supplied by tank A is equal (theoretically) to the energy supplied to tank B.

$$WH = wH_1$$

or

$$w = \frac{WH}{H_1}$$

The efficiency, η, of the ram $= \dfrac{wH_1}{WH}$, if losses in the piping system are ignored. It may be of the order of 75 percent. The efficiency of the ram is also expressed in terms of discharge called Rankine efficiency.

$$\text{Rankine efficiency} = \frac{q(H_1 - H)}{QH}$$

The ram can lift water to a height 15 times the height of the working fall.

Advantages

(*i*) Automatic and simple operation, (*ii*) Most economical as no fuel or power is required for its working, (*iii*) It is durable and has long life, (*iv*) Suitable where no source of power is available, (*v*) Air vessel helps in reducing fluctuations in flow, (*vi*) Particularly suitable for small irrigation schemes fed by a natural source like rivulet, (*vii*) Once started no attention is needed.

Disadvantages

(*i*) Large amount of water is wasted and as such limits its use for locations where natural water supply is available in abundance, (*ii*) Its working creates considerable noise,. (*iii*) Obsolete type of pump as water is becoming scarce.

25.6 HYDRAULIC LIFT

It is a device commonly used for carrying persons and/or loads from one floor of a building, more than two storey high, to another. The hydraulic lifts are of two types; (*i*) Direct acting lift, and (*ii*) Suspended lift.

1. Direct Acting Hydraulic Lift: It consists of a cage, housing standing persons, and/or loads, attached to the top end of a ram sliding in a cylinder (Fig. 25.5). High pressure water is admitted, through a pipe connection, raising the ram up and the cage fixed to it. On withdrawing the water from the cylinder, the cage moves down.

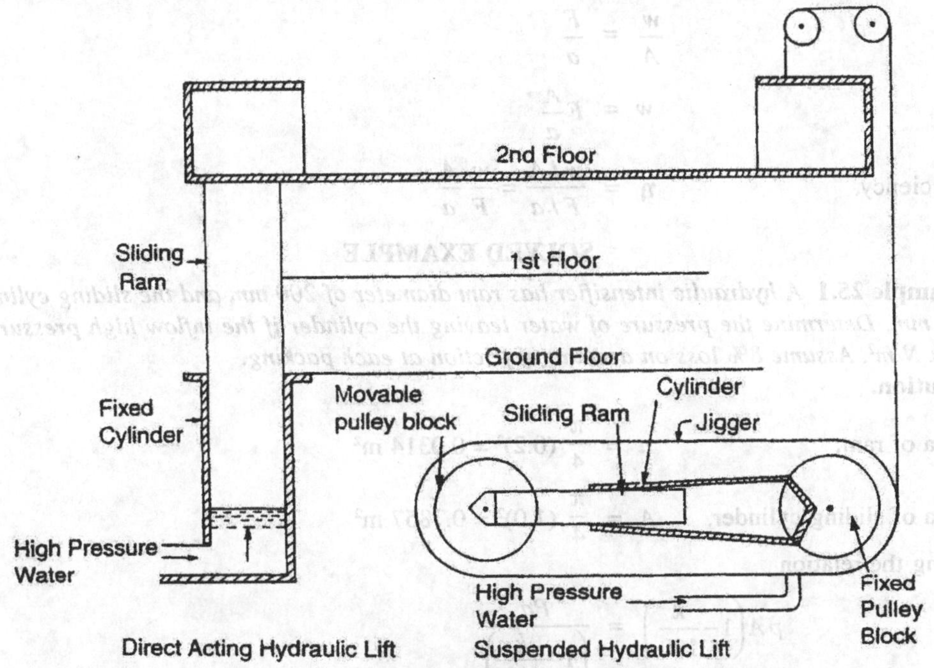

Fig. 25.5 Hydraulic lifts

2. Suspended Hydraulic Lift: It is another modified variant of the direct acting hydraulic lift. It comprises a jigger, similar to that of hydraulic lift, cage suspended by ropes, sliding rim, movable and suspended pulley blocks. High pressure water is admitted, through pipe connection, to the cylinder forcing the sliding ram outward which in turn transmits the outward movement to the pulley block. The cage gets lifted due to increased distance between the two pulley blocks and pulling of the wire rope (Fig. 25.5). The cage runs between guides suspended by four lifting ropes to support the load. The weight of cage is balanced by sliding balance weights.

The hydraulic lifts have given way to electric lifts except where danger due to fire is apprehended. But finds use as freight elevators, elevators for factories, garages etc.

25.7 HYDRAULIC PRESS

Hydraulic press is a device by means of which heavy loads are lifted by the application of a much smaller force. It works on the principle of Parcal's law according to which intensity of pressure is transmitted equally in all directions through a mass of liquid at rest. It essentially consists of a plunger sliding in a small diameter cylinder and another cylinder, larger in diameter, connected through a chamber to the smaller one and in which a plunger works (Fig. 25.6). The chamber is filled with fluid. *A* small force *F* acts on the plunger in downward direction, pressing the liquid under it. This pressure of the liquid raises the large diameter ram and lifting the load placed on it. The mechanical advantage obtained is equal to the ratio of the areas of the ram and the plunger, as elaborated below:

Let *p* is the intensity of pressure of the liquid, and *A* and *a* are the areas of ram and plunger respectively. Since the intensity of pressure is the same throughout the chamber,

$$F = pa \tag{i}$$

$$w = pA \tag{ii}$$

Equating equations (*i*) and (*ii*)

$$\frac{w}{A} = \frac{F}{a}$$

or

$$w = F\frac{A}{a}$$

Efficiency,

$$\eta = \frac{w/A}{F/a} = \frac{w}{F}\frac{A}{a}$$

SOLVED EXAMPLE

Example 25.1 *A hydraulic intensifier has ram diameter of 200 mm and the sliding cylinder of 1000 mm. Determine the pressure of water leaving the cylinder if the inflow high pressure is 20,000 k N/m². Assume 8% loss on account of friction at each packing.*

Solution.

Area of ram,

$$a = \frac{\pi}{4}(0.2)^2 = 0.0314 \text{ m}^2$$

Area of sliding cylinder,

$$A = \frac{\pi}{4}(1.0)^2 = 0.7857 \text{ m}^2$$

Using the relation

$$pA\left(1-\frac{k}{100}\right) = \frac{Pa}{\left(1-\frac{k}{100}\right)}$$

$$p\times 0.7857\left(1-\frac{8}{100}\right) = \frac{20{,}000\times 0.0314}{1-\frac{8}{100}}$$

or

$$p = 944 \text{ k N/m}^2 \quad \textbf{Ans.}$$

Example 25.2 *A hydraulic intensifier has a capacity of 0.030 m³ and stroke length of 1.5 m. The water pressure on the leaving cylinder and supply cylinder is 20 MPa (= 20 × 10⁵ N/m²) and 50 MPa respectively. Determine diameter of the ram and sliding cylinder.*

Solution.

Capacity of intensifier = area of ram × stroke length

$$0.030 = A_2 \times 1.5$$

or

$$A_2 = 0.02 \text{ m}^2$$

∴ Diameter of ram, $D_2 = \sqrt{\dfrac{0.02 \times 4}{\pi}}$

$= 0.16$ m **Ans.**

Considering equilibrium of cylinder

$$P_1A_1 = P_2 \times A_2$$

$$20 \times 10^5 \times A_1 = 50 \times 10^5 \times 0.02$$

∴ $A_1 = 0.05\ \text{m}^2$

∴ $D_1 = \sqrt{\dfrac{0.05 \times 4}{\pi}} = 0.25$ m **Ans.**

Example 25.3 *An hydraulic accumulator has a ram of 150 mm diameter and a lift of 10 m. Water is supplied at 50 bar. Determine (i) capacity of accumulator, (ii) load on ram.*

Solution.

Area of ram $= [\pi (0.15)^2]/4 = 0.0177\ \text{m}^2$

(*i*) Capacity of accumulator $= \text{P.A.L.} = 50 \times 10^5 \times 0.0177 \times 10$

$= 8.85 \times 10^5$ Nm

$= \dfrac{8.85 \times 10^5}{1000 \times 60 \times 60} = 0.246$ kWh **Ans.**

(∵ 1 kwh = 1000 × 60 × 60 Nm)

(*ii*) Load on ram, $W = p \times A = 50 \times 10^5 \times 0.0177 = 88.5$ kN **Ans.**

Example 25.4 *A hydraulic crane has ram of 200 mm diameter. Water is supplied with pressure of 8000 kN/m² at inlet end through a 50 mm diameter pipe, 500 m long. The pressure required on the ram is 500 kN/m². The ratio between the movement of load and ram is 5 : 1. Assume coefficient of friction for pipe as 0.01. Find out:*

(*i*) *Maximum speed required to lift 20 kN*

(*ii*) *Load and speed of lifting at maximum power from the crane.*

Solution.

(*i*) Pressure head at inlet end of pipe = 8000/9.81 = 815.5 m ($w = 9.81$ kN/m²)

Friction loss in pipe $= \dfrac{flv^2}{d \times 2g} = \dfrac{0.01 \times 500 V^2}{0.05 \times 2 \times 9.81} = 5.10\ V^2$

Pressure to overcome mechanical loss = 500/9.81 = 51 m

Pressure head on ram = Pressure at inlet – friction loss in pipe – mechanical loss

$= (815.5 - 5.10\ V^2 - 51.0)$ m (*i*)

Pressure head required on ram

$$= \frac{20 \times 5}{\frac{\pi}{4}(0.20)^2 \times 9.81} \quad (ii)$$

From (*i*) and (*ii*)

$$815.5 - 5.10\ V^2 - 51.0 = 324.3$$

or $V = 9.29$ m/s

Discharge of ram = Discharge of pipe

$$V_r \times \frac{\pi}{4}(0.20)^2 = \frac{\pi(0.05)^2}{4} \times 9.29$$

or $V_r = 0.58$ m/s

∴ Maximum speed of load $= 0.58 \times 5 = 2.9$ m/s **Ans.**

(*ii*) Since pressure loss in pipe is 1/3rd available head to obtain maximum transmission of power

$$\frac{1}{3} \times 815.5 = 5.10\ V^2$$

or $$v = 7.3 \text{ m/s}$$

Speed of lifting head $$= \left(\frac{0.05}{0.20}\right)^2 \times 5 \times 7.3$$

$$= 2.28 \text{ m/s} \quad \textbf{Ans.}$$

Head on ram $$= \left(815.5 - 51.0 - \frac{815.5}{3}\right) = 493 \text{ m}$$

$$= 493 \times 9.81 = 4836 \text{ kN/m}^2$$

Load as can be lifted $$= \frac{4836}{5}\left[\frac{\pi}{4}(0.20)^2\right]$$

$$= 30.40 \text{ kN} \quad \textbf{Ans.}$$

Example 25.5 *A hydraulic lift has a lift capacity of 50 kN against a height of 10 m, once in 100 seconds. The speed of lift is 0.5 m/s. Find out, (i) Power required to run the lift, (ii) working period of lift, and idle period of lift.*

Solution.

(*i*) Work done in lifting load $= w \times H = 50 \times 10 = 500$ kNm in 100 seconds

Work done/s $= 500/100 = 5$ kNm/s

Power required to run lift, $P =$ 5 kW **Ans.**

(*ii*) Working period of lift $= H/V = 10/0.5 = 20$ seconds

(*iii*) Idle period of lift $=$ Total time − working period

$= 100 - 20 = 80$ second **Ans.**

QUESTIONS

1. Explain the working of an air lift pump. (*AMIE s'01*)
   A double acting actuator (jack/hydraulic cylinder) is made to extend. The cylinder dia. is 5 cm and rod 2 cm. If a positive displacement pump provides a flow of 40 cumes:
   (*i*) What is the velocity of the piston and flow rate out of the actuator? If the pump flow is applied to the annular side :
   (*ii*) What is the flow out of the full area side and the cosequent velocity of reaction. (*AMIE s'2000*)
2. In a hydranlic lift the piston area ratio is 100:1, the longer piston (ram) dia. is 150 mm and the pumping piston (plunger) must be moved up and down 400 times in order to lift a load of 10 kN on the large piston through a distance of 140 mm.
   Calculate (*i*) pump (plunger) stroke length, (*ii*) force required on the plunger to raise the load. (*AMIE s'2000*)
3. A laboratory hydraulic ram has the following data: Speed head 3m, delivery head 17m, total water supplied 900 l/minute, discharge 90 l/minute. Determine Rankine efficiency. (*AMIE s'0*)
4. By means of a neat sketch explain the working of a differential hydraulic accumulator. (*AMIE s'93*)
5. The sliding ram of an accumulator is 40 cm dia. and total moving load is 5000 newtons. If frictional resistance against ram movement is 4000 newtons, find the pressure of water when: (*i*) loaded ram is moving up with uniform velocity, and (*ii*) loaded ram is moving down with uniform velocity. (*AMIE s'93*)

6. (*a*) With the help of a neat sketch, explain the working of a hydraulic torque convertor. Draw its performance characteristics.

   (*b*) A hydram uses 4 cu.m/minute of water under a head of 3m and pumps 0.25 cu m/minute to a tank 24m high above 100 mm dia. pipe. Take Darcy's *f* for pipe as 0.04 and calculate the efficiency of the ram. (*AMIE s*'01)

7. (*a*) Describe with neat sketch the working of a hydraulic press.

   (*b*) Derive an expression each for: (*i*) unit power, (*ii*) specific power (*AMIE s*'99)

   (*c*) Explain the working of an air lift pump, with the help of neat sketch. (*AMIE w*'99)

8. (*a*) By means of a neat sketch explain the working of an air lift pump.

   (*b*) How does torque convertor driffer from fluid coupling? Explain one of them with a neat sketch. (*AMIE s*'97)

   Write note on jet pump. (*AMIE w*'95)

9. By means of a sketch explain the working of an air lift pump. Where are such pumps used? (*AMIE s*'94)

10. (*a*) Explain with sketches the working of a hydraulic ram and define the efficiency of the ram.

    (*b*) Explain with a neat sketch the principle and working of an air-lift pump. (*AMIE w*'93)

11. What is cavitation and what are its causes? How will you prevent the cavitation in hydraulic machines? (*AMIE w*'93)

APPENDIX–I

DETERMINATION OF NUMBER AND TYPE OF UNITS

As illustration consider a power plant to be commissioned in two stages:

Stage I : Firm power 100 MW. With 60% load factor, the working requirement of the plant is 166.6 MW.

Stage II : Ultimate stage of development with 300 cumecs water supplyfirm power of 200 MW. With 60% load factor, working requirement is 333.33 MW to be installed.

1. NUMBER OF UNITS

Depending upon latest trend of bigger unit sizes, there can be possibilities of : (*i*) 1 units × 166.66 MW, (*ii*) 2 units × 83.33 MW, (*iii*) 3 units × 55.50 MW, and (*iv*) 4 units × 41.66 MW.

Stage I:

(*i*) If other considerations of transport limitations, penstock size etc. do not limit the unit size, only 1 unit of 166.66 MW is selected, then overall efficiency of the machine will be too low. Also generally a standby unit is provided in all installations which in the case of a single unit will mean doubling the installed capacity.

(*ii*) 2 units of 88.33 MW installation will have to be run on a very much reduced part load for 100 MW firm power and a standby unit will mean 1.5 times the installed capacity.

(*iii*) Further, a unit of 166.66 MW and 83.33 MW sizes can not be installed economically due to the following reasons:

(*a*) In this moderate head water development ranging from minimum net head of 35 m to maximum head of 50 m, the discharge for such bigger units becomes tremendous and it means extraordinary bigger diameter penstock, difficult to install and fabricate, as well as bigger intake opening, bigger high head gate and raised minimum drawdown level requiring more height of dam. Even for unit sizes of 55.50 MW, economical diameter comes to approximately 7 m which matches well with the maximum size of penstocks installed in the country.

(*b*) Castings especially of runner and inlet valve door preferably to be transported in one piece becomes very big than the limitations on transport by Indian Railways and split construction will impair strength and will be very much uneconomical also.

(*c*) Due to perhaps above reasons, the biggest unit for head less than 60 m is of 75 MW at Ukai dam (Gujarat); the other being as given in Table 1.

Table 1. Typical Indian installations

Name of project	*Units installed No. × MW*	*Head (m)*	*Penstock dia. (m)*
Rihand (U.P.)	6 × 50	68.5	5.20
Ukai (Gujarat)	4 × 75	47.8	7.01
Pong (Himachal)	6 × 60	66.5	7.28, bifurcated into 5.025 m
Ramganga (U.P.)	3 × 68	54.9	7.92, bifurcated into 3.87 m
Ramapartap Sagar (Rajasthan)	4 × 43	54	6.10
Kadana (Gujarat)	4 × 60	47.5	5.5
Gandhi Sagar (U.P.)	5 × 23	48.47	4.75

(*d*) 2 units of 83.33 MW or 1 unit of 166.66 MW will have to be on significantly reduced part load for firm power of 100 MW resulting in appreciably reduced overall efficiency.

(*iv*) Next preferable choice lies with unit size of 55.50 MW.

No. of units required in 1st stage = 166.66/55.50 = 3 units.

Total units, keeping 1 standby = 4 units.

Stage II

Total units, with 1 standby in stage II = 333.33/55.50 + 1 = 7 units.

4 units of 55.50 MW suit for 1st stage considering the load factor of 60%, standby provision, total cost of power house, transport feasibility, penstock diameter requirement and overall annual generation efficiency.

Hence 4 units of 55.50 MW for stage I and 7 units in ultimate stage development are selected.

The tabulation of turbine efficiency output and load variation requirements for 55.50 MW size units at maximum, design and minimum head is as given in Table 2:

Table 2. Efficiency Vs head

Head (m)	*% of design head*	*% Efficiency head*	*% H.P. of best efficiency H.P. at design head*	*Maximum out put (MW)*	*% discharge of at design head*	*No. of 55.50 MW required at % load in Ist stage*	
						Peak requirement	*Firm power*
Design head 44 in (wt. av. net head)	100	0.87 or 87	100	55.5	100	3 machines full load	2 machines 90% load
Min. net head 37 m	84	87	78	43.33	92	4 machines full load	3 machines 78% load
Max. net head 50 m	114	86	120	66.60 (20% overloading)	106	3 machines 85% load	2 machines 75% load

In view of reasonable overall efficiency, part load running requirement, economic and feasible diameter of penstock, feasibility of transport of single casting runner and load factor on the power house, 55.50 MW size of the units selected is alright.

Justification for one Standby Unit

The turbines and generators require three types of maintenance viz.,

1. **Capital Maintenance:** This means complete overhauling of turbine and generator, which is usually done once in five year period and takes 3-4 months for unit size of 55.50 MW.

2. **Annual Maintenance:** This includes turbine maintenance, inspection, painting, setting etc. and takes about 1 month for unit size of 55.50 MW.

3. **Emergency Maintenance:** This is done as and when emergency warrants.

Generally 10% standby units can cope up the situation when the units are deployed in rotation, hence one standby unit for stage I and similarly stage II development will suffice.

2. TYPES OF UNITS

1. **Specific Speed:** Specific speed is a fundamental index for the selection of unit. Various empirical formulae give a specific speed for the turbine design head.

From graph of design head Vs specific speed (N_s) as per Fig. 10.8 it is evident that within the latest trend, specific speed values for the given design head, the selection of unit is a marginal case between Francis, Kalpan and propeller. Propeller type turbine can, however, be straightaway ruled out which is installed up to 30 m for nearly constant head, discharge and load.

Choice between Francis and Kaplan Turbines

The choice between Francis and Kaplan turbines, in this marginal and overlapping case, is governed by techno economic considerations of the generating equipment, power house cost and relative benefits of power generation, briefly discussed as under:

1. Initial Cost: The dimensions of Kaplan turbine for the same head and rated power in the overlapping region are generally bigger than the Francis turbine, the higher specific speed values notwithstanding. It means that rock excavation, structural works, equipment cost and installation cost is higher than Francis turbine unit of same size power for the same head due to the bigger turbine dimensions, larger centre to centre spacing of units and overall more size of power house.

The Kaplan turbines are to be set deeper with regard to minimum tail water level. The cavitation coefficient for the kaplan turbine, in general, is much higher than that for Francis turbine necessitating relatively deeper submergence and higher excavation cost.

2. Efficiency: Fall of efficiency of Kaplan turbines with variations in head is much less in comparison to Francis turbines and as such in overlapping head ranges, the selection of the type of turbine is affected by head variation existing at site. The head variation, in order to ensure a normal runner life with minimum cavitation and to achieve reasonable efficiency levels., in the case of Francis and Kaplan turbines is restricted to 125 to 65 percent, as per IS: 12837 : 1989 (refer Table 10.4). Thus with the range of permitted head variation both Francis and Kaplan turbines suit the requirement.

Francis turbine has better peak efficiency (93%) at full load and is nearly the same as of Kaplan turbine (92%). But in the higher specific speed zone, Francis turbine shows better part load efficiencies only than propeller but is not as good as Kaplan turbine. The fall of efficiency at part load for Francis is much steeper in comparison to that for Kaplan turbine, necessity of operating turbines at part loads for long time influences choice of turbine in overlapping head ranges.

Though even a small drop in efficiency costs a large revenue loss but in this power station of base load having 7 units of 55.5 MW each, part load running requirements of individual prime movers are minimised. Further, with the interlinking of this station within integrated power grid, part load working of the units may not be required very often. As such presently Francis turbines are being opted more leading to a broadening of their field of application, partially invading those that were exclusively considered previously to Kaplan turbines.

3. Hydraulic Thrust: The thrust bearing of an axial flow runner (Kaplan units) is more heavily loaded than that of corresponding Francis runner and therefore, Kaplan is expensive and thrust bearing heavily stressed.

Pressure Rise and Speed Rise: The permissible pressure rise and speed rise for the two types of turbines, as under, show that these two parameters are limited in Francis turbine than in Kaplan turbine.

Turbine	*Pressure rise (%)*	*Speed rise (%)*
Francis	30–35	35–55
Kaplan	30–50	30–65

The generator units are designed to withstand full runway speed which proportionately adds to the generator cost. It is observed that the little economy due to higher operating speed of Kaplan turbine is offset by a higher runaway speed.

4. Minimum Output: Normal range for continuous operation for the two types of turbines, as under, indicates higher minimum output for continuous operation in favour of Francis turbine :

Turbine	*Minimum output for continuous operation (%)*
Francis	50
Kaplan	30–40

Maintenance: Both Francis and Kaplan turbines enable replacement of all the runner blades without dismanling generator. For Francis runners usually in situ repair is done. However, as per experience on existing installations recent trend is to favour Francis turbines rather than Kaplan turbines in the overlapping region. Control, operation and maintenance of Kaplan units is found to be more troublesome and expensive than Francis turbines.

The control gear of Kaplan is more complicated and requires additional maintenance compared to Francis type. Recent practice is to change the old runners with, Francis type in overlapping regions.

Considering the above techno economic factors, Francis turbines of vertical shaft setting are selected which has several advantages over horizontal setting such as:

(*i*) No large loads act on the radial bearing,

(*ii*) The scroll casing can be conveniently arranged in the power house,

(*iii*) The runner can be arranged below tailwater level in order to improve its cavitation characteristics.

(*iv*) The generating set is easy to install, dismantle and service in the course of operation.

APPENDIX–II

Presented in International Conference on Accelerated hydro power development and pumped storage development, 1-3 December 1997, New Delhi

ACCELERATED HYDRO POWER DEVELOPMENT – A CASE STUDY

R.K. Sharma

Consultant, Water Resources

SYNOPSIS

Energy in the form of electricity provides the basic infrastructure input to national economy, its productivity improvement means higher national productivity. Of all the sources of energy, hydro power is the cheapest, pollution free, resources saving, non-consumptive and self replenishing. In India whereas large hydel potential is awaiting exploitation, critical shortage of energy is hampering all aspects of the country's development which underlines the urgent need for adding hydro power in preference to thermal power for rapid and ecofriendly advance of the country to enter the 21st century. Hydro projects known for long gestation period can now be constructed expeditiously. The paper highlights the factors for accelerated development of hydro power based on the experience of modified and expeditious methods which enabled completion of Pong Power plant on Beas river in a period of 39 months.

1. INTRODUCTION

Electricity, one of the prime movers of the economy, is the most versatile and preferred form of energy. The generation of energy is synonymous with the generation of a nation's prosperity. The per capita consumption of electricity is a reliable index of a nation's progress and development. Though statistically the performance of India in this field is impressive yet per capita consumption is very low as compared to developed and a number of a developing countries. Also considering the size, population and rapidly increasing demand for power for large scale industrialisation, modernisation of agriculture, urbanisation, high living standard etc., the electricity consumption doubles in every ten years and as such generation too needs to be doubled to maintain the pace of the development. The country has yet to go a long way to enter the era of 21st century.

In the present era of shortage of energy and consequent search for commercial exploitation of non-conventional sources of power which is yet in development stage, the country has necessarily to depend on two conventional sources, that is the hydro electric and thermal power and to a smaller extent on nuclear power because of the technological constraints.

2. HYDRO-THERMAL MIX

In the power production, optimum mix of hydro, thermal and nuclear power is determined by system analysis studies taking into consideration the percentage of each type of source depending on availability of natural resource and operation of various plants to meet the demand of load with overall economy in view. The hydel-thermal mix which was 40:60 at the end of Fifth Plan has went down to undesirable value of 28:72 which needs correction keeping into consideration environment, health of power system and its economics because in overall perspective hydro power means development at the least environment cost and highly economical. An ideal hydro-thermal-nuclear mix for an healthy sustainable development is 41:56:13. To achieve this mix, rapid increase in hydro power generating capacity is imperative.

3. HYDRO POWER DEVELOPMENT

Hydro power generation has assumed all the more importance in the recognition of its distinct and clear advantages over thermal being inexhaustible, perpetual and renewable energy source, more reliable because of its non-dependence on costly and at times unreliable supply of coal, imported oil etc., and transportation problems associated therewith and above all the cheapest and pollution free energy source, simple in construction and maintenance, immune to inflation and threat of shortage of coal due to long distance transportation and strikes etc. India has every conceivable type of hydro generation. The hydro power potential of the country is quite large-84,044 MW at 60% load factor. The available capacity and proportion developed are shown in Table 1.

Table 1 Hydro Power Potential and Utilisation

River Basin	*Power potential (MW)*	*Utilisation (%)*
Indus	19,988	12.90
Ganga	10,715	17.07
Central Indian rivers	2,740	19.00
West flowing peninsular	6,149	51.84
East flowing peninsular	9,532	37.00
Brahmaputra	34,920	1.06
Total	84,044	

The non-conventional hydro power potential too is quite large, briefed in Table 2.

Table 2 Unconventional Hydro Power Potential

Pumped storage	96,524
Small hydro	5,000
Tidal energy	10,000
Total	1,11,524 MW

The hydro power development in the country commenced in 1897 with commissioning of a 130 kW hydro power station near Darjeeling in West Bengal state and has by now graduated to large power grids. However, the installed capacity of 20,967 MW is only 25% of the utilisable potential thereby leaving additional capacity to be created as more than 60,000 MW which requires quantum jump in the efforts for hydro power development.

Hydro power has several advantages in comparison to thermal power. Hydro power station presents cleaner environments. Furthermore, it is estimated that every 1000 MW of hydro power releases about three million tonnes of coal annually besides saving recurring cost of Rs. 300 crores per annum on coal and transport. The socio-economic benefits of the hydro power projects in the remote areas where they are generally located also can not be lost sight of. The hydro-electric projects in their useful multipurpose concept are also greatly instrumental in the development of agriculture, besides the benefits of drinking water, flood control, pisciculture, recreation facilities etc. The remarkable development which has taken place in Punjab, Haryana and Rajasthan is largely due to the tremendous amount of water and power released from Bhakra and Beas complex.

4. CONSTRAINTS IN HYDRO DEVELOPMENT

The main disadvantages attributed to hydro-electric power stations are high initial investment, longer gestation period and dependence on nature but these are not true as discussed hereinafter:

4.1 Higher Initial Investment

Hydro power plants are considered to require relatively larger capital investment but thermal plants too need huge expenditure on the development of coal mines etc., briefed as under:

(*i*) The additional investment is required for the development of new mines to feed the large thermal power plants. Besides, large amount of ash has to be disposed of daily at the power plant site, and investment on rehabilitation of the ecological environment disturbed in mining area and in close proximity of the thermal plant.

(*ii*) Investment on rolling stock, new rail lines, improvement of existing rail tracks and signalling facilities etc., are also required for setting up of thermal plants.

(*iii*) The auxiliary consumption of a thermal power house is 10-12%. The useful capacity of the station, and therefore, the capital investment per KW of the useful thermal power thus becomes 10-12% more.

(*iv*) According to recent survey the investment on thermal plant is double that of the hydro equivalent. Furthermore, the average cost of thermal power generation is about three times that of the hydel power.

4.2 Longer Gestation Period

The longer gestation period attributed to the hydro power is on account of the inclusion of the period of investigation and surveys of hydro electric projects whereas the gestation for the activities associated with thermal plants such as exploration and development of coal mines etc., is not taken into account. Hydel projects can now be constructed expeditiously with expertise and experience by now developed in the country. The advance planning, technology updating, better administrative and management techniques can help reduction in gestation period and real costs besides improving quality of the work. As a result future hydel projects can be completed in 5 to 7 years as in other countries as also like the thermal projects. Some of the factors which have emerged for cutting down gestation period of hydel projects are as follows:

(*i*) Investigation and surveys with latest tools and techniques such as satellite and aerial photography for topographical survey, multi-spectral remote sensing imageries for geological mapping and gammaray logging and three dimensional shear wave technique for geophysical exploration.

(*ii*) Early provision of infrastructure and proper strengthening of the construction agencies in terms of organisation, residential accommodation, communication, construction techniques and skilled man power.

(*iii*) Proper construction planning with matching annual budgets. Inadequate funding is the single major factor for the overrun in time and cost.

(*iv*) Development of optimal equipment both qualitatively and quantitatively besides satisfactory performance of indigenous equipment and to avoid delayed and disorderly supply.

(*v*) The Chief Executive of the project is vested with adequate administrative and financial powers and authorisation for taking on the spot decisions to facilitate speedy completion of the project.

(*vi*) The future hydro power projects sites have necessarily to be in very remote and inhospitable areas. The engineers, technicians and workers deserve necessary motivation and incentive like those offered to Army, Border roads organisation etc.

(*vii*) Modified and expeditious construction methods like adoption of modern mechanical techniques. Likewise, improved rock supporting methods and concrete lining of tunnels as practised in many foreign countries need to be adopted.

(*vii*) Use of modern management techniques and close monitoring of the construction work in progress for timely corrective measures.

(*ix*) With proper planning and construction techniques interim benefits of generation may even be derived before raising of the dam to full designed height.

(*x*) Standardisation of the project components such as shape and size of tunnels, shafts, turbines etc., can help reduction in design and construction time.

Pong power plant at Pong dam on Beas river in Himachal Pradesh is a case where in its first unit with installed capacity of 60 MW was commissioned in 39 months despite delay of nearly 6 months on account of late supply of embedded parts from the suppliers and manufacturing defects noticed in some parts during installation and testing as against 4-5 years for similar other plants in the country. This achievement was made possible by adopting unconventional construction methods and close monitoring of the day to day progress.

4.3 Dependence on Nature

It is true to some extent but it has to be considered that the hydro power stations are designed on percentage availability of run off of a river. Generation from hydro plant is less only in the case of continued dry spell and it is for such contingencies that the adequate back up thermal capacity is required.

5. INNOVATIVE EXPERIENCE OF PONG POWER PLANT CONSTRUCTION

5.1 Salient Features of Pong Power Plant

Pong power plant is a multistorey reinforced concrete framed structure 33 m wide, 43 m high and 111 m long and is located in stilling basin of penstock tunnels, earlier used for dissipation of energy of water during diversion stage. It has six units of 60 MW each, six unit bays for installation of machines, service bay for assembly of equipment and housing auxiliaries, service bay extension for accommodating miscellaneous equipment and underground rooms towards left for installation of compressors. Unit bays and service bays comprise 4 and 5 floors respectively. Each bay is separated from the adjoining bay by 12 mm thick expansion joint. Vertical shaft Francis turbines with rated output of 85,000 BHP, speed 166.7 r.m.p. and rated head of 65.6 m and an Umbrella type generator supplied by M/s BHEL Bhopal have been installed.

5.2 Construction Period

In conventional method, the power house is raised to its top level including EOT crane beams and roof, and the erection of electrical equipment is taken in hand after erection of the EOT crane. The minimum period for different activities from the start of concreting to the commencement of generation in conventional method of construction even after neglecting the avoidable delay is as under:

	Description	*Minimum period required (Months)*
(*i*)	First stage concreting-service bay and unit bay	24
(*ii*)	Erection of EOT cranes	3
(*iii*)	Erection of draft tube, speed ring and scroll casing	6
(*iv*)	Second stage concreting around draft tube and scroll casing	4
(*v*)	Erection of generating equipment and commissioning	12
	Total	49 months

The experience gained at Pong power plant, indicates that hydro power stations in this country can be constructed in a period of about three years by modifying the conventional construction techniques. Some of the modified methods adopted for expediting its completion are summarised as under:

(*i*) Reduction in the total construction period was made possible by planning simultaneous operations, by identifying the areas of interdependency of civil and electrical works right at the start of the construction. Once the substructure was ready, the work of erection of draft tube and scroll casing were started with independent handling facility for the equipment. A small EOT gantry comprising a 10 tonne electric hoist available on the project and supported on steel beams and columns was installed in the pit. Erection of scroll casing was thus taken up earlier by a period of about 16 months.

(*ii*) After installation of scroll casing and second stage concreting, the next activity was the installation of turbine which had to be first assembled in the service bay floor El. 344 m with the help of the EOT crane. The erection of the crane itself takes about three months. To match with the programmes of erection of generating equipment, the civil structures of service bay and the first unit bay were thus substantially completed upto El. 353 m to enable timely erection of EOT crane in about 18 months time which considering the quantum of work involved was almost impossible to achieve with the conventional method of construction. To enable early erection of crane, construction methods were modified and gantry columns and walls of service bay were raised first without floors which expedited the erection of the crane.

(*iii*) In many power plants, EOT cranes are erected after the completion of service bay roof. In view of the difficulties in handling of the various parts of the crane during erection, this method takes more time in erection of crane. At Pong power plant EOT cranes were erected first and roof which comprised precast slabs was completed last which expedited the erection of crane.

(*iv*) The stator erection can not be exposed to wind and rain and roof is thus required to be provided before taking up this work. In conventional method the erection of reinforcement for the columns is generally done at site. To expedite this power plant construction reinforcement for the columns was pre-assembled and erected at site with the help of crane. Only joints of reinforcement were thus welded at site which reduced the concreting period.

(*v*) The concreting of portal roof beams and slabs in conventional design and construction present considerable difficulties and consumes a lot of time in erecting scaffold and shuttering. At this plant, the reinforcement of roof beams was designed to self-support the frame work without any propes. The reinforcement and shuttering of beams was completed on the ground and then placed at site with the help of crane. The precast slabs used for the roof were cast in advance and placed immediately after the portal beams had been cured. The prefabricaton and precast techniques adopted for the roof beams and slabs expedited the construction.

(*vi*) Hydro power station construction involved multiplicity of activities involving different agencies for civil and electrical works to be accomplished within a time bound programme. The conventional method of monitoring progress through bar charts based on quality basis serves only limited purpose and adoption of new techniques of monitoring progress such as PERT/CPM is a must. For this plant construction, the inter-relationship of activities of both civil and electrical works were identified with the help of CPM network. The duration time of each activity was fixed on the basis of the earliest time for similar activities taken on other projects. Civil works programme was drawn up to match with the erection schedule of electrical equipment which was based on the scheduled deliveries by M/s BHEL. Where the constraint of civil work was felt, the construction methods were modified to achieve the critical level to suit the erection programme of electrical equipment. The detailed liftwise schedule within the laid target of critical level was prepared and day to day progress was watched. At times when there was a tendency of slipping of schedule, immediate corrective action was taken to put more resources. Thus, by close monitoring of the progress, the first stage civil works schedule was achieved without any slippage.

(*vii*) At this plant, the installation of unit transformers was dependent upon the erection of complicated wye, combined reducing bends. In order to expedite construction of deck for erection of transformer, the penstocks were installed from both upstream and downstream ends. This method expedited the construction of deck and the transformers were thus installed as per required dates.

(*viii*) In order to accomplish the activities on the critical path such as welding of penstock liners, placing of concrete, assembly of scroll casing, rotor etc., incentives were offered to the groups of workmen engaged on the job which expedited the progress on the critical items.

6. CONCLUSION

India has enormous hydel power potential but development of hydro energy is rather slow. Avoidable delay in the non-utilisation of this potential represents wastage as the energy loss is non-recoverable. The present trend in favour of thermal is not justified due to various factors highlighted in the paper. The gestation period of the hydro power generation can be cut down by the adoption of suitable modifications in design and construction methods by using better modern techniques in this era of rapid technological development. The eco-friendly hydro power even permits storage of the energy in the form of pumped storage.

REFERENCES

1. Basur J.C. et. al Pong Power Plant some suggessions.
2. Naidu B.S.K. (1992). Planning and Management of hydro power resources in India. CBIP publication No. 226.
3. Sharma R.K. Hydro Power (1985) BBMB News Bulletin.
4. Sharma R.K. (1997) Hydrology and water resources engineering, Dhanpat Rai Publications, Daryaganj, New Delhi - 110002.
5. Sharma R.K. (1990), Irrigation Engineering, Vol. I, II, III M/s Oxford and IBH publishing company, 66 Janpath, New Delhi-110001.

APPENDIX–III

DESIGN OF SURGE SHAFT

Differential type surge shaft at Beas sutlej link project (Himachal Pradesh) is the largest in Asia. Provided at the end of Sundernagar Sutlej tunnel (Fig. 1), it has 22.86 m inside diameter and about 125 m high reinforced concrete shaft with 7.62 m diameter RCC riser placed inside and monolithic with the main shaft (Fig. 2). A by-pass tunnel 6.71 m inside diameter and about 300 m long, with centre line at El. 789.43 m takes off from the riser shaft which also serves as a gallery for reducing the downsurge. The main shaft is connected to the Sundernager sutlej tunnel through three port openings 3.66 × 1.07 m each in a RCC diaphragm slab with bottom El. 773.58 m placed over the bottom chamber wherefrom three 4.88 m diameter penstock headers at centre line El. 766.57 m take off to lead water to Dehar power plant. The top of surge shaft is kept at El. 885.44 m. The bottom of the surge shaft is depressed by 1.85 m to serve as a trap for any material coming from Sundernagar sutlej tunnel or from the top of the surge shaft. For recording water levels in surge shaft and riser due to fluctuations two 100 mm diameter pipes are embedded at El. 776.33 m in the surge tank lining and taken through header tunnel of penstock to surge indicator room located near the penstock valve chamber.

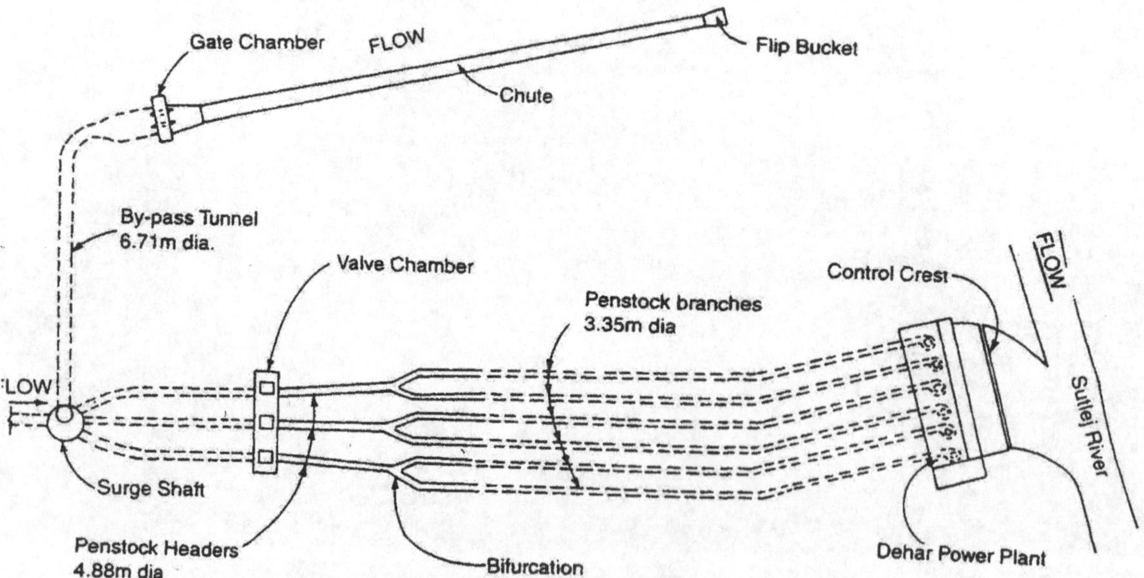

Fig. 1 General layout of surge shaft penstock by-pass and Dehar power plant

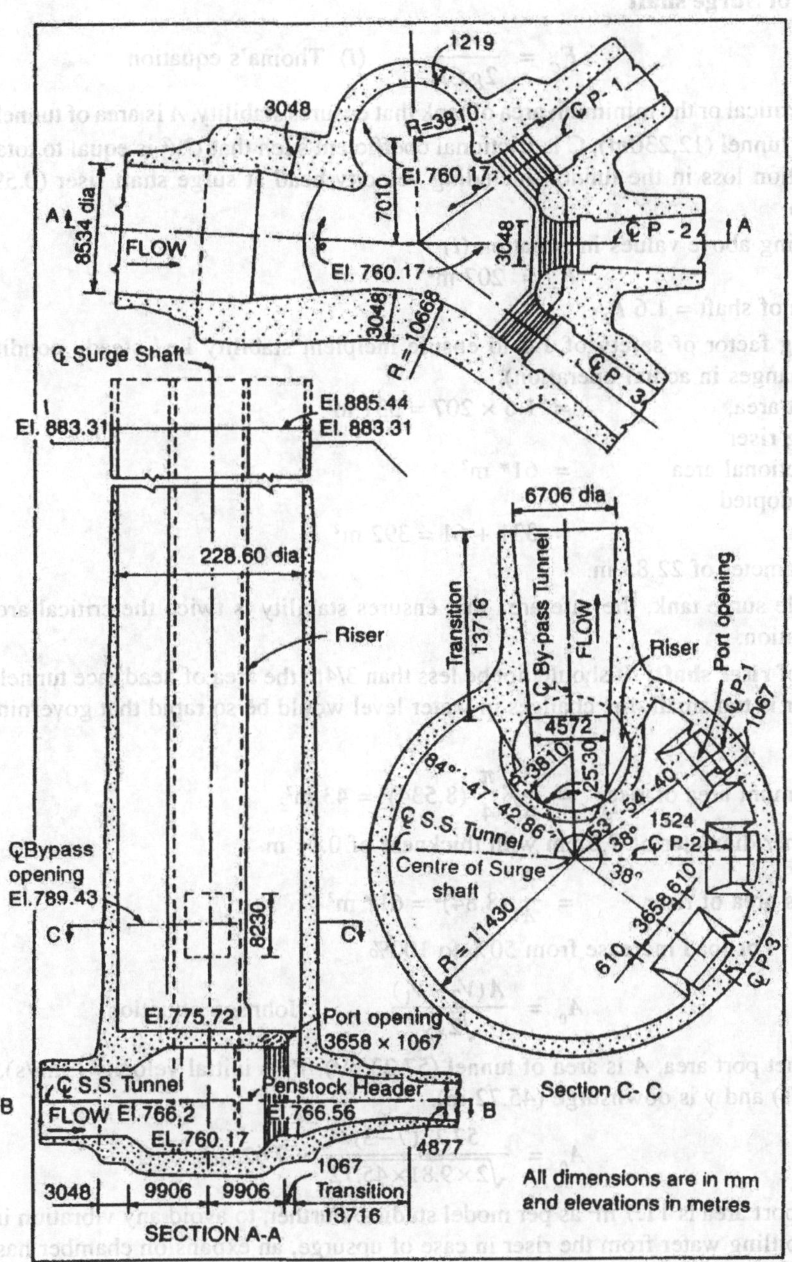

Fig. 2 Details of Dehar surge shaft.

HYDRAULIC DESIGN

Loading Conditions

Upsurge: (*i*) Complete load rejection, and

(*ii*) Increase of load from 75% to 100% followed by complete load rejection.

Downsurge: (*i*) 50% increase of load from half to full.

(*ii*) Complete load rejection followed by 50% load acceptance.

1. Area of Surge shaft

$$F_c = \frac{AL}{2g\,CH} \qquad (i) \text{ Thoma's equation}$$

Where, F_c is critical or the minimum area of tank that ensures stability, A is area of tunnel (57.23m^2), L is length of tunnel (12,230m), C is frictional coefficient such that CV^2 is equal to total minimum expected friction loss in the tunnel, including velocity head at surge shaft riser (0.59), H is net head 286.5 m.

Substituting above values in equation (i)

$$F_c = 207 \text{ m}^2$$

Safe area of shaft = 1.6 F_c

(assuming factor of safety of 1.6 to ensure incipient stability i.e., steady conditions under slight load changes in actual operation).

∴ Safe shaft area, excluding riser = 1.6 × 207 = 331 m^2

Gross sectional area of riser adopted = 61* m^2

∴ Total area = 331 + 61 = 392 m^2

Adopt diameter of 22.85 m.

For simple surge tank, the safe area that ensures stability is twice the critical area given by Thoma's equation.

2. Area of riser shaft: It should not be less than 3/4th the area of head race tunnel (8.534 m). If area of riser is too small, the changes in water level would be so rapid that governing becomes difficult.

∴ Minimum area of riser $= \frac{3}{4}\times\frac{\pi}{4}(8.534)^2 = 43 \text{ m}^2$.

Adopt inner diameter of 7.62 m with thickness of 0.61 m.

∴ Gross area of riser $= \frac{\pi}{4}(8.84)^2 = 61^* \text{ m}^2$

Port area : For load increase from 50% to 100%

$$A_0 = \frac{A(V_2 - V_1)}{\sqrt{2gy}} \quad \text{.... Johnson equation}$$

Where, A_0 is net port area, A is area of tunnel (57.23 m^2), V_1 is initial velocity (3 m/s), V_2 is final velocity (7 m/s) and y is downsurge (45.72 m).

$$\therefore \quad A_0 = \frac{57.23(7-3)}{\sqrt{2\times 9.81\times 45.72}} = 7.62 \text{ m}^2.$$

Adopted port area is 11.7 m^2 as per model studies. Further, to avoid any vibration in the surge shaft due to spilling water from the riser in case of upsurge, an expansion chamber has also been provided on the top of the riser segregated from the main shaft.

3. Height of surge shaft: To determine top and bottom elevations of the surge shaft, upsurge and downsurge have been calculated. The upsurge is calculated both for 1st stage and 2nd stage with coefficient of rugosity n as 0.011 and 0.012 respectively. The downsurge is calculated for second stage only with coefficient of rugosity n as 0.014. The coefficient of discharge for ports for upsurge and downsurge is assumed as 0.7 and 0.9 respectively as confirmed on model. The upsurge and downsurge are calculated by Johnson's equation, results thus obtained checked by graphical and arithmetic integration. The studies by arithmetic integration are done on an electronic computer.

Top of surge shaft is kept at El. 885.44 m to cate for upsurge during the 1st stage only and a spilling chamber at top bench is provided to be suitably extended for 2nd stage. The bottom of main shaft is kept at El. 775.72 m to cater for downsurge during 2nd stage.

4. Hydraulic Stability: Stability considerations require that the cross section of surge shaft is such that it prevents increase of oscillation of surges from partial load changes.

The plant is to work in big power grid with the whole inter-connected system having an installed capacity of the order of 3000 MW.

Thomas area of surge shaft in the plant in a grid;
$$F = \frac{AL}{2gCH}\left[1-\frac{3}{2}(1-k)\right]$$

Where, k is ratio of load taken by the power plant to the total load of the grid.

If $k = {}^1/_3$, $F = 0$.

Thus, when a station supplies less than one third of the load in a system, its surge shaft is always stable. Therefore, so long as the plant works in an integrated power grid its hydraulic stability is ensured. In addition, computer studies are carried out for conditions of full load rejection and load acceptance from 50% to 100% (Fig. 3). These studies are carried out for three successive half cycles of surge oscillation which indicates dampening of the surges.

The empirical rule to limit total fluctuations in head from upsurge to downsurge to a value not greater than 5% under a sudden load change of 20% is also satisfied, assuming a coefficient of rugosity of the order of 0.014.

Upsurge is determined for loading condition of full load rejection by Johnson's equation:

$$100\,N_r = 100\,VC\sqrt{\frac{2gF}{Al}}$$

where, N_r is constant, V is velocity in tunnel (23.56 ft/s or 7.18 m/s), C is a friction coefficient such that CV^2 is equal to total minimum expected frictional loss in the tunnel including velocity head at surge tank riser (0.196 and 0.59 in MKS), F is area of surge tank, A is area of tunnel, L is length of tunnel.

Substituting these values in the equation.

$$100\,N_r = 52.8$$

N_r is a constant corresponding to 100 N_r from curves by Johnson = 2.2

Upsurge $= K_r CV^2$

$= 2.2 \times 0.2345 \times 23.56^2 = 285$ ft. (86.87 m)

Maximum pond level = 2760 ft. (El. 841.25 m)

Loss in tunnel ≐ 130.2 ft (39.68 m)for $N = 0.012$

Upsurge level $= 2760 - 130.2 + 285$

= 2915 ft. (888.49 m)

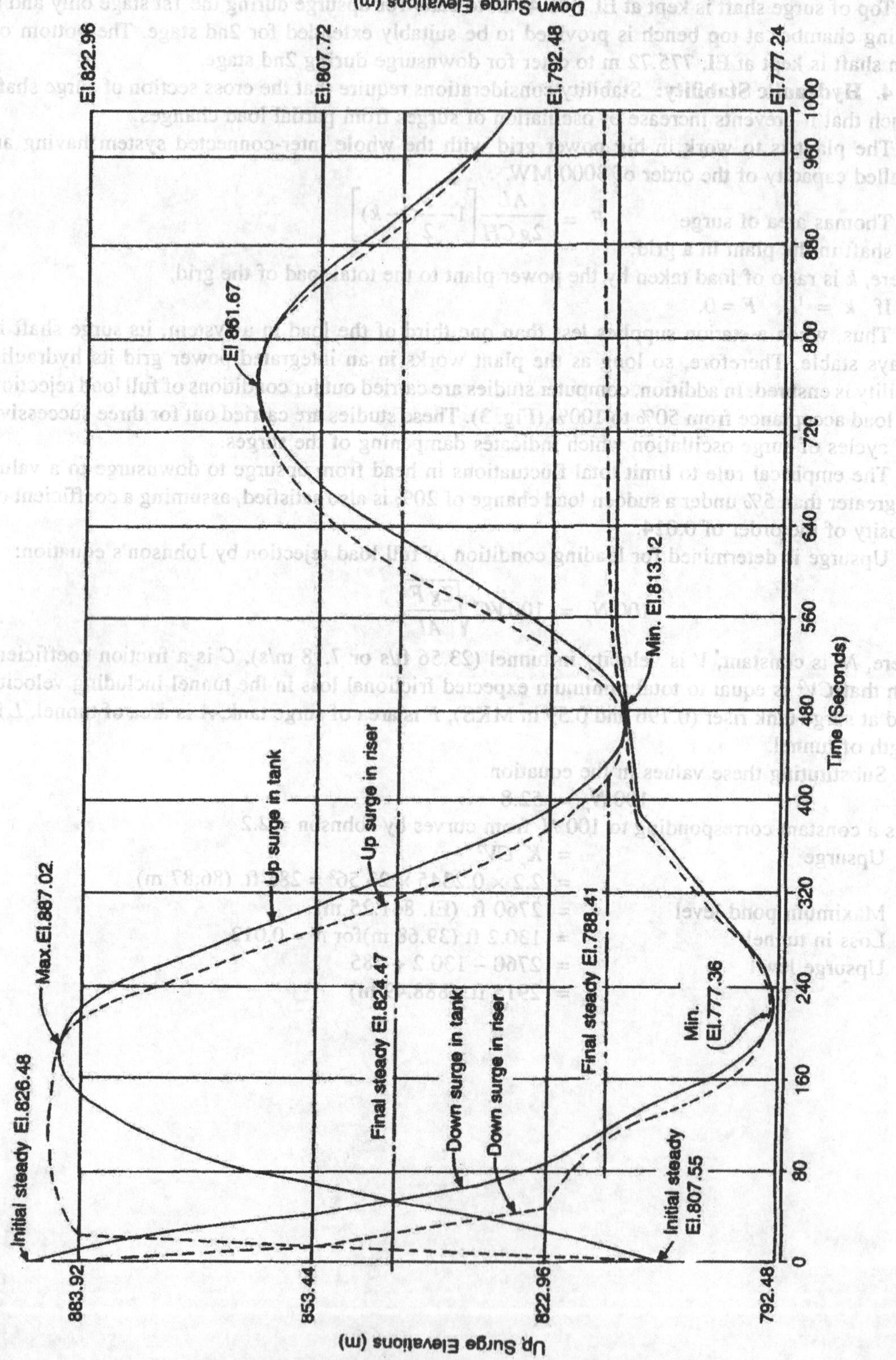

Fig. 3 Mass oscillation curves for upsurge and downsurge.

APPENDIX–IV

OBJECTIVE TYPE QUESTIONS

1. WATER POWER ENGINEERING

1. The primary power at a hydro plant is determined considering:
   (*a*) minimum stream flow (*b*) available pondage
   (*c*) probable load curve (*d*) all of these (*d*)
2. Maximum of primary power is obtained by operation of storage by:
   (*a*) yearly-use method (*b*) safe yield method (*b*)
3. Indicate method of studying effect of storage on a power plant situated:
   (*a*) at reservoir (*b*) at some distance from reservoir.
   (*a*) mass curve method
   (*b*) hydrograph method with accompanying reservoir depletion curve.
4. Wheel capacity and available power is best determined by:
   (*a*) flow-duration curve (*b*) hydrograph
   (*c*) mass curve (*a*)
5. For economical location penstock is laid:
   (*a*) on grade varied as per topography
   (*b*) on as flat grade as possible consistent with keeping below hydraulic gradient for greater part of its length and sudden drop near power plant.
   (*c*) on as steep grade as possible consistent with keeping below hydraulic gradient and sudden drop near power plant (*b*)
6. The simplest method to eliminate positive water-hammer pressure is to provide–to take the rejected flow:
   (*a*) simple surge tank at upper end of penstock
   (*b*) pressure relief valve at turbine
   (*c*) none of them (*d*) both of them (*d*)
7. Pressure relief is not effective for:
   (*a*) positive water hammer (*b*) negative water hammer
   (*c*) none of these (*d*) both of these (*b*)
8. A high ground is available close to a power house for the location of surge tank. Indicate the type of tank you will prefer:
   (*a*) simple surge tank (*b*) tower type tank (*c*) restricted tank
   (*d*) tank resting on ground containing penstock and an orifice for inflow and outflow provided on top of conduit (*d*)
9. Substructure of a power house should provide space for:
   (*a*) turbine governors (*b*) generators for horizontal units
   (*c*) switch board (*d*) low tension switches and other auxiliaries
   (*e*) all (*e*)
10. In out door type power plant:
    (*a*) substructure is omitted (*b*) superstructure is omitted
    (*c*) intermediate structure is omitted (*d*) none is omitted (*b*)

11. Substructure for an impulse turbine is relatively:
(*a*) simple (*b*) complicated
(*a*) because turbines are small in size and draft tubes are not there to be provided.
12. Transmission line is a line meant for:
(*a*) transmission of water from reservoir to power plant
(*b*) transport of power from generation source to load centre
(*c*) transmission of power from one load centre to another.
(*b*) and (*c*)
13. For large blocks of power and long distances, the cost per kW transmitted is less at:
(*a*) lower voltage (*b*) higher voltage
(*c*) not related to voltage (*d*) maximum number of circuits (*b*)
14. High voltage for transmission is used to:
(*a*) transmit large amount of power over long distances
(*b*) transmit large amount of power over short distances
(*c*) transmit small amount of power over long distances
(*d*) large terminals on tap-off stations are involved (*a*)
15. Load on a turbine is increased suddenly or a great amount at a time:
(*a*) usually (*b*) seldom (*b*)
16. The function of surge tank is to:
(*a*) avoid flow in reverse direction (*b*) smoothen the flow
(*c*) act as a reservoir for emergency condition
(*d*) relieve the pipe line of excessive pressure transients. (*d*)
17. In the determination of area of surge tank factor of safety of 2.0 and 1.6 is applicable to:
(*a*) simple (*b*) restricted (*c*) differential 2.0 (*a*), 1.6 (*b*) & (*c*)
18. Classify reservoir type power intakes with respect to head :
(*a*) 50 m (*b*) 15-30 m (*c*) 15 (*a*) high (*b*) medium (*c*) low head
19. Indicate type of hydro plant:
(*a*) up to 100 kW (*b*) up to 2 MW
(*c*) up to 25 MW (*a*) micro (*b*) mini (*c*) small
20. A small hydro plant has firm discharge as average of the lean period inflows in a dependable year:
(*a*) 100% (*b*) 90% (*c*) 80% (*d*) 75% (*d*)
21. Tidal plant can give only ________ supply of power. (*intermittent*)
22. Tidal plant is not suitable for ________ area. (*plain*)
23. In a hydro power project, reservoir is allowed to be depleted up to:
(*a*) dead storage level (*b*) minimum drawdown level
(*c*) flood storage level (*d*) normal reservoir level (*b*)
24. Dependability of live storage in a power project is:
(*a*) 100% (*b*) 90% (*c*) 75% (*d*) 60% (*b*)
25. Dependability of a run-of-river power project is:
(*a*) 100% (*b*) 90% (*c*) 75% (*d*) 60% (*b*)
26. Storage capacity of a reservoir is determined by:
(*a*) mass curve method (*b*) ratio method
(*c*) calendar year method (*a*)
27. Trap efficiency of a reservoir indicates:
(*a*) water trapped (*b*) sediment retained
(*c*) sediment inflow (*d*) sediment outflow (*b*)
28. Trap efficiency is low for a reservoir with:
(*a*) small capacity-inflow ratio (*b*) large capacity-inflow ratio (*a*)

29. Provision for sedimentation (ha m/yr/sq km of C.A) made as per ISI recommendation is:
(*a*) 0.01 to 0.02 (*b*) 0.1 to 0.2
(*c*) 0.15 to 0.25 (*d*) 0.25 to 0.35 (*a*)

30. Evaporation from reservoir surface is usually the highest in the month of ---------, give reason.
(*a*) April (*b*) May
(*c*) June (*d*) July
(*b*) because in June water spread is the lowest, being the pre-filling month.

31. V-shaped narrow valley is suitable for:
(*a*) Rockfill dam (*b*) Overflow concrete dam
(*c*) Arch dam (d) Earthen dam (*c*)

32. Indicate classification of the dams:
(*a*) > 15 m high (*b*) 12 m high, crest length > 500 m.
(both high dams)

33. Surge tank is necessarily provided in:
(*a*) long penstocks (*b*) short length penstocks
(*c*) surface penstocks (*d*) embedded penstocks (*a*)

34. Surge tank provision is necessary in tailrace consisting of:
(*a*) open channel (*b*) long pressure tunnel
(*c*) short length pressure tunnel (*d*) long free flow tunnel (*b*)

35. Water hammer effect vis-a-vis turbine operation is headrace and tailrace tunnels is:
(*a*) indentical (*b*) reverse (*b*)

36. Pressure release valve can cater for:
(*a*) turbine gate closure and consequent rejecting surplus water
(*b*) gate opening and supplying additional water
(*c*) negative water pressure (*d*) positive water pressure (*a*) and (*d*)

37. Location of surge tank is the best where ground:
(*a*) drops rapidly towards power house (*b*) rises rapidly towards power house
(*c*) midway in rising and dropping ground (*a*)

38. Function of a surge tank is to:
(*a*) Store water on load rejection
(*b*) furnish additional water during increased load damand
(*c*) both to store and supply water (*d*) none (*c*)

39. Maximum pressure occurs in full flowing-power tunnel at the time of:
(*a*) load rejection (*b*) load acceptance (*a*)
(*a*) load rejection (*b*) load acceptance (*b*)

40. Minimum pressure occurs in full flowing power tunnel at the time of:
(*a*) load rejection (*b*) load acceptance (*b*)
(*c*) Headrace (*d*) Tailrace (*a*)

41. In hydraulic tunnels located near mountain slopes, the governing criteria is:
(*a*) lateral cover (*b*) vertical cover (*a*)

42. Indicate tunnel sections excavated through:
(*a*) hard rock (*b*) loose and weak rock
(*a*) arched roof
(*b*) circular or horse shoe section

43. Indicate tunnel shape from construction point of view for
(*a*) smaller tunnel (*b*) large tunnel
(*a*) D–shape
(*b*) Horse shoe

44. Maximum flow occurs in–when ratio of depth to diameter is:
Circular section (a) 0.90 (b) 0.92 (c) 0.93 (d) 0.94 (c)
Horse shoe section (a) 0.90 (b) 0.92 (c) 0.93 (d) 0.94 (d)

45. Usual velocity (m/s) in–power tunnels is:
(a) 3 (b) 4 (c) 5 (d) 6 (d)
(a) 6 (b) 7 (c) 8 (d) 9 (d)
(a) 5 (b) 6 (c) 7 (d) 8 (d)

46. The load factor for the peak day of the year determines the required:
(a) generating capacity (b) water storage (c) pondage (a)

47. The requirement of peak week or month determines the amount of:
(a) water storage (b) generating capacity (c) none of the two (a)

48. Load factor for a power system serving highly industrial area vis-a-vis residential area is:
(a) same (b) high (c) low (b)

49. Load factor for a power system serving residential area compared to highly industrial area is:
(a) high (b) same (c) low (c)

50. If load factor is high, the unit cost of energy is comparatively:
(a) high (b) low (c) same (b)

51. With high load factor the power system is operating at:
(a) near capacity (b) near best capacity (c) idle for large part of time (a) & (b)

52. During high and low stream flow, the functions of thermal and hydro plants are sometimes:
(a) the same (b) reversed (b)

53. The optimum point of operation in a combined system usually occurs when water power capacity is somewhere between ________ of total capacity.
(a) one-half to one-third (b) one-third to two-third
(c) one-half to three-fourth (b)

54. A system supplying hydro power alone is at a disadvantage if its carry-over storage is _____ to meet demand of a severe drought
(a) adequate (b) inadequate (c) optimum (d) surplus (b)

55. The peak load of a 40 MW capacity power plant is 30MW. The loads having maximum demands of 25, 10, 5 and 7 MW are connected to the system. The diversity factor of the plant is ____________ . (1.57)

56. In (3) if the load factor for the year is 50%, the energy supplied by the system is ____________ kWh/year. (205.86×10^6

57. The fact that a conductor carries more current on the surface as compared to core is known as fault.
(a) skin effect (b) corona
(c) permeability (d) unsymmetrical fault (a)

58. Presence of ozone as a result of corona is harmful because:
(a) it gives bad odour (b) it corrodes the material
(c) it transfers energy to the ground (d) reduces power factor (b)

59. The function of steel wire in a ACSR conductor is:
(a) take care of surges (b) prevent corona
(c) reduce inductance (d) provide additional mechanical stength (d)

60. A 30 km transmission line carrying power at 33 kV is known as:
(a) short transmission line (b) medium transmission line
(c) high power line (d) ultra high voltage line (b)

61. The permissible voltage variation in a distribution system at consumer end is:
(*a*) ± 5% (*b*) ± 6%
(*c*) ± 7% (*d*) ± 10%

62. The sag of a transmission line is least affected by:
(*a*) self weight of conductors (*b*) temperature of surrounding air
(*c*) current through conductor (*d*) ice deposited on conductor (*c*)

63. A pumped storage hydroelectric plant is a:
(*a*) high head plant (*b*) run-off river plant
(*c*) base load plant (*d*) peak load plant (AMIES' 01) (*d*)

64. The correct sequence, in the direction of flow of water, of installations in a hydro power plant is:
(*a*) reservoir, penstock, turbine, surge tank
(*b*) reservoir, surge tank, penstock, turbine
(*c*) reservoir, penstock, surge tank, turbine
(*d*) reservoir, surge tank, turbine, penstock (AMIES '01) (*b*)

65. Surge tank in hydropower plants are provided for:
(*a*) controlling surge in river (*b*) controlling water hammer in penstock pipes
(*c*) constant generation of power (*d*) removal of silt from reservoir(AMIES '98) (*b*)

66. In hydel installation power generation is calculated on the basis of:
(*a*) gross head (*b*) rated head
(*c*) design head (*d*) net head (AMIES '97) (*d*)

67. In a hydro power scheme, the pipe between tank and power house is known as .... (AMIES '01) (Penstock)

68. Run off river plant is one which:
(*a*) has a limited pondage (*b*) has no pondage at all
(*c*) is a high head plant (*d*) is a pumped storage plant (AMIES '99) (*a*)

69. A medium head hydroelectric scheme is one where:
(*a*) gross head is between 15 to 60 m
(*b*) net head operating on turbine is between 15 to 60 m
(*c*) total loss of head in penstock is between 15 to 60 m
(*d*) difference in level of tailwater and centre line of turbine is between 15 to 60 m. (*b*)
(AMIE W'95)

70. Match the following:

(*i*) spiral casing — (*a*) are employed to regulate flow into the runner and to direct the water onto the runner at an appropriate angle.
(*ii*) Guide vanes — (*b*) transforms the major part of the exit kinetic energy into pressure energy and thereby improves the turbine efficiency
(*iii*) Draft tubes — (*c*) transforms hydraulic energy into mechanical energy
(*iv*) Runner — (*d*) distributes the flow over the periphery of the runner so that the entrance velocity is same at inlet tips.
(*v*) Penstock — (*e*) the flow direction is changed from radial to axial direction.
(*vi*) Whirl chamber — (*f*) Water Conveying Conduit

(AMIE W' 01)

FLUID POWER ENGINEERING

1. HYDRAULIC TURBINES

1. Indicate which is applicable to a reaction turbine:
   (*a*) all the energy of water is converted into velocity before entering wheel
   (*b*) pressure of water is atmospheric
   (*c*) wheel must not run full (*d*) wheel must run full (*d*)
2. Indicate which are the characteristics of an impulse turbine:
   (*a*) all the energy of water is converted into velocity before entering the wheel
   (*b*) pressure of water is atmospheric
   (*c*) water is admitted only over the whole circumference
   (*d*) water is admitted over part of the circumference (*a*), (*b*) and (*d*)
3. Specific speed of a turbine is:
   (*a*) the speed which develops unit power under a unit head
   (*b*) the speed which develops unit power under a unit discharge (*a*)
4. Indicate type of turbine suitable for a specific speed of 1000 and head 50 m
   (*a*) Francis (*b*) Pelton (*c*) Kaplan (*d*) Bulb (*c*)
5. Indicate type of turbine suitable for a specific speed of 1000 and head 10 m.
   (*a*) Francis (*b*) Pelton (*c*) Kaplan (*d*) Bulb (*d*)
6. What type of turbine is suitable for specific speed of 30 and head 2000 m
   (*a*) Francis (*b*) Pelton (*c*) Kalpan (*d*) Bulb (*b*)
7. Under part load condition indicate the type of turbine which has minimum efficiency:
   (*a*) Francis (*b*) Pelton (*c*) Kaplan (*d*) Propeller (*d*)
8. The specific speed of a Francis turbine ranges between:
   (*a*) 50 – 250 (*b*) 70 – 350 (*c*) 80 – 400 (*d*) 100 – 700 (*c*)
9. Cavitation in a turbine causes:
   (*a*) low efficiency (*b*) blade surface is damaged
   (*c*) vibration and noise (*d*) all (*d*)
10. Watertight casing is required in–turbine:
    (*a*) Francis (*b*) Pelton (*c*) Kalpan (*d*) Propeller (*b*)
11. Which turbine has the following characteristics:
    (*a*) water enters over a part or whole of the circumference of the wheel
    (*b*) flow regulation is done by needle valve
    (*c*) installed above tail race level (*d*) watertight casing is required (*impulse*)
12. The power obtained from a turbine shaft–power supplied by water at entry to turbine is:
    (*a*) mechanical losses (*b*) hydraulic losses
    (*c*) mechanical + volumetric losses (*d*) hydraulic + mechanical losses (*d*)
13. The turbine with the following characteristics is called ________
    (*a*) water is admitted over whole circumference of the wheel
    (*b*) water tight casing is not required
    (*c*) unit may be installed above or below the tail race level
    (*d*) flow regulation is done by a guide vane assembly (*Reaction*).
14. The effective head at a turbine is:
    (*a*) gross head-head loss in penstock
    (*b*) gross head-head loss in penstock - velocity head at turbine exit
    (*c*) none of the above (*b*)
15. Specific speed of a turbine is given by the relation:
    (*a*) $\frac{N\sqrt{P}}{H^{5/4}}$ (*b*) $\frac{N\sqrt{P}}{H^{3/4}}$ (*c*) $\frac{N\sqrt{P}}{H^{1/4}}$ (*d*) $\frac{N\sqrt{P}}{H^{1/2}}$ (*a*)

16. The power in specific speed equation is:
(*a*) water power (*b*) power in kW
(*c*) shaft power (*d*) none of these (*c*)
17. Draft tube in a reaction turbine serves to:
(*a*) Reconvert kinetic energy into flow energy
(*b*) increase flow rate
(*c*) safety to turbine (*d*) none of these (*d*)
18. At part load–operation, the turbine most efficient is
(*a*) Francis (*b*) Pelton (*c*) Propeller (*d*) Kaplan (*d*)
19. At part gate, the turbine with high relative efficiency efficient is, give reason
(*a*) Francis (*b*) Pelton (*c*) Propeller (*d*) Kaplan
(*b*) due to absence of shock losses
20. Unit power in a turbine is:
(*a*) $\frac{P}{H^{1/2}}$ (*b*) $\frac{P}{H}$ (*c*) $\frac{P}{H^{3/2}}$ (*d*) $\frac{P}{H^{3/4}}$ (*c*)
21. The water flow in an inward flow reaction turbine:
(*a*) enters at outer periphery of wheel and then flows towards the centre of the wheel
(*b*) enters at centre of wheel and then flows towards outer periphery of the wheel
(*c*) none of these (*a*)
22. State to which type of turbine it is applicable
(*a*) simple in construction
(*b*) not so likely to become clogged in operating
(*c*) load efficiency curve has a relatively sharp peak
(*d*) economical (*e*) total weight is considerably less
(Fixed-blade propeller)
23. Indicate the type of installation is it:

	Head (m)	*Speed ratio*	N_S
(*a*)	475	0.44	15
(*b*)	10	6	1000

(*a*) High head-impulse turbine (*b*) Low head–Bulb turbine
24. Indicate which is not relevant to economical (low cost) installation:
(*a*) relatively high head and small discharge
(*b*) penstock location ideal and quick drop near power plant site
(*c*) few large wheel units
(*d*) long transmission lines
(*d*) high load factor, operating in large power system (*d*)
25. Indicate whether the installation is high cost or low cost:
(*a*) large discharge (*b*) relatively low head
(*c*) variable flow (*d*) poor dam foundation
(*e*) large number of units (*f*) long transmission line (*High cost*)
26. Indicate type of turbine which has:
(*a*) entire flow from headwater to tailwater takes place in a closed conduit system
(*b*) only part of available head is converted into kinetic energy, a substantial part remaining in pressure head
(*c*) flow resembles that in a closed conduit pipe (*Reaction*)
27. Higher the sepecific speed of a hydraulic turbine for a given head and horse power :
(*a*) larger is the size of turbine (*b*) smaller is the size of turbine
(*c*) higher the cost of installation (*d*) lower the cost of installation (*b*) and (*d*)

28. **Higher the specific speed of a hydraulic turbine for a given head and horse power:**
    (*a*) smaller is the size of generator (*b*) lower is the cost installation
    (*c*) larger is the size of generator (*d*) higher is the cost of installation (*a*) and (*b*)
29. **The by-pass of flow from turbine to tailrace is done by means of:**
    (*i*) Impulse turbine (*a*) pressure relief valve needle nozzle supplemented
    (*ii*) Reaction turbine (*b*) by jet deflection or auxiliary nozzle
    [(*i*) – *b*] [(*ii*) – *a*]
30. **Reaction turbines are equipped with spiral or volute casings made of True or false**
    Higher head cast steel
    Moderate heads steel plate
    Low heads concrete
    Lowest heads open flume setting True
31. **Reaction turbine may have flow through runner**
    (*a*) inward (*b*) outward
    (*c*) axial (*d*) all of these (*d*)
32. **Indicate type of turbine in which energy:**
    (*a*) all available head is converted into kinetic or velocity head
    (*b*) water flows with free surface like that in an open channel
    (*c*) water is in contact with air
    (*d*) housing provided is to prevent splashing and guide the discharge
    (*e*) housing contains air at substantially atmospheric pressure (*Impulse turbine*)
33. **Which turbine is used for:**
    (*a*) high-head installation (*b*) water contains silt
    (*c*) maintain good efficiency (*d*) low range of specific speed (*impulse*)
34. **Pressure in penstock varies from minimum to maximum :**
    (*a*) at junction with scroll casing to upstream end
    (*b*) at upstream end junction with scroll casing
    (*c*) at sharp drop to turbine end (*b*)
35. **Indicate the unit of specific P in speed of turbine:**
    (*a*) $n_s = \dfrac{n\sqrt{P \times 1.358}}{H^{5/4}}$ (*b*) $n_s = \dfrac{n\sqrt{P}}{H^{5/4}}$ (*a*) kW, (*b*) H.P.
36. **Indicate the type of casing used for gross head:**
    (*a*) up to 40 m (*b*) above 30 m (*a*) concrete (*b*) steel
37. **In Hydel water conductor system, turbine is essentially provided with pressure relief valve to limit pressure rise in:**
    (*a*) Kaplan (*b*) Francis (*c*) Pelton
    (*a*) and (*b*)
38. **Cavitation and serious damage is comparatively much serious in the turbine:**
    (*a*) Kaplan (*b*) Francis (*c*) Pelton (*a*) and (*b*)
39. **Which is quicker for replacement, spear and nozzle or guide vanes in turbines:**
    (*a*) Pelton (*b*) Kaplan (*c*) Francis (*a*)
40. **Runner is essentially transported as one piece of turbine:**
    (*a*) Pelton (*b*) Francis (*c*) Kaplan (*d*) Propeller (*a*)
41. **In a power plant turbine setting comes to be low for cavitationfree runner at a given specific speed. What remedy do you suggest so that construction of the plant does not turn out to be uneconomical:**
    (*a*) specific speed be reduced by decreasing speed of rotation
    (*b*) specific speed be increased by increasing speed of rotation (*a*)

42. Indicate type of joint provided in power plant structure:
(*a*) to eliminate tensile stresses due to shrinkage
(*b*) to eliminate compressive stresses
(*c*) to prevent transferring of stresses (*a*) contraction (*b*) separation
43. Necessity of limiting pressure rise with long water conductor system by using pressure release valve is not required in:
(*a*) Kaplan turbine (*b*) Francis turbine
(*c*) Pelton turbine (*d*) none of these (*c*)
44. Pick up correct answer
Specific speed of a Pelton wheel turbine is:
(*a*) below 60 (*b*) above 60
(*c*) above 60 but below 100 (*d*) above 100 but below 300 (AMIE)
(*c*) below 65 m (As per Bis)
45. Kaplan turbine is an axial flow turbine which requires water head:
(*a*) below 50 m (*b*) below 50 m but above 40 m
(*c*) below 100 m but above 60 m (*d*) below 300 m but above 100 m (AMIE)
(*a*) 10-60 m
46. Efficiency of a Pelton turbine at constant head________with increase in power.
(*a*) remains constant (*b*) decreases
(*c*) increases (*d*) first increases and then decreases (*c*)
47. Maximum efficiency of a centrifugal pump is invariant with:
(*a*) power (*b*) head
(*c*) discharge and speed (*d*) none of the above.
(*a*) Since maximum efficiency at a given speed occurs at a some fixed discharge for a given speed of operation there is a particulars discharge (or head) where efficiency is maximum.

Examine the statement and indicate whether True or False :

48. The governing of turbine is the process of regulating flow rate.
49. A braking jet directs a jet of water at the front centre of the bucket.
50. A draft tube increases the efficiency of a reaction turbine by increasing flow rate.
51. In a reaction turbine no work is done by guide vavies.
52. A mixed flow reaction turbine has partly radial and partly axial flow.
(*i*) True, (*ii*) False; Bracking directs jet of water on the back of buckets, (*iii*) False; draft tube reconverts kinetic energy to flow energy to increase efficiency of the turbine, (*iv*) False; gruide vanes are provided around the runner to regulate the water flowing through the turbine, (*v*) True, flow through Francis runner is at first inwards in the radial direction, gradually changing to axial.
53. The best r.p.m. for highest efficiency of a turbine approximately corresponds to a setting ___________of gate opening:
(*a*) 75% (*b*) 90% (*c*) 95% (*d*) 100% (*a*)
54. The full-load point of a turbine corresponds to a point of:
(*a*) maximum efficiency (*b*) best speed
(*c*) maximum efficiency under best speed (*c*)
55. Opening of wheel gate beyond full-load point under normal speed means development of:
(*a*) less power (*b*) more power (*c*) no change (*b*)
56. Full gate opening is defined as the point of:
(*a*) maximum power under maximum speed
(*b*) maximum power under normal speed
(*c*) optimum power under optimum speed (*b*)

57. High head installation is well adapted for a wheel of:
(*a*) high specific speed (*b*) low specific speed
(*c*) medium specific speed (*b*)

58. Low head installation is well adapted for a turbine of:
(*a*) high speed runner (*b*) low speed runner (*c*) medium speed runner (*a*)

59. Power plant used to take peak load has:
(*a*) high load factor (*b*) low load factor (*c*) average load factor (*b*)

60. For the power plant run to its full capacity for maximum period of operation to give high load factor, the production cost of electric energy is:
(*a*) high (*b*) low (*c*) moderate (*d*) exhorbitant (*b*)

61. In hydel installations power generation is calculated on the basis of:
(*a*) gross head (*b*) rated head (*c*) design head (*d*) net head (*d*)

62. Match the sentences of column A to that in column B

Column A	Column B
(*a*) hydraulic jump	(*A*) spillway
(*b*) seepage	(*B*) hydroelectric project
(*c*) penstock	(*C*) earth dam

(*a*) → (*A*), (*b*) → (*C*), (*c*) → (*B*)

63. Match in pairs:

(*i*) impulse turbine	(*a*) positive displacement
(*ii*) centrifugal pump	(*b*) spear governor
(*iii*) gear pump	(*c*) a constant rate discharge pump
(*iv*) reciprocating pump	(*d*) flow character
(*v*) Reynolds number	(*e*) air vessel

(*i*) → (*b*), (*ii*) → (*c*), (*iii*) → (*a*), (*iv*) → (*e*), (*v*) → (*d*)

64. Fill in the blanks:

(*i*) The efficiency of a pelton turbine is ________ than that of a kaplan turbine. (less)

(*ii*) The bulb turbines are used for ________ head plants. (low)

(*iii*) In an impulse turbine ________ is employed. (velocity or impulse)

(*iv*) In a reaction turbine ________ is employed. (pressure or water head)

(*v*) Francis turbine is a ________ turbine. (reaction)

(*vi*) The ratio of shaft power to ________ is known as overall efficiency of a turbine. (water power)

(*vii*) The speed of a turbine working under ________ is known as unit speed. (in head)

(*viii*) Impulse is equal to change of ________ (momentum)

(*ix*) The specific speed of turbines is ________ in different systems of units. (different)

(*x*) With increase in unit speed, the discharge in case of Francis turbine ________. (decreases)

(*xi*) Kaplan turbine is a ________ turbine. (reaction)

(*xii*) Hydraulic turbine is a device in which fluid power is converted to ________ power. (mechanical)

(*xiii*) Kaplan turbines are axial-flow turbines with ________ blades. (movable)

(*xiv*) High velocity in runner. Means rapid wear, high friction losses and high pressure making castings difficult to fabricate.

(*xv*) A Francis turbine is not suitable for high heads, because ________.

(*xvi*) A turbine produces 300 kW under a head of 30 m. Its discharge with an overall efficiency of 80% is ________. (1.27 cumecs)

65. Indicate the appropriate choice:
(*a*) specific speed indicates (*b*) type and shape of turbine
(*c*) speed of rotation of turbine (*d*) size of the machines (*a*) – (*b*), (*c*) – (*d*)
66. The specific speeds of Kaplan, Francis and Pelton turbines are in:
(*a*) the increasing order (*b*) the decreasing order (*b*)
(*c*) neither increasing nor decreasing order
67. Higher specific speed of turbine (300 to 1000) indicates that turbine is
(*a*) Pelton (*b*) Kaplan (*c*) Francis
(*d*) any one of above (*e*) none of the above (*b*)
68. An impulse turbine:
(*a*) operates submerged (*b*) requires draft tube
(*c*) is most suited for low head applications
(*d*) Operates by initial complete conversion to kinetic energy (*d*)
(*e*) is not exposed to atmosphere
69. Runaway speed of a hydraulic turbine is:
(*a*) the speed if the runner of the turbine is allowed to revolve freely without any load and with the wicket gates wide open
(*b*) the speed at which the turbine runner is damped
(*c*) the speed at which maximum overload
(*d*) full load speed (*a*)
70. Francis, kaplan and propeller turbines fall under the category of:
(*a*) impulse turbine (*b*) reaction turbine
(*c*) impulse reaction combined (*d*) axial flow (*b*)
71. For low head and high discharge the suitable turbine is:
(*a*) Pelton (*b*) Francis (*c*) Kaplan (*d*) none of these (*c*)
72. Runaway speed of a pelton turbine means the speed at:
(*a*) no load (*b*) full load
(*c*) no load with no governor mechanism (*d*) none of the above (*c*)
73. Operating characteristic curves of a turbine means the curves drawn at constant
(*a*) speed (*b*) efficiency (*c*) head (*d*) none of these (*d*)
74. The units of specific speed of turbine are:
(*a*) RPM (*b*) no units
(*c*) rev./minutes kwm$^{1/2}$ (*d*) none of these (*a*)
75. For part load operation, the turbine recommended is
(*a*) Kaplan (*b*) Francis
(*c*) Pelton (*d*) Propeller (*a*)
76. The value of speed ratio for Francis turbine is about:
(*a*) 2 (*b*) 1 (*c*) 0.5 (*d*) 0.9 (*d*)
77. For maximum hydraulic efficiency of a Pelton impulse turbine, the speed ratio should be:
(*a*) < 0.5 (*b*) $= 0.5$ (*c*) > 0.5 (*d*) none of these (*a*)
78. The use of draft tube in a reaction turbine helps to:
(*a*) increase the flow rate (*b*) provide safety to turbine
(*c*) transport water to downstream without eddies
(*d*) reconvert residual kinetic energy to pressure energy (*d*)
79. High specific speed (300 to 1000) and low head (below 30 m) indicate that the turbine is:
(*a*) Pelton (*b*) Franics (*c*) Propelles (*d*) Kaplan (*d*)
80. The net available head in a Pelton turbine installation is the:
(*a*) kinetic energy of the jet issuing from the nozzle
(*b*) difference in elevation between forebay water level and the nozzle outlet
(*c*) head at the base of the nozzle

(*d*) difference in level between water levels at the forebay and the tailwater level (*a*)

81. A hydro turbine rotates at N r.p.m. operating under a net head H and having a discharge Q while developing an output power P. The specific speed is expressed as:
(*a*) $\frac{NH^{5/4}}{P^{1/2}}$ (*b*) $\frac{NP^{1/2}}{\sigma H^{5/4}}$ (*c*) $\frac{NP^{1/2}}{H^{5/4}}$ (*d*) None of these (*c*)

82. Pelton turbine is essentially a:
(*a*) mixed flow turbine (*b*) radial flow turbine
(*c*) axial flow turbine (*d*) tangential flow turbine (*c*)

83. With increase in load on the generator, the speed of the turbine should:
(*a*) increase (*b*) decrease (*c*) remain same (*b*)

84. A turbine produces 7500 kW at 125 r.p.m. under a head of 25 m. If head is reduced to 20 m, it will run at:
(*a*) 168.8 r.p.m. (*b*) 111.8 r.p.m. (*c*) 120.7 r.p.m. (*d*) 150 r.p.m.

85. Specific speed indicates:
(*a*) type and shape of turbine (*b*) speed of rotation of turbine
(*c*) size of machine (*a*)

86. For the same power, the size of a turbine:
(*a*) increase with speed (*b*) remains same with the speed
(*c*) decreases with speed (*c*)

87. Higher specific speed of turbine (300 to 800) indicates turbine is:
(*a*) Pelton (*b*) Kaplan (*c*) Francis
(*d*) any one of above (*d*) none of above

88. An adjustable blade propeller turbine is called a:
(*a*) Pelton turbine (*b*) Kaplan turbine
(*c*) Francis turbine (*d*) higher head turbine (*e*) none of them. (*b*)

89. Regulation of a Pelton turbine is done by changing:
(*a*) head available at nozzle (*b*) area of nozzle
(*c*) velocity of flow from nozzle (*d*) blade angle
(*e*) length of the nozzle (*b*)

90. Unit power *Pu* of a turbine developing a power *P* under a head *H* is:
(*a*) $P/H^{5/2}$ (*b*) $P/H^{1/2}$ (*c*) $PH^{3/2}$ (*d*) $P/H^{3/2}$ (*d*)

91. Kaplan turbine is
(*a*) low head axial flow turbine (*b*) higher head axial flow turbine
(*c*) high head mixed flow turbine (*d*) mixed flow impulse turbine (*a*)

92. Francis turbine is Essentially a:
(*a*) mixed flow turbine (*b*) radial flow turbine
(*c*) axial flow turbine (*d*) tangential flow turbine (*a*)

93. Reaction turbines are used for:
(*a*) low head only (*b*) high head only
(*c*) high head and low discharge (*d*) low head and high discharge (*d*)

94. A draft tube is used with:
(*a*) Pelton wheel only (*b*) Francis turbine only
(*c*) Francis and Kaplan turbines (*d*) any turbines (*c*)

95. In general sense, water turbines may be put in the following decreasing order of specific speed as:
(*a*) Propeller, Francis, Pelton (*b*) Pelton, Francis, Francis
(*c*) Kaplan, Impulse, Francis (*d*) Francis, Kaplan, Pelton (*a*)

96. The turbines in which the general direction of flow is parallel to the turbine axis are called:
(*a*) axial flow turbines (*b*) radial flow turbines
(*c*) tangential flow turbines (*d*) none of the above (*a*)

97. A turbine produces 7500 kW at 135 r.p.m. under a head of 25 m. If head is reduced to 20m, it will run at
(*a*) 168.8 r.p.m. (*b*) 108 r.p.m. (*c*) 120.7 r.p.m. (*d*) 150.9 r.p.m.
$[135\,(20/25)^{1/2}]$ (*c*)

98. Specific speed of a pelton wheel turbine is:
(*a*) below 60 (*b*) above 60
(*c*) above 60 but below 100 (*d*) above 100 but below 300 (*a*) below 65

99. Kaplan turbine is an axial flow turbine which requires water head:
(*a*) below 50 m (*b*) below 50 but above 40 m
(*c*) below 100 but above 60 m (*d*) below 300 but above 100 m (*a*) below 60

100. An impulse turbine
(*a*) always operates submerged (*b*) makes use of draft tube
(*c*) is most suited for low head installations
(*d*) operates by initial complete conversion to kinetic energy (*d*)

101. Francis turbine is:
(*a*) an impulse turbine (*b*) an axial flow turbine
(*c*) a reaction turbine (*d*) a tangential flow turbine (*c*)

102. Unit power of a turbine is given by:
(*a*) $\frac{P}{H^{1/2}}$ (*b*) $\frac{P}{H^{1/4}}$ (*c*) $\frac{P}{H^{3/4}}$ (*d*) $\frac{P}{H^{3/2}}$ (*d*)

103. Francis, Kaplan and propeller turbines fall under the category of:
(*a*) impulse turbine (*b*) reaction turbine
(*c*) impulse reaction combined (*d*) axial flow (*b*)

104. The maximum head of a Kaplan turbine is limited to:
(*a*) 25 m (*b*) 70 m (*c*) 125m (*b*) 60 m

105. Unit speed Nu is given by
(*a*) N/H (*b*) N/H^2 (*c*) $N/H^{1/2}$ (*c*)

106. Match the Pairs :
1. (*i*) Impulse turbine (*a*) Positive displacement
(*ii*) Centrifugal pump (*b*) Spear governor
(*iii*) Gear pump (*c*) A constant rate discharge pump
(*iv*) Reciprocating pump (*d*) Flow character
(*v*) Reynolds number (*e*) Air vessel
[(*i*) → (*b*)], [(*ii*) → (*c*)] [(*iii*) → (*a*)] [(*iv*) → (*e*)] [(*v*) → (*d*)]

107. Indicate True or False :
(*i*) The govering of turbine is the process of regulating flow rate. (True)
(*ii*) A braking jet directs a jet of water at the front centre of the bucket. (False)
(*iii*) A draft tube increases efficiency of a reaction turbine by increasing flow rate. (False)
(*iv*) In a reaction turbine no work is done by the guide vanes. (False)
(*v*) A mixed flow reaction turbine has partly radial and partly axial flow. (True)
(*vi*) Pelton turbine is a reaction turbine. (False)
(*vii*) Kaplan turbine is an axial flow turbine. (True)
(*viii*) Cavitation is defined as the phenomenon of formation of vapour bubbles and sudden collapsing of the vapour bubbles. (True)

(*ix*) Francis turbine is outward radial flow turbine. (False)
(*x*) Ratio of jet of water diameter to Pleton turbine diameter is known as jet ratio. (True)
(*xi*) Pelton is tangential flow Turbine. (True)
(*xii*) Pressure head is converted into kinetic energy in impulse turbine. (True)
(*xiii*) Draft tubes are employed in impulse turbines. (False)
(*xiv*) Pelton wheel turbine is a reaction turbine. (False)
(*xv*) Francis is inward mixed flow reaction turbine. (True)
(*xvi*) A high head turbine has high specific speed and speed of rotation. (False)

108. The specific speed as is usually expressed in turbine practice is a dimensionless number. (False)
109. In Kaplan turbine blade angles are adjusted according to the load on the turbine. (True)
110. The speed of a turbine working under ______________ is known as unit speed. (1 m Head)
111. Kaplan turbines are axial-flow turbines with ____________ blades. (Adjustable)
112. In Francis turbine, the discharge ____________ with increase of speed. (Increase)
113. A turbine is called reaction turbine if at the inlet of turbine the total energy is:
(*a*) kinetic (*b*) pressure
(*c*) kinetic and pressure (*d*) none of the above (*c*)
(AMIE Turbomachinery W' 01)
114. Jet ratio (*m*) is defined as the ratio of:
(*a*) Diameter of jet of water to diameter of Pelton wheel
(*b*) Velocity of vane to velocity of jet of water
(*c*) Velocity of flow to velocity of jet of water
(*d*) Diameter of pelton wheel to diameter of jet (*a*)
115. For high head and low discharge, the suitable turbine is:
(*a*) Pelton (*b*) Francis
(*c*) Kaplan (*d*) Propeller (*a*)
116. Cavitation will take place if the pressure of the flowing fluid at any point is:
(*a*) equal to vapour pressure of the fluid
(*b*) less than the vapour pressure of the fluid
(*c*) more than the vapour pressure of the fluid
(*d*) None of the above (*a*)
117. Turbine run at constant speed. The speed is kept constant by varying:
(*a*) load (*b*) discharge
(*c*) net head (*d*) inlet valve (AMIE W '01) (*b*)
118. A turbine produces 500 kW under net head of 30 m. The discharge in cumecs, when the overall efficiency is 85% is:
(*a*) 1 (*b*) 2
(*c*) 3 (*d*) 4 (AMIE W '01) (*b*)
119. In .... turbines, all available energy is converted into kinetic energy. (Impulse)
120. For maximum operating efficiency it is essential to properly maintain:
(*a*) turbine
(*b*) runner seal clearances within allowable tolerances
(*c*) runner so that it is not pitted
(*d*) guide vane clearances at vane ends to reduce wicket-gate leakage.
(*e*) turbine and generator systems are in alignment
(*f*) generator is not oil leaking and no mechanical damage to insulation
(*g*) trashrack is clear of debris, ice etc.
(*h*) all of these (*h*)

121. Hydraulic turbines testing for their performance/efficiency includes net head measurement at wheel case, turbine discharge measurement, turbine, output measurement when the turbines are:
(a) new (b) at intervals during their life
(c) none of them (d) both of them (d)

2. CENTRIFUGAL PUMPS

1. The size of foot valve and reflux valve should be equal to the size of:
(a) delivery pipe (b) suction pipe (b)
2. The size of suction and delivery pipes should be selected so that friction head loss does not exceed ________ percent of total equivalent length of piping system up to delivery pipe
(a) 5% (b) 10% (c) 15% (d) 20% (b)
3. The pump should operate at maximum pump efficiency during peak demand in the range of head and discharge in ______ season:
(a) winter (b) summer (c) rainy (b)
4. If delivery offset distance is more than 3 m, the size of pipe should be ______ to reduce friction losses
(a) smaller (b) larger (b)
5. Foot valve is not required in the case of:
(a) submersible pump (b) centrifugal pump (a)
6. In a centrifugal pump, non-return valve is required on delivery side, if total head is more than:
(a) 10.3 m (b) 15m (c) 20m (d) 25m (d)
7. Suction pipe is not required in the case of:
(a) submersible pump (b) centrifugal pump
(c) reciprocating pump (a)
8. While computing frictional losses, an additional pipe length of ______ m is provided over and above pipe length equivalent to static head to take into account submergence of suction pipe, offset pipe length of suction and delivery sides and bends
(a) 5 (b) 10 (c) 12 (d) 15 (b)
(Its components are submergence of suction pipe (1 m), offset pipe-length of suction and delivery side (5 m) and 2 bends (4 m equivalent pipe length).
9. Final total head is multiplied by ____________ in the case of old/rusted G.I. pipes
(a) 1.0 (b) 1.05 (c) 1.10 (d) 1.15 (b)
10. Indicate type of pump which has the following characteristics:
(a) economical in capital cost, installation and maintenance
(b) fewer moving parts (c) operated at very high speed
(d) compact and less space required *(Reciprocating pump)*
11. For lifting muddy water, the pumpset suitable is:
(a) reciprocating (b) centrifugal (b)
12. For lifting muddy water, the impeller used in a centrifugal pump is:
(a) open (b) closed (c) semi-open (a)
13. In a centrifugal pump a regulating valve is provided to regulate the flow in:
(a) suction pipe (b) delivery pipe (b)
14. Casing in a centrifugal pump serves to:
(a) guide water to and from impeller (b) as part of suction pipe
(c) as part of delivery pipe (a)
15. Electric motor of a centrifugal pump is started with delivery valve:
(a) open (b) closed (b)

16. In a centrifugal pump when impeller attains normal speed and water is drawn in, the delivery valve is:
    (*a*) open (*b*) closed (*a*)
17. Specific speed of a centrifugal pump is the speed of a centrifugal pump which delivers:
    (*a*) unit quantity of water against a unit head
    (*b*) unit discharge against a unit power (*a*)
18. The suction lift of a centrifugal pump can be up to
    (*a*) 10.28 m (*b*) 12 m (*c*) 15 m (none)
19. The centrifugal pumps are arranged in series to:
    (*a*) deliver large quantity of water against low load
    (*b*) deliver small quantity of water against very high head
    (*c*) deliver large quantity of water against high head
    (*d*) deliver small quantity of water against low head (*b*)
19. A centrifugal pump can be arranged to deliver large quantity of water against relatively small head:
    (*a*) in series (*b*) in parallel (*b*)
20. A centrifugal pump does not work to its optimum capacity and pressure if:
    (*a*) not properly primed (*b*) leakage of air into the pump (*b*)
21. A centrifugal pump has low-efficiency if:
    (*a*) speed is very high (*b*) speed is very low
    (*c*) head is very high (*a*)
22. Indicate appropriate choice:
    (*i*) The low specific speed (10 m) indicate that the pump is:
    (*a*) axial flow (*b*) radial type (*c*) mixed flow type (*b*)
    (*ii*) An axial flow pump is started with discharge valve:
    (*a*) fully closed (*b*) fully open (*c*) partly closed (*b*)
    (*iii*) Higher specific speed (161 to 500) of centrifugal pump indicates that the pump is:
    (*a*) axial flow (*b*) radial type (*c*) mixed flow type
    (*d*) any one of the above (*e*) none of the above (*a*)
    (*iv*) If pump NPSH requirements are not satisfied:
    (*a*) it will not develop head (*b*) it will cavitate
    (*c*) efficiency will be low (*d*) will consume excessive power
    (*e*) no flow will take place (*b*)
    (*v*) The pressure at which separation takes place is known as separation pressure. For water, the limiting value of separation pressure head is:
    (*a*) 2.5 m (abs.) (*b*) 7.5 m (abs.) (*c*) 10.3 m (abs.) (*d*) 13.3 (abs.) (*c*)
23. If speed N of centrifugal pump is increased, the power requirement:
    (*a*) remains constant (*b*) $\propto$ N (*c*) $\propto N^2$ (*d*) $\propto N^3$ (*d*)
24. When a piping system is made up primairty of friction head and very little of vertical lift, then the pump characteristics should be ________ .
25. With increase in pump speed, the minimum NPSH ________ . (Decreases)
26. Ideally a centrifugal pump may be set at a maximum height of ________ m above the sump.
27. For an ideal contrifugal pump:
    (*a*) head varies inversely as speed (*b*) power varies as the square of speed
    (*c*) discharge varies directly with speed (*d*) discharge varies as square of speed. (*c*)
28. A gear pump is:
    (*a*) constant discharge pump (*b*) a variable discharge machine

(*c*) a true version of centrifugal pump (*d*) an air vessel (*a*)

29. A centrifugal pump has:
(*a*) constant head (*b*) constant speed (*c*) constant discharge (*c*)
(*d*) constant power requirement at all speeds

30. Starting of a centrifugal pump requires:
(*a*) priming (*b*) delivery valve fully open
(*c*) an air vessel (*d*) a surge tank (*a*)

31. Flood water is pumped by a:
(*a*) gear pump (*b*) reciprocating pump with multistaging
(*c*) sluice gate (*d*) centrifugal pump (*d*)

32. Specific speed of a pump is defined as the speed of a unit of:
(*a*) unit size with unit discharge at unit head
(*b*) such a size that it requires unit power for unit head
(*c*) such size that it delivers unit discharge at unit power

33. Cavitation is caused by:
(*a*) high velocity (*b*) low barometric pressure
(*c*) low pressure (*d*) high pressure (*c*)

34. Water is to be lifted by a net head of 180m. Idetical pumps with specific speed of 30 and rotative speed of 1450 r.p.m. capable of discharge 200 l/s are available. The number of pumps required is:
(*a*) 1 (*b*) 2 (*c*) 3 (*d*) 4 (*c*)

35. Higher specific speed (161 to 500) of centrifugal pump indicates that the pump is:
(*a*) radial type (*b*) axial flow (*c*) mixed flow type
(*d*) any one of the above (*e*) none of the above (*b*)

36. Cavitation in centrifugal pump can be reduced by:
(*a*) reducing discharge (*b*) reducing suction head
(*c*) increasing flow velocity (*d*) all the above methods (*b*)

37. In general, vanes of a centrifugal pump are:
(*a*) curved forward (*b*) curved backward (*c*) radial (*d*) twisted (*b*)

38. The operating characteristic curves of a centrifugal pump show that at constant speed, the mononetic head ________ as discharge increased. (Decreased)

39. Hydraulic similarities include geometric similarity, dynamic similarly and ________ similarity. (Kinematic)

40. Multistage centrifugal pumps are used to produce high ________. (Head or Discharge)

41. A centrifugal pump requires 10 kW when running at 1000 r.p.m. This pump when running at 2000 r.p.m. will require: ________
(*a*) 80 kW (*b*) 32 kW (*c*) 64 kW (*d*) $16\sqrt{2}$ kW (*b*)

42. A centrifugal pump has:
(*a*) constant head (*b*) constant speed (*c*) constant discharge
(*d*) constant power required at all speed

43. A fast centrifugal pump impeller will have:
(*a*) forward facing blades (*b*) radial blades
(*c*) backward facing blades (*d*) propeller type blades

44. Cavitation parameter σ is defined in terms of NPSH and net head H as σ =
(*a*) $\frac{NPSH}{\sqrt{H}}$ (*b*) $\frac{H}{NPSH}$ (*c*) $\frac{H}{\sqrt{NPSH}}$ (*d*) $\frac{NPSH}{H}$ (*d*)

45. A pump is required to 150 l/s at a head of 45 m when running at 1750 r.p.m. The specific speed of the pump is:
(*a*) 17.4 (*b*) 39 (*c*) 89 (*d*) 1233 (*b*)

46. Starting of a centrifugal pump requires:
(*a*) priming (*b*) delivery valve fully open
(*c*) air vessel (*d*) a surge tank (*a*)

47. Flood water is pumped by a:
(*a*) gear pump (*b*) reciprocating pump with multistage
(*c*) slince gate (*d*) centrifugal pump (*d*)

48. Flow through a centrifugal pump working at a constant head ______ with speed
(*a*) decreases (*b*) increases (*c*) no effect (*b*)

49. Viscerous fluid is pumped by:
(*a*) reciprocating pump (*b*) centrifugal pump
(*c*) screw pump (*d*) jet pump (*c*)

50. For pumping water from deep wells, which pump is usually employed:
(*a*) air lift pump (*b*) centrifugal pump
(*c*) reciprocating pump (*d*) jet pump (*a*)

51. In general vanes of a centrifugal pump are:
(*a*) curved forward (*b*) curved backward (*c*) radial (*d*) twisted (*b*)

52. For same efficiency, if the speed of a centrifugal pump is doubled, its power requirement increases by:
(*a*) 2 times (*b*) 4 times (*c*) 6 times (*d*) 8 times (*d*)

53. A centrifugal pump requires 16 kW when running at 1200 r.p.m. This pump when running at 2400 r.p.m. will require:
(*a*) 128 kW (*b*) 32 kW (*c*) 64 kW (*d*) $16\sqrt{2}$ kW (*a*)

54. Water is to be lifted to a height of 180 m. Identical centrifugal pump of specific speed 30, rotational speed of 1450 r.p.m. and discharging 200 lt/s are available. The number of pumps required will be:
(*a*) 2 (*b*) 4 (*c*) 1 (*d*) 3 (*d*, in series)

55. Maximum efficiency of a centrifugal pump is invariant with:
(*a*) power (*b*) head
(*c*) discharge and speed (designed) (*d*) none of the above (*c*)

56. A radial flow pump is started with the discharge valve:
(*a*) fully closed (*b*) fully open (*c*) partly closed (*a*)

57. Fill in the blanks:
(*i*) Vane pumps have ________ rotors. (external or internal)
(*ii*) Radial-flow pumps ________ be started with their delivery valves closed. (are to)
(*iii*) Blades of centrifugal pumps are-mostly ________ curved. (backward)
(*iv*) The upper limit of supply pressure of a positive-displacement pump is controlled by a ________ valve, attached to the pump. (sluice)
(*v*) Slip factor of a centrifugal pump is minimum when the vanes are ________. (surved backward)
(*vi*) The monometric efficiency of a centrifugal pump is given by ________. (H_{man} / head imparted by impeller to liquid).

58. State whether True or False:
(*i*) Work done by the centrifugal pump is proportional to the area of the indicator diagram. (False)
(*ii*) In a multistage pump, the discharge is as many times the number of stages. (True)

59. Specific speed of a pump is the speed at which a pump runs when:
(*a*) head developed is unity and discharge is one cubic metre
(*b*) head developed is unity and shaft horse power is unity

(c) discharge is one cubic metre and shaft horse power is unity
(d) none of the above (AMIE Turbomachinery W '01)

60. If a high head is to be developed, the impellers are connected in .... (AMIE W '01) (Series)

61. A centrifugal impeller with backward curved vanes has a constant head discharge characteristic. (False)

3. RECIPROCATING PUMPS

1. Indicate whatever is not applicable to a reciprocating pump
(a) low efficiency (b) discharge is dependent on its speed
(c) adapted to small capacities and high heads
(d) used in drilling operations (a)

2. Reciprocating pump is suitable for:
(a) high operating speeds (b) large capacities and large heads
(c) constant discharge without regard to pressure fluctuations
(d) large capacities and low heads (c)

3. Indicate whatever is applicable to a reciprocating pump:
(a) water in the beginning of delivery stroke, is retarded
(b) water at the end of delivery stroke is accelerated
(c) absolute pressure head at the end of delivery stroke is not less than vapour pressure to avoid separation
(d) absolute pressure head at the beginning of suction stroke is not below vapour pressure to avoid separation (c) and (d)

4. Air vessel is provided in a reciprocating pump to:
(a) obtain constant discharge at a uniform rate
(b) run pump at low speed to avoid separation
(c) increase work done
(d) economise in power needed to drive pump (a) and (d)

5. Discharge equation for a double-acting reciprocating pump is:
(a) $\frac{LAN}{60}$ (b) $\frac{2LAN}{60}$ (c) $\frac{3LAN}{60}$ (d) $\frac{4LAN}{60}$ (b)

Choose the appropriate answer:

6. The work saved by fitting an air vessel to a single acting reciprocating pump is:
(a) 28.9% (b) 39.2% (c) 48.8% (d) 84.8% (d)

7. If two pumps identical in all respects and each capable of delivering a discharge Q against a head H are connected in series, the resulting discharge:
(a) 2Q against a head 2H (b) 2Q against a head H
(c) Q against a head 2H (d) Q against a head H (c)

8. The pumps should be connected in ________ for discharging a large quantity of liquid. (Parallel)

9. The slip in case of reciprocating pumps is:
(a) +ve (b) −ve (c) zero (d) +ve or −ve (d)

10. Reciprocating pump is best suited
(a) where operating speeds are much higher
(b) for pumping large quantity of fluid at smaller head
(c) where constant head is required regardless of pressure fluctuations.
(d) where constant discharge is required regardless of pressure (d)

11. The speed of a reciprocating pump is determined from the consideration of:
(a) vibration (b) friction (c) flow separation (d) none of the above. (c)

12. Air vessels are used in reciprocating pumps mainly to:
   (*a*) reduce flow fluctuation (*b*) reduce the chance of separation
   (*c*) reduce energy lost due to friction
   (*d*) increase maximum speed without separation (*all*)
13. Air vessels are used in a reciprocating pump to:
   (*a*) obtain continuous supply of water at a uniform a rate
   (*b*) run the pump at higher speed without separation taking place
   (*c*) economise on power in driving it (*d*) all of the above (*d*)
14. Air vessel is provided in a reciprocating pump to:
   (*a*) increase delivery head (*b*) reduce suction head
   (*c*) reduce acceleration head (*d*) all of above (*c*)
15. Reciprocating pump is adapted to:
   (*a*) large discharge and large head (*b*) small discharge and small head
   (*c*) small discharge and higher head (*d*) large discharge and low head (*c*)
16. A rotary pump is a positive displacement pump with:
   (*a*) a circular motion (*b*) centrifugal motion (*c*) both of them (*a*)
17. Motion pump which imparts velocity to the water:
   (*a*) centrifugal pump (*b*) rotary pump (*a*)
18. The air vessel is placed in a reciprocating pump on:
   (*a*) discharge side (*b*) suction side (*c*) connected to cylinder
   (*d*) discharge and delivery sides close to cylinder (*d*)
19. The usual shape of an air veseel is a reciprocating pump is like:
   (*a*) a cylinder (*b*) cone (*c*) elongated circle (*d*) none of these (*b*)
20. The volume of air vessel in a reciprocating pump is–times the displacement of single direct acting pump
   (*a*) 4 to 8 (*b*) 5 to 8 (*c*) 6 to8 (*d*) 5 to 10 (*c*)
21. The air vessel is connected to the discharge pipe with neck down to:
   (*a*) reduce surface area of water in contact with the air
   (*b*) absorb more amount of air by water
   (*c*) absorb minimum amount of air by water
   (*d*) provide more area for air (*a*) and (*c*)
   (with absorption the vessel gradually fills up with water and is rendered ineffective in cushioning shock.)
22. In a reciprocating pump the valves usually provided in discharge pipe are:
   (*a*) check valve (*b*) gate valve (*c*) relief valve (*d*) all of them (*d*)
23. What are the factors which reduce the output of a fast running pump:
   (*a*) great speed (*b*) slip (*a*) and (*b*)
   (If speed is too great, water cannot flow through inlet valve fast enough to completely fill the cylinder).
24. The capacity of a reciprocating pump having voluntary opening valves is limited by:
   (*a*) pistion speed
   (*b*) number of times the valves will open and close smoothly and quietly
   (*c*) none of these (*b*)
25. The total head on a pump is:
   (*a*) head (*b*) lift (*c*) head & lift (*c*)
26. The discharge pipe of a rotary pump is carried through a riser–times the diamter:
   (*a*) 5 (*b*) 10 (*c*) 7 (*d*) 3 (*a*)
27. The discharge pipe is raised at discharge end to:
   (*a*) provide free flow (*b*) make flow visible

(*c*) prevent gas or air pockets in the pump and act as a seal in high vacuum service
(*d*) all of these (*c*)

28. Fill in the blanks:
(*i*) The saving in a work in a reciprocating pump by fitting air vessel is due to smoothening of ________ head. (acceleration/friction/both)
(*ii*) Percentage power saved by fitting air vessels is ________ in a single acting reciprocating pump than in a double acting pump. (more)
(*iii*) for cavitation free operation of a centrifugal pump, the available NPSH of the pump must be
(*a*) = required NPSH (*b*) < required NPSH (*c*) > required NPSH (*c*)
(*iv*) In two pumps identical in all respects, and each capable of delivering a discharge, Q against a head H are connected in parallel, the resulting discharge is:
(*a*) 2Q against a head 2H (*b*) 2Q against a head H
(*c*) Q against a head 2H (*d*) 2Q against a head $\sqrt{2}$ H (*b*)
(*v*) Manometric head of a centrifugal pump is given by:
(*a*) static head (*b*) static head + losses
(*c*) static head + losses + kinetic head at exit (*d*) none of these (*c*)
(*vi*) Discharge (theoretical) to run a reciprocating pump is given by:
(*a*) $\frac{wQ(H_s + H_d)}{60}$ (*b*) $\frac{wQ(H_s + H_d)}{75}$ (*c*) $\frac{wQ\,H_d}{75}$ (*d*) $\frac{wQ(H_s - H_d)}{75}$ (*b*)
(*vii*) A reciprocating pump is a:
(*a*) high speed pump (*b*) low speed pump
(*c*) low head pump (*d*) high discharge pump (*b*)

29. True or False :
(*i*) A gear pump can operate at higher pressures than an axial flow pump. (True)
(*ii*) To obtain a continuous supply of liquid at a uniform rate, air vessel is used in reciprocating pumps. (True)
(*iii*) To run a reciprocating pump at a higher speed without separation taking place, the delivery pipe should be
(*a*) horizontal from the pump and then vertical
(*b*) vertical from the pump and then horizontal (*a*)
(*iv*) In a reciprocating pump, the area of the indicator diagram is proportional to:
(*a*) discharge of pump (*b*) power to drive pump
(*c*) work done by pump (*d*) none of the above (*c*)

30. An air vessel used in a reciprocating pump is filled with:
(*a*) atmosphere air (*b*) compressed air (*b*)
(*c*) water (*d*) any compressed gas (AMIE W '01)

31. Examine the statement and indicate whether True or False:
A centrifugal impeller with backward curved vanes has a constant head-discharge characteristic. (AMIE W '01) (False)

4. MISCELLANEOUS HYDRAULIC MACHINES

1. A hydraulic ram can lift:
(*a*) large quantity of water to a great height (*b*) small quantity of water to a great height
(*c*) any quantity of water to any height (*d*) only heavy weight and not water
(AMIE) (*b*)
2. Hydraulic crane is a device which
(*a*) raises or transfers head loads (*b*) pumps water to a great height
(*c*) acts as a fire extinguisher (*d*) none of the above (AMIE) (*a*)

3. Hydraulic press is a machine for:
   (*a*) lifting large load through application of small force
   (*b*) applications requiring tremendous pressures rather than small or medium pressures
   (*c*) lifting smaller loads through smaller distances
   (*d*) none of these (*a*) and (*b*)
4. Hydraulic accumulator is a machine used for accumulating
   (*a*) hydraulic energy to be expended intermittently
   (*b*) kinetic energy (*c*) strain energy (*d*) none of these (*a*)
5. Hydraulic intensifier is a device used for:
   (*a*) converting a low hydraulic pressure into a high pressure
   (*b*) decreasing intensity of hydraulic pressure through energy available from large amount of water at a low pressure
   (*c*) increasing intensity of hydraulic pressure through energy available from large amount of water at a low pressure
   (*d*) none of these (*a*) & (*c*)
6. Hydraulic intensifier work with:
   (*a*) water supplied at low pressure operates a piston in a cylinder which in turn operates a ram of smaller diameter in a smaller cylinder
   (*b*) water supplied at high pressure operates a ram of large diameter
   (*c*) water supplied at low pressure operates a piston of small diameter which in turn operates a ram of large diameter
   (*d*) none of these (*a*)
7. Hydraulic ram is a pump which works on the principle of:
   (*a*) centrifugal action (*b*) reciprocating action
   (*c*) water hammer (*d*) none of the above (*c*)
8. True or False :
   (*i*) When there is an intermittent demand for power, an accumulator in the hydraulic system reduces the capacity of the power source. (True)
   (*ii*) The hydraulic ram works on the principle of water hammer. (True)
   A hydraulic ram can lift:
   (*a*) large quantity of water to a great height
   (*b*) small quantity of water to a great height
   (*c*) any quantity of water to any height
   (*d*) only heavy weight and not water (*b*)
   (*iii*) Hydraulic crane is a device which
   (*a*) raises or transfers head loads (*b*) pumps water to a great height
   (*c*) acts as a fire extinguisher (*d*) none of the above (*a*)
   (*iv*) An intensifier is used for:
   (*a*) increasing velocity (*b*) increasing pressure
   (*c*) decreasing velocity (*d*) decreasing pressure (*b*)
   (*v*) Which of the machines works on the princple of Pascal's law:
   (*a*) Jet pump (*b*) Air lift pump (*c*) Hydraulic ram (*d*) Hydraulic press (*d*)
9. Fill in the blanks:
   (*i*) Hydraulic intensifier is a device to increase __________ of water. (pressure/velocity/discharge)
   (*ii*) Product of pressure and __________ is known as capacity of hydraulic accumulator. (volume of accumulator)

BIBLIOGRAPHY

A. ARTICLES AND BOOKS

Abbet, Robert W. (1956). American civil engineering practice, Vol. I to III. New York: John Willey and Sons.

Aggarwal B.K., Singh M.P. Appraisal of Hydro-Electric Projects.

Aggarwal, C.K. (1999): Baspa II Hydroelectric project stage II (300 MW). First private hydro power project under construction.

Allende, J.L. (1978). Dam drainage techniques and skills, 10th Congress Irr. and Drg. trans. Vol. IV.

Arakeri, H.R., Donahue, Roy (1984). Principles of soil conservation and water management, M/s Oxford & IBH Co. New Delhi.

Arora K.R. (1996). Irrigation, water power and water resources engineering: Standard Publishers, distributors, 1705 B, Nai Sarak, Delhi-110006.

Arora S.C. Domkundwar (1992) : A Course on Power Plant Engineering, Dhanpat Rai & Sons, New Delhi.

Arun Kumar (1999): New technology for small hydro power projects.

Bharat singh. (1975). Irrigation Engineering. Roorkee: Nem Chand and Sons.

Biswas, K. (1980). Environmental assessment of hydel power projects. Bhagirath.

Churchill, M. A. (1948). Discussion of analysis and use of reservoir sedimentation data by Gattschalk. L. C. Proc. Federal Inter-Agency Sedimentation Conference, Washington. D. C.

Comrie, J. (1961). Civil engineering reference book, Vols., I, II, III and IV, London: Butterworths.

Creager, W. P., Justin, J. D. and Hinds, J. (1961). Engineering for dams, Vols. I, II and III. New York: John Wiley and Sons.

Dandekar M.M. and Sharma, K.N. (1979): Water power engineering M/s Vikas Publishing House Pvt. Ltd., New Delhi.

Davis, C.V. and Sorenson, K.E. (1969). A handbook of applied hydraulics. New York: McGraw--Hill Book Co.

Desai, S.K. & Dadal Raju (1999): New Austrian tunneling method (NATM) tunneling in Bhutan.

Diwatra E, Chalvaraj A.S. and Murthy G.N. (1974): Manual on design, fabrication, erection and maintenance of steel penstocks: Hydel Civil Design Directorate CW & PC, New Delhi.

Doland, J. (1954). Hydropower Engineering. New York: Ronald Press.

Duggal, K. N. Soni, J. P. (1996). Elements of water resources engineering: M/s New Age International Publication, New Delhi.

Datta, A.K. (1999): Purulia pumped storage hydro electric project — a case study.

Datta O.P. & Khurana J.C. (1977) Beas Project Unit I : Dehar Surge Shaft. Indian Journal of Power & River Valley Development.

Eakin, H.A. (1939) Silting of reservoirs. Tech. Publication No. 524, U.S. Deptt. of Agriculture.

Garg, S.K. (1977). Water resources and hydrology. Delhi: Khanna Publishers.

Garg, S.K. (1978). Irrigation engineering and hydraulic strucrtures. Delhi: Khanna Publishers.

Goel, R.S. (2000). Environment management in hydro power: M/s Oxford IBH Publishing CO. New Delhi.

Goel, R.S. (2000). Environmental impacts assessment of water resources projects: M/s Oxford IBH Publishing CO. New Delhi.

Goel, R.S. (2000). Hydro power and river valley development: M/s Oxford IBH Publishing CO. New Delhi.

Goel, R.S. and Sinha, K.N. (1999): Investigation, planning and evolution of layouts for hydel projects in India — a critical review.

Goel R.S. (1980). Thesis for M.E., Project Report. XXIV Course, Vol-II, University of Roorkee, Roorkee.

Gopalakrishnan V (1997) : Economic Evaluation 'of Pumped Storage Power Plant— International Conference on Accelerated Hydro Power Development & Pumped Storage Development, New Delhi.

Gopalakrishnan, V and Subramaniapillai (1997): Planning of a pumped storage power project: Tamil Nadu Electricity Board, Chennai.

Goval, K.K. (1999): Hydro power development — complementary role of small hydro.

Graham Frank D (1943) : Audels Pumps. Hydraulics Air Compressors. Theo Audel & Co. Publishers, New York.

Gupta, V.K. (2000): Power projects of NHPC Indian Journal of Power and River Valley Development vol 50.

Golze, A.R. (1977). Handbook of dam engineering, New York: Van Nostrand Rheinhold Co.

Hague M.E. : Hydrologic Considerations in Appraisal of Water Resources Projects.

Janakiran, K.S. (1999): International and national funding of small hydro projects—financing arrangements.

Khanna, P.N. (1971). Indian practical civil engineers handbook, New Delhi: Engineer's Publishers.

Khosla, A.N. (1939). Silting of reservoirs, New Delhi: C.B.I.P. Publication.

Khurmi R.S. (2000) : A text book of hydraulics, fluid mechanics and hydraulic machines.

Lafita Raymond (1997): Hydro power development— a world perspective. International conference on accelerated hydro power development and pumped storage development, New Delhi.

Lal R.P. : Layout of Hydro Power Station.

Leliavsky, S. (1965). Design textbook in civil engineering, Vols. I, II, III, IV and V. New Delhi: Oxford & IBH Publishing Co.

Lewitt E.H. (1927) : Hydraulics : Sir Isaac Pitman & Sons Ltd. New York.

Linsley, R.K. and Franzini, J.B. (1979). Water resources engineering. New York: McGraw-Hill book Co.

Linsley, R.K., Kholer, M.A. and Paulhus, J. (1975). Hydrology for Engineers, New York: McGraw-Hill Book Co.

Linsley, R.K., Kholer, M.A. and Paulhus, J. (1949). Applied hydrology, New York: McGraw-Hill Book Co.

March Patrick and Fisher Richard, K (2000): It is not easy being green environmental techniques enhance conventional hydro power role in sustainable development. Water vision 2000 (USA).

March, P.A. (2000): Knowledge management system promotes paradigm shifts in the operation maintenance and environment performance of hydroelectric power plants. Water vision 2000 (USA).

Mathur, S.K. and Arora, B.M. (1997): 420 MW Laknwar project accelerates hydro power development on river Yamuna in U.P. International conference on accelerated hydro power development and pumped storage development, CBIP.

Mohan Vishwa Nath (1999): Approach for optimum utilisation of hydro potential.

Mohile, A.D. (1999). River valley development-A comprehensive approach.

Murthy, Y.K. (1976). Environmental impacts of water development projects, Journal of I. and P., 33, 1, pp. 17-19.

Naidu, B.S.K. (1999): Strategy for growth and environmental sustainability of hydro power projects in India.

Natarajan, M.K. (1984): Principles of fluid mechanics, M/s Oxford and IBH Publishing Co., New Delhi.

Nikolaev. B.A. (1980) Pumps, Fans, Compressors : Mir Publishers, Moscow.

Prasad, R.S. and Goel R.S. (1999): Environmental concerns and raising possible awareness for sustainable water-resources development in India.

Prasad Yogendra (2000): An overview on hydro power. Indian Journal of Power and River Valley Development vol. 50.

Priyani, V.B. (1967). The fundamental principles of irrigation and water power. Anand; Charotar Book Stall.

Punmia, B.C. and Pande, B.B.L. (1979). Irrigation and water power engineering. Delhi: Standard Publishers.

Ramaswamy Nand Iyer R.A. (1990): Planning and designing of hydro power complex for Sardar Sarovar project. Power and River Valley Development, sept. 1990.

Ramaswamy N. Shah, B.J. and Patel, M.S. (1990). Construction aspects of underground power complex for Sardar Sarovar Project. Power and River valley Development, sept. 1990.

Ramaswamy, N and Shah, B.J. (1990): Design and construction aspects of canal power house.

Rajput, R.K. (1998): Text book of hydraulic machines, M/s S. Chand and Co. New Delhi.

Rajput R.K. (1998) : A text book of Fluid Mechanics. S. Chand & Co. New Delhi.

Sahasrabudha, S.R. (1978). Irrigation engineering and hydraulic structures. Ludhiana: Katson Publishing House.

Sharma, H.D. (1995). Embankment dams: M/s Oxford & IBH Publishing Co. New Delhi.

Sharma, R.K. (1979 through 1997). A textbook of hydrology and water resources engineering New Delhi: Dhanpat Rai Publication Pvt. Ltd., Daryaganj, Delhi.

Sharma, R.K. (1984). Text book of irrigation engineering and hydraulic structures: M/s Oxford & IBH Publishing Co., New Delhi.

Sharma, R.K. (1984). Text book of irrigation engineering vol. I, II, III. (1990-1992): M/s Oxford & IBH Publishing Co., New Delhi.

Sharma, R.K. et.al. Socio-economic aspects of water management at Beas Dam — a case study, presented at 64th R&D Session CBIP, 1988.

Sharma, R.K. et.al. Accelerated hydropower development-International conference on hydro power (1998).

Sharma, R.K. and Gupta, G.D. Watershed management presented at International conference on watershed management (1998).

Sharma, R.K. Hydrological impacts of Pong reservoir presented at International conference on emerging trends in hydrology, University of Roorkee, Roorkee (1997)

Sharma, R.K. Beas water and pisciculture development-H.P. fishery deptt. (1999) Roorkee.

Sharma, R.K. Hydrometeorological instruments along Pong reservoir-CBIP, ICOLD, BBMB symposium (1997).

Sharma, R.K. et.al. Capacity evaluation of Pong reservoir-CBIP, ICOLD, BBMB seminar (1996).

Sharma, R.K. et.al. Sediment measurement at Pong reservoir-CBIP, R&D session, (1993).

Sharma, R.K. et.al. Dam safety at Pong — an appraisal-CBIP, R&D session, (1991).

Sharma, R.K. et.al. Monitoring environmental impacts of Pong reservoir-CBIP, R&D session, (1992).

Sharma, R.K. Water conservation — need of day-BBMB water day, (1991).

Sharma, R.K. et.al. Environment and reservoir silting-BBMB seminar, (1990).

Sharma, R.K. et.al. Sustainable economic development by Pong dam — a case study-Wapcos Institution of engineers, (1990).

Sharma, R.K. et.al. Water quality of Pong reservoir — a case study-National symposium, (1990).

Sharma, R.K. et.al. Sediment measurement & Control at Pong reservoir, a case study, presented in Indo-British Workshop on Sediment measurement and control, Chandigarh, India, 1988.

Sharma, R.K. et.al. Hydro power – Its expeditious development, presented in First National Water Convention, New Delhi (1987).

Sharma, R.K. Lecture on environmental impacts of Bhakra-Beas system. Wapcos, Institution of engineers (India), New Delhi (2000).

Sharma, R.K. et.al. Sedimentation in Pong reservoir — a case study, presented in 53rd R&D Session of CBIP, 1986.

Sinha, B.K. (2000): Standardisation in the field of hydro power and river valley development: A critical review and management issues.

Sinha, G.P. (1999): Hydro Power Development (prospectus, problems and strategies and action plan).

Srivastava, R.N. and Goel, R.S. (1999): Hydro Power Development in India — prospects, problems and strategies. Indian Journal of Power and River Valley Development, Vol. XLVIX July-August, 1999.

Tideman, E.M. (1998). Watershed Management — Guidelines for Indian Conditions. New Delhi: Omega Scientific Publications.

Urquhart, L. (1959). Civil engineering handbook. New York: McGraw-Hill Book Co.

Yogendra Prasad (1999): Strategies and experiences of Hydro Power Development in India.

Yogendra Singh : Design consideration for combating silt erosion in hydraulic turbines.

B. MANUALS AND REPORTS

Ministry of Irrigation, Govt. of India (1980). National perspective for water resources development in India. Water World, Vol. III.

Total catchment management (1987). CBIP, New Delhi.

Concrete Manual (1956). U.S.B.R. Publication.

Manual on barrages and weirs on permeable foundation, Vol. I and II. (1985). CBIP Publication No. 179.

USBR Design Standard No. 7 (1956): Valves, gates and steel conduits.

Corps of Engineers US Army (1960): Planning and Design of Hydroelectric Power Plant Structures.

Power Line (March 2001) : Economics of Wind Power Generation.

Power Line (March 2001) : Low Costs, High Savings.

Assessment of Small Hydro Potential, Govt. of India Publication.

C. BUREAU OF INDIAN STANDARDS SPECIFICATIONS

IS: 1893–1975 Criteria for earthquake resistant design of structures.

IS: 3872–1966 Code of practice for lining of canals with burnt clay tiles.

IS: 3873–1966 Code of practice for laying in situ cement concrete lining on canals.

IS: 4515 (1993)

IS: 4880 (Part 1 to 4): Hydraulic tunnels— general design, geometric design, hydraulic design, and concrete lining in rock.

IS: 4558–1995 Under drainage for lined canals.

IS: 4613 (1984): Radial gates.

IS: 5330 (1984) Anchors blocks.

IS: 5477 (Part–I, II and III): Fixing capacity of reservoirs.

IS: 5496 (1993): Dimensioning and layout of elbow type draft tubes.

IS: 6004–1971: Criteria for hydraulic design of sediment ejector.

IS: 7418: 1991. Criteria for design of spiral casing (concrete and steel).

IS: 8826–1978: Design of large earth and rockfill dams.

IS: 7894–1975: Code of practice for stability analysis of earth dams.

IS: 8444–1977: Design of under–seepage control measures for earth and rockfill dams.

IS: 9120:1979. Cavities in underground power stations.

IS: 9349 (1986): Medium and high head slide gates.

IS: 9751 (1995): Hydro power intakes.

IS: 11388 (1995): Trashracks.

IS: 11570 (1985): Intake structures.

IS: 12800 (part 1 to 3): Selection of turbines pumped storage power houses, small, mini and micro power houses.

IS: 12837 (1989): Hydraulic turbines for medium large power houses.

IS: 8237–1985: Protection of slope for reservoir.

IS: 8835–1978: Planning and design of surface drains.

IS: 10635–1993: Free board requirement in embankment dams.

IS: 7323–1994: Operation of reservoirs.

IS: 6572–1998: Design of solid gravity dams.

IS: 12967 (part 1) 1990: Hydraulic transients.